Mobile WiMAX

To my wife Shahrnaz and my children Roya and Nima

Mobile WiMAX
A Systems Approach to Understanding IEEE 802.16m Radio Access Technology

Sassan Ahmadi

ELSEVIER

AMSTERDAM • BOSTON • HEIDELBERG • LONDON • NEW YORK • OXFORD
PARIS • SAN DIEGO • SAN FRANCISCO • SINGAPORE • SYDNEY • TOKYO

Academic Press is an imprint of Elsevier

Academic Press is an imprint of Elsevier
The Boulevard, Langford Lane, Kidlington, Oxford, OX5 1GB, UK
30 Corporate Drive, Suite 400, Burlington, MA 01803, USA

First published 2011

Notices

Knowledge and best practice in this field are constantly changing. As new research and experience broaden our understanding, changes in research methods, professional practices, or medical treatment may become necessary.

Practitioners and researchers must always rely on their own experience and knowledge in evaluating and using any information, methods, compounds, or experiments described herein. In using such information or methods they should be mindful of their own safety and the safety of others, including parties for whom they have a professional responsibility.

To the fullest extent of the law, neither the Publisher nor the authors, contributors, or editors, assume any liability for any injury and/or damage to persons or property as a matter of products liability, negligence or otherwise, or from any use or operation of any methods, products, instructions, or ideas contained in the material herein.

British Library Cataloguing in Publication Data
Ahmadi, Sassan.
   Mobile WiMAX : a systems approach to understanding the IEEE
   802.16m radio access network.
   1. IEEE 802.16 (Standard) 2. Wireless communication
   systems. 3. Mobile communication systems.
   I. Title
   621.3'84-dc22

Library of Congress Control Number: 2010935393

ISBN: 978-0-12-810193-3

For information on all Academic Press publications
visit our website at www.elsevierdirect.com

Printed and bound in the United States

10 11 12 11 10 9 8 7 6 5 4 3 2 1

Contents

Preface

Wireless communication comprises a wide range of technologies, services, and applications that have come into existence to meet the particular needs of users in different deployment scenarios. Wireless systems can be broadly characterized by content and services offered, reliability and performance, operational frequency bands, standards defining those systems, data rates supported, bi-directional and uni-directional delivery mechanisms, degree of mobility, regulatory requirements, complexity, and cost. The number of mobile subscribers has increased dramatically worldwide in the past decade. The growth in the number of mobile subscribers will be further intensified by the adoption of broadband mobile access technologies in developing countries such as India and China with large populations. It is envisioned that potentially the entire world population will have access to broadband mobile services, depending on economic conditions and favorable cost structures offered by regional network operators. There are already more mobile devices than fixed-line telephones or fixed computing platforms, such as desktop computers, that can access the Internet. The number of mobile devices is expected to continue to grow more rapidly than nomadic and stationary devices. Mobile terminals will be the most commonly used platforms for accessing and exchanging information. In particular, users will expect a dynamic, continuing stream of new applications, capabilities, and services that are ubiquitous and available across a range of devices using a single subscription and a single identity. Versatile communication systems offering customized and ubiquitous services based on diverse individual needs require flexibility in the technology in order to satisfy multiple demands simultaneously. Wireless multimedia traffic is increasing far more rapidly than voice, and will increasingly dominate traffic flows. The paradigm shift from predominantly circuit-switched air interface design to full IP-based delivery has provided the mobile users with the ability to more efficiently, more reliably, and more securely utilize packet-switched services such as e-mail, file transfers, messaging, browsing, gaming, voice-over Internet protocol, location-based, multicast, and broadcast services. These services can be either symmetrical or asymmetrical (in terms of the use of radio resources in the downlink or uplink) and real-time or non real-time, with different quality of service requirements. The new applications consume relatively larger bandwidths, resulting in higher data rate requirements.

In defining the framework for the development of IMT-Advanced and systems beyond IMT-Advanced radio interface technologies, it is important to understand the usage models and technology trends that will affect the design and deployment of such systems. In particular, the framework should be based on increasing user expectations and the growing demand for mobile services, as well as the evolving nature of the services and applications that may become available in the future. The trend toward integration and convergence of wireless systems and services can be characterized by connectivity (provision of an information pipe including intelligence in the network and the terminal), content (information including push and pull services as well as peer-to-peer applications), and e-commerce (electronic transactions and financial services). This trend may be viewed as the integration and convergence of information technology, telecommunications, and content, which has resulted in new service delivery dynamics and a new paradigm in wireless telecommunications, where value-added services have provided significant benefits to both the end users and the service providers.

Present mobile communication systems have evolved by incremental enhancements of system capabilities, and gradual addition of new functionalities and features to baseline IMT-2000 systems. The capabilities of IMT-2000 systems have continued to steadily evolve over the past decade as

IMT-2000 technologies are upgraded and deployed (e.g., mobile WiMAX and the migration of UMTS systems to HSPA+). The IMT-Advanced and systems beyond IMT-Advanced are going to be realized by functional fusion of existing IMT-2000 system components, enhanced and new functions, nomadic wireless access systems, and other wireless systems with high commonality and seamless interworking. The systems beyond IMT-Advanced will encompass the capabilities of previous systems, as well as other communication schemes such as machine-to-machine, machine-to-person, and person-to-machine.

The framework for the development of IMT-Advanced and systems beyond IMT-Advanced can be viewed from multiple perspectives including users, manufacturers, application developers, network operators, and service and content providers. From the user's perspective, there is a demand for a variety of services, content, and applications whose capabilities will increase over time. The users expect services to be ubiquitously available through a variety of delivery mechanisms and service providers using a variety of wireless devices. From the service provision perspective, the domains share some common characteristics. Wireless service provision is characterized by global mobile access (terminal and personal mobility), improved security and reliability, higher service quality, and access to personalized multimedia services, the Internet, and location-based services via one or multiple user terminals. Multi-radio operation requires seamless interaction of systems so that the user can receive/transmit a variety of content via different delivery mechanisms depending on the device capabilities, location and mobility, as well as the user profile. Different radio access systems can be connected via flexible core networks and appropriate interworking functions. In this way, a user can be connected through different radio access systems to the network and can utilize the services. The interworking among different radio access systems in terms of horizontal or vertical handover and seamless connectivity with service negotiation, mobility, security, and QoS management are the key requirements of radio-agnostic networks.

The similarity of services and applications across different radio access systems is beneficial not only to users, but also to network operators and content providers, stimulating the current trend towards convergence. Furthermore, similar user experience across different radio interface systems leads to large-scale adoption of products and services, common applications, and content. Access to a service or an application may be performed using one system or using multiple systems simultaneously. The increasing prevalence of IP-based applications has been a key driver for this convergence, and has accelerated the convergence trend in the core network and radio air interface.

The evolution of IMT-2000 baseline systems and the IMT-Advanced systems has employed several new concepts and functionalities, including adaptive modulation and coding and link adaptation, OFDM-based multiple access schemes, single-user/multi-user multi-antenna concepts and techniques, dynamic QoS control, mobility management and handover between heterogeneous radio interfaces (vertical and horizontal), robust packet transmission, error detection and correction, multi-user detection, and interference cancellation. Systems beyond IMT-Advanced may further utilize sophisticated schemes including software defined radio and reconfigurable RF and baseband processing, adaptive radio interface, mobile *ad hoc* networks, routing algorithms, and cooperative communication.

In response to this demand, the IEEE 802.16 Working Group began the development of a new amendment to the IEEE 802.16 standard (i.e., IEEE 802.16m) in January 2007 as an advanced air interface to meet the requirements of ITU-R/IMT-Advanced for the fourth-generation of cellular systems. The 3rd Generation Partnership Project started a similar effort in 2008 to upgrade the UMTS standards and to further enhance its family of LTE technologies.

Many articles, book chapters, and books have been published on the subject of mobile WiMAX and 3GPP LTE, varying from academic theses to network operator analyses and manufacturers' application notes. By their very nature, these publications have viewed these subjects from one particular perspective, whether it is academic, operational, or promotional. A very different and unique approach has been taken in this book; a top-down system approach to understanding the system operation and design principles of the underlying functional components of 4th generation radio access networks. This book can be considered as the most up-to-date technical reference for the design of 4G cellular systems. In this book, the protocol layers and functional elements of both the IEEE 802.16m- and 3GPP LTE-Advanced-based radio access and core networks are described. While the main focus of the book (as will be understood from the title) is to provide readers with an in-depth understanding of the IEEE 802.16m radio access system design, and to demonstrate the operation of the end-to-end system; a detailed description of the 3GPP LTE Release 9 and 3GPP LTE-Advanced Release 10 systems is provided to allow readers to better understand the similarities and differences between the two systems by contrasting the protocols and functional elements. It can be concluded that, aside from the marketing propaganda and hype surrounding these technologies, the 3GPP LTE and mobile WiMAX systems are technically equivalent and a fair comparison of the two technologies and their evolutionary paths reveals a similar performance as far as user experience is concerned.

In order to ensure the self-sufficiency of the material, the theoretical background and necessary definitions of all terms and topics has been provided either as footnotes or in separate sections to enable in-depth understanding of the subject under consideration without distracting the reader, and with no impact on the continuity of the subject matter. Additional technical references are cited in each chapter for further study. Each chapter in this book provides a top-down systematic description of the IEEE 802.16m entities and functional blocks, such as state transition models and corresponding procedures, protocol structures, etc., (including similarities and differences with the legacy mobile WiMAX systems to emphasize improvements) starting at the most general level and working toward the details or specifics of the protocols and procedures. The description of corresponding 3GPP LTE/LTE-Advanced protocols and procedures are further provided to enable readers to contrast the analogous terminal and base station behaviors, protocols, and functionalities. Such contrast is crucial in the design of inter-system interworking functions and to provide better understanding of the design strengths and weaknesses of each system.

Introduction

International Mobile Telecommunications-Advanced systems are broadband mobile wireless access systems that include new capabilities and versatility that goes beyond those of IMT-2000 systems. IMT-Advanced has provided a global framework for the development of the next generation of wireless radio access networks that enable low-delay, high-speed, bi-directional data access, unified messaging, and broadband wireless multimedia in the form of new service classes. Such systems provide access to a variety of mobile telecommunication services through entirely packet-based access/core networks. The IMT-Advanced systems support low to very high mobility applications and a wide range of data rates proportional to usage models and user density. The design and operational requirements concerning the 4th generation of radio interface technologies may vary from different perspectives with certain commonalities as follows:

End User

- Ubiquitous mobile Internet access;
- Easy access to applications and services with high quality at reasonable cost;
- Easily understandable user interface;
- Long battery life;
- Large choice of access terminals;
- Enhanced service capabilities;
- User-friendly billing policies.

Content Provider

- Flexible billing;
- Ability to adapt content to user requirements depending on terminal type, location, mobility, and user preferences;
- Access to a sizable market based on the similarity of application programming interfaces.

Service Provider

- Fast, open service creation, validation, and provisioning;
- Quality of service and security management;
- Automatic service adaptation as a function of available data rate and type of terminal;
- Flexible billing.

Network Operator

- Optimization of resources in terms of spectrum and equipment;
- Quality of service and security management;
- Ability to provide differentiated services;
- Flexible network configuration;
- Reduced cost of terminals and network equipment based on global economies of scale;
- Smooth transition from legacy systems to new systems;
- Maximizing commonalities among various radio access systems including sharing of mobile platforms, subscriber identity modules, network elements, radio sites;

- Single authentication process independent of the access network;
- Flexible billing;
- Access type selection optimizing service delivery.

Manufacturer or Application Developer

- Reduced cost of terminals and network equipment based on global economies of scale;
- Access to global markets;
- Open physical and logical interfaces between modular and integrated subsystems;
- Programmable/configurable platforms that enable fast and low-cost development.

The capabilities of IMT-2000 systems have continuously evolved over the past decade as IMT-2000 technologies have been upgraded and widely deployed. From the radio access perspective, the evolved IMT-2000 systems have built on the legacy systems, further enhanced the radio interface functionalities/protocols, and at the same time new systems have emerged to replace the existing IMT-2000 radio access systems in the long-term. This evolution has improved the reliability and throughput of the cellular systems and promoted the development of an expanding number of services and applications. The similarity of services and applications across different IMT technologies and frequency bands is not only beneficial to users, but also a similar user experience generally leads to a large-scale deployment of products and services. The technologies, applications, and services associated with systems beyond IMT-Advanced could well be radically different from the present systems, challenging our perceptions of what may be considered viable by today's standards and going beyond what has just been achieved by the IMT-Advanced radio systems.

The IEEE 802.16 Working Group began the development of a new amendment to the IEEE 802.16 baseline standard in January 2007 as an advanced air interface, in order to materialize the ITU-R vision for the IMT-Advanced systems as laid out in Recommendation ITU-R M.1645. The requirements for the IEEE 802.16m standard were selected to ensure competitiveness with the emerging 4th generation radio access technologies, while extending and significantly improving the functionality and efficiency of the legacy system. The areas of improvement and extension included control/signaling mechanisms, L1/L2 overhead reduction, coverage of control and traffic channels at the cell-edge, downlink/uplink link budget, air-link access latency, client power consumption including uplink peak-to-average power ratio reduction, transmission and detection of control channels, scan latency and network entry/ re-entry procedures, downlink and uplink symbol structure and subchannelization schemes, MAC management messages, MAC headers, support of the FDD duplex scheme, advanced single-user and multi-user MIMO techniques, relay, femto-cells, enhanced multicast and broadcast, enhanced location-based services, and self-configuration networks. The IMT-Advanced requirements defined and approved by ITU-R and published as Report ITU-R M.2134 were referred to as target requirements in the IEEE 802.16m system requirement document, and were evaluated based on the methodology and guidelines specified by Report ITU-R M.2135-1. The IEEE 802.16m baseline functional and performance requirements were evaluated according to the IEEE 802.16m evaluation methodology document. The IMT-Advanced requirements are a subset of the IEEE 802.16m system requirements, and thus are less stringent than baseline requirements. Since satisfaction of the baseline requirements would imply a minimum-featured (baseline) system, any minimum performance of the IEEE 802.16m implementation could potentially meet the IMT-Advanced requirements and could be certified as an IMT-Advanced technology. The candidate proposal submitted by the IEEE to the ITU-R

(IEEE 802.16m) proved to meet and exceed the requirements of IMT-Advanced systems, and thus qualified as an IMT-Advanced technology.

In the course of the development of the IEEE 802.16m, and unlike the process used in the previous amendments of the IEEE 802.16 standard, the IEEE 802.16m Task Group developed system requirements and evaluation methodology documents to help discipline and organize the process for the development of the new amendment. This would allow system design and selection criteria with widely agreed targets using unified simulation assumptions and methodology. The group further developed a system description document to unambiguously describe the RAN architecture and system operation of the IEEE 802.16m entities, which set a framework for the development of the IEEE 802.16m standard specification. To enable a smooth transition from Release 1.0 mobile WiMAX systems to the new generation of the mobile WiMAX radio access network, and to maximize reuse of legacy protocols, strict backward compatibility was required. The author's original view and understanding of backward compatibility was similar to that already seen in other cellular systems such as the migration of $1 \times$ EV-DO Revision 0 to $1 \times$ EV-DO Revision A, to $1 \times$ EV-DO Revision B on the cdma2000 path and evolution of UMTS Release 99 to HSDPA to HSPA, and to HSPA+ on the WCDMA path. In these examples, the core legacy protocols were reused and new protocols were added as complementary solutions, such that the evolved systems maintained strict backward compatibility with the legacy systems, allowing gradual upgrades of the base stations, mobile stations, and network elements. Had it been materialized, the author's vision would have resulted in a fully backward compatible system with improvement and extension of the legacy protocols and functionalities built on top of the existing protocols as opposed to from ground up. However, the enthusiasm for the IMT-Advanced systems and the ambitious baseline requirements set by the IEEE 802.16 group resulted in deviation from the original vision and the new amendment turned into describing a new system that was built more or less from scratch. A large number of legacy physical, lower and upper MAC protocols were replaced with new and non-backward compatible protocols and functions. The co-deployment of the legacy and the new systems on the same RF carrier is only possible via time-division or frequency-division multiplexing of the legacy and new protocols in the downlink and uplink legacy/new zones. More specifically, the legacy and new zones are time division multiplexed in the downlink and are frequency division multiplexed in the uplink. Figure 1 illustrates an example where the legacy system is supported in an IEEE 802.16m system. The overhead channels corresponding to each system (i.e., synchronization, control, and broadcast channels) are duplicated due to incompatibility of the physical structures and transmission formats of these overhead channels. Although IEEE 802.16m specifies handover mechanisms to and from the legacy systems, the handover protocols, MAC messages, and triggers are different, requiring a separate protocol/software stack for dual-mode implementation of the two systems. Table 1 compares the physical layer and lower MAC features of the legacy mobile WiMAX and IEEE 802.16m. It can be seen that many important features and functions such as HARQ, subchannelization, control channels, and MIMO modes have changed in the IEEE 802.16m, making migration from legacy systems to the IEEE 802.16m systems not straightforward and also expensive. The complexity of later upgrades is similar to that of migration of UMTS/HSPA systems to 3GPP LTE systems given the non-backward compatible nature of 3GPP LTE enhancements relative to UMTS. The features and functions listed in this table will be described in Chapters 9 and 10.

As a result of extensive changes and enhancements in the IEEE 802.16m standard relative to legacy mobile WiMAX, it will not be surprising to realize that the throughput and performance of the IEEE

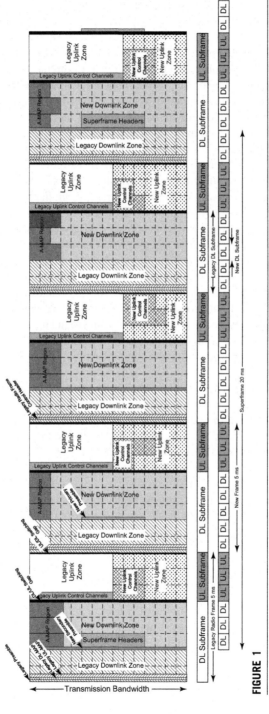

FIGURE 1

Example Sharing of Time-Frequency Resources over one Radio Frame between IEEE 802.16m and the Legacy Systems in TDD Mode

Table 1 Comparison of the Legacy Mobile WiMAX Features with IEEE 802.16m

Feature	Legacy Mobile WiMAX based on Release 1.0	IEEE 802.16m
Duplexing Scheme	TDD	TDD and FDD
Frame Structure	5 ms radio frames with flexible time-zones	5 ms radio frames with subframe-based fixed time-zones
Superframe Structure	Not supported	20 ms duration (4 consecutive radio frames)
Operating Bandwidth (MHz)	5, 7, 8.75, and 10	5, 7, 8.75, 10, and 20 (up to 100 MHz with carrier aggregation and other channel bandwidths through tone dropping)
Resource Block Size	Fixed 48 data sub-carriers	18 sub-carriers by 6 OFDM symbols physical resource units and variable number of data sub-carriers depending on the MIMO mode
Control Channel Subchannelization	Partial Usage of Sub-Channels in the downlink and uplink (distributed permutations)	Distributed logical resource units (tone-pair based distributed permutations)
Traffic Channel Subchannelization	Partial Usage of Sub-Channels in the downlink and uplink (distributed permutations)	Distributed logical resource units (distributed permutations) Sub-band logical resource units (localized permutations) Mini-band logical resource units (physical resource unit-based diversity permutations)
Permutation Zone Multiplexing	Time Division Multiplexing of different zones	Frequency Division Multiplexing in the same subframe
Pilot Design	Common (non-precoded) and dedicated (precoded) pilots depending on the permutation zone	Non-adaptive precoded pilots for distributed logical resource units, dedicated pilots per physical resource unit for sub-band and mini-band logical resource units; interlaced pilots for interference mitigation
Turbo Codes	Convolutional Turbo Codes with minimum code rate of $1/3$ and repetition coding	Convolutional Turbo Codes with minimum code rate of $1/3$ and rate matching

(Continued)

Table 1 Comparison of the Legacy Mobile WiMAX Features with IEEE 802.16m   *Continued*

Feature	Legacy Mobile WiMAX based on Release 1.0	IEEE 802.16m
Convolutional Codes	Tail-Biting Convolutional Codes with minimum code rate of ½	Tail-Biting Convolutional Codes with minimum code rate of $\frac{1}{5}$
DL HARQ	Asynchronous Chase Combining	Asynchronous Incremental Redundancy (Chase Combining as a special case)
UL HARQ	Asynchronous Chase Combining	Synchronous Incremental Redundancy
Downlink Open-loop Single-user MIMO	Space-Time Block Coding, Spatial Multiplexing; Cyclic Delay Diversity for more than two transmit antennas	Space Frequency Block Coding, Spatial Multiplexing, Non-adaptive precoding for more than two transmit antennas
Downlink Closed-loop Single-user MIMO	Sounding-based	Transformed codebook-based scheme using sub-band logical resource unit, Long-term covariance matrix or codebook based using mini-band logical resource units Sounding-based using sub-band or mini-band logical resource units
Uplink Open-loop Single-user MIMO	Not Supported	Space-Frequency Block Coding/Spatial Multiplexing, Non-adaptive precoding for more than two transmit antennas with distributed logical resource units
Uplink Closed-loop Single-user MIMO	Not Supported	Codebook-based precoding using sub-band or mini-band logical resource units
Downlink Multi-user MIMO	Not Supported	Multi-User Zero-Forcing precoding based on transformed codebook or sounding
Uplink Multi-User MIMO	Single-transmit-antenna Collaborative MIMO	Collaborative MIMO for up to four transmit antennas (codebook-based or vendor-specific precoding for more than one transmit antenna)
Uplink Power Control	Basic open-loop power control, Message-based closed-loop power control	Improved open-loop power control (SINR-based) and signaling-based closed-loop power control

Fractional Frequency Reuse	Basic Fractional Frequency Reuse		Advanced Fractional Frequency Reuse support with up to 4 frequency partitions (1 reuse-1 and 3 reuse-3), Low power transmission in other reuse-3 partitions
Downlink Control Channels	Medium Access Protocol	Compressed Medium Access Protocol/Sub-Medium Access Protocol, jointly-coded, once per frame, Time Division Multiplexed with data	Individual (user-specific) MAP, separately-coded, once per subframe, Frequency Division Multiplexed with data
	Broadcast Channel	Frame Control Header/Downlink Channel Descriptor/Uplink Channel Descriptor	Primary and Secondary Superframe Headers
	Synchronization Channel	Full bandwidth, 114 codes, once per frame	Primary preamble in 5 MHz bandwidth once per superframe. Secondary preamble in full bandwidth, 768 codes, 2 times per superframe
	Midamble	Not Supported	Full bandwidth, once per frame, used for PMI/CQI feedback
	Channel Quality and Precoding Matrix Feedbacks	4-bit/6-bit CQI	Primary and Secondary Fast Feedback Channel for CQI/PMI feedback
Uplink Control Channels	Bandwidth Request	Reuse of initial ranging structure and sequence; 5-step access	3 uplink 6 × 6 tiles, regular (5-step) and fast (3-step) contention-based access
	Sounding	One OFDM symbol in the uplink subframe, CDM and FDM for mobile station multiplexing	One OFDM symbol in the uplink subframe, CDM and FDM for mobile station and antenna multiplexing

802.16m surpasses that of the legacy system, resulting in extended capabilities to support a variety of existing and future services and applications with high quality and capacity. Table 2 compares the throughput of the two systems under selected test scenarios that were specified in the IMT-Advanced evaluation methodology document.

In Table 2, a TDD system with 10 MHz bandwidth and frequency reuse 1, as well as a DL:UL ratio of 29:18 was assumed for both systems. The legacy system employs a 4×2 single-user MIMO configuration and sounding-based beamforming in the downlink, along with a 1×4 collaborative MIMO in the uplink. The IEEE 802.16m uses a 4×2 multi-user MIMO in the downlink in addition to a 2×4 collaborative MIMO in the uplink with codebook-based beamforming for both links. There are up to four multi-user MIMO users in the downlink and up to two multi-user MIMO users in the uplink.

A common confusion arises concerning the terminologies used for mobile and base stations compliant with different versions of the IEEE 802.16 standard and mobile WiMAX system profile. The IEEE 802.16-2009 standard specifies a large number of optional features and parameters that may define various mobile station and base station configurations. One of the possible implementation variants was selected and specified by the WiMAX Forum as Release 1.0 of the mobile WiMAX system profile. The latter configuration was chosen by the IEEE 802.16m as the reference for backward compatibility. Consequently, when referring to a mobile station and base station in different amendments of the IEEE 802.16 standard, as well as mobile WiMAX profiles, one must make sure that a consistent reference is made, and that backward compatibility and interoperability can be maintained. Unlike the IEEE 802.16m specification that refers to the new IEEE 802.16 entities as "advanced mobile station," "advanced base station," and "advanced relay station" to differentiate them from their counterparts in the IEEE 802.16-2009 and IEEE 802.16j-2009 standards specifications, we refer to these entities as mobile station, base station, and relay station, assuming that the reference system is compliant with Release 1.0 of the mobile WiMAX system profile and that the extended functions and protocols corresponding to IEEE 802.16m can be distinguished from their legacy counterparts by the reader.

Similar to the IEEE, the 3GPP initiated a project on the long-term evolution of UMTS radio interface in late 2004 to maintain 3GPP's competitive edge over other cellular technologies. The

Table 2 Comparison of the Throughput of the Legacy Mobile WiMAX and IEEE 802.16m Systems

	Downlink Spectral Efficiency (bits/s/Hz/cell)		Uplink Spectral Efficiency (bits/s/Hz/cell)	
	IMT-Advanced Urban Microcell Test Environment (3 km/h)	IMT-Advanced Urban Macrocell Test Environment (30 km/h)	IMT-Advanced Urban Microcell Test Environment (3 km/h)	IMT-Advanced Urban Macrocell Test Environment (30 km/h)
Legacy Mobile WiMAX based on Release 1.0	2.02	1.44	1.85	1.70
IEEE 802.16m	3.22	2.45	2.46	2.25

evolved UMTS terrestrial radio access network substantially improved end-user throughputs, and sector capacity, and reduced user-plane and control-plane latencies, bringing a significantly improved user experience with full mobility. With the emergence of the Internet protocol as the protocol of choice for carrying all types of traffic, the 3GPP LTE provides support for IP-based traffic with end-to-end quality of service. Voice traffic is supported mainly as voice over IP, enabling integration with other multimedia services. Unlike its predecessors, which were developed within the framework of UMTS architecture, 3GPP specified an evolved packet core architecture to support the E-UTRAN through a reduction in the number of network elements and simplification of functionality, but most importantly allowing for connections and handover to other fixed and wireless access technologies, providing network operators with the ability to deliver seamless mobility experience. Similar to the IEEE 802.16, 3GPP set aggressive performance requirements for LTE that relied on improved physical layer technologies, such as OFDM and single-user and/or multi-user MIMO techniques, and streamlined Layer 2/Layer 3 protocols and functionalities. The main objectives of 3GPP LTE were to minimize the system and user equipment complexities, to allow flexible spectrum deployment in the existing or new frequency bands, and to enable coexistence with other 3GPP radio access technologies. The 3GPP LTE has been used as the baseline and further enhanced under 3GPP Release 10 to meet the requirements of the IMT-Advanced. A candidate proposal based on the latter enhancements (3GPP LTE-Advanced) was submitted to the ITU-R and subsequently qualified as an IMT-Advanced technology. However, concurrent with the 3GPP LTE standard development, the operators were rolling out HSPA networks to upgrade their 2G and 2.5G, and early 3G infrastructure, thus they were not ready to embrace yet another paradigm shift in radio access and core network technologies. Therefore, 3GPP has continued to improve UMTS technologies by adding multi-antenna support at the base station, higher modulation order in the downlink, multi-carrier support, etc., to extend the lifespan of 3G systems. It is anticipated that the new releases of 3GPP standards (i.e., LTE/LTE-Advanced) will not be commercially available worldwide on a large scale until current operators' investments are properly returned.

A comparison of 3GPP LTE-Advanced and IEEE 802.16m basic and advanced features and functionalities reveals that the two systems are very similar and may perform similarly under the same operating conditions. Therefore, there is effectively no technical or performance distinction between the two technologies. It will be shown throughout this book that the two radio access technologies are practically equivalent as far as user experience is concerned. Table 3 summarizes the major differences between IEEE 802.16m and 3GPP LTE-Advanced physical layer protocols. The features and functions listed in this table will be described in Chapters 9 and 10.

In the course of design and development of the IEEE 802.16m standard, the author decided to write a book and to take a different approach than was typically taken in other books and journal articles. The author's idea was to take a top-down systems approach in describing the design and operation of the IEEE 802.16m, and to contrast the 3GPP LTE/LTE-Advanced and IEEE 802.16m/mobile WiMAX algorithms and protocols to allow readers to better understand both systems. The addition of the 3GPP LTE/LTE-Advanced protocols and system description further expanded the scope of the book to a systems approach to understanding the design and operation of 4th generation cellular systems. There has been no attempt anywhere in this book to compare, side-by-side, the performance and efficiency of the mobile WiMAX and 3GPP LTE systems and to conclude that one system outperforms the other, rather, it is left to the reader to arrive at such a conclusion. In addition to a top-down systems approach, another distinction of this book compared to other publications in the literature is the

Table 3 Major Differences between IEEE 802.16m and 3GPP LTE-Advanced Physical Layers

Feature	3GPP LTE-Advanced	IEEE 802.16m
Multiple Access Scheme	Downlink: OFDMA Uplink: SC-FDMA	Downlink: OFDMA Uplink: OFDMA
Control Channel Multiplexing with Data	Time Division Multiplex (Resource occupied by control channel in units of OFDM symbols)	Frequency Division Multiplex (Resource occupied by control channel in physical resource block units)
Channel State Information (CSI) Feedback	Long-term CSI and Short-term CSI (e.g., sounding)	Base codebook with long-term channel covariance matrix and Sounding
Scheduling Period	Per Transmission Time Interval (TTI) scheduling and Persistent scheduling	Short and long TTI scheduling and Persistent scheduling
Physical Resource Block Size	12 sub-carriers $\times$ 14 OFDM/SC-FDMA Symbols = 168 Resource elements	18 sub-carriers $\times$ 6 OFDM symbols = 108 Resource elements
Usable Bandwidth at 10 MHz	600 sub-carriers $\times$ 15 kHz (sub-carrier spacing) = 9 MHz (Spectrum Occupancy = 90%)	864 sub-carriers $\times$ 10.9375 kHz (sub-carrier spacing) = 9.45 MHz (Spectrum Occupancy = 94.5%)
Usable OFDM/SC-FDMA Symbols per 5 ms	70 OFDM/SC-FDMA symbols (FDD) 56 OFDM/SC-FDMA symbols (TDD)	51 OFDM symbols (FDD) 50 OFDM symbols (TDD)
Usable Resource Elements per 5 ms	42000 Resource Elements (sub-carriers)	44064 Resource Elements (sub-carriers)
Modulation and Coding Scheme Levels	27 Levels	32 Levels
Downlink Antenna Configuration for IMT-Advanced Scenarios	$4 \times 2/8 \times 2$	4×2
Uplink Antenna Configuration for IMT-Advanced Scenarios	$1 \times 4/1 \times 8/2 \times 4$	2×4
Multi-antenna Schemes for IMT-Advanced Scenarios	Single-user MIMO, Multi-user MIMO/Beamforming, Coordinated Multipoint Transmission	Multi-user MIMO/Beamforming
Number of Users Paired in Downlink Multi-user MIMO	Up to 2 users paired in self-evaluation	Up to 4 users paired in self-evaluation
L1/L2 Overhead	Statically Modeled Number of OFDM symbols L = 1 (18%) Number of OFDM symbols L = 2 (24%) Number of OFDM symbols L = 3 (31%)	Dynamically Modeled Example: IMT-Advanced Urban Macrocell Scenario TDD = 11% (Control channel) + 11% (Pilot) $\approx$ 22% FDD = 14% (Control channel) + 11 % (Pilot) $\approx$ 25%

inclusion of the theoretical background or a description of uncommon terminologies and concepts in each chapter, so that readers can understand the subject matter without getting distracted with additional reading in the citations and references. In each chapter the design criteria and justification for modifications and extensions relative to the legacy systems have been described.

The present book begins with an introduction to the history of broadband mobile wireless access and an overview of the IEEE and 3GPP standards and standardization processes in Chapter 1. The approach taken in this book required the author to review the network architecture and to examine each and every significant network element in mobile WiMAX and 3GPP LTE networks. Since the WiMAX Forum has yet to update the WiMAX Network Architecture specification to support the IEEE 802.16m standard, the latest revision of the WiMAX Network Architecture document which is publicly available from the WiMAX Forum has been used. It is expected that the early deployment of IEEE 802.16m would rely on the legacy network architecture until network upgrades become available. Once the access network and core network aspects of the system are described, we turn our attention to the reference model and protocol structure of IEEE 802.16m and 3GPP LTE/LTE-Advanced, and discuss the operation and behavior of each entity (base station, mobile station, and relay station), as well as functional components and their interactions in the protocol stack. The remaining chapters of this book are organized to be consistent with the protocol layers, starting from the network layer and moving down to the physical layer. The overall operation of the mobile station, relay station, and base station and their corresponding state machines are described in Chapter 4. Perhaps this chapter is the most important part of the book, as far as understanding the general operation of the system is concerned. Chapter 5 describes the interface with the packet data network. Chapters 6 and 7 describe the medium access control layer protocols. Due to the size of content, the medium access control and physical layer chapters (Chapters 6, 7, 9 and 10) have been divided into two parts. The security aspects of the systems under consideration are described in Chapter 8. The additional functional components, algorithms, and protocols which have been introduced by the 3GPP LTE-Advanced are emphasized so that they are not confused with the legacy components. The multi-carrier operation of the IEEE 802.16m and 3GPP LTE-Advanced are described in Chapter 11. The performance evaluation of the IEEE 802.16m and 3GPP LTE-Advanced against the IMT-Advanced requirements has been described in Chapter 12, where all the performance metrics are defined and link-level and system-level simulation methodologies and parameters are elaborated.

The existing mobile broadband radio access systems will continue to evolve and new systems will emerge. The vision, service and system requirements for systems beyond IMT-Advanced will be defined as soon as the IMT-Advanced standardization process winds down. While it is not exactly clear what technologies will be incorporated into the design of such systems and whether the existing radio access technologies will converge into a single universal radio interface, it is envisioned that the future radio interfaces will rely on distributed antenna systems, low-power emission, distributed computing, seamless connectivity, software defined radio, cognitive radio systems, multi-resolution wireless multimedia, and cooperative communication concepts, as well as reconfigurable RF and baseband circuitry in order to provide a higher quality of user experience, higher capacities, and a wider range of services with minimal cost and complexity.

Acknowledgements

The author would like to acknowledge and sincerely thank his colleagues at Intel Corporation, ZTE Corporation, Samsung Electronics, Motorola, LG Electronics, the IEEE 802.16, and the 3GPP RAN groups for their contributions, consultation, and assistance in proofreading and improving the quality and content of the chapters of this book.

The author would like to sincerely thank Academic Press (Elsevier) publishing and editorial staff for providing the author with the opportunity to publish this book and for their assistance, cooperation, patience, and understanding throughout the past two years.

Finally, the author would like to thank his wife (Shahrnaz) and his children (Roya and Nima) for their unwavering encouragement, support, patience, and understanding throughout this long and challenging project.

Abbreviations

Abbreviation	Description
1xEV-DO	1× Evolution Data Only (Air Interface)
3-DES	Triple Data Encryption Standard
3G	3rd Generation (of Cellular Systems)
3GPP	3rd Generation Partnership Project
3GPP2	3rd Generation Partnership Project 2
4G	4th Generation (of Cellular Systems)
AAA	Authentication, Authorization, and Accounting
AAI	Advanced Air Interface
AAS	Adaptive Antenna System
ABS	Advanced Base Station
ACID	HARQ Channel Identifier
ACK	Acknowledgement
ACLR	Adjacent Channel Leakage Ratio
ACM	Account Management
ACS	Adjacent Channel Selectivity
AES	Advanced Encryption Standard
AGC	Automatic Gain Control
AGMH	Advanced Generic MAC Header
aGPS	Adaptive Grant Polling Service
AI_SN	HARQ Identifier Sequence Number
AK	Authorization Key
AKID	Authorization Key Identifier
AM	Acknowledged Mode
A-MAP	Advanced Medium Access Protocol
AMBR	Aggregate Maximum Bit Rate
AMC	Adaptive Modulation and Coding
AMS	Advanced Mobile Station
AoA	Angle of Arrival
A-Preamble	Advanced Preamble
ARFCN	Absolute Radio-Frequency Channel Number
ARP	Allocation and Retention Priority
ARQ	Automatic Repeat reQuest
ARS	Advanced Relay Station
AS	Access Stratum
ASA	Authentication and Service Authorization
ASN	Access Service Network
ASN.1	Abstract Syntax Notation One

(Continued)

Abbreviation	Description
ASN-GW	Access Service Network Gateway
ASP	Application Service Provider
ASR	Anchor Switch Reporting
ATDD	Adaptive Time Division Duplexing
ATM	Asynchronous Transfer Mode
AuC	Authentication Center
AWGN	Additive White Gaussian Noise
BCC	Block Convolutional Code
BCCH	Broadcast Control Channel
BCH	Broadcast Channel
BE	Best Effort
BER	Bit Error Ratio
BLER	Block Error Rate
BPSK	Binary Phase Shift Keying
BR	Bandwidth Request
BS	Base Station
BSID	Base Station Identifier
BSN	Block Sequence Number
BSR	Buffer Status Report
BTC	Block Turbo Code
BW	Bandwidth
BWA	Broadband Wireless Access
C/I	Carrier-to-Interference Ratio
C/N	Carrier-to-Noise Ratio
CA	Certification Authority
CAZAC	Constant Amplitude Zero Auto-Correlation
CBC	Cell Broadcast Center
CBC	Cipher Block Chaining
CBC-MAC	Cipher Block Chaining Message Authentication Code
CC	Confirmation Code
CC	Component Carrier
CC	Convolutional Code
CCDF	Complementary CDF
CCE	Control Channel Element
CCH	Control Subchannel
CCI	Co-Channel Interference
CCM	CTR Mode With CBC-MAC
CCO	Cell Change Order
CCS	Common Channel Signaling
CCV	Clock Comparison Value

Abbreviation	Description
CDD	Cyclic Delay Diversity
CDF	Cumulative Distribution Function
CDMA	Code Division Multiple Access
CDR	Conjugate Data Repetition
Cell_ID	Cell Identifier
ChID	Channel Identifier
CID	Connection Identifier
CINR	Carrier to Interference-plus-Noise Ratio
CIR	Channel Impulse Response
CLC	Collocated Coexistence
CLP	Cell Loss Priority
CLRU	Contiguous Logical Resource Unit
CM	Cubic Metric
CMAC	Cipher-Based Message Authentication Code
CMAS	Commercial Mobile Alert Service
CMC	Connection Mobility Control
CMI	Codebook Matrix Index
CMIP	Client Mobile IP
COBRA	Common Object Requesting Broker Architecture
Co-MIMO	Collaborative MIMO
CoMP	Coordinated Multi-Point Transmission
CoRe	Constellation Re-Arrangement
CP	Cyclic Prefix
C-Plane	Control Plane
CPS	Common Part Sublayer
CQI	Channel Quality Indicator
CQICH	Channel Quality Indicator Channel
CRC	Cyclic Redundancy Check
CRID	Context Retention Identifier
C-RNTI	Cell RNTI
CRU	Contiguous Resource Unit
CRV	Constellation Rearrangement Version
CS	Convergence Sublayer
CSA	Common Subframe Allocation
CSCF	Centralized Scheduling Configuration
CSCH	Centralized Scheduling
CSG	Closed Subscriber Group
CSI	Channel State Information
CSM	Collaborative Spatial Multiplexing
CSMA/CA	Carrier Sense Multiple Access with Collision Avoidance

(*Continued*)

Abbreviation	Description
CSMA/CD	Carrier Sense Multiple Access with Collision Detection
CSN	Connectivity Service Network
CTC	Convolutional Turbo Code
CTR	Counter Mode Encryption
DAMA	Demand Assigned Multiple Access
DARS	Digital Audio Radio Satellite
dBi	Decibels (Relative to Isotropic Radiator)
dBm	Decibels (Relative to 1 mW)
DC	Direct Current
DCAS	Downlink Contiguous Resource Unit Allocation Size
DCCH	Dedicated Control Channel
DCD	Downlink Channel Descriptor
DCI	Downlink Control Information
DCR	Deregistration with Content Retention
DES	Data Encryption Standard
DFS	Dynamic Frequency Selection
DFTS	DFT Spread (OFDM)
DHCP	Dynamic Host Configuration Protocol
DID	Deregistration Identifier
DIUC	Downlink Interval Usage Code
DL	Downlink
DLFP	Downlink Frame Prefix
DLRU	Distributed Logical Resource Unit
DOCSIS	Data over Cable Service Interface Specification
DP	Decision Point
DPF	Data Path Function
DRB	Data Radio Bearer
DRS	Demodulation Reference Signal
DRU	Distributed Resource Unit
DRX	Discontinuous Reception
DSA	Dynamic Service Addition
DSAC	Downlink Sub-band Allocation Count
DSC	Dynamic Service Change
DSCH	Distributed Scheduling
DSCP	Differentiated Services Code-Point
DSD	Dynamic Service Deletion
DSx	Dynamic Service Addition, Change, or Deletion
DTCH	Dedicated Traffic Channel
D-TDoA	Downlink Time Difference of Arrival
DTX	Discontinuous Transmission

Abbreviation	Description
DwPTS	Downlink Pilot Time Slot
EAP	Extensible Authentication Protocol
EBB	Entry Before Break
EC	Encryption Control
ECB	Electronic Code Book
ECGI	E-UTRAN Cell Global Identifier
E-CID	Enhanced Cell-ID (Positioning Method)
ECM	EPS Connection Management
ECRTP	IP-Header-Compression CS PDU Format
EDE	Encrypt-Decrypt-Encrypt
EDGE	Enhanced Data Rates for GSM Evolution
EESM	Exponential Effective SINR Mapping
EESS	Earth Exploratory Satellite System
EH	Extended Header
EIK EAP	Integrity Key
EIRP	Effective Isotropic Radiated Power
EKS	Encryption Key Sequence
e-LBS	Enhanced Location Based Services
eMBMS	Enhanced Multimedia Broadcast Multicast Service
EMM	EPS Mobility Management
eNB	E-UTRAN NodeB (Base Station)
EP	Enforcement Point
EPC	Evolved Packet Core
ePDG	Evolved Packet Data Gateway
EPRE	Energy per Resource Element
EPS	Evolved Packet System
E-RAB	E-UTRAN Radio Access Bearer
ESM	Effective SINR Mapping
ETS	Emergency Telecommunications Service
ETWS	Earthquake and Tsunami Warning System
EUI-48	48-bit IEEE Extended Unique Identifier
E-UTRA	Evolved UTRA
E-UTRAN	Evolved UTRAN
EVM	Error Vector Magnitude
FA	Frequency Assignment
FA	Foreign Agent
FBSS	Fast Base Station Switching
FC	Fragmentation Control
FCAPS	Fault, Configuration, Account, Performance and Security Management
FCH	Frame Control Header

(*Continued*)

Abbreviation	Description
FDD	Frequency Division Duplex
FDM	Frequency Division Multiplexing
FEC	Forward Error Correction
FER	Frame Error Rate
FFR	Fractional Frequency Reuse
FFSH	Fast-Feedback Allocation Sub-header
FFT	Fast Fourier Transform
FHDC	Frequency Hopping Diversity Coding
FID	Flow Identifier
FMT	Feedback Mini-Tile
FP	Frequency Partition
FPC	Frequency Partition Configuration
FPC	Fast Power Control
FPCT	Frequency Partition Count
FPEH	Fragmentation and Packing Extended Header
FPS	Frequency Partition Size
FPSC	Frequency Partition Sub-band Count
FSH	Fragmentation Sub-header
FSN	Fragment Sequence Number
FSS	Fixed Satellite Service
FTP	File Transfer Protocol
FUSC	Full Usage of Subchannels
GBR	Guaranteed Bit Rate
GERAN	GSM EDGE Radio Access Network
GF	Galois Field
GGSN	Gateway GPRS Support Node
GKEK	Group Key Encryption Key
GMH	Generic MAC Header
GMSH	Grant Management Sub-header
GNSS	Global Navigation Satellite System
GP	Guard Period
GPCS	Generic Packet Convergence Sublayer
GPI	Grant and Polling Interval
GPRS	General Packet Radio Service
GPS	Global Positioning System
GRA	Group Resource Allocation
GRE	Generic Routing Encapsulation
GS	Guard Symbol
GSM	Global System for Mobile Communication
GTEK	Group Traffic Encryption Key

Abbreviation	Description
HA	Home Agent
HARQ	Hybrid Automatic Repeat Request
HCS	Header Check Sequence
H-CSN	Home CSN
HE	Horizontal Encoding
HEC	Header Error Check
HeNB	Home eNB
H-FDD	Half-Duplex Frequency Division Duplex
HFN	Hyper-Frame Number
HHO	Hard Handover
HMAC	Hashed Message Authentication Code
HMT	HARQ Mini-Tiles
H-NSP	Home NSP
HO	Handover
H-PURDA	Hard Public Use Reservation by Departure Allocation
HRPD	High Rate Packet Data
HSDPA	High Speed Downlink Packet Access
HSPA	High Speed Packet Access
HSS	Home Subscriber Server
HT	Header Type
HTTP	Hypertext Transfer Protocol
IANA	Internet Assigned Numbers Authority
ICIC	Inter-Cell Interference Coordination
ICV	Integrity Check Value
IDFT	Inverse Discrete Fourier Transform
IDL	Interface Description Language
IE	Information Element
IEEE	Institute of Electrical and Electronics Engineers
IEEE-SA	IEEE Standards Association
IETF	Internet Engineering Task Force
IFDMA	Interleaved Frequency Division Multiple Access
IFFT	Inverse Fast Fourier Transform
IMAP	Internet Message Access Protocol
IMM	Idle Mode Management
IMS	IP Multimedia Subsystem
IMT	International Mobile Telecommunications
IoT	Interference over Thermal
IP	Internet Protocol
IPCS	Internet Protocol Convergence Sublayer
IPSec	IP Security

(*Continued*)

Abbreviation	Description
IPv4	Internet Protocol version 4
IPv6	Internet Protocol version 6
IR	Incremental Redundancy
ISO	International Standardization Organization
ITU	International Telecommunication Union
ITU-R	ITU Radio-communication Sector
ITU-T	ITU Telecommunication Standardization Sector
IV	Initialization Vector
IWF	Interworking Function
KEK	Key Encryption Key
L1	Layer 1 (Physical Layer)
L2	Layer 2 (Medium Access Control Layer)
L3	Layer 3 (Network Layer in OSI and RRC Sublayer in 3GPP)
LAN	Local Area Network
Layer 1	Layer 1 (Physical Layer)
Layer 2	Layer 2 (Medium Access Control Layer)
Layer 3	Layer 3 (Network Layer in OSI and RRC Sublayer in 3GPP)
LB	Load Balancing
LBS	Location Based Services
LCG	Logical Channel Group
LCR	Low Chip Rate
LDM	Low Duty Mode
LDPC	Low-Density Parity Check Code
LFSR	Linear Feedback Shift Register
LLC	Logical Link Control
LOS	Line-of-Sight
LPPa	LTE Positioning Protocol Annex
LR	Location Register
LRU	Logical Resource Unit
LSB	Least Significant Bit
LTE	Long Term Evolution
MAC	Medium Access Control
MAK	Multicast and Broadcast Service Authorization Key
MAN	Metropolitan Area Network
MBMS	Multimedia Broadcast Multicast Service
MBR	Maximum Bit Rate
MBS	Multicast and Broadcast Service
MBSFN	Multimedia and Broadcast Multicast Service over Single Frequency Network
MCCH	Multicast Control Channel
MCE	Multi-Cell/Multicast Coordination Entity

Abbreviation	Description
MCEH	MAC Control Extended Header
MCH	Multicast Channel
MCID	Multicast Connection Identifier
MCS	Modulation and Coding Scheme
MDHO	Macro Diversity Handover
MDS	Multipoint Distribution Service
MEF	MIMO Encoder Format
MEH	Multiplexing Extended Header
MGTEK	Multicast and Broadcast Service Group Traffic Encryption Key
MI	Mutual Information
MIB	Master Information Block
MIB	Management Information Base
MIC	Message Integrity Check
MIH	Media Independent Handover
MIHF	Media Independent Handover Function
MIMO	Multiple Input Multiple Output
MIP	Mobile IP
MISO	Multiple-Input Single-Output
ML	Maximum-Likelihood
MLRU	Minimum Advanced Medium Access Protocol Logical Resource Unit
MMDS	Multi-channel Multipoint Distribution Service
MME	Mobility Management Entity
MMIB	Mean Mutual Information per Bit
MMSE	Minimum-Mean Square Error
MPDU	MAC Protocol Data Unit
MPEG	Moving Pictures Experts Group
MRC	Maximal Ratio Combining
MS	Mobile Station
MSA	MCH Subframe Allocation
MSB	Most Significant Bit
MSDU	MAC Service Data Unit
MSI	MCH Scheduling Information
MSID	Mobile Station Identifier
MSK	Master Session Key
MSP	MCH Scheduling Period
MTCH	Multicast Traffic Channel
MU	Multi-User
MU-MIMO	Multi-User Multiple Input Multiple Output
NACC	Network Assisted Cell Change
NACK	Negative Acknowledgement

(Continued)

Abbreviation	Description
NAI	Network Access Identifier
NAP	Network Access Provider
NAS	Non-Access Stratum
NAS	Network Access Server
NCC	Next Hop Chaining Counter
NCFG	Network Configuration
NCMS	Network Control and Management System
NCMS-N	Network Control and Management System at the BS Side (Network Side)
NCMS-E	Network Control and Management System at the MS Side
NEM	Network Entry Management
NENT	Network Entry
NGN	Next Generation Networks
NH	Next Hop Key
NIP	Normalized Interference Power
NLoS	Non-Line-of-Sight
NLRU	Mini-band Logical Resource Unit
NNI	Network-to-Network Interface (or Network Node Interface)
NNSF	NAS Node Selection Function
NR	Neighbor-cell Relation
NRM	Network Reference Model
NRT	Neighbor Relation Table
nrtPS	Non-Real-Time Polling Service
NS/EP	National Security/Emergency Preparedness
NSP	Network Service Provider
NS-RCH	Non-Synchronized Ranging Channel
OFDM	Orthogonal Frequency Division Multiplexing
OFDMA	Orthogonal Frequency Division Multiple Access
OID	Object Identifier
OSG	Open Subscriber Group
OSI	Open System Interconnection
OTA	Over-the-Air
OTDoA	Observed Time Difference of Arrival (Positioning Method)
OUI	Organizationally Unique Identifier
PA	Persistent Allocation
PA	Power Amplifier
PA	Paging Agent
PAK	Primary Authorization Key
PAN	Personal Area Network
PAPR	Peak-To-Average Power Ratio
PA-Preamble	Primary Advanced Preamble

Abbreviation	Description
PAR	Project Authorization Request
PARC	Per-Antenna Rate Control
PBCH	Physical Broadcast Channel
PBR	Prioritized Bit Rate
PBR	Piggyback Request
PC	Paging Controller
PCCH	Paging Control Channel
PCEP	Policy and Charging Enforcement Point
PCFICH	Physical Control Format Indicator Channel
PCH	Paging Channel
PCI	Physical Cell Identifier
PCID	Paging Controller Identifier
PCRF	Policy and Charging Rule Function
PDCCH	Physical Downlink Control Channel
PDCP	Packet Data Convergence Protocol
PDSCH	Physical Downlink Shared Channel
PDU	Protocol Data Unit
PER	Packet Error Rate
PER	Packet Encoding Rules
PFBCH	Primary Fast Feedback Channel
PG	Paging Group
PGID	Paging Group Identifier
P-GW	PDN Gateway
PHICH	Physical Hybrid-ARQ Indicator Channel
PHS	Payload Header Suppression
PHSF	Payload Header Suppression Field
PHSI	Payload Header Suppression Index
PHSM	Payload Header Suppression Mask
PHSS	Payload Header Suppression Size
PHSV	Payload Header Suppression Valid
PHY	Physical Layer
PKM	Privacy Key Management
PLMN	Public Land Mobile Network
PMCH	Physical Multicast Channel
PMD	Physical Medium Dependent
PMI	Precoding Matrix Index
PMIP	Proxy Mobile IP
PMK	Pair-wise Master Key
PMP	Point-to-Multipoint
PN	Packet Number

(Continued)

Abbreviation	Description
POP3	Post Office Protocol version 3
PPP	Point-to-Point Protocol
PPRU	Permuted Physical Resource Unit
PRACH	Physical Random Access Channel
PRB	Physical Resource Block
PRBS	Pseudo-Random Binary Sequence
P-RNTI	Paging RNTI
PRU	Physical Resource Unit
PS	Physical Slot
PSAP	Public Safety Answering Point
PSC	Packet Scheduling
PSC	Power Saving Class
P-SFH	Primary Superframe Header
PSH	Packing Sub-header
PSI	Pilot Stream Index
PTI	Payload Type Indicator
PUCCH	Physical Uplink Control Channel
PUSC	Partial Usage of Subchannels
PUSC-ASCA	Partial Usage of Subchannels – Adjacent Sub-Carrier Allocation
PUSCH	Physical Uplink Shared Channel
PVC	Permanent Virtual Circuit
PWS	Public Warning System
QAM	Quadrature Amplitude Modulation
QCI	QoS Class Identifier
QoS	Quality of Service
QPSK	Quadrature Phase-Shift Keying
RAC	Radio Admission Control
RACH	Random Access Channel
RADIUS	Remote Authentication Dial-in User Service
RA-ID	Random Access Identifier
RA-RNTI	Random Access RNTI
RAT	Radio Access Technology
RB	Radio Bearer
RBC	Radio Bearer Control
RBIR	Received Bit Mutual Information Rate
RCH	Ranging Channel
RCP	Ranging Cyclic Prefix
RD	Relative Delay
REG	Resource Element Group
RF	Radio Frequency

Abbreviation	Description
RFMT	Reordered Feedback Mini-Tile
RHMT	Reordered HARQ Mini-Tile
RI	Rank Indication
RIM	RAN Information Management
RIT	Radio Interface Technology
RIV	Resource Indication Value
RLAN	Radio Local Access Network
RLC	Radio Link Control
RNC	Radio Network Controller
RNG	Ranging
RNL	Radio Network Layer
RNTI	Radio Network Temporary Identifier
ROHC	Robust Header Compression
RP	Ranging Preamble
RP	Reference Point
RRA	Radio Resource Agent
RRC	Radio Resource Control
RRCM	Radio Resource Control and Management
RRM	Radio Resource Management
RS	Reed–Solomon Code
RS	Relay Station
RSS	Receive Signal Strength
RSSI	Receive Signal Strength Indicator
RTD	Round Trip Delay
RTG	Receive/Transmit Transition Gap
RTP	Real-time Transport Protocol
rtPS	Real-Time Polling Service
RU	Resource Unit
RUIM	Removable User Identify Module
RX	Receiver
RxDS	Receiver Delay Spread Clearing Interval
S1-MME	S1 for the Control-plane
S1-U	S1 for the User-plane
SA	Security Association
S-ABS	Serving ABS
SAC	Sub-band Allocation Count
SAE	System Architecture Evolution
SAID	Security Association Identifier
SAP	Service Access Point
SA-Preamble	Secondary Advanced Preamble

(Continued)

Abbreviation	Description
SC-FDMA	Single Carrier – Frequency Division Multiple Access
SCH	Synchronization Channel
SCID	Sleep Cycle Identifier
SCM	Spatial Channel Model
SDF	Service Data Flow
SDMA	Spatial Division Multiple Access
SDU	Service Data Unit
SeGW	Security Gateway
SF	Service Flow
SFBC	Space-Frequency Block Code
SFBCH	Secondary Fast-Feedback Channel
SFH	Superframe Header
SFID	Service Flow Identifier
SFM	Service Flow Management
SFN	System Frame Number
SFN	Single Frequency Network
SGSN	Serving GPRS Support Node
S-GW	Serving Gateway
SHA	Secure Hash Algorithm
SI	System Information
SIB	System Information Block
SIC	Successive Interference Cancellation
SIM	Subscriber Identity Module
SIMO	Single-Input Multiple-Output
SINR	Signal-to-Interference plus Noise Ratio
SIQ	Service Information Query
SI-RNTI	System Information RNTI
SISO	Single-Input Single-Output
SLA	Service Level Agreement
SLRU	Sub-band Logical Resource Unit
SM	Spatial Multiplexing
SMMSE	Successive MMSE
SMR	Specialized Mobile Radio
SN	Sequence Number
SNMP	Simple Network Management Protocol
SNR	Signal-to-Noise Ratio
SOHO	Small Office Home Office
SON	Self-Organizing Networks
SORTD	Spatial Orthogonal-Resource Transmit Diversity
SPID	Subscriber Profile Identifier for RAT/Frequency Priority

Abbreviation	Description
SPID	Sub-Packet Identifier
SPMH	Short-Packet MAC Header
SR	Scheduling Request
SRB	Signaling Radio Bearer
S-RCH	Synchronized Ranging Channel
SRIT	Set of Radio Interface Technologies
SRS	Sounding Reference Signal
SS	Subscriber Station
S-SFH	Secondary Superframe Header
SSID	Subscriber Station Identification (MAC Address)
SSM	Subscriber Station Management
SSTG	Subscriber Station Transition Gap
STC	Space-Time Coding
STID	Station Identifier
STTD	Space-Time Transmit Diversity
SU	Single-User
SU	Scheduling Unit
SU-MIMO	Single-User Multiple Input Multiple Output
SVC	Switched Virtual Circuit
SVD	Singular Value Decomposition
TA	Tracking Area
T-ABS	Target ABS
TAC	Type Approval Code
TB	Transport Block
TBCC	Tail-Biting Convolutional Code
TCM	Trellis Coded Modulation
TCP	Transport Control Protocol
TCS	Transmission Convergence Sublayer
TDD	Time Division Duplex
TDM	Time Division Multiplexing
TDMA	Time Division Multiple Access
TDoA	Time Difference of Arrival
TEK	Traffic Encryption Key
TFT	Traffic Flow Template
TFTP	Trivial File Transfer Protocol
THP	Tomlinson–Harashima Precoding
TLV	Type/Length/Value
TM	Transparent Mode
TNL	Transport Network Layer
ToA	Time of Arrival

(*Continued*)

Abbreviation	Description
TSTID	Temporary Station Identifier
TTG	Transmit/Receive Transition Gap
TTI	Transmission Time Interval
TUSC	Tile Usage of Subchannels
TX	Transmitter
UCAS	Uplink Contiguous Resource Unit Allocation Size
UCD	Uplink Channel Descriptor
UDP	User Datagram Protocol
UE	User Equipment
UE-AMBR	UE Aggregate Maximum Bit Rate
UEP	Unequal Error Protection
UFPC	Uplink Frequency Partition Configuration
UGS	Unsolicited Grant Service
UIUC	Uplink Interval Usage Code
UL	Uplink
UM	Unacknowledged Mode
UMD	Unacknowledged Mode Data
UMTS	Universal Mobile Telecommunication System
UNI	User-To-Network Interface (or User-Network Interface)
U-NII	Unlicensed National Information Infrastructure
UPE	User-Plane Entity
U-plane	User-Plane
UpPTS	Uplink Pilot Time Slot
USAC	Uplink Sub-Band Allocation Count
USIM	Universal SIM
UTC	Universal Coordinated Time
U-TDoA	Uplink Time Difference of Arrival
UTRA	Universal Terrestrial Radio Access
UTRAN	Universal Terrestrial Radio Access Network
UWB	Ultra Wideband
VC	Virtual Channel
VCI	Virtual Channel Identifier
V-CSN	Visited CSN
VE	Vertical Encoding
VLAN	Virtual Local Area Network
V-NSP	Visited NSP
VoIP	Voice over Internet Protocol
VP	Virtual Path
VPI	Virtual Path Identifier
VPN	Virtual Private Network

Abbreviation	Description
VRB	Virtual Resource Block
WAN	Wide Area Network
WCDMA	Wideband CDMA
WiMAX	Worldwide Interoperability for Microwave Access
WLAN	Wireless Local Area Network
WRC	World Radio Conference
X2-C	X2-Control Plane
X2-U	X2-User Plane
XOR	Exclusive-OR (Logical Operation)
ZF	Zero-Forcing

Introduction to Mobile Broadband Wireless Access

1

INTRODUCTION

The last two decades have witnessed a rapid growth in the number of subscribers and incredible advancement in technology of cellular communication from simple, all-circuit-switched, analog first generation systems with limited voice service capabilities, limited mobility, and small capacity to the third generation systems with significantly increased capacity, advanced all-digital packet-switched all-IP implementations that offer a variety of multimedia services. With the increasing demand for high-quality wireless multimedia services, the radio access technologies continue to advance with faster pace toward the next generation of systems. The general characteristics envisioned for the fourth generation of the cellular systems include all-IP core networks, support for a wide range of user mobility, significantly improved user throughput and system capacity, reliability and robustness, seamless connectivity, reduced access latencies, etc.

In this chapter we discuss the current status of broadband wireless access technologies and the efforts that are made by prominent standardization organizations to materialize the vision and to fulfill the objectives for the next generation of broadband radio access systems. Presently, the most important activities in this area are conducted by the Institute of Electrical and Electronics Engineers and 3rd Generation Partnership Project. These two organizations have historically contributed to the development and advancement of fixed and mobile broadband systems such as the IEEE 802.16, IEEE 802.11, IEEE 802.3, and the UMTS family of standards. Both organizations have already taken significant steps toward the next generation of fixed and mobile broadband wireless access technologies also known as IMT-Advanced systems.

There is a great amount of commonality and similarity between the latest generations of wireless access system standards that started with similar system requirements and has further continued with similar functional blocks, protocols, and baseband processing, resulting in the notion of ultimate convergence in the 4th or later generations of broadband wireless access technologies. An attempt will be made to provide the background information and justification for this viewpoint throughout this chapter, while adhering to a systematic and structured approach.

1.1 MOBILE BROADBAND WIRELESS ACCESS TECHNOLOGIES

Wireless broadband technologies provide ubiquitous broadband access to mobile users, enabling consumers with a broad range of mobility and a variety of wireless multimedia services and applications. Broadband wireless access technologies provide broadband data access through wireless media to consumer and business markets. The most common example of broadband wireless access is

Mobile WiMAX. DOI: 10.1016/B978-0-12-374964-2.10001-3

1

wireless local area network. There have been continued efforts to deliver ubiquitous broadband wireless access by developing and deploying advanced radio access technologies such as 3GPP UMTS and LTE, as well as mobile WiMAX systems. The broadband wireless access is also an attractive option to network operators in geographically remote areas with no or limited wired network. The advantages in terms of savings in speed of deployment and installation costs are further motivation for broadband wireless access technologies.

There are various types of broadband wireless access technologies that are classified based on the coverage area and user mobility as follows:

1. Personal Area Network (PAN) is a wireless data network used for communication among data devices/peripherals around a user. The wireless PAN coverage area is typically limited to a few meters with no mobility. Examples of PAN technologies include Bluetooth or IEEE 802.15.1 [1] and Ultra Wideband (UWB) technology [2].

2. Local Area Network (LAN) is a wireless or wireline data network used for communication among data/voice devices covering small areas such as home or office environments with no or limited mobility. Examples include Ethernet (fixed wired LAN) [3] and Wi-Fi or IEEE 802.11 [4] (wireless LAN for fixed and nomadic users).

3. Metropolitan Area Network (MAN) is a data network that connects a number of LANs or a group of stationary/mobile users distributed in a relatively large geographical area. Wireless infrastructure or optical fiber connections are typically used to link the dispersed LANs. Examples include the IEEE 802.16-2004 (fixed WiMAX) [5] and Ethernet-based MAN [3].

4. Wide Area Network (WAN) is a data network that connects geographically dispersed users via a set of inter-connected switching nodes, hosts, LANs, etc., and covers a wide geographical area. Examples of WAN include the Internet [3] and cellular networks such as 3GPP UMTS [6], 3GPP LTE [7], and mobile WiMAX or IEEE 802.16-2009 [8].

The user demand for broadband wireless services and applications are continually growing. In particular, users expect a dynamic, continuing stream of new applications, capabilities, and services that are ubiquitous and available across a range of devices using a single subscription and a single or unique identity. Offering customized and ubiquitous services based on diverse individual needs through versatile communication systems will require certain considerations in the technology design and deployment.

A number of important factors are accelerating the adoption of wireless data services. These include increased user demand for wireless multimedia services, advances in smart-phone technologies, and global coverage of broadband wired and wireless access. In the meantime, application and content providers are either optimizing their offerings or developing new applications to address the needs and expectations of fixed and mobile users.

Wireless multimedia applications are growing far more rapidly than voice, and are increasingly dominating network traffic. There has been a gradual change from predominantly circuit-switched to packet-based and all-IP networks since the beginning of this millennium [9]. This change will provide the user with the ability to more efficiently utilize multimedia services including e-mail, file transfers, IP TV, VoIP, interactive gaming, messaging, and distribution services. These services are either symmetrical or asymmetrical and real-time or non real-time. They require wider frequency bandwidths, lower transmission and processing latencies, and higher data throughputs.

It is envisioned that within the next decade a large number of the world population would have access to advanced mobile communication devices. The statistics suggest that the number of broadband wireless service subscribers can exceed two billion in the next few years [9]. There are already more portable handsets than either fixed line telephones or wired line equipment such as desktops that can access the Internet, and the number of mobile devices is expected to continue to grow more rapidly than fixed line devices. Mobile terminals will be the most commonly used devices for accessing and exchanging information as well as e-commerce [10]. This trend is viewed as the integration and convergence of information technology, telecommunications, and content. This trend has resulted in new service delivery dynamics and a paradigm shift in telecommunications that will benefit both end users and service providers [10].

The following general requirements are applied to telecommunication services and applications, noting that the requirements may be different from one service offering to another:

- Seamless and continuous connectivity, as well as seamless handover across heterogeneous networks to support a wide range of user mobility from stationary to high speed. This includes mobility management and inter-system interoperability when users are in multi-mode service [11].
- Low power consumption in multi-mode devices through complexity and size reduction.
- Application scalability and quality of service to maintain services despite changes of radio channel condition by adapting the data rate and/or the error tolerance of the application.
- Security and data integrity for multimedia and e-commerce applications. In the latter, authentication of user information integrity and protection of user information are required to support high security services and prevent security breaches.
- Prioritization for applications with urgency such as emergency/disaster. Such applications require higher priority than other applications and support of prioritization of access to network resources.
- Location determination capability and accuracy to enable certain location-dependent applications. An important aspect of this capability is the ability to protect the privacy information of the user.
- Broadcast and multicast and efficient support for point-to-multipoint transmission is required because broadcast and multicast services are expected to be an important part of an operator's service offering in the future.
- Presence to allow a set of users to be informed about the availability, willingness, and means of communication of the other users in a group.
- Usability and interactiveness of applications to allow easy and convenient use of services. The usability may include voice recognition and user-friendliness of human-to-machine interfaces. Good user experience plays a crucial role in the acceptance and proliferation of services.

In defining the framework for development of IMT-Advanced, and systems beyond IMT-Advanced, it is important to understand the user demands and technology trends that will affect the development of such systems. In particular, the framework should be based on increasing user expectations and the growing demand for mobile services, as well as the evolving nature of the services and applications that may become available.

Figure 1-1 shows four service classes (conversational, interactive, streaming, and background services) and their characteristics in terms of reliability, bit rate, and latency [12]. We will further discuss these requirements and characteristics in the next sections. In this figure, BER denotes bit error rate which is a measure of reliability of communication link, and is the ratio of the number of incorrectly-received information bits to the total number of information bits sent within a certain time interval.

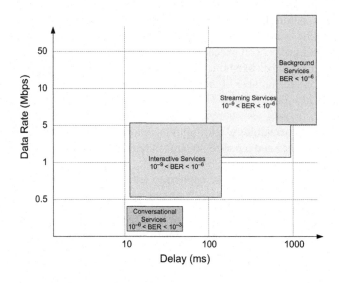

FIGURE 1-1

Service classes and their characteristics

Prominent standards developing organizations such as the 3rd Generation Partnership Project (3GPP),[i] the 3rd Generation Partnership Project 2 (3GPP2),[ii] and the Institute of Electrical and Electronics Engineers Standards Association (IEEE-SA)[iii] have actively contributed to the design, development, and proliferation of broadband wireless systems in the past decade. A number of broadband wireless access standards for fixed, nomadic, and mobile systems have been developed by these standardization groups and deployed by a large number of operators across the globe [9,13,14].

1.1.1 The 4th Generation of Mobile Broadband Wireless Access Technologies

International Mobile Telecommunications-Advanced (IMT-Advanced) or alternatively 4th Generation (4G) cellular systems are mobile systems that extend and improve upon the capabilities of the IMT-2000 family of standards. Such systems are expected to provide users with access to a variety of advanced IP-based services and applications, supported by mobile and fixed broadband networks, which are predominantly packet-based. The IMT-Advanced systems can support a wide range of data rates, with different quality of service requirements, proportional to user mobility conditions in multi-user environments. The key features of IMT-Advanced systems can be summarized as follows [11,15]:

- Enhanced cell and peak spectral efficiencies, and cell-edge user throughput to support advanced services and applications;
- Lower airlink access and signaling latencies to support delay sensitive applications;

---

[i]http://www.3gpp.org
[ii]http://www.3gpp2.org
[iii]http://standards.ieee.org

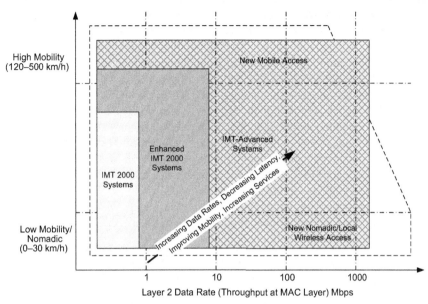

FIGURE 1-2

Illustration of the capabilities and evolution of IMT-2000 systems [10]

- Support of higher user mobility while maintaining session connectivity;
- Efficient utilization of spectrum;
- Inter-technology interoperability, allowing worldwide roaming capability;
- Enhanced air–interface–agnostic applications and services;
- Lower system complexity and implementation cost;
- Convergence of fixed and mobile networks;
- Capability of interworking with other radio access systems.

These features enable IMT-Advanced systems to accommodate emerging applications and services. The capabilities of IMT systems have been continuously enhanced proportional to user demand and technology advancements in the past decade. However, the framework and overall objectives of the systems beyond IMT-2000 are considered to be a paradigm shift in the design and development of radio air interfaces [10]. Present mobile communication systems have evolved by incremental addition of more capabilities and enhancement of features to the baseline systems. Examples include evolution of the UMTS family of standards in 3GPP. The systems beyond IMT-2000 have been realized by combining the existing components of IMT-2000 systems, with enhanced and newly developed functions, nomadic wireless access systems, and other wireless systems with high commonality and seamless interworking.

The capabilities and evolution of IMT-2000 systems is illustrated in Figure 1-2. As it appears from the figure, the services and performance of the systems noticeably increase as the systems evolve from one generation to another.

The International Telecommunication Union (ITU)[iv] framework for the future development of IMT-2000 and systems beyond IMT-2000 encompasses both the radio access, i.e., IMT-Advanced systems initiative, and the core network, i.e., Next Generation Networks (NGN)[v] project. However, it is recognized that, in the future, the evolution of technologies and redistribution of traditional functions between radio access networks and core networks in practical systems may blur this distinction.

Table 1-1 shows the service requirements for IMT-Advanced systems [11]. There are four user experience classes where each is further divided into a number of service classes based on the intrinsic characteristics of corresponding services, such as required throughput and latency to ensure the QoS requirements for each class are met. The service class requirements can be translated into system requirements, which are directly mapped to data transport over a wireless network. As a result, a limited number of QoS attributes, such as data throughput, packet delay and/or delay variations (often referred to as jitter), bit/packet error rate are defined.

Examples of applications corresponding to each service class are provided in Table 1-1. In this table, the interactive gaming services mainly involve data transferred between multiple users that are connected to a server, or directly between the equipment of multiple users. Real-time communication with low delay and low jitter is required for interactive gaming. Multimedia refers to media that uses multiple forms of information content and information processing (e.g., text, audio, graphics, animation, video, and interactivity) to inform or entertain users. Wireless multimedia is an essential element of various application services described in this section which must be supported by IMT-Advanced systems. Furthermore, location-based services, which depend on the present location of a user, enable users to find other people, vehicles, resources, services, or machines. Video conferencing is a full-duplex, real-time audiovisual communication between or among end users. Remote collaboration is sharing of files and documents in real-time among users that are members of a project. It mainly involves data transferred between multiple users that are connected to a server or directly between the user terminals. This includes facilities for a virtual office that is a personal online office, where the data and files can be shared in real-time. Mobile commerce service is the buying and selling of goods and services through wireless terminals. It mainly involves data transferred between user equipment and financial servers connected with secured databases. This service also enables the real-time sharing and management of information on products, inventory, availability, etc. This service requires a high level of reliability. Mobile broadcasting is a point-to-multipoint transmission of multimedia content over one or multiple radio access networks. This further includes interactive content or IP-TV, which requires the ability to interact with an audio/video program by exchanging multimedia information. The IP television is a system where a digital television service is delivered using a broadband IP network infrastructure.

1.1.2 Requirements of 4G Mobile Broadband Wireless Access Systems

The service and application requirements have been translated to design requirements for the next generation of mobile broadband wireless access systems. The design requirements encompass a wide range of system attributes, such as data and signaling transmission latency over the airlink, system data

---

[iv]http://www.itu.int/ITU-R/

[v]A Next Generation Network (NGN) is a packet-switched access network capable of providing telecommunication services through use of multiple broadband, QoS-enabled transport technologies where service-related functions are independent of underlying transport-related technologies. It offers unrestricted access by users to different service providers. It supports generalized mobility which will allow consistent and ubiquitous provision of services to users [16].

Table 1-1 IMT-Advanced Service Classification Requirements [11]

User Experience Class	Description	Service Class	Service Parameters (Numerical Values)		Example Services [12]
Conversational	The basic conversational service class comprises basic services that are dominated by voice communication characteristics. The rich conversational service class consists of services that mainly provide synchronous communication enhanced by additional media such as video, collaborative document viewing, etc. Conversational low delay class comprises real-time services that have very strict delay and delay jitter requirements.	Basic conversational service	Throughput	<20 kbps	Voice telephony (including VoIP) Emergency calling Push-to-talk
			Delay	50 ms	
		Rich conversational service	Throughput	5 Mbps	Video conference High-quality video telephony Remote collaboration e-Education (e.g., video call to teacher) Consultation (e.g., video interaction with doctor) Mobile commerce
			Delay	20 ms	
		Conversational low delay	Throughput	150 kbps	Interactive gaming Consultation Priority service
			Delay	10 ms	
Streaming	The differentiating factor of these service classes is the live or non-live nature of the content transmitted. In case of live content, buffering possibilities are very limited, which makes the service very delay-sensitive. In the case of non-live (i.e., pre-recorded) content, play-out buffers at the receiver side provide a high robustness against delay and jitter.	Streaming live	Throughput	2–50 Mbps	Emergency calling Public alerting e-Education (e.g., remote lecture) Consultation (e.g., remote monitoring) Machine-to-machine (e.g., surveillance) Mobile broadcasting/multicasting, Multimedia
			Delay	100 ms	
		Streaming non-live	Throughput	2–50 Mbps	Mobile broadcasting/multicasting, e-Education (e.g., education movies), Multimedia, Mobile commerce, Remote collaboration
			Delay	1 s	

(Continued)

Table 1-1 IMT-Advanced Service Classification Requirements [11] *Continued*

User Experience Class	Description	Service Class	Service Parameters (Numerical Values)		Example Services [12]
Interactive	In the interactive user experience class, two service classes are distinguished. Interactive services that permit relatively high delay which usually follow a request-response pattern (e.g., web browsing, database query, etc.). In such cases, response times in the order of a few seconds are permitted. Interactive services requiring significantly lower delays are remote server access (e.g., IMAP)[i] or remote collaboration.	Interactive high delay	Throughput Delay	500 kbps 200 ms	e-Education (e.g., data search) Consultation (e.g., data search) Internet browsing Mobile commerce Location-based services ITS-enabled services
		Interactive low delay	Throughput Delay	500 kbps 20 ms	Emergency calling e-mail (IMAP server access) Remote collaboration (e.g., desktop sharing) Public alerts (e.g., with feedback) Messaging (instant messaging) Mobile broadcasting/multicasting (mobile interactive personalized TV) Interactive gaming
Background	The background class only contains delay-insensitive services, so that there is no need for further differentiation.	Background	Throughput Delay	5–50 Mbps < 2 s	Messaging Video messaging Public alerts E-mail (transfer RX/TX, e.g., POP[ii] server access) Machine-to-machine File transfer/download e-Education (file download/upload) Consultation (file download/upload) Internet browsing, Location-based service

[i]The Internet Message Access Protocol (IMAP) is one of the prevalent IP protocols for e-mail retrieval. Many e-mail clients and servers support this protocol as a means of transferring e-mail messages from a server to a client. Once configured, the client's use of this protocol remains transparent to the user [3,17].
[ii]The Post Office Protocol version 3 (POP3) is an application-layer IP protocol used by local e-mail clients to retrieve e-mail from a remote server over a TCP/IP connection [3,18].

rates and user cell-edge, and average throughputs, etc. The requirements presented in Table 1-2, defined by ITU-R Working Party 5D, are for the purpose of consistent definition, specification, and evaluation of the candidate IMT-Advanced proposals in conjunction with the development of recommendations and reports such as the framework and key characteristics, and the detailed specifications of IMT-Advanced.

The intention of these requirements is to ensure that IMT-Advanced technologies are able to fulfill the objectives of the 4th generation of cellular systems and to set a specific level of minimum performance that each proposed technology needs to achieve in order to be considered by ITU-R for IMT-Advanced. These requirements are not intended to restrict the full range of capabilities or performance that candidate technologies for IMT-Advanced might achieve, nor is it intended to describe how the IMT-Advanced technologies might perform in actual deployments under operating conditions that could be different from those presented in ITU-R Recommendations and Reports on IMT-Advanced. Satisfaction of these requirements are verified through link- and system-level simulations (for VoIP capacity, spectral efficiencies, and mobility), analytical (for user and control-plane latencies, as well as handover interruption time), and inspection (for service requirements and bandwidth scalability) [19]. The methodology, guidelines, and common configuration parameters are specified in reference [19].

The requirements for system and user data rates are described in the form of frequency-normalized spectral efficiency, to make them independent of bandwidth and practical spectrum allocations. Some of the requirements noted in Table 1-2 are evaluated via system-level simulations, and the mobility requirement is evaluated through link-level simulation. Note that in Table 1-2, a cell or equivalently a sector is defined as a physical partition of the coverage area of a base station. The base station coverage area is typically partitioned into three or more sectors/cells to mitigate interference effects.

In the above table, the cell spectral efficiency is defined as the aggregate throughput; i.e., the number of correctly received bits delivered at the data link layer over a certain period of time, of all users divided by the product of the effective bandwidth, the frequency reuse factor, and the number of cells. The cell spectral efficiency is measured in bit/sec/Hz/cell.

The peak spectral efficiency is the highest theoretical data rate normalized by bandwidth (assuming error-free conditions) assignable to a single mobile station when all available radio resources for the corresponding link are utilized, excluding radio resources that are used for physical layer synchronization, reference signals, guard bands, and guard times (collectively known as Layer 1 overhead). Bandwidth scalability is the ability to operate with different bandwidth allocations using single or multiple RF carriers.

The normalized user throughput is defined as the number of correctly received bits by a user at the data link layer over a certain period of time, divided by the total spectrum. The cell edge user spectral efficiency is defined as 5% point of Cumulative Distribution Function (CDF) [22] of the normalized user throughput. Control-plane latency is defined as the transition time from idle-state to connected-state. The transition time (assuming downlink paging latency and core network signaling delay are excluded) of less than 100 ms is required for IMT-Advanced systems. The user-plane latency (also known as transport delay) is defined as the one-way transit time between a packet being available at the IP layer of the origin (user terminal in the uplink or base station in the downlink) and the availability of this packet at IP layer of the destination (base station in the uplink or user terminal in the downlink). User-Plane packet delay includes delay introduced by associated protocols and signaling, assuming the user terminal is in the active mode. The IMT-Advanced systems are required to achieve a transport

Table 1-2 Comparison of the Requirements of Prominent Mobile Broadband Wireless Systems

Requirements	IMT-Advanced [15]	IEEE 802.16m [21]	3GPP LTE-Advanced [20]
Peak spectrum efficiency (bit/sec/Hz) (system-level)	DL: 15 (4 × 4) UL: 6.75 (2 × 4)	DL: 8.0/15.0 (2 × 2/4 × 4) UL: 2.8/6.75 (1 × 2/2 × 4)	DL: 30 (8 × 8) UL: 15 (4 × 4)
Cell spectral efficiency (bit/sec/Hz/sector) (system-level)	DL: (4 × 2) = 2.2 UL: (2 × 4) = 1.4 (Base coverage urban)	DL: (2 × 2) = 2.6 UL: (1 × 2) = 1.3 (Mixed mobility)	DL: (4 × 2) = 2.6 UL: (2 × 4) = 2.0
Cell-edge user spectral efficiency (bit/sec/Hz) (system-level)	DL: (4 × 2) = 0.06 UL: (2 × 4) = 0.03 (Base coverage urban)	DL: (2 × 2) = 0.09 UL: (1 × 2) = 0.05 (Mixed mobility)	DL: (4 × 2) = 0.09 UL: (2 × 4) = 0.07 (Base coverage urban)
Latency	C-plane: 100 msec (idle to active) U-plane: 10 msec	C-plane: 100 msec (idle to active) U-plane: 10 msec	C-plane: 50 msec (idle/camped state to connected) 10 msec (dormant state to active state) U-plane: 10 msec
Mobility bit/sec/Hz at km/h (link-level)	55 at 120 km/h 25 at 350 km/h	Optimal performance up to 10 km/h Graceful: degradation up to 120 km/h Connectivity up to 350 km/h Up to 500 km/h depending on operating frequency	Optimal performance up to 10 km/h Up to 500 km/h depending on operating frequency
Handover interruption time (msec)	Intra frequency: 27.5 Inter frequency: 40 (in a band) 60 (between bands)	Intra frequency: 27.5 Inter frequency: 40 (in a band) 60 (between bands)	Not specified
VoIP capacity (Active users/sector/MHz) (system-level)	40 (4 × 2 and 2 × 4) (Base coverage urban)	60 (DL: 2 × 2 and UL: 1 × 2)	VoIP capacity should be improved relative to that of E-UTRA for all antenna configurations

Antenna configuration	Not specified	DL: 2 × 2 (baseline), 2 × 4, 4 × 2, 4 × 4, 8 × 8 UL: 1 × 2 (baseline), 1 × 4, 2 × 4, 4 × 4	DL: 2 × 2 (baseline), 2 × 4, 4 × 2, 4 × 4, 8 × 8 UL: 1 × 2 (baseline), 1 × 4, 2 × 4, 4 × 4
Cell range and coverage	Not specified	Up to 100 km optimal performance up to 5 km	Requirements for coverage in Release 8 E-UTRA are applicable for Advanced E-UTRA
Multicast and broadcast service (MBS) (system-level)	Not specified	4 bit/sec/Hz for ISD 0.5 km 2 bit/sec/Hz for ISD 1.5 km	Not specified
MBS channel reselection interruption time	Not specified	1.0 sec (intra-frequency) 1.5 sec (inter-frequency)	Not specified
Location based services (LBS)	Not specified	Location determination latency <30 sec MS-based position determination accuracy <50 m Network-based position determination accuracy <100 m	Not specified
Operating bandwidth	Up to 40 MHz (with band aggregation)	5–20 MHz (up to 100 MHz through band aggregation)	Up to 100 MHz through band aggregation
Duplex scheme	Not specified	TDD, FDD (support for H-FDD terminals)	TDD, FDD (support for H-FDD terminals)
Operating frequencies (MHz)	IMT bands 450–470 698–960 1710–2025 2110–2200 2300–2400 2500–2690 3400–3600	IMT bands 450–470 698–960 1710–2025 2110–2200 2300–2400 2500–2690 3400–3600	450–470 698–862 790–862 2300–2400 3400–4200 4400–4990

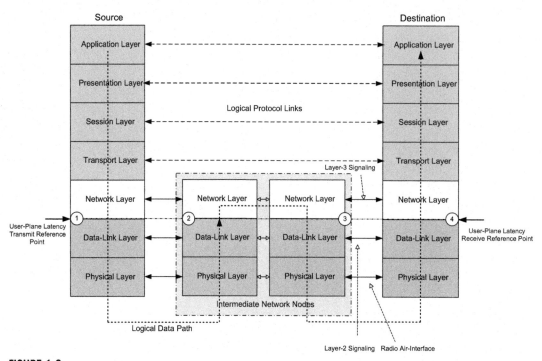

FIGURE 1-3

Graphical interpretation of user-plane latency and protocol layer interconnections

delay of less than 10 ms in unloaded conditions (i.e., single user with single data stream) for small IP packets (e.g., 0 byte payload + IP header) for both downlink and uplink, as hypothetically shown in Figure 1-3. In Figure 1-3, the user-plane latency is measured from point 1 to point 2, or from point 3 to point 4.

The following mobility classes are defined in IMT-Advanced systems:

- Stationary: 0 km/h;
- Pedestrian: 0 to 10 km/h;
- Vehicular: 10 to 120 km/h;
- High speed vehicular: 120 to 350 km/h.

A mobility class is supported if the traffic channel link-level data rate, normalized by bandwidth, on the uplink, is as shown in Table 1-2, when the user is moving at the maximum speed in that mobility class in each of the test environments.

The handover interruption time is defined as the time duration during which a user terminal cannot exchange user-plane packets with any base station. The handover interruption time includes the time required to execute any radio access network procedure, radio resource signaling protocol, or other message exchanges between the user terminal and the radio access network. For the purposes of determining handover interruption time, interactions with the core network are not considered. It is also assumed that all necessary attributes of the target base station are known at initiation of the handover from the serving base station.

Voice-over-IP (VoIP) capacity is calculated assuming a 12.2 kbps codec with 50% speech activity factor, such that the percentage of users in outage is less than 2% where a user is defined to have experienced a voice outage if less than 98% of the VoIP packets have been delivered successfully to the user within a one way radio access delay bound of 50 msec. The VoIP capacity is the minimum of the calculated capacity for downlink or uplink divided by the effective bandwidth in the respective link direction.

For comparison purposes, Table 1-2 further provides the system requirements for the 3GPP LTE-Advanced [20] and the IEEE 802.16m [21] which both have currently targeted certification as IMT-Advanced technologies. There is a great amount of similarity between the requirements, features, and target performance of the two technologies.

As shown in Figure 1-2 and considering the requirements in Table 1-2, the IMT-Advanced systems will provide significant improvements relative to IMT-2000 systems in terms of capacity and performance (user throughputs and access latencies), and variety of unicast and/or multicast based applications and services, as well as more spectrum utilization and deployment flexibility.

1.1.3 Convergence of Mobile Broadband Wireless Access Technologies

As is evident from the previous section, the new mobile broadband wireless access systems are being developed based on similar system and service requirements using similar (if not the same at least conceptually) technologies. This would make one believe that a convergence in mobile broadband wireless technology is imminent in the next decade. While the previous attempts to harmonize the 3rd generation of mobile access technologies have failed due to incompatibility of core components, political and regional issues, competition in the industry, etc., the current trend in the prominent standardization organization combined with more pressure from the network operators due to frag-mentation of the market is compelling and/or persuading the major proponents of the existing tech-nologies to converge.

The convergence of core network technologies has already been initiated and currently mobile WiMAX radio access technology can interwork with 3GPP core network and *vice versa* (see the interworking of trusted networks in 3GPP literature [23]). The inter-technology interworking func-tions, handover between heterogeneous technologies, as well as use of unified core network functions such as user authentication, authorization, and accounting for various radio access technologies, are examples of such a convergence trend. However, the convergence of radio interfaces has proved to be more difficult, due to the use of incompatible protocol structure, multiple access schemes, and baseband processing.

Figure 1-4 illustrates a transport and service layered model that is agnostic of the radio access network while ensuring seamless mobility and service/session continuity. The converged system shown in Figure 1-4 comprises an all-IP common core network and application layer that commu-nicates to user terminals through a set of heterogeneous radio access networks. Some variants of this operation model are currently deployed and being used by some network operators. However, the cost and complexity of multi-radio implementations, the coexistence of multiple radio access technologies collocated on the user terminal, and inter-system interference issues adversely impact the long-term viability of this approach. The operators now desire convergence of radio access networks to ensure

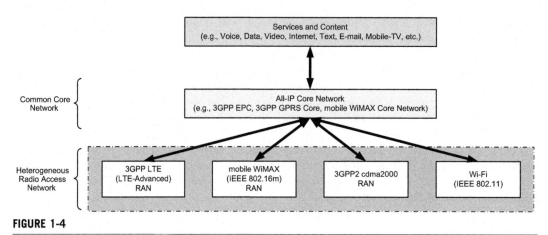

FIGURE 1-4

Example of a RAN-agnostic transport and service layered model

lower cost of deployment and simplicity of future upgrades, as well as global roaming and service continuity across different operators' networks. While the market fragmentation may help lower the royalties and foster innovative competing solutions, it does increase the operating cost and system complexity.

The use of common functional elements and protocols in the new generation of cellular systems is an indication that these systems are moving toward ultimate convergence, resulting in a unified and scalable radio access technology that can be deployed in various deployment scenarios (e.g., indoor as a femto-cell, outdoor as a micro- or macro-cell, and in high-speed rural environments) by appropriately reconfiguring the system parameters to match the radio propagation characteristics of that environment.

1.2 INTRODUCTION TO THE IEEE 802.16 STANDARDS

The development of standards in IEEE was assigned to the IEEE Standards Association. The activities of the IEEE-SA are administered by an elected board of governors. The development and maintenance of the standards are overseen by the IEEE-SA Standards Board, which manages the process, approves new projects, and further approves the balloted drafts as the IEEE standards. The IEEE 802 LAN/MAN Standards Committee (also known as IEEE 802) has been in operation since March 1980. The objective of this group has been to develop LAN and MAN standards and protocols that are mapped to the lower layers (i.e., physical and data link layers) of the seven-layer Open System Interconnection (OSI) reference model[vi] (see Figure 1-3). The IEEE 802 splits the OSI data link layer into two sub-layers, namely Logical Link Control (LLC) and Medium Access Control (MAC) [24]. The most widely used standards are for the Ethernet family [25], Token Ring [26], Wireless LAN [27], Wireless

---

[vi]The Open Systems Interconnection reference model (OSI) is an abstract description for layered communications and computer network protocol design which was developed as part of the Open Systems Interconnection initiative. The OSI model divides generic network architecture into seven layers which, from top to bottom, are defined as *Application*, *Presentation*, *Session*, *Transport*, *Network*, *Data-Link*, and *Physical* layers.

PAN [28], Wireless MAN [29,30], Bridging and Virtual Bridged LANs [31]. There are several active Working Groups under IEEE 802 that focus on different methods of broadband wireless and wired access. Table 1-3 shows the active IEEE 802 Working Groups and their charter. These groups operate under the IEEE 802 Executive Committee.

As illustrated in Figure 1-5, the process for developing standards in IEEE 802 starts with submission of a proposal consisting of the *Project Authorization Request* (PAR), which is the charter for the project, and *Five Criteria*, which is the basis for determining whether to consider a PAR, addressing *Broad Market Potential, Compatibility, Distinct Identity, Technical Feasibility*, and *Economic Viability* matters. The process further includes a *Call for Interest*, which is a brief meeting to outline the proposed topic and determine whether there is interest in investigating a possible project, and a *Study Group* that is a group formed to investigate a project and produce the PAR and Five Criteria. The proposal is submitted to the IEEE 802 EC, and once approved is forwarded to the IEEE-SA Review Committee (RevCom) and the IEEE-SA Standards Board for final approval. On final approval of the proposal a working group or task group is formed to develop the standard specifications.

The task group will develop a draft standard, and once completed approves the draft for Letter Ballot within the Working Group. The draft will be refined through several recirculations until all outstanding comments against the draft by Working Group members are resolved based on consensus. The final output of the Letter Ballot process will be forwarded to Sponsor Ballot conducted by a group of individuals affiliated with industry and academia. The draft standard in the Sponsor Ballot may be further revised and refined through several recirculations. The final draft standard is reviewed and approved by the IEEE 802 EC, the New Standards Committee (NesCom), and the IEEE-SA Standards Board, and is published as an IEEE standard. This process is shown graphically in Figure 1-5.

An IEEE standard has a validity period of five years from the date of approval by the IEEE-SA Standards Board. Before that validity period expires, the sponsor must initiate the reaffirmation process. This process affirms that the technical content of the standard is still valid and the document is reaffirmed for another five year period. During this five year validity period, amendments (i.e., addition of new features and functionalities) and corrigenda (i.e., fixing errors and minor updates) may need to be developed that offer minor revisions to the base standard. If there is a need to significantly update the base standard or any of its published amendments or corrigenda, the sponsor may consider the revision process. It is important to note that at the end of the five year validity period, an IEEE standard has to be revised, reaffirmed, or withdrawn.

The development of the IEEE 802.16 standards, along with associated amendments and revisions, is the responsibility of the IEEE Working Group 802.16. The IEEE 802.16 Working Group is a member of the IEEE 802 group for Wireless Metropolitan Area Networks, in particular Part 16: Air Interface for Broadband Wireless Access Systems. The working group develops standards and recommended practices to support the development and deployment of fixed and mobile broadband wireless access systems. The IEEE 802.16 activities were initiated in August 1998. The Working Group's initial interest was the 10–66 GHz range. The 2–11 GHz amendment project that led to the IEEE 802.16a was approved in March 2000. The 802.16a project primarily involved the development of new physical layer specifications, with supporting enhancements to the basic data link layer. In addition, the Working Group completed the IEEE 802.16.2 standard (Recommended Practice for Coexistence of Fixed Broadband Wireless Access Systems) to address 10–66 GHz coexistence and, through the amendment, project 802.16.2a, expanded its recommendations to include licensed bands from 2 to 11 GHz. The mobility management capabilities were added to the

Table 1-3 IEEE 802 Working Groups

	Working Group	Charter and Scope	Link to Resources
Executive Committee (EC)	802.1 Bridging and Architecture	The IEEE 802.1 Working Group develops standards and recommended practices in the areas of 802 LAN/MAN architecture, internetworking among 802 LANs, MANs and other wide area networks, 802 Security, 802 overall network management, and protocol layers above the MAC and LLC layers.	http://grouper.ieee.org/groups/802/1/
	802.3 Carrier Sense Multiple Access/Collision Detect – Wired Ethernet	The IEEE 802.3 Working Group develops standards for Ethernet based LANs	http://grouper.ieee.org/groups/802/3/
	802.11 Wireless LAN	The IEEE 802.11 Working Group develops standards for Wireless LANs	http://grouper.ieee.org/groups/802/11/
	802.15 Wireless Personal Area Network	The 802.15 Working Group develops consensus on standards for Personal Area Networks or short distance wireless networks and further addresses wireless networking of portable and mobile computing devices such as PCs, Personal Digital Assistants, peripherals, cell phones, pagers, and consumer electronics; allowing these devices to communicate and interoperate with one another. The goal of this group is to publish standards, recommended practices, or guides that have broad market applicability and deal effectively with the issues of coexistence and interoperability with other wired and wireless networking solutions	http://grouper.ieee.org/groups/802/15/
	802.16 Broadband Wireless Access	The IEEE 802.16 Working Group on Broadband Wireless Access Standards develops standards and recommended practices to support the development and deployment of broadband Wireless Metropolitan Area Networks.	http://grouper.ieee.org/groups/802/16/
	802.17 Resilient Packet Ring	The IEEE 802.17 Resilient Packet Ring Working Group develops standards to support the development and deployment of Resilient Packet Ring networks in Local, Metropolitan, and Wide Area Networks for resilient and efficient transfer of data packets at rates scalable to many gigabits per second. These standards build upon existing Physical Layer specifications, and will develop new physical layers where appropriate.	http://grouper.ieee.org/groups/802/17/

802.18 Radio Regulatory-Technical Advisory Group	The mandate of this Working Group monitoring of, and active participation in, ongoing radio regulatory activities, at both the national and international levels, are an important part of IEEE 802 work.	http://grouper.ieee.org/groups/802/18/
802.19 Coexistence Technical Advisory Group	The IEEE 802.19 Coexistence Technical Advisory Group develops and maintains policies defining the responsibilities of 802 standards developers to address issues of coexistence with existing standards and other standards under development. It offers assessments to the Sponsor Executive Committee regarding the degree to which standards developers have conformed to those conventions. The TAG may also develop coexistence documentation of interest to the technical community outside 802.	http://grouper.ieee.org/groups/802/19/
802.20 Mobile Broadband Wireless Access	The 802.20 Working Group specifies physical and medium access control layers of an air-interface for interoperable mobile broadband wireless access systems, operating in licensed bands below 3.5 GHz, optimized for IP-data transport, with peak data rates per user in excess of 1 Mbps. It supports various vehicular mobility classes up to 250 km/h in a MAN environment and targets spectral efficiencies, sustained user data rates and numbers of active users that are all significantly higher than achieved by existing mobile systems.	http://www.ieee802.org/20/
802.21 Media Independent Handover	The 802.21 Working Group develops standards to enable handover and interoperability between heterogeneous network types including both 802 and non-802 networks.	http://www.ieee802.org/21/
802.22 Wireless Regional Area Networks	The 802.22 Working Group develops standard for a cognitive radio-based PHY/MAC/air-interface for use by license-exempt devices on a non-interfering basis in spectrum that is allocated to the TV Broadcast Service.	http://www.ieee802.org/22/

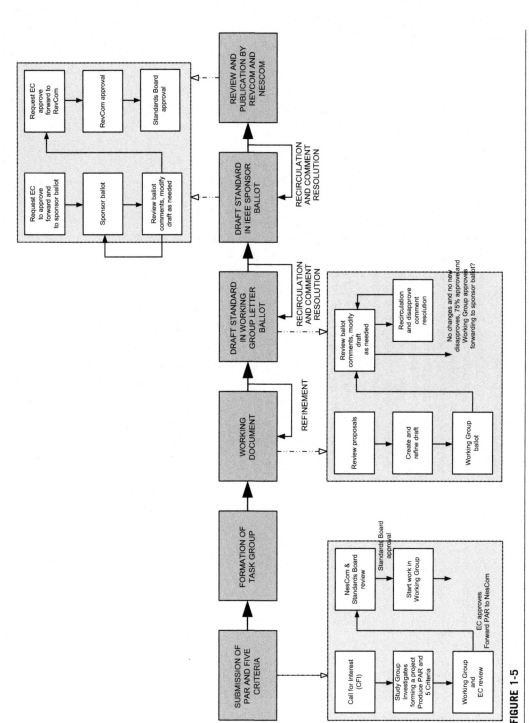

FIGURE 1-5

The IEEE 802 standardization process [32]

IEEE 802.16 standard in December 2005, when the IEEE 802.16e project was accomplished and was refined and improved through a series of corrigenda and amendments (see Table 1-4). Therefore, the IEEE 802.16 standards have evolved from line-of-sight (LOS) single-carrier fixed-wireless technology to Non-LOS (NLOS) multi-carrier mobile broadband wireless technology over the past decade.

1.2.1 Evolution of the IEEE 802.16 Standards

The amendments and revisions of the IEEE 802.16 standard have preserved the essence of their base standards. Various physical layers were combined with the original IEEE 802.16 standard medium access control protocols (i.e., single carrier, OFDM 256, and OFDMA physical layers); however, depending on the capabilities of the physical layer, some changes in the data link layer protocols were made (e.g., handover and power management schemes were added to support mobility in IEEE 802.16e). The principles of IEEE 802.16 data link layer protocols were inherited from DOCSIS[vii] standard. While this philosophy might have worked well for the fixed versions of the IEEE 802.16 standard, it has caused some inefficiency in support of mobility in the later amendments/revisions of the standard. The evolved IEEE 802.16 standards have not necessarily maintained backward compatibility and interoperability with their legacy base standards (e.g., IEEE 802.16e-2005 was not backward compatible with IEEE 802.16-2004).

Table 1-4 provides some useful information on the evolution of the IEEE 802.16 standards (already released or under development by the IEEE 802.16 Working Group as of September 2010) with hyperlinks to additional information sources. The IEEE 802.16-2009 standard is the second revision of the IEEE 802.16 standard (the first revision was released as IEEE Std. 802.16-2004) that encompasses the previous amendments and corrigenda released by this working group. This revision serves as the base standard for IEEE 802.16m, the advanced air interface, which is currently under development in the IEEE 802.16 Working Group whose release is expected in March 2011.

The IEEE 802.16-2009 standard [8] contains some feature enhancements relative to the IEEE 802.16e-2005, including Frequency Division Duplex (FDD) mode enablement and bug fixes, Half-Duplex FDD terminal operation, persistent scheduling, support of 20 MHz bandwidth, improved multi-antenna transmission and processing schemes, and enhancement of multicast and broadcast services, multi-radio coexistence, location-based services, as well as load balancing. The use of complementary grouping of the mobile stations and use of two resource allocation medium access protocols per radio frame resulted in higher VoIP capacity and lower airlink transmission latency.

Relaying and cooperative communication have emerged as important research topics in wireless communication in the past few years to improve performance and coverage of wireless links. In May 2006, the IEEE 802.16 Working Group assigned a task group to incorporate relay capabilities in the IEEE 802.16e-2005 air interface standard. The IEEE 802.16j task group finalized the multi-hop relay specification in 2009. Although this amendment is fully compatible with 802.16e-2005 mobile stations over the access link (i.e., the link between the relay and mobile stations), an IEEE 802.16j compliant

---

[vii]Data over Cable Service Interface Specification (DOCSIS) is an international standard developed by a consortium and defines the interface requirements for a data over cable system. It enables transfer of high-speed data over an existing cable TV system. It is employed by many cable television operators to provide Internet access over their existing hybrid fiber or coaxial infrastructure.

Table 1-4 IEEE 802.16 Standards

Standard/ Project	Description	Release Date	Link to Resources
IEEE P802.16-2004/Cor2	Corrigendum to IEEE Std 802.16-2004, Air Interface for Fixed and Mobile Broadband Wireless Access Systems	May 2007 (Terminated)	http://ieee802.org/16/pubs/80216_Cor2.html
IEEE Project P802.16d	Amendment to IEEE Std 802.16, Air Interface for Fixed Broadband Wireless Access Systems – Amendment 3: Detailed System Profiles for 2-11 GHz	August 2003 (Obsolete)	http://ieee802.org/16/pubs/P80216d.html
IEEE Standard 802.16/ Conformance02-2003	IEEE Standard for Conformance to IEEE 802.16 – Part 2: Test Suite Structure and Test Purposes for 10–66 GHz WirelessMAN-SC	February 2004 (Withdrawn)	http://ieee802.org/16/pubs/80216_Conf02-2003.html
IEEE Standard 802.16/ Conformance01-2003	IEEE Standard for Conformance to IEEE 802.16 – Part 1: Protocol Implementation Conformance Statements for 10–66 GHz WirelessMAN-SC Air Interface	August 2003 (Withdrawn)	http://ieee802.org/16/pubs/80216_Conf01-2003.html
IEEE Standard 802.16-2001	IEEE Standard for Local and metropolitan area networks – Part 16: Air Interface for Fixed Broadband Wireless Access Systems	April 2002 (Superseded)	http://ieee802.org/16/pubs/80216-2001.html
IEEE Standard 802.16a-2003	Amendment to IEEE Std 802.16, Air Interface for Fixed Broadband Wireless Access Systems – Amendment 2: Medium Access Control Modifications and Additional Physical Layer Specifications for 2–11 GHz	April 2003 (Superseded)	http://ieee802.org/16/pubs/80216a-2003.html
IEEE Standard 802.16c-2002	Amendment to IEEE Std 802.16, Air Interface for Fixed Broadband Wireless Access Systems – Amendment 1: Detailed System Profiles for 10–66 GHz	January 2003 (Superseded)	http://ieee802.org/16/pubs/80216c-2002.html
IEEE Standard 802.16.2-2001	IEEE Recommended Practice for Local and Metropolitan Area Networks – Coexistence of Fixed Broadband Wireless Access Systems	September 2001 (Superseded)	http://ieee802.org/16/pubs/802162-2001.html

Standard	Description	Date (Status)	URL
IEEE Std 802.16k	Amendment of IEEE Std 802.1D (as previously amended by IEEE Std 802.17a), Standard for Local and Metropolitan Area Networks: Media Access Control (MAC) Bridges – Bridging of 802.16	August 2007 (Superseded)	http://ieee802.org/16/pubs/80216k.html
IEEE 802.16/Conformance04	Standard for Conformance to IEEE Standard 802.16 – Part 4: Protocol Implementation Conformance Statement (PICS) Proforma for Frequencies below 11 GHz	January 2007 (In force)	http://ieee802.org/16/pubs/80216_Conf04-2006.html
IEEE Standard 802.16/Conformance03-2004	IEEE Standard for Conformance to IEEE 802.16 – Part 3: Radio Conformance Tests (RCT) for 10–66 GHz WirelessMAN-SC Air Interface	June 2004 (In force)	http://ieee802.org/16/pubs/80216_Conf03-2004.html
IEEE Standard 802.16.2-2004	Revision of IEEE Std 802.16.2-2001, IEEE Recommended Practice for Local and Metropolitan Area Networks – Coexistence of Fixed Broadband Wireless Access Systems	March 2004 (Superseded)	http://ieee802.org/16/pubs/802162-2004.html
IEEE Std 802.16-2004/Cor1	Corrigendum to IEEE Std 802.16-2004, Air Interface for Fixed Broadband Wireless Access Systems	February 2006 (Superseded)	http://ieee802.org/16/pubs/80216_Cor1.html
IEEE Std 802.16e-2005	Amendment to IEEE Std 802.16, Air Interface for Fixed Broadband Wireless Access Systems – Physical and Medium Access Control Layers for Combined Fixed and Mobile Operation in Licensed Bands	February 2006 (Superseded)	http://ieee802.org/16/pubs/80216e.html
IEEE Std 802.16f	Amendment to IEEE Std 802.16, Air Interface for Fixed Broadband Wireless Access Systems – Management Information Base	December 2005 (Superseded)	http://ieee802.org/16/pubs/80216f.html
IEEE Project P802.16g	Amendment of IEEE Std 802.16, Air Interface for Fixed Broadband Wireless Access Systems – Management Plane Procedures and Services	December 2007 (Superseded)	http://ieee802.org/16/pubs/80216g.html
IEEE Standard 802.16-2004	Revision of IEEE Std 802.16 (including IEEE Std 802.16-2001, IEEE Std 802.16c-2002, and IEEE Std 802.16a-2003) Air Interface for Fixed Broadband Wireless Access Systems	October 2004 (Superseded)	http://ieee802.org/16/pubs/80216-2004.html

(Continued)

Table 1-4 IEEE 802.16 Standards *Continued*

Standard/ Project	Description	Release Date	Link to Resources
IEEE Std 802.16-2009	IEEE Standard for Local and Metropolitan Area Networks – Part 16: Air Interface for Broadband Wireless Access Systems Revision of IEEE Std 802.16-2004, developed by Maintenance Task Group under the project draft title "P802.16Rev2." This work resulted in the second revision of IEEE Std 802.16, following IEEE Std 802.16-2001 and IEEE Std 802.16-2004. It consolidates and obsoletes IEEE Standards 802.16-2004, 802.16e-2005 and 802.16-2004/Cor1-2005, 802.16f-2005, and 802.16g-2007.	May 2009	http://ieee802.org/16/pubs/80216-2009.html
IEEE Std 802.16j	Amendment of IEEE Std 802.16, Air Interface for Fixed and Mobile Broadband Wireless Access Systems – Multi-hop Relay Specification	June 2009	http://ieee802.org/16/pubs/80216j.html
IEEE Project P802.16i	Project to amend IEEE Std 802.16, Air Interface for Fixed Broadband Wireless Access Systems – Management Plane Procedures and Services	March 2008 (Withdrawn)	http://ieee802.org/16/pubs/80216i.html
IEEE Std 802.16h-2010	Amendment of IEEE Std 802.16, Air Interface for Fixed Broadband Wireless Access Systems – Improved Coexistence Mechanisms for License-Exempt Operation	May 2010	http://ieee802.org/16/pubs/80216h.html
IEEE Project P802.16m	Project to amend IEEE Std 802.16-2009, Air Interface for Fixed and Mobile Broadband Wireless Access Systems – Advanced Air Interface	March 2011	http://ieee802.org/16/tgm/index.html

base station is required for relays to operate over the relay links (i.e., the link between relay and base stations). Furthermore, due to the incorporation of various inconsistent optional features and functionalities, a relay system profile (i.e., a set of consistent functional components to form a working system) needs to be developed outside the IEEE to enable industry-wide implementation and deployment.

Since January 2007, the IEEE 802.16 Working Group has embarked on the development of a new amendment of the IEEE 802.16 standard (i.e., IEEE 802.16m) as an advanced air interface to meet the requirements of ITU-R/IMT-Advanced for 4G systems, as well as the next generation mobile network operators. Depending on the available bandwidth and multi-antenna mode, IEEE 802.16m systems will be capable of over-the-air data transfer rates in excess of 1 Gbit/sec and support of a wide range of high-quality and high-capacity IP-based services and applications while maintaining full backward compatibility with the existing mobile WiMAX systems (to preserve investments and continuing support for the first generation products). The IEEE 802.16m will be suitable for both green-field and mixed deployments with legacy mobile stations and base stations. The backward compatibility feature would allow upgrades and evolution paths for existing deployments. It will enable roaming and seamless connectivity across IMT-Advanced and IMT-2000 systems through the use of appropriate interworking functions. The IEEE 802.16m systems further utilize multi-hop relay architectures for improved coverage and performance.

1.3 INTRODUCTION TO WiMAX FORUM MOBILE SYSTEM PROFILES

The IEEE 802 standards only define the physical and medium access control layers (i.e., Layers 1 and 2 of the OSI reference model). This approach has worked well for technologies such as Ethernet and Wi-Fi, relying on other bodies such as the IETF to set the standards for higher layer protocols, such as TCP/IP, SIP, VoIP, and IPsec. In the mobile wireless domain, standards bodies such as 3GPP and 3GPP2 specify standards encompassing radio air interface, network interfaces, and signaling protocols, because they require not only air link interoperability, but also inter-vendor inter-network interoperability for roaming, multi-vendor access, and billing. Vendors and operators have recognized this issue and have formed additional forums to develop standard network reference models for open inter-network interfaces. The WiMAX Forum is an industry-led and non-profit organization formed to develop system profiles, certify, and promote the compatibility and interoperability of broadband wireless products based on the harmonized IEEE 802.16-2004 and its amendments and/or revisions. The ultimate goal of the WiMAX Forum is to promote and accelerate the introduction of cost-effective broadband wireless access services into the marketplace. Standards-based, interoperable solutions enable economies of scale that, in turn, drive price and performance levels unachievable by proprietary solutions, making WiMAX Forum-certified products the most competitive at delivering broadband services on a wide scale. WiMAX Forum-certified products are fully interoperable and support broadband fixed, portable, and mobile services. Along these lines, the WiMAX Forum works closely with service providers and regulators to ensure that WiMAX Forum-certified systems meet customer and government requirements.

WiMAX Forum has been developing the mobile WiMAX system profiles that define the mandatory and optional features of the IEEE 802.16 standard that are necessary to build a mobile WiMAX compliant air interface which can be certified by WiMAX Forum [33]. The mobile

WiMAX system profile enables mobile systems to be configured based on a common feature set, thus ensuring basic functionality for terminals and base stations that are fully interoperable. Some elements of the base station profiles are specified as optional to provide additional flexibility for regional deployments based on specific usage models that may require different configurations that are either capacity-optimized or coverage-optimized. Release 1.0 of the mobile WiMAX system profiles covers 5, 7, 8.75, and 10 MHz channel bandwidths for licensed spectrum allocations in the 2.3, 2.5, 3.3, and 3.5 GHz frequency bands. The term WiMAX (Worldwide Interoperability for Microwave Access) has been used generically to describe wireless systems based on the WiMAX certification profiles, and the IEEE 802.16-2004 air interface standard and its amendments or revisions. Fixed WiMAX is used to describe IEEE 802.16-2004-based systems and mobile WiMAX is used to describe IEEE 802.16-2009 or IEEE 802.16m based systems. The WiMAX Forum further develops higher-level networking specifications for mobile WiMAX systems beyond that defined by the IEEE 802.16 air interface standard (see Chapter 2).

The combined effort of the IEEE 802.16 Working Group and the WiMAX Forum help define the end-to-end system solution for a mobile WiMAX network. A certification profile is defined as a particular instantiation of a system profile (i.e., a predefined set of parameters extracted from the system profiles) where the operating frequency, channel bandwidth, and duplexing mode are also specified. The WiMAX equipments are certified for interoperability according to a particular certification profile. Furthermore, Protocol Implementation Conformance Specification (PICS) is a questionnaire to be completed by vendors that specifies which features have been implemented in the product for conformance testing [14].

Release 1.5 of the mobile WiMAX system profiles, based on the latest revision of the IEEE 802.16 standard [8], provides a short-term migration of WiMAX systems that include additional features relative to Release 1.0, such as FDD support and Time Division Duplex (TDD) enhanced operation, new band classes, support of multicast and broadcast and location-based services, and femto-cells [33]. Release 1.5 systems are fully backward compatible with Release 1.0. The WiMAX Forum is currently working toward Release 2 of mobile WiMAX system profiles based on the IEEE 802.16m standard whose completion is expected in 2011 [34].

Figure 1-6 illustrates the organization of the WiMAX Forum and provides a brief description of the charter of each working group. The WiMAX Forum working groups operate under a Technical Steering Committee (TSC). The TSC functions include overseeing technical development of WiMAX Forum specifications and certifications, ensuring integrated, timely and predictable technical programs within WiMAX Forum, planning and approving roadmaps, work items, and specifications, receiving, reviewing, and acting upon reports from working groups.

1.4 INTRODUCTION TO 3GPP STANDARDS

The original scope of 3GPP was to produce technical specifications and technical reports for 3G mobile systems based on evolved Global System for Mobile communication (GSM) core networks [35] and the radio access technologies that they support; i.e., Universal Terrestrial Radio Access (UTRA) with both FDD and TDD duplexing modes. The scope was subsequently amended to include the maintenance and development of GSM technical specifications and technical reports, including evolved radio access technologies; e.g., General Packet Radio Service (GPRS) and Enhanced Data rates for GSM

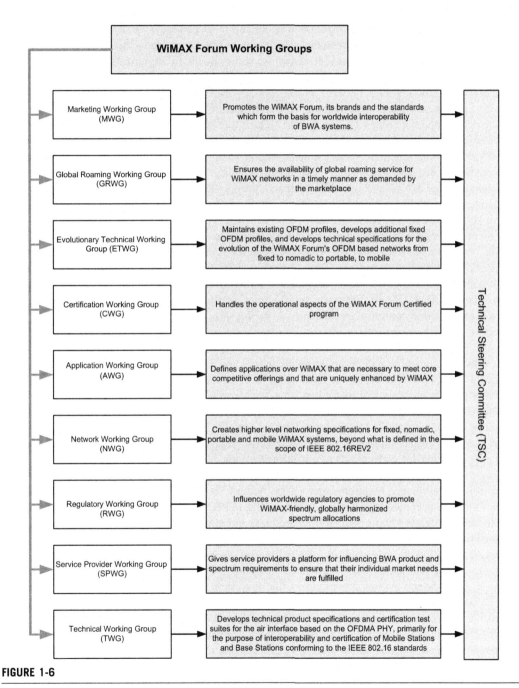

FIGURE 1-6

The organization and working groups of the WiMAX Forum [14]

Evolution (EDGE). The 3GPP organization was created in December 1998 by approval of the 3rd Generation Partnership Project Agreement. The latest 3GPP Scope and Objectives document has evolved from this original agreement [36]. Table 1-5 shows the 3GPP organization and the charter of the technical specification groups and their subcommittees.

The standardization process in 3GPP is substantially different from that of the IEEE. The 3GPP standardization process starts with developing a requirement document where the targets to be achieved are determined (i.e., Stage 1). The system architecture is defined where the underlying building blocks and interfaces are decided (i.e., Stage 2). Various working groups are engaged in developing the detailed specifications where every functional component, interface, and signaling protocol is specified (i.e., Stage 3). The compliance and interoperability testing and verification follow upon completion of Stage 3. Note that in 3GPP standards, the data link layer is divided into four sub-layers; i.e., Medium Access Control (MAC), Radio Link Control (RLC), and Packet Data Convergence Protocol (PDCP), as well as Radio Resource Control (RRC) [7].

As mentioned earlier, the growing demand for mobile Internet and wireless multimedia applications has motivated development of broadband wireless access technologies in recent years. As a result, 3GPP initiated work on Long-Term Evolution (LTE) in late 2004. The 3GPP LTE will ensure 3GPP's competitive edge over other cellular technologies. The Evolved UMTS Terrestrial Radio Access Network (E-UTRAN) substantially improves end-user throughputs, sector capacity, and reduces user-plane and control-plane latencies, bringing significantly improved user experience with full mobility. With the emergence of Internet Protocol as the protocol of choice for carrying all types of traffic, the 3GPP LTE is expected to provide support for IP-based traffic with end-to-end Quality of Service (QoS). Voice traffic will be supported mainly as voice over IP, enabling integration with other multimedia services. Initial deployments of 3GPP LTE are expected in late 2010 and commercial availability on a larger scale will likely happen a few years later.

Unlike its predecessors, which were developed within the framework of Release 99 UMTS architecture, 3GPP has specified the Evolved Packet Core (EPC) architecture to support the E-UTRAN through reduction in the number of network elements and simplification of functionality, but most importantly allowing for connections and handover to other fixed and wireless access technologies, providing the network operators the ability to deliver a seamless mobility experience [37]. 3GPP has set aggressive performance requirements for LTE that rely on improved physical layer technologies such as Orthogonal Frequency Division Multiplexing (OFDM) and Single-User and/or Multi-User Multiple-Input Multiple-Output (MIMO) techniques, and streamlined data link layer protocols and functionalities. The main objectives of LTE are to minimize the system and user equipment (UE) complexities, to allow flexible spectrum deployment in the existing or new frequency bands, and to enable coexistence with other 3GPP radio access technologies.

The 3GPP long-term evolution symbolizes the migration of the Universal Mobile Telecommunication System (UMTS) family of standards from systems that supported both circuit-switched and packet-switched voice/data communications to an all-IP, packet-only system. The development of the LTE air interface is closely coupled with 3GPP System Architecture Evolution (SAE) project to define the overall system architecture and evolved packet core network [37]. It has been shown that the 3GPP standards have evolved toward higher performance and data rates, lower access latencies, and increasing capability to support emerging wireless applications. To achieve higher downlink and uplink data rates, UMTS operators today are upgrading their 3G networks with High Speed Downlink Packet Access (HSDPA), which was specified in 3GPP Release 5, and High Speed Uplink Packet

Table 1-5 3GPP Organization[iii]

Project Coordination Group (PCG)

This group is the ultimate decision making body in 3GPP responsible for final adoption of 3GPP Technical Specification Group work items, to ratify election results, and the resources committed to 3GPP.

GSM/EDGE Radio Access Network (GERAN)

This group is responsible for the specification of the radio access part of GSM/EDGE, more specifically RF front-end, Layer 1, 2 and 3, internal and external interfaces, conformance test specifications for all aspects of GERAN base stations and terminals, and GERAN specifications for the nodes in the GERAN.

Radio Access Network (RAN)

This group is responsible for the definition of the functions, requirements and interfaces of the UTRA/E-UTRA network in its two modes, FDD and TDD.

GERAN WG1

This group is responsible for RF aspects of GERAN, Internal GERAN interface specifications, Specifications for GERAN radio performance and RF system aspects, conformance test specifications for testing of all aspects of GERAN base stations, etc.

GERAN WG2

This group is responsible for protocol aspects of GERAN. It specifies the data link layer protocols and the interfaces between these layers and the physical layer

GERAN WG3

This group is responsible for conformance test specifications for testing of all aspects of GERAN terminals liaising with other technical groups to ensure overall coordination.

RAN WG1

This group is responsible for the specification of the physical layer of the radio interface for UE, UTRAN, Evolved UTRAN, and beyond, covering both FDD and TDD modes of radio interface. It is also responsible for handling physical layer related UE capabilities and physical layer related parameters used in UE test developed in the RAN working group.

RAN WG2

This group is in charge of the radio interface architecture and protocols, the specification of the radio resource control protocol, the strategies of radio resource management, and the services provided by the physical layer to the upper layers.

RAN WG3

This group is responsible for the overall UTRAN/E-UTRAN architecture, the specification of inter-base station protocols, and the interface between the mobile station and base station.

RAN WG4

This group works on the radio aspects of UTRAN/E-UTRAN. It performs simulations of diverse radio system scenarios and derives the minimum requirements for transmission and reception parameters. Once these requirements are set, the group defines the test procedures that will be used to verify them (only for base stations).

(Continued)

Table 1-5 3GPP Organization[iii] *Continued*

Service and System Aspects (SA)

The Service and System Aspects group is responsible for the overall architecture and service capabilities of systems based on 3GPP specifications, and has a responsibility for inter-group coordination.

RAN WG5

This group works on the specification of conformance testing at the radio interface for the user equipment. The test specifications are based on the requirements defined by other groups, such as RAN WG4 for the radio test cases, and RAN WG2 and CT WG1 for the signaling and protocols test cases.

SA WG1 (Services)

This group works on the services and features for 3G. The group sets high-level requirements for the overall system and provides this in a Stage 1 description in the form of specifications and reports.

SA WG2 (Architecture)

This group is in charge of developing Stage 2 of the 3GPP network, and identifies the main functions and entities of the network, how these entities are linked to each other, and the information they exchange. The group has a system-wide view, and decides on how new functions integrate with the existing network entities.

SA WG3 (Security)

This group is responsible for the security of the 3GPP system, performing analyses of potential security threats to the system, considering the new threats introduced by the IP based services and systems, and setting the security requirements for the overall 3GPP system.

SA WG4 (Codec)

This group deals with the specifications for speech, audio, video, and multimedia codecs, in both circuit-switched and packet-switched environments.

SA WG5 (Telecom Management)

This group specifies the management framework and requirements for the management of the 3G system, delivering architecture descriptions of the telecommunication management network and coordinating across other working groups all work pertinent to the 3G system telecom management.

CT WG1

This group is responsible for the 3GPP specifications that define the user equipment – core network Layer 3 radio protocols and the core network side of the reference points.

CT WG3

This group specifies the bearer capabilities for circuit- and packet-switched data services, the necessary interworking functions towards both, and the user equipment and terminal equipment in the external network.

CT WG4

This group defines stage 2 and stage 3 aspects within the core network focusing on supplementary services, basic call processing, mobility management within the core network, bearer independent architecture, GPRS between network entities, transcoder free operation, etc.

CT WG6

This group is responsible for the development and maintenance of specifications and associated test specifications for the 3GPP smart card applications, and the interface with the mobile terminal.

Core Network and Terminals (CT)

This group is responsible for specifying terminal interfaces, terminal capabilities, and the core network part of 3GPP systems.

ⁱⁱⁱ*http://www.3gpp.org/Specification-Groups*

Access (HSUPA), which was specified in 3GPP Release 6. The HSDPA and HSUPA, collectively known as HSPA, have been further upgraded in Releases 7 and 8 with additional features such as higher-order modulations in the downlink and uplink, downlink open-loop MIMO, improvements of data link layer protocols, and continuous connectivity for packet data users.

The Release 8 of 3GPP LTE and SAE were completed in March 2009, followed by the user equipment conformance testing specifications. The 3GPP is currently working toward development of an advanced standard specification to address the IMT-Advanced requirements and services for the 4th generation of cellular systems [15]. The 3GPP LTE-Advanced is part of 3GPP Release 10 and its completion is expected by March 2011. The IMT-Advanced compliant technologies may be available for commercial deployment beyond 2012. The requirements for LTE-Advanced are provided in reference [20]. The salient features of LTE-Advanced include multi-carrier support, spatial multiplexing, extension of LTE downlink spatial multiplexing to eight layers, and uplink spatial multiplying up to four layers, coordinated multiple point transmission and reception, relaying functionality to improve the coverage and throughput, group mobility, temporary network deployment, and to provide coverage in new areas.

References

[1] Bluetooth wireless technology. <https://www.bluetooth.org/apps/content/>.
[2] Ultra-wideband (UWB) technology. <http://www.intel.com/technology/comms/uwb/>.
[3] Andrew S. Tanenbaum, Computer Networks, fourth ed. Prentice Hall, 2002.
[4] Matthew Gast, 802.11 Wireless Networks: The Definitive Guide, second ed., O'Reilly Media, Inc., 2005.
[5] Carl Eklund, et al., WirelessMAN: Inside the IEEE 802.16 Standard for Wireless Metropolitan Area Networks, IEEE Standards Information Network/IEEE Press, 2006.
[6] Harri Holma, Antti Toskala, WCDMA for UMTS: HSPA Evolution and LTE, fourth ed., John Wiley & Sons, 2007.
[7] Erik Dahlman, et al., 3G Evolution: HSPA and LTE for Mobile Broadband, second ed., Academic Press, 2008.
[8] IEEE Std 802.16-2009, IEEE Standard for Local and Metropolitan Area Networks – Part 16: Air Interface for Broadband Wireless Access Systems, May 2009.
[9] 3G Americas, Mobile Broadband Evolution 3GPP Release 8 and Beyond June 2008. <http://3gamericas.org/>.
[10] Recommendation ITU-R M.1645, Framework and Overall Objectives of the Future Development of IMT-2000 and Systems Beyond IMT-2000, June 2003.
[11] Recommendation ITU-R M.1822, Framework for Services Supported by IMT, October 2007.
[12] WINNER Project, Usage Scenarious, D13 version 1.0, June 2005. <http://www.ist-winner.org>.
[13] CDMA Development Group (CDG). <http://www.cdg.org>.
[14] WiMAX Forum. <http://www.wimaxforum.org>.
[15] REPORT ITU-R M.2134, Requirements Related to Technical Performance for IMT-Advanced Radio Interface(s), October 2008.
[16] ITU-T Next Generation Networks Global Standards Initiative. <http://www.itu.int/ITU-T/ngn/>.
[17] Internet Message Access Protocol, Wikipedia. <http://en.wikipedia.org/wiki/Imap>.
[18] Post Office Protocol, Wikipedia. <http://en.wikipedia.org/wiki/Pop3>.
[19] Report ITU-R M.2135–1, Guidelines for Evaluation of Radio Interface Technologies for IMT-Advanced, December 2009.

[20] 3GPP TR 36.913, Requirements for Further Advancements for Evolved Universal Terrestrial Radio Access (E-UTRA) (LTE-Advanced), March 2009.

[21] IEEE C80216m–07/002r10, IEEE 802.16m System Requirements, January 2010.

[22] Athanasios Papoulis, Probability, Random Variables, and Stochastic Processes, fourth ed., McGraw Hill, 2002.

[23] 3GPP TS 23.402, Architecture Enhancements for Non-3GPP Accesses, March 2009. (ftp://ftp.3gpp.org/specs/).

[24] IEEE 802.2 Logical Link Control (LLC). <http://grouper.ieee.org/groups/802/2/>.

[25] IEEE 802.3 ETHERNET. <http://grouper.ieee.org/groups/802/3/>.

[26] IEEE 802.5 Token Ring standards. <http://www.ieee802.org/5/www8025org/>.

[27] IEEE 802.11 Wireless Local Area Networks. <http://grouper.ieee.org/groups/802/11/>.

[28] IEEE 802.15 Wireless PAN. <http://grouper.ieee.org/groups/802/15/>.

[29] IEEE 802.16 Broadband Wireless Access Standards. <http://grouper.ieee.org/groups/802/16/>.

[30] IEEE 802.20 Mobile Broadband Wireless Access. <http://www.ieee802.org/20/>.

[31] Bridging and Virtual Bridged LANs IEEE 802.1. <http://grouper.ieee.org/groups/802/1/>.

[32] IEEE Standards Association. <http://standards.ieee.org/>.

[33] WiMAX Forum Mobile System Profile Specification Release 1.5, February 2009. <http://www.wimaxforum.org/resources/documents/technical/release>.

[34] WiMAX Forum, Requirements for WiMAX Air Interface System Profile Release 2.0, March 2009. <http://www.wimaxforum.org/resources>.

[35] K. Vijay Garg, E. Joseph, Wilkes, Principles and Applications of GSM, Prentice Hall, 1999.

[36] 3rd Generation Partnership Project Agreement, <http://www.3gpp.org/>.

[37] 3GPP TR 23.882, 3GPP System Architecture Evolution: Report on Technical Options and Conclusions, September 2008. (ftp://ftp.3gpp.org/specs/).

WiMAX Network Architecture

INTRODUCTION

The IEEE 802.16-2009 standard only defines physical and medium access control layer protocols. This approach might have worked well for technologies such as Ethernet and Wi-Fi, relying on other bodies such as the Internet Engineering Task Force (IETF)[i] to set the standards for higher layer protocols such as TCP/IP [1], SIP [2], and Ipsec [3], it cannot be generalized to cellular standards. Standard developing organizations such as 3GPP and 3GPP2 specify end-to-end wireless access systems including a wide range of interfaces and network protocols because practical systems require not only the air link interoperability but also inter-vendor and inter-network interoperability for roaming, multi-vendor access networks, and inter-operator billing. WiMAX vendors and operators have recognized this issue and have established additional industry groups to develop standard network reference models for open network interfaces. Examples of these groups include WiMAX Forum's Network Working Group,[ii] which is focused on creating higher-level networking specifications for fixed, nomadic, portable, and mobile WiMAX systems beyond what is defined in the IEEE 802.16 standards, and the Service Provider Working Group,[iii] which defines requirements and priorities according to operators' needs.

The WiMAX network architecture is a non-hierarchical end-to-end all-IP framework for mobile WiMAX systems that is based on maximal use of non-proprietary standard IP protocols, and is compatible with external service enablers such as IP Multimedia Subsystem [4]. A distinctive feature of WiMAX network architecture is decoupling of access, connectivity, and service networks to allow combination of multi-vendor implementations of physical network entities, as long as they comply with the normative protocols and procedures across applicable interfaces that are defined in the WiMAX network specification.

This chapter provides a top-down systematic description of WiMAX network architecture, starting at the most general level and working toward details or specifics of the network components and their interconnections. We further provide an overview of 3GPP evolved network architecture to enable the readers to contrast the corresponding procedures and functionalities.

2.1 DESIGN PRINCIPLES OF WiMAX NETWORK ARCHITECTURE

In this section, the principles based on which the WiMAX network architecture has been designed are described. The tenets are divided into different categories and are applicable to all releases of the WiMAX network architecture [5].

---

[i]http://www.ietf.org
[ii]http://www.wimaxforum.org/about/network-working-group-nwg
[iii]http://www.wimaxforum.org/about/service-provider-working-group-spwg

Mobile WiMAX. DOI: 10.1016/B978-0-12-374964-2.10002-5

The architecture and network reference model (NRM) accommodate all usage models provisioned for WiMAX systems [6]. The packet-switched architecture is based on the IEEE 802.16 standard and its amendments or revisions, as well as the use of appropriate IETF protocols and IEEE Ethernet standards. This framework permits decoupling of access network and corresponding entities from connectivity IP services, and allows network elements of the Connectivity Serving Network (CSN) to be agnostic of the IEEE 802.16 air interface. The architecture is based on functional decomposition principles (i.e., decomposition of features into functional entities across interoperability reference points without specific implementation assumptions, including the network entities and interfaces). Such a framework is modular and sufficiently flexible to accommodate a broad range of deployment options such as [5]:

- Scalable WiMAX networks (sparse to dense radio coverage and capacity);
- Diverse radio propagation environments (e.g., urban, suburban, or rural);
- Licensed and/or licensed-exempt spectrum;
- Hierarchical, non-hierarchical, or mesh topologies, and their variants;
- Coexistence of fixed, nomadic, portable, and mobile usage models;
- Interworking and integration with non-WiMAX core networks (e.g., 3GPP EPC) [7].

The architecture exploits the IEEE 802.16-2009 procedures and logical separation between such procedures and IP addressing, routing, and connectivity management procedures, and protocols to enable the use of the access architecture primitives in standalone and interworking deployment scenarios and it further supports sharing of Access Service Networks (ASN) by multiple Network Access Providers (NAP). The network design supports discovery and selection of accessible Network Service Providers (NSP) by an MS, and allows NAPs that employ one or more ASN topologies. Network Access Provider is a business entity that provides WiMAX radio access infrastructure to one or more WiMAX Network Service Providers. A NAP implements WiMAX network infrastructure using one or more ASNs. Network Service Provider is a business entity that provides IP connectivity and WiMAX services to WiMAX subscribers compliant with the service level agreement (SLA) established with WiMAX subscribers. To provide these services, an NSP sets up contractual agreements with one or more NAPs. Additionally, an NSP also establishes roaming agreements with other NSPs, and contractual agreements with third-party application providers for providing WiMAX services to subscribers. From a WiMAX subscriber perspective, an NSP can be classified as Home NSP (H-NSP) or Visited NSP (V-NSP).

The WiMAX network architecture supports access to operator's services through internetworking functions and specifies open and standard reference points between various groups of network functional entities, in particular between an MS, ASN, and CSN to enable multi-vendor interoperability. It is sufficiently flexible to accommodate future enhancements/extensions to the IEEE 802.16 suite of standards, and can accommodate documented geo-specific constraints.

The WiMAX network architecture supports evolution paths between the various usage models subject to reasonable technical assumptions and constraints. It does not preclude multi-vendor implementations based on different combinations of functional network entities as long as these implementations comply with the normative protocols and procedures across applicable reference points. The structural design supports a simple usage model where a single operator can deploy an

ASN together with a limited set of CSN functions, so that the operator can offer basic Internet access without consideration for roaming or interworking. The service and application design requirement for mobile WiMAX network architecture are as follows:

1. Support voice, multimedia services, and other mandated regulatory services such as emergency services and lawful interception;
2. Enable access to a variety of independent Application Service Provider (ASP) networks;
3. Support mobile communications using VoIP in applicable roaming scenarios, further supporting inter-operator policy definition, distribution and enforcement as needed for voice communications. The architecture supports the following capabilities:
   a. SLA-based resource management for subscribers;
   b. Multiple voice session for a particular subscriber;
   c. Simultaneous voice and data sessions;
   d. Prioritization of emergency voice calls and high-priority data sessions;
4. Interface with various interworking and media gateways, allowing delivery of legacy services translated over IP to WiMAX access networks;
5. Support delivery of IP broadcast and multicast services over WiMAX access networks.

The following are the design considerations for WiMAX network security:

1. The security framework is agnostic to the operator type and ASN topology, and applies consistently across mobile WiMAX networks and internetworking deployment models and usage scenarios.
2. The architecture supports robust MS authentication based on the IEEE 802.16 standards security framework.
3. An MS is able to support all commonly deployed authentication mechanisms and authentication in home and visited operator network scenarios, based on a consistent and extensible authentication framework. Furthermore, an MS is able to select between various authentication methods based on NSP type.
4. Data integrity, replay protection, and confidentiality using applicable key lengths within the WiMAX access network are supported.
5. The use of MS initiated or terminated security mechanisms such as Virtual Private Networks (VPNs) is accommodated [8].
6. Standard secure IP address management mechanisms between the MS and its home or visited NSP are supported [3].
7. The privacy of MS and host's specific states, such as authentication state, IP host configuration, and service provisioning is guaranteed.
8. Group communications are exclusively limited to authorized group membership.

Mobility and handover design considerations are as follows:

1. Inter-technology handovers, e.g., Wi-Fi, 3GPP, and 3GPP2, when such capability is enabled in multi-mode MS are supported.
2. Both IPv4- [9] or IPv6-based [10] mobility management are accommodated in the reference network design. Furthermore, WiMAX network supports an MS with multiple IP addresses and simultaneous IPv4 and IPv6 connections.

3. The network design does not preclude roaming between NSPs. It allows a single NAP to serve multiple MSs using different private and public IP domains owned by different NSPs. The NSP may be one operator or a group of operators.

4. Seamless and robust handover at different user mobility conditions are supported by dynamic and static home address configurations.

5. The WiMAX network allows dynamic assignment of the mobile IP Home Agent (HA) in the service provider network as a form of routing optimization, as well as in the home IP network as a form of load balancing.

6. The architecture allows dynamic assignment of the HA in Home CSN (H-CSN) or Visited CSN (V-CSN) based on policies.

The Quality of Service (QoS) considerations for network operation are as follows:

1. The network design supports differentiated levels of QoS, admission control,[iv] and bandwidth assignment in order to flexibly support simultaneous use of a diverse set of IP services.

2. The network protocols support implementation of policies as defined by various operators for QoS based on their SLAs, including policy enforcement per user and user group, etc., where the QoS policies may be coordinated among operators depending on subscriber SLAs.

3. The reference design employs standard IETF mechanisms for managing policy definition and policy enforcement between operators.

Other considerations for network operation including scalability, extensibility, coverage, etc., can be summarized as follows.

The architecture enables a user to manually or automatically select from available NAPs and NSPs. The architecture enables ASN and CSN system designs that readily up-scale or down-scale in terms of coverage, range, or capacity while accommodating a variety of ASN designs. It further facilitates a variety of backhaul links, both wireline and wireless with different latency and throughput characteristics, as well as incremental infrastructure deployment. WiMAX network supports phased deployment of IP services that in turn scale with increasing number of active users and concurrent IP services per user.

The architecture supports the integration of base stations of varying coverage and capacity – for example, pico, micro, and macro base stations – and allows flexible decomposition and integration of ASN functions in ASN network deployments, in order to enable use of load balancing schemes for efficient use of radio spectrum and network resources. Inter-working with existing wireless networks (e.g., 3GPP, 3GPP2) or wireline networks (e.g., DSL) is supported where the interworking function is based on standard IETF and IEEE protocol suites.

The WiMAX network supports global roaming across WiMAX operator networks including support for credential reuse, consistent use of Authentication, Authorization, and Accounting (AAA) protocols for accounting and charging, and consolidated/common billing and settlement. A variety of

---

[iv]Admission control is the ability to admit or control admission of a user to a network based on the user's service profile and network performance parameters, such instantaneous load and average delay. If a user requests access to network services but the incremental resources required to provide the grade of service specified in the user's service profile are not available, the admission control function rejects the user's access request. Note that admission control is implemented to ensure service quality and is different from authentication and authorization, which are also used to admit or deny network access.

user authentication credential formats, such as username/password, digital certificates, Subscriber Identity Module (SIM), Universal SIM (USIM), and Removable User Identify Module (RUIM) are supported. Furthermore, WiMAX network accommodates a variety of online and offline client-provisioning, enrollment, and management schemes based on open, broadly deployable, industry standards. Over-the-Air (OTA) services for MS provisioning and software upgrades are accommodated.

The network protocols accommodate use of header compression/suppression and/or payload compression for efficient use of WiMAX radio resources. The architecture supports mechanisms that enable maximum possible enforcement and fast re-establishment of established QoS SLAs due to handover impairments.

The WiMAX network supports interoperability between equipment from different manufacturers within an ASN and across ASNs. Such interoperability includes interoperability between BS and backhaul equipment within an ASN, interoperability between various ASN elements (possibly from different vendors) and CSN, with minimal or no degradation in functionality or capability of the ASN.

The IEEE 802.16-2009 standard defines multiple convergence sub-layers. The network architecture further supports the following convergence sub-layer (CS) types: Ethernet CS and IPv4/IPv6 over Ethernet CS; IPv4 CS; and IPv6 CS.

2.2 NETWORK REFERENCE MODEL

The Network Reference Model is a logical representation of the network architecture. The NRM identifies functional entities and reference points over which interoperability is achieved. Figure 2-1 illustrates the WiMAX NRM, consisting of the following logical entities: mobile station (MS); ASN; and CSN, which are described in the following sections. The interfaces R1–R8 are normative reference points [5].

Each of the entities MS, ASN, and CSN represent a group of functions. Each of these functions can be realized in a single physical entity or may be distributed over multiple physical entities. The reference model is used to ensure interoperability among different implementations of functional entities in the network. Interoperability is verified based on the definition of logical interfaces in order to achieve an end-to-end functionality, e.g., security or mobility management. Thus, the functional entities on either side of a reference point (RP) represent a collection of control or bearer-plane end-points.

The mobile station is a communication device providing radio connectivity between a user terminal and a WiMAX Base Station (BS) that is compliant with WiMAX Forum mobile system profiles [11]. The ASN is defined as a complete set of network functions required to provide radio access to a terminal. The ASN provides the following functions:

- Layer 2 connectivity with the MS;
- Transfer of AAA messages to the subscriber's Home Network Service Provider (H-NSP) for authentication, authorization, and session accounting for subscriber sessions;
- Network discovery and selection of the subscriber's preferred NSP;
- Relay functionality for establishing Layer 3 connectivity or IP address assignment to an MS;
- Radio resource management (RRM).

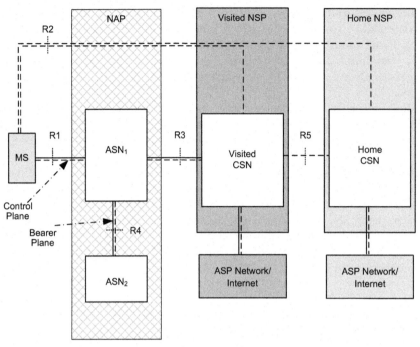

FIGURE 2-1

The WiMAX network reference model

In addition to the above functions, and to support mobility, an ASN supports the following functions:

- ASN-anchored mobility;
- CSN-anchored mobility;
- Paging and Idle State operation;
- ASN–CSN tunneling.

The ASN comprises network elements such as one or more base stations and one or more ASN Gateways (ASN-GW). An ASN may be shared by more than one Connectivity Service Network (CSN). The radio resource control functions in the BS would allow Radio Resource Management (RRM) within the BS. The CSN is defined as a set of functions that provide IP connectivity to user terminals. In the following sections, more detailed description of network elements and their interconnections will be provided.

2.2.1 Access Service Network (ASN)

The ASN is defined as a logical set of functional entities and corresponding signaling and message flows. The ASN represents an interface for functional interoperability with user terminals, on the one hand, and connectivity service functions, on the other hand. The functional decomposition of ASN is shown in Figure 2-2. Within an ASN, a BS may be logically connected to more than one ASN-GW for different MSs. For a given MS, a BS is connected to a single ASN-GW.

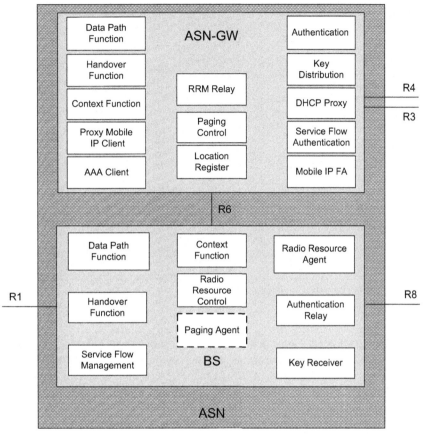

FIGURE 2-2

The functional decomposition of ASN

The ASN reference model containing multiple ASN-Gateways (ASN-GW) is illustrated in Figure 2-3. An ASN shares R1, R3, and R4 reference points with an MS, a CSN, and with another ASN, respectively. The ASN consists of one or more base stations and one or more ASN-GWs (as shown in Figure 2-3). A BS is logically connected to one or more ASN-GWs. The R4 reference point provides control and bearer-planes for interconnection and interoperability between similar or heterogeneous ASNs. When ASN is composed of multiple ASN-GWs, intra ASN mobility may involve R4 control messages and bearer-plane establishment.

2.2.2 Access Service Network Gateway (ASN-GW)

The base stations within an ASN-GW group (i.e., all base stations connected to the same ASN-GW) are connected through R8 reference point. Furthermore, the base stations within an ASN-GW group are separately connected to the ASN-GW via R6 reference point, as shown in Figure 2-3.

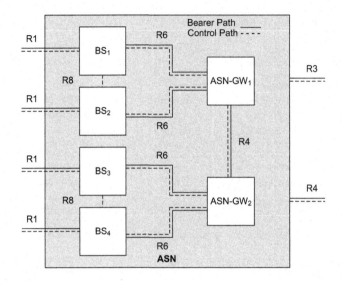

FIGURE 2-3

The ASN reference model containing multiple ASN-GW

The base station is a logical entity that implements a full instance of Medium Access Control (MAC) and Physical Layer (PHY) protocols, as specified by the IEEE 802.16-2009 standard [12] and includes one or more access functions. A BS instance represents one sector with one frequency assignment. It incorporates scheduler functions for uplink and downlink radio resources, which are typically vendor specific.

The ASN-GW is a logical entity that represents an aggregation of control functions that are either paired with a corresponding function in the ASN (e.g., BS instance), a resident function in the CSN, or a function in another ASN. The ASN-GW may also perform bearer-plane routing or bridging functions. A BS is associated with a default ASN-GW; however, the ASN-GW functions may be distributed among multiple ASN-GWs located in one or more ASNs.

The ASN functions in an ASN-GW can be divided into two groups, i.e., the Decision Point (DP) and the Enforcement Point (EP) functions. The EP category includes bearer-plane functions and the DP category includes non-bearer-plane functions. If such functional split is implemented in ASN-GW, the EP and DP functional groups would include bearer-plane and non-bearer-plane functions, respectively, and are interfaced using R7 reference point.

2.2.3 Reference Points

A Reference Point (RP) is a logical interface between two groups of functions or protocols. All protocols associated with a reference point may not always terminate in the same functional entity, i.e., two protocols associated with an RP originate and terminate in different functional entities [5]. A reference point only becomes a physical interface when the functional entities on either side of it are contained in different physical modules.

Reference Point R1

Reference point R1 consists of the protocols and procedures over the air interface between MS and ASN as specified by the IEEE 802.16-2009 standard [12] and mobile WiMAX system profile [11]. The reference point R1 further includes additional protocols related to the management plane.

Reference Point R2

The logical reference point R2, i.e., no direct protocol interface between MS and CSN, consists of protocols and procedures between the MS and CSN associated with authentication, authorization, and IP host configuration management. The authentication protocols in reference point R2 work between the MS and CSN, and are managed by the home NSP; however, under certain conditions, the ASN and CSN operated by the visited NSP may process these protocols. The reference point R2 may further support IP Host Configuration Management running between the MS and the CSN.

Reference Point R3

Reference point R3 consists of a set of control-plane protocols between the ASN and CSN to support AAA, policy enforcement, and mobility management capabilities. It also encompasses the bearer-plane protocols, including tunneling[v] to transfer user data between the ASN and the CSN.

Reference Point R4

Reference point R4 comprises a set of control and bearer-plane protocols originating and/or terminating in various functional entities of an ASN that coordinate MS mobility between ASNs and ASN-GWs, ensuring interoperability between similar or heterogeneous ASNs.

Reference Point R5

Reference point R5 consists of the set of control and bearer-plane protocols for inter-networking between the CSN operated by the home NSP, and is operated by a visited NSP.

Reference Point R6

Reference point R6 comprises a set of control and bearer-plane protocols for communication between the BS and the ASN-GW within a single ASN. The bearer-plane consists of intra-ASN data transfer between the BS and ASN-GW. The control-plane includes protocols for data plane establishment, modification, and release control corresponding to the MS mobility.

Reference Point R7

Reference point R7 consists of an optional set of control protocols such as AAA and policy coordination in the ASN-GW, as well as other protocols for coordination between the two groups of functions identified in R6. This reference point has been removed from mobile WiMAX network architecture specification Release 1.5 [12].

---

[v]Tunneling refers to a mechanism that enables disjoint packet networks to exchange data or packets via intermediate networks, while concealing the protocol details from the intermediate networks. Tunneling is generically implemented by encapsulating an end-to-end network protocol within payloads that are natively carried over the intermediate networks. Tunneling is alternately referred to as encapsulation.

Reference Point R8

Reference point R8 within an ASN consists of a set of control messages exchanged between the base stations to ensure fast and seamless handover. The control-plane consists of the inter-BS communication protocol consistent with the IEEE 802.16-2009 standard [12] and mobile WiMAX system profile [11], as well as additional protocols for data forwarding between base stations.

2.2.4 Connectivity Service Network (CSN)

Connectivity Service Network is defined as a set of network functions that provide IP connectivity to user terminals. A CSN can provide the following functions [5,13–15]:

- MS IP address and endpoint parameter allocation for user sessions;
- Internet access;
- AAA proxy/server;
- Policy and admission control based on user profiles;
- ASN-CSN tunneling support;
- Subscriber billing;
- Inter-CSN tunneling for roaming;
- Inter-ASN mobility management and mobile IP home agent functionality;
- Network services such as connectivity for peer-to-peer services, provisioning, authorization and/or connectivity to IP multimedia services, and to enable lawful interception;
- Connectivity to Internet and managed services such as IP Multimedia Subsystem (IMS),[vi] Location-Based Services (LBS),[vii] Multicast and Broadcast Services (MBS), etc.;
- Over-the-air activation and provisioning of mobile WiMAX terminals.

The CSN may further comprise network elements such as routers, AAA proxy/servers, user databases, and interworking functions. The CSN reference model is depicted Figure 2-4.

2.3 AUTHENTICATION, AUTHORIZATION, AND ACCOUNTING (AAA)

AAA refers to a framework based on IETF protocols, Remote Authentication Dial-in User Service (RADIUS) [16] or Diameter [17], which specify the procedures for authentication, authorization, and accounting associated with the user terminal's subscribed services across different access technologies. As an example, AAA includes mechanisms for secure exchange and distribution of authentication credentials and session keys for data encryption. The AAA protocols provide the following services:

- Authentication including device, user, or combined device and user authentication;
- Authorization including delivery of information to configure the session for access, mobility, QoS, and other applications;
- Accounting including delivery of billing information and other information that can be used to audit session activity by both the H-NSP and V-NSP.

[vi]IP Multimedia Subsystem (IMS) is an architectural framework for delivering IP multimedia services [4].
[vii]A location-based service is a service provided to a subscriber based on the current geographic location of the MS.

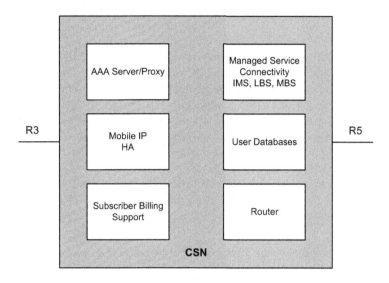

FIGURE 2-4

The CSN reference model

The AAA framework supports global roaming across operator networks, including support for reuse of credentials and consistent use of authorization and accounting. It further supports roaming between H-NSP and V-NSP. The AAA framework is based on use of RADIUS or Diameter in ASN and CSN. The AAA framework accommodates both Mobile IPv4 [9] and Mobile IPv6 [10] Security Association (SA) management. It further accommodates various network operation scenarios from fixed to full mobility. The AAA framework provides support for deploying MS authorization, user and mutual authentication between MS and the NSP, based on Privacy Key Management (PKMv2).[viii] [18] In order to ensure interoperability, the AAA framework supports Extensible Authentication Protocol (EAP)-based authentication mechanisms that include passwords, Subscriber Identity Module [19], Universal Subscriber Identity Module [20], Universal Integrated Circuit Card [21], Removable User Identity Module [22], and X.509 digital certificates [23]. The AAA framework is capable of providing the V-CSN or ASN with a temporary identifier that represents the user without revealing the user's identity.

The NAP may deploy an AAA proxy between two NASs in ASN and the AAA in CSN in order to provide security and enhanced manageability. The AAA proxy will also allow the NAP to regulate the AAA attributes received from the visited CSN, and to add additional AAA attributes that may be required by the NASs in the ASN. Note that the CSN hosts the AAA server, whereas the ASN hosts one or more NASs. The PKMv2 protocol is used to perform over-the-air user authentication. The PKMv2 transfers EAP messages over R1 reference point (i.e., the IEEE 802.16-2009 air interface or its evolution) between the MS and the BS in ASN.

[viii]Privacy Key Management (PKM) Protocol: a client/server model between the base station and mobile station that is used to secure distribution of keying material.

2.4 MOBILE IP

Mobility support within WiMAX network architecture is based on mobile IP framework. Mobile IP is an IETF protocol that allows mobile users to move from one network to another while maintaining their IP address [1,24]. The two versions of mobile IP protocols,; i.e., Mobile IPv4 and Mobile IPv6, are described in references [9] and [10], respectively. The mobile IP protocol allows transparent routing of IP datagrams on the Internet. Each mobile node is identified by its home address, irrespective of its current location in the Internet. When away from home, a mobile node is associated with a care-of address, which provides information about its current location. The mobile IP protocol specifies how a mobile node registers with its Home Agent (HA) and how the HA routes datagrams to the mobile node through a tunnel. Using mobile IP, nodes may change their point-of-attachment to the Internet without changing their IP address, allowing the application and transport-layer protocols to seamlessly maintain connection while moving. The general characteristics of mobile IP can be summarized as follows:

- Transparency of user mobility to the transport and application layer protocols;
- Interoperability with stationary hosts running conventional IP protocols;
- Scalability across the Internet;
- Security by preventing an attacker from impersonating a mobile host;
- Macro mobility by ensuring long-term connection while away from home agent.

Node mobility is realized without propagating host-specific routes throughout the Internet. Using mobile IP, a mobile device will have two addresses, i.e., a primary or permanent home address and a secondary or temporary care-of address, which is associated with the network that the mobile node is visiting. There are two types of entities in mobile IP:

- A home agent that stores information about mobile nodes whose permanent home address is in the home agent's network;
- A foreign agent (FA) which stores information about mobile nodes visiting its network. Foreign agents also advertise care-of addresses.

A node that wishes to communicate with the mobile node uses the permanent home address of the mobile node as the destination address for out-bound packets. Since the home address logically corresponds to the network associated with the home agent, conventional IP routing mechanisms forward these packets to the home agent. Instead of forwarding these packets to a destination that is physically in the same network as the home agent, the home agent redirects these packets towards the foreign agent. The home agent looks for the care-of address in the mobility binding table [25], and then tunnels the packets to the mobile node's care-of address by appending a new IP header to the original IP packet, which preserves the original IP header. The packets are detected at the end of the tunnel by removing the IP header added by the home agent and are delivered to the mobile node.

The mobile node directly sends packets to the other communicating node through the foreign agent without involvement of the home agent, using its permanent home address as the source address for the IP packets, i.e., triangular routing. The foreign agent can utilize reverse-tunneling by sending the mobile node's packets to the home agent, which forwards them to the communicating node. This mechanism is needed in networks whose gateway routers have ingress filtering enabled, and hence the source IP address of the mobile host needs to belong to the subnet of the foreign network; otherwise, the packets would be discarded by the router.

The mobile IP protocol defines an authenticated registration procedure through which a mobile node informs its home agent of its care-of address, router discovery (which allows mobile nodes to discover prospective home agent and foreign agents), and the rules for routing packets to and from mobile nodes, including the specification of one mandatory and several optional tunneling mechanisms.

Figure 2-5 depicts the protocol stack for IP Convergence Sub-layer (IP CS), i.e., a protocol layer that provides an interface between the IEEE 802.16-2009 MAC with a network layer over the data-plane [12], with routed ASN. Routing over ASN is done using IP-in-IP encapsulation protocols such as generic routing encapsulation (GRE). The GRE is a tunneling protocol that can encapsulate different network layer protocol packet types inside IP tunnels, creating a virtual point-to-point link to routers at remote points over an IP network [26]. IP tunneling is a method of connecting two disjoint IP networks with no native routing path to each other via a communication channel, i.e., the IP tunnel, which uses encapsulation techniques across an intermediate network. In IP tunneling, every IP packet with its source and destination IP network addresses is encapsulated within another packet format native to the transit network.

In order to maintain an IP session when the host IP address is changed due to mobility, the Client Mobile IP (CMIP) protocols defined by IETF can be used [27]. The CMIP allows a mobile terminal to maintain its transport connection and to continue to be reachable while moving across a network. The CMIP also provides a common IP layer mobility across different access technologies. This would be quite attractive for mobile operators operating radio access networks of different types, such as mobile WiMAX or 3GPP Long Term Evolution (LTE), etc. While CMIP ensures seamless mobility for the IP session, it introduces some disadvantages such as signaling overhead over the air interface and additional MS complexity to support the client IP mobility protocols. The Proxy Mobile IP (PMIP) defined by IETF, which is a network-based mobility management scheme, was introduced to eliminate signaling overhead and to reduce complexity/cost, as well as eliminating the requirement for a network

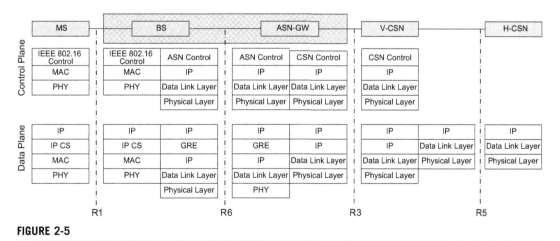

FIGURE 2-5

The protocol stack for the IP convergence sub-layer with routed ASN [14]

interface to change IP address when the mobile node changes to a new router [28]. The WiMAX network architecture supports both the CMIP and PMIP mobility schemes.

2.5 RADIO RESOURCE MANAGEMENT (RRM)

Efficient utilization of radio resources within an access network is performed by the radio resource management entity. The mobile WiMAX RRM is based on a generic architecture. The RRM defines mechanisms and procedures to share radio resource related information between BS and ASN-GW. The RRM procedures allow different BSs to communicate with each other or with a centralized RRM entity residing in the same or a different ASN to exchange information related to measurement and management of radio resources. Each BS performs radio resource measurement locally based on a distributed RRM mechanism. It is also possible to deploy RRM in an ASN using base stations with RRM function, as well as a centralized RRM entity that does not reside in the BS and collects and updates radio resource indicators such as choice of target BS, admission or rejection of service flows, etc., from several BSs. The RRM procedures facilitate the following WiMAX network functions:

- MS admission control and connection admission control, i.e., whether the required radio resources are available at a candidate target BS prior to handover;
- Service flow admission control, i.e., creation or modification of existing/additional service flows for an existing MS in the network, selection of values for admitted and active QoS parameter sets for service flows;
- Load balancing by managing and monitoring system load and use of counter-measures to enable the system back to normal loading condition;
- Handover preparation and control for improvement/maintenance of overall performance indicators (for example, the RRM may assist in system load balancing by facilitating selection of the most suitable BS during a handover).

The RRM is composed of two functional entities, i.e., radio resource agent (RRA) and radio resource control (RRC). The radio resource agent is a functional entity that resides in the BS. Each BS includes a radio resource agent. It maintains a database of collected radio resource indicators. An RRA entity is responsible for assisting local radio resource management, as well as communicating to the RRC to collect and measure radio resource indicators from the BS and from a plurality of mobile terminals served by the BS using MAC management procedures as specified by the IEEE 802.16 specifications. It also communicates RRM control information over the air interface to the MS, as defined by the IEEE 802.16 specifications. An example of such RRM control information is a list of neighbor BSs and their parameters. It further performs signaling with RRC for radio resource management functions, as well as controlling the radio resources of the serving BS, based on the local measurements and reports received by the BS and information received from the RRC functional entity. The local resource control includes power control, monitoring the MAC and PHY functions, modifying the contents of the neighbor advertisement message, assisting the local service flow management function and policy management for service flow admission control, making determinations and conducting actions based on radio resource policy, assisting the local handover functions.

The radio resource control functional entity may reside in BS, in ASN-GW, or as a standalone server in an ASN, and is responsible for collection of radio resource indicators from associated RRAs.

The RRC can be collocated with RRA in the BS. The RRC functional entity may communicate with other RRCs in neighboring BSs which may be in the same or different ASN. The RRC may also reside in the ASN-GW and communicate to other RRAs across R6 reference point. When the RRC is located in the ASN, each RRA is associated with exactly one RRC. The RRC relay functional entity may reside in ASN-GW for the purpose of relaying RRM messages. The RRC relay cannot terminate RRM messages, but only relays them to the final destination RRC. Standard RRM procedures are required between RRA and RRC, and between RRCs across network interfaces to ensure interoperability. These procedures are classified into two types: information reporting procedures for delivery of BS radio resource indicators from RRA to RRC; and between RRCs and decision support procedures from RRC to RRA for communicating recommendations on aggregated RRM status (e.g., in neighboring BSs) for various purposes.

The RRM primitives can be used either to report radio resource indicators (i.e., from RRA to RRC or between RRCs) or to communicate decisions from RRC to RRA. The former type of primitive is called information reporting primitive and the latter is called decision support primitive. The available radio resource information provided by the RRAs to RRC is used by RRC for load balancing. The RRC may interact with the handover controller to ensure load balance.

2.6 MOBILITY MANAGEMENT

The WiMAX network architecture supports two types of mobility: ASN-anchored mobility (intra-ASN) and CSN-anchored mobility. ASN-anchored mobility refers to a scenario where a mobile terminal moves between two base stations belonging to the same ASN while maintaining the same foreign agent at the ASN. The handover in this case utilizes R6 and R8 reference points. The CSN-anchored mobility refers to an inter-ASN mobility scenario where the mobile station moves to a new anchor foreign agent and the new FA and CSN exchange signaling messages to establish data forwarding paths. The handover in this case is performed via R3 reference point with tunneling over R4 to transfer undelivered packets.

Figure 2-6 illustrates three different mobility scenarios supported in WiMAX networks. When the mobile station moves from positions 1 to 2, or 1 to 3, an ASN-anchored mobility through R8 or R6 reference points, respectively, is implied, whereas moving from position 1 to 4 involves a CSN-anchored mobility scheme though R3 reference point.

2.6.1 ASN-anchored Mobility

ASN-anchored mobility management is defined as mobility scenario not involving a care-of address update (i.e., an MIP re-registration). The procedures described for ASN-anchored mobility management also apply for mobility in networks that are not based on mobile IP. The functional requirements for ASN-anchored mobility management include support of different mobility classes, minimization of packet loss and handover latency while maintaining packet ordering, compliance with the security architecture defined in the IEEE 802.16-2009 specification and IETF EAP protocols, support of private addresses allocated by the H-NSP or V-NSP, support of Macro-Diversity Handover (MDHO) and Fast Base Station Switching (FBSS) handover mechanisms, support of MS in active, idle, and sleep modes (see Chapter 6 for details), minimization of the number of roundtrip signaling between BS and

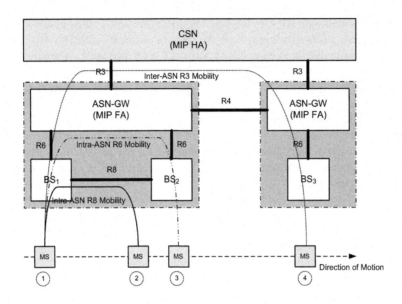

FIGURE 2-6

Mobility scenarios supported in the WiMAX network architecture

intra-ASN mobility anchor point to execute handover, independence of handover control primitives, and data path enforcement control primitives such that it allows separation of handover control and data path enforcement control, and consideration for QoS.

The ASN-anchored mobility management is defined by data path function (DPF), which manages the data path setup and includes procedures for data packet transmission between two functional entities, handover function, which controls overall handover decision and signaling procedures related to handover, and context function, which performs the exchanges required in order to set-up any state or retrieve any state in network elements. Each of these functions is viewed as a peer-to-peer inter-action corresponding to the function.

The DPF manages data-plane establishment between the two peers. This includes set-up of any tunnels and/or additional functionality that may be required for handling the bearer-plane. The DPF is used to set-up the bearer-plane between base stations, other entities such as gateways, or between gateways and base stations. Any additional requirements such as support of multicast or broadcast can be also handled by this function. The DPF further supports the use of packet sequence number.

Each DPF is responsible for instantiating and managing bearer path with another DPF, and for selecting the payload traversing the established bearer path. There are two types of data path functions:

1. **DPF Type 1** is used for IP or Ethernet packet forwarding with Layer 2 or Layer 3 transport. For DPF Type 1, the IP or Ethernet packets are forwarded using Layer 2 bridging (e.g., Ethernet protocols) or Layer 3 routing (e.g., GRE or IP-in-IP protocols) between two DPFs. Additional semantics can be applied to the transport header and payload to handle scenarios such as header compression and sequenced delivery.

2. **DPF Type 2** is a forwarding scheme with Layer 2 or Layer 3 transport. For DPF Type 2, data path bearer is also typically a generic Layer 3 tunneling protocol such as IP-in-IP or GRE, or a Layer 2 bridging protocol such as Ethernet. The payload is a Layer 2 data packet which is defined as an IEEE 802.16-2009 MAC Service Data Unit (MSDU) or part of it appended with additional information such as Connection Identifier (CID) of the target BS, Automatic Repeat reQuest (ARQ) parameters, etc. In DPF Type 2, Layer 2 session state (e.g., ARQ state) is anchored in the Anchor DPF.

The data path function can be further classified based on its role in handover and initial entry procedures as follows:

- **Anchor DPF** is a DPF located at one end of the data path, which anchors the data path associated with the MS during handovers. This function forwards the received data packet to the serving DPF using either of the aforementioned types. This function buffers the data packets from the network and maintains some state information related to bearer for MS during handovers.
- **Serving DPF** is a DPF located at other end of a data path, which currently has association with the serving PHY/MAC functions and takes charge of transmission of all messages associated with the corresponding MS. This DP function, associated with a serving BS, communicates with the Anchor DP function through Type-1 or Type-2 data path, to forward/receive MS data packets.
- **Target DPF** is a DPF which has been selected as the target for the handover. This DPF, which is associated with a target BS, communicates with the anchor DPF to prepare a data path to replace the current data path following the completion of the handover. Upon successful completion of handover, it will assume the role of serving DPF.
- **Relaying DPF** is a DPF which helps transfer information between the serving, target, and anchor DPFs.

The HandOver (HO) function makes the handover decisions and performs the signaling procedures related to handover. The handover function supports both mobile-initiated and network-initiated handover mechanisms. The handover function can be further classified by its role in handover operation as follows:

- **Serving HO Function** which controls overall handover decision operation and signaling procedures related to handover. It signals the target handover function via zero or more relaying handover functions to prepare for handover, and sends the result to the MS.
- **Relaying HO Function** relays handover-related control messages between the serving and target handover functions. A relaying handover function may modify the content of handover messages and further impact handover decisions.
- **Target HO Function** selects the target or a potential target for the handover.

There is an MS-related context in the network or network-related context in the MS that needs to be either transferred and/or updated due to intra-NAP mobility. More specifically, the MS specific context in the context function associated with the serving or anchor handover function needs to be updated. Furthermore, the MS specific context in the context function associated with the serving handover function needs to be transferred to the context function associated with the target handover function. This will also require some network specific context in the MS to be updated. The WiMAX network architecture defines the primitives between peer context functions that are used to transfer MS specific context between a context function acting as context-server and a context function acting as context-client. The information transfer concerning a specific MS can be triggered to populate the context corresponding to the MS at the target base station, to inform the network regarding the idle mode behavior of the MS, and to inform the network of initial network entry of a specific MS. The context function can be further classified into *Context-Server*, i.e., the context function is the repository of the most updated session context information for the MS, *Context-Client*, i.e., the context function associated with the functional entity physical air interface link, and *Relaying Context Function*, i.e., the context function which relays information between the context server and the context client.

2.6.2 CSN-anchored Mobility

The CSN-anchored mobility refers to mobility across different ASNs (see Figure 2-6) or alternatively to mobility across different IP subnets, and thereby requires network layer mobility management. The mobile IP protocols are used to manage mobility across IP subnets [29], and to enable CSN-anchored mobility. This section describes mobile IP based macro-mobility between the ASN and CSN across R3 reference point. In the case of IPv4, this implies re-anchoring of the current FA to a new FA, and the consequent binding updates (or MIP re-registration) to update the upstream and downstream data forwarding paths. In CSN-anchored mobility, the anchor mobile IP FA of the MS is changed. The new FA and CSN exchange messages to establish a data forwarding path. The CSN-anchored mobility management is established between ASN and CSN that are in the same or different administrative domains. The mobility management may further extend to handovers across ASNs in the same administrative domain. The procedures for CSN-anchored mobility management and the change of MS point of attachment to the ASN may not be synchronized. In this case, the procedures may be delayed relative to the completion of link layer handover by the MS.

In an intra-NAP R3 mobility scenario, an MS is moving between FAs within a single NAP domain. The R3 mobility event results in a handover between two FAs, thereby relocating the ASN R3 reference anchor point in the NAP. Note that R3 mobility does not automatically terminate or otherwise interfere with idle/sleep operation of the MS. The CSN-anchored mobility accommodates the scenario in which the MS remains in idle state or sleep mode until it is ready to transmit uplink traffic or is notified of downlink traffic by the serving BS. In all non-roaming scenarios, the HA is located in the CSN of H-NSP. For roaming scenarios, the HA is located in the CSN of either the H-NSP or V-NSP, depending on roaming agreement between H-NSP and V-NSP, user subscription profile and policy in H-NSP. The CSN-anchored mobility within a single NAP administrative domain does not introduce significant latency and packet loss. A make-before-break handover operation (i.e., when a data path is established between the MS and target BS before the data path with the serving BS is broken) is feasible within the same NAP administrative domain. To accomplish this procedure, the previous anchor FA maintains data flow continuity while signaling to establish the data path to a new anchor FA. The PMIP procedures do not require additional signaling over-the-air or additional data headers to perform CSN-anchored mobility. The CSN-anchored mobility activities are transparent to the MS. The MS uses Dynamic Host Configuration Protocol (DHCP) for IP address assignment and host configuration. DHCP is a network application protocol used by devices to obtain configuration information for operation in an IP network. This protocol reduces system administration workload, allowing devices to be added to the network with minimal user intervention [30].

2.7 PAGING AND IDLE STATE OPERATION

Paging refers to procedures used by the network to notify mobile stations (in the coverage area of a predefined set of base stations identified as a paging group) that are in the idle mode of pending downlink traffic. In addition, location update refers to procedures to obtain location update by an MS in the idle mode. Paging procedures are implemented using MAC management messages exchanged between the MS and the BS controlled through higher-layer paging management functions. The Idle State refers to a mobile station state where the MS can become periodically available for downlink broadcast traffic without registration with a specific BS. The Idle State comprises two separate modes, paging available mode and paging unavailable mode. During Idle State, the MS may attempt power saving by switching between Paging Available mode and Paging Unavailable mode. In the Paging Available mode, the MS may be paged by the BS. If the MS is paged, it performs network re-entry procedures. The MS performs location update procedure during Idle State. As shown in Figure 2-7, the paging reference model can be decomposed into three separate functional entities:

- **Paging Controller** (PC) is a functional entity that administers the activity of an MS in Idle State in the network. It is identified by a PC identifier [12], and can either be collocated with the BS or separate from the BS across R6 reference point. There are two types of PCs: *Anchor PC* that maintains the updated location information of the MS; and *Relay PC* that participates in relaying of paging and location management messages between paging agent and the anchor PC.

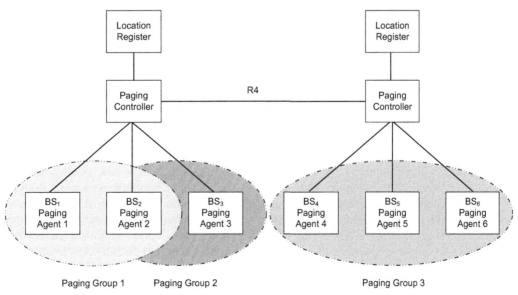

FIGURE 2-7

The paging network reference model

- **Paging Agent (PA)** is a functional entity that manages the interaction between the PC and the IEEE 802.16-2009 standard specified paging-related functionality implemented in the BS. A PA is collocated with the BS.
- **Paging Group (PG)**, as defined in the IEEE 802.16-2009 standard, consists of one or more Paging Agents. A Paging Group resides entirely within a NAP boundary and is managed and provisioned by the network operator.
- **Location Register (LR)** is a distributed database with each instance corresponding to an Anchor PC. Location registers contain information about mobile stations in Idle State. The information for each MS includes current Paging Group ID (PGID), paging cycle, paging offset, last reported BS Identifier (BS ID), last reported Relay PC ID, etc [5].

There are typically multiple paging groups inside an operator's network. A BS and its corresponding collocated PA can be part of more than one paging group. The paging controller can be collocated with the BS. The paging control messages between PCs are exchanged across R4 reference point.

The LR consists of a location database which is accessible through PC and tracks the current paging group of each mobile station in Idle State within the network. It also stores the context information required for paging. When the MS moves across paging groups, location update occurs across PCs via R6 and/or R4 reference points, and information is updated in the LR that is associated with the anchor PC. The LR entry is created when an MS enters Idle State. It is required to perform location update when an Idle State MS moves and crosses the boundary of its current paging group.

2.8 OVERVIEW OF 3GPP EVOLVED PACKET CORE NETWORK ARCHITECTURE

The Third Generation Partnership Project (3GPP) initiated work on Long-Term Evolution (LTE) in late 2004. The Evolved UMTS Terrestrial Radio Access Network (E-UTRAN) has substantially improved end-user throughputs, sector capacity, and has reduced user-plane and control-plane latencies, bringing significantly improved user experience with full mobility. With the emergence of Internet protocol as the protocol of choice for carrying all types of traffic, 3GPP LTE provides support for IP-based traffic with end-to-end Quality of Service (QoS). Voice traffic is supported mainly as voice-over-IP, enabling integration with other multimedia services. Initial deployments of 3GPP LTE are expected by 2011, and commercial availability on a larger scale will likely happen a few years later.

Unlike its predecessors, which were developed within the framework of Release 99 UMTS architecture, 3GPP has specified the Evolved Packet Core (EPC) architecture to support the E-UTRAN through reduction in the number of network elements and simplification of functionality, but most importantly allowing for connections and handover to other fixed and wireless access technologies, providing the network operators with the ability to deliver a seamless mobility experience. The main objectives of 3GPP LTE were to minimize the system and User Equipment (UE) complexities, to allow flexible spectrum deployment in the existing or new frequency bands, and to enable coexistence with other 3GPP radio access technologies. Some general principles taken into consideration in the design of E-UTRAN architecture, as well as the E-UTRAN interfaces, are as follows [31]:

- Signaling and data transport networks are logically separated;
- E-UTRAN and EPC functions are fully separated from transport functions. Addressing schemes used in E-UTRAN and EPC are not associated with the addressing schemes of transport

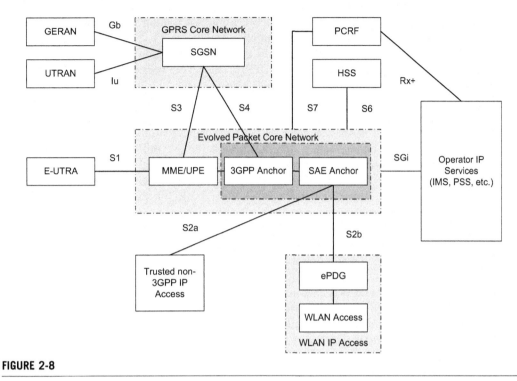

FIGURE 2-8

The 3GPP evolved packet core network reference model [32,33]

functions. The fact that some E-UTRAN or EPC functions reside in the same equipment as some transport functions does not make the transport functions part of the E-UTRAN or the EPC;

- The mobility for RRC connection is fully controlled by the E-UTRAN;
- The interfaces should be based on a logical model of the entity controlled through this interface;
- One physical network element can implement multiple logical nodes.

Figure 2-8 illustrates the EPC network reference model. The functional entities and their corresponding interfaces are defined as follows [32,33]:

Functional Entities

- **Mobility Management Entity (MME)** manages and stores UE idle state context (e.g., UE/user identities, UE mobility state, and user security parameters). It generates temporary identities and allocates them to UEs. It checks authorization as to whether the UE may camp on the Tracking Area (TA) or on the Public Land Mobile Network (PLMN).[ix] It also authenticates the user.
- **User-Plane Entity (UPE)** terminates the idle-state UE, downlink data path and triggers/initiates paging when downlink data is available for the UE. It manages and stores UE context

---

[ix]A Public Land Mobile Network (PLMN) is a network established and operated by an administration or by a recognized operating agency for the specific purpose of providing land/mobile telecommunications services to the public.

(e.g., parameters of the IP bearer service or network internal routing information). It performs replication of the user traffic in case of interception.

- **Evolved Packet Data Gateway** (**ePDG**) secures the access of a UE to the EPC (for non-trusted non-3GPP access) by means of an IP Security (IPSec)[x] tunnel between itself and the UE.
- **3GPP Anchor** manages mobility for 2G/3G and LTE systems.
- **SAE Anchor** manages mobility for non-3GPP radio access technologies.
- **Serving GPRS Support Node** (**SGSN**) provides connections for GSM EDGE Radio Access Network (GERAN) and UTRAN networks, and is responsible for the delivery of data packets from and to the mobile stations within its geographical service area. Its tasks include packet routing and transfer, mobility management (i.e., attachment and location management), logical link management, and authentication and charging functions. The location register of the SGSN stores location information and user profiles of all GPRS users registered with this SGSN.
- **Policy and Charging Rule Function** (**PCRF**) manages QoS aspects and encompasses the following functions for IP Connectivity Access Network: flow-based charging including charging control and online credit control and policy control (e.g., gating control, QoS control, QoS signaling, etc.) [32].
- **Home Subscriber Server** (**HSS**) is a master user database that supports the IMS network entities and contains the subscriber profiles. It performs authentication and authorization of the user, and can provide information about the subscriber's location and IP information. It is similar to the GSM Home Location Register (HLR) and Authentication Center (AUC).

Reference Points

- **S1** provides access to E-UTRA radio resources for the transport of user-plane and control-plane traffic/signaling. The S1 reference point enables MME and UPE separation, and further allows deployments of combined MME and UPE components.
- **S2a** provides the user-plane related control and mobility support between a trusted non-3GPP IP access and the System Architecture Evolution (SAE) Anchor.
- **S2b** provides the user-plane related control and mobility support between ePDG and the SAE Anchor.
- **S3** enables user and bearer information exchange for inter-3GPP access system mobility in idle and/ or active state. It is based on Gn reference point as defined between Serving GPRS Support Nodes (SGSN). The General Packet Radio Service (GPRS) is a packet-switched mobile data service available to Global System for Mobile communications (GSM) users, as well as in the 3G systems.
- **S4** provides the user-plane with related control and mobility support between GPRS Core and the 3GPP Anchor, and is based on Gn reference point as defined between SGSN and Gateway GPRS Support Node (GGSN).
- **S5a** provides the user-plane with related control and mobility support between MME/UPE and 3GPP anchor.

---

[x]IP Security (IPsec) is a set of protocols for securing IP-based communications by authenticating and encrypting each IP packet of a data stream. IPSec also includes protocols for establishing mutual authentication between agents at the beginning of the session and negotiation of cryptographic keys to be used during the session. IPSec can be used to protect data flows between a pair of hosts, between a pair of security gateways, or between a security gateway and a host [3].

- **S5b** provides the user-plane with related control and mobility support between 3GPP Anchor and SAE Anchor.
- **S6** enables transfer of subscription and authentication data for authenticating/authorizing user access to the evolved system (i.e., AAA interface).
- **S7** provides transfer of QoS policy and charging rules from Policy Charging Rule Function (PCRF) to Policy and Charging Enforcement Point (PCEP).
- **SGi** is the reference point between the EPC and the packet data network. Packet data network is an operator's external public/private network or an intra-operator packet data network for provision of IMS services.
- **Iu** and **Gb** are interfaces with UTRAN and GERAN, respectively.

The E-UTRAN architecture is shown in Figure 2-9. The E-UTRAN comprises Evolved NodeB (eNB), or equivalently E-UTRA base stations, providing the E-UTRA user-plane and control-plane protocol terminations towards the user equipment. Note that the Radio Network Controller (RNC) functions have been included in the eNB to reduce the architectural complexity and to further reduce the latency across the network. The eNBs are interconnected with each other through X2 interface [31]. The eNBs are also connected by means of the S1 interface to the EPC, or more specifically to the MME through the S1-MME reference point and to the Serving Gateway (S-GW) via the S1-U interface [32]. The S1 interface supports a multipoint connection among MMEs/S-GWs and eNBs. The E-UTRAN overall architecture is described in references [31-33].

The following functions are typically performed by eNB [31]:

- Radio resource management that includes radio bearer control, radio admission control, connection management, dynamic allocation of resources to UEs in both uplink and downlink (i.e., scheduling);
- Header compression and encryption of user payloads;

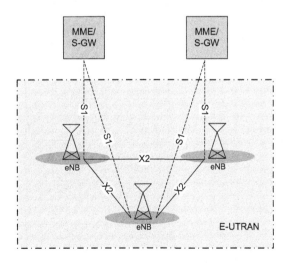

FIGURE 2-9

E-UTRAN architecture [31]

- Selection of an MME at UE attachment when no routing to an MME can be determined from the information provided by the UE;
- Routing of U-plane data toward S-GW;
- Scheduling and transmission of paging messages (originated from the MME);
- Scheduling and transmission of broadcast information (originated from the MME);
- Measurement and reporting for support of mobility and scheduling.

Figure 2-10 illustrates the functional decomposition of E-UTRAN and EPC [31]. The MME is the main control node for the 3GPP LTE access network. It is responsible for idle-mode UE tracking and paging procedure including re-transmissions. The MME functions include Non-Access Stratum (NAS) signaling and NAS signaling security, inter-core-network signaling for mobility between 3GPP access networks, idle mode UE accessibility (including control and execution of paging re-transmission), tracking area list management (for UE in idle and active mode), Packet Data Network (PDN) gateway and S-GW selection, MME selection for handovers with MME change, roaming, authentication of the users, and bearer management functions including dedicated bearer establishment [31]. The NAS signaling terminates at the MME and it is also responsible for generation and allocation of temporary identities to the UEs. It verifies the authorization of the UE to camp on the service provider's Public Land Mobile Network (PLMN) and enforces UE roaming restrictions.

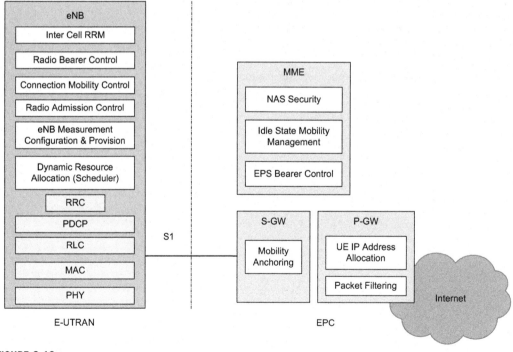

FIGURE 2-10

The functional split between E-UTRAN and EPC [31]

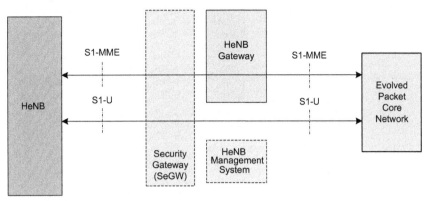

FIGURE 2-11

E-UTRAN HeNB logical architecture [31]

The Serving Gateway (S-GW) routes and forwards user data packets, while also acting as the mobility anchor for the user-plane during inter-eNB handovers and as the anchor for mobility between LTE and other 3GPP technologies. The S-GW functions include local mobility anchor point for inter-eNB handover, mobility anchoring for inter-3GPP mobility, E-UTRAN idle mode downlink packet buffering and initiation of network triggered service request procedure, lawful interception, packet routing and forwarding, and transport-level packet marking in the uplink and the downlink [31].

The Packet Data Network Gateway (P-GW) provides connectivity of the UE to external packet data networks by being the point of exit and entry of traffic for the UE. A UE may have simultaneous connectivity with more than one P-GW for accessing multiple packet data networks. The P-GW functions include per-user packet filtering, lawful interception, UE IP address allocation, and transport-level packet marking in the downlink [31].

The baseline 3GPP network architecture can be further extended to include femto-cells.[xi] [31] Figure 2-11 shows a logical network reference model for the support of femto-cells, or alternatively Home eNB (HeNB) that includes a set of S1 interfaces to connect the HeNB to the EPC. The configuration and authentication entities as shown here are common between HeNBs and Home NBs (HNB).

The HeNB GW serves as a focal point for the control-plane, particularly the S1-MME interface. The S1-U interface from the HeNB may be terminated at the HeNB GW, or a direct logical user-plane connection between HeNB and S-GW may be used. Release 9 of 3GPP specifications do not support X2 interface between HeNBs. The extended definition of S1 reference point includes the interface between the following entities:

- HeNB GW and the core network;
- HeNB and the HeNB GW;

[xi]A femto-cell is a small cellular base station typically designed for use in residential or small office environments. It connects to the operator's network via a broadband connection. A femto-cell allows the operators to extend service coverage indoors, especially where network access would otherwise be limited or unavailable. The femto-cell incorporates reduced functionalities of a typical base station and allows a simpler and self-organized deployment.

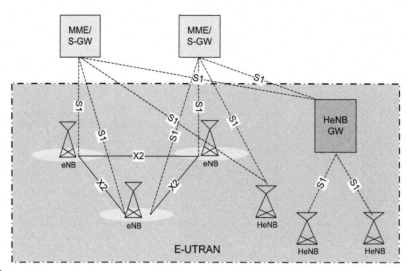

FIGURE 2-12

Overall E-UTRAN architecture including HeNBs [31]

- HeNB and the core network;
- eNB and the core network.

As shown in Figure 2-12, the HeNB GW appears to the MME as an eNB, and to the HeNB as an MME. The S1 interface between the HeNB and the EPC is the same regardless of whether the HeNB is connected to the EPC through a HeNB GW. The HeNB GW is connected to the EPC in a way that inward and/or outward movement toward the cells served by the HeNB GW does not necessarily require inter-MME handovers. The functions supported by the HeNB are similar to those supported by an eNB (with the exception of Non Access Stratum Node Selection Function [31]) and the protocols between a HeNB and the EPC are the same as those between an eNB and the EPC.

The HeNB performs the same functions as an eNB, as described earlier with the following considerations in case of connection to the HeNB GW:

- Discovery of a suitable serving HeNB GW;
- A HeNB only connects to a single HeNB GW at one time;
- If the HeNB is connected to a HeNB GW, it will not simultaneously connect to another HeNB GW or another MME;
- The Type Approval Code (TAC) and PLMN identifier used by the HeNB is also supported by the HeNB GW [31];
- When the HeNB connects to a HeNB GW, selection of an MME at UE attachment is hosted by the HeNB GW instead of the HeNB;
- HeNBs may be deployed without network planning. An HeNB may be moved from one geographical area to another, and therefore it may need to connect to different HeNB GWs depending on its location.

The HeNB GW performs the following functions:

- Relaying UE-associated S1 messages between the MME serving the UE and the HeNB serving the UE;
- Terminating non-UE associated S1 procedures towards the HeNB and towards the MME. Note that when an HeNB GW is deployed, non-UE associated procedures are run between HeNBs and the HeNB GW, and between the HeNB GW and the MME and optionally terminating S1-U interface with the HeNB and with the S-GW;
- Supporting TAC and PLMN identifier used by the HeNB.

In addition to functions listed earlier, the MME provides access control for UEs that are members of Closed Subscriber Groups (CSG) in HeNB deployments: a CSG HeNB is only accessible to a group of UEs which are members of the CSG. Access would be denied to the UEs that are not members of CSG, except for emergency services.

References

[1] Douglas E. Comer, Internetworking with TCP/IP Vol. 1: Principles, Protocols, and Architecture, fifth ed., Prentice Hall, 2005.
[2] IETF RFC 3261, J. Rosenberg, H. Schulzrinne, G. Camarillo, A. Johnston, J. Peterson, et al., SIP: Session Initiation Protocol, June 2002.
[3] Internet Protocol Security (IPsec), <http://en.wikipedia.org/wiki/Ipsec>.
[4] Miikka Poikselka, Georg Mayer, IMS: IP Multimedia Concepts and Services, third ed., John Wiley & Sons, 2009.
[5] WiMAX Forum Network Architecture Stage 2: Architecture Tenets, Reference Model and Reference Points Release 1.5, September 2009. <http://www.wimaxforum.org/resources/documents/technical/release>.
[6] Recommendations and Requirements for Networks based on WiMAX Forum Certified Products, Release 2.0, May 2009 <http://www.wimaxforum.org/resources/documents/technical/release>.
[7] 3GPP TS 23.401, General Packet Radio Service (GPRS) Enhancements for Evolved Universal Terrestrial Radio Access Network (E-UTRAN), March 2009. (ftp://ftp.3gpp.org/specs/).
[8] Virtual Private Network (VPN). <http://en.wikipedia.org/wiki/Virtual_private_network>.
[9] IETF RFC 3344, in: C. Perkins (Ed.), IP Mobility Support for IPv4, August 2002.
[10] IETF RFC 2460, S. Deering, R. Hinden, Internet Protocol, Version 6 (IPv6) Specification, December 1998.
[11] WiMAX Forum Mobile System Profile Specification Release 1.5, February 2009. <http://www.wimaxforum.org/resources/documents/technical/release>.
[12] IEEE Std 802.16-2009, IEEE Standard for Local and Metropolitan Area Networks – Part 16: Air Interface for Broadband Wireless Access Systems, May 2009.
[13] Kamran Etemad, "Overview of Mobile WiMAX Technology and Evolution", IEEE Communications Magazine, October 2008.
[14] Jeffrey G. Andrews, Arunabha Ghosh, Rias Muhamed, Fundamentals of WiMAX: Understanding Broadband Wireless Networking, first ed., Prentice Hall, 2007.
[15] Loutfi Nuaymi, WiMAX: Technology for Broadband Wireless Access, first ed., John Wiley & Sons Ltd, 2007.
[16] IETF RFC 2138, C. Rigney, A. Rubens, W. Simpson, S. Willens, Remote Authentication Dial in User Service (RADIUS), April 1997.
[17] IEFT RFC 3588, P. Calhoun, J. Loughney, E. Guttman, G. Zorn, J. Arkko, Diameter Base Protocol, September 2003.

[18] Yan Zhang, Jun Zheng, Miao Ma, Handbook of Research on Wireless Security, Information Science Publishing 2008.

[19] 3GPP TS 11.11, Specification of the Subscriber Identity Module – Mobile Equipment (SIM-ME) Interface, June 2003. (ftp://ftp.3gpp.org/specs/).

[20] Jae Hyung Joo, Jeong-Jun Suh, Young Yong Kim, "Secure Remote USIM (Universal Subscriber Identity Module) Card Application Management Protocol for W-CDMA Networks", International Conference on Consumer Electronics 2006. ICCE '06, January 2006.

[21] UICC (Universal Integrated Circuit Card). <http://en.wikipedia.org/wiki/UICC>.

[22] 3GPP2 CS0023-C v2.0, Removable User Identity Module for Spread Spectrum Systems (RUIM), October 2008. <http://www.3gpp2.org/Public_html/Specs/>.

[23] ITU-T X.509, Information technology – Open Systems Interconnection – The Directory: Public-key and attribute certificate frameworks, November 2008. <http://www.itu.int/rec/T-REC-X.509/en>.

[24] Mobile IP <http://en.wikipedia.org/wiki/Mobile_ip>.

[25] "Cisco Home Agent Redundancy and Load Balancing", White Paper, Cisco Systems, 2009. <http://www.cisco.com/en/US/prod/collateral/wireless/wirelssw/ps5940/prod_white_paper0900aecd802921f0.html>.

[26] IETF RFC 4023, in: T. Worster, Y. Rekhter, E. Rosen (Eds.), Encapsulating MPLS in IP or Generic Routing Encapsulation (GRE), March 2005.

[27] IETF RFC 4332, K. Leung, A. Patel, Cisco' s Mobile IPv4 Host Configuration Extensions, December 2005.

[28] IETF RFC 5213, Gundavelli, Ed., K. Leung, V. Devarapalli, K. Chowdhury, B. Patil, S. Proxy Mobile IPv6, August 2008.

[29] IP subnet <http://en.wikipedia.org/wiki/Subnetwork>.

[30] IETF RFC 2131, R. Droms, Dynamic Host Configuration Protocol, March 1997.

[31] 3GPP TS 36.300, Evolved Universal Terrestrial Radio Access (E-UTRA) and Evolved Universal Terrestrial Radio Access Network (E-UTRAN); Overall description; Stage 2, March 2010. (ftp://ftp.3gpp.org/specs/).

[32] 3GPP TR 23.882, 3GPP System Architecture Evolution: Report on Technical Options and Conclusions, September 2008. (ftp://ftp.3gpp.org/specs/).

[33] 3GPP TS 23.402, Architecture Enhancements for Non-3GPP Accesses, March 2009. (ftp://ftp.3gpp.org/specs/).

IEEE 802.16m Reference Model and Protocol Structure

INTRODUCTION

The IEEE 802.16-2009 standard defines a generic reference model where major functional blocks (i.e., physical layer, security sub-layer, MAC common part sub-layer, and service specific convergence sub-layer) and their interfaces, the premises of IEEE 802.16 entity, and a general network control and management system are specified. The IEEE 802.16m has modified this reference model by further classifying the MAC common part sub-layer functions into two functional groups, resulting in a more structured approach to characterizing the data link layer functions and their interoperation.

The earlier revisions and/or amendments of the IEEE 802.16 standard did not explicitly define any detailed protocol structure; rather, the functional elements in the specification were implicitly classified as convergence sub-layer, MAC common part sub-layer, security sub-layer, and physical layer. While each of these layers and/or sub-layers comprises constituent functions and protocols, no perspective was provided on how various components were interconnected and interoperated from a system standpoint. In fact, the IEEE 802.16 standards have never been developed with a system engineering approach; rather, they specify components and building blocks that can be integrated (obviously various combinations are potentially possible) to build a working and performing system. An example is the mobile WiMAX system profiles [1], where a specific set of IEEE 802.16-2009 features were selected to form a mobile broadband wireless access system. In an attempt to improve the clarity of the previous IEEE 802.16 standards and to take a systematic approach in development of the advanced air interface, IEEE 802.16m has defined a protocol structure and the functional components are classified into different layers and sub-layers, as well as differentiated based on data-plane or control-plane categories.

The protocols and functional elements defined by the IEEE 802.16 standard correspond to the physical and data link layers of the Open System Interconnection (OSI) seven-layer network reference model as shown in Figure 3-1.

In the context of protocol structure, we will frequently use the terms "service" and "protocol." It must be noted that services and protocols are distinct concepts. A service is a set of primitives or operations that a layer provides to the layer(s) with which it is interfaced [2]. The service defines what operations a layer performs without specifying how the operations are implemented. It is further related to the interface between two adjacent layers. A protocol, in contrast, is a set of rules presiding over the format and interpretation of the information/messages that are exchanged by peer entities within a layer. The entities use protocols to implement their service definitions. Thus, a protocol is related to the implementation of a service.

Mobile WiMAX. DOI: 10.1016/B978-0-12-374964-2.10003-7

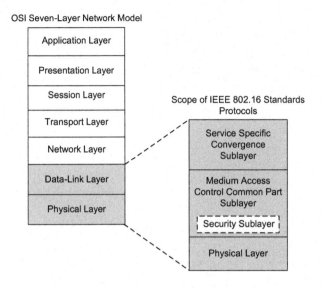

FIGURE 3-1

The mapping of IEEE 802.16 protocol layers to an OSI seven-layer network model

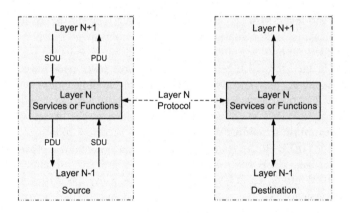

FIGURE 3-2

An illustration of service, protocol, PDU, and SDU concepts [2]

As shown in Figure 3-2, a Protocol Data Unit (PDU) is a packet exchange between peer entities of the same protocol layer located at the source and destination. On the downward direction, the PDU is the data unit generated for the next lower layer. On the upward direction, it is the data unit received from the previous lower layer. A Service Data Unit (SDU), on the other hand, is a data unit exchanged between two adjacent protocol layers. On the downward direction, the SDU is the data unit received from the previous higher layer. On the upward direction, it is the data unit sent to the next higher layer.

This chapter provides a top-down systematic description of IEEE 802.16m reference model and protocol structure, starting at the most general level and working toward details or specifics of the protocol layers, their functional constituents and interconnections. An overview of 3GPP LTE protocol structure is further provided to enable readers to contrast the corresponding protocols and functionalities.

It must be noted that while the IEEE 802.16 standard does define a generic network reference model (or a network abstraction model), the mobile WiMAX systems use the specific network reference model and system architecture that were described in Chapter 2 to achieve interoperability. Therefore, the network reference model and associated components and interfaces described in Section 3.1 are only informative, and they should not be interpreted as normative for implementation and deployment of the IEEE 802.16m systems.

3.1 THE IEEE 802.16M REFERENCE MODEL

Figure 3-3 illustrates the IEEE 802.16 reference model [3]. The data link layer of IEEE 802.16 standard comprises three sub-layers. The service-specific convergence sub-layer (CS) provides any

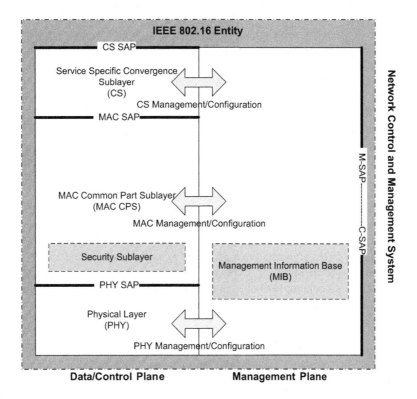

FIGURE 3-3

The IEEE 802.16 reference model [3]

transformation or mapping of network-layer data packets into MAC SDUs. On the transmitter side, the CS receives the data packets through the CS Service Access Point (SAP) and delivers MAC SDUs to the MAC Common Part Sub-layer (MAC CPS) through the MAC SAP. This includes classifying network-layer SDUs and associating them with the proper MAC Service Flow Identifiers (SFID) and Connection Identifiers (CID). The convergence sub-layer also includes payload header suppression function to compress the higher-layer protocol headers. Multiple CS specifications are provided for interfacing with various network-layer protocols such as Asynchronous Transfer Mode (ATM)[i] and packet-switched protocols such as IP or Ethernet. The internal format of the CS payload is unique to the CS, and the MAC CPS is not required to understand the format of or parse any information from the CS payload.

The MAC CPS provides the core MAC functionality of system access, bandwidth allocation, connection establishment, and connection maintenance. It can receive data from the various convergence sub-layers, through the MAC SAP classified into particular MAC connections. An example of MAC CPS service definition is given in reference [3]. The Quality of Service (QoS) is further applied to the transmission and scheduling of data over the physical layer.

The MAC also contains a separate security sub-layer providing authentication, secure key exchange, and encryption. The user data, physical layer control, and statistics are transferred between the MAC CPS and the Physical Layer (PHY) via the PHY SAP which is implementation-specific.

The IEEE 802.16 physical layer protocols include multiple specifications, defined through several amendments and revisions, each appropriate for a particular frequency range and application. The IEEE 802.16 compliant devices include mobile stations or base stations. Given that the IEEE 802.16 devices may be part of a larger network, and therefore would require interfacing with entities for management and control purposes, a Network Control and Management System (NCMS) abstraction has been introduced in the IEEE 802.16 standard as a "black box" containing these entities [3]. The NCMS abstraction allows the physical and MAC layers specified in the IEEE 802.16 standard to be independent of the network architecture, the transport network, and the protocols used in the backhaul, and therefore would allow greater flexibility. The NCMS entity logically exists at both BS and MS sides of the radio interface. Any necessary inter-BS coordination is coordinated through the NCMS entity at the BS. An IEEE 802.16 entity is defined as a logical entity in an MS or BS that comprises the physical and MAC layers on the data, control, and management planes.

The IEEE 802.16f amendment (currently part of IEEE 802.16-2009 standard [3]) provided enhancements to IEEE 802.16-2004 standard, defining a management information base (MIB), for the physical and medium access control layers and the associated management procedures. The management information base originates from the Open Systems Interconnection Network Management Model and is a type of hierarchical database used to manage the devices in a communication network [5,6]. It comprises a collection of objects in a virtual database used to manage entities such as routers and switches in a network.

---

[i]Asynchronous Transfer Mode (ATM) is a packet switching protocol that encodes data into small fixed-sized cells and provides data link layer services that run over OSI layer 1, differing from other technologies based on packet-switched networks such as IP or Ethernet, in which variable-sized packets are used. ATM exploits properties of both circuit-switched and small packet-switched networks, making it suitable for wide area data networking, as well as real-time media transport. ATM uses a connection-oriented model and establishes a virtual circuit between two end-points before the actual data exchange begins [4].

The IEEE 802.16 standard describes the use of a Simple Network Management Protocol (SNMP),[ii] i.e., an IETF protocol suite, as the network management reference model. The standard consists of a Network Management System (NMS), managed nodes, and a service flow database. The BS and MS managed nodes collect and store the managed objects in the form of WirelessMAN Interface MIB and Device MIB that are made available to network management system via management protocols, such as SNMP. A Network Control System contains the service flow and the associated Quality of Service information that have to be provided to BS when an MS enters into the network. The Control SAP (C-SAP) and Management SAP (M-SAP) interface the control and management plane functions with the upper layers. The NCMS entity presents within each MS. The NCMS is a layer-independent entity that may be viewed as a management entity or control entity. Generic system management entities can perform functions through NCMS and standard management protocols can be implemented in the NCMS. If the secondary management connection does not exist, the SNMP messages, or other management protocol messages, may go through another interface in the customer premise or on a transport connection over the air interface. Figure 3-4 describes a simplified network reference

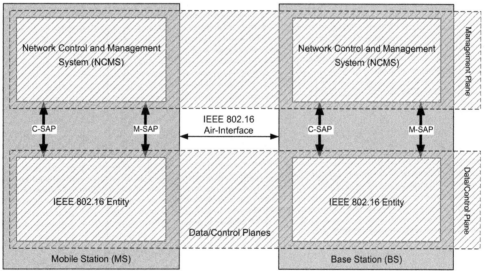

FIGURE 3-4

The IEEE 802.16 generic network reference model [3]

[ii]An SNMP-managed network consists of three key components: (1) a managed device; (2) an agent; and (3) a network management system. A managed device is a network node that contains an SNMP agent and resides in a managed network. Managed devices collect and store management information and make this information available to NMSs using SNMP. Managed devices, sometimes called network elements, can be any type of device including, but not limited to, routers, access servers, switches, etc. An agent is a network-management software module that resides in a managed device. An agent has local knowledge of management information and translates that information into a form compatible with SNMP. A network management system executes applications that monitor and control managed devices. The NMSs provide the processing and memory resources required for network management. One or more NMSs may exist on any managed network [7,8].

model. Multiple mobile stations may be attached to a BS. The MS communicates to the BS over the air interface using a primary management connection, basic connection or a secondary management connection [3]. The latter connection types have been replaced with new connection types in IEEE 802.16m standard [12].

3.1.1 The MS and BS Interface

The MAC management PDUs that are exchanged over the primary management connection[iii] can trigger, or are triggered by, primitives that are exchanged over either the C-SAP or the M-SAP, depending on the particular management or control operation. The messages that are exchanged over the secondary management connection can trigger or are triggered by primitives that are exchanged over the M-SAP. This interface is a set of SAP between an IEEE 802.16 entity and NCMS as shown in Figure 3-5. It consists of two parts: the M-SAP is used for delay-tolerant management-plane primitives; and the C-SAP is used for delay-sensitive control-plane primitives that support handovers,

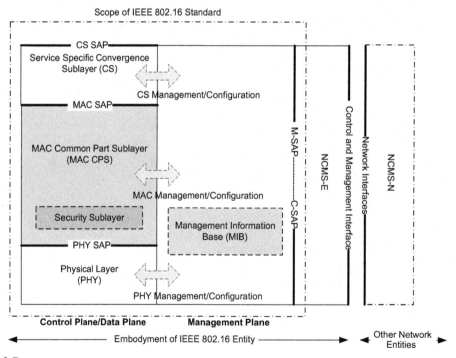

FIGURE 3-5

Partitioning of the IEEE 802.16 network control and management system [3]

[iii]A management connection is used for transporting MAC management messages or standards-based messages. The primary management connection is established during network entry and is used to transport delay-tolerant MAC management messages. The secondary management connection may be established during MS registration that is used to transport standards-based messages; e.g., SNMP, DHCP messages.

security context management, radio resource management, and low power operations such as idle mode and paging functions.

3.1.2 Network Control and Management System

The Network Control and Management System is not part of the IEEE 802.16 standards, and is treated as a "black box." It may be distributed with components residing on different nodes in a network. Part of the NCMS may be physically collocated with the IEEE 802.16 entity referred to as NCMS-E. The remaining part of the NCMS may be physically distributed across one or more network entities. This part of the NCMS is referred to as NCMS-N. Figure 3-5 shows the partitioning of the NCMS into NCMS-E and NCMS-N. The NCMS-E may have its own software platform and network protocol implementation, allowing it to communicate with external entities in the NCMS-N. The NCMS-E may provide an SNMP Agent compliant to IETF RFC3418 [13] and the SNMP/TCP/IP protocol stack, to allow for interactions with an SNMP manager. The NCMS-E may provide an Object Request Broker and implement a protocol stack to interact with components on other network entities within NCMS-N based on the CORBA architecture.[iv] The messages available to a manager in the NCMS-N are specified using Interface Description Language (IDL).[v] These messages encapsulate the interactions with the MIB. The IEEE 802.16 entity can be managed through Web Services.[vi] [11]

The decomposition of Network Control and Management System is depicted in Figure 3-6. These entities may be centrally located or distributed across the network. The exact functionality of these entities and their services is outside the scope of the IEEE 802.16 standard, but is shown here for illustration purposes and to allow description of the management and control procedures. The NCMS service manifestations on the MS and BS may have different configurations and functions.

The IEEE 802.16m reference model is very similar to that of the IEEE 802.16-2009 standard, with the exception of soft classification of MAC common part sub-layer into radio resource control and management and medium access control functions. As shown in Figure 3-7, this functional partitioning is logical, i.e., no SAP is required between the two classes of functions and no additional sub-headers are appended to the SDUs. Furthermore, the functional elements on the data and control paths are explicitly classified into data- and control-plane functions. While similar functionalities exist in the IEEE 802.16-2009 standard, the functions and protocols are not explicitly categorized in the legacy standard except explicit separation of PHY, MAC CPS, and CS functions in the specification [3].

The categorization of the functions based on functional characteristics and relative position in the data/signaling processing path would ease analogy, and contrast with other radio access technologies

---

[iv]The Common Object Requesting Broker Architecture (CORBA) is a standard defined by the Object Management Group that enables software components written in multiple computer languages and running on multiple computers to work together [9].

[v]An Interface Description Language (IDL) is a specification language used to describe a software component's interface. IDLs describe an interface in a language-independent way, enabling communication between software components that do not share a language; e.g., between components written in C++ and Java [10].

[vi]A Web Service is a software system designed to support interoperable machine-to-machine interaction over a network. It has an interface described in a machine-processable format. Other systems interact with the web service in a manner prescribed by its description using SOAP-messages (Simple Object Access Protocol is a lightweight protocol intended for exchanging structured information in a decentralized, distributed environment) typically conveyed using HTTP with an XML serialization in conjunction with other web-related standards.

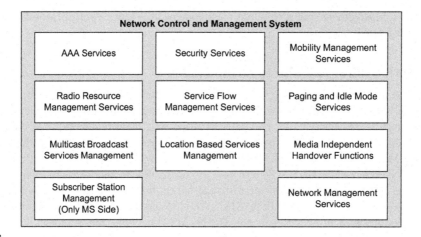

FIGURE 3-6

Decomposition of the network control and management system [3]

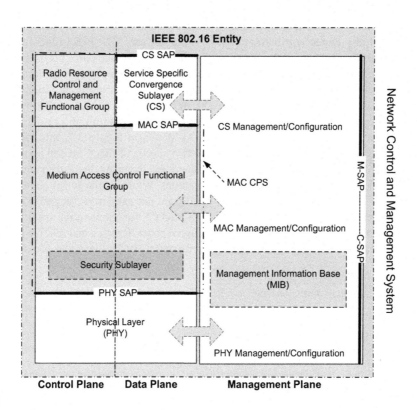

FIGURE 3-7

The IEEE 802.16m reference model [12]

such as 3GPP LTE/LTE-Advanced that have been designed based on similar structured protocol design methodology. Furthermore, the structured functional/protocol design in IEEE 802.16m would eliminate the inherent complexity and ambiguity of studying, understanding, and implementing the legacy standard.

It must be noted that there are new, modified or extended functions and protocols that are classified under generic classes of PHY, MAC CPS, and CS in IEEE 802.16m, where there are no counterparts in the legacy standard. Therefore, similarity of the reference models should not be interpreted as functional compatibility at the service access points. The backward compatibility of the IEEE 802.16m with the legacy standard ensures that non-compatible functions/protocols are not utilized in the time intervals where legacy base stations and mobile stations are supported in the network.

3.1.3 Data-Plane

The MAC and PHY functions of the IEEE 802.16m can be classified into three categories namely data-plane, control-plane, and management-plane. The data-plane (alternatively known as user-plane) comprises functions in the user data processing path, such as service flow classification and header compression, as well as MAC and PHY data packet processing and encryption functions. As shown in Figure 3-8, the IEEE 802.16m data-plane entity comprises the service specific Convergence Sub-layer,

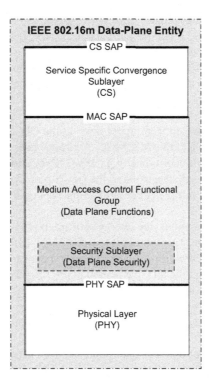

FIGURE 3-8

The IEEE 802.16m data-plane entity [12]

MAC functional group, security, and physical layer protocols corresponding to data-plane user packet processing. The MAC and PHY SAPs, while conceptually the same as those specified by IEEE 802.16-2009 standard, have different manifestation due to the new and/or modified MAC and PHY features introduced in the IEEE 802.16m standard. There is no significant change in CS SAP relative to that specified in the IEEE 802.16-2009 standard [3].

The CS provides a mapping of external network data formats (e.g., IP layer packets, ATM cells) received through CS SAP into MAC SDUs that are delivered to the MAC CPS via MAC SAP. This includes classifying external network SDUs and associating them with a proper Service Flow Identifier and Connection Identifier. A Service Flow (SF) is a MAC transport service that provides unidirectional transport of packets in the downlink or uplink [14]. It is identified by a 32-bit SFID. A service flow is characterized by a set of QoS parameters, i.e., a parameter set associated with a service flow identifier containing traffic parameters which define scheduling behavior of uplink or downlink service flows associated with transport connections. An admitted and active service flow is uniquely mapped to a CID. Note that the IEEE 802.16 standard supports two phase activation model. i.e., the resources for a service are first admitted or reserved and once the BS and MS negotiations are completed, the resources are activated [3].

The Generic Packet CS (GPCS) is a network layer protocol-agnostic packet convergence sub-layer that supports multiple network protocols over IEEE 802.16 air interface. The GPCS provides a generic packet convergence sub-layer. This layer uses the MAC SAP and exposes a SAP to GPCS applications. The GPCS does not redefine or replace other convergence sub-layers. Instead, it provides a SAP that is not protocol specific. With GPCS, packet parsing occurs above GPCS. The results of packet parsing are classification parameters provided to the GPCS SAP for parameterized classification; however, upper layer packet parsing is left to the GPCS application. With GPCS, the upper layer protocol that is immediately above the IEEE 802.16 GPCS is identified by a parameter known as GPCS protocol type. The GPCS protocol type is included in service flow management primitives and connection establishment messages. The GPCS defines a set of SAP parameters as the result of upper layer packet parsing. These are passed from upper layer to the GPCS in addition to the data packet. The SAP parameters include SFID, the MS MAC Address, data, and length. The GPCS allows multiplexing of multiple layer protocol types (e.g., IPv4, IPv6, and Ethernet) over the same IEEE 802.16 MAC connection. It is outside the scope of the GPCS protocol to specify how the upper layer multiplexes and de-multiplexes multiple protocol data packets over an IEEE 802.16 connection or service flow [3].

In multimedia streaming applications, the overhead of Internet Protocol (IP) [15], User Datagram Protocol (UDP) [16], and Real-time Transport Protocol (RTP) [17,18] payload headers are 40 bytes for IPv4 (or 60 bytes for IPv6 [19]). For voice-over-IP, this corresponds to approximately 60% of the total amount of encoded voice data (e.g., the RTP payload of 3GPP Adaptive Multi-Rate 12.2 kbps full-rate codec consists of 33 bytes [20]). Such large overheads may be tolerable in wired links where capacity is often not an issue, but are excessive for wireless systems where bandwidth is scarce. The IEEE 802.16 standard defines a native header compression algorithm that is part of the convergence sub-layer. The Payload Header Suppression (PHS) defined in the IEEE 802.16 standard compresses the repetitive or redundant parts of the payload header received from network layer. The PHS operation is based on the PHS rules, which provide all the parameters corresponding to header suppression of the SDU. Other standard header compression algorithms, such as Robust Header Compression (RoHC) defined by IETF [21], are also supported.

The RoHC scheme may be used as an alternative to PHS to compress the RTP/UDP/IP header of an IP packet. When RoHC is enabled for a service flow, the service flow constitutes what in IETF RFC 3095 is referred to as a RoHC channel [21,3]. Two service flows cannot share an RoHC channel, and two RoHC channels cannot share the same service flow. On a service flow for which RoHC has been enabled, all of the IP packet passes through the RoHC compressor on the transmitter side and the decompressor on the receiver side. The support of RoHC is negotiated between the BS and MS during capability negotiation [3].

The data-plane part of MAC CPS includes functions such as Automatic Repeat reQuest (ARQ), Packet Fragmentation/Packing, MAC PDU formation and encryption. The scheduler on the BS side allocates radio resources and multiplexes the users, and selects the appropriate MIMO mode, modulation and coding scheme based on the measurement reports that are received from the mobile stations. The ARQ is an error control mechanism at data link layer where the receiver may request the transmitter to resend a block of data that was erroneously detected or not received. An ARQ block is a distinct unit of data that is carried on an ARQ-enabled connection. Such a data unit is assigned a sequence number and is managed as a distinct entity by the ARQ state machines. The ARQ block size is a parameter that is negotiated during connection establishment. The ARQ mechanism may be disabled for some delay sensitive applications such as VoIP. Fragmentation is a process in which a MAC SDU is divided into one or more MAC SDU fragments. Packing is a process where multiple MAC SDUs are packed into a single MAC PDU payload. Both processes may be initiated by either a BS for a downlink connection or an MS for an uplink connection. Several MAC PDUs may be concatenated into a single transmission in the downlink or uplink.

The MAC PDUs containing user data are processed by the physical layer for over-the-air transmission. It must be noted that the physical layer processing of the user traffic can be different in terms of the permissible MIMO modes or modulation and coding schemes that are used.

3.1.4 Control-Plane

A set of Layer 2 control functions are needed to support various radio resource configuration, coordination, signaling, and management. This set of functions is collectively referred to as control-plane functions. The IEEE 802.16m control-plane entity comprises Radio Resource Control and Management (RRCM), MAC functional group, and physical layer protocols corresponding to control path. The RRCM functional class includes all control and management functions such as network entry/re-entry management, paging and idle mode management, multicast and broadcast service, etc. This group of functions is also known as Radio Resource Control (RRC) in other air interface standards such as 3GPP LTE. The MAC functional group consists of functions that perform physical layer control and signaling, scheduling services, QoS, etc. This functional group corresponds to Radio Link Control (RLC) and MAC layers in other air interface standards such as 3GPP LTE.

Figure 3-9 illustrates the IEEE 802.16m control-plane entity. As shown in this figure, the RRCM performs control and management of lower-layer functions. The control information is communicated with the mobile station via MAC management messages. The underlying functional elements of the RRCM sub-layer will be described in Section 3.2.

The security sub-layer in Figure 3-9 is shown with dotted line, since the IEEE 802.16m selectively encrypts and protects unicast MAC management messages. If the selective

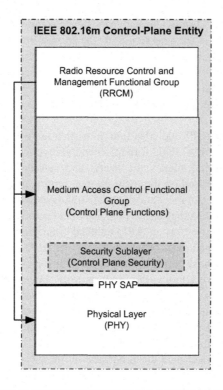

FIGURE 3-9

The IEEE 802.16m control-plane entity [12]

confidentiality protection is utilized, the negotiated keying materials and cipher suites are used to encrypt the management messages. There are three levels of selective confidentiality protection applied to MAC management messages in the IEEE 802.16m: (1) no protection where the MS and BS have no shared security context or protection is not required, then the management messages are neither encrypted nor authenticated. Management messages before the authorization phase also fall into this category; (2) cipher-based message authentication code[vii] (CMAC) integrity protection protects the integrity of the entire MAC management message; and (3) advanced encryption standard-based[viii] (AES-CCM) authentication/encryption protects the integrity of payload and MAC header [3].

[vii]A cipher-based MAC (CMAC) is a block cipher-based message/code authentication algorithm. It may be used to provide assurance of the authenticity and integrity of user payloads. AES-CMAC provides stronger assurance of data integrity than a checksum or an error-detecting code. The verification of a checksum or an error-detecting code detects only accidental modifications of the data, while CMAC is designed to detect intentional, unauthorized modifications of the data, as well as accidental modifications.

[viii]The counter with CBC-MAC (CCM), as defined in IETF RFC 3610, is a generic authenticated encryption block cipher mode. CCM is only defined for use with 128-bit block ciphers, such as AES. It is an authenticated encryption algorithm designed to provide both authentication and privacy.

3.1.5 Management-Plane

A management-plane is also defined for external management and system configuration. Therefore, all management and protocol configuration entities, as well as management information base, fall into the management-plane category. Definition of management information bases are out of scope of the IEEE 802.16m standard. As shown in Figure 3-10, the management entity and the management information bases contained in the management-plane, configure and manage the functional entities in the data- and control-plane protocol layers. The IEEE 802.16 specification includes control and management SAPs as part of the management-plane that expose control-plane and management-plane functions to upper layers. Management-plane primitives and the C-SAP are used for more time sensitive control-plane primitives that support handovers, security context management, radio resource management, and low power system operations.

In addition, under the IEEE 802.16 standard, a user can be associated with a number of service flows, each characterized with different QoS parameters. This information is provisioned in a subscriber management system (e.g., AAA database) or a policy server. There are two service models: (1) static service model, where the subscriber station is not allowed to change the parameters of provisioned service flows or create new service flows dynamically; and (2) dynamic service model, where an MS or BS may create, modify or delete service flows dynamically. In the latter case,

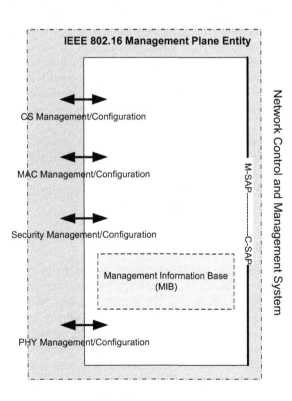

FIGURE 3-10

The IEEE 802.16 management-plane entity [3]

a dynamic service flow request is evaluated against the provisioned information to decide whether the request could be authorized. More precisely, the following steps are provisioned in the IEEE 802.16 specification for dynamic service flow creation [22]:

1. Permitted service flows and associated QoS parameters are provisioned for each subscriber via the management plane (management entity).
2. A service flow request initiated by the MS or BS is evaluated against the provisioned information, and the service flow is created if permissible.
3. A service flow thus created transitions to an admitted, and finally to an active, state due to BS action (this is possible under both static and dynamic service models). Transition to the admitted state involves the invocation of admission control in the BS and (soft) resource reservation, and transition to the active state involves actual resource assignment for the service flow. The service flow can directly transit from provisioned state to active state without going through admitted state.
4. A service flow can also transition in the reverse from an active to an admitted to a provisioned state.
5. A dynamically created service flow may also be modified or deleted.

As mentioned earlier, management information bases are collections of various network objects that are operated with the use of a Simple Network Management Protocol or SNMP [8,13]. The exact structure of the objects included in the management information base will depend on the configuration of the particular SNMP. However, additional extensions can allow for the addition of new objects outside the initial structure. Both the initial management information base and its extensions can be related to specific functions within a network. Some MIBs may be related to the definition of the domain name system, while other extensions may be associated with network objects like the fiber distributed data interface. While the initial management information base is usually defined as part of the SNMP, the extensions are generally set up as part of the basic management information base.

The Subscriber Station Management Primitives are a set of primitives to manage the status of mobile station. A management entity in the NCMS can change the status of mobile terminal. Those primitives are also used to notify the NCMS of information or events which are related to the status of the mobile terminal. The NCMS is a layer-independent entity that may be viewed as a management entity or control entity.

3.1.6 Service Access Point

A Service Access Point is defined as a reference point in a protocol stack where the services of a layer are available to its immediately neighboring layer. In other words, a SAP is a mapping between services of two neighboring layers. There are a number of SAPs in the IEEE 802.16 reference model (see Figure 3-7) that interface the adjacent protocol layers including PHY, MAC, and CS SAPs. The Management SAP may include primitives related to System configuration, Monitoring statistics, Notifications/Triggers, and Multi-mode interface management. The NCMS interacts with the MIB through the M-SAP. The Control SAP may include, but is not limited to, primitives related to handovers (e.g., notification of handover request from MS), idle mode mobility management (e.g., mobile station entering idle mode), subscriber and session management (e.g., mobile station requesting session set-up), radio resource management, AAA server signaling (e.g., *Extensible Authentication Protocol*[ix] payloads), Media Independent Handover (MIH)[x] services, and location detection and reporting capability. Unlike 3GPP LTE, the IEEE 802.16m does not explicitly define logical, transport, and physical channels (although the functionalities exist in the air interface protocols). In that case, the mapping between logical to transport and transport to physical channels would determine the SAPs between the corresponding layers.

3.1.7 Media-Independent Handover Reference Model for IEEE 802.16

The IEEE 802.21-2008 standard provides link-layer intelligence and other related network information to upper layers to optimize handovers between heterogeneous networks [23]. This includes media types specified by 3GPP, 3GPP2, and both wired and wireless media in the IEEE 802 family of standards. In the IEEE 802.21-2008 standard, media refers to the method or mode of accessing a telecommunication system (e.g., cable, radio, satellite), as opposed to sensory aspects of communication (e.g., audio, video). The standard addresses the support of handovers for both mobile and stationary users. For mobile users, handovers can occur when wireless link conditions change due to the users' movement. For the stationary user, handovers become imminent when the surrounding network environment changes, making one network more attractive than another. As an example, when making a network transition during a phone call, the handover procedures should be executed in such a way that any perceptible interruption to the conversation will be minimized. The standard supports cooperative use of information available at the mobile node and within the network infrastructure. The mobile node is well-positioned to detect available networks. The network infrastructure is well-suited to store overall network information, such as neighborhood cell lists, location of mobile nodes, and higher layer service availability. Both the mobile node and the network make decisions about connectivity. In general, both the mobile node and the network points of attachment (such as base stations and access points) can be multi-modal (i.e., capable of supporting multiple radio standards and simultaneously supporting connections on more than one radio interface).

Figure 3-11 shows the Media Independent Handover Function (MIHF), i.e., a function that realizes MIH services, for IEEE 802.16 based systems. The M-SAP and C-SAP are common between the MIHF and Network Control and Management System. The M-SAP specifies the interface between the MIHF and the management plane and allows MIHF payload to be encapsulated in management messages (such as MOB_MIH-MSG defined in the IEEE 802.16-2009 standard [3]). The primitives specified by M-SAP are used by a mobile node to transfer packets to a base station, both before and after it has completed the network entry procedures. The C-SAP specifies the interface between the MIHF and control-plane. M-SAP and C-SAP also transport MIH messages to peer MIHF entities. The CS-SAP is used to transfer packets from higher layer protocol entities after appropriate connections have been established with the network. The MIH-SAP specifies the interface of the MIHF with other higher layer entities such as transport layer, handover policy engine, and Layer 3 mobility protocols. In this model, C-SAP and M-SAP provide link services defined by MIH-LINK-SAP; C-SAP provides services before network entry; while CS-SAP provides services over the data-plane after network entry.

[ix]Extensible Authentication Protocol (EAP), as defined by IETF RFC 3748 and updated by IETF RFC 5247, is a universal authentication framework commonly used in wireless networks and point-to-point connections [24].
[x]Media Independent Handover (MIH) is a standard developed by the IEEE 802.21 to enable handover and interoperability between heterogeneous network types including both 802 and non-802 networks [23].

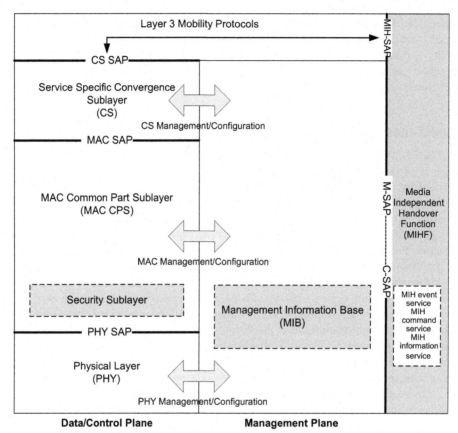

FIGURE 3-11

MIH reference model for the IEEE 802.16 standard [23]

3.2 **THE IEEE 802.16m PROTOCOL STRUCTURE**

In this section, we further examine the functional elements of each protocol layer and their interactions. The 802.16m MAC common part sub-layer functions are classified into radio resource control and management functional group and medium access control functional group. The control-plane functions and data-plane functions are also separately classified. This would allow more organized, efficient, and structured method for specifying the MAC services in the IEEE 802.16m standard specification. As shown in Figure 3-12, the radio resource control and management functional group comprises several functional blocks including:

- Radio resource management block adjusts radio network parameters related to the traffic load, and also includes the functions of load control (load balancing), admission control, and interference control;

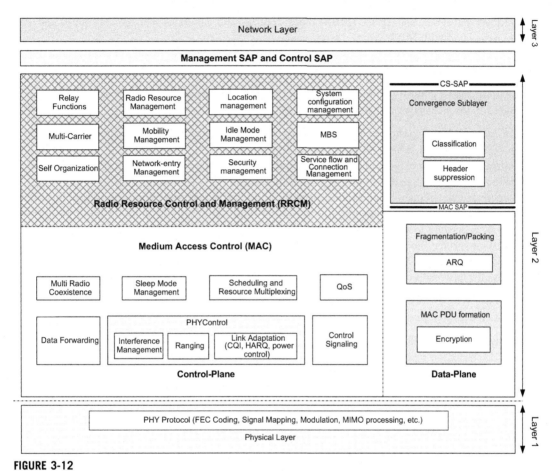

FIGURE 3-12

The IEEE 802.16m general protocol stack [12]

- Mobility management block scans neighbor BSs and decides whether MS should perform handover operation;
- Network-entry management block controls initialization and access procedures and generates management messages during initialization and access procedures;
- Location management block supports location based service (LBS), generates messages including the LBS information, and manages location update operation during idle mode;
- Idle mode management block controls idle mode operation, and generates the paging advertisement message based on paging message from paging controller in the core network;
- Security management block performs key management for secure communication. Using managed key, traffic encryption/decryption and authentication are performed;
- System configuration management block manages system configuration parameters, and generates broadcast control messages such as superframe headers;

- Multicast and broadcast service (MBS) block controls and generates management messages and data associated with MBS;
- Service flow and connection management block allocates Station Identifier (STID) and Flow Identifiers (FIDs) during access/handover service flow creation procedures.

The medium access control functional group, on the control plane, includes functional blocks which are related to physical layer and link controls such as:

- PHY control block performs PHY signaling such as ranging, channel quality measurement/ feedback (CQI), and HARQ ACK or NACK signaling;
- Control signaling block generates resource allocation messages such as advanced medium access protocol, as well as specific control signaling messages;
- Sleep mode management block handles sleep mode operation and generates management messages related to sleep operation, and may communicate with the scheduler block in order to operate properly according to sleep period;
- Quality-of-service block performs rate control based on QoS input parameters from connection management function for each connection;
- Scheduling and resource multiplexing block schedules and multiplexes packets based on properties of connections.

The MAC functional group on the data-plane includes functional blocks such as:

- Fragmentation/packing block performs fragmentation or packing of MAC Service Data Units (MSDU) based on input from the scheduling and resource multiplexing block;
- Automatic Repeat Request block performs MAC ARQ function. For ARQ-enabled connections, a logical ARQ block is generated from fragmented or packed MSDUs of the same flow and sequentially numbered;
- MAC protocol data unit formation block constructs MAC PDU (MPDU) such that BS/MS can transmit user traffic or management messages into PHY channels.

The IEEE 802.16m protocol structure is similar to that of the IEEE 802.16, with some additional functional blocks in the control-plane for new features including the following:

- Relay functions enable relay functionalities and packet routing in relay networks.
- Self organization and self-optimization functions enable home BS or femto-cells and plug-and-play form of operation for indoor BS (i.e., femto-cell[xi]).
- Multi-carrier functions enable control and operation of a number of adjacent or non-adjacent RF carriers (i.e., virtual wideband operation) where the RF carriers can be assigned to unicast and/ or multicast and broadcast services. A single MAC instantiation will be used to control several physical layers. The mobile terminal is not required to support multi-carrier operation. However, if it does support multi-carrier operation, it may receive control and signaling, broadcast, and synchronization channels through a primary carrier and traffic assignments (or services) via the secondary carriers.

[xi]Femto-cells are low-power wireless access points that operate in licensed spectrum to connect standard mobile devices to a mobile operator's network using residential DSL or cable broadband connections [25,26].

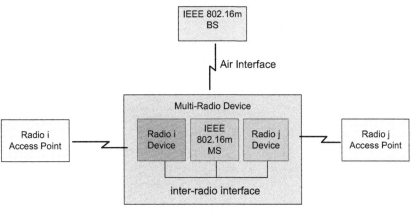

FIGURE 3-13

A generic multi-radio coexistence model [12]

- Multi-radio coexistence functions in IEEE 802.16m enable the MS to generate MAC management messages in order to report information on its collocated radio activities, and enable the BS to generate MAC management messages to respond with the appropriate actions to support multi-radio coexistence operation. Furthermore, the multi-radio coexistence functional block at the BS communicates with the scheduler functional block to assist proper scheduling of the MS according to the reported collocated coexistence activities. The multi-radio coexistence function is independent of the sleep mode operation to enable optimal power efficiency with a high level of coexistence support. However, when sleep mode provides sufficient collocated coexistence support, the multi-radio coexistence function may not be used (see Figure 3-13).
- Interference management functions are used to manage the inter-cell/sector interference effects. The procedures include MAC layer functions (e.g., interference measurement/assessment reports sent via MAC signaling and interference mitigation by scheduling and flexible frequency reuse), and PHY functions (e.g., transmit power control, interference randomization, interference cancellation, interference measurement, transmit beamforming/precoding). The inter-BS coordination functions coordinate the operation of multiple base stations by exchanging information, about interference statistics between the base stations via core-network signaling.

3.2.1 Data-Plane and Control-Plane Functions in Base Stations and Mobile Stations

Figure 3-14 shows the user data processing path at the BS and MS. As shown in the figure, the user data traverses the path from network layer to physical layer and *vice versa*. In the transmitter side, a network layer packet is processed by the convergence sub-layer, the ARQ function (if enabled), the fragmentation/packing function, and the MAC PDU formation function, to form the MAC PDU to be sent to the physical layer for processing. In the receiver side, a physical layer SDU is processed by MAC PDU formation function, the fragmentation/packing function,

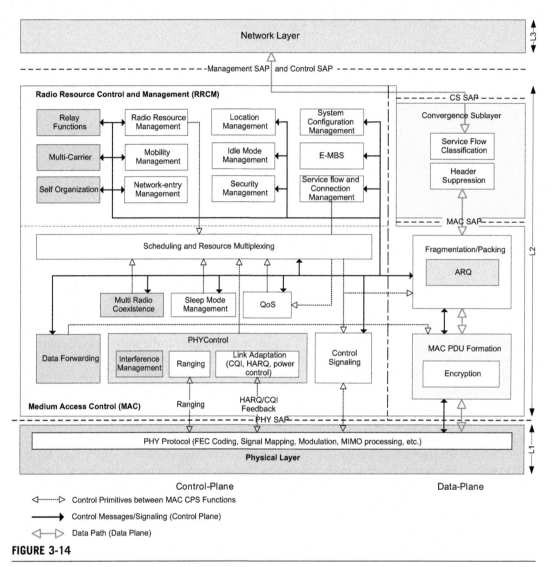

FIGURE 3-14

Signal flow graph in data- and control-planes [12]

the ARQ function (if enabled), and the convergence sub-layer function, to form the network layer packets. The control primitives between the MAC CPS functions and between the MAC CPS and PHY that are related to the processing of user traffic data are also shown in Figure 3-14.

The control-plane signaling and processing flow graph at the BS and the MS are shown in Figure 3-14. In the transmitter side, the flow of control primitives from control-plane functions to data-plane functions and processing of control-plane signals by data-plane functions in order to construct MAC management messages and MAC header/sub-headers, to be transmitted over the air interface, are illustrated. In the receiver side, the arrows show the processing of the MAC control messages through data-plane functions and the reception of the corresponding control-plane signals by control-plane functions. The dotted arrows show the control primitives between MAC CPS functions and between MAC CPS and physical layer functions that are related to the processing of control-plane signaling. The control primitives to/from M-SAP/C-SAP define the network related functionalities, such as inter-BS interference management, inter/intra RAT mobility management, etc., as well as management-related functionalities, such as location management, system configuration, etc.

Figure 3-15 illustrates the IP packet processing in an IEEE 802.16m base station transmitter and mobile station receiver. The main functional components of each layer and their interconnections are identified. It is further shown how the MAC scheduler in the base station, based on the periodic reports and measurements provided by each mobile station, generates appropriate control signals to select the best modulation and coding scheme, re-transmission method and number of re-transmissions, MIMO mode, and antenna configuration according to the channel conditions that the mobile station is experiencing.

3.2.2 Data-Plane and Control-Plane Functions in Relay Stations

Multi-hop relay is an optional entity that may be deployed in conjunction with base stations to provide additional coverage or performance improvements in a radio access network. In relay-enabled networks, the BS may be replaced by a multi-hop relay BS (i.e., a BS that supports relay capability over the relay links) and one or more relay stations (RS). The traffic and signaling between the mobile station and relay-enabled BS are relayed by the RS, thus extending the coverage and performance of the system in areas where the relay stations are deployed. Each RS is under the control of a relay-enabled BS [27].

In a multi-hop relay system, the traffic and signaling between an access RS and the BS may also be relayed through intermediate relay stations. The RS may either be fixed in location or it may be mobile. The mobile station may also communicate directly with the serving BS. The various relay-enabled BS features defined in the IEEE 802.16j-2009 standard allow a multi-hop relay system to be configured in several modes. The air interface protocols, including the mobility features on the access link (i.e., RS-MS link), remain unchanged.

The IEEE 802.16j-2009 standard specified a set of new functionalities on the relay link to support the RS–BS communication. Two different modes; i.e., centralized and distributed scheduling modes, were specified for controlling the allocation of bandwidths for an MS or an RS. In centralized scheduling mode the bandwidth allocation for subordinate mobile stations of an RS is determined at the serving BS. On the other hand, in distributed scheduling mode the bandwidth allocation of the subordinate stations is determined by the RS, in cooperation with the BS. Two different types of RS are defined, namely transparent and non-transparent. A non-transparent RS can operate in both centralized and distributed scheduling mode, while a transparent RS can only operate in centralized scheduling mode. A transparent RS communicates with the base station and subordinate mobile stations using the same carrier frequency. A non-transparent RS may communicate with the base station and the subordinate mobile stations via the same or different carrier frequencies.

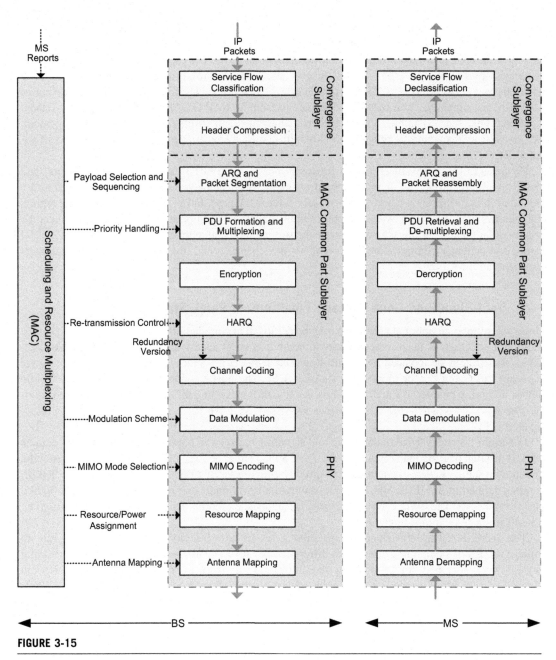

FIGURE 3-15

IP Packet processing and retrieval in the BS and MS

Relaying in the IEEE 802.16m system is performed using a decode-and-forward paradigm and supports TDD and FDD duplex modes. In TDD deployments, the relay stations operate in time-division transmit and receive (TTR) mode,[xii] whereby the access and relay link communications are multiplexed using time division multiplexing over a single RF carrier. In the IEEE 802.16m system, the relay stations operate in non-transparent mode, which essentially means that the relay stations compose and transmit the synchronization channels, system information, and the control channels for the subordinate stations. In any IEEE 802.16m deployment supporting relay functionality, a distributed scheduling model is used where each infrastructure station (BS or RS) schedules the radio resources on its subordinate links. In the case of a relay station, the scheduling of the resources is within the radio resources assigned by the BS. The BS notifies the relay and mobile stations of the frame structure configuration. The radio frame is divided into access and relay zones. In the access zone, the BS and the RS transmit to, or receive from, the mobile stations. In the relay zone, the BS transmits to the relay and the mobile stations, or receives from the relay and mobile stations. The start times of the frame structures of the BS and relay stations are aligned in time. The BS and relay stations transmit synchronization channels, system information, and the control channels to the mobile stations at the same time.

The MAC layer of a relay station includes signaling extensions to support functions such as network entry of an RS and of an MS through an RS, bandwidth request, forwarding of PDUs, connection management, and handover. Two different security modes are defined in the IEEE 802.16j-2009 standard: (1) a centralized security mode that is based on key management between the BS and an MS; and (2) a distributed security mode which incorporates authentication and key management between the BS and a non-transparent access RS, and between the access-RS and an MS. An RS may be configured to operate either in normal CID allocation mode, where the primary management, secondary, and basic CIDs are allocated by the BS, or in local CID allocation mode where the primary management and basic CID are allocated by the RS.

The IEEE 802.16m RS uses the same security architecture and procedures as an MS to establish privacy, authentication, and confidentiality between itself and the BS on the relay link. The IEEE 802.16m relay stations use a distributed security model. The security association is established between an MS and an RS during the key exchange similar to a macro BS. The RS uses a set of active keys shared with the MS to perform encryption/decryption and integrity protection on the access link. The RS runs a secure encapsulation protocol with the BS based on the primary security association. The access RS uses a set of active keys shared with the BS to perform encryption/decryption and integrity protection on the relay link. The MAC PDUs are encapsulated within one relay MAC PDU and are encrypted or decrypted by primary security association, which is established between the RS and the BS. The security contexts used for the relay link (between a BS and an RS) and the access links (between an RS and an MS) are different and are maintained independently. The key management is the same as that performed by a macro BS.

Figure 3-16 shows the IEEE 802.16m relay station protocol stack. An RS may consist of a subset of the protocol functions shown in Figure 3-16; however, the ingredients of each subset of functions depend on the type or category of the RS, as well as other deployment requirements. The functional

---

[xii]Time-division transmit and receive is a relay mechanism where transmission to subordinate station and reception from the super-ordinate station or transmission to the super-ordinate station and reception from the subordinate station is separated in time.

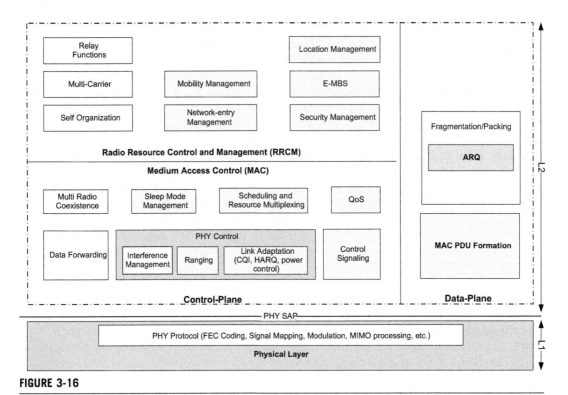

FIGURE 3-16

Protocol structure of the IEEE 802.16m relay stations [12]

blocks and the definitions provided in this section do not imply their support in all IEEE 802.16m RS implementations. The IEEE 802.16m relay capabilities are partially (and conceptually) based on the functionalities and features specified in the IEEE 802.16j-2009 standard [27]. The IEEE 802.16j-2009 standard does not define a system profile; therefore, many non-interoperable realizations of the relay stations can be considered. The WiMAX Forum technical working group started (and later abandoned) an initiative to define a relay system profile to facilitate certification and deployment of interoperable relay stations in mobile WiMAX systems. The non-transparent relay stations perform the same functions as a regular base station; however in some usage models, some functionalities of the regular base station may not be implemented in the relay station, resulting in less complexity and lower cost of implementation and deployment.

The IEEE 802.16m RS MAC CPS is divided into two sub-layers: (1) Radio Resource Control and Management sub-layer; and (2) Medium Access Control sub-layer. The RS RRCM sub-layer includes the following functional blocks that are related to the RS radio resource management functions:

- Mobility management;
- Network-entry management;
- Location management;
- Security management;

- Multicast and broadcast service;
- Relay functions;
- Self organization;
- Multi-carrier operation.

In Figure 3-16, the mobility management block supports the MS handover operation in cooperation with the BS. The network-entry management block performs RS/MS initialization procedures, as well as the RS network entry/attachment procedure to the BS. The network-entry management block may generate management messages needed during RS/MS initialization procedures and performing the network entry. The location management block supports location-based services including positioning data at the RS and reporting location information to the BS. The security management block performs the key management functions for the RS. Since an IEEE 802.16m relay uses a distributed security model, there are two sets of security protocols on the access and the relay links.

The enhanced multicast and broadcast service block is responsible for coordination, scheduling, and distribution of the E-MBS content to the subscribed users in the relay coverage area. The relay functional block includes procedures to maintain relay paths. The self-organization block performs functions to support the RS self-configuration and the RS self-optimization mechanisms which are coordinated by the BS. These functions include procedures to request the relay stations or mobile stations to report measurements for self-configuration and self-optimization, receive measurements from the relay stations or mobile stations, and report measurements to the BS. These functions also include procedures to adjust the RS parameters and configuration for self-configuration and/or optimization with or without coordination with the BS.

The multi-carrier operation block enables a common MAC entity to control a physical layer that may span over multiple frequency channels at the RS. The RS MAC sub-layer includes the following functional blocks which are related to the physical layer and link control:

- Physical layer control;
- Control signaling;
- Sleep mode management;
- Quality of service;
- Scheduling and resource multiplexing;
- ARQ function;
- Fragmentation/packing;
- MAC PDU formation;
- Data forwarding;
- Multi-radio coexistence.

As shown in Figure 3-16, the physical layer control block manages signaling schemes such as ranging, measurement, reporting, and HARQ feedback at the RS. Based on CQI and HARQ feedback, the physical layer control block estimates channel conditions of RS/MS and performs link adaptation. The control signaling block performs the RS resource allocation and generates control messages. The sleep mode management block manages sleep mode operation of mobile stations serviced by the RS in coordination with the BS. The QoS block performs rate control according to QoS parameters. The scheduling and resource multiplexing block which resides in the RS is used to support distributed scheduling, and schedules the transmission of MAC PDUs. The ARQ block assists MAC ARQ functions between the BS and RS over the relay link and between MS and RS over the access link. The MAC SDUs may be fragmented or augmented depending on the size of the payloads. The fragmentation/packing block in the RS side includes the unpacking and repacking of data fragments that have been received for relaying, in order to adapt the size of MAC PDUs to the estimated channel quality of the outbound link.

The MAC PDU formation block constructs MAC PDUs which contain user traffic or management messages. User traffic is assumed to have originated at either the BS or MS. The MAC PDU construction block may add or modify MAC PDU control information (e.g., MAC header). The data forwarding block performs routing functions on the link between the BS and the RS or MS. The data forwarding block may work in conjunction with other blocks such as scheduling and resource multiplexing and MAC PDU formation.

The interference management block at the RS performs inter-cell and inter-RS interference management. This function includes reception of interference level measurements and selection of transmission format used for the mobile stations attached to the RS. The control functions can be divided among the serving BS and the relay stations using a centralized model or a distributed model. In a centralized model, the serving BS generates control signals, and the relay stations communicate the control information between the BS and MS. In a distributed model, the RS generates control signals for the subordinate mobile stations and makes the BS aware of that control information. The determination of whether a particular control function should be centralized or distributed is made independently for each control function.

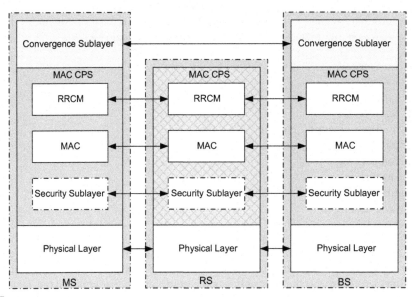

FIGURE 3-17

Protocol termination in a relay-enabled network

The multi-radio coexistence block within the RS coordinates collocated multi-radio operation of the subordinate mobile stations in coordination with the BS. Based on the earlier description of the functions and protocols performed by a relay station, the control- and data-plane protocol termination is shown in Figure 3-17. The convergence sub-layer protocols are terminated at the MS and the BS. However, some of the MAC CPS protocols are terminated at the RS. Due to the use of a distributed security model, the security functions including encryption and packet validation are terminated at the RS on the relay and access links.

3.2.3 Protocol Structure for Support of Multi-Carrier Operation

The generic protocol structure for support of multi-carrier operation is illustrated in Figure 3-18. A single MAC instance controls a number of physical layers spanning over multiple frequency bands. Some MAC messages transmitted over one RF carrier may also apply to other RF carriers. The RF channels may be of different bandwidths (e.g., 5, 10, and 20 MHz), and can be contiguous or non-

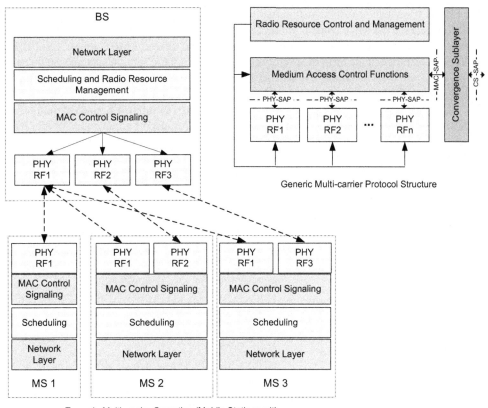

FIGURE 3-18

Multi-carrier operation using a single MAC instantiation [12]

contiguous in frequency. The RF channels may support different duplexing schemes, e.g., frequency division duplex (FDD) mode, time division duplex (TDD) mode, or a combination of multicast and/or unicast RF carriers [12]. As shown in Figure 3-18, the MAC entity can provide simultaneous service to mobile stations with different bandwidth capabilities, such as operation over one RF channel at a time or aggregation across contiguous or non-contiguous frequency bands.

3.2.4 Protocol Structure for Support of Multicast and Broadcast Services

Multicast and broadcast service is a point-to-multipoint communication scheme where data packets are transmitted simultaneously from a single source to multiple destinations. The term broadcast refers to the ability to deliver content to all users. Multicast, on the other hand, refers to distribution of content among a specific group of users that are subscribed to those services. The multicast and broadcast content is transmitted over a geographical area referred to as a zone. An MBS zone is a collection of one or more base stations transmitting the same content. Each BS capable of MBS service may belong to one or more MBS zones. Each MBS zone is identified by a unique zone identifier [3].

An MS can receive the MBS content within the MBS zone in connected state or idle state. A BS may provide multicast and broadcast services corresponding to different MBS zones. The MBS data bursts may be transmitted in the form of several sub-packets, and these sub-packets may be transmitted in different time intervals to allow the MS to combine the sub-packets without transmission of acknowledgement. The mobile stations in an MBS zone are assigned a common multicast station identifier. The IEEE 802.16m supports two types of MBS access: (1) single-BS; and (2) multi-BS. The single-BS access is implemented over multicast and broadcast transport connections within one BS, while multi-BS access is realized by transmitting MBS data through multiple base stations. The MBS PDUs are transmitted by all base stations in the same MBS zone. That transmission is supported either in the non-macro diversity mode or macro diversity mode. An MBS zone may be formed by only one

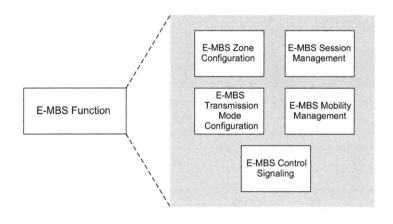

FIGURE 3-19

Breakdown of the E-MBS function (control-plane) [12]

BS. The MS may support both single-BS and multi-BS access. The MBS service may be delivered via a dedicated RF carrier or a mixed unicast, multicast, and broadcast RF carrier.

The IEEE 802.16m Enhanced Multicast and Broadcast Service (E-MBS) consists of MAC and PHY protocols that define interactions between the mobile stations and the base stations. While the basic definitions of IEEE 802.16m E-MBS are consistent with that of the IEEE 802.16-2009 standard [3], some enhancements and extensions are incorporated to provide improved functionality and performance [12]. The breakdown of E-MBS MAC function into constituent components is shown in Figure 3-19. In the control-plane, E-MBS MAC function operates in conjunction with other unicast service MAC functions. The unicast MAC functions may operate independently from E-MBS MAC function. The E-MBS MAC function may operate differently depending on whether it is operating in active mode or idle mode [12].

The E-MBS MAC function consists of the following sub-blocks:

- E-MBS Zone Configuration: this function manages the configuration and advertisement of E-MBS zones. A BS may belong to multiple E-MBS zones.
- E-MBS Transmission Mode Configuration: this function describes the transmission mode in which E-MBS is delivered over the air interface such as single-BS and multi-BS transmission.
- E-MBS Session Management: this function manages E-MBS service registration and deregistration and session start, update, or termination.
- E-MBS Mobility Management: this block manages the zone update procedures when an MS crosses the E-MBS zone boundary.
- E-MBS Control Signaling: this block broadcasts the E-MBS scheduling and physical channel mapping to facilitate E-MBS reception and power saving.

3.3 3GPP LTE/LTE-ADVANCED PROTOCOL STRUCTURE

In this section, we provide an overview of 3GPP LTE/LTE-Advanced protocol structure. Figures 3-20 and 3-21 illustrate the user-plane (U-plane) and control-plane (C-plane) protocol stacks, respectively. In the C-plane, the Non-Access Stratum (NAS) functional block is used for network attachment, authentication, setting up bearers, and mobility management. All NAS messages are ciphered and

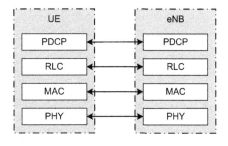

FIGURE 3-20

The user-plane protocol stack [29]

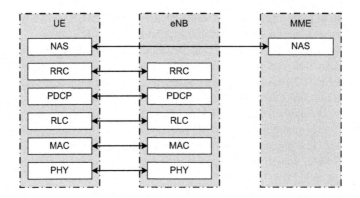

FIGURE 3-21

The control-plane protocol stack [29]

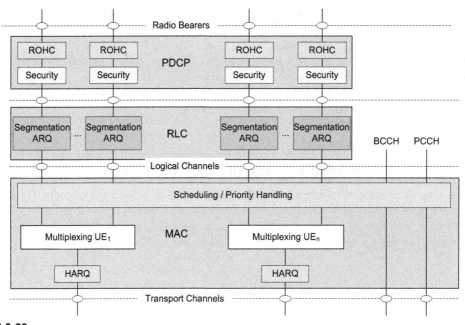

FIGURE 3-22

The 3GPP LTE Layer 2 structure in the downlink [29]

integrity protected by the Mobility Management Entity (MME) and User Equipment (UE), i.e., 3GPP LTE mobile station [28].

The Layer 2 functions in 3GPP LTE are classified into the following categories: Medium Access Control (MAC); Radio Link Control (RLC); and Packet Data Convergence Protocol (PDCP) functions [29]. Figures 3-22 and 3-23 illustrate the structure of Layer 2 protocols in 3GPP LTE downlink and

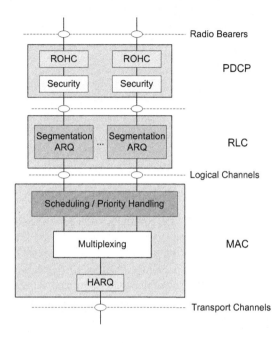

FIGURE 3-23

The 3GPP LTE Layer 2 structure in the uplink [29]

uplink. The SAP for peer-to-peer communication is marked with circles at the interface between the sub-layers. The SAP between the physical layer and the MAC sub-layer provides the transport channels. The SAP between the MAC sub-layer and the RLC sub-layer provide the logical channels. The multiplexing of several logical channels (i.e., radio bearers) on the same transport channel (i.e., transport block) is performed by the MAC sub-layer. Each logical channel is defined by type of information that is transferred. The logical channels are generally classified into two groups: (1) control channels (for the transfer of control-plane information); and (2) traffic channels (for the transfer of user-plane information).

As shown in Figure 3-21, the RRC layer in the evolved NodeB (eNB), i.e., 3GPP LTE base station, makes handover decisions based on neighbor cell measurements reported by the UE, performs paging of the users over the air interface, broadcasts system information, controls UE measurement and reporting functions such as the periodicity of channel quality indicator reports, and further allocates cell-level temporary identifiers to active users. It also executes transfer of UE context from the serving eNB to the target eNB during handover, and performs integrity protection of RRC messages. The RRC layer is responsible for setting up and maintenance of radio bearers. Note that RRC layer in 3GPP protocol hierarchy is considered as Layer 3. The main services and functions of the RRC sub-layer include [29,30]:

- Broadcast of system information;
- Paging;

- Establishment, maintenance, and release of a RRC connection between the UE and E-UTRAN, including allocation of temporary identifiers between UE and E-UTRAN and configuration of signaling radio bearers for RRC connection;
- Security functions, including key management;
- Establishment, configuration, maintenance, and release of point-to-point radio bearers;
- Mobility functions, including UE measurement reporting and control of the reporting for inter-cell and inter-RAT mobility, handover, UE cell selection and reselection, control of cell selection and reselection, context transfer at handover;
- Establishment, configuration, maintenance, and release of radio bearers for Multimedia Broadcast Multicast Service (MBMS);
- QoS management functions.

The 3GPP LTE RRC consists of the following states:

- RRC_IDLE is a state where a UE specific Discontinuous Reception (DRX) may be configured by upper layers. In the idle mode, the UE conserves power and does not inform the network of each cell change. The network knows the location of the UE to the granularity of a few cells, called the Tracking Area (TA). The UE monitors a paging channel to detect incoming traffic, performs neighboring cell measurements and cell selection/reselection, and acquires System Information.
- RRC_CONNECTED is a state where transfer of unicast data to/from UE is performed and the UE may be configured with a UE specific DRX or Discontinuous Transmission (DTX). The UE monitors control channels associated with the shared data channel to determine if data is scheduled for it, provides channel quality and feedback information, performs neighboring cell measurements and measurement reporting, and acquires System Information.

In the U-plane, the PDCP layer is responsible for compressing or decompressing the headers of IP packets using robust header compression to enable efficient use of air interface resources. This layer also performs ciphering of both user-plane and control-plane traffic. Because the NAS messages are carried in RRC, they are effectively double ciphered and integrity protected, once at the MME and again at the eNB. Therefore, the services and functions provided by the PDCP layer in the U-plane include header compression and decompression, transfer of user data between NAS and RLC layer, sequential delivery of upper layer PDUs and duplicate detection of lower layer SDUs at handover for radio link layer acknowledged mode, re-transmission of PDCP SDUs at handover for radio link layer acknowledged mode, and ciphering. The services and functions provided by the PDCP for the C-plane include ciphering and integrity protection, and transfer of control-plane data where PDCP receives PDCP SDUs from RRC and forwards them to the radio link control layer [29,31].

The RLC layer is used to format and transport traffic between the UE and the eNB. The RLC provides three different reliability modes for data transport, i.e., acknowledged mode (AM), unacknowledged mode (UM), and transparent mode (TM). The unacknowledged mode is suitable for transport of real-time services since such services are delay-sensitive and cannot tolerate delay due to re-transmissions. The acknowledged mode is appropriate for non-real-time services such as file transfers. The transparent mode is used when the size of packet data units are known in advance, such as for broadcasting system configuration information. The RLC layer also provides sequential delivery of service data units to the upper layers and eliminates duplicate packets from being delivered to the upper layers. It may also segment the service data units. Furthermore, there are two levels of re-transmissions for providing reliability, the HARQ at the MAC layer, and ARQ at the RLC layer. The

ARQ is required to handle residual errors that are not corrected by HARQ, and is kept simple by the use of a single-bit feedback mechanism. An N-process stop-and-wait HARQ is employed that has asynchronous re-transmissions in the DL and synchronous re-transmissions in the UL. In practice, multiple stop-and-wait HARQ processes are operated in parallel, i.e., when one HARQ process is waiting for an acknowledgment another process can use the channel to send sequentially-ordered sub-packets, thus improving the throughput. The synchronous HARQ means that the re-transmissions of HARQ sub-packets occur at predefined periodic intervals. Hence, no explicit signaling is required to indicate to the receiver the re-transmission schedule. Asynchronous HARQ offers the flexibility of scheduling re-transmissions based on air interface conditions (i.e., scheduling gain) [29,32]. The services and functions provided by the MAC layer can be summarized as follows [32]:

- Mapping between logical channels and transport channels;
- Multiplexing/de-multiplexing of RLC protocol data units corresponding to one or different radio bearers into/from transport blocks delivered to/from the physical layer on transport channels;
- Traffic volume measurement reporting;
- Error correction through HARQ;
- Priority handling between logical channels of one UE;
- Priority handling between UEs through dynamic scheduling;
- Transport format selection.

E-UTRA provides ARQ and HARQ functionalities. The ARQ functionality provides error correction by re-transmissions in acknowledged mode at Layer 2. The HARQ functionality ensures delivery between peer entities at Layer 1. The HARQ within the MAC layer is characterized by an N-process stop-and-wait protocol and re-transmission of transport blocks upon failure of earlier transmissions. A total of eight HARQ processes are supported [29].

The 3GPP LTE-Advanced system extends the capabilities of 3GPP LTE Rel-8 with support of carrier aggregation, where two or more component carriers are aggregated in order to support wider transmission bandwidths up to 100 MHz and for spectrum aggregation. A user terminal may simultaneously receive or transmit one or multiple component carriers depending on its capabilities. From the UE perspective, the Layer 2 aspects of HARQ are similar to those of Rel-8. There is one transport block (in absence of spatial multiplexing, up to two transport blocks in case of spatial multiplexing) and one independent HARQ entity per scheduled component carrier. Each transport block is mapped to

Table 3-1 Summary of the Differences between MAC and RRC Control [33]

	MAC Control		**RRC Control**
Control Entity	MAC		RRC
Signaling Type	Physical control channel	MAC control PDU	RRC message
Signaling Reliability	$\sim 10^{-2}$ (No HARQ re-transmission)	$\sim 10^{-3}$ (HARQ re-transmissions)	$\sim 10^{-6}$ (ARQ re-transmissions)
Control Latency	Very short	Short	Longer
Extensibility	None	Limited	High
Security	No integrity protection No ciphering	No integrity protection No ciphering	Integrity protected Ciphering

a single component carrier on which all HARQ re-transmissions may take place. A UE may be scheduled over multiple component carriers simultaneously, but at most one random access procedure will be ongoing at any time. Whenever a UE is configured with only one component carrier, the 3GPP LTE Rel-9 DRX is the baseline. In other cases, the same DRX operation will be applied to all configured component carriers. Therefore, the Layer 2 structure of the 3GPP LTE-Advanced is similar to that of 3GPP LTE Rel-8, except for the addition of the multi-carrier functionality; however, the multi-carrier nature of the physical layer is only exposed to the MAC layer through transport channels, where one HARQ entity is required per component carrier [34].

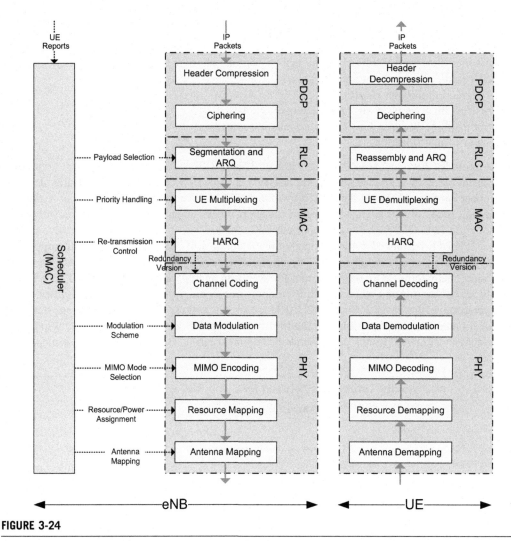

FIGURE 3-24

User data processing and signaling protocols in 3GPP LTE eNB and UE

The main difference between MAC and RRC control lies in the signaling reliability. The signaling corresponding to state transitions and radio bearer configurations should be performed by RRC sub-layer due to signaling reliability. The different characteristics of MAC and RRC control are summarized in Table 3-1.

The physical layer provides information transfer services to MAC and higher layers. The physical layer transport services are described by how and with what characteristics data is transferred over the radio interface. This should be clearly separated from the classification of what is transported, which relates to the concept of logical channels at MAC sub-layer.

Figure 3-24 illustrates the IP packet processing in the eNB and UE transmitter and receiver sides, respectively. The main functions of each protocol layer and their interconnections have been identified. One of the noticeable differences with IEEE 802.16m data processing is the location of encryption or ciphering of the user payload. Unlike the IEEE 802.16m, 3GPP LTE encrypts the packets in the PDCP layer before delivering the service data units to the RLC and MAC layers. The base station scheduler generates appropriate control signals for RLC, MAC, and PHY layers based on periodic reports and measurements received from each UE to ensure robustness and reliability of the connections, given varying radio channel conditions.

References

[1] WMF-T23-001/002/003-R015v01, WiMAX Forum Mobile System Profile Specification: Release 1.5, August 2009, <http://www.wimaxforum.org/resources/documents/technical/release>.
[2] Andrew S. Tanenbaum, Computer Networks, fourth ed., Prentice Hall, 2002.
[3] IEEE Std 802.16-2009, IEEE Standard for Local and Metropolitan Area Networks – Part 16: Air Interface for Broadband Wireless Access Systems, May 2009.
[4] Asynchronous Transfer Mode, Wikipedia, <http://en.wikipedia.org/wiki/ATM_(Asynchronous_Transfer_Mode>.
[5] Open Systems Interconnection Reference Model, Wikipedia, <http://en.wikipedia.org/wiki/OSI_model>.
[6] IETF RFC 1155, M. Rose, Structure and Identification of Management Information for TCP/IP-based Internets, May 1990.
[7] Simple Network Management Protocol, Wikipedia, <http://en.wikipedia.org/wiki/SNMP>.
[8] IETF RFC 1157, J. Case, A Simple Network Management Protocol (SNMP), May 1990.
[9] Common Object Requesting Broker Architecture (CORBA), Wikipedia <http://en.wikipedia.org/wiki/CORBA_architecture>.
[10] Interactive Data Language (IDL), Wikipedia, <http://en.wikipedia.org/wiki/IDL_(programming_language)>.
[11] Web Service, Wikipedia, <http://en.wikipedia.org/wiki/Web_Services>.
[12] IEEE 802.16m–08/003r9, IEEE 802.16m System Description Document, May 2009, <http://ieee802.org/16/tgm/index.html>.
[13] IETF RFC 3418, R. Presuhn, Management Information Base (MIB) for the Simple Network Management Protocol (SNMP), December 2002.
[14] Loutfi Nuaymi, WiMAX: Technology for Broadband Wireless Access, first ed., John Wiley & Sons, 2007.
[15] IETF RFC 791, Internet Protocol, DARPA Internet Program Protocol Specification, September 1981.
[16] IETF RFC 768, J. Postel, User Datagram Protocol, August 1980.
[17] IETF RFC 3550, H. Schulzrinne, RTP: A Transport Protocol for Real-Time Application, July 2003.
[18] Colin Perkins, RTP: Audio and Video for the Internet, first ed., Addison-Wesley, 2003.

[19] IETF RFC 2460, S. Deering, Internet Protocol, Version 6 (IPv6) Specification, December 1998.

[20] IETF RFC 3267, J. Sjoberg, Real-Time Transport Protocol (RTP) Payload Format and File Storage Format for the Adaptive Multi-Rate (AMR) and Adaptive Multi-Rate Wideband (AMR-WB) Audio Codecs, June 2002.

[21] IETF RFC 3095, C. Bormann, RObust Header Compression (ROHC): Framework and four profiles: RTP, UDP, ESP, and uncompressed, July 2001.

[22] WiMAX Forum Network Architecture Release 1.5 Version 1 –Stage 2, Architecture Tenets, Reference Model and Reference Points, November 2009, <http://www.wimaxforum.org/resources/documents/technical/release>.

[23] IEEE Std 802.21-2008, IEEE Standard for Local and Metropolitan Area Networks – Part 21: Media Independent Handover Services, January 2009.

[24] Extensible Authentication Protocol (EAP), Wikipedia, <http://en.wikipedia.org/wiki/Extensible_Authentication_Protocol>.

[25] Femto Forum, <http://femtoforum.org>.

[26] Femtocell, Wikipedia, <http://en.wikipedia.org/wiki/Femto_cell>.

[27] IEEE Std 802.16j-2009, IEEE Standard for Local and Metropolitan Area Networks Part 16: Air-interface for Fixed and Mobile Broadband Wireless Access Systems, Multi-hop Relay Specification, July 2009.

[28] E. Dahlman, et al., 3G Evolution: HSPA and LTE for Mobile Broadband, second ed., Academic Press, October 2008.

[29] 3GPP TS 36.300, Evolved Universal Terrestrial Radio Access (E-UTRA) and Evolved Universal Terrestrial Radio Access Network (E-UTRAN); Overall description; Stage 2, March 2010.

[30] 3GPP TS 36.331, Evolved Universal Terrestrial Radio Access (E-UTRA); Radio Resource Control (RRC); Protocol specification, March 2010.

[31] 3GPP TS 36.323, Evolved Universal Terrestrial Radio Access (E-UTRA); Packet Data Convergence Protocol (PDCP) specification, March 2010.

[32] 3GPP TS 36.321, Evolved Universal Terrestrial Radio Access (E-UTRA); Medium Access Control (MAC) protocol specification, March 2010.

[33] 3GPP TS 36.322, Evolved Universal Terrestrial Radio Access (E-UTRA); Radio Link Control (RLC) protocol specification, March 2010.

[34] 3GPP TR 36.912, Feasibility study for Further Advancements for E-UTRA (LTE-Advanced), March 2010.

IEEE 802.16m System Operation and State Diagrams

INTRODUCTION

This chapter provides a detailed description of the operation of IEEE 802.16m entities (i.e., mobile station, base station, femto base station, and relay station) through use of state diagrams and call flows. An attempt has been made to characterize the behavior of IEEE 802.16m systems in various operating conditions such as system entry/re-entry, cell selection/reselection, intra/inter-radio access network handover, power management, and inactivity intervals. This chapter describes how the IEEE 802.16m system entities operate and what procedures or protocols are involved, without going through the implementation details of each function or protocol. The detailed algorithmic description of each function and protocol will be provided in following chapters. Several scattered call flows and state diagrams were used in reference [1] to demonstrate the behavior of the legacy mobile and base stations, making it difficult to coherently understand the system behavior. The IEEE 802.16 standards have not generally been developed with a system-minded view; rather, they specify components and building blocks that can be integrated to build a working and performing system. An example is the mobile WiMAX profiles where a specific set of IEEE 802.16-2009 standard features (one out of many possible configurations) were selected to form a mobile broadband wireless access system. Detailed IEEE 802.16m entities' state transition diagrams comprising states, constituent functions, and protocols within each state, and inter-state transition paths conditioned to certain events would help the understanding and implementation of the standards specification [2–5]. It further helps to understand the behavior of the system without struggling with the distracting details of each constituent function.

State diagrams are used to describe the behavior of a system. They can describe possible states of a system and transitions between them as certain events occur. The system described by a state diagram must be composed of a finite number of states. However, in some cases, the state diagram may represent a reasonable abstraction of the system. There are many forms of state diagrams which differ slightly and have different semantics. State diagrams can be used to graphically represent finite state machines (i.e., a model of behavior composed of a finite number of states, transitions between those states, and actions). A state is defined as a finite set of procedures or functions that are executed in a unique order. In the state diagram, each state may have some inputs and outputs, where deterministic transitions to other states or the same state happen based on certain conditions. In this chapter, the notion of mode is used to describe a sub-state or a collection of procedures/protocols that are associated with a certain state. The unique definition of states and their corresponding modes and

Mobile WiMAX. DOI: 10.1016/B978-0-12-374964-2.10004-9

protocols, and internal and external transitions, is imperative to the unambiguous behavior of the system. Also, it is important to show the reaction of the system to an unsuccessful execution of a certain procedure. The state diagrams described in the succeeding sections are used to characterize the behavior of IEEE 802.16m system entities.

This chapter provides a top-down systematic description of IEEE 802.16m entities' state transition models and corresponding procedures, starting at the most general level and working toward the details or specifics of the protocols and transition paths. An overview of 3GPP LTE/LTE-Advanced states and user equipment state transitions is further provided to enable readers to contrast the corresponding terminal and base station behaviors, protocols, and functionalities. Such contrast is crucial in the design of inter-system interworking functions.

4.1 IEEE 802.16M MOBILE STATION STATE DIAGRAMS

A mobile state diagram (i.e., a set of states and procedures between which the mobile station transits when operating in the system to receive and transmit data or control signals) for the IEEE 802.16-2009 standard reference system based on common understanding of its behavior was established (see Figure 4-1) as a basis for characterization of IEEE 802.16m systems. A mobile station compliant with IEEE 802.16-2009 standard [1] operates in four distinct states,[i] as follows:

- **Initialization State**: initialization is a state where a mobile station without any connection performs cell search and cell selection by scanning and synchronizing to a cell and acquiring the cell identification, as well as the system configuration information. The mobile station returns to the scanning step if it fails to properly perform the required procedures in the next steps (shown with gray-colored arrows in Figure 4-1).
- **Access State**: access is a (transient) state where the mobile station performs network entry with the selected base station. The Access State comprises the following procedures: (1) initial ranging and uplink synchronization; (2) basic capability negotiation; (3) authentication, authorization, and key exchange; (4) registration with the BS; and (5) service flow establishment (IP connection). The mobile station receives specific user identification as part of Access State procedures. The IP address assignment may follow using appropriate procedures.
- **Connected State**: the Connected State includes procedures that the mobile station performs to transmit or receive control signals and/or data in the Downlink (DL) and Uplink (UL), while measuring the received RF signal strength of the neighboring base stations and conserving power. The Connected State consists of the following modes:
  - Active Mode: In this mode, downlink or uplink transmissions at appropriate intervals can be scheduled for the MS by the serving BS. Channel quality measurements based on downlink reference signals may be conducted and reported to the BS by the MS in order to assist the BS scheduler in properly adapting to the link characteristics and selecting the most appropriate transmission formats for data and control signals sent to (or received from) the MS. The MS may transit from Active Mode to Sleep Mode during sleep intervals (with

---

[i]The access is not designated as an independent state in the legacy standard, rather it is considered as a set of messages that are exchanged between the mobile station and the base station [1]. However, we categorize those messages and procedures as Access State in this book. The Access State can be considered as a transient state.

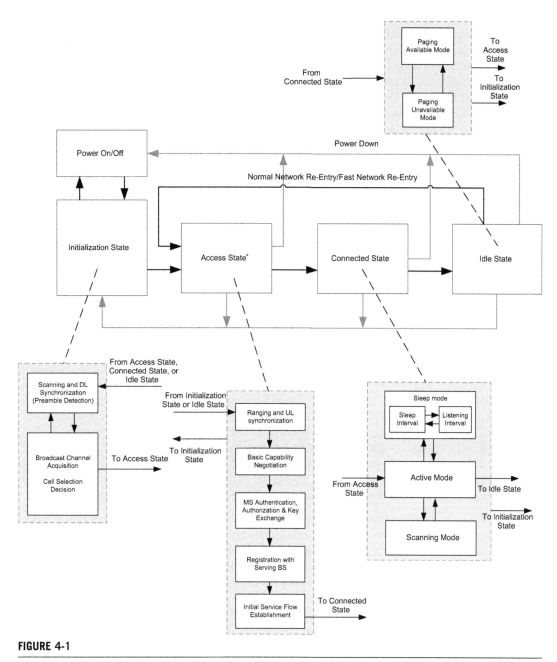

FIGURE 4-1

IEEE 802.16m mobile station state transition diagram [3]

intermittent DL/UL traffic) or to Idle State (when it has no DL/UL traffic and following deregistration with the BS) in order to conserve power.

- Sleep Mode: this mode is intended to minimize MS power usage and decrease usage of serving BS radio resources by pre-negotiated periods of absence from the serving BS air interface while connected to the BS. The MS in Active Mode transitions to Sleep Mode through Sleep Mode MAC management messages. The MS does not transmit and receive any information to/from its serving BS during the sleep interval. A traffic indication message from the serving BS can trigger an MS in the Sleep Mode to transit to Active Mode.
- Scanning Mode: In this mode, the MS performs the scanning operation (i.e., detecting and measuring the signal strength of neighboring base stations) and may temporarily be unavailable to the serving BS. The MS can be instructed to transit to Scanning Mode via explicit MAC signaling.

During the Connected State, the MS maintains at least one transport connection as established during Access State; however, the MS and BS may establish additional transport connections. It must be noted that each connection has unique Quality of Service parameters and is associated with a service flow. In addition, in order to reduce power consumption, either MS or BS can request transition to Sleep Mode. Also, the MS can scan neighbor base stations and perform cell reselection which provides more robust and reliable connections. The scanning interval is defined as a time period intended for the mobile station to monitor neighbor base stations and to determine the suitability of those base stations as targets for handover.

- **Idle State**: Idle State allows an MS to become periodically available for downlink broadcast messages without registration at a specific BS as the MS traverses across the network populated by multiple base stations, and thus allows the MS to conserve power and operational resources. The Idle State consists of two separate modes, Paging Available Mode and Paging Unavailable Mode. During Idle State, the MS may attempt power saving by switching between Paging Available and Paging Unavailable modes. In Paging Available Mode, the MS may be paged by the BS. If the MS is paged, it transitions to the Access State for network re-entry. The MS may perform location update procedure during Idle State. In the Paging Unavailable Mode, the MS is not required to monitor the downlink channel in order to reduce its power consumption.

The IEEE 802.16m mobile station state transition diagram comprises four distinct operational states and is similar to that of the legacy system (as shown in Figure 4-1) with the exception that intra-state procedures and protocols have been simplified to reduce the network entry/re-entry latency, and to enable fast cell selection or reselection. As an example, the location of the system configuration information has been fixed in IEEE 802.16m standard so that on successful downlink synchronization, the superframe headers containing the system configuration information can be located and acquired. This would enable the MS to make decisions for attachment to the BS without acquiring and decoding the legacy downlink control channel and the downlink and uplink channel descriptors that were transmitted as MAC management messages in the legacy system. This modification would further result in power saving in the MS, due to shortening and simplification of the initialization procedure. In the next sections, the detailed procedures corresponding to each state are described.

Although both normal and fast network re-entry processes are shown as transition from the Idle State to the Access State in Figure 4-1, there are differences that distinguish the two processes which

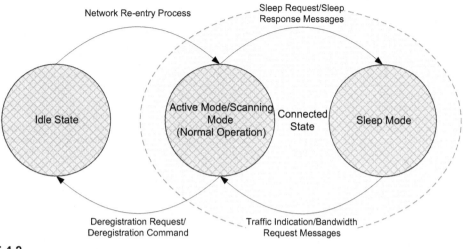

FIGURE 4-2

Steady-state behavior of IEEE 802.16 mobile stations [1]

will be discussed in the next sections. The steady-state behavior of IEEE 802.16 mobile stations is shown in Figure 4-2. The latter figure applies to both IEEE 802.16m [2] and IEEE 802.16-2009 standard [1] compliant mobile stations, although the details of procedures might be different in the two systems. Once the MS completes the initial network entry, it starts normal operation in the Active Mode while periodically scanning the neighboring base stations for handover. It may transition to the Idle State through deregistration messages or exit the Idle State and enter the Active Mode by performing network re-entry procedures. The MS may transition to Sleep Mode after negotiating the sleep intervals with the serving BS, and it may exit Sleep Mode on receiving a traffic indication message or availability of uplink traffic.

4.1.1 Initialization State

Figure 4-3 provides a decomposition of the Initialization State. It is shown that the Initialization State consists of two steps: (1) cell search; and (2) cell selection. In the first step, the MS scans for the candidate base stations to which the MS can attach and detect the Cell Identification (Cell_ID). The Cell_ID is a unique code or a synchronization sequence that is assigned to each cell (i.e., a BS or a BS sector) in order to identify the cell exclusively.[ii] The scanning is performed by measuring an appropriate metric corresponding to the received RF signal strength. In the second step, the MS acquires the system configuration parameters and concludes cell selection.

Upon power ON/OFF or after any loss of RF signal, the MS acquires a downlink RF channel. The MS is required to use nonvolatile storage in which the last operational parameters are stored and therefore it may first try to acquire the previous downlink RF channel. If the MS fails to acquire the last

---

[ii]It must be clarified that Cell_ID or physical layer identifier is not unique throughout the operator's network. In practice, the number of Cell_ID sequences are limited, thus they are reused as long as the cells that are assigned the same Cell_ID are sufficiently far apart (geographically) such that there is no confusion in the signals received from those cells.

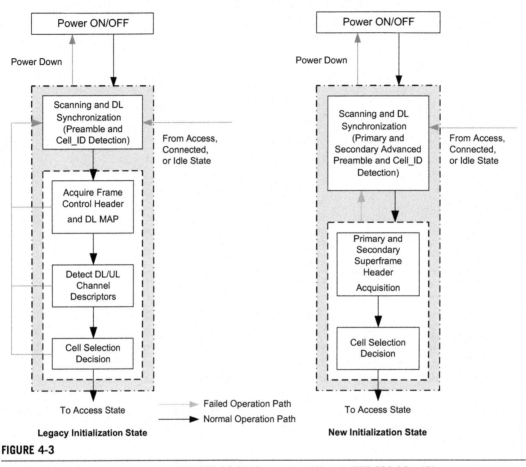

FIGURE 4-3

Initialization state procedures in the IEEE 802.16-2009 standard [1] and IEEE 802.16m [3]

RF channel that was used, it begins to scan the available downlink RF channels, as specified in the band class specifications by the WiMAX Forum [8], until it finds a valid downlink signal. Once the downlink synchronization has been achieved and Cell_ID has been detected, the MS attempts to acquire the system configuration information in the second step. The legacy system configuration information consists of all essential parameters, such as RF carrier center frequency and bandwidth, DL/UL radio (in TDD systems), resource partitioning and multi-antenna parameters, etc., that a mobile station must acquire for successful initial network entry and cell selection, as well as handover. The system information can be classified into static and dynamic sets where the static parameters remain unchanged over a long period of time and dynamic parameters change over time. The system parameters also include information about native neighbor cells (or alternative radio access networks that might be accessible by the mobile station) to facilitate inter-RAT or intra-RAT handover operation.

The initialization procedures in the legacy and IEEE 802.16m systems are different. In order to reduce the latency due to downlink scanning and initial network entry (or network re-entry), the initialization procedures have been simplified in IEEE 802.16m. As shown in Figure 4-3 for the legacy system, following successful non-hierarchical synchronization[iii] using the preamble, the Cell_ID is detected. The system configuration information is acquired through successful completion of the following steps [6,7,15]:

1. Detection and decoding of the frame control header, which follows the preamble symbol in every radio frame and contains information about the transmission format and length of the DL control channel;
2. Detection and decoding of DL control channel, i.e., DL MAP which appears in every radio frame, in order to locate the DL and UL channel descriptors through a broadcast information element;
3. Acquisition of the system configuration information through the DL and UL channel descriptors.

If the above actions are successfully performed, the mobile station will be able to select a suitable base station and proceed to the Access State; otherwise, the mobile station will have to repeat to initialization procedures and find a new candidate base station. Note that failure in any of the above steps would result in restarting of the initialization procedures.

The problem with the legacy initialization procedures is that the location of the DL/UL channel descriptors is not known to the mobile station *a priori*, consequently there is a need to locate the channel descriptors through detection and decoding of the DL control channel (i.e., DL MAP). The DL MAP contains a broadcast information element which can be received by all mobile stations in the coverage area of the base station that provides the location of the system configuration information. Another problem is that the DL/UL channel descriptor messages are usually transmitted periodically over long periods of time (i.e., every few hundred milliseconds and up to 10 seconds [1]) and this would delay the cell selection decision. In addition, the number of unique Cell_IDs specified in the legacy standard was 114 [1], which was not sufficient to support dense deployments and femto-cell operation. Therefore, the synchronization and system information broadcast schemes were redesigned to resolve these issues.

The IEEE 802.16m has simplified the initialization process by: (1) fixing the location of the system configuration information at a predetermined location, thereby dissociating the system information acquisition from the DL control channel detection; (2) reducing the number of system configuration parameters; (3) classifying the system configuration parameters based on their timing sensitivity and necessity for network entry, handover, paging, etc., thus transmitting system information sub-packets with different periodicity; and (4) increasing the number of unique Cell_IDs from 114 to 768 and changing the synchronization scheme from non-hierarchical to hierarchical [3].

Figure 4-3 illustrates the initialization procedures in IEEE 802.16m. The cell search starts with scanning for the available downlink RF channels. The DL synchronization is achieved by successful acquisition of the primary advanced preamble. The primary advanced preamble carries information about system bandwidth (e.g., 5, 10, 20 MHz) and multi-carrier configuration (i.e., fully configured or partially configured RF carrier). Once the primary advanced preamble is detected, the MS proceeds to acquisition of the secondary advanced preamble. The secondary advanced preamble carries a set of

---

[iii]In a non-hierarchical synchronization (i.e., a single step synchronization scheme), the preamble consists of one or more cell-specific waveforms and is used for time synchronization and frame alignment. In a hierarchical synchronization (i.e., a multi-step synchronization scheme), two or more unique set of signals are used for time synchronization and detection of the Cell_ID, and other relevant information for the cell or network.

768 distinct Cell_IDs that have been partitioned into a number of subsets where each subset corresponds to a certain type of base station (e.g., closed/open subscriber group femto base stations or macro base stations).

Upon successful acquisition of system timing and cell identification, the MS attempts to detect and decode the system configuration information. This information is carried via the superframe headers. The superframe headers, comprising primary and secondary superframe headers, are control elements that are periodically (while a large part of this information remains unchanged over a long period of time, some parts may change more frequently) broadcast using a robust and reliable transmission format to ensure the information can be correctly detected by all mobile stations in the coverage area of a base station. The correct and timely detection of the system information is essential for successful network entry/re-entry and handover. The secondary superframe header content is divided into three sub-packets where sub-packets carry essential information for various system processes, such as initial network entry, network re-entry, Idle-State operation, etc., according to their respective timing sensitivity.

Once the system parameters are successfully acquired, the cell selection decision can be made by taking certain considerations into account. For example, the mobile station may have a preference in selecting a specific type of base station (e.g., a femto-cell in an indoor environment), even though other types of the base station may be available or the MS may not be authorized to access a group of base stations despite the fact that their received RF signal strength might be good. Also, the IEEE 802.16m standard defines a system parameter called *Cell Bar* bit that is carried in the secondary superframe header. If *Cell Bar* bit is set to 1, the cell does not allow any new initial entry due to excess load in the cell [2].

4.1.2 Access State

Once the mobile station selects a suitable cell, it proceeds to the Access State to complete the network entry/re-entry procedures. Figure 4-4 shows the Access State procedures. The first step is to perform the initial ranging and uplink synchronization with the base station. The initial ranging is the process of acquiring the correct timing-offset, frequency-offset, and transmit-power adjustments so that the MS uplink transmissions are aligned with the BS uplink frame. The physical layer processing delays are relatively constant and the variations are accounted for in the guard time. The initial contention-based ranging consists of the following steps [2]:

- The MS selects a ranging opportunity using random backoff based on a binary truncated exponential algorithm.[iv] After selecting the ranging opportunity, the MS chooses a ranging sequence (alternatively known as ranging preamble) from the set of permissible initial ranging sequences. The selected ranging sequence is sent to the BS within the selected ranging opportunity.

[iv]The binary exponential backoff or truncated binary exponential backoff refers to an algorithm used to space out repeated re-transmissions of the same block of data. As an example, the re-transmission of frames in Carrier Sense Multiple Access with Collision Avoidance (CSMA/CA) and Carrier Sense Multiple Access with Collision Detection (CSMA/CD) networks, where this algorithm is part of the channel access method used to send data over these networks. The re-transmission is delayed by an amount of time derived from the slot time and the number of attempts to re-transmit. After n collisions, a random number of slot times between 0 and $2^n - 1$ is chosen. For the first collision, each sender might wait 0 or 1 slot times. After the second collision, the senders might wait 0, 1, 2, or 3 slot times, and so forth. As the number of re-transmission attempts increases, the number of possibilities for delay increases. The "truncated" simply means that after a certain number of increases, the exponentiation stops, i.e., the re-transmission timeout reaches a ceiling, and thereafter does not increase any further [9].

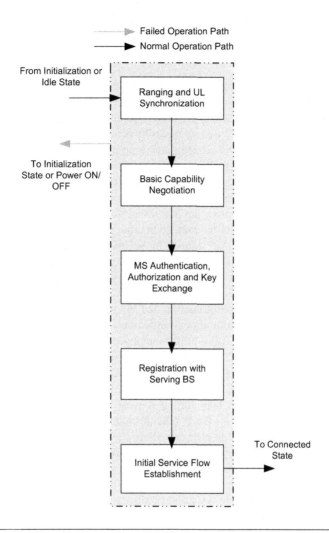

FIGURE 4-4

Access state procedures in the IEEE 802.16-2009 standard and IEEE 802.16m [3]

- The BS responds with ranging acknowledgement MAC control message within a time interval from the frame where at least one initial ranging preamble code is detected. This message provides responses to all successfully received and detected ranging preamble codes in all the ranging opportunities in a frame indicated by the frame identifier. If all the detected ranging preamble codes indicate "*success*" status, the BS provides uplink bandwidth allocations for each detected ranging preamble code within a certain time interval. If the MS receives neither the ranging acknowledgement message (corresponding to the initial ranging preamble code and opportunity selected by the MS) nor uplink resource allocation, it assumes that its initial ranging request has failed and restarts the initial ranging procedure.

- The initial ranging attempt by the MS may result in three possible outcomes that are signaled via the ranging acknowledgement message (i.e., the message contains ranging status indications from the BS to the MS) as follows:
  - The initial ranging preamble code was not successfully detected by the BS or the BS has determined that uplink transmission adjustments are necessary; therefore, the BS provides the required time, power, and possibly frequency adjustments to the MS ("*continue*" status).
  - The initial ranging preamble code was successfully detected by the BS and the BS may provide the MS with time/frequency/power adjustments or uplink bandwidth allocation in order for the MS to send a ranging request message ("*success*" status).
  - The initial ranging preamble code may or may not have been successfully detected by the BS and the BS requests the MS to abort the ranging process ("*abort*" status).
- Based on the outcome ranging response, the MS takes the following actions:
  - Upon receiving a "*continue*" status indication and adjustment parameters in the ranging acknowledgement message, the MS adjusts its uplink transmission parameters according to the instructions by the BS and continues the ranging process by sending a ranging preamble code randomly chosen from the initial ranging code-set (using random selection rather than random backoff) in the initial ranging opportunity.
  - Upon receiving a "*success*" status notification, the MS waits for uplink bandwidth allocation. If the MS does not receive bandwidth allocation within a certain time interval after sending the ranging preamble code, it restarts the initial ranging procedure, or it may return to downlink synchronization stage again. If the MS receives uplink bandwidth allocation, it sends a ranging request message. If the granted bandwidth allocation cannot accommodate the entire ranging request message, the message may be fragmented in order to fit to the allocated resources, and the MS may further request additional resources for the remaining message fragments. In response to the bandwidth request, the BS provides uplink allocation via a downlink control information element where the MS is identified by the same Random Access Identifier (RA-ID) that was used for the previous allocation. This RA-ID is used until completion of the ranging response. If the MS does not receive bandwidth allocation, or the ranging request and response message exchange are not completed within 128 frames, the MS re-sends the ranging preamble code and reinitiates the initial ranging procedure.
- The BS assigns temporary STID (TSTID) to the MS via the ranging response message upon successful ranging. The initial ranging process is concluded after receiving the ranging response message. The TSTID is used until permanent STID is assigned following successful registration.

In the case where the IEEE 802.16m mobile and base stations are interfaced with a legacy ASN, the actual MAC address is included in the ranging request message similar to that of the legacy systems [1], because the IEEE 802.16m station identifiers are not recognized by the legacy ASN. Immediately after completion of the initial ranging, the MS informs the BS of its capabilities by transmitting a capability negotiation request message with its capabilities set to "on." [1] The BS responds with a capability negotiation response message with the intersection of the mobile station's and the base station's capabilities set to "on." The problem with this way of capability negotiation in IEEE 802.16-2009 standard is that the length of the messages can grow excessively large without the possibility of fragmentation (this problem has been fixed for some MAC management messages that can be fragmented in the latest revision of the standard [1]), resulting in increasing error probability and reliability issues for the cell-edge users. Furthermore, the BS and the MS are unnecessarily required to negotiate some capabilities and parameters that are required for normal operation of the system, resulting in inefficient use of radio resources, increased network entry/re-entry latency, and compromising reliability.

The IEEE 802.16m simplifies the capability negotiation step by assuming that the MS, by default, is expected to support the baseline capabilities. A *Capability Class* is defined as a unique set of functions, configuration parameters, air interface protocol revision, and/or services that can uniquely describe a mobile station implementation or configuration while operating in the network. Each *Capability Class* is identified by its corresponding *CAPABILITY_INDEX*. If the MS is capable of supporting higher revisions of physical layer or medium access control layer protocols or further wishes to use enhanced features, it sends a capability negotiation request message to the BS indicating the highest *CAPABILITY_INDEX* that it can support. The *CAPABILITY_INDEX = 0* indicates the default capability index and basic feature set or configuration parameters and may not need to be signaled. Upon receipt of the capability negotiation request message containing the *CAPA-BILITY_INDEX* from the MS, the BS determines whether it could allow or could support the requested feature set or the MAC and/or PHY protocol revisions. If the BS does support or can allow the use of enhanced features, it responds with a capability negotiation response message to inform the MS of its decision. The BS may also respond with a *CAPABILITY_INDEX* which is numerically different than that requested by the MS. The higher the numerical values of the *CAPABILITY_INDEX* parameter, the more enhanced features or higher protocol revisions are used. The *CAPABILITY_INDEX* values range from 0 to N_{max}, where N_{max} denotes the maximum value of the *CAPABILITY_INDEX* parameter. In the case of failure at any stages of operation, the MS and BS fall back to baseline capability and restart negotiations for a new *Capability Class*, if necessary [2].

If a privacy key management function is enabled, the BS and MS perform authorization and key exchange procedures. The security procedures provide the users with privacy, authentication, and confidentiality over the air interface. This is done by applying cryptographic transforms to MAC PDUs carried across connections between the MS and BS.

Registration is the process by which the MS is granted entry to the network and a managed MS (see Chapter 3 for more information) receives its *Secondary Management CID*,[v] and thus becomes manageable. To register with a BS, the MS sends a registration request message to the BS. The BS responds with a registration response message that includes the secondary management CID for a managed MS. In IEEE 802.16m, the BS will allocate and transfer a Station Identifier (STID) to the MS through an encrypted registration response message (if management message encryption is supported) and the temporary STID, which was allocated during initial ranging procedure, is discarded.

In IEEE 802.16m on successful registration with the BS, downlink and uplink Flow Identifiers (FID) are assigned to the MS without using the dynamic service allocation procedure, in order to activate one pre-provisioned service flow for each downlink and uplink FID which can be used for upper layer signaling (e.g., DHCP). Note that the 16-bit legacy CID has been divided into a 12-bit Station ID (that identifies the MS) and a 4-bit FID (that identifies each of the assigned service flows to

---

[v]Secondary management is a connection that may be established during the registration step which can be used to transport standard messages such as Simple Network Management Protocol (SNMP) or Dynamic Host Configuration Protocol (DHCP).

the MS). The temporary STID and STID are assigned to the MS during initial entry while in Connected State, to avoid explicit transmission of the MS MAC Address[vi] over the air interface that could potentially compromise the privacy and security of the user. On failing to complete any one of the steps of network entry, the MS repeats the Initialization State procedures; otherwise, it proceeds to Connected State for downlink or uplink transmissions.

During the registration procedure, the MS and BS negotiate supported IP protocol versions and may further negotiate host configuration parameters. The MS is required to notify the network, through the BS, of IP versions (IPv4 and/or IPv6) it supports by including this information in the registration request message. After the network selects one of the IP versions, the BS informs the MS of the selected IP service type (e.g., IPv4 only, IPv6 only, or IPv4/IPv6 dual mode) by including this information in the registration response message. The MS may indicate its capability of configuring host parameters (host address or home network prefix) and may request the additional parameters such as DNS server address to the network through the BS using registration request message. If the MS indicates the capability of configuring host parameters and the BS can support the requested host configuration parameters, the BS includes either the IPv4-Host-Address information element, the IPv6-Home-Network-Prefix information, or both, and the requested host configuration parameters in the registration response message; otherwise, the MS will be configured by upper layer protocols.

There are two types of connections specified in IEEE 802.16m; control connections and transport connections. Control connections are used to carry MAC control messages. Transport connections are used to carry user data including upper layer signaling messages such as DHCP, etc., as well as data-plane signaling such as ARQ feedback. A MAC control message is never transferred over a transport connection, and user traffic (except short message service using ranging request and response messages) is never transferred over a control connection. One pair of bi-directional (DL/UL) unicast control connections are automatically established when an MS performs initial network entry where a unicast or broadcast control FID value is appropriately selected and assigned to the connection. Once the TSTID is allocated to the MS, the control connections are established automatically. The FIDs for the control connections are never changed during handover or network re-entry.

4.1.3 Connected State

As shown in Figure 4-5, the mobile station in the Connected State operates in one of the three modes: Sleep; Active; or Scanning Mode. During the Connected State, the MS maintains two connections that were established during Access State (i.e., a control connection for management messages and a transport connection for data transmission). The MS and the BS may establish additional transport connections. The MS may remain in the Connected State during handover process (i.e., exchanging data/control with the serving BS while signaling with the target BS). The BS can ask the MS to transition from the Connected State to the Idle State after deregistration. The Idle State initiation may begin only after MS deregistration. Failure to maintain the connections at any time prompts the MS to transition to the Initialization State.

[vi]The MS, RS, and BS are identified by a globally unique 48-bit IEEE Extended Unique Identifier (EUI-48™) based on the 24-bit Organizationally Unique Identifier (OUI) value administered by the IEEE Registration Authority (see http://standards.ieee.org/regauth/index.html for more information).

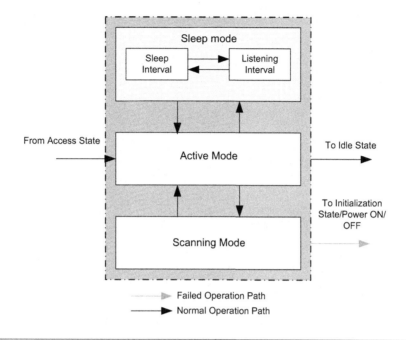

FIGURE 4-5

Connected state procedures [3]

Active Mode

An MS in Active Mode can be scheduled for downlink or uplink transmissions at appropriate intervals by the serving BS. The MS may conduct channel quality measurements based on downlink reference signals and report the measurements to the BS to assist the BS scheduler in properly adapting to the link characteristics and selecting the most appropriate transmission formats for data and control signals sent to (or received from) the MS. The MS may transit from Active Mode to Sleep Mode during sleep intervals (with intermittent DL/UL traffic) or to Idle State (when it has no DL/UL traffic and following deregistration with the BS) in order to conserve power.

Sleep Mode

Sleep is a mode in which an MS becomes unavailable to the serving BS during pre-negotiated intervals while connected to the BS. The Sleep Mode can only be activated when an MS is in the Connected State. When Sleep Mode is activated, the MS is provided with a series of sleep cycles that typically consists of a listening window followed by a sleep window (as shown in Figure 4-6). A sleep cycle is the sum of the sleep and listening intervals. During the sleep window, the BS does not transmit any downlink unicast MAC PDU to the MS; therefore, the MS may power down its transmission chain or perform other activities that do not require communication with the BS. During the listening window, the MS may receive downlink transmissions in the same way as in normal operation. The MS must ensure that it has the most up-to-date system information. The downlink synchronization and system information acquisition may be performed by the MS, if it wakes up at superframe intervals (i.e., every

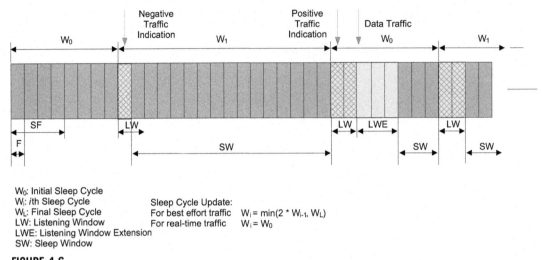

W₀: Initial Sleep Cycle
W_i: ith Sleep Cycle
W_L: Final Sleep Cycle   Sleep Cycle Update:
LW: Listening Window   For best effort traffic $W_i = min(2 * W_{i-1}, W_L)$
LWE: Listening Window Extension For real-time traffic $W_i = W_0$
SW: Sleep Window

FIGURE 4-6

An illustration of sleep mode operation [27]

20 ms) prior to its listening window to verify whether the Superframe Number and System Config-uration Description Change Count parameters are as expected. Upon wakeup, if the synchronization with the serving BS is lost or the MS detects base stations other than the serving BS, then the MS exits the Sleep Mode and performs (handover and) network re-entry procedures with the target BS, as described earlier.

The length of successive sleep cycles may remain constant or may be adaptive based on traffic conditions. The sleep and listening windows may also be dynamically adjusted for the purpose of data transmission and MAC control signaling. The MS and BS may keep up to 16 previously used sleep cycle patterns and associated sleep cycle identifiers. The sleep cycle pattern can be changed by sending a sleep request management message to the serving BS. The BS responds using a sleep response management message either approving the request or suggesting another sleep pattern. A single sleep cycle setting, identified by a sleep cycle identifier, is applied per MS across all the active connections of the MS. The Sleep Mode entry can be initiated either by the MS or the BS. When the MS is in Active Mode, the Sleep Mode parameters are negotiated between the MS and the serving BS.

The sleep cycle is measured in units of frames. The start of the listening window is aligned at the frame boundary. The MS must ensure that it has the latest system information for proper operation. If the system information is not up-to-date, the MS will not transmit in the listening window until it receives the current system information. The BS transmissions are aligned with the MS at the boundaries of the sleep and listening windows.

During the sleep window, the MS is unavailable to receive any downlink control and data trans-mission from the serving BS. The IEEE 802.16m standard provides a framework for dynamically adjusting the duration of sleep windows. If the MS has data or signaling to transmit to the BS during the sleep window, the MS can interrupt the sleep window and request bandwidth for uplink trans-mission with or without terminating the Sleep Mode. The MS can also send or receive control/data during the listening window. The listening window is measured in units of frames. After termination of a listening window, the MS may go back to sleep for the remainder of the current sleep cycle, if there is no scheduled downlink/uplink activity.

During the listening window, the BS may transmit the traffic indication message intended for one or multiple terminals. It indicates if there is traffic addressed to one or multiple mobile stations. The traffic indication message is transmitted at predefined locations. A negative traffic indication in the traffic indication message can trigger the MS to go back to Sleep Mode for the rest of the current sleep cycle. The listening window duration can be dynamically adjusted based on traffic availability or signaling in the MS or BS. In addition, the listening window can be extended to the end of the current sleep cycle. The Sleep Mode termination can be initiated either by the MS or the BS.

Scanning Mode

The scanning interval is a time period designated for the mobile station to monitor neighbor base stations and to determine the most suitable candidates for handover. A MAC management message may be transmitted by the MS to request a scanning interval for this purpose while in the Connected State. The scanning procedure provides an opportunity for the MS to perform neighbor-cell measurements for handover decisions. The MS may use any interval not used for communication with the serving BS to perform scanning. In addition, the MS may perform scanning procedure without interrupting its communication with the serving BS if the MS supports such capability, which requires dual or multiple radios.

The MS selects the scanning candidate base stations by information obtained from the BS or information stored in the MS. The BS or MS may prioritize the neighbor base stations to be scanned based on various metrics, such as cell type, loading, received signal strength, and geographical location. As part of the scanning procedure, the MS measures the received RF signal strength from the selected base stations and reports the measurement results back to the serving BS. The measurements may be used by the MS or the network to determine the best target BS for handover. The measurements in IEEE 802.16m are a superset of those specified in the IEEE 802.16-2009 standard [1,2]. The serving BS defines triggering conditions and rules for the MS for sending the scanning report.

Cell reselection refers to a process where an MS scans and/or associates with one or more base stations in order to determine their suitability, along with other performance considerations, as a handover target. The MS may incorporate information acquired from a neighbor advertisement message while searching neighbor base stations for cell reselection. The serving BS may schedule scanning intervals or sleep intervals to conduct cell reselection activity. Such procedure does not involve termination of existing connection to a serving BS.

4.1.4 **Idle State**

The Idle State is intended as a mechanism to allow the MS to become periodically available for downlink broadcast messages without registration at a specific BS as the MS traverses across different cells, typically over a large geographical area. The Idle State benefits the MS by relaxing the requirement for handover and other normal operation signaling requirements in the Connected State. By restricting the MS activity to scanning in discrete intervals, the Idle State allows the MS to conserve power and operational resources.

The Idle State further benefits the network and the BS by providing a simple method for alerting the MS to pending downlink traffic, and additionally by eliminating airlink and network handover signaling for an inactive MS. The base stations are divided into logical clusters called paging groups. The purpose of these logical groupings is to create a contiguous coverage area where an MS without uplink transmission can be paged in the downlink on availability of any traffic in the BS queue for that MS. The size of paging groups should be large enough so that most terminals will remain within the same paging group most of the time and small enough so that the paging overhead is reasonable. A BS may be a member of one or more paging groups. Thus, there are different groupings of base stations with varying paging cycles and paging offsets. The concept of different paging cycles, paging offsets, and listening intervals for mobile stations is illustrated in Figures 4-7 and 4-8. It is shown that different mobile stations may have negotiated different values of paging offset and paging cycle with the serving base station. The grouping of the base stations provides support for the geographical requirements of Idle State operation and additionally may support differentiated and dynamic QoS requirements and scalable load-balancing. The paging groups are identified by a unique 16-bit Paging Group ID (PGID) in the network [6,7]. Figure 4-9 shows an example of four paging groups defined over multiple base stations arranged in a hexagonal grid.

The location management, i.e., the process of identifying and tracking a mobile station in the network, consists of location update and paging procedures. The location update is a procedure in which the MS periodically informs the network of its current location. The location is defined as an area that is served by a group of base stations belonging to the same paging group. The network maintains the track of the MS by updating the MS location profile in a central database. If there is pending traffic for an idle MS, the MS is paged within the paging group that was last reported. It must be noted that the greater the number of base stations within a paging group, the higher the paging resources required in the network. Network operators should make tradeoffs between using resources for location update signaling versus paging over a large geographical area.

As shown in Figure 4-10, the Idle State consists of two separate modes, Paging Available Mode and Paging Unavailable Mode. During Idle State, the MS may attempt power saving by switching between Paging Available and Paging Unavailable modes. The MS monitors the paging message during listening intervals. The start of the mobile station's paging listening interval is derived based on paging cycle and paging offset. Paging offset and paging cycle are defined in terms of number of superframes [2]. The BS transmits the list of PGIDs at the predetermined location within the radio frame. The PGID information should be received during mobile stations' paging listening interval.

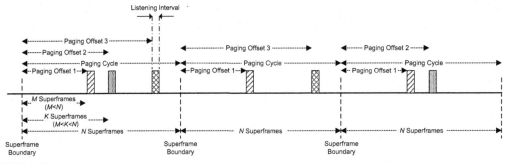

FIGURE 4-7

An illustration of paging cycle, paging offset, and listening interval concepts (example)

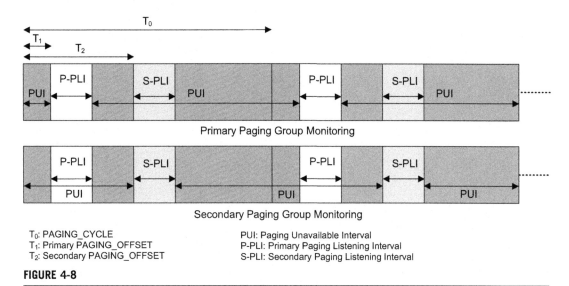

FIGURE 4-8

Multiple paging group idle state operation (example) [27]

Paging message is an MS notification which either indicates the presence of downlink traffic pending for the specific MS or it is intended to poll an MS and request a location update without requiring a full network entry, or to request an MS to perform measurements for location-based services. In addition, an emergency alert indicator is included in the paging message to notify the idle mobile stations of any emergency information.

In IEEE 802.16m, the temporary identifier, paging cycle, and paging offset are used uniquely to identify an idle MS in a particular paging group. The temporary identifier is assigned during Idle State entry or location update due to paging group change. Such identifier remains valid as long as the MS stays in the same paging group. The temporary identifier and paging cycle are used in paging messages to identify the MS. The temporary identifier, together with the paging cycle and paging offset, is used by the MS to identify itself during its network re-entry procedure as a response to paging or location update when paging group is not changed. An MS may be assigned one or more paging groups. In that case, it may be assigned multiple paging offsets within a paging cycle where each paging offset corresponds to a different paging group. The MS is not required to perform location update when it moves within its assigned paging groups. The assignment of multiple paging offsets to an MS allows the MS to monitor paging message at different paging offsets when the MS is located in one of the paging groups. If an MS is assigned to more than one paging group, one of the paging groups is called Primary Paging Group and others are known as Secondary Paging Groups. If an MS is assigned to one paging group, that paging group is considered the Primary Paging Group. When different paging offsets are assigned to an MS, the Primary Paging Offset is shorter than the Secondary Paging Offsets. The distance between two adjacent paging offsets should be sufficiently long so that the MS paged in the first paging offset can inform the network before the next paging offset in the same paging cycle, so that the network avoids unnecessary paging of the MS in the next paging offset. An Idle State MS (while in paging listening interval) wakes up at its primary paging offset and looks for primary PGID

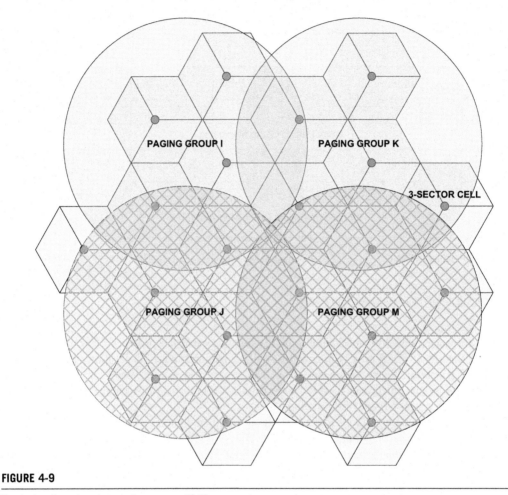

FIGURE 4-9

Example of overlapping paging groups [6,7]

information. If the MS does not detect the primary PGID, it will wake up during its secondary paging offset in the same paging cycle. If the MS can find neither primary nor secondary PGIDs, it will perform a location update. The paging message contains identification of the mobile stations to be notified of pending traffic or location update. The MS determines the start of the paging listening interval based on the paging cycle and paging offset. During the paging available interval, the MS monitors the superframe header and if there is an indication of any change in system configuration information, the MS will acquire the latest system information at the next instance of superframe header transmission (i.e., next superframe header). To provide location privacy, the paging controller may assign temporary identifiers to uniquely identify the mobile stations in the idle mode in a particular paging group. The temporary identifiers remain valid as long as the MS stays in the same paging group. The MS in the primary paging group monitors the paging message transmitted at the

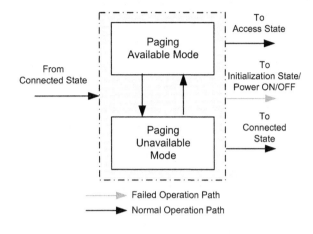

FIGURE 4-10

Idle state procedures [3]

associated paging offset. If the MS leaves the primary paging group and enters a secondary paging group, the MS will monitor the paging messages at the paging offset associated with the secondary paging group.

Within a paging listening interval, the frame that contains the paging message for one or a group of idle mobile stations is known to the idle terminals and the paging base stations. The BS should not transmit any downlink traffic or paging advertisement to the MS during a Paging Unavailable interval. During a Paging Unavailable interval, the MS may power-down, scan neighbor base stations, reselect a preferred cell, conduct ranging, or perform other activities during which the MS will not be available to any BS for downlink traffic. The MS calculates the start of the paging listening interval based on the paging cycle and paging offset. At the beginning of paging listening interval, the MS scans and synchronizes with the synchronization sequences periodically transmitted by its preferred BS. The MS detects and decodes the superframe headers and confirms whether it is located in the same paging group as it has most recently been by acquiring the PGID parameter.

In the Paging Available Mode, the MS monitors the superframe headers. If superframe headers indicate a change in system information (e.g., a change in system configuration count) then the MS should acquire the latest system information at the predetermined time when the system information is broadcast by the BS. The MS decodes the paging message at the predetermined location. If the MS decodes a paging message that contains its identification, the MS performs network re-entry or location update depending on the notification in the paging message. Otherwise, the MS returns to Paging Unavailable Mode.

The MS calculates the beginning of the paging listening interval based on the paging cycle and paging offset. The paging listening interval includes the superframe whose superframe number satisfies $N_{superframe} \bmod PAGING\_CYCLE = PAGING\_OFFSET$. The length of the paging listening interval is one superframe per paging cycle. At the beginning of the paging listening interval, the MS scans and synchronizes with the downlink synchronization channels of its preferred BS and decodes the primary superframe header of the BS. As shown in Figure 4-11, the BS transmits the PGID_Info MAC control message at a predetermined location in the paging listening interval in order to advertise the paging

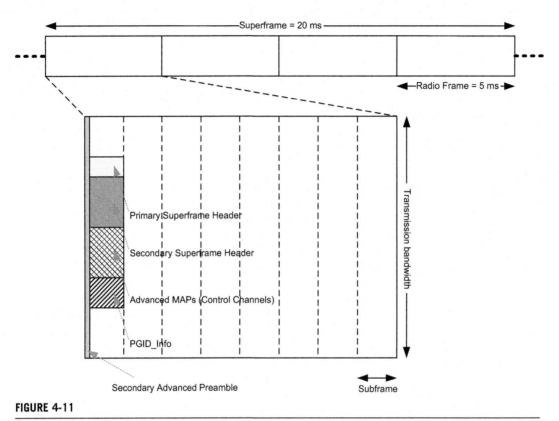

FIGURE 4-11

Transmission of PGID_Info MAC control messages [2]

groups that are supported by the BS or to which the BS is associated. The PGID_Info is transmitted by the BS regardless of any notifications for the mobile stations. The BS transmits the PGID_Info immediately after the superframe header and advanced MAPs in the first subframe of the superframe and during mobile station's paging listening interval, as shown in Figure 4-11. The PGID_Info is transmitted before any paging advertisement message in the superframe. Using the PGID_Info, the MS can determine whether it is located in the same paging group as its preferred BS or the paging group that it most recently belonged to. If none of the PGIDs to which the MS belongs are detected, the MS determines that the set of paging groups has changed and performs Idle Mode location update. The MS monitors a predetermined frame for paging message. The predetermined frame $N_{paging-frame}$ for an MS is given as $N_{paging-frame} = MS\ Deregistration\ Identifier$ mod m ($m = 1, 2, 3,$ or 4).

The MS or the serving BS can initiate the Idle State transition using procedures defined in IEEE 802.16m [2] (or alternatively corresponding procedures in reference [1] for the legacy systems). In order to reduce signaling overhead and provide location privacy, a temporary identifier is assigned to uniquely identify the idle mode mobile stations in a particular paging group. The mobile station's temporary identifier remains valid as long as the MS stays in the same paging group. The temporary identifier assignment may happen during Idle State entry or during location update due to paging group change. The temporary identifier may be used in paging messages or during the mobile station's network re-entry.

An MS exits the Idle State by invoking procedures defined in IEEE 802.16m [2] (or alternatively corresponding procedures in reference [1] for the legacy systems). For termination of the Idle State, the MS performs network re-entry with its preferred BS. The network re-entry procedure can be streamlined by the target BS possession of the MS context. An MS in the Idle State will perform a location update process based on any of the following triggers:

- Paging group location update;
- Timer-based location update;
- Power-down location update;
- Multicast and Broadcast Service (MBS) location update.

During a paging group location update, timer-based location update, or MBS location update, the MS may update its temporary identifier, paging cycle, and paging offset. If an MS updates its location, depending on the security association, it informs its preferred BS using either a secure or unsecure location update process. The location update consists of conditional evaluation and location update signaling. The MS performs the location update process when it detects a change in paging group. The change of paging group is detected by monitoring the Paging Group IDs, which are transmitted by the BS. The MS periodically performs a location update process prior to the expiration of the Idle State timer. At each location update, including paging group location update, the Idle State timer is reset and restarted. The MS attempts to complete a location update once as part of its regular power-down procedure. For an MS receiving MBS data in the Idle State and during MBS zone transition, the MS may perform the MBS location update process to acquire the MBS zone information for continuous reception of MBS data.

The transition to Idle State requires the MS to deregister from the serving BS. The MS deregistration is a process originated by shutdown or some failed conditions where the MS is deregistered from the network and its context is deleted. The following entities may start MS deregistration process [4,5]:

- MS, when initiates shutdown;
- ASN, as result of shutdown or failure triggers in the network;
- Home AAA server located in CSN also is able to trigger MS deregistration.

The MS deregistration procedure covers different scenarios as follows:

- MS deregistration as a result of MS shutdown;
- MS deregistration from the serving BS (and probably re-initialization in other BS/ network);
- Enforcing MS to halt any transmissions (including MAC management messaging);
- Enforcing MS to halt traffic transmissions;
- Erasing MS context in the ASN entities when radio link with the MS has been lost.

Deregistration signaling over the air interface is performed using IEEE 802.16m [2] (or alternatively corresponding procedures in reference [1] for the legacy systems) MAC control/management messages as follows:

- Deregistration command used by the BS to signal deregistration command to MS. It may be unsolicited or in response to an MS-initiated deregistration request.
- Deregistration request that the MS sends to the BS to request deregistration.

In order to send or receive short message service (SMS) during the Idle Mode without performing network re-entry, the SMS data may be included in a ranging request/response message. The maximum size of the short message payload is 140 bytes [2]

4.2 NETWORK ENTRY

This section describes the network entry/re-entry procedures from the perspective of the mobile and base stations. Network entry is a set of procedures that an MS must follow in order to enter the operator's network and to obtain network services. The network entry procedures that follow the cell search and cell selection include initial ranging and uplink synchronization, basic capability negotiation, authorization, authentication, key exchange, registration with the network, IP connectivity establishment, and transfer of operational parameters [1,2]. It must be noted that, depending on the current operational state of the mobile station, the network entry procedures can be different.

A mobile station which is just powered on must perform an initial network entry which, as described earlier, includes successful completion of the Initialization and Access State procedures to establish data- and control-planes and to enable any data transfer in the downlink or uplink. The initial network entry procedures in the mobile station and base station are illustrated in Figures 4-12 and 4-13. There are other access- and core-network entities such as ASN-GW/Authenticator and Home AAA Server involved in the initial network entry process which are not shown in the figures. Those network entities are involved in authentication, authorization, and key exchange, as well as in the registration procedures with the BS [5].

The MAC management messages in IEEE 802.16m are distinguished from their legacy counterparts by an "AAI" prefix denoting "Advanced Air Interface" messaging. There are some timers associated with execution of each procedure that are denoted by T_i. These timers are used to control the amount of permissible time to enact or execute a specific procedure that once expired, forces the restart of the same or a different procedure. The MAC management messages that are shown in Figures 4-12 and 4-13 will be introduced and discussed in Chapter 6. However, the procedures such as initial ranging (AAI_RNG-REQ/RSP), basic capability negotiation (AAI_SBC-REQ/RSP), and registration with the BS (AAI_REG-REQ/RSP) have already been discussed in the description of IEEE 802.16m states.

Network re-entry is a sequence of procedures similar to network entry, where the MS attempts to transition from the Idle State to the Connected State or during handover and prior to establishment of data-plane with the target BS. The MS may exit the Idle State and re-enter the network either in response to a page by the BS (i.e., network-initiated), or to start communication with the BS (i.e., MS-initiated). The network re-entry steps may differ depending on whether the serving ASN has the MS context. If the serving ASN does not have the MS context when the MS is trying to re-enter the network from Idle State, the entire context has to be retrieved from the Anchor Paging Controller. In other words, the MS tries to re-enter the network when the *management resource holding timer*[vii] has

---

[vii]When the MS enters Idle State, the BS in the serving ASN starts a timer called "Management Resource Holding Timer" to keep the time that the BS maintains connection information of the MS after the BS sends the deregistration command to the MS. The BS retains the R1 context as well as the R4 and R6 data paths for the MS until the timer expires or until the context is revoked by the Anchor PC. When located in the same ASN, the Anchor PC sends a control message to the serving BS to revoke the MS context if the MS has entered the network at a different BS before the management resource holding timer at the serving BS expires [4,5].

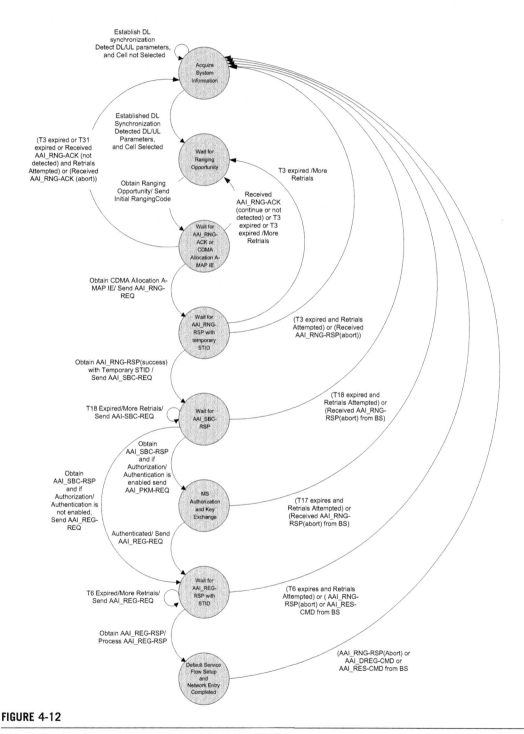

FIGURE 4-12

Initial network entry procedures in the mobile station [1,2]

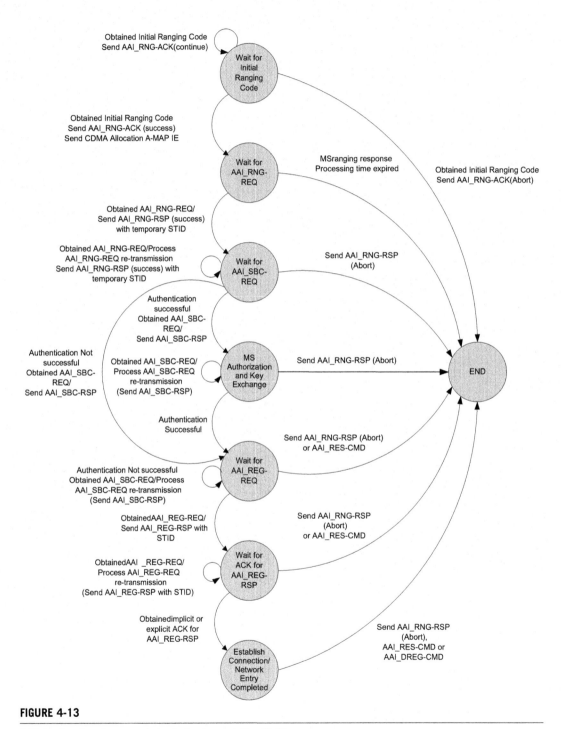

FIGURE 4-13

Initial network entry procedures in the base station [1,2]

expired in the network, and consequently the MS must perform the normal network re-entry procedures.

If the serving ASN has the MS context or the context in the serving BS is not revoked before the *management resource holding timer* expires, the procedures for the MS to exit Idle State can be further simplified. The serving BS releases the MS context and the data paths for this MS only upon expiration of this timer. In this case, if the MS has maintained downlink synchronization with the BS during Idle State and if the system information at the MS is up-to-date, the MS may skip Initialization State procedures and directly transition to Access State and perform a light ranging, re-authentication, and re-registration with the serving BS. This form of network re-entry is referred to as fast network re-entry.

Note that in the IEEE 802.16 standard, an MS is deregistered with the serving BS prior to entering the Idle State. In addition, during handover, the MS must search and select a target BS and perform network re-entry process in order to establish a data-plane with the target BS and detach from the serving BS. Since the MS context may remain in the network when the MS transitions to the Idle State for a limited time, there is no need for a full authentication and registration on transition from the Idle State.

Cell reselection is the process where an MS scans and associates with one or more base stations in order to determine their suitability, along with other performance considerations, as a handover target. The MS may use neighbor advertisement messages from the serving BS or may negotiate scanning intervals with the BS to search and select candidate base stations for handover. The cell reselection is performed in the Scanning Mode and as part of the Connected State.

4.2.1 **Normal Network Re-Entry**

As shown in Figure 4-1, normal network re-entry is performed when an MS in the Idle State transitions to the Connected State (through Access State) as a result of paging or availability of scheduled uplink data or messages for transmission. The Initialization State procedures may be skipped, depending on whether system information at the MS is up-to-date and downlink synchronization with the cell has been maintained during idle period. Also during handover, the MS must perform a successful network re-entry with the target BS to establish data-plane and start or resume data transmission with the new BS.

During network re-entry, the MS sends a ranging request message that includes a *Ranging Purpose Indication* parameter and *Paging Controller ID*. If the target BS has not already received the MS information over the backhaul, then the target BS may request the information from the paging controller over the core network. Regardless of having received the MS information from the paging controller, the target BS may request the MS information from another network management entity via the backhaul.

The network re-entry is similar to network entry, except it may be shortened by the target BS's possession of MS information obtained from paging controller or other network entity over the backbone. In order to notify an MS seeking network re-entry from Idle State about the procedural steps that may be omitted during its network re-entry attempt (due to the availability of the MS service and operational context), the target BS includes a handover process optimization parameter in the ranging response message to the MS, indicating which re-entry management messages may be skipped [1]. Furthermore, if the target BS does not share a current and valid security context with the MS, or in the event that the BS has opted to instruct the MS to use *Unsecure Location Update*, the MS initiates network re-entry from the Idle State.

On completion of network re-entry, the target BS (now the new serving BS) provides the MS with uplink allocation for transmission of *Sequence Number Report* signaling header containing the least significant bits of the sequence number of ARQ block or virtual MAC SDU number for continuation of the interrupted data transmission due to handover. After reception of this signaling header, the BS resumes transmission of the data of the corresponding downlink service flow starting from MAC SDUs pointed by the sequence number [1].

4.2.2 Fast Network Re-Entry

Fast network re-entry is a process used to accelerate the transition from the Idle State to the Connected State. As mentioned earlier, some of the Access State procedures may be skipped or simplified if the MS context has not been revoked or expired, and the BS can obtain this information from the previous serving BS. Note that the IEEE 802.16 standard does not support Idle State handover (i.e., the MS cannot perform or be instructed to perform a handover during an idle period). Therefore, when the MS moves from the coverage area of the previous serving BS (i.e., the BS which was serving the MS prior to transition to the Idle State) to the coverage area of a new BS while in Idle State, the MS must perform the network re-entry process.

For the purpose of expediting network re-entry of the MS with the target BS, the serving BS may negotiate with the target BS allocation of a non-contention-based ranging opportunity for the MS, i.e., an unsolicited uplink allocation for transmission of ranging request message. The agreed time takes into account the *Handover Indication Readiness Timer* (i.e., the minimum time that the MS may require to process handover request or handover response messages from the BS) and the *BS Switching Timer* (i.e., the minimum time the MS requires between transmission of handover indication message at the serving BS until it is able to receive fast ranging information element at the target BS) [1]. The serving BS indicates the time of the fast (i.e., non-contention-based) ranging opportunity negotiated with the potential target base stations via an information field in the handover request/response message. The network re-entry procedures are illustrated in Figure 4-14. Some of the steps may be skipped or shortened as described earlier. The MS ID* shown in Figure 4-14 is a hash value (scrambled) of the real MS ID (MAC address of the mobile station) that is temporarily used prior to establishment of a secure connection between the MS and the BS, in order to avoid revealing the user identity over the air interface [2].

Deregistration with Content Retention (DCR) is a mode in IEEE 802.16m where an MS can deregister from the network while its context is maintained in a network entity until the *Context Retention Timer* is valid to expedite the mobile station's network re-entry. In that case, the network assigns a 72-bit Context Retention Identifier (CRID) to each MS during network entry. The MS is identified by the CRID in coverage loss recovery and DCR mode, where the CRID allows the network to retrieve MS context. The network may assign the MS a new CRID if necessary. The CRID is used to uniquely identify the DCR mode mobile stations. The DCR mode may only be terminated on MS re-entry to the network or on expiration of the *Context Retention Timer.*

4.3 STATE TRANSITIONS AND MOBILITY

Handover (HO) is defined as the migration of a mobile station between the air interfaces of different base stations or cells. The BS associated with the MS before the handover is called the serving BS,

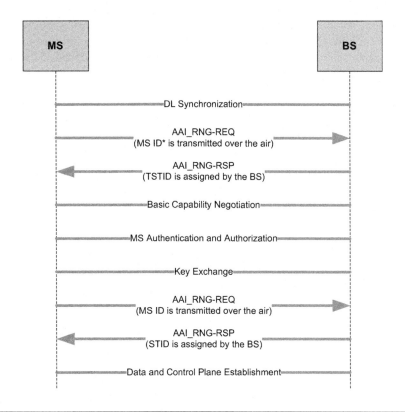

FIGURE 4-14

Summary of network entry procedures in IEEE 802.16m [2]

whereas the BS associated with the MS after the handover is denoted as the target BS. The handover process involves a number of interactions and message exchanges between the MS and neighboring base stations for the purpose of scanning, ranging, parameter negotiation, and information exchange. In this process, which may take tens of milliseconds, the MS connection to the serving BS may be temporarily interrupted (i.e., handover interruption time). The handover process consists of the following five steps:

1. Cell reselection by scanning neighboring base stations and compiling a list of candidate base stations. The MS may use neighbor BS information acquired from a decoded neighbor advertisement message or may request to schedule scanning intervals or sleep intervals to scan (and possibly to perform ranging with) neighbor base stations for the purpose of evaluating MS interest in handover to a potential target BS. The cell reselection process does not need to occur in conjunction with any specific contemplated handover decision.
2. Handover decision that may be originated from the MS (mobile-initiated handover) or the serving BS (base station-initiated handover). The handover decision is carried out with a notification of MS or BS intent for handover through handover request messages.
3. Downlink synchronization with the target BS (if necessary).

4. Handover ranging[viii], uplink synchronization, and the subsequent network re-entry procedures with the target BS (i.e., basic capability negotiation, authorization, authentication, registration, and service flow establishment), some of which may be skipped or simplified if the target BS can obtain the MS context from the serving BS over the backbone.

5. Termination of the MS context at the serving BS that includes information in the queues, ARQ state machine, counters, timers, header compression information, etc.

The above handover procedures are illustrated in Figure 4-15. The normal operation in the figure corresponds to the Active Mode. The cell reselection process shown in the figure is part of the Scanning Mode, and the network re-entry process corresponds to the Initialization and Access States. However, as mentioned earlier, some of the network re-entry steps may be skipped and the process may be further streamlined depending on the conditions that were discussed in the previous sections.

Figure 4-16 shows the IEEE 802.16m mobile-initiated handover process. In IEEE 802.16m, the handover procedure may be initiated by either the MS or BS. In MS-initiated handover, the MS sends a handover initiation message to the serving BS. The BS responds to the handover initiation message by sending a handover command message to the MS. In BS-initiated handover, the BS sends a handover command message to the MS. In both cases, the handover command message includes one or more target base stations. If the handover command message includes only one target BS, the MS executes the handover procedures as directed by the BS. An MS may send a handover indication message to the serving BS before the expiration of *Disconnect Timer* (i.e., the time interval before cessation of downlink/uplink services to the MS by the serving BS once the handover process is initiated). The serving BS stops downlink traffic transmission and uplink allocations to the MS after expiration of the disconnect timer or after reception of a handover indication message.

If the handover command message includes more than one target base station, the MS selects one of these targets and informs the BS of its selection by sending a handover indication message to the serving BS before the expiration of the disconnect timer. The network re-entry procedures with the target BS may be accelerated (i.e., fast network re-entry) if the MS information is available to the target BS as explained in the previous sections. Therefore, it can be concluded that the data transmission to and from an MS initiating handover (or being instructed to handover) will be interrupted for a period of time starting from a handover indication message transmission till re-establishment of the data-plane with the target BS. During handover interruption time, the MS may exit the Connected State and perform (normal or fast) network re-entry procedures with the target BS. The MS will transition to the Active Mode once the data-plane is established with the target BS (see Figure 4-15).

4.4 STATE TRANSITIONS IN RELAY STATIONS

Optional multi-hop relay architecture may be used to provide extended coverage or performance improvement in a radio access network [10–12]. In multi-hop relay networks, the BS may be replaced

---

[viii]An MS that wishes to perform handover ranging follows a process similar to that defined for initial ranging, except that random selection is used instead of random backoff, and the handover ranging code is used instead of the initial ranging code. Note that the set of physical codes for the ranging process are divided into three subsets: handover ranging; initial ranging; and periodic ranging. Alternatively, if the target BS is notified in advance of the MS handover, it may provide bandwidth allocation information to the MS using a fast ranging information element to send a ranging request message (i.e., non-contention-based ranging) [1,2].

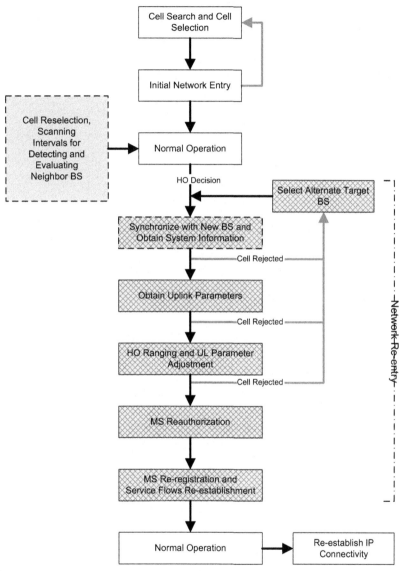

FIGURE 4-15

The handover process and network re-entry [1]

by a relay-enabled BS (i.e., a BS with relay functionalities) and one or more Relay Stations (RS). Traffic and signaling between the MS and the BS can be relayed by the RS, thereby extending the coverage and improving the performance of the system in areas where relay stations are deployed. Each RS is under the supervision of a relay-enabled BS. In a multi-hop system, traffic and signaling

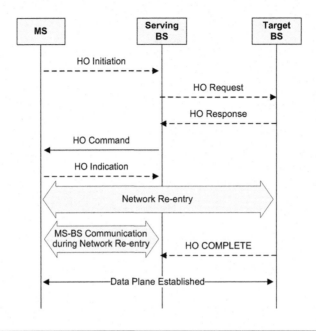

FIGURE 4-16

The IEEE 802.16m handover process (mobile-initiated handover) [2]

between an access RS and the BS may also be relayed through intermediate relay stations. The RS may be fixed in location or mobile. The MS may also communicate directly with the BS. Figure 4-17 illustrates the state transition diagram of an IEEE 802.16m relay station. The relay station state diagram consists of three states: (1) Initialization State; (2) Access State; and (3) Operational State. Note that the IEEE 802.16m standard currently supports two-hop relaying only; therefore, some of the procedures concerning multi-hop relaying may not apply.

4.4.1 Initialization State

In the Initialization State, the relay station performs cell selection by scanning and synchronizing to a base station (in the case of two-hop relay network topology) or another relay station (in the case of multi-hop relay network topology) to acquire the system timing and configuration information prior to entering the Access State. If the relay station successfully acquires system timing, decodes the configuration information, and selects a target access station, it transitions to the Access State; otherwise, it repeats the scanning and synchronization procedures (see Figure 4-18).

The RS follows the same scanning and synchronization procedures as described for an MS. In addition, the RS may store preamble indices and corresponding received RF signal strength values in order to report this information to the serving BS after registration, i.e., by sending a relay station neighbor measurement report during the neighbor station measurement and reporting stage. The BS indicates a request for this information through the ranging response message during initial ranging.

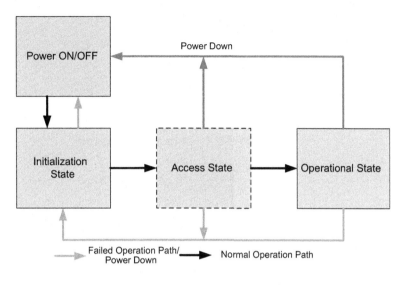

FIGURE 4-17

State transition diagram of IEEE 802.16m relay station [3]

According to the IEEE 802.16j-2009 standard specification, the BS and the RS may exchange messages to indicate support for the first stage access station selection in the multi-hop relay network [10]. An RS attempting network entry may obtain the system information transmitted by neighboring relay and base stations to select an access station. The RS then proceeds to the rest of the network entry process with the selected access station (i.e., the Access State procedures shown in Figure 4-19). The BS may request the RS to perform neighbor measurements during network entry or re-entry using the RS network entry optimization parameter in the ranging response message. The neighbor station measurement report can include the signal strength and preamble index of neighbor non-transparent stations with unique BS identifications, as well as signal strength and preamble indices of neighbor transparent or non-transparent relay stations with shared BS identifications.

After registration, if the neighbor measurement is not required, the RS can skip the neighbor station measurement reporting and second stage access station selection phases and directly proceed to the next stage, as indicated by the RS network entry optimization parameter in the ranging response message. If the RS is a mobile RS and the relay-link channel descriptor[ix] provides the list of preamble indices reserved for mobile relay stations, the mobile RS must report the received signal strengths and the corresponding preamble indices. Once the RS is informed of the monitoring scheme, the BS does not change the frame configuration of the super-ordinate station of the RS before the RS enters the Operational State. The RS utilizes the neighbor measurement mechanisms that are specified in the standard [10]. The RS then sends the measurements to the BS using the relay neighbor measurement

---

[ix]The relay link channel descriptor defines the characteristics of the relay link between an RS and its super-ordinate station. The channel descriptor is transmitted to subordinate relay stations using the RS primary management CID or multicast management CID on an event-driven basis when one of the parameters relating to the characteristics of the relay link changes [10].

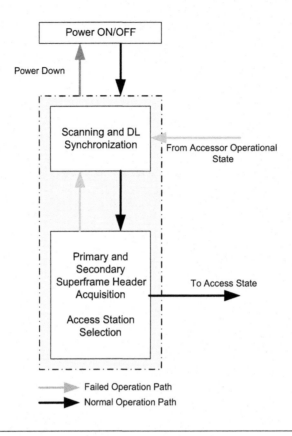

FIGURE 4-18

Procedures in the initialization state of an IEEE 802.16m relay

report. This measurement report is used by the BS during the second stage access station selection to select a new access station (or to continue with the current access station). The BS may assign to the RS a preamble index based on the report from the RS. During this phase, the BS determines the RS operational parameters and sends a relay link channel descriptor to inform the RS of the relay link characteristics and configures the parameters at the RS. The RS configuration message contains all information that is required for the RS to transition to the Operational State, as well as parameters for proper RS operation. The remaining steps prior to transition to the Operational State for different RS types are as follows:

- A time-division transmit and receive RS detects and decodes the relay zone control channels using the relay configuration message sent by the BS.
- A transparent or simultaneous transmit and receive RS receives the relay control channels from the access station in the access zone or in the relay zone, if the RS coexists with a time-division transmit and receive RS, in order to retain synchronization of relay control channels.
- A time-division transmit and receive RS may maintain its synchronization by listening to the relay preamble transmitted by its super-ordinate station.

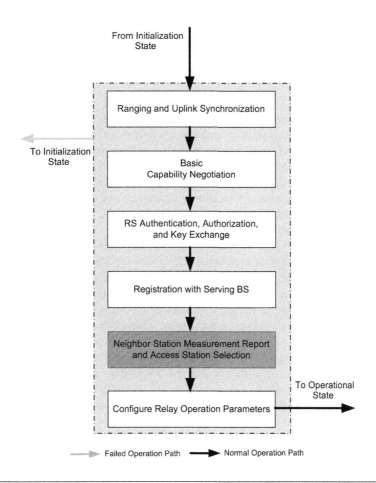

FIGURE 4-19

Procedures in the access state of an IEEE 802.16m relay

- A simultaneous transmit and receive RS may maintain its synchronization by listening to the frame start/relay preamble of its super-ordinate station.
- A transparent RS may maintain its synchronization by listening to the preamble transmission from its super-ordinate station.

According to the procedures specified in [10], in the Operational State, a non-transparent RS transmits its own frame start preamble at the frame indicated by the *Action Frame Number* parameter in the RS configuration message [10]. The frame number used by the non-transparent RS with unique BS identification takes into account the *RS Frame Offset* parameter [10]. If an *RS Frame Offset* is not provided, the RS uses the same frame number as its super-ordinate station, i.e., the RS considers the *RS Frame Offset* is zero. The relay frame configuration may be changed during the Operational State in a quasi-static manner using the RS configuration message, or in a dynamic manner using relay frame control header. The *Frame Configuration* parameter may be transmitted in RS configuration message to indicate the mechanism that must be followed for determining the frame structure in the upcoming frame(s). During the configuration phase of the network entry procedure, this parameter is included in the RS configuration message.

4.4.2 Access State

As shown in Figure 4-19, a relay station performs network entry with the target BS while in the Access State. The network entry is a multi-step process consisting of ranging, capability negotiation, authentication and authorization, registration, neighbor station measurement and access station selection (not applicable to the case of two-hop relay paradigm), and configuration of the relay station operational parameters. The relay station receives a temporary identifier and transitions to the Operational State for normal operations. On failure to complete any of the steps of network entry, the relay station transitions to the Initialization State. In Figure 4-19, some of the procedures can be skipped during RS network re-entry [3].

In the case of multi-hop relaying, the optional second-stage access station selection is performed after the neighbor station measurement reporting and before the RS operational parameter configuration phases. During this operation, the BS determines the path (i.e., access station) for this RS based on the RS neighbor station measurement reports and other information such as path loading. If the current access station is not changed, the RS continues network entry with this station and skips the second-stage access selection phase. If the current access station is changed and is located in the same cell, the BS sends the RS access station selection request message to indicate the preamble index of the selected access station. The RS responds with the acknowledgement message. The BS and the RS then perform network re-entry as described earlier, and the BS releases resources allocated to the corresponding RS. If initial network entry is required for the RS due to failure in the network re-entry with the selected access station, the original access station is the first candidate for network entry.

4.4.3 Operational State

During the Operational State, the relay station performs the tasks that are required to transmit and/or receive user data in the access (i.e., the radio link between the mobile station and the relay station) or relay link (i.e., the path between the relay station and the base station). Figures 4-16 and 4-17 show the state machines for the relay-enabled base station and relay station compliant with the IEEE 802.16j-2009 standard. It must be noted that the legacy mobile stations are not relay-aware, and IEEE 802.16j-2009 standard compliant relays assume that the access link protocols between the MS and RS remain identical to those specified by the IEEE 802.16-2009 standard to avoid interoperability issues. However, the IEEE 802.16m-compliant devices are relay-aware, (i.e., different cell identifier code subsets are assigned to various cell types) facilitating the discovery and association with the relay stations, as well as unifying the relay and access link protocols based on the IEEE 802.16m standard.

The RS follows the same scanning and synchronization procedure as that specified for the MS. In addition, the RS may store preamble indices and corresponding signal strength information in order to report the stored values to the serving relay-enabled BS[x] after registration by sending a relay neighbor measurement report during the neighbor station measurement phase and on request. The relay-enabled BS indicates the request for this information through the ranging response message during initial ranging.

---

[x]It must be noted that not all base stations are relay-enabled, as relay functionality is optional.

The second "access station selection" procedure defined in the IEEE 802.16j-2009 standard for multi-hop relay networks may be enabled if an indication is included in the RS network entry optimization in the ranging response message. The alternative access station selection may take place after neighbor station measurement reporting stage and before the configuration of RS operational parameters. During this process, the relay-enabled BS determines the path (i.e., access station) for this RS, based on the RS neighbor station measurements and other information such as path loading. If the current access station is not changed, the RS continues network entry with the current access station and skips the second access station selection phase. If the current access station is changed and is located in the same cell, the relay-enabled BS sends a relay station selection request message to the RS to indicate the preamble index of the selected access station. The RS responds with a generic acknowledgement message. Subsequently, the relay-enabled BS and the RS perform network re-entry, as described earlier. The relay-enabled BS releases resources allocated to the corresponding RS. If initial network entry is required for the RS due to failure in the network re-entry with the selected access station, the original access station would be the first candidate for network entry. Figures 4-20 and 4-21 illustrate these procedures. The timers denoted by T_i in the figures are used for measuring the permissible time per process and upon expiration of corresponding timers, other procedures are enacted. The timers and the MAC management messages shown in these figures are defined in reference [10]. Note that the uplink relay zone configuration action frame indicates the effective action time of the uplink relay zone configuration defined by the relay control channels on the relay link. If the effective action time is defined as 0, the uplink relay configuration defined by the relay control channels is effective in the current frame; if the value is set to N, the configuration defined by the relay control channels in frame i is effective in frame $i + N$. Note that the IEEE 802.16m standard has not specified a detailed state-machine for the relay station and Figures 4-20 and 4-21 are depicted based on the procedures specified in the IEEE 802.16j-2009 standard for completeness of the subject and for reader's understanding of the relay station behavior.

4.5 OPERATIONAL STATES OF FEMTO BASE STATIONS

A femto BS or femto access node is a base station with a low transmit power, typically installed by a subscriber in home or small office/home office environment to provide access to a closed or open group of users as configured by the subscriber and/or the access provider. A femto BS is connected to the service provider's network via a broadband wired (or wireless) connection and typically operates in licensed spectrum, and may use the same or different frequency as macro base stations. Its coverage may overlap with that of a macro BS. A femto BS may be intended to serve open subscriber groups or closed subscriber groups [16]. Figure 4-22 illustrates the femto BS state diagram. The state diagram consists of the Initialization and Operational States.

The Initialization State includes procedures such as configuration of radio interface parameters and time synchronization. After successful attachment to the network, a femto BS enters the Operational State. The Operational State consists of two modes: Normal Mode and Low-Duty Mode. In Low-Duty Mode, the femto BS will be periodically unavailable to reduce the adjacent/co-channel interference to neighboring femto or macro base stations. Femto base stations are required to synchronize with the network timing to the extent that synchronization error does not disrupt network operations. They may use different schemes to achieve synchronization with the network depending on the deployment

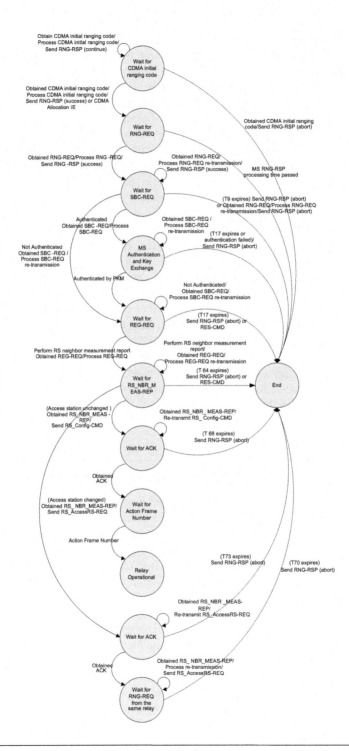

FIGURE 4-20

Network entry state machine of the relay-enabled BS [10]

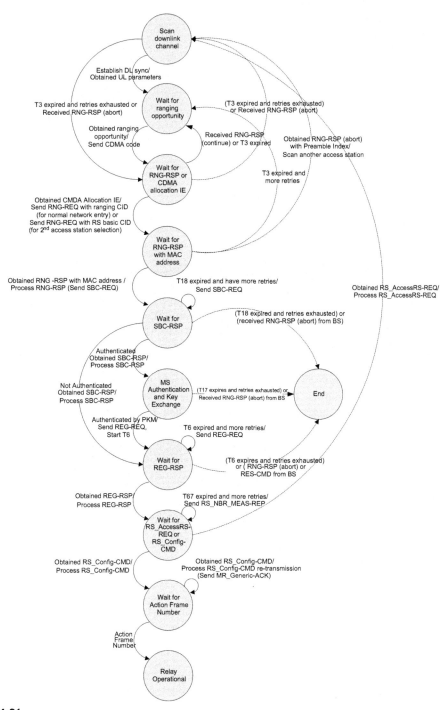

FIGURE 4-21

Network entry state machine of the relay station [10]

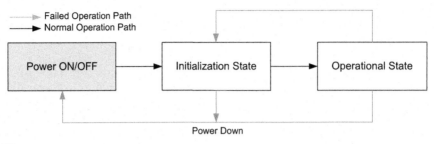

FIGURE 4-22

State transition diagram of a femto base station

scenario. A femto BS may obtain time and frequency synchronization by using schemes such as GPS,[xi] IEEE1588,[xii] etc.

Femto base stations are plug-and-play devices that, once connected to the operator's core network, are automatically configured and are self-optimized and self-organized while in operation. Since femto BS has limited coverage, it is typically used as an overlay with the macro BS (Although it may operate in remote and non-overlay cases, as well). The femto BS and macro BS may operate on the same or different frequency assignment (FA) or frequency band. Femto BS may belong to an open (accessible to all subscribers) or closed (accessible to authorized subscribers) group.

Since the femto base stations are usually densely deployed, the use of the typical cell search and cell selection methods, which are utilized in initial network entry/re-entry or handover in macro base stations, would be cumbersome and would result in extreme overhead in system operation, MS complexity, and power consumption. Furthermore, private femto base stations may be overloaded with signaling with unauthorized mobile stations trying to select them as target base stations for handover or system entry/re-entry.

As mentioned earlier, the femto base stations may belong to closed subscriber groups (CSG) or open subscriber groups (OSG). In CSG, the access and services are restricted to authorized mobile stations. The credentials or electronic certificates may be provided to the mobile station by the cellular system operator at the time of subscription. The serving macro BS may have the knowledge (through MS context generated following session set-up) of the CSGs where the MS may have access. The OSGs, on the other hand, are publicly accessible and no specific authorization is required.

[xi]The Global Positioning System (GPS) is a global navigation satellite system developed by the United States Department of Defense and managed by the United States Air Force. It is the only fully functional satellite navigation system in the world. It can be used freely by anyone, unless the system is technically restricted. These restrictions can be applied to specific regions by the US Department of Defense. The GPS can be used almost anywhere near the earth, and is often used by civilians for navigation purposes. An unobstructed line of sight to four satellites is required for non-degraded performance. The GPS location determination accuracy is about 15 meters (50 ft). The GPS uses a constellation of between 24 and 32 medium Earth orbit satellites that transmit precise radio signals, which allow GPS receivers to determine their current location, the time, and their velocity [13].

[xii]The Precision Time Protocol is a time-transfer protocol defined in the IEEE 1588-2002 standard that allows precise synchronization of networks (e.g., Ethernet). Accuracy within the nanosecond range can be achieved with this protocol when using hardware generated timestamps [14].

During normal operation and as part of handover, the MS typically measures the received RF signal strength from the serving and the neighboring base stations and selects the most viable candidate base stations as target for handover. The measurements are usually conducted on the synchronization sequences that are unique to each base station and further carry the Cell_ID. The type of the cell may also be communicated via the synchronization sequences if a hierarchical synchronization scheme is utilized. Since the number of bits that can be carried through the synchronization sequences are limited, additional information on the cell type and other configuration restrictions are broadcast as part of system configuration information.

In the IEEE 802.16m, there are two steps of downlink synchronization. The downlink synchronization is achieved by successful acquisition of the primary advanced preamble. The primary advanced preamble carries information about system bandwidth (e.g., 5, 10, 20 MHz) and multi-carrier configuration (i.e., fully configured or partially configured RF carrier). Once the primary advanced preamble is detected, the MS proceeds to acquisition of the secondary advanced preambles. The secondary advanced preamble carries a set of 768 distinct Cell_IDs that have been partitioned into a number of subsets where each subset corresponds to a certain type of base station (e.g., closed/open subscriber group femto base stations or macro base stations). Normally the MS has to acquire the synchronization sequences and to detect the Cell_ID followed by detection of the broadcast channel to complete the cell selection. If the MS realizes that the cell is a non-accessible femto BS or access node, it has to restart the cell search and to select another cell. This effort would prolong the initial network entry/re-entry and handover, and may involve a great number of unsuccessful trials.

On successful acquisition of system timing and cell identification, the MS attempts to detect and decode the system configuration information. This information is carried via the Superframe Headers (SFH) in IEEE 802.16m. The superframe headers, comprising Primary and Secondary Superframe Headers (P-SFH and S-SFH), are control elements that are periodically (while a large part of this information remains unchanged over a long period of time, some parts may change more frequently) broadcast using a robust and reliable transmission format to ensure the information can be correctly detected by all mobile stations in the coverage area of a base station. The correct and timely detection of the system information is essential for successful network entry/re-entry and handover. The S-SFH content is divided into three sub-packets (SP1, SP2, and SP3) where sub-packets carry essential information for various system processes such as initial network entry, network re-entry, Idle-State operation, etc., according to their respective timing sensitivity. The S-SFH SP1 and SP2 carry the 24-bit Operator ID and 24 least significant bits of the BS-ID. Once the system parameters are successfully acquired, the cell selection can be made by taking certain considerations into account. For example, the mobile station may have a preference in selecting a specific type of the base station (e.g., a femto-cell in indoor environment) even though other types of base stations may be available, or the MS may not be authorized to access a group of base stations despite the fact that their received RF signal strength might be good. The MS may or may not be femto-aware. If the MS is not femto-aware or is using an older version of air interface protocols, the legacy network entry/re-entry or handover procedures are utilized.

Femto base stations are distinguished from macro base stations by the use of different secondary preamble sequences, in order to enable early distinction of femto BS from the macro BS which helps the MS to avoid unnecessary network (re)entry and handovers to/from a femto BS. A large number of femto base stations may be configured within the same CSG, which has the same group of authorized mobile stations. A common identifier may be assigned to all CSG femto base stations which are part of the same CSG. An MS may use this identifier for accessibility checking for the CSG femto base stations. A common identifier known as CSG ID is used to identify the base stations belonging to the same CSG. The CSG ID is unique within the same operator ID. The CSG ID, as a part of the BS ID, may be derived from the full 48-bit BS ID, or may be provided to the CSG femto BS during initial network entry in the registration response, or may be pre-provisioned by the network. The mobile station's CSG White List (i.e., a local table in the MS containing the identities of all the CSG femto base stations to which the MS is subscribed and is authorized to access)[xiii] may contain the identifiers of allowable femto base stations.

During network entry or re-entry, the MS begins scanning of the neighbor base stations through RF measurements. The detection of the Cell_ID helps categorizing the BS type and, depending on the preference of the MS, a macro or femto candidate is selected. Failure in any stage of the cell search and cell selection will result in repeating the scanning and downlink synchronization steps. The Operator-ID (i.e., the 24 most significant bits of BS-ID) and the least significant bits of the BS-ID will help the MS to determine whether it is authorized to access to the target BS. The BS-ID in Figure 4-23 refers to the least significant bits of the 48-bit BS-ID parameter. The MS must be subscribed to the operator identified by the Operator-ID and must have the BS-ID in its White List The algorithm illustrated in Figure 4-23 provides the procedures which the MS follows to complete cell selection/re-selection when femto and macro base stations are deployed on the same or a different FA. If the femto base stations are deployed in a different FA (inter-FA) the same algorithm is applicable, except that the MS scans a different frequency band and conducts RF measurements during scanning in that frequency band. The other procedures remain intact. If the femto and macro base stations are deployed in the same FA (intra-FA), the same algorithm is used and the MS conducts RF measurements during scanning in the same frequency band. Combination of the inter-FA and intra-FA scanning is also possible where the neighbor macro base stations operate in the same frequency band and femto base stations operate in a different frequency band(s).

Note that there is no uplink transmission during execution of this algorithm, and all signals and identifiers are received and decoded via downlink transmissions that will happen regardless of the MS scanning. Therefore, no additional signaling overhead will be imposed on the target femto base stations. During handover and Scanning Mode, the serving BS broadcasts mobile neighbor advertisement management messages at a periodic interval to identify the network and define the characteristics of neighbor BS to potential MS seeking initial network entry or handover. Since the femto base stations that belong to closed subscriber groups are not accessible to all mobile stations in the cell, and considering the broadcast nature of the neighbor advertisement message (although in some cases a unicast neighbor advertisement message may be used in IEEE 802.16m during handover), the macro BS should not broadcast the information of CSG femto base stations. This helps reduce the size of the mobile neighbor advertisement message.

The scanning interval is defined as the time during which the MS scans for available base stations. The mobile scanning interval request management message sent by the MS in the Connected State

---

[xiii]The CSG White List contains the list of femto base stations to which the MS is subscribed and which it can access. These femto base stations are identified based on a common identifier. The mobile station's local White List may contain the allowable BS-IDs or common identifiers of CSGs and relevant information to help derivation of the actual BS-IDs from the common identifier. In addition, the White List may include absolute or relative location information of CSG femto BS, such as GPS information and overlay macro BS identification.

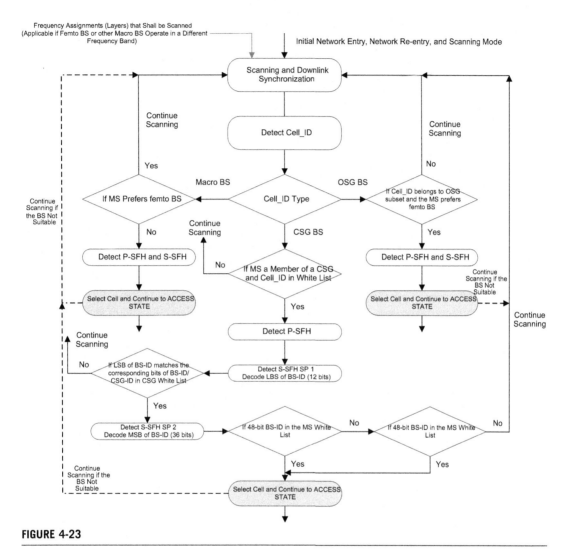

FIGURE 4-23

Discovery and association process of femto-aware mobile stations [2]

contains a group of neighbor base stations for which scanning and association are requested. This message may include the BS-ID of the CSG femto base stations to which the MS is subscribed, if the MS is a femto-preferred terminal, which means it prefers to be associated with a femto BS even though macro stations may be available in its neighborhood. Note that this is a unicast MAC management message, based on which the serving BS may update the MS context to include the CSG femto BS list that the MS may belong to. On reception of a mobile scanning interval request message by the serving BS, the BS responds with a mobile scanning interval response message granting or refusing the MS scanning interval request or a different interval and further containing the list of recommended neighbor base stations by the serving BS. Following receipt of the mobile scanning interval response message granting the request, the MS scans for one or more neighbor base stations and may attempt to select the cell and associate with that cell through completion of the Access State procedures. The MS scans all the recommended neighbor base stations identified in mobile scanning interval response message and sends a report to the serving BS at the end of the scanning interval via a mobile scanning interval report management message. The OSG femto base stations operate similar to macro base stations when paging an MS. Depending on network topology, a femto-cell or a macro BS may be assigned to one or more paging groups. The overlay macro BS and the CSG femto BS may share the same paging group identifications.

In addition to the normal operation mode, femto base stations may support low duty operation mode in order to reduce interference in neighbor cells. As shown in Figure 4-24, the low duty operation mode consists of available and unavailable intervals. During an available interval, the femto BS may become active on the air interface for synchronization and signaling purposes such as paging, ranging, or for data transmission opportunities for the mobile stations.

During the available interval, the femto base station may become active on the air interface for activities such as paging, transmitting system information and synchronization signals (see Figure 4-24), ranging, or user data transmission. During the unavailable interval, the femto base station does not transmit on the air interface (while this is the case in the current version of IEEE 802.16m specification, this may have some practical implications on the implementation and operation of femto base stations, constraining and prolonging network discovery and association of the mobile stations). An unavailable interval may be used for synchronization with the overlay macro BS or for measuring the interference from neighboring cells. The femto BS may enter low duty mode if there are no attached terminals and there are no terminals in the process of network entry. Figure 4-24 provides an example of normal and low duty mode operation where the femto BS transmits overhead channels such as preambles and the system information during low-duty mode to allow terminals to discover this femto BS and perform network entry or re-entry, if necessary and authorized. A sequence of availability and unavailability intervals forms a low duty mode (LDM) pattern. The default LDM pattern is the periodic repetition of one available interval and one unavailable interval. An available interval for the default LDM pattern begins with the frame including the primary preamble. The default LDM pattern parameters include available interval (in units of superframes), unavailable interval (in units of superframes), and the start superframe offset. There may be one or more LDM patterns in a femto BS deployment. The parameters of the default LDM pattern can be pre-provisioned or directly sent to the MS during initial network entry with the femto BS using the registration response message [2].

4.6 3GPP LTE USER EQUIPMENT STATES AND STATE TRANSITIONS

The Radio Resource Control (RRC) sub-layer in 3GPP LTE performs control-plane functions such as broadcast of system information related to access stratum and non-access stratum, paging, establishment, maintenance and release of an RRC connection between the User Equipment (UE) and E-UTRAN, signaling radio bearer management, security handling, mobility management including UE measurement reporting and configuration, active mode handover, idle mode mobility control, Multimedia Broadcast Multicast Service (MBMS) notification services and radio bearer management for MBMS, QoS management and NAS direct message transfer between NAS and UE. A 3GPP LTE

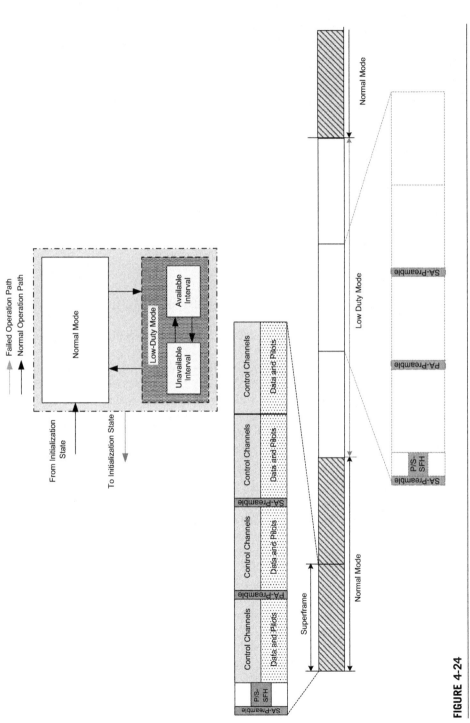

FIGURE 4-24

Femto BS operational state procedures (assuming that femto BS transmits overhead channels during low duty mode)

compliant UE has two steady-state operational states: RRC_CONNECTED and RRC_IDLE. A UE is in RRC_CONNECTED when an RRC connection or control-plane has been established. If this is not the case, i.e., no RRC connection is in place, the UE is in RRC_IDLE state. The 3GPP UE states only correspond to the steady-state behavior, and the transient procedures for cell search and selection, as well as network entry/re-entry, are not designated as states. The 3GPP UE steady-state states should be compared with the IEEE 802.16 counterparts in Figure 4-2 in order to better understand the similarities and differences of the two systems. The RRC states can further be characterized as follows.

- RRC_IDLE:
  - UE specific Discontinuous Reception (DRX[xiv]) is configured by upper layers;
  - UE controlled mobility is performed;
  - the UE monitors a paging channel to detect incoming calls;
  - system information change is tracked, and for Earthquake and Tsunami Warning System (ETWS) capable UEs, ETWS notification is monitored;
  - UE performs neighboring cell measurements and cell selection/re-selection and acquires system information.
- RRC_CONNECTED:
  - Transfer of unicast data to/from UE;
  - At lower layers, the UE may be configured with a UE specific DRX (this type of DRX is similar to IEEE 802.16m sleep cycle);
  - Network controlled mobility, i.e., handover and cell change order with optional Network Assisted Cell Change (NACC) to GERAN.
  - The UE monitors a paging channel and/or System Information Block Type 1 contents to detect system information change, and for ETWS capable UEs, ETWS notification. The UE further monitors control channels associated with the shared data channel to determine if data is scheduled for it, and provides channel quality and feedback information. The UE also performs neighboring cell measurements and measurement reporting, and acquires system information.

Figure 4-25 illustrates the RRC states in E-UTRA and shows the mobility support between E-UTRAN, UTRAN, and GERAN, where a Cell Change Order (CCO) message is used to command the UE to change to another radio access technology [17].

The mobility support between E-UTRAN, cdma2000 1xRTT [20,21,23], and cdma2000 High Rate Packet Data (HRPD) or cdma 1xEV-DO [22,23] radio access technologies are shown in Figure 4-26.

The inter-RAT handover procedures support signaling, conversational, and non-conversational services. The mobility between E-UTRA and non-3GPP systems other than cdma2000 has also been studied [23–25]. In addition to the state transitions shown in Figures 4-25 and 4-26, there is support for connection release with redirection information from E-UTRA RRC_CONNECTED to GERAN, UTRAN, and cdma2000 (HRPD Idle/ 1xRTT Dormant modes).

As a general principle, in RRC_IDLE to RRC_CONNECTED state transitions, RRC protection keys and user-plane protection keys are generated, while keys for non-access stratum protection as well as higher layer keys are assumed to be already available in the MME. On the other hand, in

---

[xiv]Note: The DRX cycle in 3GPP LTE idle mode is conceptually similar to Paging Available/Unavailable cycles in IEEE 802.16m idle mode.

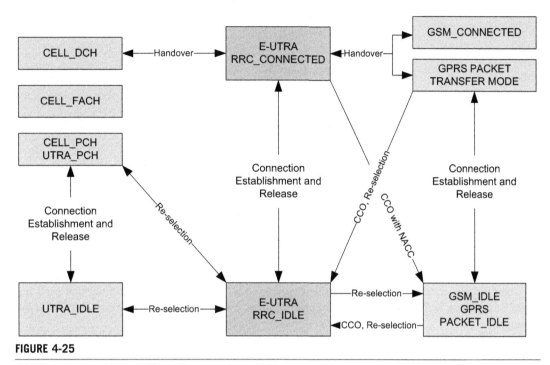

FIGURE 4-25

E-UTRA states and Inter-RAT mobility procedures [18]

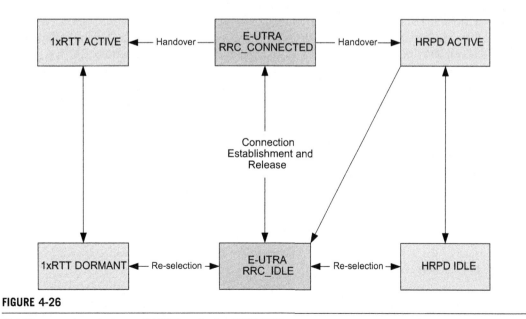

FIGURE 4-26

Mobility procedures between E-UTRA and cdma2000 [18]

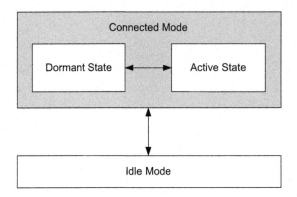

FIGURE 4-27

State transitions in 3GPP LTE-Advanced UE [26]

transitions from RRC_CONNECTED to RRC_IDLE, the eNBs delete the keys they store, such that state for idle mode UEs only has to be maintained in MME. It is also assumed that eNB no longer stores state information about the corresponding UE and deletes the current keys from its memory.

In 3GPP LTE-Advanced, the requirement for UE transition from the Idle Mode, with allocated IP address, to the Connected Mode is less than 50 ms, including the establishment of the user-plane and excluding the S1 transfer delay. The transition requirement from Dormant State (Sleep Mode counterpart in IEEE 802.16m) to Active State (corresponding to Active Mode in IEEE 802.16m) in Connected Mode is less than 10 ms. These requirements are stricter than those for 3GPP LTE. In the Dormant State, the UE has an established RRC connection and radio bearers. The UE has already been identified at cell level, but it may be in DRX to save power during inactivity periods. The UE may be either synchronized or unsynchronized (see Figure 4-27).

Although 3GPP LTE already fulfills the latency requirements of IMT-Advanced systems, several new techniques are being utilized in 3GPP LTE-Advanced to further reduce the user-plane and control-plane latencies as follows [26]:

- Combined RRC Connection Request and NAS Service Request allows those two messages to be processed in parallel at the eNB and MME, respectively, reducing overall transition latency from Idle Mode to Connected Mode by approximately 20 ms.
- Reduced processing delays in different nodes form the major part of the delay (about 75% for the transition from Idle Mode to Connected Mode assuming a combined request); thereby any improvement has a large impact on the overall latency
- Reduced Random Access Channel (RACH) scheduling period from 10 ms to 5 ms results in decreasing of the average waiting time for the UE to initiate the procedure to transit from Idle Mode to Connected Mode by 2.5 ms.

Furthermore, a shorter cycle of Physical Uplink Control Channel (PUCCH) would reduce the average waiting time for a synchronized UE to request radio resources in the Connected Mode, expediting the transition from the Dormant State in the Connected Mode.

4.6.1 Acquisition of System Information

System information in 3GPP LTE is divided into the Master Information Block (MIB) and a number of System Information Blocks (SIBs). The MIB includes a limited number of most essential and most frequently transmitted parameters that are needed to acquire other information from the cell and is transmitted in the broadcast channel [19]. The SIBs other than *System Information Block Type1* are carried in System Information (SI) messages and mapping of SIBs to SI messages is flexibly configurable by the *scheduling information list* included in *System Information Block Type 1* with the following restrictions: (1) each SIB is contained only in a single SI message; (2) only SIBs having the same scheduling requirement (periodicity) can be mapped to the same SI message; and (3) *System Information Block Type 2* is always mapped to the SI message that corresponds to the first entry in the list of SI messages in the *scheduling information list*. There may be multiple SI messages transmitted with the same periodicity. *System Information Block Type 1* and all SI messages are transmitted on the downlink shared channel [17].

The change of system information (other than for ETWS) only occurs at specific radio frames. The system information may be transmitted a number of times with the same content within a modification period, as defined by its scheduling. When the network changes some of the system information, it notifies the UEs about this change. In the next modification period, the network transmits the updated system information, as illustrated in Figure 4-28 where different colors indicate different system information. On receiving a change notification, the UE acquires the new system information immediately from the start of the next modification period. The UE applies the previously acquired system information until the UE acquires the new system information.

The paging message is used to inform UEs in RRC_IDLE and in RRC_CONNECTED about a system information change. If the UE receives a paging message indicating a system information modification, the UE by default expects that the system information will change at the next modification period boundary. Although the UE may be informed about changes in system information, no further details are provided, e.g., regarding which system information will change. The *System Information Block Type 1* includes a value tag that indicates if a change has occurred in the SI messages. The UEs may use the tag on return from out-of-coverage to verify whether the previously stored SI messages are still valid. Furthermore, the UE considers stored system information to be invalid after three hours from the moment it was successfully confirmed as valid, unless otherwise specified.

E-UTRAN may not update the tags on change of some system information, such as ETWS information or regularly changing parameters like cdma2000 system time. Similarly, E-UTRAN may

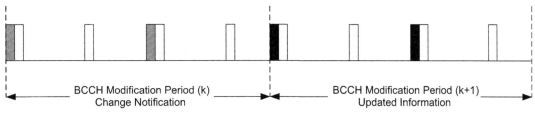

FIGURE 4-28

Change of system information in time [18]

not include the system information modification within the paging message on change of some system information. The UE verifies that stored system information remains valid by either checking the tags in *System Information Block Type 1* after the modification period boundary, or by attempting to find the system information modification indication during the modification period in case no paging is received. If no paging message is received by the UE during a modification period, the UE may assume that no change of system information will occur at the next modification period boundary. If the UE in RRC_CONNECTED during a modification period receives one paging message, it may assume from the presence/absence of system information modification whether a change of system information other than ETWS information will occur in the next modification period.

The UE performs the system information acquisition procedure (shown in Figure 4-29) to acquire the access stratum and non-access stratum system information that is broadcast by E-UTRAN. The procedure applies to UEs in RRC_IDLE and in RRC_CONNECTED. The UE applies the system information acquisition procedure on cell selection and re-selection, after handover completion, after entering E-UTRA from another RAT, on return from out-of-coverage condition, on receiving a notification that the system information has changed, on receiving an indication about the presence of an ETWS notification, on receiving a request from cdma2000 upper layers, and on exceeding the maximum validity duration. The system information acquisition procedure overwrites any stored system information.

If the UE is in RRC_IDLE state, it must ensure it has a valid version of the *Master Information Block* and *System Information Block Type 1*, as well as *System Information Block Type 2* through *System Information Block Type 8*, depending on support of the other radio access technologies. If the UE is in RRC_CONNECTED state, it must ensure possession of the *Master Information Block, System Information Block Type 1*, and *System Information Block Type 2*, as well as *System Information Block Type 8*, depending on support of cdma2000 [18].

In 3GPP LTE, the system information is classified into five categories as follows: (1) information valid across multiple cells; (2) information needed at cell search; (3) information needed prior to cell camping; (4) information needed before cell access; and (5) information needed while camping on a cell. From the UE perspective, the information that is needed at cell selection and prior to camping

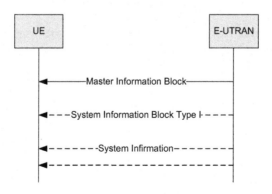

FIGURE 4-29

System information acquisition flow graph [18]

are very similar. Before a UE can camp on a cell, it needs to know if access is allowed in that cell. Thus, it would be very beneficial to know all access restrictions already at cell search phase. In order to support full mobility within the serving frequency layer, the UEs need to perform cell search periodically, and thus it is important that the information needed for cell search is readily available to reduce cell search latency and minimize UE power consumption.

Before a UE can camp on a cell, it needs to know any access-related parameters in order to avoid camping on cells where access is restricted. Thus, prior to camping on a cell, a UE needs to know any cell access restriction parameters such as Tracking Area (TA) identity, cell barring status (UE is not allowed to camp on a barred cell), and cell reservation status, i.e., the UE needs to know whether the cell is barred or reserved in order to avoid camping on a barred cell. Also barring time might be needed in order to ensure that the UE does not have to poll barring time frequently from the system information. Another option is that barring status is also indicated in the neighbor cell list, and radio access limitation parameters, i.e., any radio condition parameters that limit the access to the cell [18].

When a UE has camped on a cell, it needs to continue measuring the neighboring cells in order to stay camped. In order for the UE to start mobility procedures, it needs to receive parameters such as reporting periods, reporting event parameters, time to trigger, etc. UEs in RRC_IDLE state need cell reselection parameters. The UEs in RRC_CONNECTED state need parameters of the neighbor cells for handover and for error recovery cases. Neighbor cell lists are needed to start neighbor cell measurements. UEs in different states may use different sets of neighbor cell lists. 3GPP LTE system information can be classified into two distinctive groups: static and flexible. The static part is sent more frequently, e.g., once per frame, in the cell and has a limited capacity for information transfer. The flexible part has a flexible amount of scheduled resources available, and thus encompasses most of the SI information. The flexible part has different types of *Information Elements* which require independent scheduling in order to allow sufficiently fast reception and efficient resource utilization.

4.6.2 Connected Mode Mobility

In RRC_CONNECTED state, the network controls the UE mobility, i.e., the network decides when and where the UE should initiate handover process. For network-controlled mobility in RRC_CONNECTED state, the network triggers the handover process based on radio conditions or network load. To facilitate this process, the network may configure the UE to perform measurement reporting (including the configuration of measurement gaps). The network may also initiate handover blindly, i.e., without having received measurement reports from the UE. Before sending the handover message to the UE, the source eNB prepares one or more target cells. The target eNB generates the message that is used to perform the handover, i.e., the message including the access stratum configuration to be used in the target cell. The source eNB transparently (i.e., it does not modify values or content) forwards the handover information received from the target to the UE. When appropriate, the source eNB may initiate data forwarding for (a subset of) the data radio bearers.

After receiving the handover message, the UE attempts to access the target cell at the first available random access channel opportunity according to random access resource selection, i.e., the handover is asynchronous. Consequently, when allocating a dedicated preamble for the random access in the target cell, E-UTRA ensures it is available from the first random access opportunity that the UE may use. On successful completion of the handover, the UE sends a message to confirm the handover. After the successful completion of handover, PDCP SDUs may be re-transmitted in the target cell. This only applies for data radio bearers using RLC-AM mode [17]. The source eNB should, for some time, maintain a context to enable the UE to return in case of handover failure. After having detected handover failure, the UE attempts to resume the RRC connection, either in the source or in another cell, using the RRC re-establishment procedure. This connection resumption succeeds only if the accessed cell is prepared, i.e., a cell of the source eNB or of another eNB which has performed handover preparation.

References

[1] IEEE Std 802.16-2009, IEEE Standard for Local and Metropolitan Area Networks – PART 16: Air Interface for Broadband Wireless Access Systems, May 2009.

[2] P802.16m/D6, IEEE Standard for Local and Metropolitan Area Networks – Part 16: Air Interface for Broadband Wireless Access Systems – Advanced Air Interface, May 2010.

[3] IEEE 802.16m–09/0034r3, IEEE 802.16m System Description Document. <http://ieee802.org/16/tgm/index.html>, May 2010.

[4] WiMAX Forum Network Architecture Release 1.5 Version 1 – Stage 2, Architecture. Tenets, Reference. Model. and. Reference. Points. <http://www.wimaxforum.org/resources/documents/technical/release>, September 2009.

[5] WiMAX Forum Network Architecture Release 1.5 Version 1 – Stage 3, Detailed Protocols and Procedures. <http://www.wimaxforum.org/resources/documents/technical/release>, September 2009.

[6] Loutfi Nuaymi, WiMAX: Technology for Broadband Wireless Access, first ed., John Wiley & Sons, 2007.

[7] Jeffrey G. Andrews, Arunabha Ghosh, Rias Muhamed, Fundamentals of WiMAX: Understanding Broadband Wireless Networking, first ed., Prentice Hall, 2007.

[8] DRAFT-T23-005-R015v04-B_RSP, WiMAX Forum Mobile Radio Specification document

[9] Truncated binary exponential backoff, Wikipedia, <http://en.wikipedia.org/wiki/Binary_exponential_backoff>.

[10] IEEE Std 802.16j-2009, Amendment to IEEE 802.16-2009 standard, IEEE Standard for Local and Metropolitan Area Networks – Part 16: Air Interface for Broadband Wireless Access Systems, Multiple Relay Specification, June 2009.

[11] W. StevenPeters, Robert W. Heath Jr., "The Future of WiMAX: Multi-hop Relaying with IEEE 802.16j," IEEE Communications Magazine, January 2009.

[12] V. Genc, S. Murphy, Y. Yu, J. Murphy, "IEEE 802.16j Relay-Based Wireless Access Networks: An Overview," IEEE Wireless Communications, October 2008.

[13] Global Positioning System (GPS), Wikipedia, <http://en.wikipedia.org/wiki/GPS>.

[14] IEEE 1588, Precision Time Protocol, Wikipedia, <http://en.wikipedia.org/wiki/IEEE_1588>.

[15] Yan Zhang, Hsiao-Hwa Chen (Eds.), Mobile WiMAX: Toward Broadband Wireless Metropolitan Area Networks, first ed., Auerbach Publications, December 2007.

[16] Marcos Katz, Frank Fitzek (Eds.), WiMAX Evolution: Emerging Technologies and Applications, John Wiley & Sons Ltd, March 2009.

[17] 3GPP TS 36.300, Evolved Universal Terrestrial Radio Access (E-UTRA) and Evolved Universal Terrestrial Radio Access Network (E-UTRAN); Overall description; Stage 2, <http://www.3gpp.org/ftp/Specs/html-info/36-series.htm>, March 2010.

[18] 3GPP TS 36.331, Evolved Universal Terrestrial Radio Access (E-UTRA); Radio Resource Control (RRC); Protocol specification. <http://www.3gpp.org/ftp/Specs/html-info/36-series.htm>, March 2010.

[19] Erik Dahlman, et al., 3G Evolution: HSPA and LTE for Mobile Broadband, second ed., Academic Press, 2008.

[20] 3GPP2 CS0003-E v1.0, Medium Access Control (MAC) Standard for cdma2000 Spread Spectrum Systems. <http://www.3gpp2.org/Public_html/specs/tsgc.cfm>, June 2009.

[21] 3GPP2 CS0005-E v1.0, Upper Layer (Layer 3) Signaling Standard for cdma2000 Spread Spectrum Systems. <http://www.3gpp2.org/Public_html/specs/tsgc.cfm>, June 2009.

[22] 3GPP2 CS0024-B v2.0, cdma2000 *High Rate Packet Data Air Interface Specification*. <http://www.3gpp2.org/Public_html/specs/tsgc.cfm>, April 2007.

[23] 3GPP2 CS0087-0 v1.0, E-UTRAN – cdma2000 Connectivity and Interworking: Air Interface Specification. <http://www.3gpp2.org/Public_html/specs/tsgc.cfm>, May 2009.

[24] 3GPP2 CS0086-0 v1.0, WiMAX–HRPD Interworking: Air Interface Specification. <http://www.3gpp2.org/Public_html/specs/tsgc.cfm>, May 2009.

[25] 3GPP TR 36.938, Evolved Universal Terrestrial Radio Access Network (*E- UTRAN*); Improved Network Controlled Mobility Between E-UTRAN and 3GPP2/Mobile WiMAXR Technologies. <http://www.3gpp.org/ftp/Specs/html-info/36-series.htm>, March 2008.

[26] 3GPP TR 36.912, Feasibility Study for Further Advancements for E-UTRA (LTE-Advanced). <http://www.3gpp.org/ftp/Specs/html-info/36-series.htm>, March 2010.

[27] R.Y. Kim, S. Mohanty, "Advanced Power Management Techniques in Next-Generation Wireless Networks," IEEE Communications Magazine, Vol. 48, Issue 5, May 2010.

The IEEE 802.16m Convergence Sub-Layer

INTRODUCTION

This chapter provides a detailed description of the functional components and protocols associated with the IEEE 802.16m service-specific Convergence Sub-layer (CS). The convergence sub-layer, as shown in Figure 5-1, is located on top of the IEEE 802.16 MAC sub-layer and interfaces the MAC sub-layer with the network layer protocols and further performs the following functions:

- Accepting Protocol Data Units (PDUs) from the network layer;
- Performing classification of higher layer PDUs;
- Processing the higher layer PDUs based on the classification (i.e., payload header compression);
- Delivering CS PDUs to the MAC Service Access Point (SAP);
- Receiving CS PDUs from the peer entity.

The convergence sub-layer is part of Open System Interconnection Data-Link Layer (alternatively known as Layer 2) protocol class and is interfaced with network layer and MAC sub-layer through CS SAP and MAC SAP, respectively.

The convergence sub-layers of the IEEE 802.16m and IEEE 802.16-2009 standard have very similar behavior; the only differences are in the assignment and use of connection identifiers in the two standards, as well as exclusion of some unused legacy protocols. The Internet Protocol CS (IPCS) and Generic Packet CS (GPCS) are two types of the service-specific CS that are supported by IEEE 802.16m, which are used to transport packet data over the air interface. When using GPCS, the classification is performed in protocol layers above the CS, and the relevant information for performing classification is transparently provided during connection set-up or change. The Asynchronous Transfer Mode CS (ATM CS)[i] (see Section 5.1 of reference [1]) and Ethernet CS (see Section 5.2.4 of reference [1]) variants that were specified in the IEEE 802.16-2009 standard are no longer supported in IEEE 802.16m due to a lack of industry interest. Other air interface standards such as 3GPP LTE also use such logical interfaces between their Layer 2 service access points and the network layer protocols. The Packet Data Convergence Protocol (PDCP) in 3GPP LTE, among other functions listed above, performs ciphering and encryption of the MAC PDUs. This is an important difference between the MAC functions of IEEE 802.16 and 3GPP LTE. As shown in Figure 5-1, in IEEE 802.16m, the MAC PDUs are encrypted and integrity protected in a security

---

[i]The ATM CS is a logical interface that associates different ATM services with the MAC CPS SAP. The ATM CS accepts ATM cells from the ATM layer, performs classification, header compression (if provisioned), and delivers CS PDUs to the appropriate MAC SAP [1].

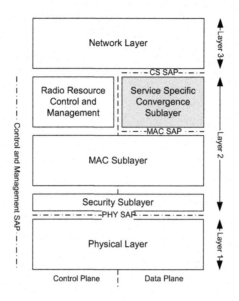

FIGURE 5-1

The location of the convergence sub-layer in the IEEE 802.16m protocol structure

sub-layer, whereas in 3GPP LTE the MAC SDUs are already encrypted and integrity protected in the PDCP layer.

This chapter provides a top-down systematic description of convergence sub-layers supported in IEEE 802.16m and their constituent protocols. An overview of 3GPP LTE PDCP layer is further provided to enable the readers to contrast the corresponding functionalities.

5.1 HEADER COMPRESSION

In some services and applications, such as voice-over-IP, interactive gaming, multimedia messaging, etc., the data payload of the IP packet is almost the same size or even smaller than the header. Over the end-to-end connection comprising multiple hops, these protocol headers are extremely important, but over a single link these headers serve no useful purpose. It is possible to compress these headers, and thus save the bandwidth and use the expensive radio resources efficiently. The header compression also provides other important benefits, such as reduction in packet loss and improved interactive response time [2]. Payload header compression is the process of suppressing the repetitive portion of payload headers at the sender and restoring them at the receiver side of a low-bandwidth/capacity-limited link. The use of header compression has a well-established history in transport of IP-based payloads over capacity-limited wireless links where more bandwidth efficient transport methods are required. The Internet Engineering Task Force (IETF) has developed several header compression protocols that are widely used in telecommunication systems.

Let's consider some examples of the compression ratios that can be achieved using header compression schemes. The IP version 4 protocol header consists of 20 bytes [3], when combined with the User Datagram Protocol (UDP)[ii] header of 8 bytes and the Real-time Transport Protocol (RTP)[iii] header of 12 bytes this results in an IPv4/UDP/RTP header size of 40 bytes. A header compression scheme typically compresses such headers to 2–4 bytes in the steady-state. Note that the RTP payload size of some commonly used voice codecs is approximately 20–40 bytes for active speech; therefore, the 40 byte IPv4/UDP/RTP protocol header size would be a relatively large overhead. Using header compression in such cases would result in major bandwidth savings. The amount of overhead for small packets when using IP version 6 (IPv6), with increased header size of 40 bytes due to increased IP address space [8], would be even larger. In low bandwidth or congested networks, the use of header compression may yield better response times due to smaller packet sizes. A small packet may further reduce the probability of packet loss [2]. It has been observed that in applications such as video transmission over wireless links, the use of header compression does not improve the video quality in spite of lower bandwidth usage. For voice transmission, the voice quality may improve while utilizing lower transmission bandwidth. In summary, header compression helps improve network transmission efficiency, quality, and speed by decreasing protocol header overhead, reducing the packet loss, decreasing interactive response time, and increasing network core users per channel bandwidth means less infrastructure deployment costs. Figure 5-2 illustrates the structure of IPv6, UDP, and RTP headers.

The IP protocol together with transport protocols such as TCP or UDP and application-layer protocols (e.g., RTP) are described in the form of payload headers. The information carried in the header helps the applications to communicate over large distances connected by multiple links or hops in the network. This information consists of source and destination addresses, ports, protocol identifiers, sequence numbers, error checksums, etc. Under nominal conditions, most of the information carried in packet headers remains the same or changes in specific patterns. By observing the fields that remain constant or change in specific patterns, it is possible either not to send them in each packet or to represent them in a smaller number of bits than would have been originally required. This process is referred to as compression. The process of header compression uses the concept of flow context, which is a collection of information about field values and change patterns of field values in the packet

---

[ii]The User Datagram Protocol is part of the Internet Protocol Suite, i.e., the set of network protocols used for the Internet. The use of UDP would allow computer applications to send datagrams to other hosts on an IP network without requiring prior signaling to set up data paths. The UDP protocol is defined in IETF RFC 768. The UDP protocol uses a simple transmission model without implicit hand-shaking dialogues for guaranteeing reliability, ordering, or data integrity. Thus, UDP provides an unreliable service and datagrams may be lost, or arrive out of order or duplicated. The UDP assumes that error checking and correction is either not necessary or is performed at the application layer, avoiding the overhead of such processing at the network interface level. Delay-sensitive applications such as VoIP and interactive gaming often use UDP because dropping packets would avoid potentially variable and long delays. If error correction schemes are needed at the transport layer, an application may use the Transport Control Protocol (TCP) which is designed for this purpose. Unlike TCP, UDP is compatible with packet broadcasting and multicasting. Common applications include media streaming, IPTV, VoIP, and interactive gaming [4,5].

[iii]The Real-time Transport Protocol defines a standardized packet format for delivering audio and video over the Internet. It is defined by IETF RFC 3550 [26]. The RTP protocol is used in communication and entertainment systems that involve media streaming, such as telephony, video teleconference applications, and web-based push-to-talk features. Relying on signaling protocols such as Session Initiation Protocol (SIP), the RTP is one of the key ingredients of voice and media transport over IP networks. The RTP protocol is usually used in conjunction with the RTP Control Protocol (RTCP). While RTP carries the media streams (e.g., audio and video), the RTCP is used to monitor transmission statistics and QoS information. When both protocols are used in conjunction, RTP is usually originated and received on even port numbers, whereas RTCP uses the next highest odd port number [6,7].

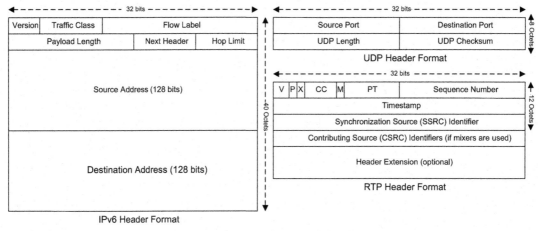

FIGURE 5-2

IPv6, UDP, and RTP header formats

header. This context is formed on the compressor and the decompressor side for each packet flow. The first few packets of a newly identified flow are used to build the context on both sides. These packets are sent without compression. The number of these first few packets, which are initially sent uncompressed, is closely related to link characteristics like bit error rate and round trip time. Once the context is established on both sides, the compressor compresses the payload headers as much as possible. By taking into account the link conditions and feedback from the decompressor, the compressed packet header sizes may vary. At certain intervals and in the case of error recovery, uncompressed packet headers are sent to reconstruct the context and revert back to normal operational mode, which is sending compressed packet headers. The header compression module is a part of the protocol stack on the devices. It is a feature which must be negotiated before it can be used on a link. Both end points must agree if they support header compression, and on the related parameters to be negotiated.

Considering the end-to-end connectivity over IP, the header compression does not introduce any changes in the data payload when it compresses and decompresses the header. The header compression is a hop-to-hop process and is not applied in end-to-end connection. At each hop in the IP network, it becomes necessary to decompress the packet to be able to perform the operations such as routing, QoS negotiation, and parameter adjustment, etc. Header compression is best suited for specific links in the network characterized by relatively low bandwidth, high bit error rates, and long round trip times. The performance of a header compression scheme can be described with three parameters [9]: compression efficiency; robustness; and compression transparency. The compression efficiency is determined by how much the header sizes are reduced by the compression scheme. The compression transparency is a measure of the extent to which the scheme ensures that the decompressed headers are semantically identical to the original headers. If all decompressed headers are semantically identical to the corresponding original headers, the transparency is considered to be 100%. Compression transparency is high when damage propagation is low. When the context of the decompressor is not consistent with the context of the compressor, decompression may fail to reproduce the original header. This condition can

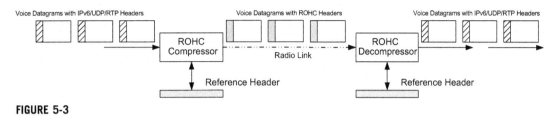

FIGURE 5-3

An example of ROHC compression/decompression of IP/UDP/RTP headers for communication over a radio link

occur when the context of the decompressor has not been initialized properly, or when packets have been lost or damaged between compressor and decompressor.

5.1.1 Robust Header Compression

The RObust Header Compression (ROHC) scheme reduces the size of the transmitted IP/UDP/RTP header by removing redundancies. This mechanism starts by classifying header fields into different classes according to their variation pattern. The fields that are classified as *inferred* are not sent. The *static* fields are sent initially and then are not sent anymore, and the fields with varying information are always sent. The ROHC mechanism is based on a context,[iv] which is maintained by both ends, i.e., the compressor and the decompressor (see Figure 5-3). The context encompasses the entire header and ROHC information. Each context has a Context Identifier, which identifies the flows. The ROHC scheme operates in one of the following three operation modes [14–19]:

1. Uni-directional mode (U): in the uni-directional mode of operation, packets are only sent in one direction, from compressor to decompressor. This mode therefore makes ROHC usable over links where a return path from decompressor to compressor is unavailable or undesirable.
2. Optimistic mode (O): the bi-directional optimistic mode is similar to the uni-directional mode, except that a feedback channel is used to send error recovery requests and (optionally) acknowledgments of significant context updates from the decompressor to compressor. The O-mode aims to maximize compression efficiency and sparse usage of the feedback channel.
3. Reliable mode (R): the bi-directional reliable mode differs in many ways from the previous two. The most important differences are a more intensive usage of the feedback channel and a stricter logic at both the compressor and the decompressor that prevents loss of context synchronization between compressor and decompressor except for very high residual bit error rates.

The U-mode is used when the link is uni-directional or when feedback is not possible. For bi-directional links, O-mode uses positive feedback packets (ACK) and R-mode use positive and negative

---

[iv]The context of the compressor is the state it uses to compress a header. The context of the decompressor is the state it uses to decompress a header. Either of these, or combinations of the two, is usually referred to as "context." The context contains relevant information from previous headers in the packet stream, such as static fields and possible reference values for compression and decompression. Moreover, additional information describing the packet stream is also part of the context, for example information about how the IP identifier field changes and the typical inter-packet increase in sequence numbers or timestamps [9].

feedback packets (ACK and NACK). ROHC always starts header compression using U-mode even if it is used in a bi-directional link. ROHC does not attempt re-transmission when an error occurs and the erroneous packet is dropped. The ROHC feedback is used only to indicate to the compressor side that there was an error and probably the context is damaged. After receiving a negative feedback, the compressor always reduces its compression level. The ROHC compressor has three compression states as follows [10]:

1. Initialization and Refresh (IR), where the compressor has just been created or reset and full packet headers are sent;
2. First Order (FO) where the compressor has detected and stored the static fields such as IP addresses and port numbers on both sides of the connection;
3. Second Order (SO) where the compressor is suppressing all dynamic fields such as RTP sequence numbers, and sending only a logical sequence number and partial checksum to cause the other side to generate based on prediction, and verify the headers of the next expected packet.

Each compression state uses a different header format in order to send the header information. The IR compression state establishes the context, which contains static and dynamic header information. The FO compression state provides the change pattern of dynamic fields. The SO compression state sends encoded values of *Sequence Number* (SN) and *Timestamp* (TS), forming the minimal size packets (see Figure 5-4). Using this header format, all header fields can be generated at the other end of the radio link using the previously established change pattern. When some updates or errors occur, the compressor returns to upper compression states. It only transitions to the SO compression state after re-transmitting the updated information and re-establishing the change pattern in the decompressor.

In the U-mode, the feedback channel is not used. To increase the compression level, an optimistic approach is used for the compressor to ensure that the context has been correctly established at the decompressor side. This means that the compressor uses the same header format for a number of packets. Since the compressor does not know whether the context is lost, it also uses two timers to be able to return to the FO and IR compression states. The decompressor works at the receiving end of the link and decompresses the headers based on the header fields' information of the context. Both the compressor and the decompressor use a context to store all the information about the header fields. To

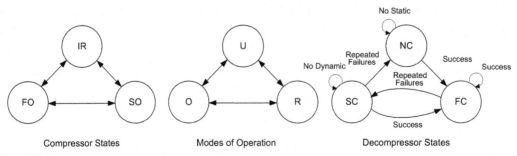

FIGURE 5-4

ROHC state machines [9]

ensure correct decompression, the context should be always synchronized. The decompressor has three states as follows: (1) No Context (NC), where there is no context synchronization; (2) Static Context (SC), where the dynamic information of the context has been lost; and (3) Full Context (FC), when the decompressor has all the information about header fields. In FC state, the decompressor transitions to the initial states as soon as it detects corruption of the context. The decompressor uses the "*k out of n*" rule by looking at the last *n* received packets. If *k* CRC failures have occurred, it assumes the context has been corrupted and transitions to an initial state (SC or NC). The decompressor also sends feedback according to the operation mode (see Figure 5-4).

The values of the ROHC compression parameters that determine the efficiency and robustness are not defined in ROHC specification and are not negotiated initially, but are stated as implementation dependent. The values of these parameters stay fixed during the compression process. Those compression parameters are as follows [10]:

- *L*: in U-mode and O-mode the ROHC compressor uses a confidence variable *L* in order to ensure the correct transmission of header information.
- *Timer_1* (*IR_TIMEOUT*): in U-mode, the compressor uses this timer to return to the IR compression level and periodically resends static information.
- *Timer_2* (*FO_TIMEOUT*): the compressor also uses another timer in U-mode; this timer is used to go downward to FO compression level if the compressor is working in SO compression level.
- *Sliding Window Width* (*SWW*): the compressor, while compressing header fields like *Sequence Number* (SN) and *Timestamp*, uses Window-based Least Significant Bit (*W_LSB*) encoding that uses a Sliding Window of width equal to *SWW*.

W_LSB encoding is used to compress those header fields whose change pattern is known. When using this encoding, the compressor sends only the least significant bits. The decompressor uses these bits to construct the original value of the encoding fields.

- *k* and *n*: the ROHC decompressor uses a "*k out of n*" failure rule, where *k* is the number of packets received with an error in the last *n* transmitted packets. This rule is used in the state machine of the decompressor to assume the damage of context and move downwards to a state after sending a negative acknowledgment to the compressor, if a bi-directional link is used. The decompressor does not assume context corruption and remains in the current state until *k* packets arrive with error in the last *n* packets.

5.2 SERVICE FLOW CLASSIFICATION AND IDENTIFICATION

A service flow is a uni-directional flow of packets with a particular set of QoS parameters and is identified by a Service Flow Identifier (SFID). The QoS parameters may include traffic priority, maximum sustained traffic rate, maximum burst rate, minimum tolerable rate, scheduling type, ARQ type, maximum delay, tolerated jitter, service data unit type and size, bandwidth request mechanism to be used, transmission PDU formation rules, etc. Service flows may be provisioned through a network management system or created dynamically via defined signaling mechanisms in the standard. The base station is responsible for assigning the SFID and mapping it to unique transport connections.

5.2.1 Service Flow Attributes

A service flow is characterized by the following attributes [1,12]:

- Service Flow Identifier: a 32-bit SFID is assigned to each existing service flow. The SFID serves as the main identifier for the service flow in the mobile station. A service flow has at least one SFID and the associated direction.
- Station Identifier (STID): the IEEE 802.16m BS assigns a 12-bit-long STID to the MS during network entry/re-entry that uniquely identifies the MS within the domain of the BS. Each IEEE 802.16m MS registered in the network has an assigned STID. Certain STID values are reserved for broadcast, multicast, emergency alert, and ranging. The STID is the twelve most significant bits of 16-bit Connection Identifier (CID) defined in the IEEE 802.16-2009 standard. The CID of the transport connection exists only when the service flow is admitted or active. The relationship between SFID and Transport CID, when present, is unique. An SFID is never associated with more than one transport CID, and a transport CID is never associated with more than one SFID.
- Flow Identifier (FID): each IEEE 802.16m MS connection is assigned a 4-bit FID that uniquely identifies the connection within the MS. The FID is the four least significant bits of a 16-bit CID, and identifies management connections and transport connections. Certain values of FID may be reserved.
- *ProvisionedQoSParamSet*: a QoS parameter set is provisioned via means outside of the scope of the IEEE 802.16 standard, such as the network management system.
- *AdmittedQoSParamSet*: a set of QoS parameters for which the BS (and possibly the MS) are reserving resources. The main resource to be reserved is bandwidth, but this also includes any other memory- or time-based resource required to subsequently activate the flow.
- *ActiveQoSParamSet*: a set of QoS parameters (associated with the service flow) defining the service that is actually provided. The packet associated with an active service flow can only be forwarded.
- Authorization Module: a logical function within the BS that approves or denies every change to QoS parameters and classifiers associated with a service flow. As such, it limits the possible values of the *AdmittedQoSParamSet* and *ActiveQoSParamSet* parameters.

5.2.2 Service Flow Types

There are three basic types of service flows defined in the standard [1]; however, other types may be supported. The basic service flow types are as follows [1,13]:

- **Provisioned service flows**: a service flow may be provisioned but not immediately activated. In other words, the description of any such service flow contains an attribute that provisions but defers activation and admission. The network reserves an SFID for such a service flow. The BS may also require an exchange with a policy module prior to admission. As a result, the MS may choose to activate a provisioned service flow by passing the SFID and the associated QoS parameter sets to the BS in the dynamic service change request message. If the MS is authorized and resources are available, the BS responds by mapping the service flow to a CID (or alternatively a STID+FID combination for an IEEE 802.16m MS). The BS may choose to activate a service flow by passing the SFID as well as the CID and the associated QoS parameter sets to the MS in the dynamic service change request message. Such a provisioned service flow may be activated and deactivated many times through exchange of MAC management messages. In all cases, the original SFID is used when reactivating the service flow.

- **Admitted service flows**: the standard supports a two-phase activation model that is often utilized in cellular telephony applications. In the two-phase activation model, the resources for a call are first admitted, and once the end-to-end negotiation is completed, the resources are activated. The two-phase model serves the following purposes: (1) conserving network resources until a complete end-to-end connection has been established; (2) performing policy checks and admission control on resources as quickly as possible, and in particular, before informing the other side of a connection request; and (3) preventing several potential theft-of-service scenarios.
- **Active service flows**: a service flow that has a non-null *ActiveQoSParamSet* with committed resources by the BS. An admitted service flow may be activated by providing an *ActiveQoSParamSet*, signaling the resources actually desired at the current time.

A service flow may be provisioned and immediately activated. Alternatively, a service flow may be created dynamically and immediately activated. In this case, two-phase activation can be skipped and the service flow is available for immediate use upon authorization. The provisioning of service flows is outside of the scope of the standard. During provisioning, a service flow is instantiated, i.e., an SFID and a provisioned type are assigned. For some service flows, a dynamic service addition is activated by the network entry procedure. The service flows are enabled following the transfer of the operational parameters. In this case, the service flow type may change to "admitted" or "active". Thus, the service flow is mapped onto a certain connection. Service flow encodings contain either a full definition of service attributes or a service class name. A service class name is an ASCII string, which is known to the BS and indirectly specifies a set of QoS parameters.

5.2.3 Service Flow Classification

Classification is the process by which a MAC SDU is mapped onto a particular transport connection for transmission between MAC peers. The mapping process associates a MAC SDU with a transport connection, which also creates an association with the service flow characteristics of that connection. This process facilitates the delivery of MAC SDUs with the appropriate QoS constraints. A classification rule is a set of matching criteria applied to each packet entering the IEEE 802.16 entity. It consists of some protocol-specific packet matching criteria (e.g., destination IP address), classification rule priority, and a reference to a CID (or, for an IEEE 802.16m BS or an MS, reference to a STID+FID combination). If a packet matches the specified packet matching criteria, it is then delivered to the SAP for delivery on the connection defined by the CID or STID+FID. Implementation of each specific classification capability (e.g., IPv4 based classification) is optional. The service flow characteristics of the connection provide the QoS for that packet. This is illustrated in Figure 5-5.

Several classification rules may each refer to the same service flow. The classification rule priority is used for prioritizing the mapping of classification rules to packets. Explicit ordering is necessary because the patterns used by classification rules may overlap. While the priority does not have to be unique, it must be ensured that there is no ambiguity in classification within a classification rule priority. The downlink classification rules are applied by the BS to packets it is transmitting, and the

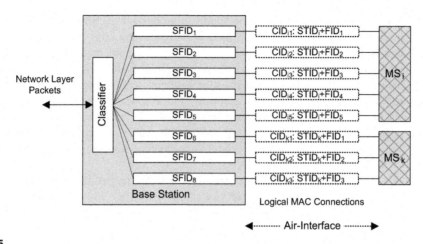

FIGURE 5-5

An illustration of service flow classification, SFID, STID, FID, and CID concepts [13]

uplink classification rules are applied at the MS. It is possible for a packet to fail to match the set of defined classification rules. In this case, the CS discards the packet.

Figure 5-5 shows the relationship between the classifier, SFID, STID, FID, and CID. Note that each mobile station is assigned a unique station identifier and a number of flow identifiers associated with the active connections with the mobile station. Despite the fact that the most significant bits of CID remained unchanged for the active connections between the BS and the MS in the legacy standard, the entire CID was communicated through MAC headers and messages over the airlink, resulting in inefficient use of radio resources. In IEEE 802.16m, the CID has been split into STID and FID, and only the flow identifiers are signaled over the air interface once the connections are established.

5.3 PACKET CONVERGENCE SUB-LAYER

The packet CS is a variant of the convergence sub-layer that interfaces the IP network with the IEEE 802.16m MAC CPS. The packet CS performs the following functions by utilizing the services of the MAC layer:

- Classification of the higher layer PDUs into the appropriate transport connections;
- Payload header compression;
- Delivery of the CS PDUs to the MAC SAP for transport to the peer MAC SAP at the receive side;
- Receipt of the CS SDU from the peer MAC SAP;
- Rebuilding of any compressed payload header information.

The CS on the transmitter side is responsible for delivering the MAC SDUs (or alternatively CS PDUs) to the MAC SAP. The MAC is responsible for delivery of the MAC SDUs to peer MAC SAP in accordance with the QoS, fragmentation, concatenation, and other transport functions associated with the connection's service flow characteristics. The receiving CS is responsible for accepting the MAC PDUs (or alternatively CS SDUs) from the peer MAC SAP and delivering them to the higher layer entity.

5.3.1 **Packet CS Payload Header Suppression**

The IEEE 802.16-2009 standard convergence sub-layer specifies a native Payload Header Suppression (PHS) scheme which can be optionally used to compress the IP/UDP/RTP payload headers [1]. A comparison of the performance and robustness of the native PHS algorithm and standard ROHC confirms the superiority of the latter. It is expected that all IEEE 802.16m implementations will utilize ROHC as the header compression scheme for all types of traffic.

In PHS, the repetitive portion of the payload headers of the higher layers is suppressed in the MAC SDU by the sending entity and restored by the receiving entity. In the uplink, the sending entity is the MS and the receiving entity is the BS. In the DL, the sending entity is the BS and the receiving entity is the MS. If PHS is enabled for a MAC connection, each MAC SDU is prefixed with a PHS Index (PHSI), which references the PHS Field (PHSF), i.e., a string of bytes representing the header portion of a PDU in which one or more bytes are suppressed. The sending entity uses classification rules to map packets to a service flow. The classification rule uniquely maps packets to its associated PHS rule. The receiving entity uses the CID or STID+FID and the PHSI to restore the PHSF. Once a PHSF has been assigned to a PHSI, it will not change. To change the value of a PHSF on a service flow, a new PHS rule must be defined; the old rule is removed from the service flow, and the new rule is added. When all classification rules associated with the PHS rule are deleted, then the PHS rule can also be deleted. The PHS scheme has a PHS Valid (PHSV) option to verify the payload header before suppressing it. The PHS scheme has also a PHS Mask (PHSM) option to allow certain bytes not to be suppressed. The PHSM facilitates suppression of header fields that remain static, while enabling transmission of fields that change from packet to packet.

The BS assigns the PHSI values in the same way that it assigns the CID or STID+FID values. Either the sending or the receiving entity specifies the PHSF and the PHS Size (PHSS). This provision allows for preconfigured headers or for higher level signaling protocols outside the scope of the standard to establish cache entries. It is the responsibility of the higher layer service entity to generate a PHS rule that uniquely identifies the suppressed header within the service flow. It is also the responsibility of the higher layer service entity to guarantee that the byte strings that are being suppressed are constant from packet to packet for the duration of the active service flow.

The operation of the PHS is illustrated in Figure 5-6. The format of the IP CS PDU is shown in Figure 5-7 for two cases: (1) when header suppression is enabled at the connection, but not applied to the CS PDU; or (2) with header suppression. In the case where PHS is not enabled, the PHSI field is omitted.

The ROHC may be used instead of PHS to compress the IP headers. The MS and the BS signal the enabling and support of ROHC protocol by setting the seventh bit of Request/Transmission Policy parameter to zero (the value of this parameter would allow specification of certain attributes for the associated service flow). When ROHC is enabled for a service flow, the service flow constitutes that which in IETF RFC 3095 is referred to as a ROHC channel. Feedback sent on an ROHC channel consists of one or more concatenated feedback elements. When an ROHC compressor has transformed original packets into ROHC packets with compressed headers (see Figure 5-3), these ROHC packets are sent to the corresponding decompressor through a logical point-to-point connection dedicated to that traffic. Such a logical channel, which only has to carry data in this single direction from compressor to decompressor, is referred to as an ROHC channel. An ROHC compressor instance is a logical entity that performs header compression according to one or several ROHC profiles. There is a one-to-one

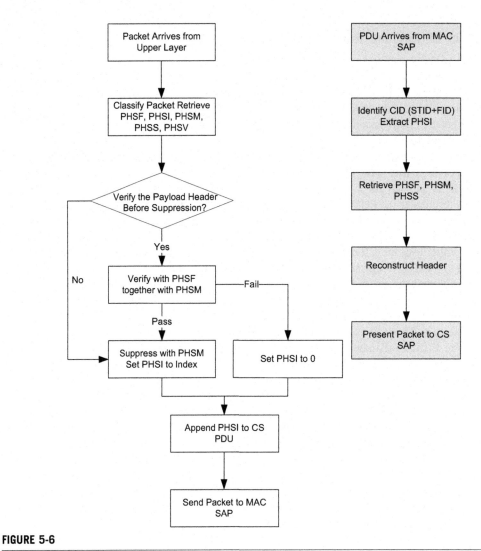

FIGURE 5-6

Packet CS payload header suppression flow graph [1]

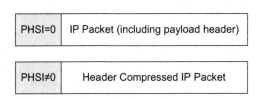

FIGURE 5-7

IP CS PDU format with and without header suppression [1]

correspondence between an ROHC compressor instance and an ROHC channel, where the ROHC compressor is located at the input end of the ROHC channel. Similarly, an ROHC decompressor instance is a logical entity that performs header decompression according to one or several ROHC profiles. There is a one-to-one correspondence between an ROHC decompressor instance and an ROHC channel, where the ROHC decompressor is located at the output end of the ROHC channel.

A channel is essentially a logical point-to-point connection between the IP interfaces of two communicating network elements. By that definition, the channel characterizes a logical connection that is needed to make header compression generally applicable, and further the channel properties control whether compression can operate in a uni-directional or a bi-directional manner. An ROHC channel has the same properties as a channel, with the difference that an ROHC channel is always uni-directional. An ROHC channel therefore has one single input endpoint, connected to a single ROHC compressor instance, and one single output endpoint, connected to a single ROHC decompressor instance. An ROHC channel must be logically dedicated to one ROHC compressor and one ROHC decompressor, also referred to as ROHC peers, creating a one-to-one mapping between an ROHC channel and two ROHC compressor/decompressor peers.

5.4 GENERIC PACKET CONVERGENCE SUB-LAYER

The Generic Packet CS (GPCS) is a protocol-independent packet convergence sub-layer that can support multiple network-layer protocols (e.g., IPv4, IPv6, or IEEE 802.3 Ethernet) over the IEEE 802.16 air interface. The GPCS provides a generic packet convergence layer. This layer uses the MAC SAP and exposes a SAP to GPCS applications. It does not redefine or replace other convergence sub-layers. Instead, it provides a SAP that is not protocol specific. With GPCS, packet parsing happens above the convergence sub-layer, resulting in parameters that are passed to the GPCS SAP for classification. Upper layer packet parsing is left to the GPCS application. For the MS and BS, the upper layer protocol that is immediately on top of the IEEE 802.16 GPCS is identified by a TLV parameter and GPCS protocol type [1]. The GPCS protocol type is included in Control Service Flow Management (C-SFM) primitives and Dynamic Service Addition messages during connection establishment. The GPCS protocol defines a set of SAP parameters as the result of upper layer packet parsing. These are passed from the upper layer to the GPCS in addition to the data packet. For the MS and BS, the SAP parameters in the IEEE 802.16-2009 standard include SFID, MS MAC Address, data, and length [1]. In IEEE 802.16m, the MAC address is replaced by STID for user privacy protection.

The GPCS allows multiplexing of multiple layer protocol types (e.g., IPv4, IPv6, IEEE 802.3 Ethernet) over the same IEEE 802.16 connection using appropriate upper layer protocol that supports multiplexing. This capability is signaled by the IEEE 802.16-2009 standard compliant mobile and base stations using the *GPCS_PROTOCOL_TYPE* parameter and by the IEEE 802.16m-compliant mobile and base stations using the Dynamic Service Addition/Change/Deletion messages to indicate that multiple protocols are supported for a connection/service flow. The multiplexing and de-multiplexing of multiple protocol data packets over an IEEE 802.16 connection/service flow is outside the scope of GPCS; however, interoperability between peer protocol layers must be ensured.

The GPCS can further transparently support IEEE 802.1D, bridging over the IEEE 802.16 air interface, since it requires the upper layer to provide the MS MAC Address and SFID with every packet, where the MS MAC Address and SFID can represent a port, and a port is either a unicast port or

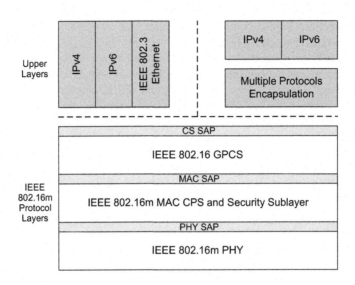

FIGURE 5-8

GPCS protocol layering model [1]

a broadcast port [11]. The PHS algorithm defines rules for how packets with suppressed fields are reconstructed, based on the PHSI and the associated PHS rule. This reconstruction method can also be applied on packets transferred over the GPCS. The GPCS protocol layering model is shown in Figure 5-8.

The GPCS uses the GPCS SAP, an instance of the logical CS SAP. The GPCS SAP parameters enable the upper layer protocols to generically pass information to the GPCS so that the GPCS does not need to interpret upper layer protocol headers in order to map the upper layer data packets into proper IEEE 802.16m MAC connections. Since the SAP parameters are explicit, the parsing portion of the classification process is the responsibility of the upper layer. The parameters are relevant for SAP data path primitives, *GPCS_DATA.request*, and *GPCS_DATA.indication* as described in reference [1]. The parameters SFID, MS MAC Address, STID (for mobile and base stations compliant with IEEE 802.16m), FID (for mobile and base stations compliant with IEEE 802.16m), length, and data are described in reference [12]. Note that SFID and MS MAC Address are not transferred over the IEEE 802.16 air interface. The GPCS maps the CID or STID+FID to SFID and MS MAC Address, and then passes them to the upper layer of the GPCS through *GPCS_DATA.indication*, where the CID or STID+FID are provided in MAC SAP.

5.5 THE 3GPP LTE PACKET DATA CONVERGENCE PROTOCOL

The 3GPP Packet Data Convergence Protocol is part of Layer 2 protocols that provide the following services and functions to other protocol layers in the user-plane [20]:

- Header compression and decompression using ROHC;
- Transfer of user data;
- In-sequence delivery of upper layer PDUs at PDCP re-establishment procedure for RLC AM;

- Duplicate detection of lower layer SDUs at PDCP re-establishment procedure for RLC AM;
- Re-transmission of PDCP SDUs at handover for RLC AM;
- Ciphering and deciphering;
- Timer-based SDU discarding in the uplink.

The main services and functions of the PDCP in the control-plane include Ciphering and Integrity Protection, as well as Transfer of control-plane data. Figure 5-9 illustrates one possible structure for the PDCP sub-layer (without restricting other implementations), and is based on the radio interface protocol architecture defined in reference [21]. A PDCP PDU is a bit string that is byte aligned.

As shown in Figure 5-9, each Radio Bearer (RB) is associated with one PDCP entity. Each PDCP entity is associated with one or two (one for each direction) RLC entities depending on the RB characteristic (i.e., uni-directional or bi-directional) and RLC mode [22]. The PDCP entities are located in the PDCP sub-layer and are configured by upper layers [21]. The Control SAP (C-SAP) is the PDCP interface with the RRC layer.

Several PDCP entities may be defined for a UE. Each PDCP entity carrying user-plane data may be configured to use header compression. Each PDCP entity is carrying the data of one radio bearer. The 3GPP LTE supports the ROHC protocol as specified by reference [9]. Each PDCP entity uses one ROHC instance. A PDCP entity is associated with either the control-plane or the user-plane, depending on which radio bearer is carrying data. Figure 5-10 shows the functional decomposition of the PDCP entity in PDCP sub-layer. The figure is based on the radio interface protocol architecture defined in reference [20]. The PDCP provides services to the RRC sub-layer and upper layers in the user-plane at

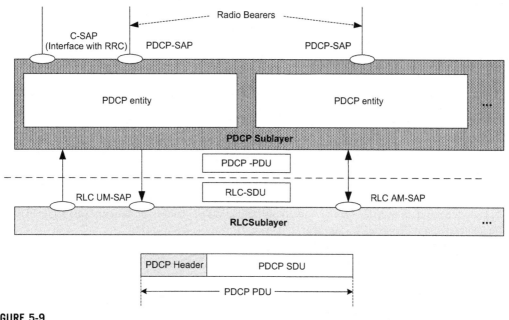

FIGURE 5-9

PDCP sub-layer and structure of PDCP PDU [21]

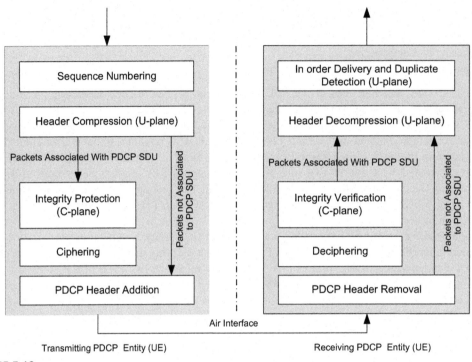

FIGURE 5-10

Functional decomposition of PDCP entities [21]

the UE, or to the relay at the evolved Node B (eNB). The following services are provided by PDCP to upper layers:

- Transfer of user-plane and control-plane data;
- Payload header compression;
- Ciphering;
- Integrity protection.

The maximum supported size of a PDCP SDU is 8188 octets [21]. The following services are provided by lower layers to PDCP sub-layer:

- Acknowledged data transfer service including indication of successful delivery of PDCP PDUs;
- Unacknowledged data transfer service;
- In-sequence delivery, except at re-establishment of lower layers;
- Duplicate discarding, except at re-establishment of lower layers.

As mentioned earlier, the PDCP sub-layer performs the following functions:

- Header compression and decompression of IP data flows using the ROHC protocol;
- Transfer of data (user-plane or control-plane);
- Maintenance of PDCP Sequence Numbers (SN);

- In-sequence delivery of upper layer PDUs at re-establishment of lower layers;
- Removal of duplicate SDUs at re-establishment of lower layers for radio bearers mapped on RLC AM;
- Ciphering and deciphering of user-plane data and control-plane data;
- Integrity protection and integrity verification of control-plane data;
- Timer-based discard;
- Duplicate discarding.

The PDCP uses the services provided by the RLC sub-layer. The PDCP is used for SRBs (Signaling Radio Bearer carrying control-plane data) and DRBs (Data Radio Bearer carrying user-plane data) mapped on Dedicated Control Channel (DCCH) and Dedicated Traffic Channel (DTCH) type of logical channels. The PDCP is not used for other types of logical channels.

The ROHC protocol is utilized for header compression in 3GPP LTE. There are multiple header compression algorithms, called profiles, defined for the ROHC protocol. Each profile is specific to the particular network layer, transport layer or upper layer protocol combination, e.g., TCP/IP or RTP/UDP/IP. The multiplexing of different flows (with or without header compression) over the ROHC channels, as well as association of a specific IP flow with a specific context state during initialization of the compression algorithm for that flow, are described in reference [25]. The implementation of ROHC functionality and the supported header compression profiles are not specified in 3GPP LTE specifications. In 3GPP LTE specifications, the support of the following ROHC profiles is described in Table 5-1.

Table 5-1 Supported Header Compression Protocols and Profiles [21]

Profile Identifier	Usage	Reference
0 × 0000	No compression	RFC 4995
0 × 0001	RTP/UDP/IP	RFC 3095, RFC 4815
0 × 0002	UDP/IP	RFC 3095, RFC 4815
0 × 0003	ESP[j]/IP	RFC 3095, RFC 4815
0 × 0004	IP	RFC 3843, RFC 4815
0 × 0006	TCP/IP	RFC 4996
0 × 0101	RTP/UDP/IP	RFC 5225
0 × 0102	UDP/IP	RFC 5225
0 × 0103	ESP/IP	RFC 5225
0 × 0104	IP	RFC 5225

[j] Encapsulating Security Payload (ESP) is a key component of the IPsec protocol suite. The ESP protocol provides confidentiality and integrity by encrypting data to be protected and placing the encrypted data in the data portion of the IP ESP. Depending on the user's security requirements, this mechanism may be used to encrypt either a transport-layer segment (e.g., TCP, UDP) or the entire IP datagram. Encapsulating the protected data is necessary to provide confidentiality for the entire original datagram. ESP also supports encryption-only and authentication-only configurations, but using encryption without authentication is not recommended because of insecurity. Unlike Authentication Header (AH), ESP does not protect the IP packet header. However, in the Tunnel Mode, where the entire original IP packet is encapsulated with a new packet header added, ESP protection is applied to the entire inner IP packet (including the inner header) while the outer header remains unprotected. ESP operates directly on top of IP using IP protocol number 50 [24].

PDCP entities associated with DRBs can be configured by upper layers to use header compression. RFC 4995 has configuration parameters that are mandatory and that must be configured by upper layers between compressor and decompressor peers. Those parameters define the ROHC channel. As described earlier, the ROHC channel is a uni-directional channel, i.e., there is one channel for the downlink and one for the uplink. Therefore, there is one set of parameters for each channel and the same values are used for both channels belonging to the same PDCP.

The ciphering function includes both ciphering and deciphering and is performed in the PDCP sub-layer. Note that, unlike IEEE 802.16m where the CS PDUs are not encrypted and the encryption is performed on MAC PDUs, in 3GPP LTE the RLC SDUs are ciphered. For the control-plane, the data unit that is ciphered is the data part of the PDCP PDU and the Message Authentication Code for Integrity (MAC-I), i.e., a 32-bit field that carries message authentication code. For the user-plane, the data unit that is ciphered is the data part of the PDCP PDU; ciphering is not applicable to PDCP control PDUs. The ciphering algorithm and key to be used by the PDCP entity are configured by upper layers, and the ciphering method is applied according to Security Architecture of 3GPP System Architecture Evolution [23]. The ciphering function is activated by upper layers. After security activation, the ciphering function is applied to all PDCP PDUs indicated by upper layers for the downlink and uplink transmissions. The parameters that are required by PDCP for ciphering are defined in reference [23] and are input to the ciphering algorithm. The required inputs to the ciphering function include the *COUNT* value and *DIRECTION* (direction of the transmission).The parameters required by PDCP which are provided by upper layers are *BEARER* (defined as the radio bearer identifier) and *KEY* (the ciphering keys for the control-plane and for the user-plane) [21].

The integrity protection function includes both integrity protection and integrity verification, and is performed in PDCP for PDCP entities associated with SRBs. The data unit that is integrity protected is the PDU header and the data part of the PDU prior to ciphering. The integrity protection algorithm and key to be used by the PDCP entity are configured by upper layers and the integrity protection method is applied according to Security Architecture of 3GPP System Architecture Evolution [23]. The integrity protection function is activated by upper layers. Following security activation, the integrity protection function is applied to all PDUs including and subsequent to the PDU indicated by upper layers for the downlink and uplink transmissions. As the RRC message which activates the integrity protection function is itself integrity protected with the configuration included in that RRC message, the message must be decoded by RRC before the integrity protection verification could be performed for the PDU in which the message was received. The parameters that are required by PDCP for integrity protection are defined in reference [23] and are input to the integrity protection algorithm. The required inputs to the integrity protection function include the *COUNT* value and *DIRECTION* (direction of the transmission). The parameters required by PDCP which are provided by upper layers are *BEARER* (defined as the radio bearer identifier) and the *KEY*.

During transmission, the UE computes the value of the MAC-I field and, at reception, it verifies the integrity of the PDCP PDU by calculating the X-MAC, i.e., Computed MAC-I, based on the input parameters. If the calculated X-MAC corresponds to the received MAC-I, integrity protection is verified successfully. When a PDCP entity receives a PDCP PDU that contains reserved or invalid values, the PDCP entity discards the received PDU. The PDCP Data PDU is used to transport PDCP SDU SN and user-plane data containing an uncompressed PDCP SDU, user-plane data containing a compressed PDCP SDU, or control-plane data as well as MAC-I field for SRBs.

The PDCP Control PDU is used to convey the PDCP status report identifying missing PDCP SDUs following PDCP re-establishment and header compression control information, e.g., ROHC feedback.

References

[1] IEEE Std 802.16-2009, IEEE Standard for Local and Metropolitan Area Networks – Part 16: Air Interface for Fixed Broadband Wireless Access Systems, May 2009.

[2] IETF RFC 2507, M. Degermark, et al., "IP Header Compression," February 1999. <http://www.rfc-editor. org/rfc/rfc2507.txt>.

[3] IEFT RFC 791, Internet Protocol, September 1981. <http://www.ietf.org/rfc/rfc0791.txt>.

[4] User Datagram Protocol (UDP), Wikipedia <http://en.wikipedia.org/wiki/User_Datagram_Protocol>.

[5] Douglas E. Comer, Internetworking with TCP/IP Vol. 1: Principles, Protocols, and Architecture, fifth ed., Prentice Hall, 2005.

[6] Real-time Transport Protocol (RTP), Wikipedia <http://en.wikipedia.org/wiki/Real-time_Transport_ Protocol>.

[7] Colin Perkins, RTP: Audio and Video for the Internet, Addison-Wesley Professional, June 2003.

[8] IEFT RFC 2460, S. Deering, R. Hinden, "Internet Protocol, Version 6 (IPv6) Specification" December 1998.

[9] IETF RFC 3095, C. Bormann, et al., "RObust Header Compression (ROHC): Framework and four profiles: RTP, UDP, ESP, and uncompressed," July 2001. <http://www.ietf.org/rfc/rfc3095.txt>.

[10] A. Minaburo, et al., "Proposed Behavior for Robust Header Compression over a radio link", 2004 IEEE International Conference on Communications, Vol. 7, June 2004.

[11] Bridging and Virtual Bridged LANs IEEE 802.1, <http://grouper.ieee.org/groups/802/1/>.

[12] P802.16m/D6, IEEE Standard for Local and Metropolitan Area Networks – Part 16: Air Interface for Broadband Wireless Access Systems, Advanced Air Interface May 2010.

[13] Loutfi Nuaymi, WiMAX: Technology for Broadband Wireless Access, first ed., John Wiley & Sons, 2007.

[14] D.E. Taylor, et al., "Robust Header Compression (ROHC) in Next-Generation Network Processors", IEEE/ACM Transactions on Networking, Vol. 13, (No. 4), August 2005.

[15] S. Ayed, et al., "Enhancing Robust Header Compression Over IEEE 802 Networks", IEEE International Conference on Wireless and Mobile Computing, Networking and Communications, (WiMob'2006) June 2006.

[16] A. Minaburo, L. Nuaymi, et al., "Configuration and Analysis of Robust Header Compression in UMTS", 14th IEEE Proceedings on Personal, Indoor and Mobile Radio Communications (PIMRC-2003), Vol. 3, September 2003.

[17] P. Fortuna, M. Ricardo, "Header compressed VoIP in IEEE 802.11", IEEE Wireless Communications, Vol. 16, June 2009.

[18] IEFT RFC 3759, L.-E. Jonsson, "RObust Header Compression (ROHC): Terminology and Channel Mapping Examples," April 2004. <http://www.ietf.org/rfc/rfc3759.txt>.

[19] IETF RFC 4815, L.-E. Jonsson, et al., "RObust Header Compression (ROHC): Corrections and Clarifications to RFC 3095," February 2007.

[20] 3GPP TS 36.300, Evolved Universal Terrestrial Radio Access (E-UTRA) and Evolved Universal Terrestrial Radio Access Network (E-UTRAN); Overall description; Stage 2 March 2010.

[21] 3GPP TS 36.323, Evolved Universal Terrestrial Radio Access (E-UTRA); Packet Data Convergence Protocol (PDCP) specification, March 2010.

[22] 3GPP TS 36.322, Evolved Universal Terrestrial Radio Access (E-UTRA); Radio Link Control (RLC) protocol specification, March 2010.

[23] 3GPP TS 33.401, 3GPP System Architecture Evolution (SAE): Security Architecture, March 2009. <ftp://ftp.3gpp.org/specs/2009-03/Rel-8/33_series/>.

[24] Encapsulating Security Payload (ESP) (http://en.wikipedia.org/wiki/IPsec#Encapsulating_Security_Payload).

[25] IETF RFC L.-E. 4995, Jonsson, et al., "The RObust Header Compression (ROHC) Framework," July 2007.

[26] IETF RFC 3550, H. Schulzrinne, et al., "RTP: A Transport Protocol for Real-Time Applications," July 2003.

The IEEE 802.16m Medium Access Control Common Part Sub-layer (Part I)

INTRODUCTION

This chapter describes the functional and operational aspects of IEEE 802.16m MAC Common Part Sub-layer (MAC CPS) on the control-plane. As shown in Figure 6-1, the MAC CPS provides an interface between the physical layer and higher protocol layers through PHY and MAC SAPs, respectively. The MAC CPS functions are classified into Radio Resource Control and Management (RRCM) and MAC functions. The soft classification of the MAC CPS into RRCM and MAC sub-layer does not require any SAP between the two classes of functions. The RRCM functions fully reside on the control-plane, whereas the MAC sub-layer functions reside on both control-plane and data-plane. The RRCM includes several functional blocks that are related to radio resource and radio access management and control, such as [5]:

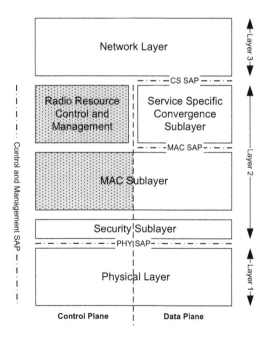

FIGURE 6-1

IEEE 802.16m control-plane MAC CPS functions [5]

Mobile WiMAX. DOI: 10.1016/B978-0-12-374964-2.10006-2
Copyright © 2011 Elsevier Inc. All rights reserved

- Radio Resource Management (RRM);
- Mobility Management;
- Network-entry Management;
- Location Management;
- Idle Mode Management;
- Security Management;
- System Configuration Management;
- Enhanced-Multicast and Broadcast Service (E-MBS);
- Service Flow and Connection Management;
- Relay Functions;
- Self Organization;
- Multi-Carrier Operation.

The control-plane part of the MAC sub-layer includes functional blocks which are related to the physical layer and link control such as:

- Physical Layer Control;
- Control Signaling;
- Sleep Mode Management;
- QoS;
- Scheduling and Resource Multiplexing;
- Multi-Radio Coexistence;
- Data Forwarding;
- Interference Management;
- Inter-BS Coordination.

The data-plane portion of the MAC sub-layer is responsible for the following functions:

- MAC SDU Fragmentation/Packing;
- ARQ;
- MAC PDU Formation.

To draw an analogy between IEEE 802.16m and 3GPP LTE Layer 2 protocols, one must note that the Layer 2 functions of the latter include MAC and RLC sub-layers and the RRC sub-layer is considered as Layer 3. The IEEE 802.16m RRCM functional group corresponds to RRC sub-layer in 3GPP LTE, and IEEE 802.16m MAC sub-layer is analogous to a combination of 3GPP LTE MAC and RLC sub-layers. The 3GPP RRC sub-layer will be described to enable the reader to understand the similarities and differences of the two standards by making heuristic correspondences between the two protocol sets.

The IEEE 802.16m MAC layer is connection-oriented. For the purpose of mapping services to varying levels of QoS at mobile stations, all data communications are manifested in the form of transport connections. Service flows may be provisioned when an MS enters into the system. Following MS registration with the serving BS, transport connections are established and associated with the service flows (one connection per service flow) to provide a reference for requesting bandwidth. Furthermore, new transport connections may be established when a user's service needs to change. A transport connection defines both the mapping between peer convergence sub-layers that utilize the MAC and a service flow. The service flow defines the QoS parameters for the PDUs that are exchanged on the connection.

The concept of service flow mapping to a transport connection is essential to the operation of the MAC CPS protocols. Service flows provide a mechanism for UL and DL QoS management. In particular, they are an integral part of the bandwidth allocation process. An MS requests UL bandwidth per connection basis by implicitly identifying the service flow. The bandwidth is granted by the serving BS to an MS as an aggregate of grants in response to per-connection requests from the MS. Transport connections, once established, may require active maintenance with requirements which may vary depending on the type of service.

In the following sections, the IEEE 802.16m MAC CPS functions are classified and described according to their location on the control-plane using a systematic approach. Some of the service control functions, such as E-MBS and location-based services, as well as some advanced features such as multi-carrier operation, relay, self-organizing networks, and security aspects will be described in separate chapters.

6.1 ADDRESSING

A 48-bit universal MAC address, as defined in IEEE 802 standard [3], uniquely defines mobile terminals over the air interface. It is used during the legacy initial ranging process to establish the appropriate connections for an MS. It is also used as part of the legacy authentication process by which the BS and MS each verify the identity of each other.

In IEEE 802.16-2009 standard, connections are identified by a 16-bit Connection Identifier (CID). In MS initialization, two pairs of management connections, basic connections (UL and DL) and primary management connections (UL and DL), are established between the MS and the BS, and a third pair of management connections (secondary management, DL and UL) may be optionally established. The three pairs of management connections correspond to three different levels of QoS for management messages between the MS and the serving BS. The basic connection is used by the BS MAC and the MS MAC to exchange short, time-urgent MAC management messages. The primary management connection is used by the BS MAC and the MS MAC to exchange longer, more delay-tolerant MAC management messages. In addition, the secondary management connection may be used by the BS and MS to transfer delay-tolerant, standards-based Dynamic Host Configuration Protocol (DHCP), Trivial File Transfer Protocol (TFTP),[i] and Simple Network Management Protocol (SNMP) messages. The messages carried on the secondary management connection may be packed and/or fragmented. Use of the secondary management connection is required only for a managed MS [1]. The CIDs for these connections are assigned in the RNG-RSP, REG-RSP, or MOB_BSHO-REQ/RSP for pre-allocation in handover (HO) scenarios. When CID pre-allocation is used during HO, a primary management CID may be derived based on basic CID without assignment in the messages. The message exchanges provide three CID values. The same CID value is assigned to both members (UL and DL) of each connection pair.

---

[i]Trivial File Transfer Protocol (TFTP) is a file transfer protocol with a very basic form of the File Transfer Protocol (FTP) functionality that was first defined in 1980.

An IEEE 802.16m MS and its active connections are identified with global and logical addresses during operation in the network. The MS, RS, and BS are identified by the globally unique 48-bit IEEE Extended Unique Identifier (EUI-48™)[ii] based on the 24-bit Organizationally Unique Identifier (OUI) value administered by the IEEE Registration Authority [4]. There are three logical identifiers defined to recognize an active user and its associated connections as follows:

- Station Identifier (STID): the serving BS assigns a 12-bit STID to the MS during network entry or re-entry that uniquely identifies the MS within the coverage area of the serving BS. Each MS registered in the network is assigned an STID. Some specific STIDs are reserved, e.g., for broadcast and multicast and ranging.
- Temporary Station ID (TSTID): this logical identifier is used to protect the mapping between the STID and the MS MAC Address. A TSTID is assigned during the initial ranging process. During registration procedure, the BS assigns and transfers an STID to the MS using an encrypted registration response message. The serving BS discards the TSTID when the MS successfully completes the authentication procedures.
- Flow Identifier (FID): each MS connection is assigned a 4-bit FID that uniquely identifies the connections with the MS. The FIDs identify control and transport connections. The downlink and uplink transport FID values are chosen from 0b0011 to 0b1111 range. An FID that has been assigned to one DL/UL transport connection cannot be assigned to another DL/UL transport connection belonging to the same MS; therefore, the assigned FIDs in the downlink (or uplink) are unique for a particular terminal. However, an FID that has been used for a DL transport connection can be assigned to another UL transport connection associated with the same MS. Some specific FIDs may be pre-assigned. An FID value of 0b0000 designates a control FID (unicast control FID when PDU is allocated by unicast assignment A-MAP IE or broadcast control FID when PDU is allocated by broadcast assignment A-MAP IE). An FID value of 0b0001 indicates a MAC signaling header.
- Deregistration Identifier (DID): the DID uniquely identifies an idle-mode MS for paging purposes.
- Context Retention Identifier (CRID): if Deregistration with Content Retention (DCR) mode is enabled, the network assigns a 72-bit CRID to each MS during network entry or on handover to an IEEE 802.16m BS in a mixed-mode operation. The MS is identified by the CRID in coverage loss recovery and DCR mode, where the CRID allows the network to retrieve the MS context. The network may assign the MS a new CRID if necessary.
- E-MBS Identifier: a 12-bit value that is used along with a 4-bit FID to uniquely identify a specific E-MBS flow in an E-MBS zone.

In order to protect the mapping between the STID and the MS MAC address, two types of STIDs are assigned to an MS during network entry: a temporary STID and a permanent STID. A TSTID is assigned during the initial ranging process and is used until the permanent STID is prescribed. The permanent STID is assigned during the registration with the BS and after successful authentication

---

[ii]The IEEE defined 48-bit extended unique identifier (EUI-48™) is a concatenation of either a 24-bit Organizationally Unique Identifier (OUI) value administered by the IEEE Registration Authority and a 24-bit extension identifier assigned by the organization with that OUI assignment, or the concatenation of a 36-bit Individual Address Block (IAB) identifier or 36-bit Organizationally Unique Identifier (OUI-36), and a 12-bit extension identifier assigned by the organization with that IAB assignment. The IEEE administers the assignment of 24-bit OUI values and 36-bit OUI-36 or IAB identifiers. The assignments of these values are public, so that a user of an EUI-48 value can identify the manufacturer that provided the value. See http://standards.ieee.org/regauth/oui/tutorials/EUI48.html for more details.

process and is encrypted during transmission. The TSTID is released after the STID is assigned. The STID stays valid during the entire session.

6.2 MAC PDU HEADERS

6.2.1 Legacy Generic MAC Header

Each MAC PDU in IEEE 802.16-2009 standard consists of a MAC header followed by a payload (also called SDU) and an optional Cyclic Redundancy Check (CRC). The CRC consists of four octets, is based on IEEE 802.3, and is calculated on the entire MAC PDU including the header. The size of MAC header (six octets) is considered a significant overhead for certain applications such as VoIP and interactive gaming, which consist of frequent bursts of small payloads. As an example, the size of Real-time Transport Protocol (RTP) payloads corresponding to VoIP application, depending on the type of the voice codec is typically less than 40 octets; therefore, each VoIP packet encapsulated in a MAC PDU would have more than 25% overhead (typically 50% with smaller packets, e.g., 20 octets for ITU-T G.729) overhead due to the size of the MAC header and the trailing CRC. Most of the fields in the legacy MAC header are not usually used in VoIP and interactive gaming.

The structure of a legacy MAC PDU is illustrated in Figure 6-2. Each PDU begins with a fixed-length MAC header that may be followed by a payload. If the payload is present, it may consist of zero or more sub-headers and zero or more MAC SDUs and/or fragments of MAC SDUs. The payload information may vary in length; therefore, the MAC PDU may have variable size. This allows the MAC to tunnel various higher layer traffic types without knowledge of the formats or bit patterns of those messages. All reserved fields are set to zero on transmission and ignored on reception. The generic MAC header is used with all MAC SDUs and MAC management messages.

The information fields of the generic MAC header are as follows:

- Header Type (HT) is an indicator of whether the MAC PDU contains any payload. If set to 1, the MAC PDU does not contain any payload and the MAC header is a signaling header.
- CRC Indicator (CI) shows whether a CRC is suffixed to the MAC PDU.
- Encryption Control (EC) indicates if the payload is encrypted. The header is never encrypted.

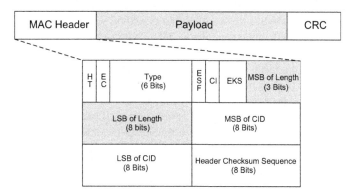

FIGURE 6-2

Structure of the legacy generic MAC header [1]

- Encryption Key Sequence (EKS) is the index of the traffic encryption key and initialization vector used to encrypt the payload. This field is only meaningful if the EC field is set to 1.
- Extended Sub-header Field (ESF) indicates whether the extended sub-header is present. If set to 1, the extended sub-header is present and follows the generic MAC header immediately. The ESF is applicable to downlink and in uplink.
- Header Checksum Sequence (HCS) is an 8-bit field that is used to detect errors in the header. The transmitter calculates the HCS value for the first five bytes of the MAC header and inserts the result into the HCS field. The HCS is the remainder of the modulo-2 division by the generator polynomial $g(D) = D^8 + D^2 + D + 1$ of the polynomial D^8 multiplied by the content of the header excluding the HCS field.

The ESF bit in the generic legacy MAC header is an indication of the presence of extended sub-headers. Using this field, a number of additional sub-headers can be used within a MAC PDU. The extended sub-header always appears immediately after the generic MAC header and before all other sub-headers. Unlike other sub-headers, extended sub-headers are not considered part of the MAC PDU payload and are not encrypted. When an IEEE 802.16-2009 standard entity transmits a MAC PDU without a payload, it sets the EC bit in the generic MAC header to 0, even if the connection on which it transmits the MAC PDU is associated with data encryption. When an entity receives a MAC PDU that does not contain a payload, it processes the MAC PDU, if the EC bit is set to 0; otherwise, it should discard this MAC PDU.

Five types of MAC sub-headers may be present in a legacy MAC PDU in conjunction with the generic MAC header; four per-PDU sub-header types and one per-SDU sub-header type [1]. The per-PDU sub-headers, i.e., extended sub-headers, Fragmentation Sub-Header (FSH), Fast-Feedback Allocation Sub-Header (FFSH), and Grant Management Sub-Header (GMSH) may be inserted in the MAC PDUs immediately following the generic MAC header. If both the FSH and GMSH are present, the GMSH takes precedence. In the DL, the FFSH always appears as the last per-PDU sub-header. The ESF bit in the generic MAC header indicates that one or more extended sub-headers are present in the PDU. The extended sub-headers always appear immediately after the generic MAC header and before all other sub-headers. The extended sub-headers are not encrypted. The only per-SDU sub-header is the Packing Sub-Header (PSH). It may be inserted before each MAC SDU, if indicated by the *Type* field. The PSH and FSH are mutually exclusive and cannot present within the same MAC PDU. When present, per-PDU sub-headers always precede the first per-SDU sub-header.

The GMSH is 2 bytes in length and is used by the MS to convey bandwidth management requests to the serving BS. This sub-header is encoded differently based on the type of UL scheduling service for the connection as given by the CID. The support of GMSH at both BS and MS is optional. When packing is used, the MAC may pack multiple SDUs into a single MAC PDU. When packing variable-length MAC SDUs, the MAC prefixes each one with a PSH. If the ARQ feedback payload bit in the MAC *Type* field is set, the ARQ feedback payload is transported. If packing is used, it is transported as the first packed payload. Note that this bit does not address the ARQ feedback payload contained inside an ARQ feedback message.

6.2.2 IEEE 802.16m MAC Headers

To mitigate the excessive overhead of the legacy MAC header for small payload applications, IEEE 802.16m has specified new MAC headers. The structure of an IEEE 802.16m MAC PDU is shown in

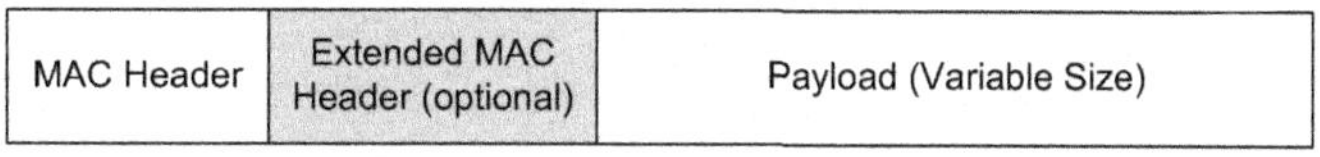

MAC Header	Extended MAC Header (optional)	Payload (Variable Size)

FIGURE 6-3

Structure of a MAC PDU [2]

Figure 6-3. Each MAC PDU consists of a MAC header and is optionally followed by one or more extended headers. The MAC PDU contains a variable-sized payload. Multiple MAC SDUs and/or SDU fragments from different unicast connections corresponding to the same MS can be multiplexed into a single MAC PDU. The multiplexed unicast connections are associated with the same security association.

The IEEE 802.16m specifies three types of MAC headers: (1) Advanced Generic MAC Header (AGMH) that is used for MAC PDUs containing either MAC management messages or user payload; (2) Short-Packet MAC Header (SPMH) that is utilized in conjunction with persistent or group allocations; and (3) Signaling MAC header. The MAC header formats are mutually exclusive and are not used simultaneously for the same connection. It must be noted that the Advanced Generic MAC header is different from that specified in the IEEE 802.16-2009 standard [1]. The Generic MAC Header (GMH) specified by the latter consists of 6 bytes that was used for all MAC PDUs containing MAC management messages or user data. While the overhead resulting from the MAC header was negligible for large user payloads, it could become a considerable overhead for the small payload applications such as VoIP and interactive gaming. It was further observed that the GMH specified by the IEEE 802.16-2009 standard contains many information fields that are not typically used. To mitigate this problem, the generic MAC header was restructured and redundant fields were removed to improve efficiency. The AGMH MAC header format is shown in Figure 6-4.

In Figure 6-4, different header fields are defined as follows:

- Flow Identifier (Flow ID or FID): this field indicates the service flow that is addressed.
- Extended Header (EH) Presence Indicator: when set to 1, this field indicates that an Extended Header is present following this AGMH.
- Length (11 bits): length of the MAC PDU in number of bytes including the AGMH and extended header (if present).

The short-packet MAC header is used for connections with persistent and group allocation (small payload applications). The structure of the short-packet MAC header is illustrated in Figure 6-5. The EH field denotes that the Extended Header when set to 1 indicates that an extended header will follow this header. The *Length* field indicates the total size of the MAC PDU in bytes including the short-packet MAC header and the extended header (if any). The *Sequence Number* denotes the MAC

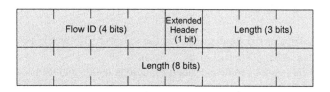

FIGURE 6-4

IEEE 802.16m advanced generic MAC header structure [2]

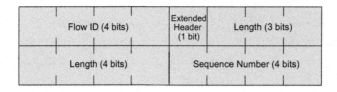

FIGURE 6-5

Structure of the short-packet MAC header [2]

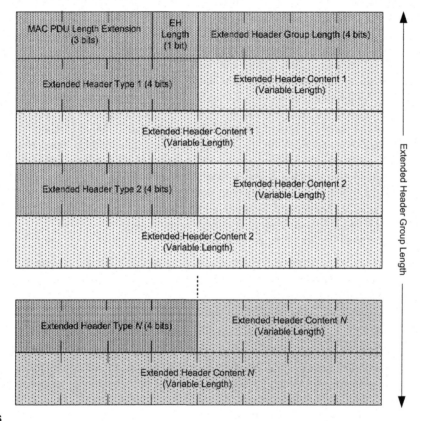

FIGURE 6-6

Structure of an extended header group [2]

PDU sequence number. The SPMH is identified by the specific FID that is provisioned statically or created dynamically using AAI_DSA-REQ/RSP messages.

The AGMH and SPMH header can be extended to include additional information fields depending on the use case and MAC PDU contents. The use of extended header is indicated by the EH flag in the MAC header. The extended header structure is illustrated in Figure 6-6. The description of the extended header fields is as follows:

- MAC PDU length extension (3 bits): These bits are appended as the most significant bits to the 11-bit *Length* field in AGMH to create a 14-bit *Length* field.
- Extended header Group Length (4/12 bits): the extended header group length field indicates the total length in bytes of the extended header group as shown in Figure 6-6.
- Extended Header Type (4 bits): this field indicates the type of extended header. There are some predefined types such as MAC SDU fragmentation or packing, MAC control, multiplexing, message acknowledgement, piggyback bandwidth request, MAC PDU length extension, ARQ feedback, and ARQ feedback polling extended headers.
- Type-specific header contents (variable): a type-specific and variable-sized field that will be elaborated on in the remaining parts of this section. An example is provided in Figure 6-7.

Table 6-1 provides a complete list of extended header types, their descriptions, use cases, and important parameters. Note that the extended header group is never encrypted.

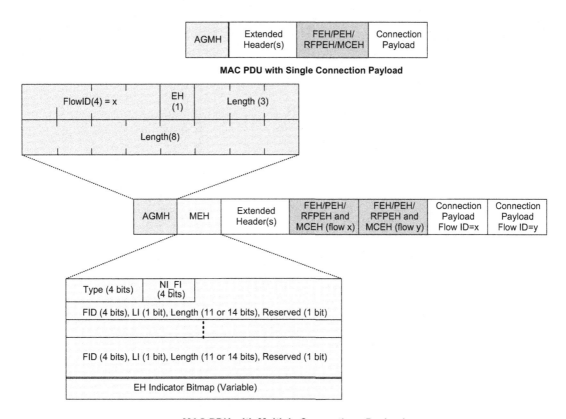

FIGURE 6-7

Example usage of FEH/PEH/RFPEH/MCEH and MEH in MAC PDU [2]

Table 6-1 Extended Header Types and Associated Parameters in IEEE 802.16m [2]

Extended Header Type	Description	Usage	Parameters
MAC SDU Fragmentation Extended Header (FEH)	The FEH is included in the MAC PDU with SPMH/AGMH headers, if the payload in the MAC PDU contains only a fragment of the MAC SDU. The FEH is also included in the MAC PDUs with AGMH header, if the payload in the MAC PDU contains a non-fragmented MAC SDU which requires a sequence number.	Fragmentation of large MAC SDUs	Fragmentation Control (FC) is a 2-bit field that describes the relationship of the MAC PDU and the MAC SDU in terms of fragmentation (see Table 6-2) Sequence Number (SN) is maintained per connection. For non ARQ connection, SN represents the MAC PDU *Payload Sequence Number* and the SN value increments by one (modulo 1024) for each MAC PDU. For ARQ connection, SN represents the ARQ block sequence number.
MAC SDU Packing Extended Header (PEH)	The PEH is included in the MAC PDUs with AGMH header, if the MAC SDUs, or MAC SDU fragments, or both are packed in the payload in the MAC PDU.	Packing of small MAC SDUs	FC and SN parameters are similar to the previous row. The 11-bit length field indicates the length of MAC SDU or MAC SDU fragment in the payload. If a MAC PDU payload consists of N MAC SDU/MAC SDU fragments, N − 1 length-fields are present to identify the length of MAC SDU/MAC SDU fragments 1 to N − 1.
MAC Control Extended Header (MCEH)	The MCEH is used on control connections. When message fragments associated with two different control messages are sent, the transmitter assigns different Control Connection Channel IDs (CCC IDs) to the MCEH of each MAC PDU.	Transmission and fragmentation of control messages	Encryption Control (EC) indicates whether the payload is encrypted (1 bit) Control Connection Channel ID (CCC ID) Channel ID to identify separate fragmentation/reassembly state machines (1 bit) Polling bit indicates whether acknowledgement is required. Fragmentation control bits (see Table 6-2) Sequence Number (6 bits) indicates the payload sequence number and is maintained per CCC ID. The SN value increments by one (modulo 64) sequentially.

Header			
Multiplexing Extended Header (MEH)	Multiplexing of different connections on the same MAC PDU	The MEH is used when data from multiple connections associated with the same security association is present in the payload contained in the MAC PDU. The AGMH header carries the FID corresponding to the payload of the first connection. The MEH carries the FIDs corresponding to remaining connections.	Number of Flow Information (N_FI) (4-bits) present in the MEH, If n connections are multiplexed, n-1 FIDs and lengths are present. (i.e., N_FI is set to n − 1). Flow Identifier (4 bits) The ith FID indicates the flow identifier of the $(i + 1)$th connection. Length Indicator (LI) indicates whether size of the payload is 11 bits or 14 bits Length of the connection payload (The ith length field indicates the length of the payload of the $(i + 1)$th connection. EH_indicator bitmap (variable size) A bit in the EH_indicator bitmap indicates the presence or absence of FEH/PEH/RFPEH/MCEH in the EH group corresponding to a connection payload. For a connection payload only one extended header from FEH, PEH, RFPEH, and MCEH are present. The FEH/PEH/RFPEH/MCEH for the connection payloads (if present) appear after the MEH in the same order as that of connection payloads.
MAC Control ACK Extended Header (MAEH)	Acknowledgement of MAC control message	This header may be used by BS and MS to indicate the reception of a specific MAC control message or BR without STID header. When a BS or an MS receives a MAC control message or MAC control message fragment with the Polling bit set to 1 in the MCEH, the BS or MS sends an MAEH or AAI_MSG-ACK message as an acknowledgement after receiving the complete message with the SN of the MAC control message PDU or the SN of the last received fragment, if fragmented.	MAEH sub-type (1 bit) indicates reception of a control message ACK_SN (6 bits) indicates that SN is retrieved from the MCEH of the MAC PDU with the Polling bit set to 1. Control Connection Channel ID (CCC ID) of the MAC control message that is received.
Piggyback Bandwidth Request Extended Header (PBREH)	Piggyback bandwidth request	PBREH is used when an MS uses a piggyback bandwidth request for one or more service flows.	Num_of_PBR (4 bits) is the number of piggyback bandwidth requests. FID (4 bits) is the flow identifier. Request type (1 bit) indicates aggregate or incremental nature of the request. BR size (19 bits) indicates the amount of requested bandwidth.

(Continued)

Table 6-1 Extended Header Types and Associated Parameters in IEEE 802.16m [2] *Continued*

Extended Header Type	Description	Usage	Parameters
MAC PDU length extended header (MLEH)	The MAC PDU length extended header is added to the MAC PDU when the MAC PDU length is larger than 2047 bytes. The MLEH would be the first extended header in the MAC PDU.	Extension of the size of MAC PDUs for large PDUs.	The length field in MLEH provides the three most significant bits of the extended length of the MAC PDU. The length field in AGMH is the 11 least significant bits of extended length of MAC PDU.
ARQ Feedback Extended Header (AFEH)	This header is used by an ARQ receiver to signal positive or negative acknowledgement.	ARQ feedback	ARQ feedback information elements.
Rearrangement Fragmentation and Packing Extended Header (RFPEH)	The RFPEH is used for ARQ enabled transport connections and is included in MAC PDU with AGMH, if the MAC PDU payload contains an ARQ sub-block which is constructed from SDU(s) or SDU fragment(s) or both are packed.	ARQ feedback for fragmented/packed MAC PDUs	FC (2 bits) fragmentation control bits (see Table 6-2). SN (10 bits) is maintained per connection and represents the ARQ block sequence number. The SN value increments by one (modulo 1024) for each MAC PDU. SSN (11 bits) is the SUB-SN of the first ARQ sub-block. LSI (1 bit) is the last ARQ sub-block indicator, indicating the last ARQ sub-block from the single ARQ block is not included in this MAC PDU or the last ARQ sub-block from the single ARQ block is included in this MAC PDU. PI (1 bit) is the Packing Indicator, indicating packing information is not present in the RFPEH or packing information is present in RFPEH. Length (11 bits) is the length of MAC SDU or MAC SDU fragment in a MAC PDU. If a MAC PDU consists of N MAC SDU/ MAC SDU fragments, N − 1 length fields are present to indicate the length of MAC SDU/ MAC SDU fragments 1 to N − 1.
ARQ Feedback Polling Extended Header (APEH)	This header is used by an ARQ transmitter to poll ARQ feedback.	ARQ feedback	Reference FID for ARQ feedback polling

Table 6-2 Fragmentation Control Information [2]

FC	Description	Usage
00	The first byte of data in the MAC PDU is the first byte of a MAC SDU. The last byte of data in the MAC PDU payload is the last byte of a MAC SDU.	One or multiple MAC SDUs are packed in a MAC PDU.
01	The first byte of data in the MAC PDU is the first byte of a MAC SDU. The last byte of data in the MAC PDU is not the last byte of a MAC SDU.	A MAC PDU with only the first fragment of a MAC SDU or a MAC PDU with one or more non-fragmented MAC SDUs followed by the first fragment of the subsequent MAC SDU.
10	The first byte of data in the MAC PDU is not the first byte of a MAC SDU. The last byte of data in the MAC PDU is the last byte of a MAC SDU.	A MAC PDU with only the last fragment of a MAC SDU or MAC PDU with the last fragment of a MAC SDU followed by one or more non-fragmented subsequent MAC SDUs.
11	The first byte of data in the MAC PDU is not the first byte of a MAC SDU. The last byte of data in the MAC PDU is not the last byte of a MAC SDU.	A MAC PDU with only the middle fragment of a MAC SDU or a MAC PDU with the last fragment of a MAC SDU followed by zero or more non-fragmented MAC SDUs followed by the first fragment of the subsequent MAC SDU.

The Fragmentation Control (FC) bits are defined in Table 6-2. An example usage of FEH/PEH/RFPEH/MCEH and MEH extended headers in a MAC PDU is illustrated in Figure 6-7.

6.3 MAC SIGNALING HEADERS

6.3.1 Legacy MAC Signaling Headers

MAC signaling headers are different types of headers that are used in MAC PDUs with no payload. They are sent as standalone or concatenated with other MAC PDUs. The IEEE 802.16-2009 standard specifies two types of signaling headers. This MAC header format is applicable to the uplink direction. The MAC header is not followed by any MAC PDU payload and CRC. The structure of Type I legacy MAC signaling headers are shown in Figure 6-8, with the description of their specific fields given in Table 6-3.

The type II MAC signaling header is also UL-specific. There is no payload following the MAC signaling header. The structure of MAC signaling header type II is illustrated in Figure 6-9. The feedback header is sent by a legacy MS either as a response to a Feedback Polling IE or as an unsolicited feedback. When sent as a response to a Feedback Polling IE, the legacy MS sends a feedback header using the assigned resource indicated in the Feedback Polling IE. When sent as unsolicited feedback, the legacy MS can either send the feedback header on currently allocated UL resource or request additional UL resource by sending an indication flag on the fast-feedback channel or the enhanced fast-feedback channel, or by sending a bandwidth request ranging code [1]. The feedback header with and without the CID field are illustrated in Figure 6-9. The feedback header with the CID field is used when the UL resource used to send the feedback header is requested through bandwidth request ranging; otherwise it is without the CID field. In Figure 6-10, the CID Inclusion

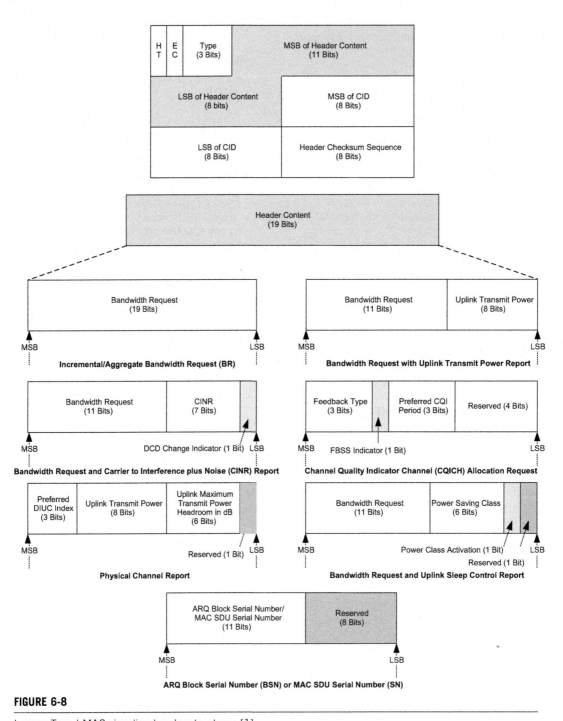

FIGURE 6-8

Legacy Type I MAC signaling header structures [1]

Table 6-3 Type Field Encodings for Legacy MAC Signaling Header Type I [1]		
Type	**MAC Header Type (HT)**	**Description of the Fields**
0b000	Incremental Bandwidth Request (BR)	The amount of requested resources by the MS in the number of bytes associated with the CID.
0b001	Aggregate Bandwidth Request	The amount of requested resources by the MS in the number of bytes associated with the CID.
0b010	Physical Channel Report	UL maximum transmit power headroom in dB for the burst that carries this header ranges from 0 to 63 in 1 dB steps. The reported value represents the difference between the maximum output power and the maximum power transmitted during the burst, index of the Downlink Interval Usage Code (DIUC) preferred by the MS, as well as UL transmit power level in dBm for this burst.
0b011	Bandwidth Request with Uplink Transmit Power Report	The amount of requested resources by the MS in the number of bytes associated with the CID and uplink power level in dBm for the burst that carries this header.
0b100	Bandwidth Request and Carrier to Interference plus Noise (CINR) Report	The amount of requested resources by the MS in the number of bytes associated with the CID. The CINR is a single value from −16.0 dB to 47.5 dB in units of 0.5 dB measured by the MS in the DL. The Downlink Channel Descriptor (DCD) change indicator is set to 1 if the DCD change count stored at the MS is not equal to that in the received DL-MAP message.
0b101	Bandwidth Request and Uplink Sleep Control Report	Power saving class identifier.
0b110	ARQ Block Serial Number (BSN) or MAC SDU Serial Number (SN)	The ARQ BSN or MAC SDU SN for the service flow addressed in this header.
0b111	Channel Quality Indicator Channel (CQICH) Allocation Request	The MS-preferred CQICH allocation period in units of frames.

Indication is set to 1 for a feedback header with the CID field, and is set to 0 for a feedback header without the CID field. The feedback type in Figure 6-9 is used as shown in Table 6-4.

The MIMO channel feedback header is used for MS to provide DL MIMO channel quality feedback to the BS. The MIMO channel feedback header can be used to provide a single or composite channel feedback.

6.3.2 IEEE 802.16m Signaling Headers

The MAC signaling header carries no payload and can be sent standalone or concatenated with other MAC PDUs in the downlink or uplink. If the MS uses an anonymously assigned uplink resource to send the signaling header, the MS is required to include the STID in the content field of the signaling header. The reserved value of Flow ID used for the MAC signaling header is 0b0001.

The structure of a MAC signaling header is shown in Figure 6-10. The *Flow ID* in the figure denotes the signaling header and the *Type* field identifies the type of signaling header. There are different types of signaling headers specified by IEEE 802.16m standard, as shown in Table 6-5. The use case, as well as important parameters of each signaling header type, are described in Table 6-5.

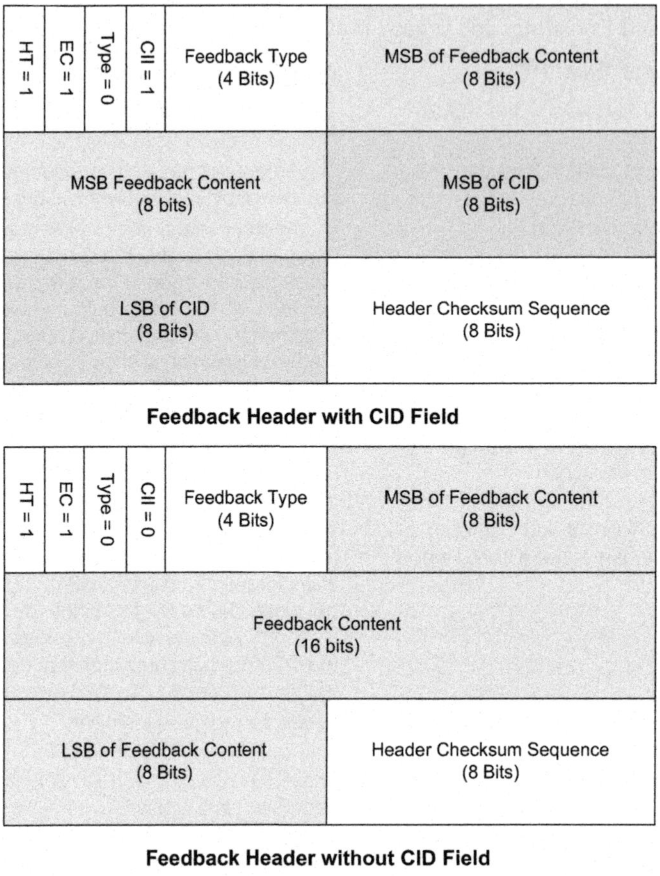

HT = 1	EC = 1	Type = 0	CII = 1	Feedback Type (4 Bits)	MSB of Feedback Content (8 Bits)
MSB Feedback Content (8 bits)					MSB of CID (8 Bits)
LSB of CID (8 Bits)					Header Checksum Sequence (8 Bits)

Feedback Header with CID Field

HT = 1	EC = 1	Type = 0	CII = 0	Feedback Type (4 Bits)	MSB of Feedback Content (8 Bits)
Feedback Content (16 bits)					
LSB of Feedback Content (8 Bits)					Header Checksum Sequence (8 Bits)

Feedback Header without CID Field

FIGURE 6-9

Legacy Type II feedback MAC signaling headers [1]

6.4 MOBILITY MANAGEMENT AND HANDOVER

6.4.1 Handover Mechanisms

The IEEE 802.16-2009 standard defines three basic mechanisms for handover [6,7,9,10]:

1. **Hard Handover (HHO):** a process that is based on Received Signal Strength Indicator (RSSI) measurements conducted on the preamble. The MS continuously measures the RSSI of the serving BS and reports the values periodically to the serving BS. When the MS notices that the RSSI is below a certain threshold, it starts scanning procedure by sending a MOB_SCN-REQ message to the serving BS. The serving BS replies by sending a MOB_SCN-RSP message that contains, among other parameters, the scanning time interval. During the scanning period, the MS measures the RSSI for all the neighbor base stations. The neighbor base stations are

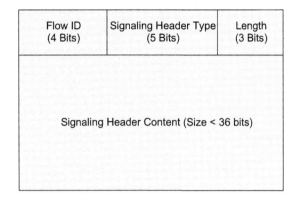

FIGURE 6-10

MAC signaling header structure [2]

Type	Feedback Content
Table 6-4 Feedback Type and Feedback Content in Legacy MAC Signaling Header Type II [1]	
0b0000	MIMO and CQI feedback
0b0001	DL average CINR of the serving or anchor BS
0b0010	MIMO coefficient feedback for up to four antennas
0b0011	Preferred DL channel DIUC feedback
0b0100	PHY channel feedback
0b0101	CQIs of up to three AMC bands
0b0111	The recommended number of frames for which the short-term precoding feedback can be used
0b1000	Multiple types of feedback
0b1001	Long-term precoding feedback
0b1010	Combined DL average CINR of all active BSs within the diversity set
0b1011	MIMO mode channel condition feedback
0b1100	CINR Feedback
0b1101	Closed-loop MIMO feedback type as follows:

0b00: Antenna grouping (Antenna grouping index)
0b01: Antenna selection (Antenna selection option index for 3 or 4 transmit antennas)
0b10: Codebook (Codebook index)
0b11: Indication of transition from closed-loop MIMO to open-loop MIMO

The 2 bit differential CQI denotes +2, +1, −1, −2 dB. The number of best bands is up to three.

A BS may send the Feedback_Polling_IE to request feedback header type (0b1101) without requesting feedback type (0b0110) to trigger the closed-loop MIMO operation with localized allocations. In this case, the MS over rides the feedback type (0b1101) with (0b0110) for the first report and may override the type (0b1101) whenever necessary. The BS may use the previously reported Precoder Matrix Indices (PMIs) and differential CQIs when the MS overrides the feedback type to (0b0110).

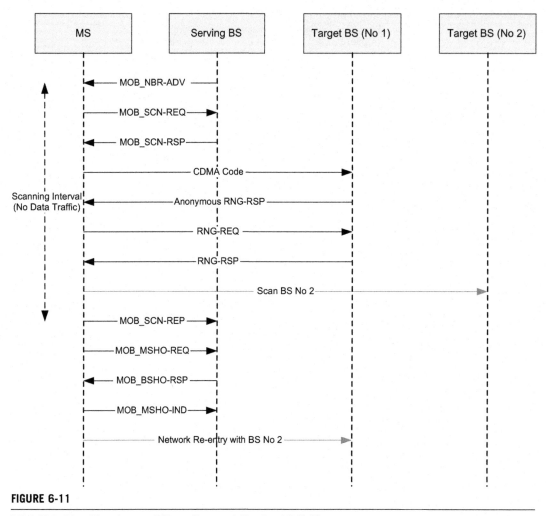

FIGURE 6-11

MS initiated hard handover call flow with association level 0 [1,9]

advertised periodically by the serving BS through the broadcast MOB_NBR-ADV message. During the scanning period, user data is not exchanged between the MS and the serving BS; instead the MS receives the preambles from each neighbor BS and calculates the RSSI. An optional procedure called association can be used to help the handover process. The association enables the MS to acquire and record ranging parameters and service availability information during scanning. Three association levels are defined in the standard as follows [1]:

a. In association level 0, the MS completes ranging during the scanning period with randomly chosen ranging Code Division Multiple Access (CDMA) code.[iii]

[iii]A unique code from an initial ranging code set that is assigned to the MS and used for dedicated ranging [1].

Table 6-5 IEEE 802.16m Signaling Header Types, their Usage, and Parameters [2]

MAC Signaling Header Type	Description	Usage	Parameters
Bandwidth Request with STID	This header is transmitted when the MS requests bandwidth through a non-dedicated uplink resource. In this case, the MS transmits this signaling header and includes its STID over the non-dedicated uplink resource.	Bandwidth Request	FID (4 bits) is the flow identifier whose value is set to 0b0001. Type (5 bits) indicates the MAC signaling header type. Length (3 bits) denotes the length of the signaling header in bytes. BR Size (19 bits) is the aggregated bandwidth request size in bytes. BR FID (4 bits) is the FID for which uplink bandwidth is requested. STID (12 bits) is the STID of the MS which requests the bandwidth.
Bandwidth Request without STID	This header is used when the MS requests bandwidth (incrementally or aggregately) through its dedicated uplink resource.	Bandwidth Request	FID (4 bits) which is set to 0b0001. Type (5 bits) denotes the MAC signaling header type. Length (3 bits) indicates the length of the signaling header in bytes. BR Type (1 bit) indicates whether the requested bandwidth is incremental or aggregate. BR Size (19 bits) is the size of the requested bandwidth in bytes. BR FID (4 bits) is the FID for which uplink bandwidth is requested.
Service Specific Scheduling Control	This header is sent either by the MS through a dedicated uplink resource or by the BS, which needs to change or acknowledge the scheduling parameters of its service flow. If the service-specific scheduling control header is sent by the MS, bandwidth request type is included to indicate the incremental or aggregated nature of the request. It may further indicate	Change or acknowledge the scheduling or QoS parameters	FID (4 bits) and is set to 0b0001. Type (5 bits) is the MAC signaling header type. Length (3 bits) indicates the length of the signaling header in bytes. BR Type (1 bit) indicates whether the requested bandwidth is incremental or aggregated. BR Size (19 bits) is the bandwidth request size in bytes. BR FID (4 bits) is the FID of service flow for which uplink bandwidth is requested.

(Continued)

Table 6-5 IEEE 802.16m Signaling Header Types, their Usage, and Parameters [2] *Continued*

MAC Signaling Header Type	Description	Usage	Parameters
	adaptation request when it is sent for an adaptive grant polling service flow. The BS may acknowledge or modify the QoS parameters for adaptive grant polling service flow by sending the service-specific scheduling control header in the downlink, which indicates adaptation response. If the Sleep Cycle Identifier (SCID) change indicator is set to 1, the BS sends an acknowledgement to confirm the change of sleep mode configuration. If the BS accepts the change of sleep mode configuration, the BS sends an MAEH extended header as acknowledgement. The BS may modify the terminal's sleep cycle setting. In this case, the BS replies with a sleep control header including a different SCID to the MS.		QoS parameter change indicator (1 bit) indicates whether there is a change in QoS parameters. SCID change indicator (1 bit). Grant Polling Interval (GPI) (6 bits) indicates new GPI (in number of frames) to use for future allocations. Minimum grant delay (6 bits) indicates minimum delay (in number of frames) of the requested grant.
Sleep Control	This header is used to carry control signaling related to sleep operation by a BS or an MS. In the case where this header is used to respond to a previously used sleep control header, the MS or the BS are required to set the response indication flag to 1 and to use the same sub-type as the previously received sleep control header. If the BS-initiated request and an MS-initiated request occur concurrently, the BS-initiated request takes precedence over the MS-initiated request.	Configuration of sleep mode operation parameters	FID (4 bits) is flow identifier and is set to 0b0001. Type (5 bits) denotes the MAC signaling header type. Length (3 bits) indicates the length of the signaling header in bytes. Sleep Control Header Sub-type (3 bits): 0b000 = Listening Window control; 0b001 = Resume Sleep Cycle Indication; 0b010 = Sleep cycle configuration switch; 0b011 = Sleep Cycle control; 0b100 = Multi-Carrier Listening Window control. Response Indication flag (1 bit). Listening Window End or Extension flag (1 bit). Last frame of Extended Listening

		Window (7 bits) indicates the frame that extended listening window is terminated. Scheduled Sleep Cycle Interruption flag (1 bit). Start Frame Offset for Scheduled. Sleep Cycle Interruption (7 bits) denotes the number of frames in the future from the frame containing this header in which the scheduled sleep cycle interruption will occur. SCID (4 bits) corresponding to the new sleep cycle setting. Start Frame Offset for new sleep cycle configuration (3 bits) are the least significant bits of the frame number in which sleep cycle setting is to be applied. Next Sleep Cycle Flag (NSCF) (2 bits). New Initial Sleep Cycle (6 bits) is used when the current sleep cycle is reset, if this value is included, the current sleep cycle is replaced by this value. Target Carrier Index Bitmap (4 bits) If the nth bit is set, it indicates that downlink data transmission on the secondary carrier whose logical carrier index is equal to (n + 1).
MS Battery Level Report	Terminal's battery level reporting	This header is used to inform the BS of the terminal's battery level. The battery level is defined as the amount of energy remaining in the battery as a percentage of full energy level.	FID (4 bits) is flow identifier and is set to 0b0001. Type (5 bits) denotes the MAC signaling header type. Length (3 bits) indicates the length of the signaling header in bytes. MS Battery Status (1 bit) indicates whether the MS is plugged into a power source. Battery Level Indication (1 bit) indicates whether the detailed battery level report is included. MS Battery Level (3 bits): 0b000: 75% < Battery Level <100%; 0b001: 50% < Battery Level <75%; 0b010: 25% < Battery Level <5%; 0b011: 5% < Battery Level <25%; 0b100: Battery level <5%. FID (4 bits) is flow identifier and is set to 0b0001.

(Continued)

Table 6-5 IEEE 802.16m Signaling Header Types, their Usage, and Parameters [2] *Continued*

MAC Signaling Header Type	Description	Usage	Parameters
Uplink Power Status Report	This header is used to inform the BS of the MS uplink power control status.	Uplink power control status reports	Type (5 bits) denotes the MAC signaling header type. Length (3 bits) indicates the length of the signaling header in bytes. Power Control Parameter Update Indicator (1 bit) indicates whether the MS has updated its power control parameters based on AAI_SCD. Change Configuration Change (4 bits). txPowerPsdBase (8 bits): The txPowerPsdBase (PSD(base)) parameter is encoded using 8 bits in 0.5 dBm steps ranging from −74 dBm to 53.5 dBm. txSirDownlink (10 bits): The txSirDownlink (SIRDL) parameter is encoded using 10 bits in 1/16 dB steps ranging from −12 dB to 51.9375 dB.
Correlation Matrix Feedback	This header may be used by the MS as a response to a Feedback Polling A-MAP IE requesting the quantized transmit correlation matrix when the BS has 2 or 4 transmit antennas.	Correlation matrix based precoding	FID (4 bits) is flow identifier and is set to 0b0001. Type (5 bits) denotes the MAC signaling header type. Length (3 bits) indicates the length of the signaling header in bytes. Entries of the quantized transmit correlation matrix.
MIMO Feedback	This header is used by the MS to send only the wideband CQI, spatial rate, or PMI information for any combinations of MIMO Feedback Modes 0, 4, 7.	MIMO feedback	FID (4 bits) is flow identifier and is set to 0b0001. Type (5 bits) denotes the MAC signaling header type. Length (3 bits) indicates the length of the signaling header in bytes. MFM bitmap (3 bits). Bitmap to indicate the MIMO Feedback Modes for which the MS is sending feedback. Wideband CQI (4 bits), Wideband STC rate (3 bits), or Wideband PMI (6 bits).
Reserved	—	—	—

b. In association level 1, the ranging CDMA code and the rendezvous time,[iv] i.e., in which frame the ranging is performed, is negotiated via BS-to-BS signaling before the scanning period. The chosen CDMA code and the rendezvous time are reported to the MS in the MOB_SCN-RSP message.

c. The association level 2 is similar to level 1, but the ranging parameters (e.g., power level) are reported to the serving BS, which collects the information from all neighbor BSs and then reports the information to the MS in a MOB_ASC-REP message.

The hard handover is triggered if the RSSI of the serving BS becomes considerably low and there is another BS with a higher RSSI value in the neighborhood. In order to avoid a ping-pong effect, the RSSI of the target BS has to exceed a certain hysteresis margin, as well. If the MS is the initiator of handover, it will send a MOB_MSHO-REQ message to the serving BS, which responds with a MOB_BSHO-RSP message. On the other hand, if BS is the initiator of the handover, it will send a MOB_BSHO-REQ to the MS. The actual handover is carried out by sending a MOB_MSHO-IND message to the serving BS. The MS context is terminated at the BS after a *Resource_Retain_Time* interval during which the MS may cancel the handover. The MS continues with the new serving BS by completing the network re-entry process. After the MS is in the Connected State, the data delivery is resumed. Figure 6-11 illustrates an example MS initiated hard handover with association level 0 [9].

2. Fast Base Station Switching (FBSS): the MS and BS both maintain a list of the base stations (i.e., *Diversity Set*) that are involved in FBSS operation. An Anchor BS, with which the MS communicates, is defined in the set. However, the MS may add or drop a BS from the list using MOB_MSHO-REQ, to which the Anchor BS responds with MOB_BSHO-RSP. The Anchor BS may be changed by using handover messages or by using fast anchor selection feedback. Figure 6-12 illustrates the exchange of management messages during the FBSS operation. If association is used, the ranging procedure is performed before adding the BS to the *Diversity Set*. The measurements are based on Carrier to Interference-plus-Noise Ratio (CINR) calculations conducted on the pilot subcarriers in DL and UL subframes.

Normal operation when MS is registered with a single BS is a particular case of FBSS with *Diversity Set* consisting of a single BS, which in this case is the Anchor BS. When operating in FBSS, the MS only communicates with the Anchor BS for UL and DL messages including management and traffic connections. Transition from one Anchor BS to another (i.e., switching) is performed without invocation of handover procedure. The BS supporting FBSS broadcasts the neighbor advertisement message that includes thresholds that may be used by the FBSS capable MS to determine whether the MOB_MSHO-REQ message should be sent to request changing the *Diversity Set*. When mean CINR of an active BS in the current *Diversity Set* is less than a certain threshold, the MS may send a MOB_MSHO-REQ message to request dropping this BS from the *Diversity Set*. When mean CINR of a neighbor BS is higher than a certain threshold, the MS may send a MOB_MSHO-REQ message to request adding this neighbor BS to the diversity set. In each case, the Anchor BS responds with a MOB_BSHO-RSP message with an updated *Diversity Set* [1].

[iv]This is the offset, measured in units of frame duration, when the BS is expected to provide a non-contention-based ranging opportunity for the MS. The offset is calculated from the frame where the RNG-RSP message is transmitted. The BS is expected to provide the non-contention-based ranging opportunity at the frame specified by the rendezvous time parameter [1].

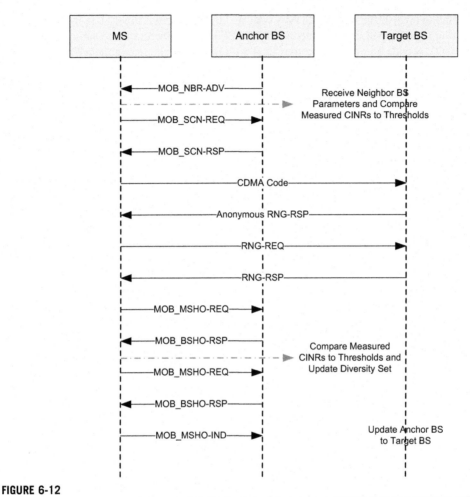

FIGURE 6-12

Fast base station switching call flow with association level 0 [1,9]

After the MS completes the initial network entry/re-entry or handover procedure, the BS automatically becomes an Anchor BS. Furthermore, the *Diversity Set* is initialized and the TEMP_BS-ID parameter of the Anchor BS is set to zero. However, the TEMP_BS-ID parameter and *Diversity Set* are maintained when Anchor BS switching occurs. As shown in Figure 6-12, the process of Anchor BS update may begin with a MOB_MSHO-REQ message from MS or a MOB_BSHO-REQ from the Anchor BS. The acknowledgement of MOB_MSHO-REQ with MOB_BSHO-RSP is required. After MS transmits MOB_MSHO-REQ, the MS does not transmit the MOB_MSHO-REQ message prior to expiration of internal timer *MS_handover_retransmission_timer*. The MS deactivates this timer on transmission of the MOB_HO-IND message or on MS receipt of a MOB_BSHO-RSP message. The process of Anchor BS update may also begin with anchor switching indication via fast-feedback channel.

If an MS that transmitted the MOB_MSHO-REQ message detects an incoming MOB_BSHO-REQ message, it will ignore the MOB_BSHO-REQ message. A BS that transmitted a MOB_BSHO-REQ message and detects an incoming MOB_MSHO-REQ or MOB_HO-IND message from the same MS ignores its own previous request. There are several conditions that are required to be met in order to enable fast BS switching between MS and a group of base stations. The base stations involved in FBSS, are required to be synchronized based on a common timing source, the frames sent by the base stations from diversity set arrive at the MS within the OFDM cyclic prefix interval, the base stations involved in FBSS must have synchronized frames and operate at same frequency channel, and the base stations involved in FBSS are also required to share or transfer MS context. The latter includes all information that MS and BS normally exchange during network entry, such as authentication state, such that an MS authenticated and registered with one of the base stations in the *Diversity Set* is automatically authenticated and registered with other base stations in the same *Diversity Set*. The context also includes a set of service flows and corresponding mapping to connections associated with the MS, current authentication, and encryption keys associated with the connections.

3. **Macro Diversity Handover** (**MDHO**): an MDHO process is initiated with a decision for an MS to transmit to and receive from multiple base stations at the same time. An MDHO can start with either a MOB_MSHO-REQ or a MOB_BSHO-REQ message. For an MS and a BS that support MDHO, the MS and the BS maintain a list of BSs that are involved in MDHO with the MS. The list is called the *Diversity Set*. Among the base stations in the *Diversity Set*, an Anchor BS is defined. The normal operation where the MS is registered with a single BS is a particular case of MDHO with *Diversity Set* consisting of a single BS; i.e., the Anchor BS. When operating in MDHO, the MS communicates with all base stations in the *Diversity Set* for UL and DL unicast messages and traffic. There are two methods for the MS to monitor DL control information and broadcast messages. In the first method, the MS monitors only the Anchor BS for DL control information and broadcast messages. In this case, the DL/UL control channels of the Anchor BS may contain burst allocation information for the other active base stations. In the second method, the MS monitors all the base stations in the *Diversity Set* for DL control information and broadcast messages. In this method, the DL/UL control channels of any active BS may contain burst allocation information for the other active base stations. The method to be used by the MS is defined during the REG-REQ and REG-RSP messaging. An MDHO process begins with a decision for an MS to transmit/receive unicast messages and traffic from multiple base stations at the same time interval. For DL MDHO, two or more base stations provide synchronized transmission of MS DL data so that macro diversity combining can be performed by the MS. For UL MDHO, the transmission from an MS is received by multiple BSs so that selection diversity on the information received by multiple base stations can be performed.

The BS supporting MDHO broadcasts the system information that includes the *H_Add* and *H_Delete* thresholds. These thresholds are used by the MDHO capable MS to determine if MOB_MSHO-REQ should be sent. When long-term CINR of an active BS in the current *Diversity Set* is lower than the *H_Delete* threshold, the MS sends a MOB_MSHO-REQ message to drop this active BS from the *Diversity Set*. When long-term CINR of a neighbor BS is greater than the *H_Add* threshold, the MS sends a MOB_MSHO-REQ to add this neighbor BS to the *Diversity Set*. The decision to update the diversity set begins with a notification by the MS through the MOB_MSHO-REQ message or by the BS through the MOB_BSHO-REQ management message.

The Anchor BS update may begin with a MOB_MSHO-REQ message from MS or a MOB_BSHO-REQ message from the Anchor BS. An acknowledgement with MOB_BSHO-RSP for this notification is required. After MS transmits MOB_MSHO-REQ, the MS will not transmit any MOB_MSHO-REQ prior to expiration of an internal timer, *MS_handover_retransmission_timer*. The MS deactivates the timer *MS_handover_retransmission_timer* on transmission of a MOB_HO-IND message by the MS or receipt of a MOB_BSHO-RSP message. The Anchor BS update may also begin with anchor switching indication via fast-feedback channel. If an MS that transmitted a MOB_MSHO-REQ message detects an incoming MOB_BSHO-REQ message, it ignores the MOB_BSHO-REQ message. Similarly, a BS that transmitted a MOB_BSHO-REQ message and detects an incoming MOB_MSHO-REQ or MOB_HO-IND message from the same MS discards its own previous request.

The base stations involved in MDHO with an MS must use the same set of CIDs for the connections that are established with the MS. The BS may assign a new set of CIDs to the MS during *Diversity Set* update through MOB_BSHO-REQ and MOB_BSHO-RSP messages. There are several conditions that are required to be met in order to enable macro diversity handover between MS and a group of active base stations. The base stations involved in MDHO are required to:

a. Be synchronized based on a common timing source (typically a GPS source);
b. Have synchronized frame structures. That is the radio frames transmitted by the base stations involved in MDHO at any time arrive at the MS within the OFDM cyclic prefix (for CP = 1/8, the arrival variation is less than 11.42 μs);
c. Have the same RF carrier frequency;
d. Use the same set of CIDs for the connections that are established with the MS;
e. Send the same MAC/PHY PDUs to the MS;
f. Share or transfer MS context that includes all information the MS and BS normally exchange during network entry, particularly authentication state, such that an MS authenticated/registered with one of base stations from the *Diversity Set* is automatically authenticated/registered with other base stations from the same *Diversity Set*. The context also includes a set of service flows and corresponding mapping to connections associated with the MS, current authentication, and encryption keys associated with the connections.

4. Seamless Handover: in addition to optimized HHO, MS and BS may perform seamless HO, which is a variant of HHO, to reduce handover latency and message overhead. The capability of seamless handover is negotiated via REG-REQ/REG-RSP management messages. An authorization policy, except *No Authorization*, is negotiated between the MS and the serving BS. The seamless handover also requires support for counter-based Traffic Encryption Key (TEK)[v] generation for handover. The seamless handover is only enabled if the MS, the serving BS, and the target base stations support seamless handover. A BS supporting seamless handover must include the connection identifier descriptor in the system information. In seamless handover, a target BS calculates primary management CID, secondary management CID, and transport CIDs for an MS by using the descriptor. During seamless HO, a serving BS must include the pre-allocated basic CID in MOB_BSHO-REQ/RSP messages for the MS. When a BS allocates a basic CID to an MS in advance, the primary management CID is allocated autonomously without explicit assignment

---

[v]Traffic Encryption Key (TEK) is a symmetric key that is used to encrypt messages. TEKs are typically changed frequently, in some systems daily and in others for every message.

in the message. Once CIDs are pre-assigned, the BS indicates whether it will perform seamless handover by including the seamless handover mode flag in MOB-BSHO-REQ/RSP management messages. When the MS receives a MOB_BSHO-REQ or MOB_BSHO-RSP message with the seamless handover mode flag set, the MS can perform seamless handover by transmitting an HO-IND message including the BS-ID of a BS among the recommended base stations that indicated support for seamless handover, i.e., a BS for which the seamless handover mode flag was set in a MOB_BSHO-REQ/RSP message. If the MS transmits a HO-IND message including the BS-ID of any BS other than the base stations which have already indicated support for seamless handover, then seamless handover is not applied to that BS. The MOB_BSHO-REQ or MOB_BSHO-RSP message may contain a specific action time. If a value for the action time is specified, pre-assigned CIDs are valid at the target BS after the expiration of the action time. A value of zero indicates that the pre-assigned CIDs are already valid and MS may initiate seamless handover at any time. During seamless handover, the target BS may allocate downlink and uplink resources for the MS before the RNG-REQ/RSP messaging, as shown in Figure 6-13.

During seamless HO, the MS is required to initiate the RNG-REQ/RSP message exchange by sending a RNG-REQ message before the deadline specified by the *Seamless HO Ranging Initiation Deadline* parameter included in the MOB_BSHO-REQ/RSP message during handover preparation. The time is measured from the instant where the MOB_BSHO-REQ/RSP message is transmitted. If the target BS does not receive a RNG-REQ message from the MS within the deadline defined by the *Seamless HO Ranging Initiation Deadline* parameter, the target BS assumes that the seamless handover has failed and stops allocating bandwidth to the MS. The MS would assume the seamless handover has failed if it does not transmit the RNG-REQ message before the deadline. If the MS transmits RNG-REQ prior to the deadline, it may still consider the handover as failed, if it does not receive a RNG-RSP within a certain period after the last transmission or re-transmission of RNG-REQ that was performed within the permissible time. When the MS considers the seamless handover has failed, it invalidates the pre-assigned CIDs. In all cases, even when the RNG-REQ/RSP message exchange has been initiated before the deadline, the seamless handover can fail if the RNG-REQ/RSP procedure does not succeed. When data packets are exchanged before the RNG-REQ/RSP procedures are completed, the receiver (MS or BS) should store the received data packets and must not forward them to the upper layers until the sender is authenticated. If the data packets belong to a service flow associated with a security architecture that supports data authentication, as indicated by the data authentication algorithm identifier in the cryptographic suite, the receiver can authenticate the sender by verifying that the cipher-text authentication code included in each data packet was produced with the TEK associated with this security architecture. If the data packets belong to a service flow associated with a security architecture that does not support data authentication the receiver can authenticate the sender when the RNG-REQ/RSP transaction completes successfully. In all cases, if the sender is authenticated, the decrypted data packets are forwarded to the upper layer in the recipient, and if the sender is not authenticated the data packets are discarded.

The RNG-REQ/RSP messaging for seamless handover is shown in Figure 6-13. The MS initiates the RNG-REQ/RNG-RSP message exchange by transmitting a RNG-REQ message to the target BS before the deadline specified by the *Seamless HO Ranging Initiation Deadline* parameter included in the MOB_BSHO-REQ/RSP message during handover preparation. The RNG-REQ message must include basic CID, CMAC_KEY_COUNT, and a valid HMAC[vi]/CMAC[vii] tuple (see Chapter 8 for security aspects). When the BS receives the RNG-REQ message, the BS responds to the RNG-REQ message by transmitting a RNG-RSP message with a valid HMAC/CMAC tuple. The RNG-RSP message includes basic CID but is not required to include any CID update TLV or SA-TEK-Update TLV for unicast connections. When MS receives the RNG-RSP message from the target BS, the target BS becomes the serving BS of the MS, and the MS transmits a BR header with a zero bandwidth request. When BS receives the BR header, the seamless handover procedure completes successfully.

6.4.2 Handover Process

The handover process consists of the following stages (as shown in Figure 6-14):

1. Cell Reselection: the MS may use neighbor BS information acquired from a decoded MOB_NBR-ADV message or may request to schedule scanning intervals or sleep intervals to scan neighbor

---

[vi]Hashed Message Authentication Code (HMAC) is a method for calculating a message authentication code involving a cryptographic hash function in combination with a secret key. As with any message authentication code, this scheme may be used to simultaneously verify both the data integrity and the authenticity of a message. Any iterative cryptographic hash function may be used in the calculation of an HMAC. The cryptographic strength of the HMAC depends on the cryptographic strength of the underlying hash function, the size of its hash output length in bits, and on the size and quality of the cryptographic key. An iterative hash function breaks up a message into blocks of a fixed size and iterates over them with a compression function [20]. The definition of HMAC requires a cryptographic hash function, which we denote by H, and a secret key K. Let's assume H denotes a cryptographic hash function where data is hashed by iterating a basic compression function on blocks of data and B denotes the byte length of such blocks, e.g., $B = 64$, and L denotes the byte length of hash outputs, e.g., $L = 16$ or 20. The authentication key K can be of any length up to B, the block length of the hash function. Applications that use keys longer than B bytes will first hash the key using H, and then use the resulting L byte string as the actual key to HMAC. Let's define two fixed and different binary strings S_{inner} and S_{outer} as follows: $S_{inner} = 0 \times 36$ repeated B times and $S_{outer} = 0 \times C$ repeated B times. To compute HMAC over the input data denoted as *information*, one should perform $H(K \oplus S_{outer}, H (K \oplus S_{inner},$ information$))$ where $\oplus$ denotes XOR operation [19].

[vii]Cipher-based Message Authentication Code (CMAC) is a block cipher-based message authentication code algorithm.It is used to validate the authenticity and the integrity of binary data. To generate an l-bit CMAC tag μ of message m using a β-bit block cipher ε and secret key k, one first generates two β-bit sub-keys k_1 and k_2 using the following algorithm. This is equivalent to multiplication by x and x^2 in a finite Galois field GF(2^β). Let $\ll$ denote a logical left-shift operator, then [21]:

1. Calculate a temporary value $k_0 = \varepsilon_k(0)$
2. If MSB(k_0) = 0 then $k_1 = k_0 \ll 1$ else $k_1 = (k_0 \ll 1) \oplus \alpha$, where α is a constant that depends only on β. The constant α is the non-leading coefficients of the lexicographically first irreducible degree-β binary polynomial with the minimal number of ones. A polynomial is called irreducible, if it cannot be factored into nontrivial polynomials over the same field. For example, in the field of rational polynomials Q[x] (i.e., polynomials $f(x)$ with rational coefficients), $f(x)$ is called irreducible, if there do not exist two non-constant polynomials $g(x)$ and $h(x)$ in x with rational coefficients such that $f(x) = g(x) h(x)$.
3. If MSB(k_1) = 0 then $k_2 = k_1 \ll 1$ else $k_2 = (k_1 \ll 1) \oplus \alpha$

As an example, assume $\beta = 4$, $\alpha = $ 0b0011, and $k_0 = \varepsilon_k(0) = $ 0b0101. Then $k_1 = $ 0b1010 and $k_2 = $ 0b0100 $\oplus$ 0b0011 $=$ 0b0111. The CMAC tag generation procedure is as follows:

1. Divide message into β-bit blocks $m = m_1 \parallel \ldots \parallel m_{n-1} \parallel m_n'$ where $m_1, \ldots, m_{n-1}$ are complete blocks
2. If m_n' is a complete block then $m_n = k_1 \oplus m_n'$ else $m_n = k_2 \oplus (m_n' \parallel$ 0b10...0)
3. Let $c_0 = $ 0b00...0
4. For $i = 1, \ldots, n$, calculate $c_i = \varepsilon_k(c_{i-1} \oplus m_i)$
5. Output $\mu = $ MSB$_\ell(c_n)$

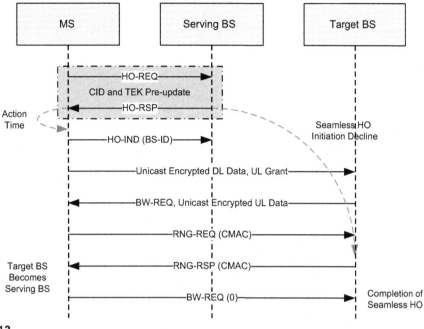

FIGURE 6-13

Seamless handover call flow [1]

base stations for the purpose of handover to a potential target BS. The cell reselection process does not need to occur in conjunction with any specific contemplated handover decision.

2. Handover Decision and Initiation: the handover process begins with a decision by an MS (or a BS) to handover from a serving BS to a target BS. The decision may originate either at the MS or at the serving BS. The handover decision is completed with a notification of MS intent to handover through a MOB_MSHO-REQ or MOB_BSHO-REQ message.

3. Downlink Synchronization: the MS synchronizes with the DL transmissions of the target BS and obtains DL and UL transmission parameters (system configuration information). If the MS had previously received a MOB_NBR-ADV message including target BS-ID, physical frequency, and other essential system parameters, this process may be simplified. If the target BS has previously received handover notification from the serving BS over the backhaul, then the target BS may allocate one or more non-contention-based initial ranging opportunities for the MS.

4. Ranging: the MS and target BS must perform initial ranging or handover ranging. If the MS RNG-REQ message includes the serving BS-ID, then the target BS may request the serving BS to provide the MS information over the backhaul. The normal network re-entry process may be shortened by target BS possession of MS information (see Chapter 4 for more information on fast network re-entry). This type of handover is referred to as optimized handover. The MS context may include static and dynamic information where static context consists of all configuration parameters that were acquired during initial network entry or via exchange of information between the BS and MS (e.g., the SBC-RSP and REG-RSP parameters, service flow encodings from DSA-REQ/RSP

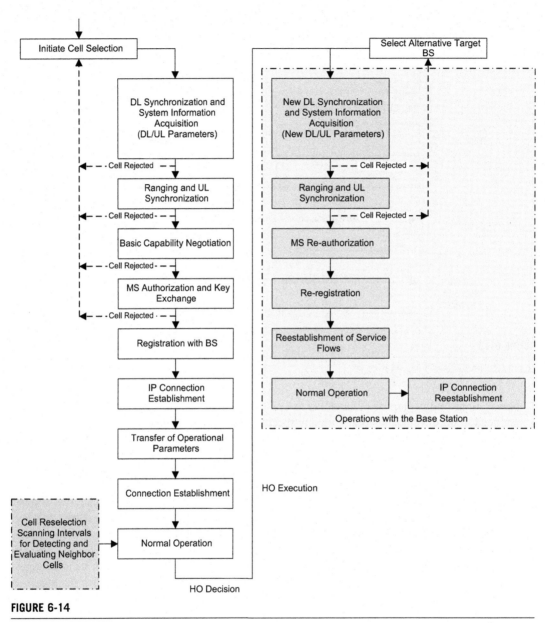

FIGURE 6-14

The handover process [1]

messages, etc.) and dynamic information consists of all counters, timers, state machine status, and data buffer contents (e.g., ARQ window). The transaction states, which may impact configuration parameters, are considered dynamic context until complete, which by then is considered static context [1]. The security information is always considered static context. Depending on the amount of information, the target BS may decide to skip one or several of the following network entry steps: (1) basic capability negotiation; (2) PKM[viii] authentication; (3) TEK establishment; (4) REG-REQ message; and (5) unsolicited REG-RSP message.

5. Termination of MS Context: the final step in handover is termination of MS context that is defined as serving BS termination of context of all connections belonging to the MS and the context associated with them (i.e., information in queues, ARQ state machine, counters, timers, header suppression information, etc., are discarded).

6. Handover Cancellation: an MS may cancel handover via a MOB_HO-IND message at any time prior to expiration of *Resource_Retain_Time* after transmission of MOB_MSHO-REQ (in case of MS-initiated handover) or MOB_BSHO-REQ (in case of BS-initiated handover).

6.4.3 IEEE 802.16m Handover Scenarios

The following handover scenarios are supported in IEEE 802.16m:

1. Handover from a legacy BS to a legacy BS;
2. Handover from a new BS to a legacy BS;
3. Handover from a legacy BS to a new BS;
4. Handover from a new BS to a new BS.

The IEEE 802.16m radio access network and mobile station use legacy handover procedures for the first scenario. A new BS periodically broadcasts the system information of the neighboring new and legacy base stations using a neighbor advertisement message. The BS generates the neighbor advertisement message based on the cell types of neighbor cells, in order to achieve overhead reduction and prioritize scanning at the MS. A broadcast neighbor advertisement message does not include information about neighbor private femto-cells. The neighbor advertisement message may include parameters required for cell selection, such as cell load and cell type.

The scanning procedure provides an opportunity for the MS to perform measurement of the neighboring cells for handover decision. In addition, the IEEE 802.16m MS may use any intervals that are not allocated by the serving BS to perform self-directed scanning. The MS may perform scanning procedure without interrupting its communication with the serving BS, if the MS supports such capability. The MS selects the candidate base stations to be scanned from information obtained from the serving BS. The BS or MS may prioritize the neighbor base stations to be scanned based on various metrics such as cell type, loading, received signal strength, or geographical location. As part of the scanning procedure, the MS conducts measurements on selected candidate base stations and reports the measurement results to the serving BS. The measurements may be used by the MS or the network to determine the best target BS for the MS. The serving BS defines triggering conditions under which the MS can send the scanning report.

In IEEE 802.16m, the handover process may be initiated by either the MS or the BS. In the case of MS initiated handover, the MS sends a handover initiation message to the serving BS (see Figure

---

[viii]Privacy Key Management (PKM) is a client/server model between the base station and mobile station that is used to secure distribution of keying material.

6-15). The serving BS responds to the handover initiation message by sending a handover command to the MS. In the case of BS-initiated handover, the serving BS sends a handover command to the MS. In both cases, i.e., MS or BS initiated handover; the handover command includes one or more target base stations. If the handover command includes only one target BS, the MS will execute the handover as directed by the BS. An MS may send a handover indication message to the serving BS before the expiration of disconnect timer. The serving BS stops sending DL data and providing UL allocations to the MS after expiration of the disconnect timer or after receiving a HO-IND message. If the handover command includes more than one target base station, the MS selects one of those target base stations and informs the serving BS of its choice by sending a handover indication message before the expiration of disconnect timer. The network re-entry procedures with the target BS may be optimized by providing the target BS with the MS context. The MS may also maintain communication with the serving BS while performing network re-entry process with the target BS, as shown in Figure 6-15.

The serving BS defines the conditions based on which the MS may decide when a target BS among those that are included in the handover command is unreachable. If all target base stations that were

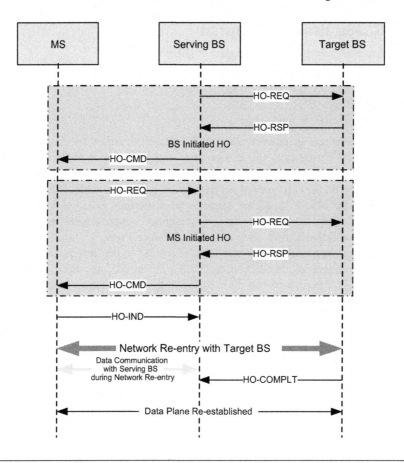

FIGURE 6-15

General handover call flow [2,5]

included in the handover command are unreachable for the MS, the MS may find a new target BS and inform the serving BS by sending a handover indication message before expiration of disconnect timer, and the MS may proceed with network re-entry with the new target BS. The MS also indicates the identity of previous serving BS to the new target BS during network re-entry. The handover procedures in IEEE 802.16m can be divided into four steps:

Handover Initiation

Handover procedure may be initiated either by the MS or the BS. When handover is initiated by the MS, the triggers and conditions are defined by the serving BS and a handover initiation message is sent by the MS to begin the handover procedure. In the case of BS-initiated handover, the handover preparation is performed before handover initiation, and the AAI_HO-CMD message is sent by the BS to initiate the handover procedure (see Figure 6-15). During handover initiation, the serving BS indicates whether the MS can maintain data communication with the serving BS while performing network re-entry with the target BS by setting the *HO_Re-entry_Mode* field in AAI_HO-CMD to 1; otherwise, the *HO_Re-entry_Mode* filed is set to 0.

Handover Preparation

During the handover preparation stage, the serving BS communicates with potential target base stations. The target BS may obtain MS context from the serving BS via core network for handover optimization. If ranging with the target BS is not performed prior to or during handover preparation, dedicated ranging resources such as ranging code or ranging opportunity with the target BS may be reserved for the MS to facilitate non-contention-based handover ranging. The information related to MS identity and security context may be updated during handover preparation. Any inconsistent system information between the MS and the target BS, if detected, may be provided to the MS by the serving BS during handover preparation.

If only one target BS is included in the handover command, the handover preparation phase is completed when the serving BS informs the MS of its handover decision via a handover command. If multiple target base stations are included in the handover command, the handover preparation phase is completed when the MS informs the serving BS about its target selection through a handover indication message. The handover command may include dedicated ranging resource allocation and resource pre-allocations for the MS at each target BS for optimized network re-entry. The handover command includes an action time for the MS to start network re-entry with each target BS, and an indication whether the MS should continue communication with the serving BS during network re-entry. The handover command further contains a disconnect timer, which indicates when the serving BS will stop sending downlink data and other regularly scheduled unsolicited uplink allocations for the MS. In the case where the MS can maintain communication with the serving BS during network re-entry, the transmission parameters for concurrent communication with both serving and target base stations are determined by the serving BS, depending on the MS capability that is negotiated between the serving and target base stations. The handover command indicates if the MS static and/or dynamic context is available at the target BS.

If the *HO_Re-entry_Mode* field of the handover command is set to 1, the serving BS will negotiate with the target BS the Entry Before Break (EBB) handover parameters (see Figure 6-16). In the case of single-carrier handover, the EBB HO parameters include *HO_Re-entry_Interleaving_Interval* and *HO_Re-entry_Iteration* for the MS to communicate with the serving BS during network re-entry. In the

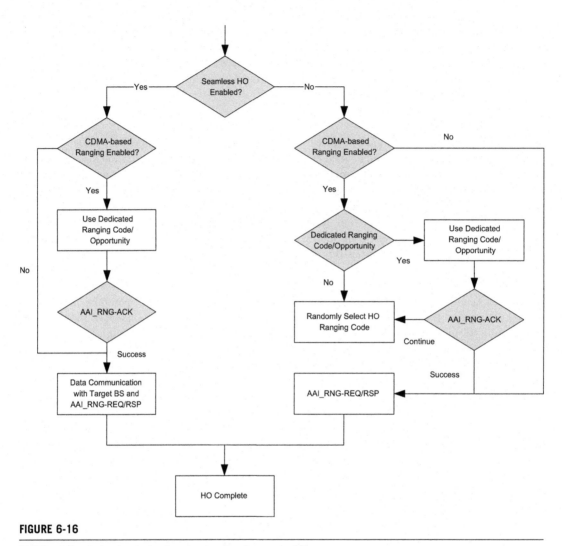

FIGURE 6-16

Mobile station behavior during handover network re-entry [2]

multicarrier handover case, the EBB HO parameters include the carrier information in the target BS for the MS performing network re-entry while continuing to communicate with the serving BS.

Handover Execution

The MS performs network re-entry with the target BS within the action time specified in the handover command. If communication is not supported between the MS and the serving BS during network re-entry, the serving BS will not allocate resources to the MS for transmission during handover interruption time. If the MS is instructed by the serving BS via a handover command, it performs network re-entry with the target BS during action time while communicating with the serving BS. However, the MS stops communication with the serving BS after network re-entry with the target BS is completed. It must be noted that the MS cannot exchange data with the target BS prior to completion of network re-entry. The network re-entry signaling with the target BS and data communication with the serving BS are time-multiplexed and the multiplexing intervals are negotiated between the MS and the serving BS.

Handover Cancellation

After handover is initiated, the handover may be canceled by the MS at any time during the handover procedure. After the handover cancellation request is processed, the MS and the serving BS can resume their normal operation. The network can advertise handover cancellation trigger conditions. When one or more of those conditions are met, the MS cancels the handover. When handover execution is complete, the MS is ready to perform network re-entry with the target BS. In addition, a handover cancellation procedure is defined to allow the MS to cancel a handover procedure. If the serving BS indicates in AAI_HO-CMD that CDMA-based handover ranging[ix] is enabled, a dedicated ranging code/opportunity is allocated to the MS by the target BS, the MS transmits the dedicated ranging code to the target BS during network re-entry. The MAC management messages in IEEE 802.16m are distinguished from their legacy counterparts by an "AAI" prefix denoting the "Advanced Air Interface" messaging. If a ranging channel is exclusively scheduled by the target BS for handover, the MS would use that ranging channel in order to avoid interference by other mobile stations attempting random access. Once the dedicated ranging code is successfully received by the target BS, it allocates uplink radio resources, allowing the MS to send an AAI_RNG-REQ message and/or UL user data. If the target BS does not receive the dedicated ranging code within *HO Ranging_Initiation_Deadline* time interval, the target BS discards the pre-assigned STID of the MS.

As shown in Figure 6-16, if CDMA-based handover ranging is enabled and the MS does not have a dedicated ranging code or a dedicated ranging opportunity with the target BS, the MS transmits a randomly-selected handover ranging code at the earliest ranging opportunity. If the MS is sufficiently synchronized with the target BS, as determined by the serving BS, the CDMA-based ranging can be skipped while performing network entry with the target BS. In that case, the MS applies independently-calculated adjustments when starting network entry with the target BS by measuring and comparing the time difference between the DL synchronization signals of the target and serving base stations. The serving BS may send an AAI_RNG-RSP message with timing/power adjustment parameters regardless of whether CDMA-based ranging is enabled. In that case, the MS must apply the adjustments to the subsequent uplink transmissions. The serving BS may also send an AAI_RNG-RSP message with ranging status parameter set to *continue* so that the MS performs CDMA-based ranging with randomly selected handover ranging code in the next ranging opportunity to properly adjust its uplink transmission timing/power.

As shown in Figure 6-17, during network re-entry, the MS initiates the AAI_RNG-REQ/RSP message by sending an AAI_RNG-REQ message before the deadline specified by the *HO Ranging Initiation Deadline* parameter contained in the AAI_HO-CMD message during handover preparation. The time is measured from the transmission instant of the AAI_HO-CMD message. If the target BS does not receive the AAI_RNG-REQ message from the MS within the time interval defined by the *HO*

[ix]The handover ranging is a process similar to the initial ranging except that in the CDMA handover ranging process, random selection is used instead of random backoff and the CDMA handover ranging code is used instead of the initial ranging code. Furthermore, if the BS is notified of an upcoming handover, it may provide bandwidth allocation information using Fast Ranging IE so that the MS can send a RNG-REQ message [1].

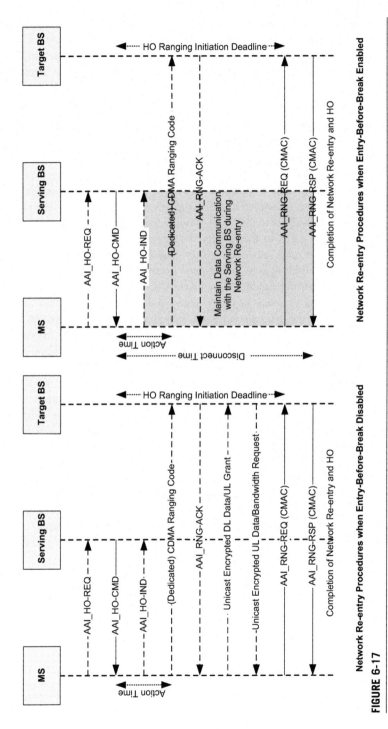

FIGURE 6-17

Handover network re-entry call flow [2]

Ranging Initiation Deadline parameter, it assumes the handover has failed and stops allocating bandwidth to the MS. If the MS transmits AAI_RNG-REQ prior to the deadline, the handover may still fail, if the MS does not receive an AAI_RNG-RSP within the permissible time.

If data packets are exchanged between the MS and the target BS before the AAI_RNG-REQ/RSP messaging is complete, the recipient (MS or target BS) stores the received data packets, but will not deliver them to the upper layers until the sender is authenticated. If the data packets belong to a service flow associated with a trusted entity that supports data authentication, as indicated by the data authentication algorithm identifier in the cryptographic suite of the IEEE 802.16m security, the receiver can authenticate the sender by verifying the cipher text authentication code included in each data packet. On the other hand, if the data packets belong to a service flow associated with a trusted entity that does not support data authentication, the receiver can authenticate the sender when the AAI_RNG-REQ/RSP messaging is successfully completed. In all cases, if the sender is authenticated, the decrypted data packets will be delivered to the upper layer in the receiving end; otherwise, the data packets are discarded. Alternatively, the sender can authenticate the receiver when the AAI_RNG-REQ/RSP messaging is successfully completed. In all cases, if the sender fails to authenticate the other end, all previously transmitted packets will be re-transmitted to the BS (or the MS) when a new connection is established.

The handover network re-entry call flow is shown in Figure 6-17. The MS initiates the AAI_RNG_REQ/RSP messaging by transmitting the AAI_RNG-REQ message to the target BS before the expiration of the *HO Ranging Initiation Deadline* time contained in the AAI_HO-CMD message. The MS MAC address or the previous serving BS-ID are not communicated over the air interface once the initial entry is completed, instead temporary and permanent station identifiers are assigned to the MS during network entry to identify the MS throughout the active session. The exclusion of the MAC address in the MAC messages was meant to improve network security. When the BS receives the AAI_RNG-REQ message including a valid CMAC tuple, the BS responds by transmitting an encrypted AAI_RNG-RSP message. The AAI_RNG-RSP message uses the STID assigned to the MS. On completion of the network re-entry with the target BS, the target BS becomes the new serving BS for the MS.

In the case of an uncontrolled handover,[x] the AAI_RNG-REQ message includes the former serving BS-ID and previously used STID, if the *Resource Retain Timer* has not expired. When the BS receives

---

[x]There are two types of handover, i.e., controlled and uncontrolled, from a QoS point of view. A controlled handover considers the following conditions [22,23]:

If the handover is MS initiated, the MS must send the BS a list of potential targets via a MOB_MSHO-REQ message.

In the case of MS-initiated handover, the network performs target selection based on the list of potential targets provided by the MS. The Anchor DPF or serving DPF may start multicasting with all potential targets.

The network sends to the MS the list of available targets for handover using MOB_BSHO-RSP or MOB_BSHO-REQ messages. If the list is void, the network refuses to accept MS handover.

The targets provided by the network to the MS should be a subset of the ones requested by the MS or reported by the MS via MOB_SCN-REP.

The MS must move to one of the targets provided by the network or reject the handover.

The MS performs the handover or rejects it by sending MOB_HO-IND.

If any of the above conditions are not met, the handover is considered an uncontrolled or un-predictive handover and the QoS cannot be guaranteed. If the MS leaves the serving BS before receiving MOB_BSHO-RSP, but it succeeds to at least send a MOB_BSHO-IND message with an indication of the target BS, this is also considered an uncontrolled handover. In the worst case, the MS may suddenly connect to the target BS without any indication given to the target BS. This is also considered as an un-predictive handover.

the AAI_RNG-REQ message including a valid CMAC tuple,[xi] the BS responds by transmitting an encrypted AAI_RNG-RSP message including an STID for the MS.

A drop is defined as a condition where an MS has stopped communication with its serving BS (either in the DL or in the UL) before the normal handover procedures with the serving BS are complete. When the MS has detected a drop during network re-entry with a target BS, it may attempt network re-entry with its preferred target BS by performing cell reselection procedures, which may include resuming communication with the serving BS.

6.4.4 Handover to and From Legacy Systems

When an IEEE 802.16m system is co-deployed with a legacy system using the same RF carrier, the cells and their radio resources may be divided between the two systems. A legacy BS advertises the system information for its neighbor base stations and the legacy zones of its neighbor new base stations. The legacy zone (LZone) is a time-frequency region (i.e., an integer number of subframes over partial or full system bandwidth) where the legacy mobile stations are supported in a new base station radio frame. A new zone (MZone) is a time-frequency region in the radio frame where the new mobile stations are supported. A new BS advertises the system information of its neighbor legacy base stations in both legacy and new zones. It further advertises the legacy zone of its neighbor new base stations, as well as system configuration of its neighbor new base stations. A new BS may indicate its legacy support through broadcast information transmitted in its legacy zone.

As soon as handover from a legacy BS to a new BS is triggered for a legacy MS, the MS is handed over from the serving legacy BS to the legacy zone of the target new BS using legacy handover signaling and procedures. A new MS may be handed over from the serving legacy BS to the legacy zone of the target new BS and then be switched to the new zones following the completion of handover procedure (i.e., zone switching). A new MS may also perform handover from a legacy BS to a new BS or directly to the new zone of the new BS, if the MS is capable of scanning the new BS prior to handover.

If handover from a new serving to a legacy target BS is triggered for a legacy MS, the MS handover is from the legacy zone of the serving BS to the target BS using legacy handover signaling and procedures [1]. If handover from a new BS to a legacy BS is triggered for a new MS, the serving BS and the MS perform handover execution using handover signaling and procedures as specified by IEEE 802.16m [2]. If necessary, the serving BS performs context mapping and protocol inter-working from the new to the legacy system. The MS performs network re-entry with target legacy BS using network re-entry signaling and procedures specified in the IEEE 802.16-2009 standard. An MS performs handover from a legacy BS to a new BS by using either zone-switching or direct handover process. The zone-switching handover is applicable to new base stations supporting coexisting legacy and new system. The direct handover is applicable to new base stations which only support new mobile stations. A new BS may also decide to keep a new MS in the legacy zone when coexisting with legacy systems.

As shown in Figure 6-18, the zone-switching handover procedure begins with a mobile station's decision to handover from the serving legacy BS to the LZone of a target new BS. The handover decision, initiation, and cancellation follow the same procedures described earlier. The zone-switching is initiated either by the MS or the new BS, and the final decision is made by the new BS. In the case of

---

[xi]Cipher-based Message Authentication Code tuple is used to protect control messages in IEEE 802.16-2009 standard. For example, if a BS intends to transmit a control signal with a CMAC tuple to an MS, the CMAC must comprise an MS authentication code for the MS to authenticate whether the control message is authentic.

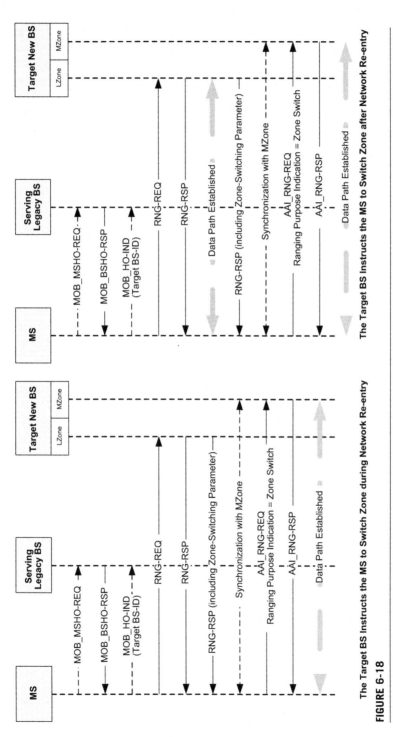

The handover procedure from legacy to new base stations [2]

FIGURE 6-18

an MS-initiated zone-switching process, the MS transmits a RNG-REQ message with the *Ranging Purpose Indicator* bit set, which implies the zone-switching request. The new BS then sends an RNG-RSP message including the zone-switching parameter. Before the zone-switching is initiated, the MS may inform the target BS whether it has already acquired the system information contained in the MZone of the target BS.

The MS performs network re-entry in the LZone of the target new BS following the procedures defined in the previous section. In addition, after the target BS is informed of the MS capability of supporting IEEE 802.16m functionalities through the MAC protocol version information field obtained either from the RNG-REQ message sent by the MS in the LZone or from the serving legacy BS over the backhaul, the new BS may instruct the MS to switch from LZone to MZone during or after LZone network re-entry. If zone-switching from Lzone to Mzone during network re-entry is required while the MS is performing network re-entry in the LZone, the BS may direct the MS to stop the LZone network re-entry and to immediately switch from the LZone to the MZone by transmitting the RNG-RSP message containing zone-switching parameter, as shown in Figure 6-18. If the BS decides to switch the MS to the MZone after the MS finishes the network re-entry in the LZone, it sends an unsolicited RNG-RSP message with zone switching parameter in LZone. The zone-switching TLV-coded parameter includes the following information:

- MZone new preamble index (i.e., the secondary preamble index that is assigned to this BS when operating as an IEEE 802.16m BS);
- Time offset between LZone and MZone (i.e., the distance in the number of subframes between the legacy and new regions);
- Zone-switching action time from LZone to MZone where the MS performs zone switching within the action time, if *HO_Reentry_Mode* is set to 0, the new BS stops all resource allocation for the MS in Lzone;
- Zone Switching Mode: if set to 1, the MS maintains its data communication with the BS in LZone while performing network re-entry with the MZone; otherwise, the MS interrupts data communication in LZone before performing network re-entry with the MZone.

Note that the security master key is shared between LZone and MZone during zone switching. The zone switching TLV coded parameter may also include the following information:

- Temporary STID to be used in the Mzone;
- HO Ranging Initiation Deadline which denotes the valid time for the temporary STID.

After receiving the zone-switching command through an RNG-RSP message in the LZone, the MS performs network re-entry with the MZone. The MS maintains its normal operation in the LZone (i.e., exchanging user data with the BS in the LZone) while performing network re-entry in the MZone, if the data path in LZone has been established before the start of zone switching and the *Zone Switching Mode* in the zone-switching TLV-coded parameter is set to 1. If synchronization with the MZone has not been accomplished, the MS starts network re-entry in the MZone by performing DL synchronization with the MZone preambles using the system information provided in the zone-switching TLV-coded parameter. The MS acquires the system information transmitted in the MZone by listening to the new BS broadcast channels.

The MS appends the CMAC tuple to the AAI_RNG-REQ message, if the security context is available for the MZone. The BS then responds with an AAI_RNG-RSP message. If the CMAC tuple in the AAI_RNG-REQ message is valid, the BS responds to the MS with the encrypted AAI_RNG-RSP message. If the CMAC in the AAI_RNG-REQ message is invalid or not included, the BS instructs the MS to initiate network entry procedure.

If an UL grant is provided in the MZone in *Zone_Switch_Action_Time*, the MS uses the grant to send an AAI_RNG-REQ message with *HO Ranging Purpose Indication* set to *Zone Switch*. Otherwise, the MS will request UL bandwidth to send the AAI_RNG-REQ by using the pre-assigned STID obtained while in the LZone. On receiving the bandwidth request, the BS provides a UL grant to the MS to send its AAI_RNG-REQ message. After receiving the AAI_RNG-REQ message, the target BS responds with the AAI_RNG-RSP message. The MS also performs basic capability negotiation during network re-entry in the MZone by exchanging AAI_REG-REQ/RSP messages. The MS context mapping from LZone to MZone is performed by the BS.

A new MS that is served by a legacy BS may scan and discover only a new BS, and may decide to directly perform handover to that BS. In this case, the MS performs legacy handover procedures with the serving legacy BS and then performs new network re-entry procedures with the target BS.

6.5 QUALITY OF SERVICE

Quality of Service (QoS) for IP networks is an industry-wide set of standards and mechanisms for ensuring high-quality and performance for user applications. By using QoS mechanisms, network administrators or operators can use existing resources more efficiently and ensure the required level of service without reactively expanding or over-provisioning their networks. Traditionally, the concept of quality in computer networks meant that all network traffic was treated equally. The result was that all network traffic received the network's best effort, with no guarantees for reliability, delay, variation in delay, or other performance characteristics. With best-effort delivery service, a single bandwidth-intensive application can result in poor or unacceptable performance for all applications. The QoS is a concept in which the service requirements of some user applications are more critical than others, requiring preferential handling of the corresponding traffic. The goal of QoS is to provide preferential service delivery for the applications that need it by ensuring sufficient bandwidth, controlling latency and jitter, and reducing data loss. The Internet Engineering Task Force (IETF) defines two major models for QoS on IP networks: Integrated Services (Intserv) and Differentiated Services (Diffserv). These models encompass several categories of mechanisms that provide preferential treatment to specified traffic. A service class defines how each QoS parameter is applied to a specific service. Table 6-6 provides the typical service classes and their characteristics.

6.5.1 Legacy QoS Classes

The IEEE 802.16-2009 standard defines several Quality of Service-related concepts, including service flow QoS scheduling, dynamic service establishment, and two-phase activation model [1]. The requirements for QoS function include the following:

1. A configuration and registration function for pre-configuring MS-based QoS service flows and traffic parameters;
2. A signaling function for dynamically establishing QoS-enabled service flows and traffic parameters;
3. Utilization of MAC scheduling and QoS traffic parameters for UL service flows;

Table 6-6 Characteristics of Service Classes [11]

Service Class Categories	Delay Requirement	Bit Rate	Bit Error Rate (BER) Margin	Use Cases
Point-to-multi-point; multi-point-to-multi-point; multi-point-to-point; highly interactive	<20 ms	1–20 Mbps	$10^{-9} \le BER \le 10^{-6}$	Tele-presence/video-conference; collaborative work navigation systems; real-time gaming; real-time video streaming
Asymmetric; interactive; low rate	20–100 ms	8–512 kbps	$10^{-9} \le BER \le 10^{-6}$	Remote control sensors; interactive geographical maps
Point to multi-point; multi-point to multi-point; multi-point to point; interactive; high rate	20–100 ms	1–50 Mbps	$10^{-6} \le BER \le 10^{-3}$	Rich data call; video broadcasting/streaming; high quality video conferencing; collaborative work
Conversational; soft BER	100–200 ms	8–512 kbps	$BER \le 10^{-3}$	Voice telephony; instant messages; multiplayer gaming; audio streaming; video telephony (medium quality); multiplayer gaming (high quality)
Conversational; symmetric QoS; tight BER	100–200 ms	1–50 Mbps	$10^{-6} \le BER \le 10^{-3}$	High quality video telephony; collaborative work; access to databases; file systems
Point-to-point unidirectional (uplink or downlink; asymmetric; delay tolerant	>200 ms	8 kbps– 50 Mbps	$10^{-9} \le BER \le 10^{-6}$	Messaging (data/voice/media); web browsing; audio on demand; internet radio; access to databases; video download/upload; peer-to-peer file sharing; video streaming

4. Utilization of QoS traffic parameters for DL service flows;
5. Grouping of service flow properties into service classes; therefore, upper-layer entities and external applications at both the MS and BS may request service flows with desired QoS parameters in a globally consistent way.

The principal mechanism for providing QoS is to associate packets transported through the MAC interface that are identified by a transport CID to a service flow. A service flow is a uni-directional flow of packets linked to a particular QoS parameter set. In other words, a service flow is a MAC transport service that provides uni-directional transport of packets either to UL packets transmitted by the MS or to DL packets transmitted by the BS. A service flow is characterized by a set of QoS parameters such as latency, jitter, and throughput requirements. The MS and BS provide the QoS according to the QoS parameter set defined for the service flow. Service flows exist in both UL and DL directions. All service flows are identified with a 32-bit Service Flow Identifier (SFID). In addition, each admitted or active service flow has a 16-bit CID. A service flow is partially characterized by the following attributes [1]:

- *Service Flow ID*: an SFID is assigned to each existing service flow. The SFID serves as the identifier for the service flow associated with an MS. A service flow has an SFID and an associated direction as a minimum.
- *CID*: the connection identifier of the transport connection exists only when the service flow is admitted or active. The relationship between SFID and transport CID, when present, is unique. An SFID is never associated with more than one transport CID, and a transport CID is never associated with more than one SFID.
- *ProvisionedQoSParamSet*: a QoS parameter set provisioned via mechanisms outside of the IEEE 802.16-2009 standard, such as the network management system.
- *AdmittedQoSParamSet*: this defines a set of QoS parameters for which the BS and possibly the MS are reserving resources. The main resource to be reserved is bandwidth, but this also includes any other memory- or time-based resource required to subsequently activate the service flow.
- *ActiveQoSParamSet*: this defines a set of QoS parameters specifying the service provided over the service flow. Note that only an active service flow may forward packets.
- *Authorization Module*: a logical function within the BS that approves or denies any changes to the QoS parameters and classifiers associated with a service flow. As such, it defines an encompassing set that limits the possible values of the *AdmittedQoSParamSet* and *ActiveQoSParamSet* attributes.

The relationship between the legacy QoS parameter sets is as shown in Figure 6-19. The *ActiveQoSParamSet* is always a subset of the *AdmittedQoSParamSet*, which is always a subset of the *AuthorizedQosParamSet* enclosure. In the dynamic authorization model, this enclosure is defined by the *Authorization Module*, also labeled as the *AuthorizedQoSParamSet*. In the provisioned authorization model, the aforementioned enclosure is specified by the *ProvisionedQoSParamSet*. There are three types of service flows, as follows [1]:

1. Provisioned: this type of service flow is provisioned by the network management system, and its *AdmittedQoSParamSet* and *ActiveQoSParamSet* attributes are both null.
2. Admitted: this type of service flow has resources reserved by the BS for its *AdmittedQoSParamSet*, but these parameters are not active (i.e., its *ActiveQoSParamSet* is null). The admitted service flows may be provisioned by other mechanisms in the network.
3. Active: this type of service flow has resources committed by the BS, and its *ActiveQoSParamSet* attribute is non-empty.

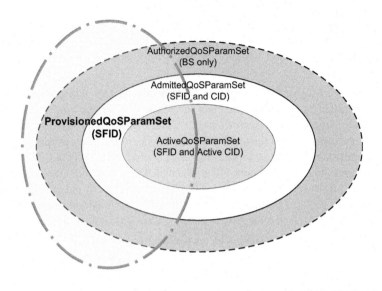

FIGURE 6-19

The relationship between the QoS parameter sets [1]

The main elements (or alternatively objects) of the legacy QoS framework and their attributes are shown in Table 6-7. Each object has a number of attributes. The optional attributes are denoted with brackets. A service flow may be associated with several PDUs, but a PDU is associated with exactly one service flow. The service flow is the central concept of the IEEE 802.16 MAC protocol. As mentioned earlier, there is a one-to-one mapping between admitted and active service flows identified by a 32-bit SFID and transport connections specified by a 16-bit CID.

The information delivered to the MAC SAP includes the CID, identifying the transport connection over which the information is delivered. The service flow for the connection is mapped to the MAC transport connection identified by the CID. A Classifier Rule uniquely maps a packet to its transport connection. The Classifier Rule is associated with zero or one PHS Rule. When creating a PHS Rule, the associated Classifier Rule Index is used as a reference. A PHS Rule is associated with a single service flow. The PHS Rules associated with the same service flow are uniquely identified by their PHSI

Object	Attribute
Table 6-7 The Main QoS Framework Elements are their Attributes [1]	
PHS Rule	PHSI; PHSS;PHSF; PHSM; PHSV
Classifier Rule	Classification rule index; Classification criteria
MAC PDU	SFID; [Service class]; CID; Payload
Service Flow	SFID; Direction; [CID]; [ProvisionedQoSParamSet]; [AdmittedQoSParamSet]; [ActiveQoSParamSet]
Transport Connection	Connection ID; QoS parameter set
Service Class	Service class name; QoS parameter set

(see Chapter 5). The Classifier Rule uniquely maps packets to its associated PHS Rule. The service class is an optional object that may be implemented at the BS and is referenced by an American Standard Code for Information Interchange (ASCII) code, which is intended for provisioning purposes. A service class is defined in the BS to include a particular QoS parameter set. The QoS parameter sets of a service flow may contain a reference to the service class name as a macro that selects all of the QoS parameters of the service class. The service flow QoS parameter sets may augment and even override the QoS parameter settings of the service class, subject to authorization by the BS.

Mobile WiMAX networks require common definition of service classes and associated *AuthorizedQoSParamSets* in order to facilitate operation and interoperability across a distributed topology [8,22,23]. Global service class names are supported to enable operation in this context. In practice, the global service class names are used as a baseline convention for communicating *AuthorizedQoSParamSet* or *AdmittedQoSParamSet*. The global service class parameters are as follows [1]:

- **Uplink/Downlink Indicator** parameter identifies service flow direction relative to the originating entity.
- **Maximum Sustained Traffic Rate** is a parameter that defines the peak information rate of the service. The rate is expressed in bits per second and pertains to the service data units at the input to the system. This parameter does not limit the instantaneous rate of the service since this is governed by the physical attributes of the entrance port. However, at the destination network interface in the UL direction, the service is regulated to conform to this parameter, on the average, over time. On the network in the DL direction, it may be assumed that the service was already regulated at the network entry. If this parameter is set to zero, then there is no explicitly mandated maximum rate. The maximum sustained traffic rate field specifies only a bound, not a guarantee that the rate is available.
- **Maximum Traffic Burst** parameter defines the maximum burst size that is accommodated for the service. Since the physical rate of input/output ports, any air interface, and the backhaul, will in general be greater than the maximum sustained traffic rate parameter for a service, this parameter describes the maximum continuous burst the system should accommodate for the service, assuming the service is not currently using any of its available resources. If the maximum traffic burst set to zero, it means that there is no maximum traffic burst reservation requirement.
- **Minimum Reserved Traffic Rate** parameter specifies the minimum rate, in bits per second, reserved for this service flow. The BS is required to satisfy the bandwidth requests for a connection up to its minimum reserved traffic rate. If less bandwidth than its minimum reserved traffic rate is requested for a connection, the BS may use the excess reserved bandwidth for other purposes. The value of this parameter excludes the MAC overhead. If the minimum reserved traffic is set to zero, it means that there is no minimum reserved traffic rate requirement.
- **Maximum Latency** is a parameter whose value specifies the maximum interval between the receipt of a packet at the convergence sub-layer of the BS or MS and the transmission of the corresponding physical layer PDU over the air interface. A value of zero for maximum latency is interpreted as no commitment.
- **SDU Indicator** is a parameter whose value specifies whether the SDUs are fixed or variable in length.
- **Paging Preference** is a single bit indicator of a mobile station's preference for the reception of paging advisory messages during the Idle State. It indicates that the BS may present paging advisory messages or other indicators to the MS when there are MAC SDUs bound for an idle mode MS.

- **Uplink Grant Scheduling Type** specifies which uplink grant scheduling service type is associated with uplink service flow. This parameter is present in the uplink direction.
- **Tolerated Jitter** is a parameter whose value specifies the maximum delay variation (jitter) for the connection. This parameter is present for a DL or UL service flow, which are associated with *Uplink Grant Scheduling Type* = UGS or ertPS.
- **Request/Transmission Policy** is a parameter whose value specifies certain attributes for the associated service flow.
- **Traffic Priority** is a parameter whose value specifies the priority of associated service flow. This parameter is present for a DL or UL service flow, which are associated with any *Uplink Grant Scheduling Types* except UGS.
- **Unsolicited Grant Interval** parameter defines the nominal interval between successive data grant opportunities for a DL service flow, which are associated with *Uplink Grant Scheduling Type* = UGS or ertPS. If this parameter is set to zero, then there is no explicitly mandated unsolicited grant interval. The maximum unsolicited grant interval field specifies only a bound, not a guarantee that the rate is available.
- **Unsolicited Polling Interval** parameter defines the maximum nominal interval between successive polling grant opportunities for an UL service flow, which are associated with *Uplink Grant Scheduling Type* = rtPS. If this parameter is set to zero, then there is no explicitly mandated unsolicited grant interval. The maximum unsolicited polling interval field specifies only a bound, not a guarantee that the rate is available.

The QoS classes specified in the IEEE 802.16-2009 standard are as follows [1,6,7]:

- **Unsolicited Grant Service (UGS)** is designed to support real-time uplink service flows that transport periodic fixed-size data packets such as VoIP without silence suppression. This service class provides fixed-size grants on a real-time periodic basis, which eliminates the overhead and latency due to MS bandwidth requests and ensures timely availability of the grants to meet the real-time characteristics of the service flow. The BS schedules the MS at periodic intervals based on the *Minimum Reserved Traffic Rate* of the service flow. The size of these grants must be sufficient to hold the fixed-length payload associated with the service flow. In order to ensure proper operation, the *Request/Transmission Policy* setting will prohibit the MS from using any contention request opportunities for this connection. The mandatory QoS parameters are *Minimum Reserved Traffic Rate, Maximum Latency, Tolerated Jitter, Uplink Grant Scheduling Type, SDU size* for fixed length SDU service flows, *Request/Transmission Policy*, and *Unsolicited Grant Interval*. If present, the *Minimum Reserved Traffic Rate* parameter has the same value as the *Maximum Sustained Traffic Rate* parameter.
- **Real-Time Polling Service (rtPS)** is designed to support real-time UL service flows that transport variable-size data packets on a periodic basis such as Moving Pictures Experts Group (MPEG) video format. This service offers real-time, periodic, and unicast request opportunities, which meet the service flow's real-time requirements and further allow the MS to specify the size of the desired grant. This service involves more overhead than UGS, but supports variable-sized grants for optimal data transport. The BS provides periodic unicast request opportunities. In order to ensure proper operation, the *Request/Transmission Policy* is set such that the MS is prohibited from using any contention-based request opportunities for this connection. The BS may render unicast request opportunities as prescribed by this class even if prior requests were not fulfilled. As a result, the

MS uses only unicast request and data transmission opportunities in order to obtain UL transmission grants. The mandatory QoS parameters corresponding to this class are *Minimum Reserved Traffic Rate, Maximum Sustained Traffic Rate, Maximum Latency, Uplink Grant Scheduling Type, Request/ Transmission Policy*, and *Unsolicited Polling Interval*.

- **Extended Real-Time Polling Service (ertPS)** is a scheduling mechanism which utilizes the advantages of UGS and rtPS. The BS provides unicast grants in an unsolicited manner similar to UGS, reducing the latency of bandwidth request. Unlike the UGS allocations, the ertPS allocations are variable-sized. The BS may provide periodic UL allocations that are used for requesting the bandwidth as well as for data transmission. By default, the size of allocations corresponds to the current value of *Maximum Sustained Traffic Rate* for the connection. The MS may request changing the size of the UL allocation either by using an *Extended Piggyback Request* field of the GMSH or the *Bandwidth Request* field of the MAC signaling headers by sending a codeword over the uplink feedback channel. The BS will not change the size of UL allocations until it receives another bandwidth change request from the MS. The mandatory QoS parameters are the *Maximum Sustained Traffic Rate*, the *Minimum Reserved Traffic Rate*, the *Maximum Latency*, the *Request/Transmission Policy*, and *Unsolicited Grant Interval*. The ertPS is designed to support real-time service flows that generate variable-size data packets on a periodic basis, such as VoIP with silence suppression.
- **Non-Real-Time Polling Service (nrtPS)** offers unicast polls on a regular basis, which ensures that the UL service flow receives request opportunities even during network congestion. The serving BS typically polls nrtPS connections every second and provides timely unicast request opportunities. The *Request/Transmission Policy* attribute is set such that the MS is allowed to use contention-based or unicast request opportunities for data transmission. The mandatory QoS parameters for this scheduling service are *Minimum Reserved Traffic Rate, Maximum Sustained Traffic Rate, Traffic Priority, Uplink Grant Scheduling Type*, and *Request/Transmission Policy*.
- **Best Effort (BE)** service is designed to support applications for which no minimum service guarantees (e.g., no rate or delay requirements) are required. The MS is allowed to use contention-based and unicast request opportunities for data transmission.

6.6 IEEE 802.16m QoS CLASSES

In addition to the legacy service class attributes, the IEEE 802.16m defines a new service class attribute called *Maximum Sustained Traffic Rate per Flow* [2]. The new attribute defines the peak information rate of the service flow. The maximum rate is denoted in bits per second and pertains to the service data units at the input of the convergence sub-layer. This parameter does not include transport, protocol or network overhead information, and does not limit the instantaneous rate of the service flow since this is governed by the physical attributes of the ingress port. However, at the destination network interface in the uplink direction, the service is regulated to ensure conformance to this parameter. The time interval that the traffic rate is averaged over is defined during service negotiation. In the downlink direction, it may be assumed that the service was already regulated at the ingress to the network. If this parameter is set to zero, then there is no explicitly mandated maximum rate. The maximum sustained traffic rate field specifies only a bound, not a guarantee that the rate is available.

In addition to the legacy scheduling services, the IEEE 802.16m supports adaptation of service flow QoS parameters. One or more QoS parameter set(s) may be defined during the initial service nego-tiation, e.g., a mandatory primary QoS parameter set per service flow and an optional secondary QoS parameter set. If multiple QoS parameter sets are defined, each of them corresponds to a specific traffic characteristic for the user data mapped to the same service flow. If the QoS requirement or traffic characteristics for UL traffic changes, the serving BS may independently perform adaptation by either changing the service flow QoS parameters or by switching among multiple service flow QoS parameter sets. The MS may also request the serving BS to perform adaptation and the BS will allocate radio resources according to the adapted service flow QoS parameters.

Scheduling services characterize the data handling mechanisms supported by the MAC scheduler for data transport on a specific connection. Each service flow is associated with a single scheduling service as in IEEE 802.16m. A scheduling service is characterized by a set of service flow QoS parameters that quantify its behavior. These parameters are established or modified using service flow management procedures.

Adaptive Grant and Polling Service (aGPS) is a new service class defined in IEEE 802.16m where the BS may grant or poll an MS periodically and may negotiate only primary QoS parameters or both primary and secondary QoS parameter sets with the MS. Initially, the BS uses QoS parameters defined in the primary QoS parameter set including primary *Grant* and *Polling Interval* (*GPI*) and primary *Grant Size*. During the service, the traffic characteristics and QoS requirement may change, for example silence-suppression-enabled VoIP alternates between talk spurt and silence period, which triggers adaptation of QoS parameters. Adaptation includes switching between primary and secondary QoS parameter sets or changing of *GPI/Grant Size* to values other than those defined in the primary or secondary QoS parameter sets when the traffic can be characterized by more than two QoS states.

Depending on the adaptation method specified during the service flow negotiation, the *Grant Size* or *GPI* can be changed by the BS on detecting certain traffic conditions, or can be triggered by explicit signaling by the MS. There are three adaptation methods [2]:

1. **Implicit**: on detecting certain traffic conditions different to those in the pre-negotiated QoS parameter set, the BS automatically changes *GPI* and/or *Grant Size*, or switches between *GPI_Primary/Grant_Size_Primary* and *GPI_Secondary/Grant_Size_Secondary*, if the secondary QoS parameter set is defined.

2. **Explicit, Sustained**: the *GPI* and *Grant Size* change is triggered by explicit signaling by the MS, such as piggyback bandwidth request, bandwidth request signaling, quick access message in bandwidth request channel, or fast-feedback channel. Such change is sustained until the next change request. If *GPI_Secondary* or *Grant_Size_Secondary* is defined, the *GPI* and *Grant Size* switches between *GPI_Primary/Grant_Size_Primary* and *GPI_Secondary/Grant_Size_Secondary*, as requested by the explicit signaling; otherwise, *GPI* and *Grant Size* changes as indicated by QoS requirement carried in the explicit signaling mechanisms.

3. **Explicit, One Time Only**: the *GPI* and *Grant Size* one-time-only change is triggered by explicit signaling by the MS, such as in piggyback bandwidth request, bandwidth request signaling, quick access message in bandwidth request channel, or the fast-feedback channel. If *GPI_Secondary/Grant_Size_Secondary* is defined, the *GPI* and *Grant Size* are switched only once from *GPI_Primary/Grant_Size_Primary* to *GPI_Secondary/Grant_Size_Secondary*; otherwise, *GPI* and *Grant Size* change as indicated by QoS requirements contained in the explicit signaling mechanisms.

The mandatory QoS parameters of aGPS are the Maximum Sustained Traffic Rate, the Request/ Transmission Policy, Primary GPI, and Primary Grant Size.

6.7 MAC MANAGEMENT/CONTROL MESSAGES

The connection-oriented MAC layers in the serving BS and the MS communicate using the MAC control messages to perform the control-plane functions. The MAC control messages are contained in MAC PDUs and are transported over broadcast, unicast or random access connections. There is a single unicast control connection. In order to improve reliability, the HARQ mechanism is used for MAC control messages that are sent over unicast control connections. Furthermore, encryption may be enabled for unicast MAC control messages. The IEEE 802.16m MAC control messages can be fragmented; however, a multiplexing extended header is not used for these messages. The IEEE 802.16m MAC management/control messages are listed in Table 6-8. The MAC management messages in IEEE 802.16m are distinguished from their legacy counterparts by an "AAI" prefix, denoting "Advanced Air Interface" messaging. Unlike the IEEE 802.16-2009 standard, the IEEE 802.16m MAC management/control messages are encoded in ASN.1 format.[xii] The encryption of MAC PDU is signaled through *Encryption Control Bit* in FEH extended header when it is set to one. As shown in Table 6-8, the MAC management messages may or may not be encrypted depending on their function. A MAC management message included in a MAC PDU whose encryption control bit value does not match the combined message type and corresponding context is discarded. Encrypted and unencrypted MAC control messages are not sent in the same PDU. The MAC management messages shown in Table 6-8 are sorted based on their usage, e.g., Network entry/re-entry, Sleep Mode operation, Idle Mode operation, etc. The IEEE 802.16-2009 standard corresponding MAC management messages (if applicable) are also included in the table to help better understanding of the similarities and differences. Note that legacy MAC management messages, depending on their types, are carried over basic or primary management connections, as well as broadcast or initial ranging connections, whereas in IEEE 802.16m, the type of connections for carrying MAC control messages (as shown in Table 6-8) are classified as unicast, broadcast, and initial ranging.

[xii]Abstract Syntax Notation One (ASN.1) is a standard for describing a message that can be sent or received in a network. ASN.1 is divided into two parts: (1) the rules of syntax for describing the contents of a message in terms of data type and content sequence or structure; and (2) how each data item is encoded in a message. ASN.1 is defined in two ISO standards for applications intended for the Open Systems Interconnection framework [16,17]. The following is an example of a message definition specified with ASN.1 notation:

Report::= SEQUENCE {author OCTET STRING, title OCTET STRING, body OCTET STRING, biblio Bibliography}

In this example, Report is the name of this type of message. SEQUENCE indicates that the message is a sequence of data items. The first four data items have the data type of OCTET STRING, meaning each is a string of bytes. The bibliography data item is another definition named Bibliography that is used.

Bibliography::= SEQUENCE {author OCTET STRING title OCTET STRING publisher OCTET STRING year OCTET STRING}

Other data types that can be specified include: INTEGER; BOOLEAN; REAL; and BIT STRING. An ENUMERATED data type is one that takes one of several possible values. Other data items can be specified as optional.

Table 6-8 IEEE 802.16m MAC Management/Control Messages [1,2]

IEEE 802.16m MAC Management Message	Message Description	Usage	IEEE 802.16-2009 Standard Equivalent Message	Security Attributes	Connection Type
AAI_SCD System Configuration Descriptor	This message is transmitted by the BS at a periodic interval to define the system configuration.	Broadcast of system information and parameters (Additional Broadcast Information not otherwise transmitted in the superframe headers).	N/A	N/A	Broadcast
AAI_LBS-ADV LBS Advertisement	AAI_LBS-ADV is a MAC control message broadcast by the BS to provide the MS with topographical information of the neighboring base stations, which can be used by the MS for triangularization or trilaterization to enable location determination. This message may also contain time and frequency information to aid satellite based, e.g., GPS, receivers for improved performance.	Broadcast of system information	N/A	N/A	Broadcast
AAI_RNG-REQ Ranging Request	A message that is transmitted by the MS at initialization and later periodically to determine network delay and to request power and/or DL burst profile change.	Network Entry/ Re-entry	RNG-REQ	Null: during initial ranging procedure when there is no security association already established or updated	Initial Ranging or Unicast

				CMAC: Other cases.	
AAI_RNG-RSP Ranging Response	A message that is transmitted by the BS in response to an AAI_RNG-REQ message. In addition, it may also be transmitted asynchronously to send corrections based on measurements that have been made on other received data or MAC messages.	Network Entry/ Re-entry	RNG-RSP	Null: during ranging procedure when there is no security association already established or updated. Encrypted/ICV: Other cases.	Initial Ranging or Unicast
AAI_RNG-ACK Ranging Acknowledgement	Aggregated CDMA Ranging Acknowledge is a message that provides responses (e.g., adjustments, ranging status, etc.) to all successfully received and decoded initial ranging requests in the initial ranging slots in a previous UL subframe, in a predefined, subsequent DL subframe.	Network Entry/ Re-entry	N/A	N/A	Broadcast
AAI_SBC-REQ MS Basic Capability Request	A message containing the maximum "Capability Class" that the MS can support. The maximum value of CAPABILITY_INDEX is denoted by N bits. A "Capability Class" is defined as a unique set of functions, configuration parameters, air interface protocol revision, and/or services that can uniquely describe a mobile station implementation or configuration while operating in the network. The MS, by default, is required to support	Network Entry/ Re-entry	SBC-REQ	Null	Unicast

(Continued)

Table 6-8 IEEE 802.16m MAC Management/Control Messages [1,2] *Continued*

IEEE 802.16m MAC Management Message	Message Description	Usage	IEEE 802.16-2009 Standard Equivalent Message	Security Attributes	Connection Type
	the basic capabilities associated with "Capability Class 0." If the MS is capable of supporting higher revisions of physical layer or medium access control layer protocols or further wishes to use enhanced features, it sends an AAI_SBC-REQ message to the BS indicating the highest "CAPABILITY_INDEX" that it can support. The CAPABILITY_INDEX = 0 indicates the default capability index and basic feature set or configuration parameters and may not need to be signaled.				
AAI_SBC-RSP MS Basic Capability Response	Upon receipt of the AAI_SBC-REQ message containing the "CAPABILITY_INDEX" from the MS, the BS determines whether it could allow or could support the requested feature set or MAC and/or PHY protocol revisions. If the BS does support or can allow the use of enhanced features, it responds with an AAI_SBC-RSP message to inform the	Network Entry/ Re-entry	SBC-RSP	Null	Unicast

MS of its decision. The BS may signal a "CAPABILITY_INDEX" which is numerically different than the one requested by the MS. The "CAPABILITY_INDEX" values range from 0 to N, where N denotes the maximum CAPABILITY_INDEX value. The features and configuration parameters included in the baseline capability class are sufficient to meet the minimum performance requirements of the standard. In case of failure in any stages of operation, the MS and BS fall back to "Capability Class 0" and restart negotiations for a new "Capability Class," if necessary.

AAI_PKM-REQ Privacy Key Management Request

This message is sent by the PKM client (MS) to PKM server (BS). PKMv3 uses AAI_PKM-REQ and AAI_PKM-RSP MAC management messages. Each message encapsulates one PKM message in the MAC control message payload. They are transmitted through unicast primary management connection. The PKM protocol messages contain the following parameters: *Code (i.e., the*

Network Entry/Re-entry

PKM-REQ

Prior to derivation of AK in network entry: NULL. Following derivation of AK in network entry and when EAP-Transfer message is enclosed: Encrypted/ICV. Following derivation of AK in network entry and when other messages are enclosed: CMAC.

Unicast

(Continued)

Table 6-8 IEEE 802.16m MAC Management/Control Messages [1,2] *Continued*

IEEE 802.16m MAC Management Message	Message Description	Usage	IEEE 802.16-2009 Standard Equivalent Message	Security Attributes	Connection Type
	type of PKM packet), PKM Identifier to match a BS response to the MS requests or an MS response to the BS requests. For retransmissions, the identifier field remains unchanged.				
AAI_PKM-RSP Privacy Key Management Response	This message is sent by the PKM server (BS) to PKM client (MS). PKMv3 uses AAI_PKM-REQ and AAI_PKM-RSP MAC management messages. Each message encapsulates one PKM message in the MAC control message payload. They are transmitted through a unicast primary management connection. The PKM protocol messages contain the following parameters: Code (i.e., the type of PKM packet); PKM Identifier to match a BS response to the MS requests or an MS response to the BS requests. For re-transmissions, the identifier field remains unchanged.	Network Entry/Re-entry	PKM-RSP	Prior to derivation of AK in network entry: NULL. Following derivation of AK in network entry and when EAP-Transfer message is enclosed: Encrypted/ICV. Following derivation of AK in network entry and when other messages are enclosed: CMAC.	Unicast
AAI_REG-REQ Registration Request	This message is sent by the MS in order to register with the BS and to indicate supported management	Network Entry/Re-entry	REG-REQ	Encrypted/ICV	Unicast

Message	Description				
	parameters, CS capabilities, IP mode, etc. It is used by the MS to negotiate general MS capabilities and perform registration during network entry. This message is encrypted and does not contain CMAC Tuple, if authentication has been completed. This message, among other parameters, includes MS MAC address to derive security keys.				
AAI_REG-RSP Registration Response	A message that is transmitted by the BS in response to an AAI_REG-REQ message during initialization to confirm registration and authentication. The AAI_REG-RSP message is encrypted and does not contain CMAC Tuple, if authentication has been completed. This message, among other parameters, includes the STID which is used for MS identification in lieu of the temporary STID which has been transferred by AAI-RNG-RSP message.	Network Entry/Re-entry	REG-RSP	Encrypted/ICV	Unicast
AAI_RES-CMD Reset Command	The AAI_RES-CMD message is transmitted by the BS to force the MS to reset itself, reinitialize its MAC, and repeat initial network entry. This message may be used if an MS is unresponsive to the BS,	Network Entry/Re-entry	RES-CMD	Encrypted/ICV	Unicast

(Continued)

Table 6-8 IEEE 802.16m MAC Management/Control Messages [1,2] *Continued*

IEEE 802.16m MAC Management Message	Message Description	Usage	IEEE 802.16-2009 Standard Equivalent Message	Security Attributes	Connection Type
	or if the BS detects continued abnormalities in the UL transmission from the MS. It contains no information except the MAC message name.				
AAI_DSA-REQ Dynamic Service Addition Request	This message is sent by an MS or BS to create a new service flow, and may contain parameters for more than one service flow. An MS or BS can generate an AAI_DSA-REQ message and include the following parameters: *Control Message Type*; Service flow parameters; Convergence sub-layer parameter encodings; and may further include the SCID in order to change sleep cycle settings, predefined BR index in order to use three-step BR procedure. The FID for the transport connection is not included in the MS-initiated AAI_DSA-REQ message. The BS assigns to the service flow a unique FID for the transport connection. The MS-initiated AAI_DSA-REQ message may use the service class name instead of the QoS	Service Flow Establishment/ Change/Deletion (Connection Management)	DSA-REQ	Encrypted/ICV	Unicast

	parameters. The BS-initiated AAI_DSA-REQ message includes the QoS parameter set associated with the service class and further include the target SAID for the service flow.				
AAI_DSA-RSP Dynamic Service Addition Response	This message is generated in response to an AAI_DSA-REQ message to approve/reject a new service flow creation. An MS or BS can generate the AAI_DSA-RSP message with the following parameters: *Control Message Type*; *Confirmation Code*; service flow parameter; as well as Convergence sub-layer parameter encodings.	Service Flow Establishment/Change/Deletion (Connection Management)	DSA-RSP	Encrypted/ICV	Unicast
AAI_DSA-ACK Dynamic Service Addition Acknowledge	This message is generated by an MS or BS in response to an AAI_DSA-RSP message and includes *Control Message Type and Confirmation Code* parameters.	Service Flow Establishment/Change/Deletion (Connection Management)	DSA-ACK	Encrypted/ICV	Unicast
AAI_DSC-REQ Dynamic Service Change Request	This message is sent by an MS or BS to dynamically change the parameters of an existing service flow. An MS or BS generate AAI_DSC-REQ message, including the following parameters: Control Message Type; Service Flow Parameters (service flow's new traffic characteristics and scheduling requirements); admitted and active QoS parameter sets currently in	Service Flow Establishment/Change/Deletion (Connection Management)	DSC-REQ	Encrypted/ICV	Unicast

(Continued)

Table 6-8 IEEE 802.16m MAC Management/Control Messages [1,2] *Continued*

IEEE 802.16m MAC Management Message	Message Description	Usage	IEEE 802.16-2009 Standard Equivalent Message	Security Attributes	Connection Type
	use by the service flow. If the DSC message is successful and it contains service flow parameters, but does not contain replacement sets for both admitted and active QoS parameter sets, the omitted sets are set to null. The service flow parameters contain an FID. Other parameters may be included in the AAI_DSC-REQ message SCID to switch sleep cycle setting, predefined BR index to be used in three-step bandwidth request procedure. An AAI_DSC-REQ message does not carry parameters for more than one service flow.				
AAI_DSC-RSP Dynamic Service Change Response	This message is generated in response to an AAI_DSC-REQ message. An MS or BS can generate an AAI_DSC-RSP message with the following parameters: Control Message Type; Confirmation Code; Service Flow Parameters (service flow's traffic characteristics and scheduling requirements). The complete specification of	Service Flow Establishment/ Change/Deletion (Connection Management)	DSC-RSP	Encrypted/ICV	Unicast

(Continued)

the service flow is included in the AAI_DSC-RSP only if it includes an expanded service class name. If a service flow parameter set contained a service class name and an admitted QoS parameter set, the AAI_DSC-RSP includes the QoS parameter set corresponding to the service class. If specific QoS parameters were also included in the service flow request, those QoS parameters are included in the AAI_DSC-RSP instead of any QoS parameters of the same type of the service class, as well as convergence sub-layer parameters such as service flow's CS-specific parameters. In response to an AAI_DSC-REQ message which contains a SCID, the SCID may be included in an AAI_DSC-RSP message as an indication of approval. If the AAI_DSC-RSP message doesn't include an SCID, the request has failed. In response to an AAI_DSC-REQ message containing a predefined BR index, the BR index may be included in the AAI_DSC-RSP message as an indication of approval. If an AAI_DSC-RSP message

Table 6-8 IEEE 802.16m MAC Management/Control Messages [1,2]   *Continued*

IEEE 802.16m MAC Management Message	Message Description	Usage	IEEE 802.16-2009 Standard Equivalent Message	Security Attributes	Connection Type
	does not include the BR index, the request has failed.				
AAI_DSC-ACK Dynamic Service Change Acknowledge	This message is generated in response to an AAI_DSC-RSP message. An MS or BS may generate the AAI_DSC-ACK message with *Control Message Type* and *Confirmation Code* parameters.	Service Flow Establishment/ Change/Deletion (Connection Management)	DSC-ACK	Encrypted/ICV	Unicast
AAI_DSD-REQ Dynamic Service Deletion Request	This message is sent by an MS or BS to delete an existing service flow. An MS or BS may generate this message with the following parameters: *Flow ID* and *Control Message Type*. The SCID may be included in the AAI_DSD-REQ message to change the sleep cycle settings.	Service Flow Establishment/ Change/Deletion (Connection Management)	DSD-REQ	Encrypted/ICV	Unicast
AAI_DSD-RSP Dynamic Service Deletion Response	This message is generated in response to an AAI_DSD-REQ message. An MS or BS may generate this message with the following parameters: *Flow ID*; *Control Message Type*; and the *Confirmation Code*. In response to the AAI_DSD-REQ message containing an SCID, the SCID may be included in the response message as an	Service Flow Establishment/ Change/Deletion (Connection Management)	DSD-RSP	Encrypted/ICV	Unicast

Message	Description	Function	Related message	Security	Type
AAI_DSX-RVD DSX Received	indication of approval. If the response message does not include an SCID, the request has failed. This message is generated by the BS to inform the MS that it has correctly received a DSX (DSA or DSC) request message. The DSX-RSP message is transmitted only after the DSX-REQ is authenticated.	Service Flow Establishment/ Change/Deletion (Connection Management)	DSx-RVD	Encrypted/ICV	Unicast
AAI_SLP-REQ Sleep Request	An MS in the Active Mode may use this message to enter the Sleep Mode. An MS in the Sleep Mode can change the Sleep Mode settings using this message. An MS in the Sleep Mode can exit from Sleep Mode using this message with a certain operation code.	Sleep Mode Operation	MOB_SLP-REQ	Encrypted/ICV	Unicast
AAI_SLP-RSP Sleep Response	This message is sent from the BS to the MS in response to an AAI_SLP-REQ message. The BS may send this message in an unsolicited manner. If the request is rejected by the BS, the MS will not retransmit the AAI_SLP-REQ message before a certain time interval signaled in the AAI_SLP-RSP message elapses.	Sleep Mode Operation	MOB_SLP-RSP	Encrypted/ICV	Unicast
AAI_TRF-IND Traffic Indication	This message is sent by the BS to the mobile stations in the first frame of mobile station's listening window.	Sleep Mode Operation	MOB_TRF-IND	N/A	Broadcast

(Continued)

Table 6-8 IEEE 802.16m MAC Management/Control Messages [1,2] *Continued*

IEEE 802.16m MAC Management Message	Message Description	Usage	IEEE 802.16-2009 Standard Equivalent Message	Security Attributes	Connection Type
	The mobile stations that have not been assigned SLPID would ignore this message. This message indicates whether there is DL traffic for mobile stations that are explicitly addressed. There are two formats for the AAI_TRF-IND message, as indicated by the format field of the message. If the MS does not find a SLPID-Group Indication bitmap or a Traffic Indication bitmap, the MS considers it as a negative indication and may go back to sleep. If the MS does not find its own SLPID in the AAI_TRF-IND message, the MS considers it as a negative indication and may go back to sleep.				
AAI_TRF_IND-REQ Traffic Indication Request	If the *Traffic Indication Message Flag* is set to 1, the MS receives an AAI_TRF-IND message in the first frame during the listening window; otherwise, the MS stays awake for the rest of the listening window. If the MS receives any unicast data during the listening window,	Sleep Mode Operation	MOB_TRF-IND	Encrypted/ICV	Unicast

Message	Description				
	then it assumes that the *Traffic Indication Message Flag* was set. If the MS receives neither the *Traffic Indication Message Flag* nor any other unicast data in the listening window, the MS would send an AAI_TRF_IND-REQ message to the BS in order to inquire about the status of *Traffic Indication Message Flag*. The BS will respond to the MS by sending an AAI_TRF_IND-RSP with *Traffic Indication Message Flag* for the MS.				
AAI_TRF_IND-RSP Traffic Indication Response	When the BS receives an AAI_TRF_IND-REQ message from an MS, the BS responds to the MS by sending the AAI_TRF_IND-RSP message with the *Traffic Indication Message Flag*. When the MS receives the AAI_TRF_IND-RSP message from the serving BS, it will update the current sleep cycle based on the *Traffic Indication Message Flag* in the AAI_TRF_IND-RSP message.	Sleep Mode Operation	N/A	Encrypted/ICV	Unicast
AAI_NBR-ADV Neighbor Advertisement	This message is transmitted by the serving BS to provide the MS with information about neighbor base stations or relay stations including the following parameters: BS type (macro, micro, femto, relay, etc.); carrier frequency; MAC	Handover Operation/Normal Operation	MOB_NBR-ADV	Null: in Unicast N/A: in Broadcast	Unicast or Broadcast

(Continued)

Table 6-8 IEEE 802.16m MAC Management/Control Messages [1,2] *Continued*

IEEE 802.16m MAC Management Message	Message Description	Usage	IEEE 802.16-2009 Standard Equivalent Message	Security Attributes	Connection Type
	version; TDD/FDD; bandwidth and cyclic prefix size; multi-carrier capability; and configuration. The BS determines and indicates the system configuration information included for each deployment type and their corresponding broadcast frequency. To allow AAI_NBR-ADV fragmentation while providing flexibility for the MS handover operation without requiring acquisition of the entire AAI_NBR-ADV message, the BS always provides the total number of deployment types and total number of recommended target base stations for each type. Each AAI_NBR-ADV fragment has corresponding indices for each deployment type and each neighbor BS. The base stations with identical type are listed in the AAI_NBR-ADV message in descending order of their cell coverage. Each AAI_NBR-ADV message carries, among other parameters, AAI_NBR-ADV change count, number				

	of total cell types, and segment information for this AAI_NBR-ADV message, system information of base stations from one or more cell types.				
AAI_SCN-REQ Scanning Interval Allocation Request	A message that is transmitted by the MS to request a scanning interval for the purpose of seeking available legacy or new base stations and determining their suitability as targets for handover. An MS may request a scanning interval during the scan interleaving interval.	Handover Operation/Normal Operation	MOB_SCN-REQ	Encrypted/ICV	Unicast
AAI_SCN-RSP Scanning Interval Allocation Response	A message that is transmitted by the BS either unsolicited or in response to an AAI_SCN-REQ message sent by an MS. The BS may transmit AAI_SCN-RSP to start MS scan reporting with or without scanning interval allocation. If the Scan Duration field contains a non-zero value, the system parameters (except cell bar information) which are carried through the superframe header are not changed during the scanning time specified in the AAI_SCN-RSP message. If the Scan Duration field contains a non-zero value, the scanning interval pattern included in the AAI_SCN-RSP message replaces the previous existing scanning interval pattern.	Handover Operation/Normal Operation	MOB_SCN-RSP	Encrypted/ICV	Unicast

(Continued)

Table 6-8 IEEE 802.16m MAC Management/Control Messages [1,2] *Continued*

IEEE 802.16m MAC Management Message	Message Description	Usage	IEEE 802.16-2009 Standard Equivalent Message	Security Attributes	Connection Type
AAI_SCN-REP Scanning Result Report	When the report mode is event-triggered in the most recently received AAI_SCN-RSP, the MS transmits an AAI_SCN-REP message to report the scanning results to its serving BS after each scanning period, if the trigger condition is met. For the periodic and one-time scan report modes, the MS reports the scanning results to its serving BS at the time indicated in the AAI_SCN-RSP message, except when it is in the scanning interval. For the periodic report mode, the MS stops reporting after all scanning intervals in the AAI-SCN-RSP message. The MS includes all available scanning results for the requested base stations specified in the AAI_SCN-RSP message. The MS may transmit an AAI_SCN-REP message to report the scanning results to its serving BS at any time.	Handover Operation/Normal Operation	MOB_SCN-REP	Encrypted/ICV	Unicast

Message	Description	Category	Message	Security	Addressing
AAI_HO-REQ MS Handover Request	In MS-initiated handover, the MS sends an AAI_HO-REQ message to the serving BS in order to initiate the handover procedure.	Handover Operation/Normal Operation	MOB_MSHO-REQ/MOB_BSHO-REQ	Encrypted/ICV	Unicast
AAI_HO-RSP	The BS transmits this message on reception of an AAI-HO-REQ message.	Handover Operation/Normal Operation	MOB_BSHO-RSP	Encrypted/ICV	Unicast
AAI_HO-CMD BS Handover Command	The serving BS sends an AAI_HO-CMD to initiate the handover procedure or to acknowledge the AAI_HO-REQ message sent by the MS.	Handover Operation/Normal Operation	HO-CMD	Encrypted/ICV	Unicast
AAI_HO-IND MS Handover Indication	The MS may send an AAI_HO-IND MAC control message during handover preparation, handover execution, and handover cancellation. If a Piggyback Extended Header is included in the AAI_HO-IND message, bandwidth request size should be transferred to the target BS.	Handover Operation/Normal Operation	MOB_HO-IND	Encrypted/ICV	Unicast
AAI_L2-XFER AAI L2 Transfer	This is a generic MAC control message which acts as a generic service container for various services including, but not limited to, device provisioning bootstrap message to the MS, GPS assistance delivery to the MS, Base stations geo-location unicast delivery to the MS, IEEE 802.21 MIH transfer, messaging service, etc. This container is also used for	Inter-RAT Operation	N/A	Encrypted/ICV	Unicast

(Continued)

Table 6-8 IEEE 802.16m MAC Management/Control Messages [1,2] *Continued*

IEEE 802.16m MAC Management Message	Message Description	Usage	IEEE 802.16-2009 Standard Equivalent Message	Security Attributes	Connection Type
	IEEE 802.16m messages that are not processed by the BS, rather they are processed by the network entities beyond the BS. The AAI_L2_XFER message is transmitted only by an authenticated MS.				
AAI_SII-ADV Service Identity Information Advertisement	This message is transmitted when the BS is ready to send the Media Independent Handover (MIH) response in a multi-RAT operation. The BS transmits this message containing the MIH response in the MAC frame	Inter-RAT Operation	N/A	N/A	Broadcast
AAI_PAG-ADV BS Paging Advertisement	This message is sent by the BS on the broadcast Idle Mode multicast connection during the BS paging interval.	Idle Mode Operation	MOB_PAG-ADV	N/A	Broadcast
AAI_DREG-REQ Deregistration Request	This message is sent by the MS to the serving BS in order to notify the BS of the MS de-registration request from the network.	Idle Mode Operation	DREG-REQ	Encrypted/ICV	Unicast
AAI_DREG-RSP Deregistration Command	This message is sent by the BS to confirm immediate termination of the service by the BS and that the MS should attempt network entry	Idle Mode Operation	DREG-RSP	Encrypted/ICV	Unicast

	with another BS later. When more than one paging groups are assigned to an MS, multiple Paging group IDs and the same number of paging offsets that correspond to the Paging group ID are included in the AAI_DREG-RSP message.		
AAI_DREG-CMD	This message is transmitted by the BS to force the MS to change its Access State. This message is either unsolicited or in response to an AAI_DREG-REQ message.	Idle Mode Operation	DREG-CMD	Encrypted/ICV	Unicast
AAI_MC-ADV Multi-carrier Advertisement	The BS broadcasts the system information on each carrier with a specific format and provides the MS with basic radio configuration for all available carriers supported in the cell through the AAI_MC-ADV message. This message is periodically broadcast by the BS and includes the multi-carrier mode and the configurations supported by the BS. The multi-carrier configuration information is relevant to all mobile stations in any multi-carrier modes or in the single carrier mode.	Multi-carrier Operation/Normal Operation	N/A	N/A	Broadcast

(Continued)

Table 6-8 IEEE 802.16m MAC Management/Control Messages [1,2] *Continued*

IEEE 802.16m MAC Management Message	Message Description	Usage	IEEE 802.16-2009 Standard Equivalent Message	Security Attributes	Connection Type
AAI_MC-REQ Multi-carrier Request	The Multi-carrier Request Message is sent by an MS to a BS to request the list of *Assigned Carriers*. This message includes information about the MS supported multi-carrier configurations that is needed by the BS to assign carriers to the MS and to activate the carriers when needed. The MS cannot send the AAI_MC-REQ message until it receives the AAI_MC-ADV message from its serving BS. The AAI_MC-REQ message includes MS multi-carrier capability parameters for carrier assignment, list of candidate assigned carriers, and support of data transmission over guard sub-carrier.	Multi-carrier Operation/Normal Operation	N/A	Null	Unicast
AAI_MC-RSP Multi-carrier Response	This message contains the BS response to the AAI_MC-REQ message to provide the MS with information about its assigned carriers including the following parameters: list of assigned carriers; support of data transmission over guard sub-carrier. While the AAI_MC-RSP message is typically sent to the MS in response to the AAI_MC-REQ	Multi-carrier Operation/Normal Operation	N/A	Null	Unicast

	message, it may also be sent by the BS to update the list of assigned carriers in an unsolicited manner.				
AAI_CM-CMD Carrier Management Command	The activation or deactivation of secondary carriers is decided by the BS based on QoS requirement, load condition of carriers, and other criteria. The BS activates and/or deactivates the secondary carrier with the AAI_CM-CMD message. The BS sends the AAI_CM-CMD message on the primary carrier and includes the following parameters: *Indication Type per DL/UL* (Activation or Deactivation); list of secondary carriers (addressed by physical carrier index); and the *Ranging Indicator* for the activated carrier. The AAI_CM-CMD message may allocate feedback channel for the activated carriers without paired UL; e.g., partially configured carrier or asymmetrically activated DL carrier.	Multi-carrier Operation/Normal Operation	N/A	Encrypted/ICV	Unicast
AAI_CM-IND Carrier Management Indication	In response to the AAI_CM-CMD message, the MS transmits the AAI_CM-IND MAC control message through the primary carrier. This message confirms that the MS has successfully	Multi-carrier Operation/Normal Operation	N/A	Encrypted/ICV	Unicast

(Continued)

Table 6-8 IEEE 802.16m MAC Management/Control Messages [1,2] *Continued*

IEEE 802.16m MAC Management Message	Message Description	Usage	IEEE 802.16-2009 Standard Equivalent Message	Security Attributes	Connection Type
	activated/deactivated the carriers listed in the AAI_CM-CMD message. In the case of activation, the confirmation is sent by the MS when the DL/UL of the newly activated carrier is ready to be used to transport data traffic.				
AAI_SingleBS_MIMO_FBK Single-BS MIMO Feedback	This message is sent by an MS to respond to Feedback Polling A-MAP IE requesting to feedback the sub-band information for MIMO Feedback Modes 2, 3, 5, or 6. It is also used to feedback the transmit correlation matrix when the BS has 8 transmit antennas.	Normal Operation	N/A	Encrypted/ICV	Unicast
AAI_MultiBS_MIMO_FBK Multi-BS MIMO Feedback	This message is sent by an MS as a response to Feedback Polling A-MAP IE requesting multi-BS MIMO feedback.	Normal Operation	N/A	Encrypted/ICV	Unicast
AAI_MultiBS_MIMO-REQ Multi-BS MIMO Request	The AAI_MULTI_BS_MIMO-REQ message is transmitted by the MS to report its preference on single BS precoding with Multi-BS MIMO coordination or multi-BS joint MIMO processing.	Normal Operation	N/A	Encrypted/ICV	Unicast

Message	Description				
AAI_CLC-REQ Collocated Coexistence Request	The MS sends this message to activate, terminate, or reconfigure one or several Type I, Type II, and/or Type III CLC classes. The AAI_CLC-REQ message is sent from the MS to the BS on the mobile station's basic connection. The MS may include AAI_CLC-INFO parameter fields.	Normal Operation	N/A	Encrypted/ICV	Unicast
AAI_CLC-RSP Collocated Coexistence Response	The BS sends the AAI_CLC-RSP message to the MS on the mobile station's basic connection in response to AAI_CLC-REQ.	Normal Operation	N/A	Encrypted/ICV	Unicast
AAI_CLC-INFO Collocated Coexistence Information	The parameters of this message are applicable to AAI_REG-RSP, AAI_RNG-REQ, AAI_RNG-RSP, AAI_CLC-REQ, AAI_SBC-REQ, and AAI_SBC-RSP messages.	Normal Operation	N/A	Encrypted/ICV	Unicast
AAI_FFR-CMD Fractional Frequency Reuse Command	This message is transmitted by the BS to instruct the MS to perform measurement over specific frequency partition.	Normal Operation	N/A	Encrypted/ICV	Unicast
AAI_FFR-REP Fractional Frequency Reuse Report	This message is sent by an MS to report the interference statistics of frequency partition.	Normal Operation	N/A	Encrypted/ICV	Unicast

(Continued)

Table 6-8 IEEE 802.16m MAC Management/Control Messages [1,2] *Continued*

IEEE 802.16m MAC Management Message	Message Description	Usage	IEEE 802.16-2009 Standard Equivalent Message	Security Attributes	Connection Type
AAI_ARQ-Feedback Stand-alone ARQ Feedback	This message is used by a receiver to inform the transmitter of the reception status of a number of ARQ blocks or ARQ sub-blocks. The ARQ feedback IE is included in this message.	Normal Operation	ARQ-Feedback	Encrypted/ICV	Unicast
AAI_ARQ-Discard ARQ Discard	This message is sent by the transmitter when skipping a number of ARQ blocks.	Normal Operation	ARQ-Discard	Encrypted/ICV	Unicast
AAI_ARQ-Reset ARQ Reset	The transmitter or receiver may send this message to reset the parent connection's ARQ transmitter and receiver state machines.	Normal Operation	ARQ-Reset	Encrypted/ICV	Unicast
AAI_SON-ADV SON Advertisement Message	A message that is used by the BS to broadcast relevant Self-Organizing Network (SON) information for action types such as BS Reconfiguration, BS Restart, BS Scanning, or BS Reliability.	Self Organization	N/A	Encrypted/ICV	Unicast

6.8 **CONNECTION AND SESSION MANAGEMENT**

A session is defined as the duration of time from the moment that an MS performs initial network entry and registers with the network and an exclusive MS context is generated in the network until the MS signs off the network and the MS context is flushed out. During this time, the MS may transit between different states (Initialization, Access, Connected, and Idle States) and may perform a number of network re-entries and re-register with the serving BS on transition from Idle to Connected State.

In the IEEE 802.16-2009 standard, a connection is defined as uni-directional mapping between base station and mobile station MAC layers. Connections are identified by a 16-bit connection identifier. There are two types of connections, i.e., management and transport connections, where the management connections can be basic, primary, or secondary type. The basic connection is used by the BS and MS MAC layers to exchange short and time-sensitive MAC control messages. The primary management connection is utilized by the BS and MS MAC layers to exchange long and delay-tolerant MAC control messages. In IEEE 802.16m, a connection is defined as a mapping between the MAC layers of a BS and one or more mobile stations. When the mapping is between a BS and an MS, the connection is called a unicast connection. Otherwise, it is a multicast or broadcast connection. Unicast connections are identified by the combination of a 12-bit STID and a 4-bit FID. Multicast and broadcast connections are identified by the reserved STIDs. In IEEE 802.16m, the connections are classified as management connections and transport connections. Management connections carry MAC control messages. Transport connections, on the other hand, are used to carry user data including upper layer signaling such as DHCP, as well as data-plane signaling such as ARQ feedback. The MAC control messages are never transferred over transport connections and user data is never transferred over management connections. One pair of bi-directional (DL/UL) unicast management/control connections are automatically established when an MS performs initial network entry. An FID with value "0" is reserved for this pair of connections. The mapping between MAC control messages and connection types is static, as shown in Table 6-8. Therefore, there is a difference between the notion of connection in IEEE 802.16m and the legacy standard. In IEEE 802.16m, connections are either management (bi-directional for unicast and uni-directional for broadcast connections) or transport (uni-directional).

Once the Temporary STID is allocated to the MS, the management connections are established automatically. The FIDs for the management connections are not changed during handover or network re-entry between IEEE802.16m compliant systems. All user data communications use transport connections. A transport connection is uni-directional and established with a unique FID that is assigned using a Dynamic Service Addition (DSA) procedure. If the *Group Create/Change* parameter is included in a DSA message, it would indicate that a group of coupled transport connections must be considered together on admission. Each transport connection is associated with an active service flow to provide various levels of QoS required by the service flow. The transport connections are established when the associated active service flows are created, and are released when the associated service flows become inactive. Once established, the FID of the transport connection is not changed during handover. The detailed mechanism for addition, change, and deletion of the transport connections can be described as follows.

Dynamic Service Addition: A set of MAC management messages for addition of a new service flow, and thereby a new transport connection

An AAI_DSA-REQ message is sent by an MS or BS to create a new service flow and may contain parameters for more than one service flow. The MS or BS generates an AAI_DSA-REQ message, including the following parameters:

- *Control Message Type* (i.e., type of the AAI_DSA-REQ message);
- *Service Flow Parameters* specify the service flow's traffic characteristics and scheduling requirements;
- *Convergence Sub-layer Parameter Encodings* specify the service flow's CS-specific parameters.

The Sleep Cycle Identifier (SCID) may be included in the AAI_DSA-REQ to change sleep cycle settings. This message may also contain the *Predefined BR Index* of the 3-step bandwidth request procedure when initiated by the BS. The FID for the transport connection is not included in the MS-initiated AAI_DSA-REQ message. At the BS, the service flow within the AAI_DSA-REQ message is assigned a unique FID for the transport connection, which will be communicated via the AAI_DSA-RSP message. The MS-initiated AAI_DSA-REQ messages may use the service class name instead of the QoS parameters. The BS-initiated AAI_DSA-REQ messages for service class names include the QoS parameter set associated with that service class, as well as the target Security Association Identifier (SAID) for the service flow. An AAI_DSA-RSP message is generated in response to a AAI_DSA-REQ message. An MS or BS generated AAI_DSA-RSP message contains the following parameters:

- *Control Message Type* (i.e., type of the AAI_DSA-RSP message);
- *Confirmation Code* for the corresponding AAI_DSA-REQ message;
- *Service Flow Parameters* to characterize service flow traffic and scheduling requirements if the procedures are executed successfully. The complete specification of the service flow is included in the AAI_DSA-RSP message, if it includes an expanded service class name;
- *Convergence Sub-layer Parameter Encodings* specify the service flow CS-specific parameters if the transaction is successful.

In response to an AAI_DSA-REQ message containing SCID, the SCID may be included in the AAI_DSA-RSP message to accept the base station or mobile station's request. If the AAI_DSA-RSP message does not include SCID, the MS would assume that the request has been denied. The base station's AAI_DSA-RSP messages for service flows that are successfully added contain an FID for the transport connection, as well as the target SCID for the service flow. If the corresponding AAI_DSA-REQ message uses the service class name to request service addition, the AAI_DSA-RSP message will contain the QoS parameter set associated with the service class name. If the service class name is used in conjunction with other QoS parameters in the AAI_DSA-REQ message, the BS accepts or rejects the AAI_DSA-REQ message using the QoS parameters in the AAI_DSA-REQ message. If these service flow encodings conflict with the service class attributes, the BS uses the AAI_DSA-REQ message values as override for those of the service class. If mobile station's AAI_DSA-RSP with successful status is sent and *Service Flow Parameters* are included, the only *Service Flow Parameters* that may be included are ARQ parameters for ARQ-enabled connections. An AAI_DSA-ACK message is generated in response to an AAI_DSA-RSP message. An MS or BS generates the AAI_DSA-ACK message containing the following parameters:

- *Control Message Type* (i.e., type of AAI_DSA-ACK message);
- *Confirmation Code* for the corresponding AAI_DSA-RSP message.

Dynamic Service Change: A set of MAC management messages for changing the parameters of an existing service flow

The MS or the BS generates the AAI_DSC-REQ message including the following parameters:

- *Control Message Type* (i.e., type of AAI_DSC-REQ message);
- *Service Flow Parameters* specify the service flow's new traffic characteristics and scheduling requirements or the admitted and active QoS parameter sets currently in use by the service flow. If the dynamic service flow change message is successful and it contains service flow parameters, but does not contain replacement sets for both admitted and active QoS parameter sets, the omitted set is set to null. The service flow parameters include the FID.

The SCID may be included in the AAI_DSC-REQ message to change sleep cycle setting. An AAI_DSC-REQ message does not carry parameters for more than one service flow. The AAI_DSC-RSP message is generated in response to the AAI_DSC-REQ message inclusive of the following parameters:

- *Control Message Type* (i.e., type of AAI_DSC-RSP message);
- *Confirmation Code* for the corresponding AAI_DSC-REQ message;
- *Service Flow Parameters* specify the service flow's traffic characteristics and scheduling requirements if the transaction is successful. Complete specification of the service flow is included in the AAI_DSC-RSP, if a service flow parameter set contained a service class name and an admitted QoS parameter set. The AAI_DSC-RSP message further includes the QoS parameter set corresponding to the service class name. If specific QoS parameters were also included in the service flow request, these QoS parameters are included in the AAI_DSC-RSP message instead of any QoS parameters of the same type of the service class name.
- *Convergence Sub-layer Parameter Encodings* specify the service flow's CS-specific parameters, if the procedures are successfully executed.

In response to an AAI_DSC-REQ message which contains SCID, the SCID may be included in an AAI_DSC-RSP message to accept the base station or mobile station's request. If the AAI_DSC-RSP message does not include SCID, the MS would assume that its request has failed. An AAI_DSC-ACK message is generated in response to a received AAI_DSC-RSP message. The MS or the BS generates the AAI_DSC-ACK message that includes the following parameters:

- *Control Message Type* (i.e., type of AAI_DSC-ACK message);
- *Confirmation Code* for the corresponding AAI_DSC-RSP message.

Dynamic Service Deletion: A set of MAC management messages for deleting an existing service flow

An AAI_DSD-REQ message is sent by an MS or BS to delete an existing service flow. The MS or BS generates an AAI_DSD-REQ message with the following parameters:

- *Flow ID* to be deleted;
- *Control Message Type* of AAI_DSD-REQ message.

The SCID may be included in the AAI_DSD-REQ to change sleep cycle setting. An AAI_DSD-RSP message is generated in response to an AAI_DSD-REQ message. The MS or BS generates an AAI_DSD-RSP message, including the following parameters:

- *Flow ID* from the AAI_DSD-REQ to which this response refers;
- *Control Message Type* of AAI_DSD-RSP message;
- *Confirmation Code* for the corresponding AAI_DSD-REQ message.

In response to an AAI_DSD-REQ message which contains SCID, the SCID may be included in the AAI_DSD-RSP message to accept the base station or mobile station's request. If the AAI_DSD-RSP message does not include SCID, the MS assumes that its request has failed.

To support Emergency Telecommunications Service (ETS)[xiii] and E-911,[xiv] the emergency service flows are given priority in admission control over the regular service flows. The default service flow parameters are defined for emergency service flows. The BS grants resources in response to an emergency service notification from the MS without going through the complete service flow set-up procedures. The MS can include an emergency service notification in initial ranging or service flow set-up requests. If a service provider wishes to support National Security/Emergency Preparedness (NS/EP) priority services, the BS would use the procedures defined by the home country regulatory body. For example, in the United States, the procedure defined by the FCC in Hard Public Use Reservation by Departure Allocation (H-PURDA) is used to support NS/EP.

6.9 MOBILITY AND POWER MANAGEMENT

The MS power conservation during normal operation is crucial to mobile user connectivity. The IEEE 802.16m provides MS power management functions including sleep and idle mode procedures to minimize MS power consumption.

[xiii]The Emergency Telecommunications Service (ITU.ETS.E106) is a successor to and generalization of two services used in the United States: Multi-Level Precedence and Pre-emption (MLPP), and the Government Emergency Telecommunication Service (GETS). Services based on these models are also used in a variety of countries throughout the world, both Public Switched Telephone Network (PSTN) and Global System for Mobile Communications (GSM)-based. Both of these services are designed to enable an authorized user to obtain service from the telephone network in times of crisis. They differ primarily in the mechanisms used and number of levels of precedence acknowledged [24].

[xiv]The FCC's wireless Enhanced 911 (E911) requirements seek to improve the effectiveness and reliability of wireless 911 services by providing the dispatchers with additional information on wireless 911 calls. The FCC's wireless E911 rules apply to all wireless licensees, broadband Personal Communications Service (PCS) licensees, and certain Specialized Mobile Radio (SMR) licensees. The FCC has divided its wireless E911 program into two phases: Phase I and Phase II. Under Phase I, the FCC requires carriers, within six months of a valid request by a local Public Safety Answering Point (PSAP), to provide the PSAP with the telephone number of the originator of a wireless 911 call and the location of the cell site or base station transmitting the call. Under Phase II, the FCC requires wireless carriers, within six months of a valid request by a PSAP, to begin providing information that is more precise to PSAPs, specifically, the latitude and longitude of the caller. This information must meet FCC accuracy standards, generally to within 50 to 300 meters, depending on the type of technology used. The deployment of E911 requires the development of new technologies and upgrades to local 911 PSAPs, as well as coordination among public safety agencies, wireless carriers, technology vendors, equipment manufacturers, and local wireline carriers (see http://www.fcc.gov/pshs/services/911-services/enhanced911).

6.9.1 **Sleep Mode Operation**

Sleep is a mode of operation in which an MS conducts pre-negotiated periods of absence from the serving BS air interface. These periods are characterized by the unavailability of the MS, as observed by the serving BS, for DL or UL traffic. Sleep mode is used to minimize MS power consumption and to decrease the use of serving BS radio resources. The Sleep Mode may also be used to support co-located multi-radio coexistence. A single power saving class for each mobile station is managed in order to operate the active connections associated with the MS. The Sleep Mode may be invoked when an MS is in the Connected State. When Sleep Mode is activated, the MS is provided with a series of alternate listening window and sleep intervals. The listening window is the time in which the MS is available to exchange control signaling, as well as data, in the uplink or downlink. The IEEE 802.16m provides a mechanism for dynamically adjusting the duration of sleep windows and listening windows based on changing traffic patterns and HARQ operations. The length of successive sleep cycles, each comprising a sleep and listening window, may remain unchanged or may change based on traffic conditions. The sleep and listening windows can be dynamically adjusted for the purpose of data transmission, as well as MAC control signaling. The MS can send and receive data and MAC control signaling without deactivating the Sleep Mode.

Sleep Mode entry is initiated either by the MS or the BS. When the MS is in Active Mode, sleep parameters are negotiated between the MS and BS. The serving BS determines when the MS can transition to Sleep Mode. MAC control signaling is used for sleep mode request and response. The sleep cycle is measured in units of frames. The start of the listening window is aligned with the frame boundaries. The MS ensures that it has the latest system information for proper operation; otherwise, the MS does not transmit in the listening window until the system information is updated. A sleep cycle is the sum of sleep and listening windows. The MS or BS may request change of sleep cycle through MAC control signaling. The BS maintains synchronization with the MS at the sleep/listening windows' boundary. The synchronization can be conducted implicitly with a predetermined procedure or explicitly using an appropriate signaling mechanism. There are 16 distinct sleep patterns or sleep cycle settings (also referred to as power saving modes in the legacy system) specified in IEEE 802.16m where each is denoted by a unique Sleep Cycle Identifier (SCID). Note that there are only three power-save modes in the legacy standard [1]. If an MS requests changes to the sleep pattern to one of the previous patterns using AAI_SLP-REQ including the associated SCID, the BS will respond with an AAI_SLP-RSP containing the same or a different SCID. There is only a single sleep cycle setting that is applied across all MS active connections.

During the sleep window, the MS is unavailable to receive any DL data and MAC control signaling from the serving BS. For the duration of the listening window, the MS can receive DL data and MAC control signaling from the serving BS. The MS can also send data, if any uplink data is scheduled for transmission. The length of the listening window is measured in units of frames. After termination of a listening window, the MS may return to sleep for the remainder of the current sleep cycle. During the listening interval, if an MS does not have any downlink or uplink traffic in certain subframes (i.e., equal-sized fractions of radio frame), the MS can turn off its RF transmit/receive circuitry until the next active subframe(s) (as shown in Figure 6-20). These short-time sleep patterns are referred to as micro-sleep cycles and help the MS to further reduce power consumption.

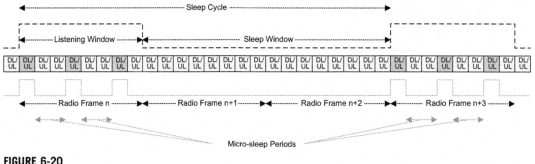

FIGURE 6-20

Example of sleep cycles and micro-sleep patterns

The serving BS may transmit the traffic indication message intended for one or multiple mobile stations during the listening window according to the Sleep Mode negotiation messages. It indicates whether there is traffic destined to one or multiple mobile stations. The traffic indication message is transmitted at a predefined location. On receiving a negative traffic indication in the traffic indication message, the MS can return to sleep for the remainder of the current sleep cycle. The duration of the listening window can be dynamically adjusted based on traffic availability or control signaling in the MS or BS. The listening window can be extended through explicit signaling. The listening window cannot be extended beyond the end of the current sleep cycle. The MS or BS can initiate the Sleep Mode termination/deactivation procedure. The BS signals the termination of the MS Sleep Mode using MAC management messages. The traffic indication signal is sent for a single MS or a group of mobile stations using the AAI_TRF-IND message. The AAI_TRF-IND message is transmitted at the first frame in the listening window of each MS. If the traffic indication is enabled for an MS with Sleep Identifier (SLPID) assigned (i.e., a unique identifier for a mobile station or a group of mobile stations in the Sleep Mode), the MS should expect a traffic indication message at predetermined intervals. The set of SLPID values range from 0 to 1023 and are divided into 32 SLPID Groups. The SLPID Groups are defined as follows: SLPID-Group#0 corresponding to SLPID values 0 … 31; SLPID-Group #1 corresponding to SLPID values 32 … 63; …; SLPID-Group#31 corresponding to SLPID values 992 … 1023. The SLPID-Group Indication Bitmap is a 32-bit field where each bit is assigned to a particular SLPID-Group. In other words, the MSB in the field is assigned to SLPID-Group#0 and subsequent bits correspond to SLPID-Group #1, 2, …, 31. Therefore, the Sleep Mode MS associated with SLPID-Group #n may return to sleep mode if the nth bit in the SLPID-Group Indication Bitmap is not set. If the corresponding bit were set, the MS would read its own Traffic Indication Bit-map in the AAI_TRF-IND message.

On receiving the traffic indication message, the MS checks whether there is a positive traffic indication (e.g., through SLPID-Group Indication Bit-map and Traffic Indication Bit-map or the SLPID assigned to it). If the MS receives a negative traffic indication, then it ends the listening window and proceeds with sleep window operation for the remainder of the sleep cycle, unless the MS has UL signaling or pending traffic for transmission. If the BS transmits a negative traffic indication to the MS, the BS will not transmit any DL traffic to the MS during the remaining part of the listening window, unless there are UL bandwidth requests or UL MAC PDUs sent from the MS which have not been processed. If a positive traffic indication is sent to a specific MS, the BS transmits at least one DL MAC PDU to the MS in the listening window. If the traffic indication message is lost or not properly detected by the MS, the MS would stay awake for the remainder of the listening window. If the MS receives any unicast data during the listening window, it assumes that the traffic indication had been positive. If neither the traffic indication message nor any unicast data in the listening window is received, the MS remains awake even after the listening window ends. The MS then sends a MAC control message (e.g., a signaling header) to inquire about the traffic indication. The BS responds to the MS by sending a unicast MAC control message containing the traffic indication. The operation of the IEEE 802.16m sleep mode with VoIP (persistent and non-persistent scheduling) and HTTP traffic models are shown in Figure 6-21 and Figure 6-22, respectively. It is shown that the MS power can be considerably conserved and the inactivity intervals of the mobile station can be dynamically adjusted for different types of user traffic [25].

6.9.2 Idle Mode Operation

Idle State[xv] provides a power saving mechanism for the MS by allowing the MS to become periodically available for downlink broadcast messages (e.g., paging message) without registration with a specific BS. The network assigns Idle Mode MS to a paging group during Idle Mode entry or location update. This allows the network to minimize the number of location updates performed by the MS and the paging signaling overhead incurred by the BS. The base stations and Idle Mode mobile stations can belong to one or multiple paging groups. Idle mode mobile stations may be assigned to paging groups of different sizes based on user mobility.

The MS monitors the paging message during tthe mobile station's paging listening interval. The start of the mobile station's paging listening interval is derived based on paging cycle and paging offset. The paging offset and paging cycle are defined in terms of the number of superframes (i.e., a collection of four radio frames). The mobile stations may be divided into logical groups in order to distribute the paging overhead.

An MS may be assigned to one or more paging groups. If an MS is assigned to multiple paging groups, it may also be assigned multiple paging offsets within a paging cycle, where each paging offset corresponds to a separate paging group. The MS is not required to perform location update when it moves within its assigned paging groups. The assignment of multiple paging offsets to an MS allows monitoring of the paging message at a different paging offset when the MS is located in one of its paging groups. When an MS is assigned to more than one paging group, one of the paging groups is called the *Primary Paging Group* and the others are known as *Secondary Paging Groups*. If an MS is assigned to one paging group, that paging group is considered the *Primary Paging Group*. When different paging offsets are assigned to an MS, the *Primary Paging Offset* is shorter than the *Secondary Paging Offset*s. The distance between two adjacent paging offsets should be long enough that the MS paged in the first paging offset can inform the network before the next paging offset in the same paging cycle, so that the network avoids unnecessary paging of the MS in the next paging offset. An Idle State MS (while in a paging listening interval) wakes up at its primary paging offset and looks for primary Paging Group Identifier (PGIDs) information. If the MS does not detect the primary PGID, it will wake

---

[xv]Idle State and Idle Mode terms are used interchangeably in this book and in the literature.

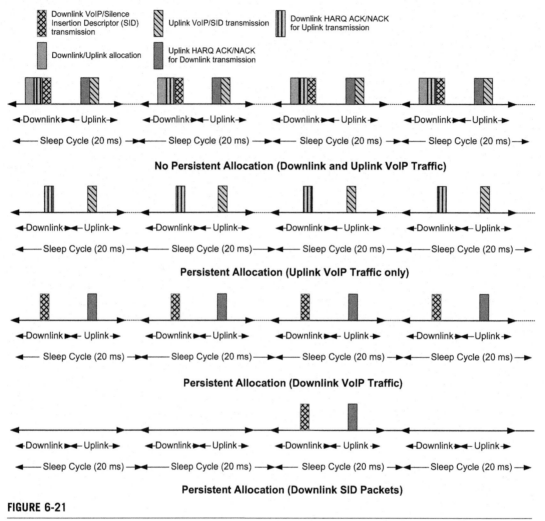

FIGURE 6-21

Sleep mode operation for VoIP (example) [25]

up during its secondary paging offset in the same paging cycle. If the MS can find neither primary nor secondary PGIDs, it will perform a location update.

The BS transmits the list of PGIDs at predetermined locations. The PGID information should be received during a mobile station's paging listening interval. The paging message includes identification of the mobile stations (i.e., temporary identifiers) to be notified of pending DL traffic or location update. The BS does not transmit any DL traffic or paging advertisement to the MS during the mobile station's paging unavailable interval. During the paging unavailable interval, the MS may power down, scan neighbor base stations, reselect a preferred cell, conduct ranging, or perform other activities which the MS would not be available for, to any BS for DL traffic. The MS derives the start of the

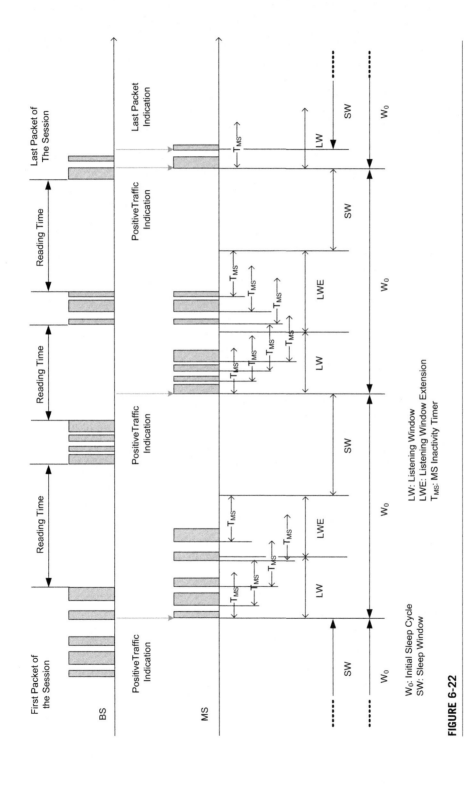

FIGURE 6-22

Sleep mode operation for HTTP traffic (example) [25]

paging listening interval based on the assigned paging cycle and paging offset. At the beginning of the paging listening interval, the MS scans and synchronizes with the downlink of its preferred BS. The MS receives and decodes the system information. The MS confirms whether it is still in the same paging group based on PGID information. During the paging listening interval, the MS monitors superframe header content. If the contents of superframe header has changed, the MS acquires the updated system information. The MS decodes the full paging message at the predetermined intervals. If the MS decodes a paging message that contains its identification, it performs network re-entry or location update depending on the notification indicated in the paging message; otherwise, the MS returns to paging unavailable interval.

An MS or serving BS initiates Idle State using procedures defined in the legacy system [1]. In order to reduce signaling overhead and to provide location privacy, a temporary identifier is assigned to uniquely identify the mobile stations in the Idle State in any paging group. The mobile station's temporary identifier remains valid as long as the MS resides in the same paging group. The temporary identifier is assigned on Idle State entry or during location update as a result of paging group change. The temporary identifier may be used in paging messages or during mobile station's network re-entry.

An MS terminates the Idle State operation using procedures defined in the legacy system [1]. For termination of Idle State, the MS performs network re-entry with its preferred BS. An MS in the Idle State performs the location update process, if any of the following location update trigger conditions are met; during paging group location update, timer based location update, or Multicast and Broadcast Service (MBS) location update, the MS may update the temporary identifier, paging cycle and paging offset. If an MS chooses to update its location, depending on the security association that the MS shares with its preferred BS, it uses either the: secure location update process or unsecure location update process.

The MS performs the location update once it detects a change in the paging group. The MS detects the change of the paging group by monitoring the PGID, which is transmitted by the BS. The MS periodically performs location update prior to the expiration of an Idle State timer. In each location update including paging group location update, the Idle State timer is reset. The MS attempts to perform a location update as part of its regular power-down process. For an MS receiving MBS data in the Idle State, during MBS zone transition, the MS may perform the MBS location update process to acquire the MBS zone information for continuous reception of MBS content.

Enhanced power saving schemes can also be utilized when the MS is in the Active Mode. In this case, the BS optimizes resource and transmission parameters to increase energy savings at the MS.

6.10 SCHEDULING SERVICES

Scheduling services demonstrate the data handling mechanisms supported by the MAC scheduler for data transport on a connection. Each connection is associated with a single scheduling service. A scheduling service is identified by a set of QoS parameters that quantify different aspects of its behavior. These parameters are managed using DSA and DSC MAC management messages. The legacy scheduling services are supported in IEEE 802.16m. In addition, IEEE 802.16m provides a specific scheduling service to support real-time non-periodic applications such as online gaming. IEEE 802.16m supports adaptation of service flow QoS parameters. One or more sets of QoS parameters are defined for one service flow. The MS and the serving BS negotiate the supported QoS parameter sets during the service flow set-up procedure. When QoS requirements/traffic characteristics for UL traffic change, the BS may change the service flow QoS parameters, such as grant/polling interval or grant size based on predefined rules. In addition, the MS may request the BS to change the service flow QoS parameter set through signaling. The BS then allocates the radio resource according to the new service flow parameter set.

To satisfy the latency requirements for network entry, handover, and state transitions, IEEE 802.16m supports fast and reliable transmission of MAC control messages. To provide reliable transmission of MAC control messages, all MAC control messages can be fragmented. The HARQ error control mechanism is applied to all unicast MAC control messages. Message timers for re-transmission are defined for all the unicast MAC control messages. The message timers may be different for various MAC control messages. If HARQ is applied during the transmission of a MAC control message, and if the HARQ process is terminated with an unsuccessful outcome before the expiration of the message timer, the MAC message management entity in the transmitter may initiate re-transmission of the entire message or the message fragment of the failed HARQ burst. When the transmitter polls a message acknowledgement, the receiver responds with a MAC layer acknowledgement. For fragmented messages, all fragments of the message must be received before the MAC layer acknowledgment is sent.

In a point-to-multi-point topology, the network operates with a centralized scheduling service located in the BS. The BS antenna system is sectorized and the BS simultaneously serves a number of independent sectors. The BS scheduler is responsible for determining the physical transmission parameters for a particular connection based on the CQI reports from the MS (channel quality), QoS requirements, and parameters of the service flow, as well as MS bandwidth requests and fairness criteria. The QoS parameters associated with service flow and the channel quality reports from the MS help the BS scheduler adjust the throughput and latency of the uplink and downlink transmissions corresponding to a certain MS (see Figure 6-23). The BS scheduler may or may not coordinate the downlink and uplink transmissions to a mobile station with other base stations. The inter-BS coordination is required in certain deployment scenarios (e.g., interference management and coordination) or MIMO schemes (e.g., multi-BS MIMO or DL/UL multi-point coordinated transmission in 3GPP LTE-Advanced). The scheduler decisions are signaled to the MS through allocation A-MAPs (DL/UL MAPs in the legacy systems) for the downlink and uplink. Figure 6-23 shows an example operation of the BS scheduler in the downlink and uplink. Depending on the type of application layer services, the service flows are classified and assigned QoS parameters, and are then mapped to downlink or uplink connections.

The IEEE 802.16 standard does not define a scheduling algorithm, and the details of scheduler implementation remain proprietary and vendor-specific. However, there are certain scheduling algorithms such as *Round Robin*, *Weighted Round Robin*, *Proportional Fair* (*PF*), *Weighted Proportional Fair*, etc., that are typically used. The details of the weighted proportional fair algorithm are provided in Appendix A of this chapter.

The MAC protocol peers communicate using a set of MAC control messages. These messages are defined based on ASN.1 syntax. The Packed Encoding Rules (PER)[xvi] are used to encode the messages

---

[xvi]Packed Encoding Rules (PER) are ASN.1 encoding conventions for producing compact transfer syntax for data structures described in ASN.1.

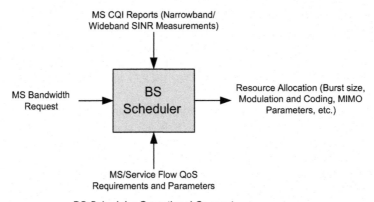

BS Scheduler Operational Concept

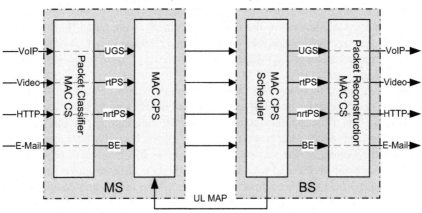

BS Scheduler Functions in Uplink

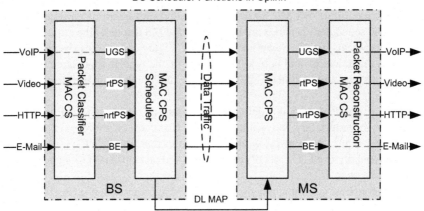

BS Scheduler Functions in Downlink

FIGURE 6-23

BS scheduler operation in downlink and uplink

for transmission over the air interface. IEEE 802.16m provides a generic MAC control message at the L2 called L2_Transfer[xvii] that acts as a generic service container for various standards-defined services, including but not limited to: device provisioning bootstrap message to MS; GPS assistance delivery to MS; base station's geo-location unicast delivery to MS; IEEE 802.21-2009 Media Independent Handover standard message transfer; etc.

6.10.1 Persistent Scheduling

Persistent scheduling is a technique used to reduce assignment overhead for connections with periodic traffic pattern and relatively fixed payload size. To allocate resources persistently to a single connection, the BS transmits separate DL or UL Persistent A-MAP IEs. The persistently allocated resource size, position, and the modulation and coding scheme, is maintained by the BS and MS until the assignment is de-allocated, changed, or an error occurs. The persistent scheduling does not include any consideration for HARQ re-transmission of data packets. The resources used for re-transmissions can be allocated one at a time as needed using a DL or UL Basic Assignment A-MAP IE.

For persistent allocation in the DL or UL, the BS transmits the DL or UL Persistent A-MAP IE, respectively. Allocation of the persistently assigned resource begins in the DL or UL subframe that is referenced by the DL or UL Persistent A-MAP IE and repeats after an allocation period that is specified in the DL or UL Persistent A-MAP IE. The attributes of the persistently scheduled resource including size, location, MIMO encoder format, and modulation and coding scheme are maintained per DL or UL Persistent A-MAP IE. The values of HARQ Channel Identifier (ACID)[xviii] field, and the number of ACIDs in the DL or UL Persistent A-MAP IE are used together to specify an implicit cycling of HARQ channel identifiers. The allocation period and number of ACIDs required for persistent operation are configured in the DL or UL Persistent A-MAP IE. In order to facilitate link adaptation and avoid resource holes, the attributes of a persistently allocated resource can be changed. To change an persistent assignment, the BS transmits the DL or UL Persistent A-MAP IEs for DL or UL reallocations. If an MS has an existing persistent allocation in a particular subframe and receives a new persistent allocation in the same subframe, the new persistent allocation replaces the original allocation. In other words, the original persistent allocation is de-allocated. When the BS sends a Persistent A-MAP IE to reallocate a persistently assigned resource, a different HARQ feedback channel must be assigned for reallocation. The reception of an ACK/NACK in the recently assigned HARQ feedback channel for the persistently assigned resource with the updated attributes will ensure that the reallocation control signaling was received correctly by the MS.

[xvii]To facilitate inter-RAT interworking functions, the IEEE 802.16m provides a MAC control message called AAI_L2_XFER that acts as a generic service container for various services. This container is also used for IEEE 802. 16m MAC control messages that are not processed by the serving BS, but rather are processed by network entities beyond the serving BS. For example, EAP transfer is processed at the authenticator entity in the access network and it is simply relayed by the BS to the authenticator. The AAI_L2_XFER is transmitted exclusively by authenticated mobile stations.

[xviii]The HARQ channel identifier is used to identify HARQ channels. Each connection can have multiple concurrent HARQ channels, each of which may have a pending encoder packet transaction.

Asynchronous HARQ re-transmission is used for downlink persistent allocations. The DL Basic Assignment A-MAP IE is transmitted to signal control information for HARQ re-transmission. Synchronous HARQ re-transmission (i.e., a HARQ scheme where the transmissions/re-transmissions are scheduled at predetermined fixed time intervals) is used for uplink persistent allocations. The UL Basic Assignment A-MAP IE is transmitted to signal control information for HARQ re-transmission. For transmissions where HARQ is enabled, an ACK or NACK is transmitted to acknowledge the success or failure of decoding of data burst. If an ACK or NACK for the data burst is detected in the assigned HARQ feedback channel, the BS would assume that the DL Persistent A-MAP IE is correctly received by the MS. If the initial data burst identified by the UL Persistent A-MAP IE is successfully decoded in HARQ period, the BS assumes that the UL Persistent A-MAP IE is correctly received. If no ACK or NACK is detected in the HARQ feedback channel, the BS assumes that the MS has not received the DL Persistent A-MAP IE and the same DL persistent allocation can be re-transmitted.

In the case of de-allocation of persistent allocations in the DL/UL, the BS transmits a HARQ feedback allocation in the DL/UL Persistent A-MAP IE. This allocation is used to identify the HARQ channel in which the ACK for the DL/UL Persistent A-MAP IE signaling the de-allocation is transmitted. In the absence of an ACK, the BS will assume that the MS has not received the DL/UL Persistent A-MAP IE, and will re-transmit the DL/UL Persistent A-MAP IE that signaled the de-allocation. If no ACK or NACK is detected in the HARQ feedback channel, the BS will assume that the MS has not received the UL Persistent A-MAP IE and re-transmit the same UL Persistent A-MAP IE.

6.10.2 Group Resource Scheduling

The Group Resource Allocation (GRA) is a scheduling mechanism that allocates resources to multiple users as a group, in order to reduce the control overhead. The users are grouped according to the commonality of channel conditions and operational parameters, such as modulation and coding scheme, MIMO mode, HARQ burst size, and resource size. A bitmap is used to represent different combinations of HARQ burst sizes and resource sizes that are used by a group. A group facilitates the dynamic link adaptation based on the limited set of MIMO mode, nominal modulation and coding rates, and HARQ data burst sizes.

The serving BS configures the MIMO mode set for each group among the predefined candidate sets for downlink and uplink. When an MS is added to the group, the configured *Group MIMO Mode Set ID* is indicated through Group Configuration MAC control message. The assigned MIMO mode to the MS in the group is chosen from the configured set as shown in Table 6-9. The BS configures a range of nominal Modulation and Coding Schemes (MCS) supported for each group by indicating the highest and lowest MCS values through Group Configuration MAC control message for both downlink and uplink. The allocation size for an MS in the group is determined based on the nominal MCS levels and HARQ burst size. The BS configures a HARQ burst size set for each group among the predetermined HARQ burst size set candidates. Those candidates are signaled to the mobile stations through broadcast channel. When an MS is added into the group, the configured *HARQ Burst Size Set ID* is signaled through Group Configuration MAC control message. The assigned HARQ burst size to an MS in the group is chosen from the configured set [2].

The addition of an MS to a group is done when group resource allocation is initialized for the MS, or when an MS in a group moves to another group. The MS must be informed in advance about group

Table 6-9 DL/UL MIMO Mode Set Candidates for Group Scheduling

ID	Link	Group MIMO Mode Set	Constrained Number of MIMO Streams
0b00	Downlink	Mode 0	N/A
0b01		Mode 0, Mode 1	2
0b10		Mode 2	1
0b11		Reserved	N/A
0b00	Uplink	Mode 0	N/A
0b01		Mode 0, Mode 1	2
0b10		Mode 2	1
0b11		Mode 3	1

information in order to enable the MS to correctly interpret resource assignment information from Group Resource Allocation MAC control message. The information is transmitted through unicast Group Configuration MAC control message. When a BS decides to use group resource allocation for an MS, the BS adds the MS to an existing group. If the existing groups are not appropriate for the MS, or if no groups currently exist, the BS may form a new group. The BS signals group configuration information via Group Configuration MAC control message, which includes the *Group ID* of the group (to which the MS is added) and the assigned *User Bitmap Index*, to the MS. Once the MS is added to the group, the resources used for initial transmission of the HARQ data burst may be allocated as part of the group until the MS is deleted from the group.

On receiving Group Configuration MAC control message, the MS obtains the required information to interpret the assigned MIMO mode, HARQ burst size, and resource size from the bitmaps in the corresponding MAC management message. Once the MS receives a Group Configuration MAC control message, it monitors its allocations until it is deleted from the designated group. The BS may delete an MS from a group based on either or both of the following conditions: (1) all the GRA-enabled connections are terminated and (2) the MIMO mode/nominal MCS/HARQ burst size suitable for the MS no longer belongs to the MIMO Mode Set/nominal MCS set/HARQ burst size set corresponding to the group. The BS may delete a number of mobile stations from a group in a subframe. The deletion information is signaled via the Group Resource Allocation MAC control message. The de-allocation information is sent by listing *De-allocated MS Index*. The *De-allocated MS Index* is determined based on the order of the MS among inactive mobile stations in the user bitmap. The length of the *De-allocated MS Index* is determined according to the number of inactive users. Figure 6-24 shows an example de-allocation with de-allocated MS index.

The deletion of an MS takes effect from the subframe in which the deletion information is sent to the MS. After sending the deletion information, the BS will for an ACK from the MS. The BS does not allocate the corresponding bitmap position to another MS until an ACK for deletion is received. After decoding the Group Resource Allocation MAC control message, if an MS realizes that it has been deleted from the group, it will not expect allocations in that group starting in the same subframe in which deletion information was sent. The MS sends an ACK to the BS to acknowledge receipt of the deletion information.

The user bitmap may be rearranged via Group Resource Allocation MAC control message in order to indicate open user bitmap positions in the current user bitmap. The open positions indicated in the

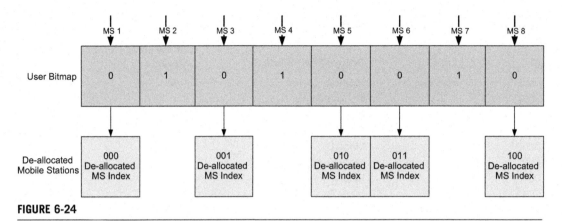

FIGURE 6-24

An example of de-allocation with a de-allocated MS index

rearranged bitmap implicitly identify the available positions for switching user bitmap. For HARQ-enabled transmissions, an ACK/NACK is transmitted to acknowledge the successful/unsuccessful decoding of a data burst. The starting HARQ feedback channel for the group is signaled in the DL Group Resource Allocation A-MAP IE and scheduled mobile stations belonging to the group will determine their individual HARQ feedback channels.

The GRA uses bitmaps to signal resource allocation information for mobile stations within a group. These bitmaps are sent in the Group Resource Allocation MAC control message. The first bitmap is the *User Bitmap* which uses one bit per user, to signal which users are scheduled in that frame. The second bitmap is *MIMO Bitmap* which is used to indicate the assigned MIMO mode, when multiple MIMO modes and their associated parameters are supported in the group. When the *MIMO Mode Set* of the group includes MU-MIMO, PSI Bitmap and Pairing Bitmap appear to determine the two mobile stations that share the same resource. The *Pairing Bitmap* is used to indicate a pair of mobile stations using different number of streams. The number of bits per pair in the *Pairing Bitmap* depends on the total number of pairs in the group. If there are N pairs in the group, the number of bits per pair is equal to $[\log_2 N]$. The third bitmap is the *Resource Allocation Bitmap* which uses N bits per MS to signal the HARQ burst size and resource size for the scheduled MS in the subframe or extended subframe that are scheduled in the frame. The scheduled mobile stations may have a different number of bits in the third bitmap when they are assigned different MIMO mode and number of streams.

6.11 BANDWIDTH REQUEST AND ALLOCATION

Bandwidth Request (BR) refers to a mechanism that a mobile station uses to inform the serving BS about the need for UL bandwidth allocation. The MS may use a contention-based random access BR indicator and an optional quick access message on BR channel, a standalone bandwidth request, a piggybacked bandwidth request carried in an extended header in the MAC PDU, or a bandwidth request using fast-feedback channel. Bandwidth requests in standalone and piggybacked schemes are made in the form of the number of bytes needed to carry the MAC PDU, excluding the physical layer overhead. The bandwidth request message from the MS indicates the size of the payload excluding any header, security, or other MAC PDU overhead that are included during transmission over the air interface. An MS requests UL bandwidth on a per-connection basis. In addition, the MS may request bandwidth for multiple connections in a piggyback scheme. Two mechanisms may be used by a network operator to impose cell access restrictions. The first mechanism is an indication of cell status and special reservations for control of cell selection and reselection. The second mechanism, referred to as access class control, prevents selected classes of users from sending initial access messages for control of emergency calls. The serving BS may advertise a minimum access class in the BR channel configuration within a DL control message. When an MS has data to send in the uplink using the contention-based random access bandwidth request mechanism, it must ensure that the priority of the information access class is higher than or equal to the minimum access class advertised by BR channel configuration within the DL control message. If this is not the case, then the MS waits until the BR channel configuration within a DL control message advertises an access class with a priority less than or equal to that requested by the MS. When the MS access class is allowed, the MS appropriately sets its internal backoff window. The bandwidth request channel and bandwidth request preambles are used for contention-based random access. Each BR channel indicates a BR opportunity. The MS decides whether to send the BR preamble sequence only or to send the BR preamble sequence together with a quick access message for the random access based BR procedure.

The three-step random-access-based BR procedure is illustrated in Figure 6-25. In step 1, the MS transmits a BR preamble sequence and a quick access message on a randomly selected BR opportunity. The BR-ACK A-MAP IE is sent in the next DL frame if the BS detects at least one BR preamble sequence in the BR opportunities in the previous frame. In this case, if the BR-ACK A-MAP IE is not sent in the next DL frame, the MS assumes an implicit NACK and may restart the BR procedure.

The BR-ACK A-MAP IE indicates the decoding status of each BR opportunity in the previous frame, all correctly received BR preamble sequences in the BR opportunities of the previous UL

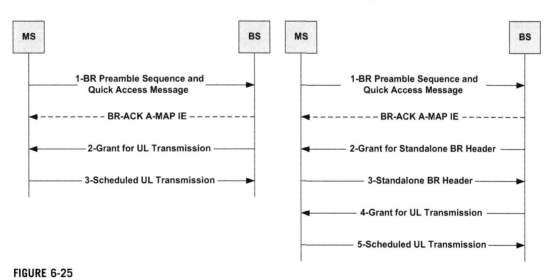

FIGURE 6-25

Three- and five-step random access bandwidth request procedures frame, and the decoding status of the quick access message for each correctly received BR preamble sequence. If the BR-ACK bitmap indicates no BR preamble sequence was detected in the BR opportunity selected by the MS or the mobile station's BR preamble sequence was not included in the selected BR opportunity in the BR-ACK A-MAP IE, the MS would consider it as a NACK. On successful decoding of the quick access message, the BS may provide a UL grant to the MS using the STID and the BR index provided in the quick access message. If the MS receives neither a UL grant nor a NACK, it starts the BR timer at the DL frame located immediately after the UL frame in which the MS has sent the BR. If the BR-ACK A-MAP IE indicates successful reception of the BR preamble sequence and quick access message, the BR timer value is set to the *Differentiated BR Timer* acquired during the DSx procedure. For all other cases, the BR timer value is fixed. The BR timer is stopped on reception of the UL grant. The MS considers the BR attempt as failed and may restart the BR process if either of the following conditions is met: (1) the MS receives a NACK or (2) the BR timer expires.

During the three-step BR procedure, if the BS is unable to decode the quick access message, the BS falls back to the five-step BR procedure illustrated in Figure 6-25. In that case, in step 2, the BS provides a UL grant to the MS using a BR-ACK A-MAP IE or CDMA Allocation A-MAP IE. In step 3, the MS transmits a standalone bandwidth request header or uses the assigned UL resource for its uplink data transmission instead of the BW-REQ message. In both cases, STID is carried in the uplink transmission. When using a five-step procedure, the MS starts the BR timer after sending the BR header in step 3. The BR timer value is set to the *Differentiated BR Timer* acquired during the DSx procedure. The MS stops the BR timer on reception of the UL grant. The BR procedure may restart after expiration of the BR timer. In a five-step random access BR procedure, an MS only sends a BR preamble sequence. If the MS receives neither UL grant nor NACK in the next DL frame following transmission of the BR preamble sequence, a fixed value timer is activated.

The BR Acknowledgement (BR-ACK) A-MAP IE indicates the decoding status of the BR opportunities in the previous UL frame. All the successfully received preamble sequences are acknowledged in ascending order. In addition, the BR-ACK A-MAP IE also includes the allocation information for the fixed-sized BR header. The UL resource and HARQ feedback are allocated to the preamble sequences whose grant indicator is set. The allocations are ordered based on the index of preamble sequences. The standalone bandwidth request header is used by the MS to send bandwidth request in step 3 of the five-step contention-based random access, or as a response to the polling from the serving BS. The MS can use any UL resource allocated to it to send the standalone BR header. The piggybacked bandwidth request is used by the MS to request bandwidth for the same or a different connection into which the user data in the MAC PDU is mapped. The bandwidth request message can also be sent from an MS to the BS using the primary fast-feedback channel. When a three-step BW-REQ procedure is used, the quick access message sent in step 1 carries 12 bits of information, including the MS addressing information with quick access sequences carrying additional 4-bit BW-REQ information. The following parameters are carried in step 1 of a three-step BW-REQ procedure:

- STID of the MS (12 bits);
- Predefined BR Index (4 bits): the mapping between the predefined BR index and BR size/QoS level is done during DSx procedure. The BR size/QoS level is determined based on the QoS parameters of the flow in the DSx messages. The BR size is in units of bytes and is mapped to the *Maximum Traffic Burst* parameter, and QoS level is mapped to the UL *Grant Scheduling Type* parameter.

When the standalone bandwidth request header is transmitted in step 3 of the five-step contention-based random access BR procedure, it includes the following parameters:

- STID of the MS;
- FID of the requesting connection;
- Aggregate bandwidth request.

When the standalone bandwidth request header is transmitted using the UL grant allocated to the MS, it may be used to request bandwidth for one or multiple flows, GPI change for aGPS, or minimum delay of the requested grant for BE QoS class, and includes the following parameters. The BS uses the size of single flow standalone bandwidth request header for polling allocation.

- FID of the requesting connection;
- Aggregate or incremental bandwidth to request one or multiple flows;
- New GPI value for aGPS or minimum delay of the requested grant for BE QoS class;
- GPI change indicator for aGPS.

A piggybacked bandwidth request includes the FID of the requesting connection and the aggregated bandwidth. Multiple requests can be included in one piggyback bandwidth request. In the bandwidth request procedure, the grant for BW-REQ message is allocated by CDMA Allocation A-MAP IE. The CDMA Allocation A-MAP IE is used for allocation of bandwidth to a user that requested bandwidth using a ranging or BR code. The MS decodes the information element and verifies the "CRC masked by the reserved STID for BR-ACK A-MAP" (MCRC) field with its specific 12-bit *Random Access Identifier* (RAID) and 4-bit *Masking Indicator*. The RA-ID is calculated by a hash function with the mobile station's random access attributes (*Frame Number Index* [4 bits], *Ranging Code/BR Code Index* [6 bits], and *Opportunity Index* [2 bits]). The *Masking Indicator* indicates the identifier used for CRC masking. The allocation is in the first UL subframe relevant to the A-MAP region, regardless of DL/UL ratio. The BS requests adjustment of timing and power by MS using *Timing Adjust* (amount of timing advance required to adjust the MS uplink transmission), *Power Level Adjust* (relative change in uplink transmission power level), and *Offset Frequency Adjust* (relative change in uplink transmission frequency) in CDMA Allocation A-MAP IE for ranging [2].

6.12 MULTI-RADIO COEXISTENCE

The IEEE 802.16m can support and facilitate collocated multi-radio operation and coexistence on a mobile platform through pre-negotiated periodic absence of the MS from the serving BS. The pattern of such a periodic absence is referred to as Collocated Coexistence (CLC) class. The following CLC class parameters are defined:

- *CLC Start Time*: the start time of a CLC class;
- *CLC Active Interval*: the time duration of a CLC class designated for collocated non-IEEE 802.16m radio activities;
- *CLC Active Cycle*: the time interval corresponding to active pattern of a CLC class repeating;
- *CLC Active Ratio*: the time ratio of CLC active intervals to CLC active cycle of a CLC class;
- *Number of Active CLC Classes*: the number of active CLC classes of the same type of an MS.

Table 6-10 Collocated Coexistence Class Parameters [2]

	CLC Active Cycle	CLC Active Interval	CLC Start Time	N_i	R_i	T_i
Type I	Microsecond	Subframe	Subframe	1	5%	8 Subframes (5 ms)
Type II	Frame	Subframe	Frame	1	30%	64 Subframes (40 ms)
Type III	Not Applicable	Superframe	Superframe	Not applicable	Not applicable	150 Superframes (3 seconds)

The IEEE 802.16m supports three CLC classes and they differ from each other in terms of the time unit of CLC start time, active cycle, and active interval, as shown in Table 6-10. Type I CLC class is recommended for non-IEEE 802.16m multi-radio activity that is low duty cycle and may not be aligned with the IEEE 802.16m radio frame boundary. Type II CLC class is recommended for scheduling flexibility. Type III CLC class is recommended for continuous non-IEEE 802.16m multi-radio activity that lasts in the order of a few seconds and has only one cycle.

The MS determines CLC active interval and cycle based on the activities of its collocated non-IEEE 802.16m multi-radios. The MS determines CLC start time only for Type I CLC class and the BS determines CLC start time for Type II and III CLC classes to ensure scheduling flexibility. The serving BS does not schedule A-MAP, user data, and HARQ feedback associated with the mobile station in the CLC active interval of an active CLC class. The downlink or uplink or both transmissions may be prohibited depending on the configuration of the CLC class, where the default action is that both DL and UL allocations are prohibited (see Figure 6-26). The support of all three types of CLC classes is mandatory for the BS and optional for the MS. The serving BS manages each type of CLC class with the following three limits:

- R_i: maximum CLC active ratio (%);
- T_i: maximum CLC active interval;
- N_i: maximum number of active CLC classes.

Where index i is set to 1, 2, or 3 to indicate Type I, II, and III CLC class, respectively. The BS may include the CLC limits in AAI_REG-RSP. The higher value of a limit indicates more support for non-IEEE 802.16m multi-radio activities. The CLC limits must exceed the default values in Table 6-10. In the example shown in Figure 6-26, the pre-allocated first re-transmission falls in a CLC active interval and the serving BS reschedules the HARQ re-transmission in the next available subframe. The static time relevance to the initial transmission is no longer valid and the serving BS sends A-MAP to notify the MS where the rescheduled HARQ feedback is located. The serving BS synchronously schedules the second re-transmission according to the rescheduled first re-transmission.

A CLC class remains active until it is deactivated by the MS in one of the Connected State modes. The MS may skip scanning operation in a scan interval, if it overlaps with a CLC active interval. The MS and the serving BS locally deactivate all CLC classes after the MS enters Idle State. During handover preparation, the new BS may obtain information about active CLC classes from the previous

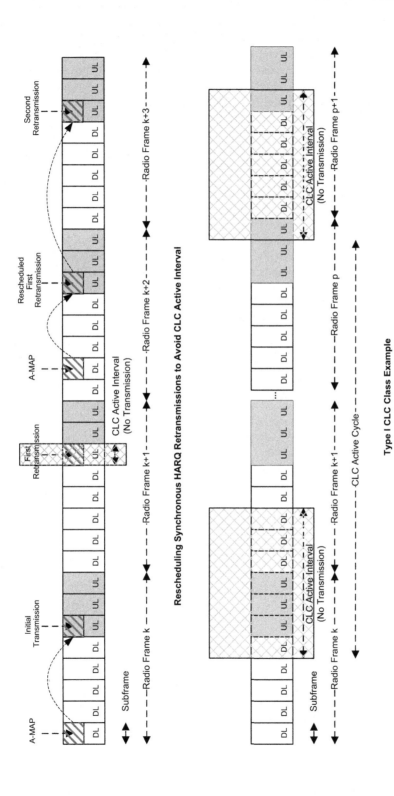

FIGURE 6-26

An illustration of system behavior during CLC active intervals [2]

BS via backhaul. The CLC active cycle and interval parameters remain the same. If the CLC active cycle is one frame, the MS uses a CLC active bitmap to configure a Type II CLC class. The bitmap is used to indicate the CLC active interval within the designated frame, where every bit in the bitmap corresponds to a subframe within the radio frame and, if set to 1, would indicate the corresponding subframe is a CLC active interval (i.e., non-IEEE 802.16m radio activity interval).

6.13 **3GPP LTE RADIO RESOURCE CONTROL FUNCTIONS**

In 3GPP LTE, a user terminal can be in two different states, as shown in Figure 6-27. The Radio Resource Control (RRC) layer in the eNB makes handover decisions based on neighbor cell measurements sent by the UE, pages for the UEs over the air-interface, broadcasts system information, controls UE measurement reporting such as the periodicity of channel quality reports and allocates cell-level temporary identifiers to active UEs. It also executes transfer of UE context from the source eNB to the target eNB during handover, and performs integrity protection of RRC messages. The RRC layer is responsible for setup and maintenance of radio bearers. A UE is in RRC_CONNECTED state when an RRC connection has been established. If no RRC connection is established, the UE is said to be in RRC_IDLE state. The RRC states can further be characterized as follows:

- RRC_IDLE is a state where a UE specific Discontinuous Reception (DRX) may be configured by upper layers. In the idle mode, the UE is saving power and does not inform the network of each cell change. The network knows the location of the UE to the granularity of a few cells, called the Tracking Area (TA). The UE monitors a paging channel to detect incoming traffic, performs neighboring cell measurements and cell selection/reselection, and acquires system information. In brief, the RRC_IDLE state is characterized as follows:
  - UE is known in EPC and has an assigned IP address;
  - UE is not known in E-UTRAN/eNB;
  - User terminal's location is known at Tracking Area level;

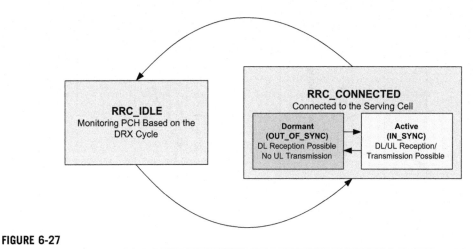

FIGURE 6-27

3GPP LTE UE states and sub-states [12]

- Unicast data transfer with UE is not possible;
- UE can be reached through paging in Tracking Areas controlled by EPC;
- UE-based cell-selection and Tracking Area update provided to EPC.
- RRC_CONNECTED is a state where transfer of unicast data to/from UE is performed and the UE may be configured with a UE specific DRX or Discontinuous Transmission (DTX). The UE monitors control channels associated with the shared data channel to determine if data is scheduled for it, provides channel quality and feedback information, performs neighboring cell measurements and measurement reporting, and acquires system information. One or multiple IP addresses are assigned to the UE, as well as a temporary identifier called Cell Radio Network Temporary Identifier (C-RNTI) that is used for signaling purposes between the UE and the network. In brief, the RRC_CONNECTED state is characterized as follows:
  - UE is known in EPC and E-UTRAN/eNB, i.e., UE context is stored in eNB;
  - UE location is known at cell level;
  - Unicast data transfer with UE is possible;
  - DRX is supported for power saving (in Dormant sub-state);
  - Mobility is controlled by the network.

According to references [12] and [15], the RRC_CONNECTED state can be further implicitly divided into sub-states: (1) Dormant (OUT_OF_SYNC), where the UE may receive DL transmission but cannot transmit in the UL, if the UE is not synchronized to the network; and (2) Active (IN_SYNC), where a network-synchronized UE can send and receive in the UL and DL, respectively. These sub-states have not been further specified in Release 9 specifications [14].

The UE monitors the Paging Channel (PCH) and/or *System Information Block Type 1* contents to detect system information change. It further monitors control channels associated with the shared data channel to determine if data is scheduled for it, provides channel quality and feedback information, performs neighbor cell measurements and measurement reporting, and acquires system information.

A signaling radio bearer (SRB) is defined as a radio bearer (RB) that is used only for the transmission of RRC and NAS messages. More specifically, the following three SRBs are defined:

- SRB0 is for RRC messages using the Common Control Channel (CCCH) logical channel;
- SRB1 is for RRC messages including piggybacked Non-Access Stratum (NAS) messages, as well as for NAS messages prior to the establishment of SRB2, all using DCCH logical channel;
- SRB2 is for NAS messages using DCCH logical channel, has a lower priority than SRB1, and is always configured by E-UTRAN after security activation.

In the downlink, the piggybacked NAS messages are used for bearer establishment/modification/release. In the uplink, the piggybacked NAS messages are used for transferring the initial NAS messages during connection set-up. The NAS messages transferred via SRB2 are also contained in RRC messages, which do not include any RRC protocol control information. Once security is established, all RRC messages on SRB1 and SRB2, including those containing NAS or non-3GPP messages, are integrity-protected and ciphered by PDCP. The NAS independently applies integrity protection and ciphering to the NAS messages. The RRC protocol offers the following services to upper layers:

- Broadcast of common control information;
- Notification of mobile terminals in RRC_IDLE, e.g., about a terminating call;
- Transfer of dedicated control information, i.e., information for one specific UE.

The following are the main services that RRC expects from lower layers:

- Integrity protection and ciphering;
- Reliable and orderly transfer of information without introducing duplicates and with support for segmentation and concatenation.

The RRC protocol includes the following main functions:

- Broadcast of system information including NAS common information, information applicable for mobile terminals in RRC_IDLE, e.g., cell (re-)selection parameters, neighbor cell information and information pertinent to mobile terminals in RRC_CONNECTED, e.g., common channel configuration information, as well as ETWS notification.
- RRC connection control including paging, establishment/modification/release of RRC connections (e.g., assignment/modification of UE identity (C-RNTI), establishment/modification/release of SRB1 and SRB2, access class barring, initial security activation (i.e., initial configuration of Access Stratum (AS) integrity protection and AS ciphering), RRC connection mobility (e.g., intra-frequency and inter-frequency handover, associated security handling, key/algorithm change, specification of RRC context information transferred between network nodes), establishment/modification/ release of radio bearers (RBs) carrying user data, radio configuration, and control (e.g., assignment/modification of ARQ configuration, HARQ configuration, DRX configuration), QoS control including assignment/modification of semi-persistent scheduling parameters for DL and UL, assignment/modification of parameters for UL rate control in the UE (i.e., allocation of a priority and a prioritized bit rate for each RB), and recovery from radio link failure.
- Inter-RAT mobility (e.g., security activation, transfer of RRC context information).
- Measurement configuration and reporting including establishment/modification/release of measurements (e.g., intra-frequency, inter-frequency and inter-RAT measurements), set-up and release of measurement gaps, and measurement reporting.
- Other functions including transfer of dedicated NAS information and non-3GPP dedicated information, transfer of UE radio access capability information, and support for E-UTRAN sharing (multiple PLMN identities).
- Generic protocol error handling.
- Support of self-configuration and self-optimization.

Random access is specified entirely in the MAC, including initial transmission power estimation. The RRC connection establishment includes establishment of SRB1. The E-UTRAN completes RRC connection establishment prior to completing the establishment of the S1 connection, i.e., prior to receiving the UE context from the EPC. Therefore, the AS security is not activated during the initial phase of the RRC connection. During the initial phase of the RRC connection, the E-UTRAN may configure the UE to perform measurement reporting. However, the UE only accepts handover messages when security has been activated. On receiving the UE context from the EPC, the E-UTRAN activates ciphering and integrity protection using the initial security activation procedure. The RRC messages to activate security are integrity-protected, while ciphering starts only after completion of the procedure. That is, the response to the message used to activate security is not ciphered, while the subsequent messages used to establish SRB2 and data RBs are both integrity-protected and ciphered.

Following activation of the initial security, the E-UTRAN establishes SRB2 and user data RBs, i.e., the E-UTRAN may perform this function prior to receiving the confirmation of the initial security activation from the UE. The E-UTRAN applies both ciphering and integrity protection for the RRC connection reconfiguration messages used to establish SRB2 and user data RBs. The E-UTRAN should release the RRC connection, if the initial security activation and/or the radio bearer establishment fails (i.e., security activation and user data RB establishment are triggered by a joint S1-procedure, which does not support partial success). For SRB2 and user data RBs, the security is always activated from the beginning, i.e., the E-UTRAN does not establish these bearers prior to activating security. The release of the RRC connections is initiated by the E-UTRAN. The procedure may be used to redirect the UE to another frequency or Radio Access Technology (RAT). In certain cases, the UE may discontinue RRC connection, i.e., it may move to RRC_IDLE without notifying the E-UTRAN.

In RRC_CONNECTED, the network controls UE mobility, i.e., the network decides when the UE must move to another cell, which may be on another frequency or RAT. For network controlled mobility in the RRC_CONNECTED state, handover is the only procedure that is defined. The network triggers the handover procedure, based on radio conditions or load. The network may configure the UE to perform measurement reporting including the configuration of measurement gaps. The network may also initiate handover without having received measurement reports from the UE. Before sending the handover message to the UE, the source eNB prepares one or more target cells. The target eNB generates the message used to perform the handover, including the AS-configuration to be used in the target cell. The source eNB transparently forwards the handover message/information received from the target to the UE. The source eNB may initiate data forwarding for some of the user data RBs. After receiving the handover message, the UE attempts to access the target cell at the first available Random Access Channel (RACH) opportunity according to random access resource selection procedure for asynchronous handover. When allocating a dedicated preamble for random access in the target cell, the E-UTRA ensures it is available from the first RACH opportunity the UE may utilize. On successful completion of the handover, the UE sends a message to confirm the handover. Following successful completion of handover, the PDCP SDUs may be re-transmitted in the target cell. This only applies for user data RBs using RLC-AM mode. Furthermore, the Sequence Number (SN) and the Hyper-Frame Number (HFN)[xix] are reset, except for user data RBs using RLC-AM mode for which both SN and HFN may continue. For a limited period of time, the source eNB maintains the UE context to allow the UE to return in case of handover failure. If handover failure is detected, the UE will attempt to resume the RRC connection, either with the source eNB or with another cell using the RRC re-establishment procedure. The connection re-establishment would succeed if the accessed cell is prepared.

The paging procedure is used to transmit paging information to a UE in RRC_IDLE and/or to inform mobile terminals in RRC_IDLE and RRC_CONNECTED states about a system information change and/or to inform about ETWS primary and/or secondary notifications. The paging information is provided to upper layers, which in response may establish an RRC connection to receive an incoming call. The E-UTRAN initiates the paging procedure by transmitting the paging message at the user equipment's paging intervals. The E-UTRAN may address multiple mobile terminals within

---

[xix]A Hyper Frame Number (HFN) is an overflow counter mechanism that is used in the eNB and UE in order to limit the actual number of sequence number bits that are needed to be sent over the radio air interface. The HFN needs to be synchronized between the UE and eNB.

a paging message by including one *PagingRecord* for each UE. The E-UTRAN may also indicate a change of system information and/or provide an ETWS notification.

The RRC connection establishment involves SRB1 establishment. The procedure is also used to transfer the initial NAS dedicated information/message from the UE to the E-UTRAN. The UE initiates the RRC connection establishment procedure when the upper layers request establishment of an RRC connection while the UE is in the RRC_IDLE state.

The RRC connection re-establishment is used to re-establish the RRC connection, including resumption of SRB1 operation and reactivation of security. A UE in RRC_CONNECTED, for which security has been activated, may initiate the procedure in order to continue the RRC connection. The connection re-establishment succeeds if the serving/target is prepared, i.e., has a valid UE context. If the E-UTRAN accepts the connection re-establishment, SRB1 operation resumes while the operation of other radio bearers are suspended. If the AS security has not been established, the UE cannot initiate the procedure and directly moves to the RRC_IDLE state.

The main difference between MAC and RRC control lies in the signaling reliability. The signaling corresponding to state transitions and radio bearer configurations should be performed by the RRC sub-layer due to signaling reliability. The different characteristics of MAC and RRC control are summarized in Table 6-11.

The 3GPP LTE RRC PDU contents are encoded using Abstract Syntax Notation One (ASN.1), as specified in ITU-T Rec. X.680[16] and X.681 [17]. Transfer syntax for RRC PDUs is derived from their ASN.1 definitions by the use of Packed Encoding Rules as specified in ITU-T Rec. X.691 [18] as shown in Figure 6-28.

6.13.1 Mobility Management and Handover

In order to support mobility, an LTE-compliant UE may conduct the following measurements (physical layer measurements include Reference Signal Received Power and E-UTRA carrier Received Signal Strength Indicator [13]):

- Intra-frequency E-UTRAN measurements;
- Inter-frequency E-UTRAN measurements;
- Inter-RAT measurements.

Table 6-11 Summary of the Differences between MAC and RRC Control

	MAC Control		RRC Control
Control Entity	MAC		RRC
Signaling Type	PDCCH	MAC Control PDU	RRC Message
Signaling Reliability	$\sim 10^{-2}$ (no re-transmission)	$\sim 10^{-3}$ (After HARQ)	$\sim 10^{-6}$ (After ARQ)
Control Latency	Very short	Short	Long
Extensibility	None	Limited	High
Security	No integrity protection No ciphering	No integrity protection No ciphering	Integrity-protected ciphering

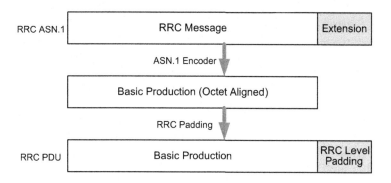

FIGURE 6-28

Creation of RRC PDUs

Measurement commands are used by E-UTRAN to order the UE to start measurements, modify measurements, or stop measurements. The reporting criteria that are used include event-triggered reporting, periodic reporting, and event-triggered periodic reporting. In E-UTRAN RRC_CON-NECTED state, network-controlled UE-assisted handovers are performed and various DRX cycles are supported. In E-UTRAN RRC_IDLE state, cell reselections are performed and DRX is supported.

The UE reports measurement information in accordance with the measurement configuration as provided by E-UTRAN. E-UTRAN provides the measurement configuration applicable for a UE in RRC_CONNECTED by means of dedicated signaling. The UE may be requested to perform the following types of measurements:

- Intra-Frequency Measurements: measurements at the downlink carrier frequency of the serving cell;
- Inter-frequency measurements: measurements at frequencies that differ from the downlink carrier frequency of the serving cell;
- Inter-RAT measurements of UTRA frequencies;
- Inter-RAT measurements of GERAN frequencies;
- Inter-RAT measurements of cdma2000 HRPD or cdma2000 1×RTT frequencies.

The measurement configuration includes the following parameters:

- **Measurement Objects**: the objects on which the UE performs the measurements. For intra-frequency and inter-frequency measurements, a measurement object is a single E-UTRA carrier frequency. Associated with this carrier frequency, E-UTRAN can configure a list of cell-specific offsets and a list of blacklisted cells. The blacklisted cells are not considered in event evaluation or measurement reporting. For inter-RAT UTRA measurements, a measurement object is a set of cells on a single UTRA carrier frequency. For inter-RAT GERAN measurements, a measurement object is a set of GERAN carrier frequencies. For inter-RAT cdma2000 measurements, a measurement object is a set of cells on a single (HRPD or 1×RTT) carrier frequency.
- **Reporting Configurations**: a list of reporting configurations where each reporting configuration consists of the following:
  - **Reporting Criterion**: the criterion that triggers the UE to send a measurement report. This can either be periodical or a single event description.

- **Reporting Format**: the quantities that the UE includes in the measurement report and associated information (e.g., number of cells to report).
- **Measurement Identities**: a list of measurement identities where each measurement identity links one measurement object with one reporting configuration. By configuring multiple measurement identities it is possible to link more than one measurement object to the same reporting configuration, as well as to link more than one reporting configuration to the same measurement object. The measurement identity is used as a reference number in the measurement report.
- **Quantity Configurations**: one quantity configuration is configured per RAT type. The quantity configuration defines the measurement quantities and associated filtering used for all event evaluation and related reporting of that measurement type. One filter can be configured per measurement quantity.
- **Measurement Gaps**: periods that the UE may use to perform measurements; i.e., no UL or DL transmissions are scheduled.

The E-UTRAN only configures a single measurement object for a given frequency, i.e., it is not possible to configure two or more measurement objects for the same frequency with different associated parameters, e.g., different offsets and/or blacklists. The E-UTRAN may configure multiple instances of the same event by configuring two reporting configurations with different thresholds. The UE maintains a single measurement object list, a single reporting configuration list, and a single measurement identities list. The measurement object list includes measurement objects that are specified per RAT type including an intra-frequency object (i.e., the object corresponding to the serving frequency), inter-frequency object(s), and inter-RAT objects. Similarly, the reporting configuration list includes E-UTRA and inter-RAT reporting configurations. Any measurement object can be linked to any reporting configuration of the same RAT type. Some reporting configurations may not be linked to a measurement object. Some measurement objects may not be linked to a reporting configuration. The measurement procedures distinguish the types of cells, i.e., the serving cell, listed cells (cells listed within the measurement objects), and detected cells (cells that are not listed within the measurement objects but are detected by the UE on the carrier frequencies indicated by the measurement objects).

For E-UTRA, the UE measures and reports on the serving cell, listed cells, and detected cells. For inter-RAT UTRA, the UE measures and reports on listed cells. For inter-RAT GERAN, the UE measures and reports on detected cells. For inter-RAT cdma2000, the UE measures and reports on listed cells. For inter-RAT UTRA and cdma2000, the UE measures and reports detected cells for the purpose of self-organization and optimization. The Closed Subscriber Group (CSG) femto-cells are not indicated within the neighbor list. Furthermore, the assumption is that for non-femto-cell deployments, the physical cell identity is unique within the area of a large macro cell.

The intra-E-UTRAN mobility support for mobile terminals in Evolved Packet System (EPS) Connection Management (ECM-CONNECTED) handles all necessary steps for relocation or hand-over procedures, such as processes that precede the final handover decision on the serving network, preparation of resources on the target network, commanding the UE to the new radio resources, and finally releasing resources on the serving network side. It contains mechanisms to transfer context data between eNBs and to update node relations on C-plane and U-plane. The UE makes measurements of attributes of the serving and neighbor cells to enable the process [13].

Depending on whether the UE needs transmission/reception gaps to perform the relevant measurements, measurements are classified as gap-assisted or non-gap-assisted. A non-gap-assisted measurement is a measurement on a cell that does not require transmission/reception gaps to allow the measurement to be performed. A gap-assisted measurement is a measurement on a cell that does require transmission/reception gaps to allow the measurement to be performed. Gap patterns are configured and activated by the RRC functional block [13]. The handover procedure is performed without EPC involvement; i.e., preparation messages are directly exchanged between the eNBs. The release of the resources at the serving eNB (or source eNB) during the handover completion phase is triggered by the eNB.

Figure 6-29 illustrates the handover procedure between eNBs within the same MME/S-GW. More information on the handover procedure can be found in reference [13]. After the downlink path is switched at the S-GW, downlink packets on the forwarding path and on the new direct path may arrive interchanged at the target eNB (see Figure 6-29). The target eNB should first deliver all forwarded packets to the UE before delivering any of the packets received on the new direct path. On handover, the serving eNB sequentially forwards all downlink PDCP SDUs that have not been acknowledged by the UE to the target eNB. The serving eNB discards any remaining downlink RLC PDUs. Correspondingly, the source eNB does not forward the downlink RLC context to the target eNB.

Measurements to be performed by a UE for intra-/inter-frequency mobility can be controlled by E-UTRAN, using broadcast or dedicated control/signaling. In the RRC_IDLE state, a UE follows the measurement parameters defined for cell reselection. In the RRC_CONNECTED state, a UE follows the measurement configurations specified by RRC managed by the E-UTRAN. Intra-frequency and inter-frequency neighbor cell measurements are defined as follows:

- Neighbor cell measurements performed by the UE are intra-frequency measurements when the current and target cell operates on the same carrier frequency.
- Neighbor cell measurements performed by the UE are inter-frequency measurements when the neighbor cell operates on a different carrier frequency compared to the current cell.

6.13.2 Scheduling and Rate Control Functions

In order to efficiently utilize the shared radio resources, a scheduling function is used in the MAC sub-layer. In this section, an overview of the scheduler is provided in terms of scheduler operation, signaling of scheduler decisions, and measurements to support scheduler operation. The MAC in eNB includes dynamic resource schedulers that allocate physical layer resources for the downlink and uplink transport channels. Different schedulers operate for the downlink and uplink shared channels. The scheduler takes into account the system load and the QoS requirements of each UE and associated radio bearers when allocating resources to mobile terminals. Per UE grants are used to allow transmission on the uplink shared channel.

The schedulers may assign resources considering the radio conditions at the UE identified through measurements conducted at the eNB and/or reported by the UE. The radio resource allocations can be valid for one or multiple Transmission Time Intervals (TTIs). The resource assignment consists of physical resource blocks, and modulation and coding schemes. Allocations for time periods longer than one TTI might also require additional information.

In the downlink, the E-UTRAN can dynamically allocate resources to UEs at each TTI via the C-RNTI (similar to STID in IEEE 802.16m) on Physical Downlink Control Channel (PDCCH). A UE always monitors the PDCCH in order to find possible allocation when its downlink reception is

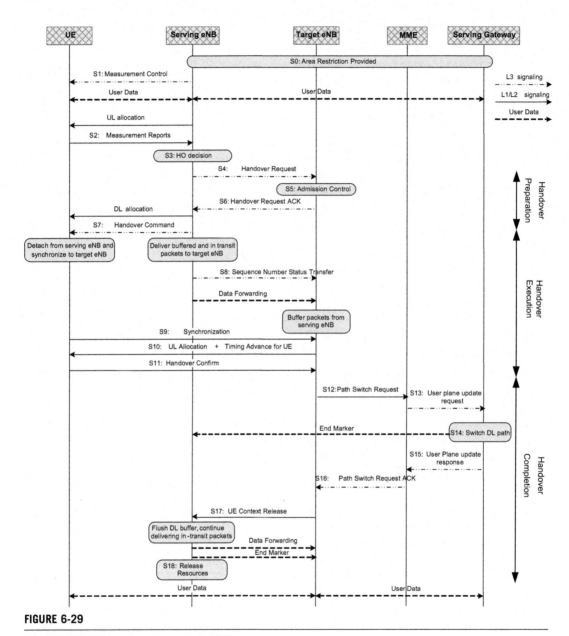

FIGURE 6-29

Intra-MME/S-GW handover procedures [13]

enabled, consistent with DRX configuration. In addition, the E-UTRAN can allocate semi-persistent downlink resources for the first HARQ transmissions to UEs where RRC defines the periodicity of the semi-persistent downlink grant and PDCCH indicates whether the downlink grant is a semi-persistent one, i.e., whether it can be implicitly reused in the following TTIs according to the periodicity defined by RRC. The re-transmissions are explicitly signaled via the PDCCH. In subframes where the UE has a semi-persistent downlink resource, if the UE cannot find its C-RNTI on the PDCCH, a downlink transmission according to the semi-persistent allocation that the UE has been assigned in the TTI is assumed. Otherwise, in the subframes where the UE has a semi-persistent downlink resource, if the UE finds its C-RNTI on the PDCCH, the PDCCH allocation overrides the semi-persistent allocation for that TTI and the UE does not decode the semi-persistent resources.

In the uplink, the E-UTRAN can dynamically allocate resources to UEs at each TTI via the C-RNTI on PDCCH. A UE always monitors the PDCCH in order to find possible allocation for uplink transmission when its downlink reception is enabled, consistent with DRX configuration. In addition, E-UTRAN can allocate a semi-persistent uplink resource for the first HARQ transmissions, and possible re-transmissions to UEs, where RRC defines the periodicity of the semi-persistent uplink grant and PDCCH indicates whether the uplink grant is a semi-persistent one, i.e., whether it can be implicitly reused in the following TTI according to the periodicity defined by RRC. In the subframes where the UE has a semi-persistent uplink resource, if the UE cannot find its C-RNTI on the PDCCH an uplink transmission according to the semi-persistent allocation that the UE has been assigned in the TTI can be made. The network performs decoding of the predefined physical resource blocks according to the predefined MCS. Otherwise, in the subframes where the UE has a semi-persistent uplink resource, if the UE finds its C-RNTI on the PDCCH, the PDCCH allocation overrides the persistent allocation for that TTI and the mobile terminal's transmission follows the PDCCH allocation. The re-transmissions are either implicitly scheduled where the UE uses the semi-persistent uplink allocation, or explicitly allocated via PDCCH where the UE does not follow the semi-persistent allocation. There is no blind decoding in uplink; therefore, when the UE does not have enough data to fill the allocated resource, padding is used.

The measurement reports are required to enable the scheduler to operate in both uplink and downlink. These include transport volume and measurements of a user terminal's radio environment. The uplink Buffer Status Report (BSR) is needed to provide support for QoS-aware packet scheduling. In E-UTRAN, uplink buffer status reports refer to the data that is buffered in for a group of radio bearers in the UE. Four radio bearer groups and two formats are used for reporting in uplink, i.e., a short format for which only one BSR is reported, and a long format for which all four BSRs are reported. Uplink buffer status reports are transmitted using MAC signaling. The time and frequency resources used by the UE to report CQI are under the control of the eNB. The CQI reporting can be either periodic or aperiodic. A UE can be configured to have both periodic and aperiodic reporting at the same time. In case both periodic and aperiodic reporting occurs in the same subframe, only the aperiodic report is transmitted in that subframe. For efficient support of localized, distributed, and MIMO transmissions, the E-UTRA supports three types of CQI reporting:

1. Wideband Type provides channel quality information over the entire system bandwidth;
2. Multi-Band Type provides channel quality information of a fraction of system bandwidth;
3. MIMO Type for open-loop or closed-loop operation (with or without PMI feedback).

When the UE is allocated Physical Uplink Shared Channel (PUSCH) resources in a subframe where a periodic CQI report is configured, the periodic CQI report is transmitted together with uplink data in PUSCH. Otherwise, the periodic CQI reports are sent in the Physical Uplink Control Channel (PUCCH). Aperiodic CQI reporting is defined as the report that is scheduled by the eNB via the PDCCH, or the report transmitted together with uplink data on PUSCH. When a CQI report is transmitted together with uplink data on PUSCH it is multiplexed with the transport block by the physical layer (i.e., the CQI report is not part of the uplink transport block). The eNB configures a set of sizes and formats of the reports. The size and format of the report depends on whether it is transmitted over PUCCH or PUSCH, and whether it is a periodic or an aperiodic CQI report.

6.13.3 Discontinuous Reception in RRC_CONNECTED State

In order to reduce UE battery consumption, E-UTRAN defines a DRX scheme per UE (similar to Sleep Mode in IEEE 802.16m) where there is no RRC or MAC sub-state to distinguish between different levels of DRX, and available DRX values are controlled by the network and start from non-DRX up to a few seconds. Measurement requirement and reporting criteria can differ according to the length of the DRX cycle, as long DRX cycles may experience more relaxed requirements. Irrespective of DRX cycle, the UE may use the first available RACH opportunity to send a UL measurement report. Immediately after sending a measurement report, the UE may change its DRX cycle (see Figure 6-30). This mechanism can be pre-configured by the eNB. HARQ operation related to data transmission is independent of DRX cycle and the UE wakes up to read the PDCCH for possible re-transmissions and/ or ACK/NACK signaling, regardless of DRX cycle. In the downlink, a timer is used to limit the time the UE can stay awake awaiting for a re-transmission. When the DRX is configured, the UE may be further programmed with an *On Duration* timer, during which the UE monitors the PDCCHs for possible allocations. When DRX is configured, periodic CQI reports can only be sent by the UE during the *Active-Time*. The RRC can further restrict periodic CQI reports so that they are only sent during the *On Duration*. A timer in the UE is used to detect need for obtaining timing advance. The following definitions apply to DRX in E-UTRAN:

- **On Duration**: time interval in downlink subframes that the UE awaits PDCCH after waking up from DRX. If the UE successfully decodes a PDCCH, the UE stays awake and starts the inactivity timer.
- **Inactivity Timer**: time interval in downlink subframes that the UE waits to successfully decode a PDCCH from the last successful decoding of a PDCCH. The UE re-enters DRX if it fails to

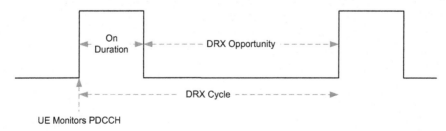

FIGURE 6-30

An illustration of the DRX cycle

decode PDCCH. The UE restarts the inactivity timer following successful decoding of PDCCH in the first transmission.

- **Active-Time**: total time duration that the UE is awake. This includes the *On Duration* of the DRX cycle, the time UE is performing continuous reception while the *Inactivity Timer* has not expired, and the time UE is performing continuous reception while waiting for a DL re-transmission. Based on the above, the minimum active time is equal to *On Duration*.

Among the above parameters, the *On Duration* and *Inactivity Timer* are of fixed lengths, while the *Active-Time* is of varying length based on scheduling decisions and UE decoding success. Only *On Duration* and *Inactivity Timer* duration are signaled to the UE by the eNB. There is only one DRX configuration used in the UE at any time. The UE applies an *On Duration* on wake-up from DRX sleep. This is also applicable in the case where the UE has only one service that is being handled through the allocation of predefined resources, allowing other signaling such as RRC to be sent during the remaining portion of the active time. New transmissions can only take place during the *Active-Time*. If PDCCH has not been successfully decoded during the *On Duration*, the UE follows the DRX configuration. This also applies to the subframes where the UE has been allocated predefined resources.

If the UE successfully decodes a PDCCH in the first transmission, the UE stays awake and starts the *Inactivity Timer*, even if PDCCH is successfully decoded in the subframes where the UE has also been allocated predefined resources, until a MAC control message instructs the UE to re-enter DRX cycle or until the *Inactivity Timer* expires. If a short DRX cycle is configured, the UE first follows the short DRX cycle and after a longer period of inactivity the UE follows the long DRX cycle; otherwise, the UE follows the long DRX cycle directly. When the DRX cycle is configured, the network detects whether the UE would remain in E-UTRAN coverage by requesting the UE to send periodic signals to the network.

6.13.4 Quality of Service

An Evolved Packet System bearer/E-UTRAN Radio Access Bearer (E-RAB)[xx] is the level of granularity for bearer-level QoS control in the EPC/E-UTRAN. As a result, service data flows mapped to the same EPS bearer receive the same bearer-level packet forwarding treatment, e.g., scheduling policy, queue management policy, rate shaping policy, RLC configuration, etc. One EPS bearer/ E-RAB is established when the UE connects to PDN and remains throughout the lifetime of the PDN connection to provide the UE with always-on IP connectivity to the PDN. This bearer is referred to as the *default* bearer. Any additional EPS bearer/E-RAB that is established to the same PDN is referred to as the *dedicated* bearer. The initial bearer-level QoS parameter values of the default bearer are assigned by the network, based on subscription information. The decision to establish or modify a dedicated bearer can only be taken by the EPC and the bearer-level QoS parameter values are always assigned by the EPC.

An EPS bearer/E-RAB is referred to as a Guaranteed Bit Rate (GBR) bearer, if dedicated network resources related to a GBR value that is associated with the EPS bearer/E-RAB are permanently allocated by the admission control function in the eNB on bearer establishment/modification.

[xx]An E-RAB uniquely identifies the concatenation of an S1 bearer and the corresponding data radio bearer. When an E-RAB exists, there is a one-to-one mapping between the E-RAB and an EPS bearer of the NAS.

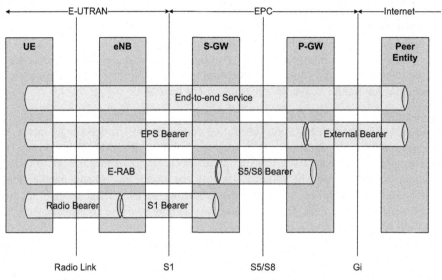

FIGURE 6-31

EPS bearer service architecture

Otherwise, an EPS bearer/E-RAB is referred to as a Non-GBR bearer. A dedicated bearer can either be a GBR or a non-GBR bearer, while a default bearer is a non-GBR bearer. The EPS bearer service architecture is depicted in Figure 6-31, where:

- A UL Traffic Flow Template (TFT) binds a service data flow to an EPS bearer in the uplink direction. Multiple service data flows can be multiplexed into the same EPS bearer by including multiple uplink packet filters in the UL TFT.
- A DL TFT in the PDN GW binds a service data flow to an EPS bearer in the downlink direction. Multiple service data flows can be multiplexed into the same EPS bearer by including multiple downlink packet filters in the DL TFT.
- An E-RAB transports the packets of an EPS bearer between the UE and the EPC. When an E-RAB exists, there is a one-to-one mapping between the E-RAB and an EPS bearer.
- A data radio bearer transports the packets of an EPS bearer between a UE and the eNB. When a data radio bearer exists, there is a one-to-one mapping between this data radio bearer and the EPS bearer/E-RAB.
- An S1 bearer transports the packets of an E-RAB between an eNB and an S-GW.
- An S5/S8 bearer transports the packets of an EPS bearer between a Serving GW and a PDN GW.
- A UE stores the mapping between an uplink packet filter and data radio bearer to create the binding between the service data flow and the data radio bearer in the uplink.
- The P-GW stores the mapping between the downlink packet filter and S5/S8a bearer to create the binding between a service data flow and the S5/S8a bearer in the downlink.
- The eNB stores a one-to-one mapping between the data radio bearer and the S1 bearer to create the binding between the data radio bearer and the S1 bearer in both the uplink and downlink.

- The S-GW stores a one-to-one mapping between the S1 bearer and the S5/S8a bearer to create the binding between the S1 bearer and the S5/S8a bearer in both the uplink and downlink.

Each EPS bearer/E-RAB (GBR and non-GBR) is associated with the following bearer-level QoS parameters:

- QoS Class Identifier (QCI): a metric that is used as a reference to access node-specific parameters that control bearer-level packet forwarding function, e.g., scheduling weights, admission thresholds, queue management thresholds, link layer protocol configuration, etc., that have been pre-configured by the network operator.
- Allocation and Retention Priority (ARP): the primary purpose of ARP is to decide whether a bearer establishment/modification request can be accepted or needs to be rejected in the case of resource limitations. In addition, the ARP can be used by the eNB to decide which bearers to drop during special resource limitations.
- Each GBR bearer is additionally associated with the bit rate that can be expected to be provided by a GBR bearer.
- Each Access Point Name (APN) access by a UE is associated with per APN Aggregate Maximum Bit Rate (APN-AMBR). An APN identifies an IP packet data network with which a mobile data user wishes to communicate. In addition to identifying a PDN, an APN may also be used to define the type of service.
- Each UE in EPS Mobility Management (EMM)-REGISTERED state is associated with per UE Aggregate Maximum Bit Rate (UE-AMBR).
- The GBR denotes bit rate of traffic per bearer, while UE-AMBR/APN-AMBR denotes bit rate of traffic per group of bearers. Each of those QoS parameters has an uplink and a downlink component.

APPENDIX A: PROPORTIONAL FAIR SCHEDULING ALGORITHM

The Proportional Fair Scheduling algorithm calculates a metric for all active users for a given scheduling interval. The user with the highest metric is allocated the resource available in the given interval, the metrics for all users are updated before the next scheduling interval, and the process repeats. To adapt this simple algorithm for OFDMA systems, the definition of scheduling interval and scheduling resource are extended to encompass a two-dimensional OFDMA time-frequency resource. Furthermore, the Proportional Fair Scheduling algorithm only applies to full-buffer traffic model and time-frequency zones which use a distributed subcarrier permutation.

For OFDMA systems, the scheduling interval is typically a radio frame and multiple users may be allocated in the same frame. Thus, some modifications can be made to the Proportional Fair Scheduling algorithm including (1) frames are equally-partitioned into regular, fixed scheduling resources that are scheduled sequentially until all available resources are assigned; (2) the metric is updated after scheduling each partition. Note that the number of resources ultimately allocated to a user depends on the metric update process, and does not preclude a single user from obtaining multiple or all the resources in a frame. For system simulations with an assumption of fixed overhead allowing for up to $N_{partition}$ resource partitions, each partition assignment should be considered as a separate packet transmission. The number of partitions, $N_{partition}$, the time constant of the filter used in the metric computation, and number of active users are simulation parameters. At any scheduling instant t, the scheduling metric $M_i(t)$ for subscriber $_i$ used by the proportional fair scheduler is given by

$$M_i(t) = \frac{T\_inst_i(t)}{[T\_average_i(t)]^{\alpha}} \tag{A6.1}$$

where $T\_inst_i(t)$ is the data rate that can be supported at scheduling instant t for subscriber i, $T\_inst_i(t)$ is a function of the CQI feedback, and consequently of the modulation and coding scheme that can meet the Packet Error Rate (PER) requirement. The $T\_average_i(t)$ denotes the throughput function smoothed by a low-pass filter at the scheduling instant t for user $_i$. The parameter α is a fairness exponent factor with default value of one. For the scheduled subscriber, $T\_average_i(t)$ is computed as:

$$T\_average_i(t) = \frac{1}{N_{PF}} * T\_inst_i(t) + (1 - \frac{1}{N_{PF}}) * T\_average_i(t - 1) \tag{A6.2}$$

and for an unscheduled subscriber,

$$T\_average_i(t) = (1 - \frac{1}{N_{PF}}) * T\_average_i(t - 1) \tag{A6.3}$$

The latency scale of the Proportional Fair scheduler, N_{PF}, is given by:

$$N_{PF} = T_{PF}N_{partitions}/T_{Frame} \tag{A6.4}$$

where T_{PF} represents the latency time scale in units of seconds and T_{Frame} is the frame duration of the system. In some implementations, the scheduler may give priority to HARQ re-transmissions.

References

[1] IEEE Std 802.16-2009, IEEE Standard for Local and Metropolitan Area Networks – Part 16: Air Interface for Fixed Broadband Wireless Access Systems, May 2009.
[2] P802.16m/D6, IEEE Standard for Local and Metropolitan Area Networks – Part 16: Air Interface for Broadband Wireless Access Systems, Advanced Air Interface, May 2010.
[3] IEEE Std 802-2001, IEEE Standard for Local and Metropolitan Area Networks: Overview and Architecture, February 2002.
[4] IEEE Registration Authority <http://standards.ieee.org/regauth/index.html>.
[5] IEEE 802.16m–0034r3, System Description Document, May 2010.
[6] Loutfi Nuaymi, WiMAX: Technology for Broadband Wireless Access,, John Wiley & Sons, 2007.
[7] Jeffrey G. Andrews, Arunabha Ghosh, Rias Muhamed, Fundamentals of WiMAX: Understanding Broadband Wireless Networking, first ed., Prentice Hall, 2007.
[8] WiMAX Forum, Mobile System Profile Release 1.5 Air Interface Specification, August 2009.
[9] Mikko Majanen et al., "Mobile WiMAX Handover Performance Evaluation," Proceedings of Fifth International Conference on Networking and Services, April 2009.
[10] Antti Makelainen, Analysis of Handoff Performance in Mobile WiMAX Networks, MS Thesis,, Helsinki University of Technology, Espoo, Finland, March 2007.
[11] IST-2003-507581 WINNER D1.3 version 1.0, Final Usage Scenarios, June 2005.

[12] Erik Dahlman, et al., 3G Evolution: HSPA and LTE for Mobile Broadband, second ed., Academic Press, 2008.

[13] 3GPP TS 36.300, Evolved Universal Terrestrial Radio Access (E-UTRA) and Evolved Universal Terrestrial Radio Access Network (E-UTRAN); Overall Description; Stage 2, March 2010.

[14] 3GPP TS 36.331, Evolved Universal Terrestrial Radio Access (E-UTRA); Radio Resource Control (RRC); Protocol Specification, March 2010.

[15] 3GPP TR 36.913, Requirements for Further Advancements for Evolved Universal Terrestrial Radio Access (E-UTRA) LTE-Advanced, March 2009.

[16] ISO 8824/ITU X.680 specifies the message syntax <http://www.itu.int/ITU-T/studygroups/com17/languages/X.680-0207.pdf>.

[17] ISO 8825/ITU X.690 specifies the basic encoding rules for ASN.1 <http://www.itu.int/ITU-T/studygroups/com17/languages/X.690-0207.pdf>.

[18] ITU-T Recommendation X.691, ASN.1 Encoding Rules: Specification of Packed Encoding Rules (PER), June 1999.

[19] IETF RFC 2104, H. Krawczyk, et al., HMAC: Keyed-Hashing for Message Authentication, February 1997.

[20] Hashed Message Authentication Code (HMAC), Wikipedia <http://en.wikipedia.org/wiki/Keyed-hash_message_authentication_code>.

[21] Cipher-based Message Authentication Code (CMAC), Wikipedia <http://en.wikipedia.org/wiki/CMAC>.

[22] WiMAX Forum Network Architecture Release 1.5 Version 1, Stage 2: Architecture Tenets, Reference Model and Reference Points, September 2009. <http://www.wimaxforum.org/resources/documents/technical/release>.

[23] WiMAX Forum Network Architecture Release 1.5 Version 1, Stage 3: Detailed Protocols and Procedures, September 2009. <http://www.wimaxforum.org/resources/documents/technical/release>.

[24] Ietf Rfc 4542, F. Baker, J. Polk, Implementing an Emergency Telecommunications Service (ETS) for Real-Time Services in the Internet Protocol Suite, May 2006.

[25] R.Y. Kim, S. Mohanty, "Advanced Power Management Techniques in Next-Generation Wireless Networks,", IEEE Communications Magazine, Vol. 48, Issue 5, May 2010.

The IEEE 802.16m Medium Access Control Common Part Sub-layer (Part II)

7

INTRODUCTION

This chapter describes the functional and operational aspects of the IEEE 802.16m MAC Common Part Sub-layer (MAC CPS) on the data-plane. As shown in Figure 7-1, the MAC CPS provides an interface between the physical layer and higher protocol layers through PHY and MAC SAPs, respectively. The MAC CPS functions are classified into Radio Resource Control and Management (RRCM), and MAC functions. The soft classification of the MAC CPS into RRCM and MAC sub-layer does not require any SAP between the two classes of functions. The RRCM functions fully reside on

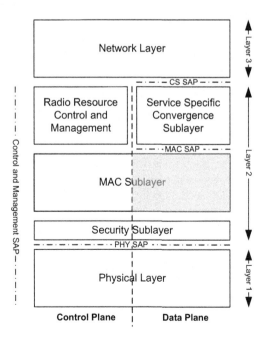

FIGURE 7-1

IEEE 802.16m data-plane MAC CPS functions

Mobile WiMAX. DOI: 10.1016/B978-0-12-374964-2.10007-4
Copyright © 2011 Elsevier Inc. All rights reserved.

the control-plane, whereas the MAC sub-layer functions reside on both control-plane and data-plane [3].

The control-plane part of the MAC sub-layer includes functional blocks which are related to the physical layer and link control such as:

- Physical Layer Control;
- Control Signaling;
- Sleep Mode Management;
- QoS;
- Scheduling and Resource Multiplexing;
- Multi-Radio Coexistence;
- Data Forwarding;
- Interference Management;
- Inter-BS Coordination.

The data-plane portion of the MAC sub-layer is responsible for the following functions:

- MAC SDU Fragmentation/Packing;
- ARQ;
- MAC PDU Formation.

To draw an analogy between the IEEE 802.16m and 3GPP LTE Layer 2 protocols, one must note that the Layer 2 functions of the latter include MAC and RLC sub-layers and the RRC sub-layer is considered Layer 3. The IEEE 802.16m RRCM functional group corresponds to RRC sub-layer in 3GPP LTE, and the IEEE 802.16m MAC sub-layer is analogous to a combination of 3GPP LTE MAC and RLC sub-layers. The 3GPP LTE MAC and RLC sub-layers will be described in this chapter to enable the reader to understand the similarities and differences of IEEE 802.16m and 3GPP LTE/LTE-Advanced demonstrating correspondences between the two protocol sets [3,9].

The IEEE 802.16m MAC is connection-oriented. For the purpose of mapping services to varying levels of QoS at mobile stations, all data communications are manifested in the form of transport connections. Service flows may be provisioned when an MS performs network entry. Following MS registration with the serving BS, transport connections are established and associated with the service flows (one connection per service flow) to provide a reference for requesting bandwidth. Furthermore, new transport connections may be established when a user's service needs to change. A transport connection defines the mapping between peer convergence sub-layers that utilize the MAC and a service flow. The service flow defines the QoS parameters for the PDUs that are exchanged on the connection.

The concept of service flow mapping to a transport connection is essential to the operation of the MAC CPS protocols. Service flows provide a mechanism for UL and DL QoS management. In particular, they are an integral part of the bandwidth allocation process. An MS requests UL bandwidth on a per connection basis by implicitly identifying the service flow. The bandwidth is granted by the serving BS to an MS as an aggregate of grants in response to per connection requests from the MS. Transport connections, once established, may require active maintenance, with requirements which may vary depending on the type of service. In the following sections, the IEEE MAC CPS functions are classified and described according to their location on the data-plane using a systematic approach.

7.1 **AUTOMATIC REPEAT REQUEST**

7.1.1 **ARQ Principles**

Automatic Repeat Request (ARQ) is an error-control mechanism for data transmission which uses acknowledgements (or negative acknowledgements) and timeouts to achieve reliable data transmission over an unreliable communication link. In an ARQ scheme, the receiver uses an error detection code, typically a Cyclic Redundancy Check (CRC), to detect whether the received packet is in error. If no error is detected in the received data, the transmitter is notified by sending a positive acknowledgement. If an error is detected, the receiver discards the packet and sends a negative acknowledgement to the transmitter, and requests a re-transmission. An Acknowledgement (ACK) or Negative Acknowledgement (NACK) is a short message sent by the receiver to the transmitter to indicate whether it has correctly or incorrectly received a data packet, respectively. Timeout is a predetermined time interval after the sender sends the packet; if the sender does not receive an acknowledgement before the timeout, it usually re-transmits the packet until it receives an acknowledgement or exceeds a pre-defined number of re-transmissions. There are three types of ARQ protocol including [4]:

- Stop-and-Wait ARQ is the basic form of ARQ protocol where the sender sends one packet at a time and then waits for an ACK or NACK signal from the receiver before sending the same or a new packet. The receiver sends an ACK signal following receipt of a good packet. If the ACK does not reach the sender before the timeout, the sender re-sends the same packet.
- Go-Back-N ARQ is a form of ARQ protocol in which the sender continuously sends a number of packets (determined by the duration of transmission window) without receiving an ACK signal from the receiver. The receiver process keeps track of the sequence number of the next packet it expects to receive, and sends the sequence number with every ACK it sends. The receiver will ignore any packet that does not have the exact sequence number it expects whether that packet is a duplicate of a packet it has already acknowledged or a packet with a sequence number higher than the one expected. Once the sender has sequentially sent all the packets in its transmission window, it will check whether all of the packets are acknowledged and will resume sequential transmission of the packets starting with the next sequence number to the one that was last acknowledged.
- Selective Repeat ARQ is a form of the ARQ protocol for transmission and acknowledgement of packets or fragments of a packet where the sending process continues to send a number of packets specified by a window size even after a packet is lost. Unlike Go-Back-N ARQ, the receiving process will continue to accept and acknowledge packets sent after an initial error. The receiver process keeps track of the sequence number of the earliest packet it has not received, and sends that number with every ACK it sends. If a packet from the sender does not reach the receiver, the sender continues to send subsequent packets until it has emptied its window. The receiver continues to fill its receiving window with the subsequent packets, replying each time with an ACK containing the sequence number of the earliest missing packet. Once the sender has sent all the packets in its window, it re-sends the packet number given by the ACKs, and then continues where it stopped.

Figure 7-2 provides an example illustration of the ARQ variants. It shows how efficiently different ARQ schemes utilize the communication channel, and use of the ARQ buffer in the transmitter and receiver. The major advantage of ARQ over Forward Error Correction (FEC) schemes is that error

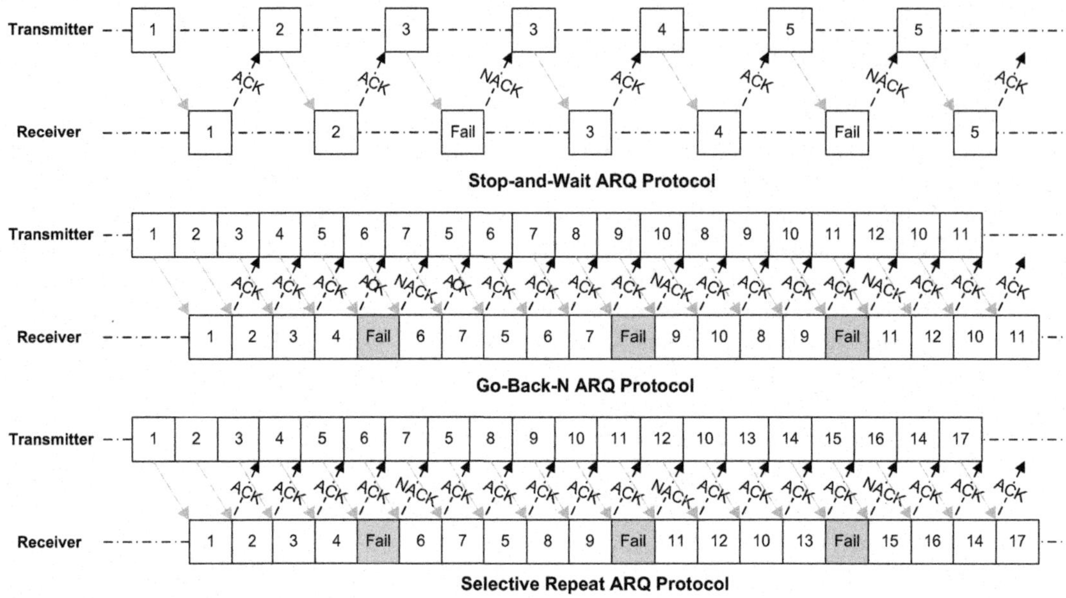

FIGURE 7-2

An example illustration of different ARQ schemes [4]

detection requires much simpler decoding mechanisms and much less redundancy than error correction. Furthermore, ARQ is adaptive, in the sense that information is re-transmitted only when errors occur. On the other hand, FEC may be desirable instead of, or in addition to, error detection, for any of the following reasons: (1) a feedback channel is not available or ARQ delay is not tolerable; (2) the re-transmission scheme is not conveniently implemented; and (3) the expected number of errors without correction would require excessive re-transmissions.

The ARQ can be used for each connection between the MS and BS. Since the use of ARQ may increase the latency due to more reliability requirements, the ARQ mechanism is usually disabled for delay-sensitive applications, such as VoIP or interactive gaming. If the ARQ mechanism is enabled, the ARQ parameters are specified and negotiated during connection set-up. A connection does not contain a mix of ARQ and non-ARQ traffic. The scope of a specific instance of ARQ is limited to one uni-directional flow.

7.1.2 IEEE 802.16m ARQ Mechanism

An ARQ block is generated from one or multiple MAC SDUs or MAC SDU fragments corresponding to the same flow. The ARQ blocks can be variable in size. An ARQ block is constructed by fragmenting MAC SDU or packing MAC SDUs and/or MAC SDU fragments. The fragmentation or packing information for the ARQ block is included in the extended header within the MAC PDU. When a MAC PDU is generated for transmission, the MAC PDU may contain one or more ARQ blocks. If the MAC PDU contains traffic from a single connection, the MAC PDU itself will be a single ARQ block. If information from multiple ARQ connections is multiplexed into one MAC PDU, the MAC PDU contains multiple ARQ blocks. The number of ARQ blocks in a MAC PDU is equal to the number of ARQ connections multiplexed in the MAC PDU. The ARQ blocks of a connection are sequentially numbered. The ARQ block Sequence Number (SN) is included in the MAC PDU using FPEH or MEH headers. The original MAC SDU ordering is maintained [2]. In the legacy system, the size of the ARQ blocks is fixed and the length of the ARQ blocks is specified by the serving BS for each connection and signaled through MAC management messages [1]. In that case, if the length of the MAC SDU is not an integer multiple of ARQ block size, the last ARQ block may be padded. The MAC SDU partitioning into ARQ blocks remains in effect until all ARQ blocks are received and acknowledged by the receiver [1].

As shown in Figure 7-3, if the initial transmission of an ARQ block fails, a re-transmission is scheduled with or without rearrangement. In the case of ARQ block re-transmission without

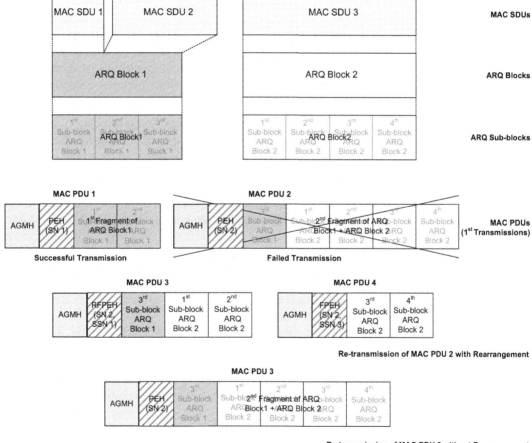

FIGURE 7-3

ARQ block initial transmission and re-transmission [2]

rearrangement, the MAC PDU contains the same ARQ block and corresponding fragmentation and packing information which was used in the initial transmission. In the case of ARQ block re-transmission with rearrangement, a single ARQ block may be fragmented into a sequence of multiple ARQ sub-blocks. A MAC PDU payload is constructed from one or more ARQ sub-blocks. The ARQ sub-blocks are sequentially numbered using ARQ block SUB_SN (SSN). The size of an ARQ sub-block is defined by ARQ_SUB_BLOCK_SIZE, which is fixed. The ARQ sub-block is maintained during re-transmission. Figure 7-3 illustrates ARQ block initial transmission and re-transmissions. Two options for re-transmission are shown, i.e., with and without rearrangements of the failed ARQ blocks.

The ARQ feedback Information Element (IE) is defined for the receiver to indicate the reception status of an ARQ block (initial transmission) and an ARQ sub-block. The ARQ feedback IE is transported either as part of an extended header (piggybacked) within a MAC PDU or a standalone MAC control message. The ARQ feedback IE supports cumulative and selective ACK. In cumulative ACK, ARQ SN or ARQ SUB_SN are reported to indicate successful reception. In selective ACK, each bit of the ACK MAP indicates the error or success of ARQ blocks.

The transmitter can request ARQ feedback poll to update the reception status of the transmitted ARQ blocks. In the downlink, an ABS may assign unsolicited bandwidth for the MS to send ARQ feedback information. The receiver sends ARQ feedback IE when these three conditions are met: (1) ARQ feedback polling request is received from the transmitter; (2) an ARQ block has been missing for a predetermined period; and (3) an ARQ discard message is received from the transmitter. The following parameters characterize an ARQ operation [2]:

- ARQ_SN_MODULUS: the number of unique ARQ sequence values equal to 1024;
- ARQ_WINDOW_SIZE: the maximum number of ARQ blocks with consecutive Block Sequence Number (BSN) in the sliding window of ARQ blocks that is managed by the receiver and the transmitter;
- ARQ_BLOCK_LIFETIME: the maximum time interval an ARQ block is managed by the transmitter ARQ state machine, once initial transmission of the block has occurred. After expiring ARQ_BLOCK_LIFETIME, the corresponding ARQ block is discarded in the ARQ window;
- ARQ_RX_PURGE_TIMEOUT: the time interval the receiver waits after successful reception of a block that does not result in advancement of ARQ_RX_WINDOW_START, before advancing ARQ_RX_WINDOW_START;
- MAX_ ARQ_BUFFER_SIZE: the maximum size of the buffer in bytes that the MS is able to allocate for the ARQ connection;
- ARQ_SYNC_LOSS_TIMEOUT: the maximum time interval ARQ_TX_WINDOW_START or ARQ_RX_WINDOW_START is allowed to remain at the same value before declaring a loss of synchronization of the sender and receiver state machines when data transfer is known to be active;
- ARQ_ERROR_DETECTION_TIMEOUT: the time interval after which the ARQ block is declared as erroneous. It is used to reorder ARQ blocks that have not arrived in order due to HARQ re-transmission.

7.1.3 ARQ State Machine

A finite-state machine can be used to model the behavior of the ARQ protocol in the transmitter and receiver sides. Each ARQ-enabled connection has an independent ARQ state machine. The ARQ state transitions are based on the status of ARQ blocks, rather than ARQ sub-blocks. Based on this model, an ARQ block may be in one of the following six states: NOT-SENT; OUTSTANDING; WAITING-FOR-RE-TRANSMISSION; DISCARD; REARRANGEMENT; or DONE. The ARQ state machine in the transmitter shown in Figure 7-4 is similar to that specified in the IEEE Standard 802.16-2009 [1]. The transmitter constructs each ARQ block using the fragmentation and packing rules. Each NOT-SENT ARQ block forms a MAC PDU and is assigned the value of the next ARQ block sequence number (ARQ_TX_NEXT_SN), which is then increased by one. The ARQ state machine variables are defined as follows:

- ARQ_TX_WINDOW_START is the lower edge of ARQ window in the transmitter;
- ARQ_TX_NEXT_SN denotes the lowest ARQ sequence number of the next ARQ block to be sent by the transmitter;
- ARQ_RX_WINODW_START is the lower edge of the ARQ window at the receiver;
- ARQ_RX_HIGHEST_SN denotes the highest sequence number of the ARQ block received incremented by one.

As shown in Figure 7-5, each ARQ block in the transmission buffer starts in NOT-SENT state before it is transmitted. When an ARQ block is initially transmitted, the ARQ_BLOCK_LIFETIME timer is set and the ARQ block transitions from NOT-SENT state to OUTSTANDING state. While an ARQ block is in OUTSTANDING state, the transmitter waits for an acknowledgement. If an acknowledgement is received, the ARQ block state transits to the DONE state. If a negative acknowledgement is received, the ARQ block state is changed to the WAITING-FOR-RE-TRANSMISSION state. When the ARQ_BLOCK_LIFETIME period expires, the ARQ block state is changed to the DISCARD state. While an ARQ block is in the WAITING-FOR-RE-TRANSMISSION state, the transmitter prepares for ARQ block re-transmission. If the ARQ block is re-transmitted, the ARQ block state is changed to OUTSTANDING. If the ARQ_BLOCK_LIFETIME period expires, the ARQ block state transits to the

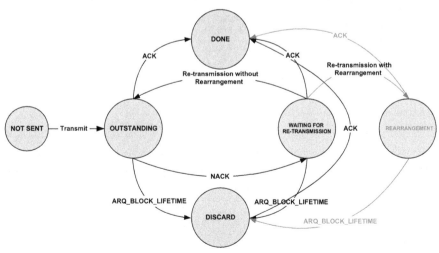

FIGURE 7-4

ARQ state machine [2]

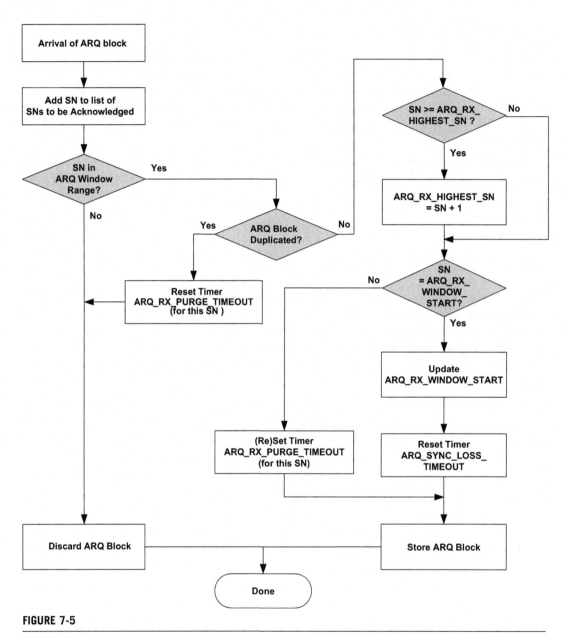

FIGURE 7-5

ARQ block reconstruction at the receiver

discard state. While the ARQ block is in the DISCARD state, the transmitter sends a discarded message and waits for acknowledgement from the receiver. If an acknowledgement for the ARQ block corresponding to the discarded message arrives, the ARQ block state transitions to the DONE state. When an ARQ block is in the DONE state, the transmitter flushes the ARQ block, and resets the timers and state variables associated with the flushed ARQ block.

When a MAC PDU is received, the receiver examines the extended header and obtains the ARQ block information. Once the receiver identifies the ARQ block sequence number and the corresponding ARQ block in the MAC PDU, the receiver state machine adds this ARQ block to the list of blocks to be acknowledged. The state machine checks whether the block sequence number falls within the ARQ window range. If sequence number is not valid, the receiver discards the corresponding ARQ block; otherwise, if the corresponding ARQ block is already received, the state machine resets ARQ_RX_PURGE_TIMEMOUT timer and discards the ARQ block. If the received ARQ block is valid and not duplicated, the receiver state machine updates the ARQ state, as shown in Figure 7-5.

The MAC SDUs are reconstructed from the received ARQ blocks and are sequentially delivered to the upper layers. The transmitter or receiver can reset the ARQ state machine, if necessary. When an ARQ reset error occurs during the ARQ reset procedure, the BS or MS may reinitialize the MAC procedures.

When an ARQ block arrives out of order, each ARQ block with an intermediate sequence number is declared as missing, and the ARQ_ERROR_DETECTION_TIMEOUT for every missing ARQ block is set. If the missing ARQ block does not arrive within the ARQ_ERROR_DETECTION_ TIMEOUT, the receiver declares the corresponding ARQ block as an error. The receiver sends feedback corresponding to each ARQ block using the ARQ feedback IE. The ARQ feedback is sent under one of the following conditions: an ARQ feedback poll is received from the transmitter; the receiver detects an ARQ block error; or when a discarded message is received from the transmitter. If all ARQ blocks in the ARQ window are received correctly, the ARQ feedback IE contains a cumulative acknowledgement. If one or more ARQ blocks in the ARQ window are in error, the ARQ feedback IE contains a selective acknowledgement to indicate the error. If the transmitter or receiver declares an ARQ synchronization loss, the transmitter or receiver may initiate the ARQ reset procedure.

7.2 HYBRID AUTOMATIC REPEAT REQUEST FUNCTIONS

7.2.1 HARQ Principles

While the ARQ error control mechanism is simple and provides high transmission reliability, the throughput of ARQ schemes drop rapidly with increasing channel error rates, and the latency due to re-transmissions could be excessively high and intolerable for some delay-sensitive applications. Systems using Forward Error Correction (FEC), on the other hand, can maintain constant throughput regardless of channel error rate. However, FEC schemes have some drawbacks. High reliability is hard to achieve with FEC and requires the use of long and powerful error correction codes that increase the complexity of implementation [4]. The drawbacks of ARQ and FEC can be overcome, if the two error control schemes are properly combined.

In order to achieve increased throughput and lower latency in packet transmission, a Hybrid ARQ (HARQ) scheme is designed to combine ARQ error-control mechanism and FEC coding. A HARQ system consists of an FEC sub-system contained in an ARQ system. In this approach, the average number of re-transmissions is reduced by using FEC through correction of the error patterns that occur more frequently; however, when the less frequent error patterns are detected, the receiver requests a re-transmission where each re-transmission carries the same or some redundant information to help packet detection. The HARQ uses FEC to correct a subset of errors at the receiver and rely on error detection to detect the remaining errors. Most practical HARQ schemes utilize CRC codes for error detection and convolutional or Turbo codes for error correction [5].

The hybrid ARQ schemes may by classified depending on the content of subsequent re-transmissions as follows [4,15]:

1. **Type I HARQ**: in type I HARQ schemes, the same data packet is transmitted in all the re-transmissions. Soft combining may be used to improve the reliability. In this type of HARQ, also referred to as Chase Combining (CC), the blocks of data, along with the CRC code, are encoded using FEC encoder before transmission. If the receiver is unable to correctly decode the data block, a re-transmission is requested. When a re-transmitted coded block is received, it is combined with the previously received block corresponding to the same information bits (using, for example, a maximal ratio combining method) and fed to the decoder. Since each re-transmission is an identical replica of the original transmission, the received E_b/N_0, i.e., the energy per information bit divided by the noise spectral power density, increases for each re-transmission, improving the likelihood of correct decoding. In chase combining HARQ, the redundancy version of the encoded bits is not changed from one transmission to the next; therefore, the puncturing pattern remains the same. The receiver uses the current and all previous HARQ transmissions of the data block in order to decode the information bits. The process continues until either the information bits are correctly decoded and pass the CRC test, or the maximum number of HARQ re-transmissions is reached. When the maximum number of re-transmissions is reached, the MAC layer resets the process and continues with fresh transmission of the same data block.

   The chase combining method is used as the mandatory HARQ scheme for data transmissions in Release 1.0 of the mobile WiMAX systems, where an N-process stop-and-wait ARQ along with FEC coding are exploited [8]. A number of parallel channels for HARQ can help improve the throughput as one process is awaiting an acknowledgement; another process can utilize the channel and transmit sub-packets. Figure 7-6 illustrates the operation of a chase combining HARQ scheme and how the re-transmission of the same coded bits change the combined energy per bit (E_b) while maintaining the effective code rate intact.

2. **Type II HARQ**: in the original type II HARQ scheme, an alternate parity/user data re-transmission scheme is used. With the proper choice of channel encoder, the user information may also be recovered by parity bits. However, most type II algorithms adopt an incremental parity re-transmission scheme, in which only additional parity bits are sent in subsequent re-transmissions. Therefore, after each re-transmission, a richer set of parity bits is available at the receiver, improving the probability of reliable decoding. In incremental schemes; however, information cannot be recovered from parity bits alone.

   In an Incremental Redundancy (IR) HARQ scheme, a number of coded bits with increasing redundancy, where each represents the same set of input bits, are generated and transmitted to the receiver when a re-transmission is requested, to assist the receiver with the decoding of the information bits. The receiver combines each re-transmission with the previously received bits

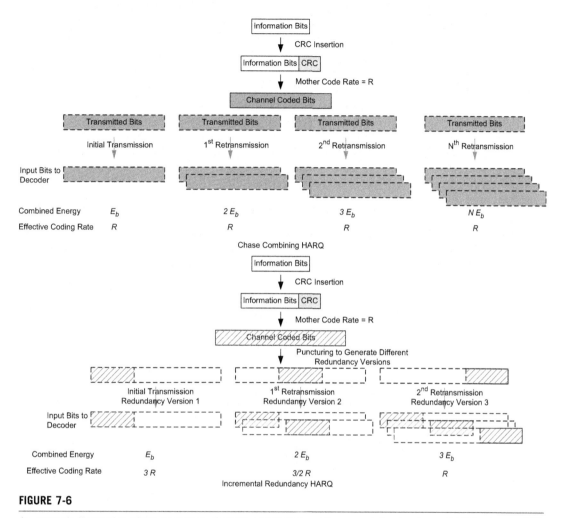

FIGURE 7-6

An example illustration of the operation of chase combining and incremental redundancy HARQ schemes [12]

belonging to the same packet. Since each re-transmission carries additional parity bits, the effective code rate is lowered by each re-transmission, as shown in Figure 7-6. The incremental redundancy is based on low-rate code and the different redundancy versions are generated by puncturing the channel coder output. In the example shown in Figure 7-6, the mother code rate is R, and 1/3 of the coded bits are transmitted in each re-transmission. Aside from increasing the received signal E_b/N_0 ratio by each re-transmission due to combining, there is a coding gain[i] as a result of each re-transmission. It must be noted that chase combining is a special case of incremental redundancy HARQ where the re-transmissions are identical copies of the original coded bits.

[i]The coding gain is defined as the difference between E_b/N_o required to achieve a given Bit Error Rate (BER) in a coded system, and the E_b/N_0 required to achieve the same BER in an uncoded system.

The incremental redundancy HARQ was an optional feature in Release 1.0 of mobile WiMAX systems [8]; however, it has been made the mandatory HARQ scheme for IEEE 802.16m systems. It is assumed that the receiver has received all previously transmitted redundancy and that each re-transmission provides some amount of information about the data packet. The puncturing pattern to be used for a given HARQ transmission is indicated by the Sub-Packet Identifier (SPID). By default, the SPID of the first transmission is always zero where all, for instance, turbo code systematic bits and some parity bits are sent. Note that only the parity bits are punctured, and each transmission by itself decodable. The SPID of the subsequent transmissions can be arbitrarily chosen by the system [7]. It must be noted that if the initial transmission with SPID of 0 is not properly received or lost, the re-transmission of only additional parity bits will not typically help decoding of the data packet, and a fresh transmission of the systematic bits might be necessary.

3. **Type III HARQ**: in a type III HARQ scheme, both user data and complementary parity bits are included in every re-transmission. The type III HARQ schemes are less efficient than incremental redundancy schemes due to repeated user data bits. Also in this case, after every re-transmission, a richer set of parity bits is available at the receiver, improving the probability of reliable decoding. Moreover, by adopting combining techniques, reliability may be further improved. In a type III incremental scheme, information can be recovered from every transmission. In a self-decodable HARQ re-transmission scheme, the user data may be recovered from every single transmission.

7.2.2 IEEE 802.16m HARQ Mechanism

The HARQ mechanism is used for all unicast data traffic in both downlink and uplink. The IEEE 802.16m HARQ scheme is based on an N-process stop-and-wait protocol. In single channel stop-and-wait, the transmitter waits after each transmission until an acknowledgement from the receiver is obtained. In the case of positive acknowledgement, a new packet is transmitted; otherwise, the same packet is re-transmitted. The N-process stop-and-wait mechanism makes use of the waiting time and transmits other sub-packets. Both BS and MS are required to maintain multiple simultaneous HARQ channels. The DL HARQ channels are identified by a HARQ Channel Identifier (ACID), whereas the UL HARQ channels are identified by both ACID and the index of the UL subframe in which the UL HARQ data burst is transmitted. Multiple UL HARQ channels in the same UL subframe are identified by different ACIDs and UL HARQ channels in different UL subframes are identified by the index of the UL subframe when they are addressed with the same ACID [3].

Generation of the HARQ sub-packets follows the channel coding procedures. The received sub-packets are combined by the FEC decoder as part of the decoding process. The use of incremental redundancy HARQ is mandatory in IEEE 802.16m compliant entities with chase combining as a special case. Each sub-packet contains part of a codeword identified by an SPID. In order to specify the start of a new transmission, a single-bit HARQ Identifier Sequence Number (AI_SN) is toggled on every new HARQ transmission attempt on the same ACID. If the AI_SN changes, the receiver will treat the corresponding HARQ transmission as a new encoder packet and discard previous HARQ transmissions associated with the same ACID.

7.2.3 **HARQ Timing and Signaling Protocols**

The IEEE 802.16m uses adaptive asynchronous HARQ in the downlink. In adaptive asynchronous HARQ, the resource allocation and transmission format for the HARQ re-transmissions may be different from the initial transmission. In the case of re-transmission, appropriate signaling is required to indicate the resource allocation and transmission format, along with other HARQ parameters [2].

On receiving a DL Basic Assignment Advanced Medium Access Protocol Information Element (A-MAP IE), the MS attempts to receive and decode the data burst. If the decoding is successful, the MS sends a positive acknowledgement to the BS; otherwise, the MS will send a negative acknowledgement to the BS. The process of re-transmissions is controlled by the BS using the ACID and AI_SN fields in the DL Basic Assignment A-MAP IE. If the AI_SN field for the ACID remains the same between two HARQ burst allocations, it indicates re-transmission. The BS may allocate different resource and transmission formats that are signaled through the DL Basic Assignment A-MAP IE for each re-transmission. If the AI_SN field for the ACID is toggled, it indicates the transmission of a new HARQ packet. The maximum number of HARQ channels per MS in the downlink is 16. The delay between two consecutive HARQ transmissions of the same data burst does not exceed the maximum [T_ReTx_Interval = 1, 2, ..., 8]. The number of re-transmissions of the same data packet does not exceed the maximum [N_MAX_ReTx = 4]. An example of HARQ operation in the downlink for TDD (DL/UL ratio of 5:3) and FDD is shown in Figure 7-7. Note that the processing time at the BS and MS is assumed to be equal to three sub-frames (a fraction of the radio frame).

The HARQ ACK/NACK timing is defined for FDD mode and for TDD mode with certain DL/UL ratios. A failed HARQ burst should be re-transmitted within the maximum re-transmission delay bound. A HARQ burst is discarded if the maximum number of re-transmissions is reached. When persistent allocation is applied to initial transmissions, HARQ re-transmissions are supported in a non-persistent manner, i.e., resources are allocated dynamically for HARQ re-transmissions. An asynchronous HARQ scheme is used in the downlink where the interval between successive transmissions/re-transmissions may vary, providing more flexibility for the downlink scheduler.

The IEEE 802.16m uses a synchronous HARQ scheme in the uplink where the interval between successive transmissions/re-transmissions is the same, resulting in a lower signaling overhead in resource assignment. In synchronous HARQ, resource allocation for the re-transmissions in the uplink can be fixed or adaptive according to control signaling. The default operation mode of HARQ in the uplink is non-adaptive, i.e., the parameters and the resource for the re-transmission are known *a priori*. The BS can signal an adaptive uplink HARQ mode.

Similar to the HARQ operation in the downlink, on receiving a UL Basic Assignment A-MAP IE, the MS transmits a HARQ sub-packet in the assigned resource. The BS attempts to decode the data packet. If the decoding is successful, the BS will send a positive acknowledgement to the MS; otherwise, the BS will send a negative acknowledgement to the MS. If re-transmission becomes necessary, and if the MS does not receive a UL Basic Assignment A-MAP IE for the failed HARQ sub-packet, the MS transmits the next sub-packet through the resources assigned in the previous sub-packet transmission with the same ACID. A UL Basic Assignment A-MAP IE may be sent to signal the re-transmission with corresponding ACID and unchanged AI_SN. On receiving the UL Basic Assignment A-MAP IE, the MS performs the HARQ re-transmission. The maximum number of HARQ channels per MS is 16 in the uplink. The maximum number of re-transmissions of the same

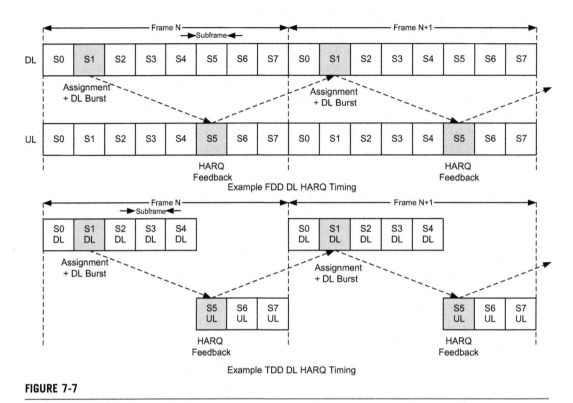

FIGURE 7-7

An example of FDD and TDD downlink HARQ timing for 5, 10, and 20 MHz channel bandwidths [2]

data packet is [N_MAX_ReTx = 4]. An example HARQ operation in the uplink for TDD (DL/UL ratio of 5:3) and FDD is shown in Figure 7-8. Note that the processing time at the BS and MS is assumed to be equal to three sub-frames.

Synchronous HARQ re-transmissions are supported when persistent allocation is utilized. With group resource allocation, the HARQ re-transmissions are allocated individually in a synchronous manner. Transmission of Assignment A-MAP IE, the downlink/uplink HARQ sub-packets, and the corresponding feedbacks are according to predefined timing. The transmission times are defined based on frame index and subframe index. The integer frame index F is between 0 and 3, since there are four radio frames per superframe (as will be described in Chapter 9). In FDD mode, the index of DL or UL subframe ranges from 0 to $SF - 1$, where SF is the number of subframes per frame, determined based on the radio frame configuration. In TDD mode, the index of DL subframe ranges from 0 to $D - 1$, where D is the number of DL subframes per frame, and the index of UL subframe is between 0 and $U - 1$, where U is the number of UL subframes per frame. The value of parameters D and U depend on specific frame configuration [2].

In FDD mode, as shown in Figure 7-9, the DL HARQ sub-packet transmission corresponding to a DL Basic Assignment A-MAP IE in the lth subframe of the ith DL frame begins in the lth subframe of the ith DL frame. Note that the DL Assignment A-MAPs point to DL allocations over the same

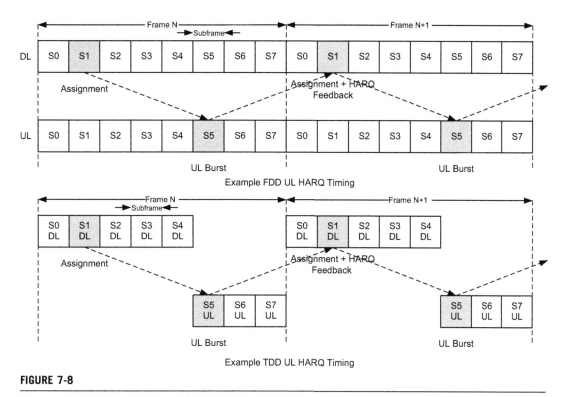

FIGURE 7-8

An example of FDD and TDD uplink HARQ timing for 5, 10, and 20 MHz channel bandwidths [2]

subframe. A HARQ feedback for the DL HARQ sub-packet is transmitted in the nth subframe of the jth UL frame. The subframe indices m, n, and frame index j are determined by using l and i according to the following equation [2]:

$$n = \lceil m + SF/2 \rceil \bmod SF$$

$$j = \left(i + \left\lceil \frac{\lceil m+SF/2 \rceil}{SF} \right\rceil + T_{offset} \right) \bmod 4$$

(7-1)

Where $\lceil . \rceil$ and $\lfloor . \rfloor$ denote *ceil* and *floor* functions, respectively, and *modulo 4* operation is due to the fact that the frame index F ranges between 0 and 3. The parameter T_{offset} is used to ensure that the HARQ feedback is not generated earlier than the minimum processing time of three subframes. This offset is set to 1 only if the time interval from completion of the HARQ sub-packet transmission to its feedback time will be shorter than the data burst processing time; otherwise it is set to zero. The value of the offset also depends on whether a short Transmission Time Interval (TTI) or a long TTI of one or four subframes is used, respectively.

In FDD mode, as shown in Figure 7-9, the UL HARQ sub-packet transmission corresponding to a UL Basic Assignment A-MAP IE in the qth subframe of the ith DL frame begins in the mth subframe

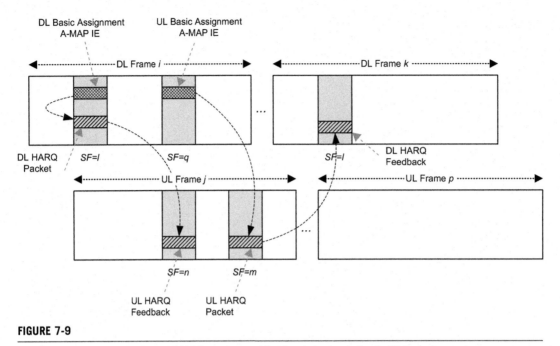

FIGURE 7-9

An illustration of downlink/uplink HARQ packet and feedback transmission in FDD mode

of the jth UL frame. A HARQ feedback for the UL HARQ sub-packet is transmitted in the qth subframe of the kth DL frame. When the UL HARQ feedback indicates a negative acknowledgement, re-transmission of the UL HARQ sub-packet is scheduled in the mth subframe of the pth UL frame. The subframe indices m, n, and frame indices j, k, and p are defined according to the following equation [2]:

$$m = n = \lceil q + SF/2 \rceil \bmod SF$$

$$j = \left(i + \left\lfloor \frac{\lceil q+SF/2 \rceil}{SF} \right\rfloor + T_{offset}\right) \bmod 4$$

$$k = \left(j + \left\lfloor \frac{\lceil q+SF/2 \rceil}{SF} \right\rfloor + T'_{offset}\right) \bmod 4$$

$$p = \left(k + \left\lfloor \frac{\lceil q+SF/2 \rceil}{SF} \right\rfloor + T''_{offset}\right) \bmod 4$$

(7-2)

The binary-valued offsets in the above equations are included to ensure the minimum processing time condition is met.

In TDD mode, as shown in Figure 7-10, the DL HARQ sub-packet transmission corresponding to a DL Basic Assignment A-MAP IE in lth DL subframe of the ith frame is scheduled in the mth DL subframe of the ith frame. A HARQ feedback for the DL HARQ sub-packet is transmitted in

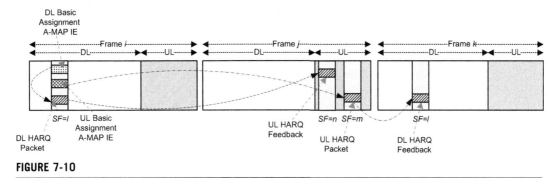

FIGURE 7-10

An illustration of downlink/Uplink HARQ packet and feedback transmission in TDD mode

the nth UL subframe of the jth frame. The subframe indices m, n, and frame index j are obtained as follows [2]:

$$m = 1$$

$$\text{for } D > U$$

$$n = \begin{cases} 0 & 0 \le m < k \\ m - k & k \le m < U + k \\ U - 1 & U + k \le m < D \end{cases} \qquad (7\text{-}3)$$

$$\text{for } D \le U \quad n = m - k$$

$$j = (i + T_{offset}) \bmod 4$$

where

$$k = \begin{cases} \lceil D - U \rceil / 2 & D \le U \\ \lfloor D - U \rfloor / 2 & D > U \end{cases} \qquad (7\text{-}4)$$

When the DL Basic Assignment A-MAP IE in the lth DL subframe of the ith frame indicates use of a long TTI and l is 0 (i.e., the first subframe of the frame), the subframe index m and the frame index j are calculated according to Equation (7-3). The DL Basic Assignment A-MAP IE in the lth ($l \ne 0$) DL subframe of the ith frame can signal the use of a long TTI. In this case, the long TTI transmission of the DL HARQ sub-packet begins in the first DL subframe of the succeeding frame. The HARQ feedback for this long TTI transmission is transmitted in the nth UL subframe of the jth frame. The subframe index n and the frame index j are derived according to Equation (7-3).

In TDD mode, as shown in Figure 7-10, the UL HARQ sub-packet transmission corresponding to a UL Basic Assignment A-MAP IE in lth DL subframe of the ith frame is scheduled in the mth UL subframe of the jth frame. The HARQ feedback is transmitted in the lth DL subframe of the kth frame. When the UL HARQ feedback indicates a negative acknowledgement, the re-transmission of the UL

HARQ sub-packet begins in the mth UL subframe of the pth frame. The subframe indices m, n, and frame indices j, k, and p are calculated as follows [2]:

$$m = \begin{cases} 0 & \text{for } 0 \le l < Q \\ l - Q & \text{for } k \le l < U + Q \\ U - 1 & \text{for } U + Q \le l < D \end{cases}$$

$$j = (i + T_{offset}) \bmod 4$$

$$k = (j + 1 + T_{offset}) \bmod 4$$

$$p = (k + T''_{offset}) \bmod 4$$

$$Q = \begin{cases} \lfloor (D - U)/2 \rfloor & D \ge U \\ \lceil (D - U)/2 \rceil & D < U \end{cases}$$

(7-5)

The binary-valued offsets in the above equations are included to ensure the minimum processing time is met.

The legacy and IEEE 802.16m frames are offset by a fixed number of subframes to accommodate new features such as the IEEE 802.16m preambles, superframe header (system configuration information), and control channels, as shown in Figure 7-11. The *Frame Offset* is a time offset between the start of the legacy frame and the start of the IEEE 802.16m frame carrying the superframe headers, defined in a unit of subframes.

The subframe configuration in the IEEE 802.16m systems that support the legacy mobile and base stations are indexed in a different manner. The DL subframe index ranges from 0 to $D - 1$, where D is the number of DL subframes dedicated to IEEE 802.16m compliant systems. Similarly, the UL subframe index ranges from 0 to $U - 1$, where U is the number of UL subframes dedicated to IEEE 802.16m compliant systems. Figure 7-11 shows an example of subframe indexing where the ratio of the total DL to UL subframes is 5 to 3. The *Frame Offset* parameter is 2. While operating the legacy and the new systems simultaneously in TDD mode, the legacy and new uplinks are frequency-division multiplexed to mitigate the uplink link budget issues of the legacy terminals. Thus, the offset parameters in Equations (7-1) to (7-5) are adjusted to accommodate the effect of *Frame Offset* in IEEE 802.16m systems with legacy support.

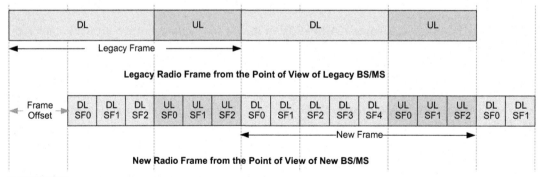

FIGURE 7-11

Relative positions of the legacy and new radio frames in TDD mode with DL/UL ratio 5:3 [3]

The Group Resource Allocation (GRA) mechanism allocates resources to multiple users as a group in order to reduce the control overhead. The mechanism takes advantage of common traffic characteristics and grouping is done based on some common parameters, such as modulation and coding scheme, MIMO mode and resource size. On receiving a DL Group Resource Allocation A-MAP IE, the scheduled MS attempts to decode the received data packet. If the decoding is successful, the MS will send a positive acknowledgement to the serving BS; otherwise, the MS will send a negative acknowledgement to the BS. With DL Group Resource Allocation, the HARQ re-transmissions are allocated individually and performed using a DL Basic Assignment A-MAP IE that carries the same ACID as the one used by the DL Group Resource Allocation A-MAP IE for the first HARQ sub-packet transmission. On receiving a UL Group Resource Allocation A-MAP IE, the MS transmits the HARQ sub-packet in the resource assigned by the UL Group Resource Allocation A-MAP IE. The serving BS will attempt to decode the data packet. If the decoding is successful, the BS will send a positive acknowledgement to the MS; otherwise, the BS will send a negative acknowledgement to the MS. With the UL Group Resource Allocation, the HARQ re-transmissions are allocated individually and performed as described earlier. In this case, the UL Basic Assignment A-MAP IE contains the same ACID as the one used by the UL Group Resource Allocation A-MAP IE for the first HARQ sub-packet transmission.

Persistent Allocation (PA) is a technique used to reduce assignment overhead for connections with a periodic traffic pattern and relatively fixed payload size. To allocate resources persistently to a single connection, the BS transmits a Persistent Allocation A-MAP IE for DL or UL. The persistently allocated resource size, resource location, and the packet modulation and coding scheme remain unchanged by the BS and MS until the persistent assignment is de-allocated, changed, or an error event occurs. The persistent scheduling does not include any special consideration for HARQ re-transmissions of data packets that were initially transmitted using persistently allocated resources. Resources for re-transmissions can be allocated one at a time as needed using a DL or UL Basic Assignment A-MAP IE. On receiving an uplink Persistent Allocation A-MAP IE, the MS periodically transmits the HARQ sub-packets through the assigned resource as specified in the UL Persistent Allocation A-MAP IE. The serving BS will attempt to decode the received data packet. If the decoding is successful, the BS will send a positive acknowledgement to the MS; otherwise, a negative acknowledgement is sent to the MS.

If the HARQ entity in the transmitter determines that the HARQ process was unsuccessfully terminated, it informs the ARQ entity in the transmitter about the failure of the HARQ packet delivery. The ARQ entity in the transmitter can then initiate re-transmission and re-segmentation of the ARQ blocks that correspond to the failed HARQ packet.

7.3 MAC PDU FORMATION

Figure 7-12 illustrates the MAC PDU construction process, input and output of each functional block, and the order in which those functions are applied for various types of connections, i.e., ARQ, non-ARQ, and control connections. The payloads associated with multiple transport connections corresponding to the same security association can be multiplexed and encrypted together in a MAC PDU. If N transport connections are multiplexed, one MEH and one FEH, PEH, RFPEH, and MCEH per connection are present in a MAC PDU. The AGMH and the MEH headers carry the information about the Flow IDs and lengths of the payloads. The FEH, PEH, and RFPEH headers carry the information

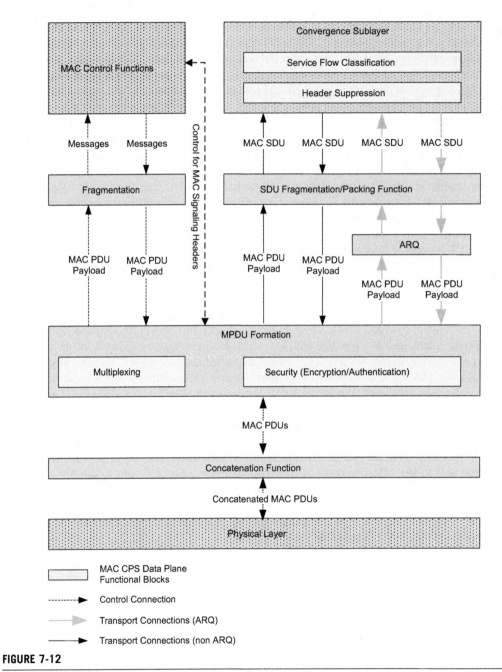

FIGURE 7-12

The MAC PDU construction process on the data-plane [1]

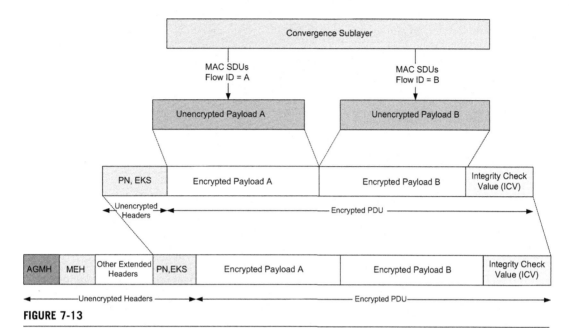

FIGURE 7-13

Multiplexing of connection payloads associated with the same security association [2]

about the transport connection payload and MCEH carries information about management connections' payload. For example, multiple connections' payloads which are encrypted using AES-CCM[ii] [13] can be multiplexed and encrypted together in a single MAC PDU. Figure 7-13 illustrates the multiplexing of two connection payloads which are associated with same security association (i.e., AES-CCM).

Multiple MAC PDUs may be concatenated into a single transmission in UL or DL directions. For the MS attached to a BS, each MAC PDU in an UL or DL PDU is uniquely identified by a Flow ID.

[ii]AES-CCM is an authenticated encryption algorithm designed to provide both authentication and privacy. Advanced Encryption Standard (AES) is a symmetric block cipher that has variable key and fixed data length. The key lengths can be independently chosen as 128, 192 or 256 bits, which result in 10, 12, and 14 rounds of operation, respectively. The data length is, however, fixed at 128 bits. The input, as well as intermediate data, can be considered as a matrix with four rows and four columns called state. Each element of the matrix is composed of eight bits, therefore enabling efficient implementation of AES on 8-bit platforms. Counter with CBC-MAC (CCM) is a mode of operation of a block cipher that combines the existing Counter with CBC-MAC modes. It uses an encryption algorithm to generate encrypted and authenticated data at the same time. The AES-CCM process requires two AES cores. In order to achieve higher throughput, two separate AES cores, one for CBC-MAC and the other for Counter Mode, are developed. In AES CBC-MAC the first AES core is working in cipher feedback mode. It compares the new input to previously encrypted data using XOR operation. The core is used to calculate the Message Integrity Code (MIC) for authenticity of data. The process starts with encrypting the first block and then successively XORs subsequent blocks and encrypts the result. The final MIC is one 128-bit block. Once the first block has been prepared, XOR-ing the current block with the previously encrypted block computes the MIC one block at a time. If the last block is not exactly 128-bits, it is padded with zeros. The final output is one 128-bit block, but the CCM requires only a 64-bit MIC, so the low order 64-bit final output is discarded. The encryption process uses the AES CBC-MAC core to generate the MIC and the AES Counter core for the encryption of data.

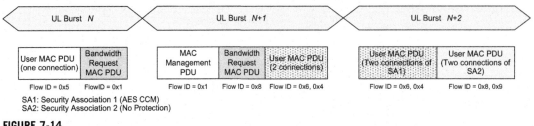

FIGURE 7-14

An example MAC PDU concatenation [2]

Figure 7-14 illustrates the concatenation concept for a UL burst transmission. Since the MAC SDUs in a MAC PDU are identified by a Flow ID in the AGMH and MEH, the receiving MAC entity is able to present the MAC SDU (after reassembling the MAC SDU from one or more MAC PDUs) to the correct instance of the MAC SAP. The MAC PDUs containing control messages, user data, and bandwidth request may be concatenated into the same transmission.

Fragmentation is a process by which a MAC SDU (or MAC control message) is divided into two or more MAC PDUs. The capabilities of fragmentation and reassembly are mandatory for IEEE 802.16m entities. For transport connections, the FPEH or MEH header in the MAC PDU provides information about the SDU fragment. The FPEH or MEH header is always present in a MAC PDU for transport connections. The SN in FPEH or MEH header is used for sequencing the SDU fragments and the fragmentation control bits in FPEH or MEH header are used to tag the SDU fragments with respect to their position in the parent SDU. For non-ARQ transport connections, the fragments are transmitted once. The sequence number assigned to each connection PDU carrying SDU fragment allows the receiver to recreate the original payload and to detect the loss of any intermediate fragments. A connection may be in only one fragmentation state at any time instant. On loss of any fragment, the receiver discards all SDU fragments on the connection until a new SDU fragment or a non-fragmented SDU is detected. For ARQ connections, the fragments are transmitted in sequence. The sequence number assigned to each ARQ PDU that carries MAC SDU fragments allows the receiver to recreate the original payload and to detect the loss of any intermediate fragments. For management connections, the FEH header in the MAC PDU provides information about the control message fragment. The sequence number in FEH is used for sequencing the control message fragments and the fragmentation control bits in the FEH header are used to tag the control message fragments with respect to their position in the parent control message.

Only one control message can be in the fragmentation state at any given time. The sequence number, assigned to each management connection PDU carrying a control message fragment, allows the receiver to recreate the original payload and to detect the loss of any intermediate fragments. On loss of any fragment, the receiver waits for the lost control message fragments until a new control message fragment or a new non-fragmented control message is detected.

Multiple MAC SDUs may be packed into a single MAC PDU. The support of packing and unpacking is mandatory in IEEE 802.16m entities; however, it is up to the transmitting side whether to pack a group of MAC SDUs into a single MAC PDU. The packing and fragmentation mechanisms for both the non-ARQ and ARQ connections are specified in the IEEE 802.16m specification [2]. A MAC PDU may contain a packed sequence of variable-length MAC SDUs. It must be noted that non-fragmented MAC SDUs and MAC SDU fragments may both be present in the same MAC PDU. The MAC appends a PEH header or an MEH header to the MAC PDU.

The fragmentation control bits are set according to the rules defined in reference [2]. The use of an PEH header for ARQ-enabled connections is similar to that for non-ARQ connections. It is up to the transmitting side to decide whether to pack a group of MAC SDUs and/or fragments in a single MAC PDU. The sequence number of the PEH header is used by the ARQ protocol to identify and re-transmit ARQ blocks. When transmitting a MAC PDU on a connection that is mapped to a security association, the sender performs encryption and data authentication on the MAC PDU payload as specified by that security association. When receiving a MAC PDU on a connection mapped to a security association, the receiver performs decryption and data authentication functions on the MAC PDU, as specified by that security association. The AGMH and the extended headers are not encrypted. The receiver determines whether the payload in the MAC PDU is encrypted based on the Flow ID in the AGMH or

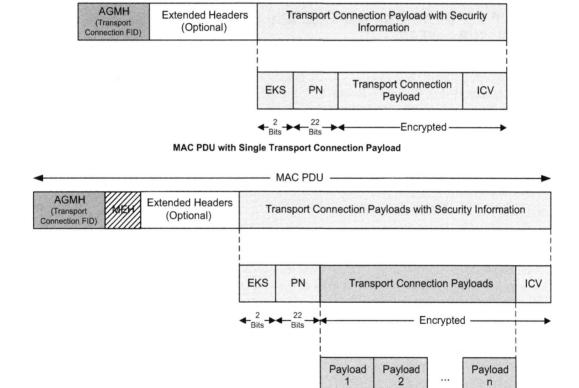

MAC PDU with Single Transport Connection Payload

MAC PDU with Multiple Transport Connection Payloads

FIGURE 7-15

MAC PDU with single/multiple (authenticated and encrypted using AES-CCM) transport connection payload(s) [1,2]

SPMH. The encryption information required to decrypt a payload at the receiving entity is present at the beginning and at the end of the connection payload. As an example, the AES CCM, Packet Number (PN), and Encryption Key Sequence (EKS) are prefixed at the beginning of the payload and an Integrity Check Value (ICV) is suffixed to the end of the in MAC PDU, as shown in Figure 7-15.

If multiple connection payloads are transmitted in the same burst and the connections are mapped to the same security association, then multiple connection payloads are multiplexed prior to encryption and the multiplexed payload is jointly encrypted. The receiver performs the decryption and data authentication on the multiplexed payload, as specified by the security association. The receiver determines whether the payload in the MAC PDU is encrypted by verifying the Flow ID in the AGMH. The encryption information needed to decrypt the multiplexed payload at the receiving station is contained in the beginning of the first connection payload and at the end of the last connection payload. As an example, the AES CCM, PN, and EKS are present in the beginning of the connection payload and the ICV is appended at the end of the connection payload in the MAC PDU, as shown in Figure 7-15.

7.4 3GPP LTE RADIO LINK CONTROL AND MEDIUM ACCESS CONTROL SUB-LAYERS

The Layer 2 functions in LTE are classified into the following categories: Medium Access Control (MAC) functions; Radio Link Control (RLC) functions; and Packet Data Convergence Protocol (PDCP) functions [9–12]. Figures 7-16 and 7-17 illustrate the structure of Layer 2 in LTE downlink and uplink. The Service Access Point (SAP) for peer-to-peer communication is marked with a small oval at the interface between the sub-layers. The SAP between the physical layer and the MAC sub-layer provides the transport channels. The SAP between the MAC sub-layer and the RLC sub-layer provide the logical channels. The multiplexing of several logical channels (i.e., radio bearers) on the same transport channel (i.e., transport block) is performed by the MAC sub-layer [9].

7.4.1 3GPP LTE MAC Sub-Layer

The services and functions provided by the MAC sub-layer can be summarized as follows:

- Mapping between logical channels and transport channels;
- Multiplexing/de-multiplexing of RLC protocol data units corresponding to one or different radio bearers into/from transport blocks delivered to/from the physical layer on transport channels;
- Traffic volume measurement reporting;
- Error correction through HARQ;
- Priority handling between logical channels of one UE;
- Priority handling between UEs through dynamic scheduling;
- Transport format selection.

The E-UTRA defines MAC entities in the UE and in the E-UTRAN. These MAC entities control the following transport channels: Broadcast Channel (BCH); Downlink Shared Channel (DL-SCH); Paging Channel (PCH); Uplink Shared Channel (UL-SCH); and Random Access Channel (RACH). The exact functions performed by the MAC entities are different in the UE and the E-UTRAN. Figure 7-18 illustrates a sample structure for the UE MAC entity.

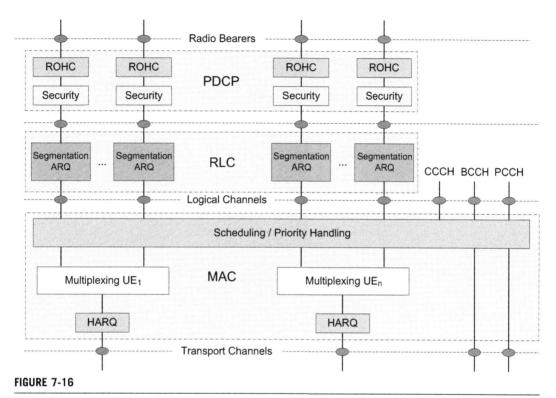

FIGURE 7-16

LTE Layer 2 structure in the downlink [9]

This MAC sub-layer provides services such as data transfer and radio resource allocation to upper layers. The physical layer provides the MAC sub-layer with data transfer services, signaling of HARQ feedback, signaling of scheduling request, and measurements. The access to the data transfer services is through the use of transport channels. The characteristics of a transport channel are defined by its transport format, which specifies the physical layer processing to be applied to the transport channel, such as channel coding and interleaving, and any other service-specific rate matching.

7.4.2 Logical and Transport Channels

The different logical and transport channels in LTE are illustrated in Figures 7-19 and 7-20, respectively. Each logical channel type is defined by what type of information is transferred. The logical channels are generally classified into two groups: (1) Control Channels (for the transfer of control-plane information) and (2) Traffic Channels (for the transfer of user-plane information), as shown in Figure 7-19 [9].

The control channels are exclusively used for transfer of control-plane information. The control channels supported by MAC can be classified as follows (see Figure 7-19):

- Broadcast Control Channel (BCCH): a downlink channel for broadcasting system control information.

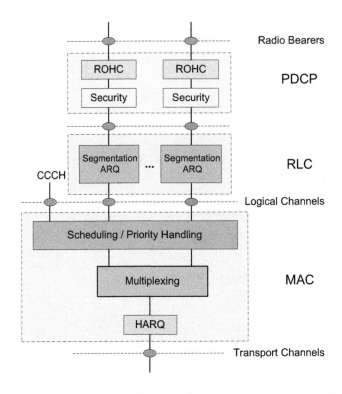

FIGURE 7-17

LTE Layer 2 structure in the uplink [9]

- Paging Control Channel (PCCH): a downlink channel that transfers paging information and system information change notifications. This channel is used for paging when the network does not know the location of the UE.
- Common Control Channel (CCCH): a channel for transmitting control information between UEs and eNBs. This channel is used for UEs having no RRC connection with the network.
- Multicast Control Channel (MCCH): a point-to-multipoint downlink channel used for transmitting MBMS control information from the network to the UE, for one or several MTCHs. This channel is only used by UEs that receive MBMS.
- Dedicated Control Channel (DCCH): a point-to-point bi-directional channel that transmits dedicated control information between a UE and the network. It is used by UEs that have RRC connection.

The traffic channels are exclusively used for the transfer of user-plane information. The traffic channels supported by MAC can be classified as follows (as shown in Figure 7-20):

- Dedicated Traffic Channel (DTCH): a point-to-point bi-directional channel dedicated to a single UE for the transfer of user information.
- Multicast Traffic Channel (MTCH): a point-to-multipoint downlink channel for transmitting traffic data from the network to the UE. This channel is only used by UEs that receive MBMS.

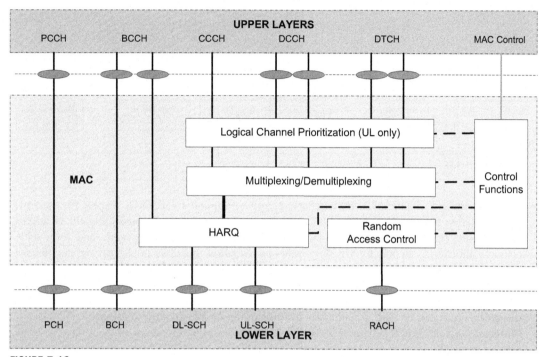

FIGURE 7-18

3GPP LTE UE MAC structure [10]

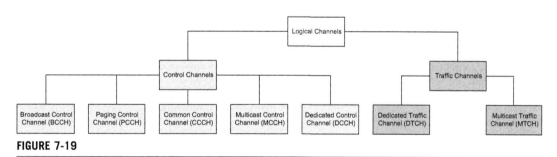

FIGURE 7-19

Classification of 3GPP LTE logical channels

The physical layer provides information transfer services to MAC and higher layers. The physical layer transport services are described by how and with what characteristics data are transferred over the radio interface. This should be clearly separated from the classification of what is transported, which relates to the concept of logical channels at the MAC sub-layer. As shown in Figure 7-20, downlink transport channels can be classified as follows:

• Broadcast Channel (BCH) characterized by fixed, pre-defined transport format and required to be broadcast in the entire coverage area of the cell;

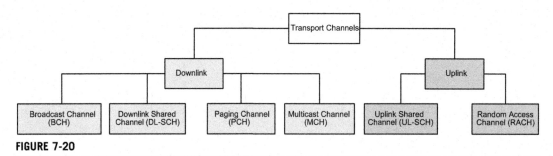

FIGURE 7-20

Classification of 3GPP LTE transport channels

- Downlink Shared Channel (DL-SCH) characterized by support for HARQ, support for dynamic link adaptation by varying the modulation, coding and transmit power, possibility for broadcast in the entire cell, possibility to use beamforming, support for both dynamic and semi-static resource allocation, support for UE discontinuous reception to enable power saving, support for MBMS transmission;
- Paging Channel (PCH) characterized by support for UE discontinuous reception in order to enable power saving, requirement for broadcast in the entire coverage area of the cell, mapped to physical resources which can be used dynamically also for traffic or other control channels;
- Multicast Channel (MCH) characterized by a requirement to be broadcast in the entire coverage area of the cell, support for macro-diversity combining of MBMS transmission on multiple cells, support for semi-static resource allocation;

The uplink transport channels are classified as follows (see Figure 7-20):

- Uplink Shared Channel (UL-SCH) characterized by the possibility to use beamforming, support for dynamic link adaptation by varying the transmit power and modulation and coding schemes, support for HARQ, support for both dynamic and semi-static resource allocation;
- Random Access Channel (RACH) characterized by limited control information and collision risk.

The mapping of the logical channels to the transport channels in the downlink and uplink is shown in Figure 7-21. As shown in Figures 7-16 and 7-17, the main services and functions provided by the RLC sub-layer include transfer of upper layer PDUs supporting AM or UM, TM data transfer, error

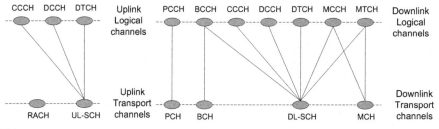

FIGURE 7-21

Mapping of logical to transport channels in the downlink and uplink [9]

correction through ARQ (since CRC check is provided by the physical layer, no CRC is needed at RLC level), segmentation according to the size of the transport block, re-segmentation of PDUs that need to be re-transmitted, concatenation of SDUs for the same radio bearer, in-sequence delivery of upper layer PDUs except during handover, duplicate detection, and protocol error detection and recovery.

The users in the 3GPP LTE system are assigned temporary identifiers to protect user privacy and confidentiality. Depending on the state of the UE, different types of temporary identifiers are used. Table 7-1 summarizes the Radio Network Temporary Identifiers (RNTI) and their usage in 3GPP LTE. For example, the Random Access RNTI (RA-RNTI) is used on the PDCCH when random access response messages are transmitted. It unambiguously identifies which time-frequency resource was utilized by the UE to transmit the random access preamble. The Msg3 acronym in Table 7-1 denotes the message transmitted on UL-SCH containing a C-RNTI MAC Control Element (CE) or CCCH SDU, submitted from an upper layer and associated with the UE Contention Resolution Identity, as part of a random access procedure [10]. The various temporary identifiers in 3GPP LTE are conceptually similar to station identifiers, as well as MAP information elements in IEEE 802.16m systems that are used to identify the users and their active connections and allocations in the downlink and uplink.

Table 7-1 Various Radio Network Temporary Identifiers and their Usage in 3GPP LTE [10]

Radio Network Temporary Identifier	Usage	Transport Channel	Logical Channel
Paging RNTI (P-RNTI)	Paging and system information change notification	PCH	PCCH
System Information RNTI (SI-RNTI)	Broadcast of system information	DL-SCH	BCCH
Random Access RNTI (RA-RNTI)	Random access response	DL-SCH	N/A
Temporary C-RNTI	Contention resolution (when no valid C-RNTI Is available)	DL-SCH	CCCH
Temporary C-RNTI	Msg3 transmission	UL-SCH	CCCH, DCCH, DTCH
Cell RNTI (C-RNTI)	Dynamically scheduled unicast transmission	DL-SCH, UL-SCH	DCCH, DTCH
C-RNTI	Triggering of PDCCH-ordered random access	N/A	N/A
Semi-Persistent Scheduling C-RNTI	Semi-persistently scheduled unicast transmission (activation, reactivation, and re-transmission)	DL-SCH, UL-SCH	DCCH, DTCH
Semi-Persistent Scheduling C-RNTI	Semi-persistently scheduled unicast transmission (deactivation)	N/A	N/A
Transmit Power Control-Physical Uplink Control Channel-RNTI (TPC-PUCCH-RNTI)	Physical layer uplink power control	N/A	N/A
Transmit Power Control-Physical Uplink Shared Channel-RNTI (TPC-PUSCH-RNTI)	Physical layer uplink power control	N/A	N/A

7.4.3 **3GPP LTE RLC Sub-Layer**

The functions of the RLC sub-layer are controlled by the RRC sub-layer and are performed by the RLC entities. For each RLC entity configured at the eNB, there is a corresponding RLC entity configured at the UE. As shown in Figure 7-22, an RLC entity receives RLC SDUs from upper layers and sends RLC PDUs to its peer RLC entity via lower layers. An RLC PDU can either be an RLC data PDU or an RLC control PDU. If an RLC entity receives RLC SDUs from the upper layer, it receives them through a single SAP between the RLC and upper layer, and after forming RLC data PDUs from the received RLC SDUs, the RLC entity delivers the RLC data PDUs to a lower layer through a single logical channel. If an RLC entity receives RLC data PDUs from a lower layer, it receives them through a single logical channel, and after forming RLC SDUs from the received RLC data PDUs, the RLC entity delivers the RLC SDUs to an upper layer through a single SAP between the RLC and the upper layer. When an RLC entity delivers RLC control PDUs to a lower layer, it delivers them via the same logical channel through which the RLC data PDUs are delivered.

An RLC entity can be configured to perform data transfer in one of the following three modes: Transparent Mode (TM); Unacknowledged Mode (UM); or Acknowledged Mode (AM). Consequently, an RLC entity is categorized as a TM RLC entity, a UM RLC entity, or an AM RLC entity, depending on the mode of data transfer that the RLC entity is configured to provide.

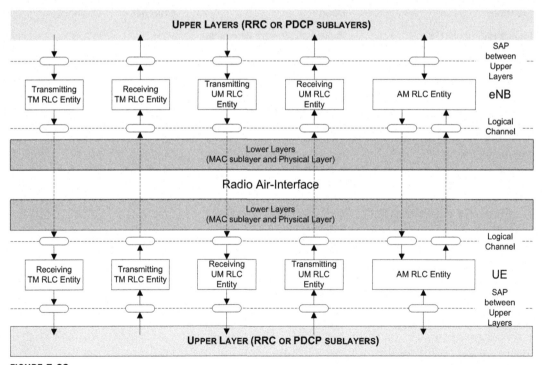

FIGURE 7-22

The structure of the 3GPP RLC sub-layer

A TM RLC entity is configured either as a transmitting TM RLC entity or a receiving TM RLC entity. The transmitting TM RLC entity receives RLC SDUs from an upper layer and sends RLC PDUs to its peer receiving TM RLC entity via the lower layers. The receiving TM RLC entity delivers RLC SDUs to the upper layer and receives RLC PDUs from its peer transmitting TM RLC entity via the lower layers. A UM RLC entity is configured either as a transmitting UM RLC entity or a receiving UM RLC entity. The transmitting UM RLC entity receives RLC SDUs from an upper layer and sends RLC PDUs to its peer receiving UM RLC entity via the lower layers. The receiving UM RLC entity delivers RLC SDUs to an upper layer and receives RLC PDUs from its peer transmitting UM RLC entity via the lower layers.

An AM RLC entity consists of a transmitting side and a receiving side. The transmitting side of an AM RLC entity receives RLC SDUs from an upper layer and sends RLC PDUs to its peer AM RLC entity via the lower layers. The receiving side of an AM RLC entity delivers RLC SDUs to an upper layer and receives RLC PDUs from its peer AM RLC entity via the lower layers. Figure 7-22 illustrates the structure of the RLC sub-layer.

The octet-aligned RLC SDUs of variable sizes are supported for all RLC entity types (i.e., TM, UM, and AM RLC entities). The RLC PDUs are formed on notification of a transmission opportunity and are delivered to the MAC sub-layer.

As shown in Figure 7-23, a TM RLC entity can be configured to deliver RLC PDUs through BCCH, DL/UL CCCH, and PCCH logical channels. When a transmitting TM RLC entity forms Transparent Mode Data (TMD) PDUs from RLC SDUs, it does not segment or concatenate the RLC SDUs, and does not include any RLC headers in the TMD PDUs. A UM RLC entity can be configured to deliver RLC PDUs through DL/UL DTCH logical channels. When a transmitting UM RLC entity constructs Unacknowledged Mode Data (UMD) PDUs from RLC SDUs, it segments and/or concatenates the RLC SDUs so that the UMD PDUs fit within the total size of RLC PDU indicated by the lower layer at the particular transmission opportunity, and includes relevant RLC headers in the UMD PDU.

When a receiving UM RLC entity receives UMD PDUs, it detects whether duplicate UMD PDUs have been received and discards duplicated UMD PDUs, it reorders the UMD PDUs, if they are received out of sequence, it detects the loss of UMD PDUs at lower layers and avoids excessive reordering delays, it reassembles RLC SDUs from the reordered UMD PDUs and delivers the RLC SDUs to the upper layer in ascending order of RLC SN, and it discards received UMD PDUs that cannot be reassembled into a RLC SDU due to loss of a UMD PDU which belonged to the particular RLC SDU at lower layers. At the time of RLC re-establishment, the receiving UM RLC entity reassembles RLC SDUs from the UMD PDUs that are received out of sequence and deliver them to the upper layer, discards any remaining UMD PDUs that could not be reassembled into RLC SDUs, initializes relevant state variables, and stops relevant timers.

As shown in Figure 7-24, an AM RLC entity can be configured to deliver RLC PDUs through DL/UL DCCH or DL/UL DTCH logical channels. When the transmitting side of an AM RLC entity constructs Acknowledged Mode Data (AMD) PDUs from RLC SDUs, it segments and/or concatenates the RLC SDUs so that the AMD PDUs fit within the total size of the RLC PDU at a particular transmission opportunity. The transmitting side of an AM RLC entity supports re-transmission of RLC data PDUs (i.e., ARQ), if the RLC data PDU to be re-transmitted does not fit within the total size of RLC PDU at a particular transmission opportunity, the AM RLC entity can re-segment the RLC data PDU into AMD PDU segments where the number of re-segmentations is unlimited.

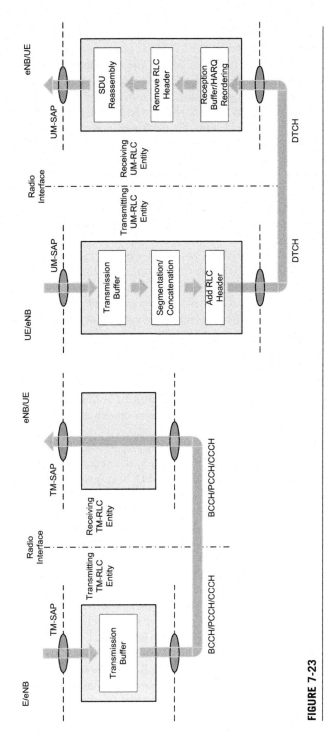

FIGURE 7-23

An illustration of 3GPP LTE TM and UM RLC entities [11]

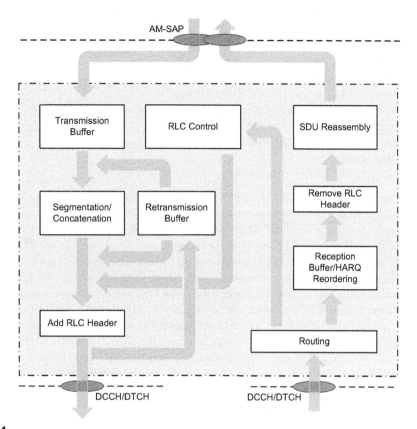

AM-SAP

Transmission Buffer

RLC Control

SDU Reassembly

Segmentation/ Concatenation

Retransmission Buffer

Remove RLC Header

Reception Buffer/HARQ Reordering

Add RLC Header

Routing

DCCH/DTCH   DCCH/DTCH

FIGURE 7-24

An illustration of 3GPP LTE AM RLC entity [11]

When the transmitting side of an AM RLC entity forms AMD PDUs from RLC SDUs received from an upper layer, or AMD PDU segments from RLC data PDUs to be re-transmitted, it includes relevant RLC headers in the RLC data PDU. The receiving side of an AM RLC entity receives RLC data PDUs, detects duplicate RLC data PDUs, and discards duplicated RLC data PDUs. It reorders the RLC data PDUs, if they are received out of sequence, detects the loss of RLC data PDUs at lower layers, and sends re-transmission requests to its peer AM RLC entity. It further reassembles RLC SDUs from the reordered RLC data PDUs and delivers the RLC SDUs to the upper layer in sequence. At the time of RLC re-establishment, the receiving side of an AM RLC entity reassembles RLC SDUs from the RLC data PDUs that are received out of sequence and delivers them to the upper layer, discarding any remaining RLC data PDUs that could not be reassembled into RLC SDUs, initializing relevant state variables, and stopping relevant timers.

The RLC sub-layer provides TM data transfer, UM data transfer, and AM data transfer, including indication of successful delivery of upper layer PDU services to RRC or PDCP. The MAC sub-layer provides data transfer and notification of transmission opportunity (along with the total size of the RLC PDUs) services to the RLC sub-layer.

The RLC sub-layer performs the following functions:

- Transfer of upper layer PDUs;
- Error correction through ARQ for AM data transfer;
- Concatenation, segmentation, and reassembly of RLC SDUs for UM and AM data transfer;
- Re-segmentation of RLC data PDUs for AM data transfer;
- Reordering of RLC data PDUs for UM and AM data transfer;
- Duplicate detection for UM and AM data transfer;
- RLC SDU discard for UM and AM data transfer;
- RLC re-establishment;
- Protocol error detection for AM data transfer.

RLC PDUs can be categorized into RLC data PDUs and RLC control PDUs. RLC data PDUs are used by TM, UM, and AM RLC entities to transfer upper layer PDUs. RLC control PDUs are used by AM RLC entity to perform ARQ procedures. The STATUS PDU is used by the receiving side of an AM RLC entity to inform the peer AM RLC entity about correctly received as well as lost RLC data PDUs.

As shown in Figure 7-25, the TMD PDU consists of only a data field, and does not include any RLC headers. The UMD PDU consists of a data field and a UMD PDU header. The UMD PDU header consists of a fixed part (i.e., fields that are present for every UMD PDU) and an extension part (i.e., fields that are present for a UMD PDU when necessary). The fixed part of the UMD PDU header itself is octet-aligned and consists of Framing Information, Extension, and Sequence Number (SN) fields. The extension part of the UMD PDU header itself is octet-aligned and includes Extension bits and Length Indicators. The AMD PDU consists of a data field and an AMD PDU header.

The AMD PDU header comprises a fixed part (i.e., fields that are present for every AMD PDU) and an extension part (i.e., fields that are present for an AMD PDU when necessary). The fixed part of the AMD PDU header itself is octet-aligned and consists of Data/Control (D/C), Re-segmentation Flag (RF), Polling bit (P), Framing Indicator (FI), Extension bit (E), and Sequence Number (SN). The extension part of the AMD PDU header itself is octet-aligned and consists of Extension and Length Indicator fields. The STATUS PDU consists of a payload and an RLC control PDU header [11]. The packet processing stages in 3GPP LTE in various sub-layers is illustrated in Figure 7-26.

7.4.4 ARQ and HARQ in LTE

The E-UTRA supports ARQ and HARQ functionalities. The ARQ functionality provides error correction by re-transmissions in the acknowledged mode at Layer 2. The HARQ functionality improves packet transmission and detection between peer entities at Layer 1. The HARQ within the MAC sub-layer is characterized by an N-process Stop-and-Wait protocol and re-transmission of transport blocks on failure of earlier transmissions. The ACK/NACK transmission in FDD mode refers to the downlink packet that was received four subframes earlier. In TDD mode, the uplink ACK/NACK timing depends on the uplink/downlink configuration. For TDD, the use of a single ACK/NACK response for multiple PDSCH transmissions is possible. A total of 8 HARQ processes are supported in FDD duplex mode [9].

An asynchronous adaptive HARQ is utilized in the downlink. The uplink ACK/NACK signaling in response to downlink (re)transmissions is sent on PUCCH or PUSCH. The PDCCH signals the HARQ process number and whether it is a fresh transmission or a re-transmission. The re-transmissions are

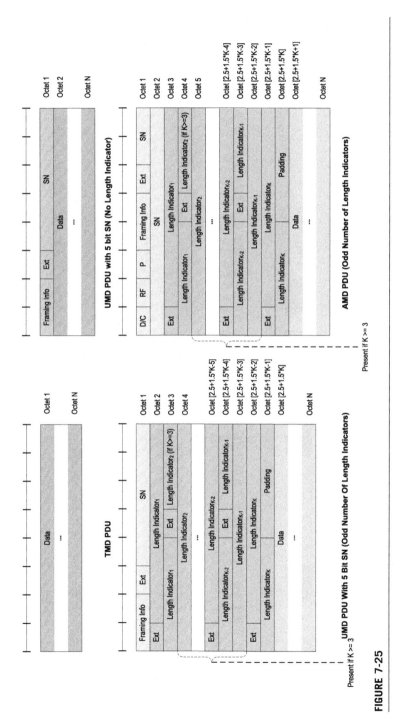

FIGURE 7-25

3GPP RLC PDU formats [11]

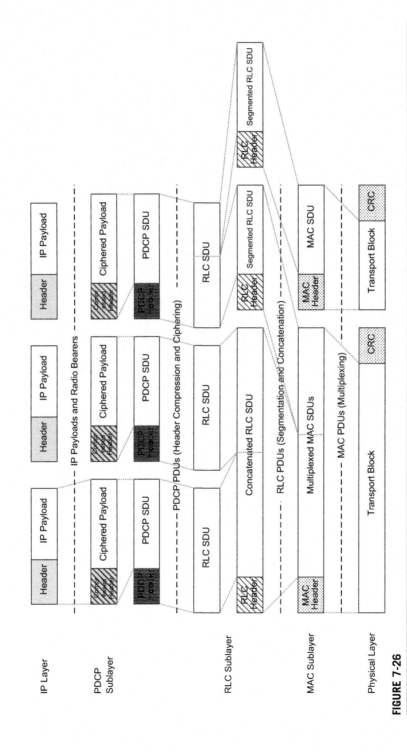

FIGURE 7-26

Packet processing in 3GPP LTE [9,12]

always scheduled through PDCCH. A synchronous HARQ scheme is supported in the uplink. The maximum number of re-transmissions can be configured per UE basis, as opposed to per radio bearer. The downlink ACK/NACK signaling in response to uplink (re)transmissions is sent on PHICH. The ARQ functionality within the RLC sub-layer is responsible for re-transmission of RLC PDUs or RLC PDU segments. The ARQ re-transmissions are based on RLC status reports and optionally based on HARQ/ARQ interactions. The polling for an RLC status report is used when needed by RLC and status reports can be triggered by upper layers [9]. When a PDCCH for the UE is correctly received, the UE follows the instructions by PDCCH, i.e., to perform a transmission or a re-transmission regardless of the content of the HARQ feedback (ACK or NACK). When no PDCCH addressed to the C-RNTI of the UE is detected, the HARQ feedback determines how the UE should perform the re-transmissions. The UE performs a non-adaptive re-transmission, i.e., a re-transmission on the same uplink resource as previously used by the same process. The UE does not perform any UL (re)transmission and maintains the data in the HARQ buffer. A PDCCH is then required to perform a re-transmission, i.e., a non-adaptive re-transmission cannot follow. Measurement gaps are of higher priority than HARQ re-transmissions; whenever a HARQ re-transmission collides with a measurement gap, the latter prevails.

The ARQ function within the RLC sub-layer re-transmits RLC PDUs or RLC PDU segments based on RLC status reports. Polling for an RLC status report is used when needed by RLC. The RLC receiver can also trigger an RLC status report after detecting a missing RLC PDU or RLC PDU segment. In addition, the E-UTRAN can allocate semi-persistent downlink resources for the first HARQ transmissions to UEs where the RRC defines the periodicity of the semi-persistent downlink grant and the PDCCH indicates whether the downlink grant is semi-persistent, i.e., if it can be implicitly reused in the following TTIs according to the periodicity defined by RRC. The re-transmissions are explicitly signaled via the PDCCH. In the subframes where the UE has a semi-persistent downlink resource, if the UE cannot find its C-RNTI on the PDCCH, a downlink transmission according to the semi-persistent allocation that the UE has been assigned in the TTI is assumed; otherwise, in the subframes where the UE has a semi-persistent downlink resource, if the UE finds its C-RNTI on the PDCCH, the PDCCH allocation supersedes the semi-persistent allocation for that TTI and the UE does not decode the semi-persistent resources. Similar principles apply to uplink scheduling. The HARQ operation related to data transmission is independent of DRX operation, and the UE wakes up to read the PDCCH for possible re-transmissions and/or ACK/NACK signaling regardless of DRX. In the downlink, a timer is used to limit the time the UE stays awake awaiting for a re-transmission.

The possibility to configure the downlink and uplink ratio results in a significant difference between TDD and FDD modes [14]. In FDD mode, there is a one-to-one correspondence between each downlink and uplink subframe; however, this is not typically the case for TDD as the number of uplink and downlink subframes in a radio frame might be different. As an example, for FDD, a fixed timing relationship has been defined such that an uplink scheduling grant received in downlink subframe n corresponds to uplink subframe $n + 4$. This provides the UE with sufficient time to prepare for the uplink transmission. Such a simple fixed timing relationship is not possible for TDD, since subframe $n + 4$ may not be an uplink subframe, and consequently other timing relationships have been defined for TDD mode. Another difference concerns the re-transmission of erroneously received data packets. For both downlink and uplink, LTE utilizes HARQ with soft combining, i.e., the terminal and the network can request re-transmission of erroneously received data packets. In case the original transmission fails, the receiver buffers the soft information and combines it with the re-transmission. The fundamental structure used is a set of parallel HARQ processes, where after a fixed time following reception of data, a positive or negative acknowledgement is transmitted. In each downlink subframe, data can be transmitted over one of the HARQ processes in the UE, while the other processes are in various stages of decoding or rescheduling. By selecting the number of processes appropriately, continuous transmission to a terminal can be achieved. In TDD mode, on the other hand, the time from reception of downlink data to transmission of an acknowledgement in the uplink depends on the uplink/downlink ratio, and the number of HARQ processes varies from 4 to 15. This results from the fact that uplink subframes are not available at all times and the feedback must be delayed at least until there is an uplink subframe.

The other main difference between 3GPP LTE FDD and TDD modes with respect to HARQ processing is related to the transmission of acknowledgements in the uplink. For frame configurations where the number of uplink subframes is greater than or equal to the number of downlink subframes, each downlink subframe has been associated with an uplink subframe in such a way that acknowledgements from at most one downlink subframe need to be transmitted in every uplink subframe. However, in configurations where the number of downlink subframes per radio frame is larger than the number of uplink subframes, reception of several downlink subframes may need to be acknowledged in a single uplink subframe. The 3GPP LTE provides two mechanisms to perform this function, either through bundling or multiplexing. The network determines which mechanism to use on a per UE basis. In any case, the amount of control channel resources in an uplink subframe increases with the increasing number of downlink subframes to be acknowledged [14]. The bundling mechanism combines the acknowledgements from multiple HARQ processes such that a positive acknowledgement will be sent only in the case where all downlink transmissions were correctly received. The advantage of this approach is that the number of acknowledgements from a UE in single subframe is minimized, which is important especially for coverage-limited terminals. When the UE cannot detect one or several downlink assignments, to avoid erroneously positive acknowledgements, and if data in the detected subframes is correctly decoded, there is an index included in the downlink assignment where the terminal can learn how many subframes have been assigned, and hence detect whether assignments have been missed.

With multiplexing, the acknowledgements from multiple HARQ processes are explicitly transmitted in an uplink subframe. This provides more detailed information about the decoding results of the different downlink transmissions, but requires a higher signal-to-noise ratio at the base station which may hence not be suitable for coverage-limited terminals. For uplink data transfers, the

TABLE 7-2 3GPP LTE Uplink HARQ Operation [9]

HARQ Feedback (UE Side)	PDCCH (UE Side)	UE Behavior
ACK or NACK	New transmission	New transmission according to PDCCH
ACK or NACK	Re-transmission	Re-transmission according to PDCCH (adaptive re-transmission)
ACK	None	No (re)transmission, retain data in HARQ buffer and a PDCCH is required to resume re-transmissions
NACK	None	Non-adaptive re-transmission

individual feedback is available and bundling is not required. Negative acknowledgements can be transmitted in a subset of subframes for asymmetric uplink/downlink configurations, and there is multiplexing in one of the downlink subframes. The same control channel formats are used for 3GPP LTE TDD and FDD modes. Nevertheless, the interpretation of transmitted bits and assignment of the resources are different. Table 7-2 summarizes the HARQ feedback and UE behavior in the uplink.

References

[1] IEEE Std 802.16-2009, IEEE Standard for Local and Metropolitan Area Networks – Part 16: Air Interface for Fixed Broadband Wireless Access Systems, May 2009.

[2] P802.16m/D6, IEEE Standard for Local and Metropolitan Area Networks – Part 16: Air Interface for Broadband Wireless Access Systems, Advanced Air Interface, May 2010.

[3] IEEE 802.16m–0034r3, System Description Document, May 2010.

[4] Shu Lin, Daniel Costello, Error Control Coding, second ed., Prentice Hall, 2004.

[5] Chris Heegard, Stephen Wicker, Turbo Coding, Springer, 1998.

[6] Loutfi Nuaymi, WiMAX: Technology for Broadband Wireless Access, John Wiley & Sons, 2007.

[7] Jeffrey G. Andrews, Arunabha Ghosh, Rias Muhamed, Fundamentals of WiMAX: Understanding Broadband Wireless Networking, first ed., Prentice Hall, 2007.

[8] WiMAX Forum, Mobile System Profile Release 1.5 Air Interface Specification, August 2009.

[9] 3GPP TS 36.300, Evolved Universal Terrestrial Radio Access (E-UTRA) and Evolved Universal Terrestrial Radio Access Network (E-UTRAN); Overall Description, Stage 2, March 2010.

[10] 3GPP TS 36.321, Evolved Universal Terrestrial Radio Access (E-UTRA); Medium Access Control (MAC), Protocol Specification, March 2010.

[11] 3GPP TS 36.322, Evolved Universal Terrestrial Radio Access (E-UTRA); Radio Link Control (RLC), Protocol Specification, March 2010.

[12] Erik Dahlman, et al., 3G Evolution: HSPA and LTE for Mobile Broadband, second ed., Academic Press, 2008.

[13] IETF RFC 4309, R. Housley, Using Advanced Encryption Standard (AES) CCM Mode with IPsec Encapsulating Security Payload (ESP), December 2005.

[14] 3G Americas, 3GPP LTE for TDD Spectrum in the Americas, November 2009.

[15] R. Comroe, D. Costello, "ARQ Schemes for Data Transmission in Mobile Radio Systems", IEEE Journal on Selected Areas in Communications Vol. 2 (Issue 4), July 1984.

The IEEE 802.16m Security Sub-Layer

8

INTRODUCTION

A wireless system uses an intrinsically open and unsecure radio channel for transmission of user signaling and traffic between the base station and mobile stations. As such, reliable and robust security and encryption procedures must be employed in order to protect confidentiality, privacy, and integrity of user traffic and credentials, and to prevent security breaches and theft of service in cellular networks.

This chapter describes the security aspects of the IEEE 802.16m standard. As shown in Figure 8-1 the security sub-layer of IEEE 802.16 is located between the MAC and the physical layers. The security functions provide users with privacy, authentication, and confidentiality by applying cryptographic transforms to MAC PDUs transported over the connections between the MS and the BS. In addition, the security sub-layer enables the operators to prevent unauthorized access to data transport services by securing the associated service flows across the network. The security sub-layer employs an authenticated client/server key management protocol in which the BS (the server) controls distribution of keying material to the MS (the client). In addition, the basic security mechanisms are reinforced by adding digital-certificate-based MS device-authentication to the key management protocol. If, during capability negotiation, the MS indicates that it does not support the IEEE 802.16m security protocols, the authorization and key exchange procedures are skipped and the MS will not be provided with any service (except emergency services). The privacy function has two component protocols: (1) an encapsulation protocol for securing packet data across the network, i.e., a set of cryptographic suites and the rules for applying those algorithms to a MAC PDU payload; and (2) a Key Management Protocol (PKM) providing the secure distribution of keying data from the BS to the MS. The MS and the BS can synchronize keying data via the key management protocol. The BS can use the protocol to enforce conditional access to network services, as well.

In IEEE 802.16m, the encryption of the user data is done after the MAC PDUs are generated. This marks a significant difference between IEEE 802.16m and 3GPP LTE, where the ciphering is performed in the PDCP sub-layer and prior to formation of the MAC PDUs. In this chapter the security functions of both standards are described to allow the readers to better understand the similarities and differences of security functions by drawing analogies [3,7–9]. Another important aspect of the IEEE 802.16m security relative to the legacy systems is the encryption of MAC management messages to protect the integrity of Layer 2 messaging and signaling over the air interface [1,2].

8.1 SECURITY ARCHITECTURE

The security architecture of IEEE 802.16m consists of the MS, the BS, and the Authenticator, as shown in Figure 8-2 [4,5].

Mobile WiMAX. DOI: 10.1016/B978-0-12-374964-2.10008-6
Copyright © 2011 Elsevier Inc.

Network Layer
CS SAP
Radio Resource Control and Management
Service Specific Convergence Sublayer
MAC SAP
MAC Sublayer
Security Sublayer
PHY SAP
Physical Layer
Control Plane
Data Plane
Control and Management SAP
Layer 3
Layer 2
Layer 1

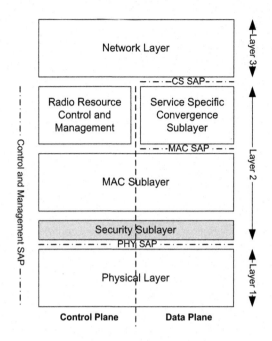

FIGURE 8-1

Location of the security sub-layer in the IEEE 802.16m protocol stack

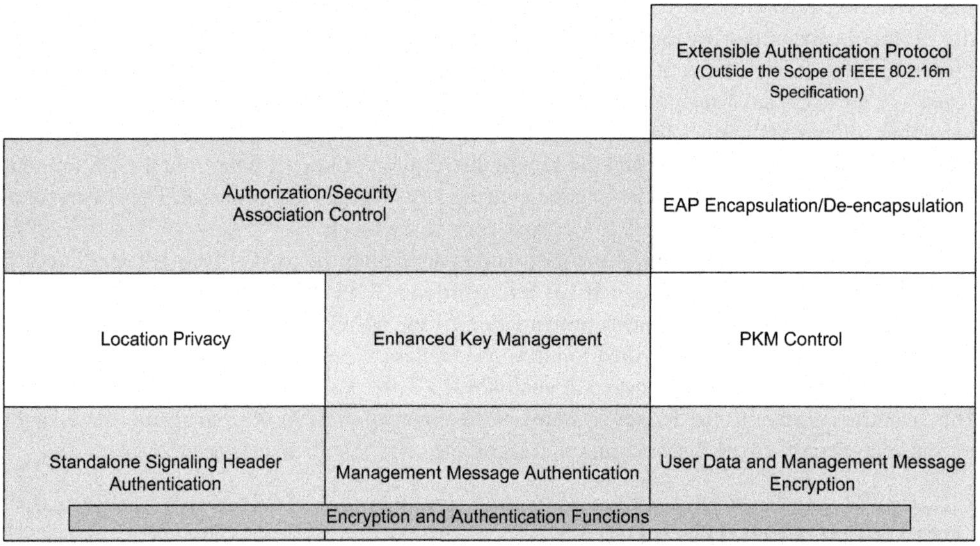

Extensible Authentication Protocol (Outside the Scope of IEEE 802.16m Specification)
Authorization/Security Association Control
EAP Encapsulation/De-encapsulation
Location Privacy
Enhanced Key Management
PKM Control
Standalone Signaling Header Authentication
Management Message Authentication
User Data and Management Message Encryption
Encryption and Authentication Functions

FIGURE 8-2

Functional blocks of IEEE 802.16m security architecture [3]

Within the MS and BS, the security functions are classified into two logical categories: (1) a security management entity; and (2) encryption and integrity. The security management entity includes the following functions [2,3]:

- Overall security management;
- Extensible Authentication Protocol (EAP) encapsulation/de-encapsulation for authentication;
- PKM control functions through key generation/derivation/distribution, and key state management;
- Authentication and Security Association (SA) control;
- Location privacy.

Encryption and integrity protection entity consists of the following functions:

- User data encryption/authentication;
- Management message authentication;
- Protection of management message confidentiality.

8.2 AUTHENTICATION

The authentication of user and device identities is conducted between the MS and BS entities using EAP. The choice of EAP methods and selection of credentials that are used during EAP-based authentication are outside the scope of the IEEE 802.16m standard. The authentication is performed during initial network entry following basic capability negotiation. The security capabilities and policies are negotiated in the authentication, authorization, and key exchange stage. The re-authentication is performed before expiration of authentication materials/credentials. The data transmission may continue during the reauthentication process by providing the MS with two sets of authentication and keying material with overlapping lifetimes. The authentication procedure is controlled by the authorization state machine, which defines the operations in each state [2,3].

8.3 KEY MANAGEMENT PROTOCOL (PKMv3)

The IEEE 802.16m uses a newer version of the PKM protocol known as PKMv3 to perform key management. The legacy standard key management procedures are based on PKMv2 [1,6]. Transparent exchange of authentication and authorization messages, key agreement, and security context exchange are among the services provided by PKMv3. The PKMv3 protocol provides mutual authentication and establishes a shared secret between the MS and the BS. The shared secret is then used to exchange or derive other keying material. This two-tiered mechanism allows frequent traffic key refreshing without additional computational complexity.

8.3.1 Key Derivation

The PKMv3 key hierarchy defines what keys are present in the system and how keys are generated. The IEEE 802.16m uses one authentication scheme based on EAP; therefore, there is only one primary source of keying material. The key that is used to protect the integrity of control messages is derived from the source keying material generated by authentication/authorization processes. The EAP-based authentication process provides the master key. All security keys are derived directly or indirectly from the master key. The master key is the shared key that is derived in the course of executing the EAP functions. The Pair-wise Master Key (PMK) is derived from the master key and is used to derive the Authorization Key (AK). The AK is used to derive other keys, including the Traffic Encryption Key (TEK) and the Cipher-based Message Authentication Code (CMAC) key.

After completing the authentication or reauthentication process, key agreement is performed to derive a PMK and an AK to verify the recently created PMK and AK, and to exchange other required security parameters. The PMK is derived by feeding parameters such as the master key, NONCE_MS,[i] NONCE_BS, etc., where NONCE_MS is a random number generated by the MS that is sent to the BS during key agreement, and NONCE_BS is a random number generated by the BS that is sent to the MS during key agreement. After EAP authentication is complete, the MS (requester), AAA, and authenticator hold a 512-bit master key, which is transferred to the authenticator from AAA using EAP attributes, and the BS starts a key agreement three-way handshake to derive a fresh PMK in both MS and authenticator [2].

8.3.2 Key Exchange

The key exchange procedure is controlled by the security key state machine, which defines the operations in each state. The security keys, such as PMK, AK, and CMAC, are locally derived by using the shared master key and other parameters during the key agreement procedure, as shown in Figure 8-3.

8.3.3 Key Usage

Each SA maintains two security keys for downlink and uplink encryption. The TEK_{DL} key is used for encryption of DL data by the BS and the TEK_{UL} key is used for encryption of UL data by the MS, the decryption is carried out according to the Encryption Key Sequence (EKS). Note that the EKS field in the legacy generic MAC header carries a 2-bit key sequence of associated TEK [1]. In certain instances where the BS derives a new TEK_{UL} and sets the TEK_{DL} to the old TEK_{UL}, the BS TEK_{DL} and MS TEK_{UL} are the same TEK with the same EKS, and both can transfer data securely until the TEK update happens from the MS side and the MS is re-syncronized with the new TEK_{UL}. The security key update is triggered either by exhausting packet numbers[ii] for the TEK_{DL} or TEK_{UL}, or by reauthentication [1].

---

[i]In security schemes, a nonce is an abbreviation for number used once. It is often a random or pseudo-random number that is issued in an authentication protocol to ensure old communications cannot be reused in replay attacks. For example, the nonces that are used in HTTP digest access authentication to calculate an MD5 digest of the password. Message-Digest algorithm 5 (MD5) is a widely used cryptographic hash function with a 128-bit hash value specified in IETF RFC 1321 [11]. The nonces are different each time the authentication challenge response code is presented, and each client request has a unique sequence number, thus making replay attacks and dictionary attacks virtually impossible. The initialization vectors are also referred to as nonces for the above reasons. To ensure that a nonce is used only once, a time-variant number such as a time stamp or a random number with a large number of bits is utilized, resulting in a statistically insignificant chance of repeating previously generated values [10].

[ii]The Packet Number (PN) associated with an SA is set to 1 when the SA is established and when a new TEK is installed. After each MAC PDU transmission, the PN is incremented by 1. On UL connections, the PN is XORed with 0×80000000 prior to encryption and transmission. On DL connections, the PN is used without such modification, which results in splitting of the PN space such that $0\times00000001 - 0\times7FFFFFFF$ values are used for the DL and $0\times80000001 - 0\times FFFFFFFF$ values are utilized in the UL, preventing a PN collision between the UL and the DL [1].

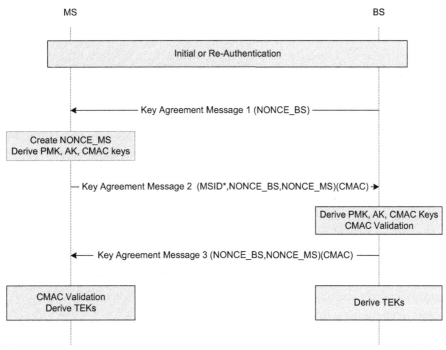

FIGURE 8-3

Initial or reauthentication key derivation and exchange [2]

The MS can request a key update when the packet number space of the TEK_{UL} is exhausted, or the TEK_{UL} is being used for downlink traffic, as well. For reauthentication, after key agreement, the AK_{old} would be valid and only one new TEK would be derived by using AK_{new}; the TEKs are updated immediately after key agreement. The MS is then resyncronized with the new TEK or another new TEK is derived from AK_{new} [2].

8.4 SECURITY ASSOCIATION MANAGEMENT

The security association (SA) is defined as the set of information required for secure communication between the BS and MS. The SA is identified using an SA Identifier (SAID). The SA is applied to the service flows once it has been established. The IEEE 802.16m only supports unicast static SA, i.e., the SA is used to provide keying context to unicast transport connections. The SA is applied to all data exchanged over the connection with the MS. Multiple connections may be mapped to the same unicast SA. However, the SA is not equally applied to all the management messages over the same management connection. Depending on the extended header indicator, the SA can be selectively applied to management connections.

If the MS and the BS decide to use *No Authorization* as their authorization policy, no security association will be established. In this case, Null SAID is used as the target SAID field in service flow creation messages. If authorization is performed but the MS and BS decide to create an unprotected service flow, the Null SAID may be used as the target SAID field in service flow creation messages.

8.5 CRYPTOGRAPHIC METHODS

The Advanced Encryption Standard (AES) is a block cipher scheme that can be used in different modes. The IETF RFC 4309 describes the use of the AES in Counter with CBC-MAC (CCM) mode with an explicit Initialization Vector (IV) as an IPsec Encapsulating Security Payload (ESP) mechanism to provide confidentiality, data origin authentication, and connectionless integrity [12]. The IEEE 802.16m standard specifies cryptographic algorithms that are used for MAC PDU protection or key encryption/decryption functions. The MS and the BS may support encryption methods and algorithms for secure transmission of MAC PDUs over the air interface. The Advanced Encryption Standard algorithm is the only supported cryptographic method in IEEE 802.16m. The AES modes specified in IEEE 802.16m are: AES-CCM mode,[iii] which provides integrity protection, and AES-CTR mode. The AES-CCM mode is supported for unicast transport and management connections and the Packet Number (PN) size is 22 bits. The AES-CTR mode is supported for unicast transport connections and the PN size is 22 bits. When some connections identified by flow identifiers are mapped to the same SA, their payloads can be multiplexed together into one MAC PDU, where the multiplexed payloads are jointly encrypted together. For example, in Figure 8-4, the payloads associated with Flow A and Flow B, which are mapped to the same SA, are jointly encrypted. The MAC header or extended headers provides the details of payloads which are multiplexed.

8.6 CONTROL-PLANE SIGNALING PROTECTION

The MAC management/control messages are selectively protected in IEEE 802.16m systems, if such capability is enabled during basic capability negotiation. If the selective confidentiality protection is enabled, the negotiated keying materials and cipher suites are used to encrypt MAC management messages. Figure 8-5 illustrates three levels of selective confidentiality protection used for MAC management/control messages in IEEE 802.16m as follows [2]:

- No protection: if the MS and BS have no shared security context (or protection is not required), the MAC management messages are neither encrypted nor authenticated. The MAC management messages before the authorization phase also fall into this category.

[iii]Advanced Encryption Standard in Counter with CBC-MAC mode is a mode of operation of a block cipher that combines the existing Counter (CTR) and CBC-MAC modes. It uses an encryption algorithm to generate encrypted and authenticated data at the same time. The AES-CCM process requires two AES cores. In order to achieve higher throughput, two separate AES cores, one for CBC-MAC and the other for Counter Mode, are developed. In AES CBC-MAC, the first AES core is working in cipher feedback mode. It XORs the new input to previously encrypted data. The core is used to calculate the MIC for the authenticity of data. The process starts with encrypting the first block and then successively XORs subsequent blocks, and encrypts the result. The final MIC is one 128-bit block. Once the first block has been prepared, XORing the current block with the previously encrypted block computes the MIC one block at a time. If the last block is not exactly 128-bit, it is padded with zeros. The final output is one 128-bit block, but the CCM requires only a 64-bit MIC, so the low order 64-bit of final output is discarded [12].

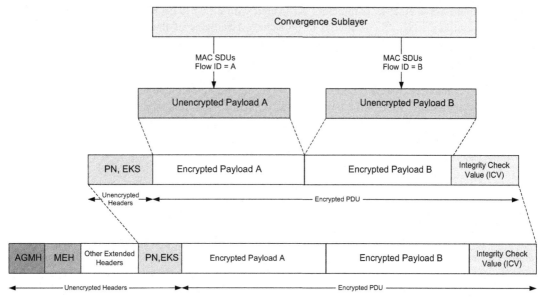

FIGURE 8-4

Multiplexed MAC PDU format [2]

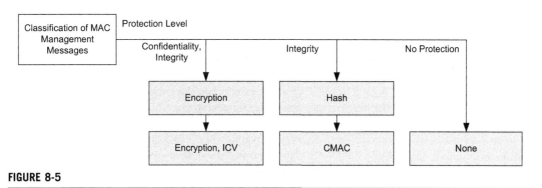

FIGURE 8-5

The IEEE 802.16m management message protection process [2,3]

- CMAC-based integrity protection: CMAC Tuple (i.e., a keyed message digest to authenticate the sender if authentication is supported) is included with the MAC management message which protects the entire MAC management/control message. The original MAC management message is plain text.
- AES-CCM based authenticated encryption: the Integrity Check Value (ICV) field is included after encrypted payload and protects the integrity of the payload, as well as the MAC header (see Figure 8-4).

The security context is a set of parameters associated with a key in each hierarchy that defines the scope while the key usage is considered to be secure. Examples of these parameters include key lifetime and counters to ensure that the same encryption will not be used more than once. When the context of the key expires, a new key is obtained to continue operation. The master key context includes all parameters associated with the master key, and is created when EAP authentication is fulfilled. This context is created on completion of successful authentication/reauthentication, and is discarded when no longer valid or if a key agreement procedure was not completed for a period of time [2]. The context is created by the MS and the authenticator when authenticating in radio frame legacy zones or new zones, and is maintained during zone switching. Other security contexts include PMK, AK, TEK, and SA contexts.

8.7 USER PRIVACY

The user identity in IEEE 802.16m systems is protected by ensuring that system hackers cannot obtain the mapping between the MS MAC address (alternatively known as MS ID) and the STID through eavesdropping the information exchange over the air interface. In order to protect the mapping between the STID and the MS MAC address, two types of STIDs are assigned to an MS during network entry, i.e., a Temporary STID (TSTID) and a permanent STID. As shown in Figure 8-6, a TSTID is assigned during the initial ranging process and is used until the permanent STID is assigned to the MS. The STID is assigned during the registration process following successful authentication, and is encrypted during transmission. The TSTID is released after the STID is securely assigned. The STID is used for all information exchange with the MS until the session is active.

The MS ID is protected by hashing the value of the real MS ID, which is denoted by MS ID* in Figure 8-6. The MS generates a new NONCE_MS and derives the MS ID* value, subsequently it sends an AAI_RNG-REQ message carrying the MS ID* to the BS. When the BS receives the AAI_RNG-REQ message, it returns an AAI_RNG-RSP message containing the temporary STID instead of the STID and the MS ID* value that the MS had sent. After the temporary STID is assigned, it is used for subsequent network entry procedures until the STID is allocated. The real MS ID is transmitted to the BS in an AAI_REG-REQ message in an encrypted manner. The STID is assigned after the authentication procedure is successfully completed. Note that the assignment message, i.e., the AAI_REG-RSP message, is encrypted. Once the MS receives the STID via the AAI_REG-RSP message, it releases the temporary STID [2].

8.8 3GPP LTE SECURITY ASPECTS

The E-UTRAN security has been designed based on the following principles. The keys used for Non-Access Stratum (NAS) and Access Stratum (AS) protection are dependent on the algorithm with which they are used. The eNB keys are cryptographically separated from the EPC keys that are used for NAS protection, making it impossible to use the eNB key to drive an EPC key. The AS and NAS keys are derived in the EPC/UE from key material that was generated by a NAS (EPC/UE) level

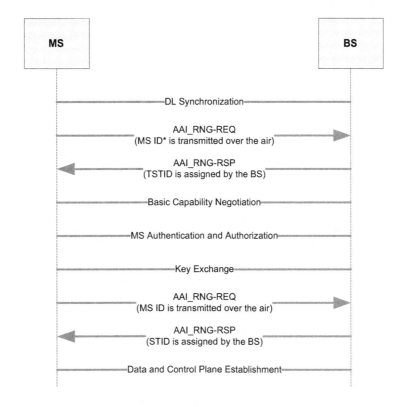

FIGURE 8-6

User identity protection during network entry [2]

AKA procedure[iv] (K_{ASME}) and is identified with a key identifier (KSI_{ASME}). The eNB key (K_{eNB}) is sent from the EPC to the eNB when the UE is entering ECM-CONNECTED state, i.e., during RRC connection or S1 context set-up. Separate AS and NAS level security mode procedures are used. The AS level security mode procedure configures AS security (i.e., RRC and user-plane) and the NAS

[iv]The Authentication and Key Agreement (AKA) mechanism is based on challenge-response mechanisms and symmetrical cryptography. The AKA scheme typically operates in a UMTS Subscriber Identity Module (USIM) or a cdma2000 Removable User Identity Module (RUIM). The third generation AKA provides substantially longer key lengths and mutual authentication compared to the second generation mechanisms, such as GSM AKA. The introduction of AKA inside EAP further enables several new applications, including the use of the AKA as a secure PPP authentication method in devices that already contain an identity module, the use of the third generation mobile network authentication infrastructure in the context of wireless LANs, as well as relying on AKA and the existing infrastructure in a seamless way with any other technology that can use EAP. In AKA, the identity module and the home environment have agreed on a secret key in advance. The "home environment" refers to the home operator's authentication network infrastructure. Furthermore, the actual authentication process starts by having the home environment produce an authentication vector, based on the secret key and a sequence number. The authentication vector contains a random part, an authenticator part used for authenticating the network to the identity module, an expected result part, a 128-bit session key for integrity check, and a 128-bit session key for encryption [13].

level security mode procedure configures NAS security. Both integrity protection and ciphering for RRC are activated within the same AS security mode command procedure, but not necessarily within the same message [7].

The user-plane ciphering is activated at the same time as RRC ciphering. The keys that are stored in the eNBs never leave a secure environment, and user-plane data ciphering/deciphering is conducted within the secure environment where the related keys are stored. Key material for the eNB keys is sent between the eNBs during ECM-CONNECTED intra-E-UTRAN mobility. A sequence number denoted as COUNT is used as an input to ciphering and integrity protection. A given sequence number must only be used once for a given eNB key, except for identical re-transmission on the same radio bearer in the same direction. The same sequence number can be used for both ciphering and integrity protection. A Hyper Frame Number (HFN), i.e., an overflow counter mechanism, is used in the eNB and the UE in order to limit the actual number of sequence number bits that need to be sent over the radio air interface. The HFN is synchronized between the UE and the eNB. If corruption of keys is detected, the UE has to restart the radio level attachment procedure, e.g., a similar radio level procedure for RRC_IDLE to RRC_CON-NECTED mode transition or initial attachment. Since SIM access is not granted in E-UTRAN, the idle mode UE that is not equipped with USIM cannot attempt to reselect the E-UTRAN. To prevent handover to E-UTRAN, the UE which is not equipped with USIM indicates E-UTRA support in UE capability signaling in other RATs. A simplified key derivation is illustrated in Figure 8-7, where [7–9]:

- K_{NASint} is a key which is only used for the protection of NAS traffic with a particular integrity algorithm. This key is derived by the UE and MME from K_{ASME}, as well as an identifier for the integrity algorithm.

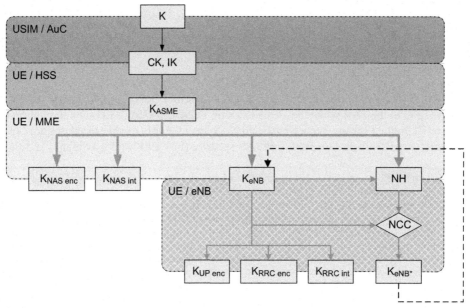

FIGURE 8-7

Key derivation procedure in 3GPP LTE [7]

- K_{NASenc} is a key which is only used for the protection of NAS traffic with a particular encryption algorithm. This key is derived by the UE and MME from $K_{ASME,}$ as well as an identifier for the encryption algorithm.
- K_{eNB} is a key derived by the UE and MME from K_{ASME}. K_{eNB} may also be derived by the target eNB from the Next Hop (NH) during handover. K_{eNB} is used for the derivation of K_{RRCint}, K_{RRCenc}, and K_{UPenc}, for the derivation of K_{eNB*} on handover.
- K_{eNB*} is a key derived by the UE and source eNB from either K_{eNB} or from a fresh NH. K_{eNB*} is used by the UE and target eNB as a new K_{eNB} for RRC and user-plane traffic.
- K_{UPenc} is a key which is only used for the protection of user-plane traffic with a particular encryption algorithm. This key is derived by the UE and eNB from K_{eNB}, as well as an identifier for the encryption algorithm.
- K_{RRCint} is a key which is only used for the protection of RRC traffic with a particular integrity algorithm. K_{RRCint} is derived by the UE and eNB from K_{eNB}, as well as an identifier for the integrity algorithm.
- K_{RRCenc} is a key which is only used for the protection of RRC traffic with a particular encryption algorithm. K_{RRCenc} is derived by the UE and eNB from K_{eNB}, as well as an identifier for the encryption algorithm.
- The Next Hop is used by the UE and eNB in the derivation of K_{eNB*} for the provision of *Forward Security*. The NH is derived by the UE and MME from K_{ASME} and K_{eNB} when the security context is established or from K_{ASME} and the previous NH.
- Next Hop Chaining Count (NCC) is a counter related to NH, i.e., the amount of key chaining that has been performed, which allows the UE to be synchronized with the eNB and to determine whether the next K_{eNB*} needs to be based on the current K_{eNB} or a fresh NH.

The AuC, HSS, and MME entities shown in Figure 8-7 are Authentication Center (i.e., the network entity that can authenticate subscribers in a mobile network), Home Subscriber Service (i.e., the main IMS database which also acts as a database in EPC that combines legacy Home Location Register and AuC functions for circuit-switched and packet-switched domains), and Mobility Management Entity, respectively [8].

In Figure 8-7, CK and IK denote keys that have been agreed during the AKA procedure. The MME invokes the AKA procedures by requesting authentication vectors to the Home Environment (HE), if no unused EPS authentication vectors have been stored. The HE sends an authentication response back to the MME that contains a fresh authentication vector, including a base key named K_{ASME}. Consequently, the EPC and the UE share K_{ASME}. From K_{ASME}, the NAS keys, and indirectly K_{eNB} keys and NH, are derived. The K_{ASME} is never transported to an entity outside of the EPC, but K_{eNB} and NH are transported to the eNB from the EPC when the UE transitions to ECM-CONNECTED. From the K_{eNB}, the eNB and UE can derive the UP and RRC keys.

The RRC and user-plane keys are refreshed during handover. K_{eNB*} is derived by the UE and source eNB from the target *Physical Cell Identifier*, target frequency, and K_{eNB}, or alternatively from the target *Physical Cell Identifier*, target frequency, and NH. The K_{eNB*} is then used as a new K_{eNB} for RRC and UP traffic at the target. If the UE transitions to ECM-IDLE state, all keys are deleted in the eNB. The reuse of COUNT is avoided for the same radio bearer identity in RRC_CONNECTED mode without K_{eNB} change, and this is left to eNB implementation. In case of HFN de-synchronization in

Table 8-1 Security Termination Points in 3GPP LTE [7]

	Ciphering	Integrity Protection
NAS Signaling	Required and terminated in MME	Required and terminated in MME
User-Plane Data	Required and terminated in eNB	Not required
RRC Signaling (AS)	Required and terminated in eNB	Required and terminated in eNB
MAC Signaling (AS)	Not required	Not required

RRC_CONNECTED mode between the UE and eNB, the UE is forced to idle mode. Table 8-1 shows the security termination points in 3GPP LTE.

Integrity protection for the user-plane is not required, and thus is not supported between the UE and serving gateway or for the transport of user-plane data between the eNB and serving gateway on the S1 interface (i.e., the interface between eNB and MME). In general, on RRC_IDLE to RRC_CONNECTED transition, RRC protection and user-plane protection keys are generated, while keys for NAS protection, as well as higher layer keys, are assumed to be already available in the MME. These higher layer keys may have been established in the MME as a result of an AKA run, or transferred from another MME during handover or idle mode mobility. On RRC_CONNECTED to RRC_IDLE transition, eNB deletes the keys that it stores, so that context for the idle mode UE only resides in the MME. The state and current keys of the UE are assumed to have been deleted at eNB on such transition. In particular, the eNB and UE delete NH, K_{eNB}, K_{RRCenc}, K_{RRCint}, K_{Upenc}, and related Next Hop Chaining Counter (NCC), while MME and UE retain K_{ASME}, K_{NASint}, and K_{NASenc}.

The RRC and user-plane keys are derived based on the algorithm identifiers and K_{eNB}, which results in new RRC and user-plane keys in each handover. The source eNB and UE independently create K_{eNB*}. The K_{eNB*} is provided to target the eNB during handover preparation time. Both the target eNB and UE consider the new K_{eNB} equal to the received K_{eNB*}. If AS keys (K_{UPenc}, K_{RRCint}, and K_{RRCenc}) need to be changed in RRC_CONNECTED, an intra-cell handover is used. Inter-RAT handover from UTRAN to E-UTRAN is only supported after activation of integrity protection in UTRAN.

References

[1] IEEE Std 802.16-2009, IEEE Standard for Local and Metropolitan Area Networks – Part 16: Air Interface for Fixed Broadband Wireless Access Systems, May 2009.

[2] P802.16m/D6, IEEE Standard for Local and Metropolitan Area Networks – Part 16: Air Interface for Broadband Wireless Access Systems, Advanced Air Interface, May 2010.

[3] IEEE 802.16m–0034r3, System Description Document, May 2010.

[4] Loutfi Nuaymi, WiMAX: Technology for Broadband Wireless Access, John Wiley & Sons, 2007.

[5] Jeffrey G. Andrews, Arunabha Ghosh, Rias Muhamed, Fundamentals of WiMAX: Understanding Broadband Wireless Networking, first ed., Prentice Hall, 2007.

[6] WiMAX Forum, Mobile System Profile Release 1.5 Air Interface Specification, August 2009.

[7] 3GPP TS 36.300, Evolved Universal Terrestrial Radio Access (E-UTRA) and Evolved Universal Terrestrial Radio Access Network (E-UTRAN); Overall description; Stage 2, March 2010.

[8]   3GPP TS 36.401, Evolved Universal Terrestrial Radio Access Network (E-UTRAN); Architecture Description, December 2009.

[9]   3GPP TS 36.323, Evolved Universal Terrestrial Radio Access (E-UTRA); Packet Data Convergence Protocol (PDCP) specification, March 2010.

[10] IETF RFC 2617, J. Franks et al., HTTP Authentication: Basic and Digest Access Authentication, June 1999.

[11] IETF RFC 1321, R. Rivest, MD5 Message-Digest Algorithm, April 1992.

[12] IETF RFC 4309, R. Housley, Using Advanced Encryption Standard (AES) CCM Mode with IPsec Encapsulating Security Payload (ESP), December 2005.

[13] IETF RFC 4187, J. Arkko and H. Haverinen, Extensible Authentication Protocol Method for 3rd Generation Authentication and Key Agreement (EAP-AKA), January 2006.

The IEEE 802.16m Physical Layer (Part I)

INTRODUCTION

This section describes the physical layer protocols and functional processing in IEEE 802.16m. As shown in Figure 9-1, the physical layer is the lowest protocol layer in baseband signal processing that interfaces with the physical media (in this case the air interface) through which the signal is transmitted and received. The physical layer receives the MAC PDUs and processes them through channel coding, interleaving, baseband modulation, multi-antenna encoding, precoding, resource and antenna mapping. The choice of an appropriate modulation and coding scheme, as well as multi-antenna transmission mode, is critical to achieve the desired reliability and system throughput in mobile wireless data communications. Typical mobile radio channels tend to be dispersive and time-variant and exhibit severe Doppler effects, multipath delay variation, intra-cell and inter-cell interference, and fading.

A good and robust design of the physical layer ensures that the system can normally operate and overcome the above deleterious effects, and can provide the maximum throughput and lowest latency under various operating conditions. The chapters on the physical layer in this book are dedicated to the systematic design of physical layer protocols and functional blocks of 4th generation cellular systems, the theoretical background on physical layer procedures, and performance evaluation of the physical layer components. The theoretical background is provided to make the book self-contained, and to ensure that the reader understands the basic theory behind the operation of various functional blocks and procedures. Additional references are provided for further study. While the focus is mainly on the techniques that were incorporated in the design of the IEEE 802.16m physical layer, the author has attempted to take a more generic and systematic approach to the design of the 4th generation cellular system physical layer, so that the reader can understand and apply the learning to the design and implementation of any OFDM-based physical layer component, irrespective of the radio access technology.

The physical layer processes both control- and data-plane signals; however, due to different design requirements and reliability and performance criteria, the procedures tend to be different. In the areas where the baseline standard has been extended or modified, comparison with the legacy system has been provided to show the improvements.

Similar to previous chapters, the description of 3GPP LTE/LTE-Advanced technologies are also provided to enable the reader to compare and understand both technologies, by making analogies between the corresponding protocols and functional blocks. The description of the physical layer procedures and measurements has been divided into two parts, where the multiple access schemes, frame structure, subchannelization schemes, coding and modulation, as well as physical layer

Mobile WiMAX. DOI: 10.1016/B978-0-12-374964-2.10009-8

335

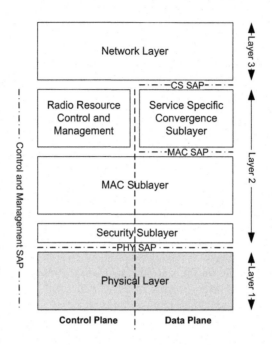

FIGURE 9-1

Location of the physical layer in IEEE 802.16m protocol stack

synchronization and broadcast information, are described in this chapter, and other physical layer topics including control structure and multi-antenna techniques are treated in the Chapter 10.

9.1 OVERVIEW OF IEEE 802.16M PHYSICAL LAYER PROCESSING

This section provides an overview of the physical layer processing in IEEE 802.16m. The processing steps and the associated control signaling are depicted in Figure 9-2. While the physical layer of the IEEE 802.16m is based on the IEEE 802.16-2009 standard, there are new or modified functional components that contribute to significantly increased performance of the IEEE 802.16m relative to the legacy standard. During the design of IEEE 802.16m, an attempt was made to identify the shortcomings of the legacy standard and to replace the inefficient protocols with new or improved procedures.

In some cases, such as frame structure, subchannelization and permutation, and control signaling, there have been great changes from the corresponding legacy frameworks which make the mixed-mode operation of the new and legacy systems more complex. However, extreme modifications were necessary to fulfill the IMT-Advanced requirements and to allow future enhancements of the radio air interface. The main criteria in the design of the new physical layer were to increase the application throughput and capacity, reduce access latency, support higher user mobility, minimize intra-cell and inter-cell interference, improve reliability of control and data channel coverage (especially at the cell edge), and reduce the complexity and signaling overhead. The overall performance evaluations provided in Chapter 12 suggest that the above goals have been fulfilled.

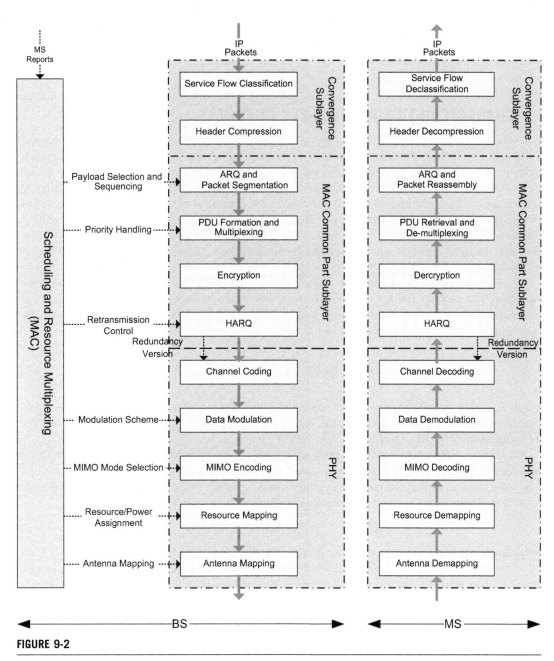

FIGURE 9-2

Physical layer processing in the downlink and uplink

In the next sections of this chapter and Chapter 10, we describe in detail the functional blocks and protocols of the physical layer and their interactions, based on the order in which the information is processed, as shown in Figure 9-2.

9.2 CHARACTERISTICS OF WIRELESS CHANNELS

In a wireless communication system, a signal can travel from the transmitter to the receiver over multiple paths. This phenomenon is referred to as multipath propagation where signal attenuation varies on different paths. This effect, also known as multipath fading, can cause stochastic fluctuations in the received signal's magnitude, phase, and angle of arrival. The propagation over different paths is caused by scattering, reflection, diffraction, and refraction of the radio waves by static and moving objects, as well as the medium. It is obvious that different propagation mechanisms results in different channel and path loss models. As a result of wave propagation over multipath fading channels, the radio signal is attenuated due to mean path loss, as well as macroscopic and microscopic fading.

In an ideal free-space propagation model, the attenuation of RF signal energy between the transmitter and receiver follows the inverse-square law. The received power expressed in terms of transmitted power is attenuated proportional to the inverse of $L_S(d)$, which is called free space loss. The received signal power can be expressed as follows:

$$P_r = P_t G_t G_r \left(\frac{\lambda}{4\pi d}\right)^2 = \frac{P_t G_t G_r}{L_s(d)} \qquad (9\text{-}1)$$

Where P_t and P_r denote the transmitted and the received signal power, G_t and G_r denote the transmitting and receiving antenna gains, d is the distance between the transmitter and the receiver, and λ is the wavelength of the RF signal.

Macroscopic fading is caused by the shadowing effects of buildings and natural obstructions, and is modeled by the local mean of a fast fading signal. The mean path loss $\overline{L}_p(d)$ as a function of distance d between the transmitter and receiver is proportional to an nth power of d relative to a reference distance d_0. In logarithmic scale, it can be expressed as:

$$\overline{L}_p(d) = L_s(d_0) + 10n \log(d/d_0) \text{ (dB)} \qquad (9\text{-}2)$$

The reference distance d_0 corresponds to a point located in the far field of the antenna, typically 1 km for large cells, 100 m for microcells, and 1 m for indoor channels. In the above equation, $\overline{L}_p(d)$ is the mean path loss which is typically $10n$ dB per decade attenuation for $d \gg d_0$. The value of n depends on the frequency, antenna heights, and propagation environment, and is equal to 2 in free space. The studies show that the path loss $L_p(d)$ is a random variable with log-normal distribution about the mean path loss $\overline{L}_p(d)$. Let $X(0, \sigma^2)$ denote a zero-mean Gaussian random variable with standard deviation σ when measured in decibels, then:

$$L_p(d) = L_s(d0) + 10n \log(d/d_0) + X \text{ (dB)} \qquad (9\text{-}3)$$

The value of X is often derived empirically based on measurements. A typical value for σ is 8 dB. The parameters that statistically describe path loss due to large-scale fading (macroscopic fading) for an arbitrary location with a specific transmitting–receiving antenna separation include the reference distance d_0, the path loss exponent n, and the standard deviation σ of X.

Microscopic fading refers to the rapid fluctuations of the received signal in time and frequency, and is caused by scattering objects between the transmitting and receiving antennas. When the received RF signal is a superposition of independent scattered components plus a line-of-sight (LoS) component, the envelope of the received signal $r(t)$ has a Rician Probability Distribution Function (PDF) and is referred to as Ricean fading. As the magnitude of the LoS component approaches zero, the Ricean PDF approaches a Rayleigh PDF. Thus:

$$f(r) = \frac{r(K+1)}{\sigma^2} e^{\left(-K-\frac{(K+1)r^2}{2\sigma^2}\right)} I_0\left(\frac{2r}{\sigma}\sqrt{K(K+1)/2}\right) \quad r \geq 0 \tag{9-4}$$

Where K and $I_0(r)$ denote the Ricean factor and zero-order modified Bessel function of the first kind, respectively. In the absence of an LoS path ($K = 0$), the Ricean PDF reduces to Rayleigh distribution.

Time-varying fading due to scattering objects or transmitter/receiver motion results in Doppler spread. The time spreading effect of small-scale or microscopic fading is manifested in the time-domain as multipath delay spread, and in the frequency-domain as channel coherence bandwidth. Similarly, the time-variation of the channel is characterized in the time-domain as channel coherence time and in the frequency-domain as Doppler spread. In a fading channel, the relationship between maximum excess delay time τ_m and symbol time τ_s can be viewed in terms of two different degradation effects, i.e., frequency-selective fading and frequency non-selective or flat-fading. A channel is said to exhibit frequency-selective fading if $\tau_m > \tau_s$. This condition occurs whenever the received multipath components of a symbol extend beyond the symbol's time duration. Such multipath dispersion of the signal results in Inter-Symbol Interference (ISI) distortion. In the case of frequency-selective fading, mitigating the distortion is possible because many of the multipath components are separable by the receiver. A channel is said to exhibit frequency non-selective or flat-fading if $\tau_m > \tau_s$. In this case, all the received multipath components of a symbol arrive within the symbol time duration; therefore, the components are not resolvable. In this case, there is no channel-induced ISI distortion, since the signal time spreading does not result in significant overlap among adjacent received symbols. There is still performance degradation because the unresolvable phasor components can add up destructively to yield a substantial reduction in signal-to-noise ratio (SNR). Also, signals that are classified as exhibiting flat-fading can sometimes experience frequency-selective distortion.

Figure 9-3 illustrates a multipath-intensity profile $\Lambda(\tau)$ versus delay τ, where the term delay refers to the excess delay. It represents the signal's propagation delay that exceeds the delay of the first signal's arrival at the receiver. For a typical wireless communication channel, the received signal usually consists of several discrete multipath components. The received signals are composed of a continuum of multipath components in some channels, such as the tropospheric channel. In order to perform measurements of the multipath intensity profile, a wideband signal; i.e., a unit impulse or Dirac delta function, is used. For a single transmitted impulse, the time τ_m between the first and last received component is defined as the maximum excess delay during which the multipath signal power typically falls to some level 10–20 dB below that of the strongest component. Note that for an ideal system with zero excess delay, the function $\Lambda(\tau)$ would consist of an ideal impulse with weight equal

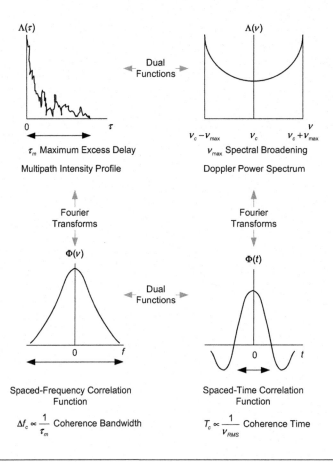

FIGURE 9-3

An illustration of the duality principle in time- and frequency-domains [3]

to the total average received signal power. In the literature, the Fourier transform of $\Lambda(\tau)$ is referred to as the spaced-frequency correlation function $\Phi(\nu)$. The spaced-frequency correlation function $\Phi(\nu)$ is the channel's response to a pair of sinusoidal signals separated in frequency by υ. The coherence bandwidth Δf_c is a measure of the frequency range over which spectral components have a strong likelihood of amplitude correlation. In other words, a signal's spectral components over this range are affected by the channel in a similar manner. Note that Δf_c and τ_m are inversely proportional, i.e., $\Delta f_c \propto 1/\tau_m$. The maximum excess delay τ_m is not the best indicator of how any given wireless system will perform over a communication channel, because different channels with the same value of τ_m can exhibit different variations of signal intensity over the delay span. Thus the delay spread is often characterized in terms of the Root Mean Square (RMS) delay spread τ_{RMS}, where:

$$\tau_{RMS} = \sqrt{\frac{\int_0^{\tau_m}(\tau - \bar{\tau})^2\Lambda(\tau)d\tau}{\int_0^{\tau_m}\Lambda(\tau)d\tau}} \tag{9-5}$$

where

$$\bar{\tau} = \frac{\int_0^{\tau_m} \tau \Lambda(\tau) d\tau}{\int_0^{\tau_m} \Lambda(\tau) d\tau} \tag{9-6}$$

A closed-form relationship between coherence bandwidth and RMS delay spread does not exist and must be derived from signal analysis of actual signal dispersion measurements in specific channels. If coherence bandwidth is defined as the frequency interval over which the channel's complex frequency transfer function has a correlation of at least 0.9, the coherence bandwidth is approximately $\Delta f_c \approx 1/(50\tau_{RMS})$. A common approximation of Δf_c corresponding to a frequency range over which the channel transfer function has a correlation of at least 0.5 is $\Delta f_c \approx 1/(5\tau_{RMS})$.

A channel is said to exhibit frequency-selective effects, if $\Delta f_c < 1/\tau_s$, where the inverse symbol rate is approximately equal to the signal bandwidth W. In practice, W may differ from $1/\tau_s$ due to filtering or data modulation. Frequency-selective fading effects arise whenever the signal's spectral components are not affected equally by the channel. This occurs whenever $\Delta f_c < W$. Frequency non-selective or flat-fading degradation occurs whenever $\Delta f_c > W$. Hence, all of the signal's spectral components will be affected by the channel in a similar manner. Flat-fading does not introduce channel-induced ISI distortion, but performance degradation can still be expected due to a loss in SNR whenever the signal is fading. In order to avoid channel-induced ISI distortion, the channel is required to exhibit flat-fading by ensuring that $\Delta f_c > W$. Hence, the channel coherence bandwidth Δf sets an upper limit on the transmission rate that can be used without incorporating an equalizer in the receiver.

Figure 9-3 shows function $\Phi(t)$, which is known as the spaced-time correlation function. It is the autocorrelation function of the channel's response to a sinusoid. This function specifies the extent to which there is correlation between the channel's response to a sinusoid sent at time t_1 and the response to a similar sinusoid sent at time t_2, where $\Delta t = t_1 - t_2$. The coherence time is a measure of the expected time duration over which the channel's response is essentially invariant. To estimate $\Phi(t)$, a sinusoidal signal is transmitted through the channel and the autocorrelation function of the channel output is calculated. The function $\Phi(t)$ and the parameter T_c provide us with information about the speed of fading channel variation. Note that for an ideal time-invariant channel, the channel's response would be highly correlated for all values of Δt and $\Phi(t)$ would be a constant function. If one ideally assumes uniformly distributed scattering around a mobile station with linearly-polarized antennas, then the Doppler power spectrum (i.e., the inverse Fourier transform of the spaced-time correlation function) $\Lambda(v)$ has a U-shaped distribution, as shown in Figure 9-3. In a time-varying fading channel, the channel response to a pure sinusoidal tone spreads over a finite frequency range of $v_c - v_{max} < v < v_c + v_{max}$, where v_c and v_{max} denote the frequency of the sinusoidal tone and the maximum Doppler spread, respectively. The RMS bandwidth of $\Lambda(v)$ is referred to as Doppler spread and is denoted by v_{RMS} that can be estimated as follows:

$$v_{RMS} = \sqrt{\frac{\int_{v_c-v_{max}}^{v_c+v_{max}} (v - \bar{v})^2 \Lambda(v) dv}{\int_{v_c-v_{max}}^{v_c+v_{max}} \Lambda(v) dv}} \tag{9-7}$$

where

$$\bar{v} = \frac{\int_{v_c-v_{max}}^{v_c+v_{max}} v \Lambda(v) dv}{\int_{v_c-v_{max}}^{v_c+v_{max}} \Lambda(v) dv} \tag{9-8}$$

The coherence time is typically defined as the time lag for which the signal autocorrelation coefficient reduces to 0.7. The coherence time is inversely proportional to Doppler spread $T_c \approx 1/\nu_{RMS}$. A common approximation for the value of coherence time as a function of Doppler spread is $T_c = 0.423/\nu_{RMS}$. It can be observed that the functions on the right side of Figure 9-3 are the dual of the functions on the left side (duality principle).

The angle spread refers to the spread in angle of arrival (AoA) of the multipath components at the receiver antenna array. At the transmitter, on the other hand, the angle spread refers to the spread in angle of departure (AoD) of the multipath components that leave the transmit antennas. If the angle spectrum function $\Theta(\theta)$ denotes the average power as a function of AoA, then the RMS angle spread can be estimated as:

$$\theta_{RMS} = \sqrt{\frac{\int_{-\pi}^{+\pi}(\theta - \bar{\theta})^2 \Theta(\theta)d\theta}{\int_{-\pi}^{+\pi}\Theta(\theta)d\theta}} \tag{9-9}$$

where

$$\bar{\theta} = \frac{\int_{-\pi}^{+\pi}\theta\Theta(\theta)d\theta}{\int_{-\pi}^{+\pi}\Theta(\theta)d\theta} \tag{9-10}$$

The angle spread causes space selective fading, which manifests itself as a variation of signal amplitude according to the location of the antennas. The space selective fading is characterized by the coherence distance D_c, which is the spatial separation at which the autocorrelation coefficient of the spatial fading reduces to 0.7. The coherence distance is inversely proportional to the angle spread $D_c \propto 1/\theta_{RMS}$.

In Figure 9-3, a duality between multipath intensity function $\Lambda(\tau)$ and Doppler power spectrum $\Lambda(\nu)$ is identified. It means that the two functions exhibit similar behavior across time domain and frequency domains. As the $\Lambda(\tau)$ function identifies expected power of the received signal as a function of delay, $\Lambda(\nu)$ identifies expected power of the received signal as a function of frequency. Similarly, the spaced-frequency correlation function $\Phi(f)$ and the spaced-time correlation function $\Phi(t)$ are dual functions. It implies that, as $\Phi(f)$ represents channel correlation in frequency, $\Phi(t)$ corresponds to the channel correlation function in time in a similar manner.

9.3 SC-FDMA AND OFDMA PRINCIPLES

The Orthogonal Frequency Division Multiplexing (OFDM) is a form of multi-carrier modulation technique that was first introduced more than four decades ago. In a conventional serial data transmission system, the information bearing symbols are transmitted sequentially, with the frequency spectrum of each symbol occupying the entire available bandwidth. Figure 9-4 illustrates an unfiltered Quadrature Amplitude Modulation (QAM) signal spectrum. It is in the form of $\sin(\pi f T_u)/\pi f T_u$ with zero crossing points at integer multiples of $1/T_u$ where T_u is the QAM symbol period. The concept of OFDM is to transmit the data bits in parallel QAM modulated sub-carriers using frequency division multiplexing. The carrier spacing is carefully selected so that each sub-carrier is located on all the other sub-carriers' zero crossing points in the frequency domain. Although there are spectral overlaps among sub-carriers, they do not interfere with each other if they are sampled at the sub-carrier

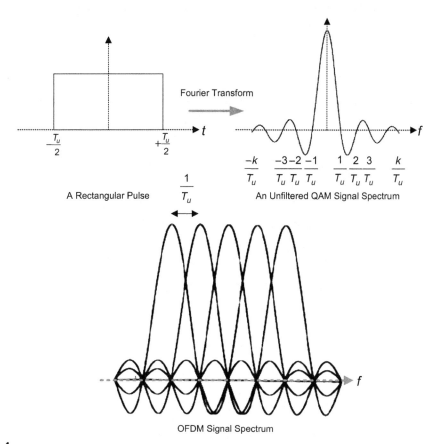

FIGURE 9-4

An illustration of OFDM Concept

frequencies. In other words, they maintain spectral orthogonality. As shown in Figure 9-4, the OFDM signal in the frequency domain is generated through aggregation of N_{FFT} parallel QAM-modulated sub-carriers, where adjacent sub-carriers are separated by sub-carrier spacing $1/T_u$. An example OFDM signal spectrum as seen on a vector signal analyzer is shown in Figure 9-5.

Since an OFDM signal consists of many parallel QAM sub-carriers, the mathematical expression of the signal in time-domain can be expressed as follows:

$$s(t) = \text{Re}\left\{ e^{j\omega_c t} \sum_{k=-(N_{FFT}-1)/2}^{(N_{FFT}-1)/2} \alpha_k e^{j2\pi k(t-t_g)/T_u} \right\} \quad mT_u \leq t \leq (m+1)T_u \quad (9\text{-}11)$$

Where $s(t)$ denotes the OFDM signal in time-domain, α_k is the complex-valued data that is QAM-modulated and transmitted over sub-carrier k, N_{FFT} is the number of sub-carriers in frequency-domain, ω_c is the RF carrier frequency, and T_g is the guard interval or Cyclic Prefix (CP). For a large number of sub-carriers, direct generation and demodulation of the OFDM signal would require arrays of coherent

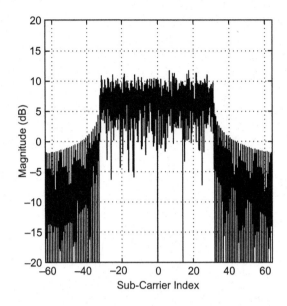

FIGURE 9-5

The OFDM signal in the frequency-domain

sinusoidal generators, which can become excessively complex and expensive. However, one can notice that the OFDM signal is actually the real part of the Inverse Discrete Fourier Transform (IDFT) of the original complex-valued data symbols α_k. It can be seen that there are $N < N_{FFT}$ sub-carriers, each carrying the corresponding data α_k. The inverse of the sub-carrier spacing $1/T_u$ is defined as the OFDM useful symbol duration, which is N_{FF} times longer than that of the original input data symbol duration.

Since IDFT is used in the OFDM modulator, the original data is defined in the frequency-domain, while the OFDM signal $s(t)$ is defined in the time-domain. The IDFT can be implemented via a computationally efficient Fast Fourier Transform (FFT) algorithm. The orthogonality of sub-carriers in OFDM can be maintained, and individual sub-carriers can be completely separated and demodulated by an FFT at the receiver when there is no ISI introduced by the communication channel. In practice, linear distortions, such as multipath delay, cause ISI between OFDM symbols, resulting in loss of orthogonality and an effect that is similar to co-channel interference. However, when delay spread is small, i.e., within a fraction of the OFDM useful symbol length, the impact of ISI is insignificant, although it depends on the order of modulation implemented by the sub-carriers. A simple solution to deal with multipath delay is to increase the OFDM effective symbol duration, such that it is much larger than the delay spread; however, when the delay spread is large, it requires a large number of sub-carriers and a large FFT size. Meanwhile, the system might become sensitive to Doppler shift and carrier instability. An alternative approach to mitigate multipath distortion is to generate a cyclically extended guard interval, where each OFDM symbol is prefixed with a periodic extension of the signal itself, as shown in Figure 9-6 where the tail of the symbol is copied to the beginning of the symbol. The OFDM symbol duration then is defined as $T_s = T_u + T_g$, where T_g is the guard interval or cyclic prefix. When the guard interval is longer than the channel impulse response or

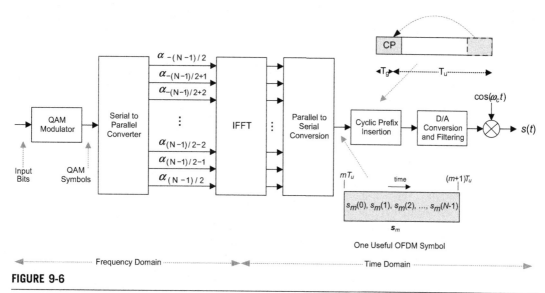

FIGURE 9-6

OFDM signal generation

the multipath delay, the ISI can be effectively eliminated. The ratio of the guard interval to useful OFDM symbol duration depends on the deployment scenario and the frequency band. Since the insertion of guard intervals will reduce data throughput, T_g is usually less than $T_u/4$. The cyclic prefix should absorb most of the signal energy dispersed by the multipath channel. The entire ISI energy is contained within the cyclic prefix if its length is greater than that of the channel RMS delay spread, i.e., $T_g > \tau_{RMS}$. In general, it is sufficient to have most of the delay spread energy absorbed by the guard interval, considering the inherent robustness of large OFDM symbols to time dispersion.

The mapping of the modulated data symbol into multiple sub-carriers also allows an increase in the symbol duration. Since the throughput on each sub-carrier is greatly reduced, the symbol duration obtained through an OFDM scheme is much larger than that of a single carrier modulation technique with a similar overall transmission bandwidth. In general, when the channel delay spread exceeds the guard time, the energy contained in the ISI will be much smaller with respect to the useful OFDM symbol energy, as long as the symbol duration is much larger than the channel delay spread, i.e., $T_s >> \tau_{RMS}$. Although large OFDM symbol duration is desirable to mitigate the ISI effects caused by time dispersion, large OFDM symbol duration can further reduce the ability to alleviate the effects of fast fading, particularly if the symbol period is large compared to the channel coherence time; then the channel can no longer be considered as time-invariant over the OFDM symbol duration and this will introduce inter-sub-carrier orthogonality loss. This can affect the performance in fast fading conditions. Hence, the symbol duration should be kept smaller than the minimum channel coherence time. Since the channel coherence time is inversely proportional to the maximum Doppler spread, the symbol duration T_s must, in general, be chosen such that $T_s << 1/\nu_{RMS}$. The large number of OFDM sub-carriers makes the bandwidth of the individual sub-carriers small relative to the overall signal bandwidth. With an adequate number of sub-carriers, the inter-carrier spacing is much smaller than the channel coherence bandwidth. Since the channel coherence bandwidth is inversely proportional to the channel delay spread τ_{RMS}, the sub-carrier separation is generally designed such that $\Delta f_c \ll 1/\tau_{RMS}$. In this case, the fading on each sub-carrier is flat and can be modeled as a complex-valued constant channel gain. The individual reception of the QAM symbols transmitted on each sub-carrier is therefore simplified to the case of a flat-fading channel. This enables a straightforward introduction of advanced MIMO schemes. Furthermore, in order to mitigate Doppler spread effects, the inter-carrier spacing should be much larger than the RMS Doppler spread $\Delta f_c \ll 1/\nu_{RMS}$. Since the OFDM sampling frequency is typically larger than the actual signal bandwidth, only a subset of sub-carriers are used to carry QAM symbols. The remaining sub-carriers are left inactive prior to the IFFT and are referred to as guard sub-carriers. The split between the active and the inactive sub-carriers is determined based on the spectral constraints, such as the bandwidth allocation and the spectral mask.

An OFDM transmitter diagram is shown in Figure 9-6. The incoming bit stream is first QAM modulated to form the complex-valued QAM symbols. The QAM symbols are converted from serial to parallel with $N < N_{FFT}$ complex-valued numbers per block, where N_{FFT} is the number of FFT/IFFT points. Each block is processed by an IFFT and the output of the IFFT forms an OFDM symbol, which is converted back to serial data for transmission. A guard interval is inserted between symbols to eliminate ISI effects caused by multipath distortion. The discrete symbols are filtered and converted to analog for RF up-conversion. The reverse process is performed at the receiver. A one-tap equalizer is usually used for each sub-carrier to correct channel distortion. The tap coefficients are calculated based on channel information. As shown in Figures 9-4 and 9-5, one can understand that the OFDM signal spectrum is approximately rectangular. Since the OFDM modulator is an IFFT processor, its physical meaning is to convert data from the frequency-domain to the time-domain and, transmit them in the time-domain over the channel. If a spectrum analyzer, which converts the input from the time-omain to the frequency-domain, is used to monitor an OFDM signal, what displayed is the original data. When there is multipath distortion, a conventional single carrier wideband transmission system suffers from frequency selective fading. A complex adaptive equalizer must be used to equalize the in-band fading. The number of taps required for the equalizer is proportional to the symbol rate and the multipath delay. For an OFDM system, if the guard interval is larger than the multipath delay, the ISI can be eliminated and orthogonality can be maintained among sub-carriers. Since each OFDM sub-carrier occupies a very narrow spectrum, in the order of a few kHz, even under severe multipath distortions, the sub-carriers are only subject to flat-fading. In other words, the OFDM converts a wideband frequency-selective fading channel to a series of narrowband frequency non-selective fading sub-channels by using the parallel multi-carrier transmission scheme. Since OFDM data sub-carriers are statistically independent and identically distributed, based on the central limit theorem, when the number of sub-carriers N_{FFT} is large, the OFDM signal distribution tends to be Gaussian.

The peak-to-average power ratio (PAPR) for a single carrier modulation signal depends on its constellation and the pulse shaping filter roll-off factor. For a Gaussian distributed OFDM signal, the cumulative distribution function (CDF)[i] of the PAPR for 99.0%, 99.9%, and 99.99% are approximately

---

[i]The CDF of the real-valued random variable X is defined as $x \rightarrow F_X(x) = P(X \le x)$, $\forall x \in i$, where the right-hand side represents the probability that the random variable X takes on a real value less than or equal to x. The CDF of X can be defined in terms of the probability density function $f(x)$ as $F(x) = \int_{-\infty}^{x} f(x)dx$. The Complementary CDF (CCDF), on the other hand, is defined as $P(X > x) = 1 - F_X(x)$.

8.3, 10.3, and 11.8 dB, respectively. Since the OFDM signal has a high PAPR, it could be clipped in the transmitter power amplifier, because of its limited dynamic range or non-linearity. Higher output back-off is required to prevent performance degradation and inter-modulation products spilling into adjacent channels. Therefore, RF power amplifiers should be operated in a very large linear region. Otherwise, the signal peaks leak into the non-linear region of the power amplifier, causing signal distortion. This signal distortion introduces inter-modulation among the sub-carriers and out-of-band emission. Thus, the power amplifiers should be operated with large power back-offs. On the other hand, this leads to very inefficient amplification and expensive transmitters. Thus, it is highly desirable to reduce the PAPR. In addition to inefficient operation of the power amplifier, a high PAPR requires a larger dynamic range for the receiver analog-to-digital (A/D) converter. To reduce the PAPR, several techniques have been proposed and used such as clipping, channel coding, temporal windowing, Tone Reservation,[ii] and Tone Injection. However, most of these methods are unable to simultaneously achieve a large reduction in PAPR with low complexity, with low coding overhead, without performance degradation, and without transmitter receiver symbol handshake. The PAPR ξ of the OFDM signal defined in Equation (9-11) is defined as follows:

$$\xi = \frac{\max|s(t)|^2}{E\{|s(t)|^2\}}\bigg|_{mT_u \leq t \leq (m+1)T_u} \tag{9-12}$$

In the above equation, $E\{.\}$ denotes the expectation operator and m is an integer. From the central-limit theorem, for large values of N_{FFT}, the real and imaginary values of $s(t)$ (from Equation (9-11) and without the Re$\{.\}$ operator) would have a Gaussian distribution. Consequently, the amplitude of the OFDM signal has a Rayleigh distribution with zero mean and a variance of N_{FFT} times the variance of one complex sinusoid. Assuming the samples are mutually uncorrelated, the cumulative distribution function for the peak power per OFDM symbol is given by:

$$P(\xi > \gamma) = [1 - (1 - e^{-\gamma})^{N_{FFT}}] \tag{9-13}$$

From the above equation, it can be seen that a large PAPR occurs only infrequently due to the relatively large values of N_{FFT} used in practice. The PAPR of an OFDM signal is depicted in Figure 9-12. Power amplifiers (PA) generally have a non-linear amplitude (AM) response where the output power is saturated for large input powers. Most applications require operation in the linear region of the PA response, where the output power is a linear function of the input. The larger the linear operation region or alternatively the larger the saturation point the more expensive the PA. Therefore, it is imperative to reduce the PAPR of the OFDM signal before processing through the PA. Figures 9-7 and 9-8 illustrate examples of PA AM/AM response and QAM signal constellation fuzziness when the OFDM signal with large PAPR is processed through a PA with low saturation point. In this example, 64 QAM modulation, 256 point IFFT with $CP = 1/4T_u$, and PA saturation point of 6 dB are assumed.

The modulation accuracy or the permissible signal constellation fuzziness is often measured via the error vector magnitude (EVM) metric. The EVM defines the average constellation error with respect to the farthest constellation point (i.e., the distance between the reference signal and the measured signal points in I–Q plane) and is calculated as follows: [2]

[ii]In the tone reservation method, the transmitter and the receiver reserve a subset of tones or sub-carriers for generating PAPR reduction signals. Those reserved tones are not used for data transmission [66].

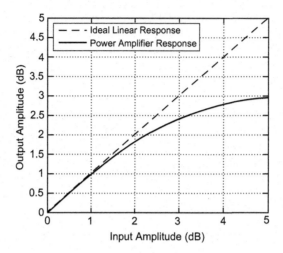

FIGURE 9-7

An example of the AM/AM response of a power amplifier (saturation point = 6 dB)

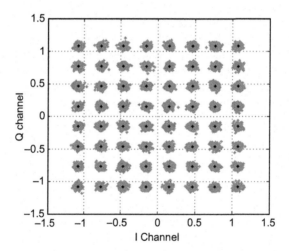

FIGURE 9-8

A 64 QAM signal constellation before and after power amplification

$$EVM = \sqrt{\frac{\displaystyle\sum_{i=0}^{N_{samples}} (\Delta I_i^2 + \Delta Q_i^2)}{N_{samples} r_{max}^2}} \tag{9-14}$$

Where $N_{samples}$ is the number of symbols used in the measurement period, ΔI_i, ΔQ_i refers to the In-phase (I) and Quadrature (Q) components of the ith error vector, and r_{max} denotes the maximum constellation amplitude. The EVM is measured over the continuous portion of a burst occupying at least one quarter of the transmission frame at maximum power settings. The permissible EVM can be estimated from the transmitter implementation margin, if the error vector is considered noise, which is added to the channel noise. The implementation margin is the excess power needed to maintain the carrier-to-noise ratio (C/N) intact when going from the ideal to the real transmitter. The EVM cannot be measured at the antenna connector, but should be measured by an ideal receiver with a certain carrier recovery loop bandwidth specified in a percentage of the symbol rate. The measured EVM include the transmitter filter accuracy, D/A-converter, modulator imbalances, untracked phase noise, and power amplifier nonlinearity.

In practice, the sub-carrier spacing is not the same among different tones due to mismatched oscillators (i.e., frequency offset), Doppler shift, and timing synchronization errors, resulting in Inter-Carrier Interference (ICI) and loss of orthogonality. If the time-domain signals corresponding to sub-carriers k and $k + m$ are represented by $x_k(t)$ and $x_{k+m}(t)$, respectively, then the ICI is defined as the inner product between $x_k(t)$ and $x_{k+m}(t)$ over the useful OFDM symbol duration T_u. Note that if the sub-carriers are ideally orthogonal, the inner product will be zero.

$$ICI_k = \sum_{m \neq k} \left| \int_0^{T_u} x_k(t) x_{k+m}^*(t) dt \right|^2$$

if

$$x_k(t) = e^{j2\pi kt/T_u}$$

$$x_{k+m}(t) = e^{j2\pi(k+m)t/T_u}$$

(9-15)

then

$$ICI_k \approx \eta(T_u \delta)^2$$

Where η is a constant, and δ denotes the frequency offset. It can be seen that the ICI increases with the increase of the OFDM symbol duration (or alternatively decrease of sub-carrier spacing) and the frequency offset. The effects of timing offset are typically less than that of the frequency offset, provided that the cyclic prefix is sufficiently large.

The impact of local oscillator phase noise on the performance of an OFDM system. It has been concluded that the oscillator phase noise may have significant effects on close sub-carrier spacing (large OFDM symbol duration in time). Long symbol duration is required for implementing a long guard interval that can mitigate long multipath delay in single frequency operation without excessive reduction of data throughput. Extensive studies suggest that phase noise in OFDM can result in two effects: a common sub-carrier phase rotation on all the sub-carriers; and a thermal-noise-like sub-carrier de-orthogonality. The common phase error, i.e., constellation rotation, on all the demodulated sub-carriers is caused by the phase noise spectrum from DC up to the frequency of sub-carrier spacing. This low-pass effect is due to the long integration time of the OFDM symbol duration. This phase error can, in principle, be corrected by using pilots within the same symbol (in-band pilots). The phase error causes sub-carrier constellation blurring rather than rotation. It results from the phase noise spectrum contained within the system bandwidth. This part of the phase noise is more crucial, since it cannot be easily corrected. In brief, an OFDM system has the following advantages:

- It is sufficiently flexible in meeting various design requirements such as low complexity, bandwidth efficiency and scalability, spectrum shaping, and low sensitivity to various impairments.

- OFDM is well-suited to multi-antenna transmission schemes.
- It requires no adaptation to instantaneous channel response for low-order modulations. It is robust to impulse interference and channel variations. Channel estimation is required for high-order QAM modulations.
- As a multi-carrier transmission technique, OFDM is less sensitive to shift in sampling time in comparison to serial transmission techniques.
- The bandwidth efficiency of an OFDM system approaches the Nyquist rate as the FFT size increases (this, however, increases system complexity and is subject to an upper-limit due to channel variations and carrier frequency offset).
- For fixed bandwidth efficiency, the complexity (multiplications per symbol) of an FFT-based OFDM system grows logarithmically with the increase in the channel multipath spread, in comparison with a linear increase of complexity for equalizers for a single carrier system.
- A properly coded and interleaved OFDM system can exceed the performance of other practical systems. Higher transmitter output back-off is required in comparison to a single carrier system, because of the high peak-to-average power ratio of an OFDM system.
- As a parallel transmission technique, OFDM is more sensitive to carrier frequency offset and tone interference than a single carrier system.

Orthogonal Frequency Division Multiple Access (OFDMA) is the multi-user variant of the OFDM scheme where multiple-access is achieved by assigning subsets of sub-carriers to different users, allowing simultaneous data transmission from several users. In OFDMA, the radio resources are two-dimensional regions over time (an integer number of OFDM symbols) and frequency (a number of contiguous or non-contiguous sub-carriers). The difference between OFDM and OFDMA is illustrated in Figure 9-9. Similarly to OFDM, OFDMA employs multiple closely spaced sub-carriers, but the sub-carriers are divided into groups of sub-carriers where each group is called a resource block. The grouping of sub-carriers into groups of resource blocks is referred to as subchannelization. The sub-carriers that form a resource block do not need to be physically adjacent. In the downlink, a resource block may be allocated to different users. In the uplink, a user may be assigned to one or more resource blocks.

Subchannelization defines subchannels that can be allocated to mobile stations depending on their channel conditions and service requirements. Using subchannelization, within the same time slot (i.e., an integer number of OFDM symbols) an OFDMA system can allocate more transmitting power to user devices with lower SNR and less power to user devices with higher SNR. Subchannelization also enables the BS to allocate higher power to sub-channels assigned to indoor mobile terminals, resulting in better indoor coverage.

The basic uplink transmission technique in 3GPP LTE is single-carrier transmission (SC-FDMA) with a cyclic prefix to achieve uplink inter-user orthogonality and to enable efficient frequency-domain equalization at the receiver side. DFT-Spread OFDM (DFT-S OFDM) a form of single-carrier transmission technique where the signal is generated in the frequency-domain similar to OFDMA, and is illustrated in Figure 9-10. In Figure 9-10, the common processing units in OFDMA and SC-FDMA are distinguished from those that are specific to SC-FDMA. This allows for a relatively high degree of commonality with the downlink OFDMA baseband processing using the same parameters, e.g., clock frequency, sub-carrier spacing, FFT/IFFT size, etc. The use of SC-FDMA in the uplink is mainly due to the relatively inferior PAPR properties of OFDMA that result in worse uplink coverage compared to

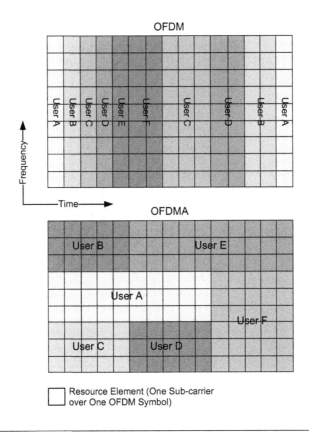

FIGURE 9-9

An illustration of the difference between OFDM and OFDMA

SC-FDMA. The PAPR characteristics are important for cost-effective design of a mobile station's power amplifiers.

The ith transmitted symbol in an SC-FDMA system without a cyclic prefix in a single transmit/receive antenna case can be expressed as a vector of length N_{FFT} samples defined by $\mathbf{y} = \mathbf{FTDx}$ where $\mathbf{x} = (x_1, x_2, \ldots, x_M)^T$ is an $M \times 1$ vector with M QAM modulated symbols (the superscript T denotes transpose operation), $\mathbf{D}$ is an $M \times M$ matrix which performs M-point DFT operation, $\mathbf{T}$ is the $N \times M$ mapping matrix for sub-carrier assignment, and $\mathbf{F}$ performs N_{FFT} point IFFT operation. After propagation through the multipath fading channel and addition of the Additive White Gaussian Noise (AWGN), removing the cyclic prefix and going through the N_{FFT} point FFT module, the received signal vector in the frequency domain can be expressed as $\mathbf{z} = \mathbf{HTDx} + \mathbf{w}$ where $\mathbf{H}$ is the diagonal matrix of channel response and w is the noise vector. Note that the maximum excess delay of the channel is assumed to be shorter than the cyclic prefix; therefore, the ISI can be mitigated by the cyclic prefix. The amplitude and phase distortion in the received signal due to the multipath channel is compensated by a Frequency Domain Equalizer (FDE) and the signal at the FDE output can be described as $\mathbf{v} = \mathbf{Cz}$ where $\mathbf{C} = \text{diag}(c_1, c_2, \ldots, c_{NFFT})$ is the diagonal matrix of FDE

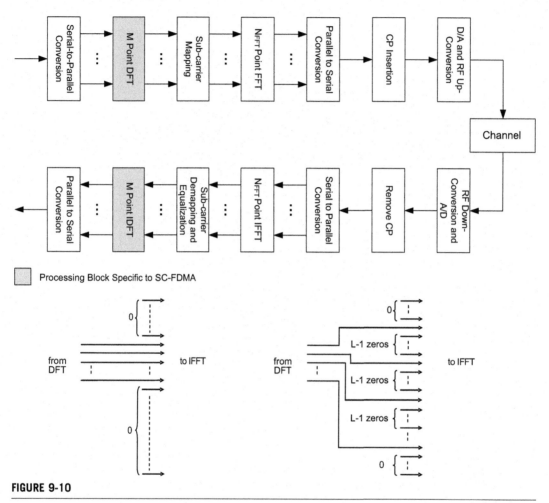

FIGURE 9-10

The transmitter structure for SC-FDMA with localized and distributed sub-carrier mapping schemes. Note that $M < N_{FFT}$

coefficients. The FDE complex coefficient can be derived using Minimum Mean-Square Error (MMSE) criterion:

$$c_k = \frac{H_k^*}{|H_k|^2 + \sigma_n^2/\sigma_s^2} \tag{9-16}$$

Where k denotes the sub-carrier index, σ_{2n} denotes the variance of the additive noise, and σ_{2s} is the variance of the transmitted pilot symbol. Following the sub-carrier demapping function and IDFT de-spreading, an $M \times 1$ vector $\hat{\mathbf{x}}$ containing M QAM modulated symbols as an estimate to the input vector $\mathbf{x}$ is obtained at the receiver. The IDFT de-spreading block in the receiver averages the noise over each sub-carrier. A particular sub-carrier may experience deep fading in a frequency selective fading channel. The IDFT de-spreading averages and spreads the fading effect, which results in a noise enhancement to all the QAM symbols. Hence, the IDFT de-spreading makes the DFT-S OFDM more sensitive to the noise.

The sub-carrier mapping in SC-FDMA determines which part of the spectrum is used for transmission by inserting a suitable number of zeros (i.e., null sub-carriers inserted between the data sub-carriers) at the upper and/or lower end as shown in Figure 9-10. Between each DFT output sample $L - 1$ zeros are inserted. A mapping with $L = 1$ (as shown in Figure 9-10) corresponds to localized transmissions, i.e., transmissions where the DFT outputs are mapped to consecutive sub-carriers. There are two sub-carrier mapping schemes (localized and distributed) that could be used in the uplink.

In order to demonstrate the similarities and differences between OFDMA and SC-FDMA operation, let's assume that one wishes to transmit a sequence of 8 QPSK symbols as shown in Figure 9-11 [92]. In the OFDMA case assuming $M = 4$, four QPSK symbols would be processed in parallel, each of them modulating its own sub-carrier at the appropriate QPSK phase. After one OFDM symbol period, a guard period or cyclic prefix is inserted to mitigate the multi-path effects. For SC-FDMA, each symbol is transmitted sequentially. With $M = 4$, there are four data symbols transmitted in one SC-FDMA symbol period. The higher-rate data symbols require four times the bandwidth and so each data symbol occupies $M \times \Delta f$ of spectrum assuming a sub-carrier spacing of Δf. After four data symbols, the CP is inserted. Note the OFDMA and SC-FDMA symbol periods are the same [92].

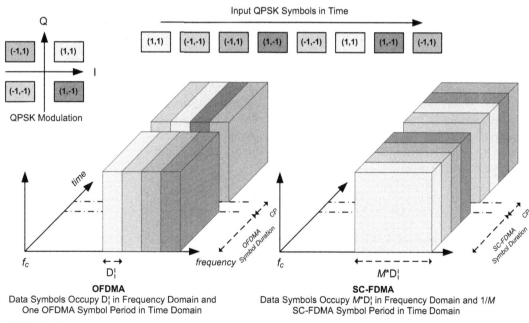

FIGURE 9-11

A comparison of OFDMA and SC-FDMA operation using QPSK modulation with $M = 4$ sub-carriers. The input QPSK symbols (shown on top), are fed into the modulator from left to right [92]

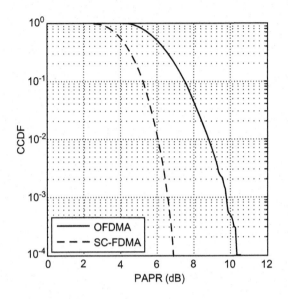

FIGURE 9-12

A comparison of OFDMA and SC-FDMA PAPR

As mentioned earlier, the PAPR of OFDMA intrinsically is inferior to SC-FDMA. Figure 9-12 shows the CCDF of OFDMA and SC-FDMA PAPRs. It can be seen that the PAPR performance of SC-FDMA is approximately 3 dB better than that of OFDMA with probability of 0.99.

A new study suggests that the PAPR of an OFDM signal does not precisely predict the PA power de-rating or power capability as accurately as the cubic metric (CM). The cubic metric has been adopted by 3GPP as a method of determining PA power de-rating, because of its accuracy over a wide range of devices and signals. This method has proven to be more accurate for WCDMA signals compared to methods that use the statistical PAPR to predict power de-rating based on the information that has been collected from several devices for a variety of signals. The cubic metric can be calculated using the following equation [121]:

$$CM = \frac{20\log\left(s^3_{norm_{RMS}}\right) - 20\log\left(s^3_{norm-reference_{RMS}}\right)}{K} \qquad (9-17)$$

Where $s_{norm_{RMS}}$ denotes the RMS value of the normalized voltage waveform of the input signal and $s_{norm-reference_{RMS}}$ is the RMS value of the normalized voltage waveform of the reference signal (i.e., signal generated using 12.2 kbps AMR speech codec [121]). The CM is used to model the impact of non-linearity in the power amplifier on the adjacent channel leakage ratio. A small value of the CM represents a small power amplifier back-off. The CM is insensitive to the highest power values of the distribution that only have very low probabilities. Thus, methods to reduce the PAPR may not have any major impact on the CM. The value of parameter K was empirically determined to be 1.85 for a set of WCDMA signals. In other words, this equation computes the cubic power of a signal $s(t)$ and compares it to a reference signal $s_{reference}(t)$ and uses the empirical slope factor K to estimate the CM value.

Table 9-1 Example CM Values Corresponding to Reference [122]

	Signal				
Type	**Modulation**	**RAW CM (dB)**	**K**	**Bandwidth Offset (dB)**	**CM (dB)**
WCDMA	Voice	1.52	–	0.00	0
OFDM	16 QAM	7.75	1.56	0.77	4.76
DFT-S OFDM	QPSK	3.44	1.56	0.77	2.00
DFT-S OFDM	16 QAM	4.85	1.56	0.77	2.90
DFT-S OFDM	64 QAM	5.18	1.56	0.77	3.11
IFDMA	QPSK	2.40	1.56	0.00	0.56
IFDMA	16 QAM	4.36	1.56	0.00	1.82
IFDMA	64 QAM	4.64	1.56	0.00	2.00

Although the signal vectors are shown as a function of time, the time scale is effectively removed in the RMS operation and is not a part of the computation. Thus, the raw CM result is not a function of symbol rate or alternatively not a function of signal bandwidth. Similarly, a change in chip rate has no affect on the PAPR. The bandwidth effects will be significant in this study since it includes signals with 3 dB bandwidths larger than the 3.84 MHz used for WCDMA. Table 9-1 shows the CM values corresponding to some example input signals, such as OFDM, DFT-S OFDM, and Interleaved Frequency Division Multiple Access (IFDMA) which is a variant of SC-FDMA. It can be seen that DFT-S OFDM has an advantage relative to other signals shown in the table. In Table 9-1, the bandwidth offset in decibel denotes a de-rating increase of 0.77 dB from the CM regression line that can be used to estimate the actual de-rating of higher bandwidth signals. The K factor is empirically derived based on measurements on various power amplifiers.

9.4 DOWNLINK AND UPLINK MULTIPLE ACCESS SCHEMES

There are various multiple access schemes that have been used in cellular systems in the past few years which allow the network to share the available radio resources (i.e., time, frequency, code, and space) among a number of users in the cell in the downlink and uplink directions. Figure 9-13 illustrates the concept of resource sharing in various multiple access schemes. As mentioned earlier, OFDMA has been a promising multiple-access scheme that has recently been used in mobile broadband wireless radio access technologies. IEEE 802.16m uses OFDMA in the downlink and uplink directions. The OFDMA parameters for IEEE 802.16m are similar to the legacy standard, to facilitate backward compatibility and interoperability between the new and the legacy systems. The IEEE 802.16m OFDMA parameters are shown in Table 9-2.

In OFDMA, an OFDM symbol is constructed of sub-carriers, the number of which is determined by the FFT size. There are several sub-carrier types: (1) data sub-carriers are used for data transmission; (2) pilot sub-carriers are utilized for channel estimation and coherent detection; and (3) null sub-carriers are not used for pilot/data transmission. The null sub-carriers are used for guard bands and DC sub-carrier. The number of sub-carriers used is always less than the FFT/IFFT size. The guard bands

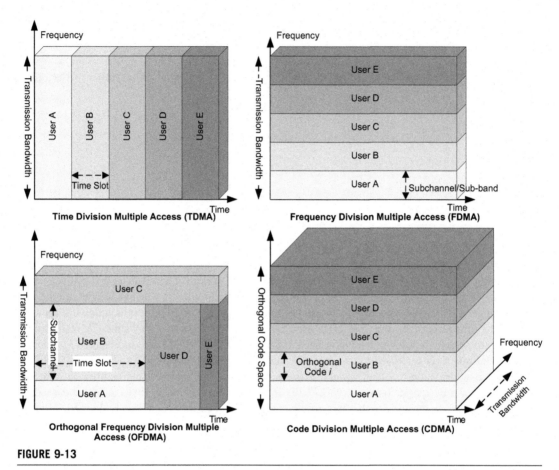

FIGURE 9-13

An illustration of different multiple access scheme concepts

are used to allow spectrum sharing, and to reduce the adjacent channel interference and out-of-band emissions. The sampling frequency is selected to be greater than or equal to the channel bandwidth. In order to ensure that the number of time samples in a 5 ms radio frame is an integer and to further simplify the design of the analog transmit filter, the sampling frequency is scaled by a factor greater than one. As shown in Table 9-2, the value of over-sampling factor n depends on the channel bandwidth and the sampling frequency $f_{sampling}$, as the function of over-sampling factor and the channel bandwidth is given as $f_{sampling} = \lfloor nBW/8000 \rfloor \times 8000$. As shown in Table 9-2, IEEE 802.16m supports a large number of channel bandwidths. The 7 and 8.75 MHz channel bandwidths are used in some legacy systems deployed in Asia and Europe. The channel bandwidths that are integer multiples of 5 MHz are more important. The regional regulatory bodies are recently allowing the network operators to deploy future mobile WiMAX systems in spectrum blocks that are integer multiples of 5 MHz multiples. The bandwidths that are not shown in Table 9-2 are supported via tone dropping, i.e., dropping the sub-carriers at the edges of the frequency band based on 10 and 20 MHz systems (e.g., for

Table 9-2 OFDMA Parameters for IEEE 802.16m [2]

Parameter			5	7	8.75	10	20
Nominal Channel Bandwidth (MHz)			5	7	8.75	10	20
Sampling Factor			28/25	8/7	8/7	28/25	28/25
Sampling Frequency (MHz)			5.6	8	10	11.2	22.4
FFT Size			512	1024	1024	1024	2048
Sub-carrier Spacing (kHz)			10.94	7.81	9.76	10.94	10.94
Useful Symbol Time T_u (µs)			91.429	128	102.4	91.429	91.429
CP $T_g = 1/8\,T_u$	Symbol Time T_s (µs)		102.857	144	115.2	102.857	102.857
	Number of Subframes per Frame		8	5	7	8	8
	FDD	Number of OFDM Symbols per 5 ms Frame	48	34	43	48	48
		Idle Time (µs)	62.857	104	46.40	62.857	62.857
	TDD	Number of OFDM Symbols per Frame	47	33	42	47	47
		TTG + RTG (µs)	165.714	248	161.6	165.714	165.714
CP $T_g = 1/16\,T_u$	Symbol Time T_s (µs)		97.143	136	108.8	97.143	97.143
	Number of Subframes per Frame		8	6	7	8	8
	FDD	Number of OFDM Symbols per Frame	51	36	45	51	51
		Idle Time (µs)	45.71	104	104	45.71	45.71
	TDD	Number of OFDM Symbols per Frame	50	35	44	50	50
		TTG + RTG (µs)	142.853	240	212.8	142.853	142.853
CP $T_g = 1/4\,T_u$	Symbol Time T_s (µs)		114.286	160	128	114.286	114.286
	Number of Subframes per Frame		7	5	6	7	7
	FDD	Number of OFDM Symbols per Frame	43	31	39	43	43
		Idle Time (µs)	85.694	40	8	85.694	85.694
	TDD	Number of OFDM Symbols per Frame	42	30	37	42	42
		TTG + RTG (µs)	199.98	200	264	199.98	199.98
Number of Guard Sub-carriers	Left		40	80	80	80	160
	Right		39	79	79	79	159
Number of Used Sub-Carriers			433	865	865	865	1729
Number of Physical Resource Units in a Subframe.			24	48	48	48	96

a 6 MHz channel bandwidth, the parameters of 10 MHz are used and the edge sub-carriers are dropped to fit the spectrum block).

9.5 IEEE 802.16M DUPLEX MODES

The term duplex refers to bi-directional communication between two devices. When unpaired spectrum or alternatively the same RF carrier is used for downlink and uplink communications, the transmit/receive functions are time-multiplexed. The Time Division Duplex (TDD) is a duplex scheme where uplink and downlink transmissions occur at different times but may share the same frequency. In other words, the downlink and uplink transmissions are multiplexed in time and are not concurrent.

When paired spectrum or alternatively two RF carriers are used for downlink and uplink communications, the transmit/receive functions are frequency-multiplexed. The Frequency Division Duplex (FDD) is a duplex scheme in which uplink and downlink transmissions occur simultaneously using different frequencies. The downlink and uplink frequencies are separated by a sufficiently large frequency separation. IEEE 802.16m supports TDD and FDD schemes with maximal commonality in baseband processing. In order to reduce the implementation complexity and cost of FDD terminals, and to further increase the reuse of baseband functional elements, a Half-Duplex FDD (H-FDD) operation is supported where the downlink and uplink transmissions are not simultaneous but occur in two different frequencies. As shown in Figure 9-14, classic H-FDD operation does not utilize the radio resources efficiently on the downlink and uplink RF carriers. The complementary grouping and scheduling of users would allow efficient use of downlink and uplink resources in an H-FDD operation. The various duplex schemes are illustrated in Figure 9-14.

9.6 FRAME STRUCTURE

As mentioned earlier, the mobile WiMAX systems are based on a subset of IEEE 802.16-2009 standard features [1]. Although the latter standard specifies a number of radio frame sizes, such as 2, 2.5, 4, 5, 8, 10, 12, and 20 ms, only the 5 ms radio frame is supported in the mobile WiMAX system profiles [6]. In order to satisfy the requirements for shorter airlink transmission latency, IEEE 802.16m transparently modified the legacy frame structure. The modifications (as shown in Figure 9-18), are transparent to the legacy devices and are only visible to the new systems. The main idea behind the modifications was to shorten the transmission time intervals and to accelerate the HARQ re-transmissions. Therefore, the legacy radio frame was partitioned into a number of subframes. In order to improve the broadcast of system information in a predefined location and at deterministic intervals, the concept of superframe and superframe header were introduced. The modified frame structure further carries new synchronization and broadcast channels to overcome some of the limitations of the legacy system, such as problems with cell selection at low Signal-to-Interference + Noise Ratios (SINR) due to the poor design of the legacy preamble and the lack of robustness at low SINRs, resulting in poor performance at the cell edge and issues with cell selection/re-selection for cell-edge users. The new superframe concept would allow prescheduled transmission of time-critical system configuration information which further simplifies the operation of the system and reduces the system overhead and scanning latency.

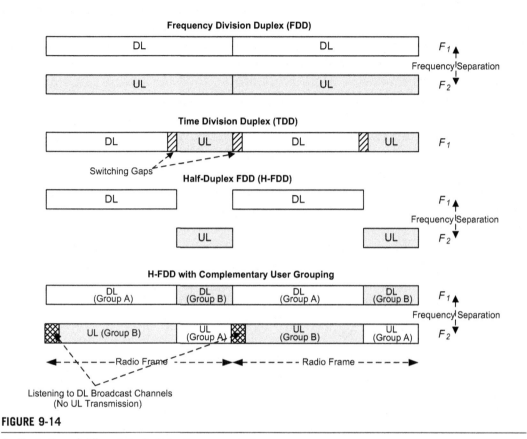

FIGURE 9-14

An illustration of different duplexing scheme concepts

The IEEE 802.16m basic frame structure is illustrated in Figure 9-15. A superframe is defined as a set of four consecutive and equally-sized radio frames. Each 20 ms superframe contains a Primary Superframe Header (P-SFH) and a Secondary Superframe Header (S-SFH). As shown in Table 9-2, the number of subframes per frame varies depending on the cyclic prefix size, the number of available OFDM symbols per frame, and the transmission bandwidth. A subframe is assigned to either DL or UL transmissions subject to the duplexing method. The basic frame structure with $CP = 1/8T_u$, due to backward compatibility with the legacy system, is of higher importance. The $CP = 1/16T_u$ is used when the sum of Round Trip Time (RTT) and the RMS delay spread of the channel is less than 5.71 µs, assuming transmission bandwidths of 5, 10, or 20 MHz. There are four types of subframes defined in the IEEE 802.16m standard: (1) type-1 subframe consists of six OFDM symbols; (2) type-2 subframe consists of seven OFDM symbols; (3) type-3 subframe consists of five OFDM symbols; and (4) type-4 subframe consists of nine OFDM symbols. The basic frame structure is applied to both FDD and TDD duplexing schemes including H-FDD operation. The number of switching points in each radio frame in TDD mode is limited to two, where a switching point is defined as a change of transmission direction from DL to UL or from UL to DL. Note that an excessive number of switching points in a radio frame

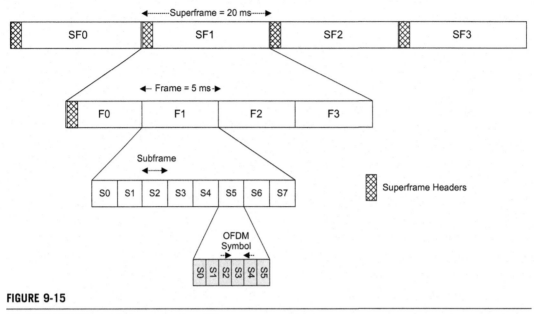

FIGURE 9-15

The IEEE 802.16m basic frame structure [2]

would result in inefficient use of radio resources due to consumption of radio resources by the switching gaps.

The frame structure from the point of view of an H-FDD mobile station is similar to the TDD frame structure, with the exception that the DL and UL transmissions occur in two separate frequency bands. Transmission gaps are required between the DL and UL subframes to allow switching of transmit/receive circuitry (i.e., a duplexer-free operation) and accommodation of maximum permissible RTT in each deployment scenario (cell size, RF carrier frequency, etc.).

For nominal channel bandwidths of 5, 10, and 20 MHz, an IEEE 802.16m frame for $CP = 1/8T_u$ has eight type-1 subframes for FDD, and seven type-1 subframes and one type-3 subframe for TDD. Figure 9-16 shows example TDD and FDD frame structures for a 5, 10, and 20 MHz channel bandwidth with a $CP = 1/8T_u$. Assuming OFDM symbol duration $T_u = 102.857$ µs and $CP = 1/8T_u$, the length of type-1 and type-3 subframes are 0.617 ms and 0.514 ms, respectively. The Transmit/Receive Transition Gap (TTG) and Receive/Transmit Transition Gap (RTG) are 105.714 µs and 60 µs, respectively. Other numerologies may result in a different number of subframes per frame and of symbols within the subframes. In FDD mode, the structure of a frame is identical for the DL and UL in each specific frame. The length of the TTG gap is selected greater than the maximum round trip time in the cell plus the switching time of the RF circuitry from transmission to reception, i.e., $t_{Gap} > 2r_{max}/c + t_{switching}$ where r_{max} denotes the maximum cell radius, c denotes the speed of light, and $t_{switching}$ represents the switching time of RF circuitry. Assuming a typical value of 30 µs for $t_{switching}$ and the value of switching gap given above, a maximum cell size of 11.4 km can be supported. Larger values of switching gaps should be used for support of larger cell sizes. The IEEE 802.16m OFDMA parameters further include $CP = 1/4T_u$ for support of large cell sizes and frequency bands of less than 1 GHz.

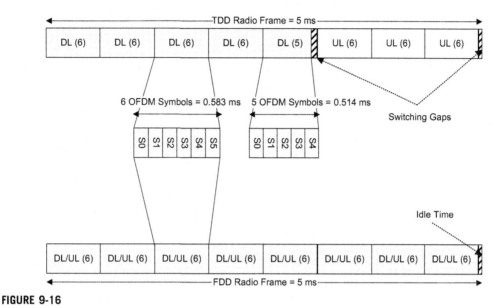

FIGURE 9-16

The frame structure with CP = 1/8 T_u for FDD and TDD (DL/UL ratio = 5:3) [2]

An IEEE 802.16m frame for a CP = $1/16T_u$ has five type-1 subframes and three type-2 subframes for FDD; six type-1 subframes and two type-2 subframes for TDD for nominal channel bandwidths of 5, 10, and 20 MHz. The subframe preceding a DL to UL switching point is a type-1 subframe.

Figure 9-17 illustrates an example of the TDD and FDD frame structure for 5, 10, and 20 MHz channel bandwidths with a CP = $1/16T_u$. With an OFDM symbol duration of 97.143 µs and a CP length of $1/16T_u$, the lengths of the type-1 and type-2 subframes are 0.583 ms and 0.680 ms, respectively. The TTG and RTG switching gaps are 82.853 µs and 60 µs, respectively. Other numerologies may result in different number of subframes per frame and symbols within the subframes.

The Transmission Time Interval (TTI) is defined as the duration of the transmission of the physical layer encoded packets over the radio air interface, and is equal to an integer number of consecutive subframes. In other words, the TTI refers to the length of an independently decodable transmission on the airlink. A data burst occupies either one subframe (i.e., the default TTI or short TTI) or multiple consecutive subframes (i.e., the long TTI). The long TTI in FDD mode contains four subframes for both the DL and UL. The long TTI in TDD mode comprises all available DL/UL subframes depending on the direction of transmission.

The legacy and IEEE 802.16m frames are offset by an integer number of subframes to accommodate new features such as the IEEE 802.16m Advanced Preamble (preamble), Superframe Header (system configuration information), and control channels, as shown in Figure 9-18. The *Frame Offset* shown in Figure 9-18 is an offset between the start of the legacy frame and the start of the IEEE 802.16m frame, defined in a unit of subframes. For uplink transmissions both Time Division Multiplexing (TDM) and Frequency Division Multiplexing (FDM) approaches can be used when supporting

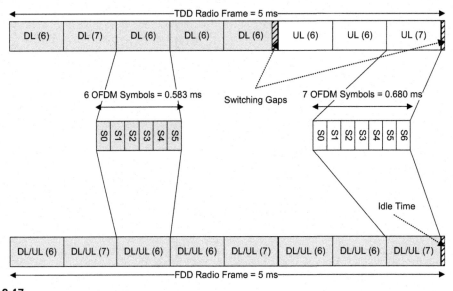

FIGURE 9-17

The frame structure with CP = 1/16 T_u for FDD and TDD (DL/UL ratio = 5:3) [2]

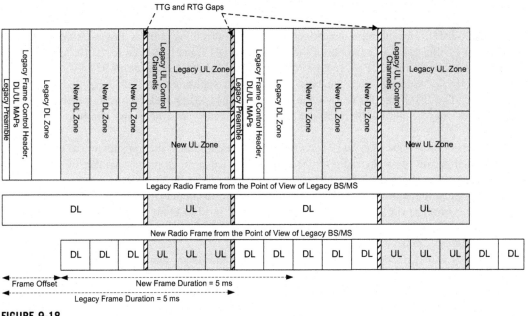

FIGURE 9-18

An example of partitioning of the radio frame and the relationship of the new and legacy frames

legacy and new mobile stations. The FDM approach is preferred to ensure minimal impact on the uplink link budget of the legacy terminals.

9.7 THE CONCEPT OF TIME ZONES AND FREQUENCY REGIONS

As shown in Figure 9-19, a time zone is defined as a time-frequency region which includes the entire transmission bandwidth extended over an integer number of consecutive subframes. Different non-overlapping time zones are time division multiplexed in time. A frequency region is a time-frequency block that occupies a fraction of the transmission bandwidth over one or more consecutive subframes in the downlink or uplink. The concept of time zones is equally applied to TDD and FDD modes. Since many physical and MAC layer features of IEEE 802.16m are new and not backward-compatible with the legacy system, the new and legacy systems are confined within time zones or frequency regions that are time or frequency multiplexed in each radio frame, respectively. The MZones are defined as time zones/frequency regions dedicated to IEEE 802.16m operation, and the LZones are time zones/ frequency regions assigned to the legacy system operation. The Mzones and LZones are time-multiplexed in the downlink. For uplink transmissions both TDM and FDM approaches are supported for multiplexing of legacy and new mobile stations. Note that downlink or uplink traffic/control for new mobile stations can be scheduled in either MZone or Lzone, depending on the operational mode with which the new MS is connected to the BS, but it cannot be scheduled in both zones at the same time, whereas the DL or UL traffic/control for the legacy mobile stations can only be scheduled in the LZones.

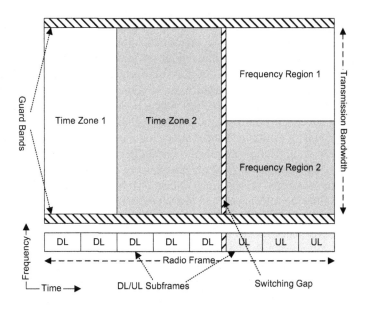

FIGURE 9-19

An illustration of the concept of time zones and frequency regions in TDD mode

In the absence of any legacy system, the LZones will disappear and the entire radio frame will be allocated to the new systems. In a mixed mode deployment of legacy and new systems, the allocation of time zones in the TDD mode is as shown in Figure 9-19. The duration of the zones may vary. Unlike the legacy standard, where the minimum size of zone was two OFDM symbols, in IEEE 802.16m the minimum size of a zone is one subframe, thus, there is less granularity compared to the legacy standard. Note that each legacy frame starts with a legacy preamble and the DL/UL MAPs, followed by a DL zone that is expected by the legacy mobile and relay stations. Similarly, in a mixed mode deployment, the UL portion of the frame in TDD mode starts with the legacy UL control channels as expected by the legacy mobile and relay stations. The coexistence is defined as a deployment where the new and legacy base stations coexist in the same frequency band and in the same or neighboring geographical areas. In a green-field deployment, the LZones can be removed. The DL to UL and UL to DL switching points must be synchronized across the network to reduce inter-cell interference. The switching points would require use of idle symbols to accommodate the switching gaps. In the case of TDD operation with the basic frame structure, the last symbol in the subframe immediately preceding a downlink-to-uplink/uplink-to-downlink switching point may be reserved for guard time and consequently not transmitted. An example frame partitioning is shown in Figure 9-18.

9.8 SUBCHANNELIZATION AND PERMUTATION

The IEEE 802.16-2009 standard defines a permutation zone as a number of contiguous OFDM symbols in the DL or the UL that use the same permutation formula. The DL or UL portions of the radio frame may contain more than one permutation zone. One of the drawbacks of the sub-channelization schemes specified in the IEEE 802.16-2009 standard is that the distributed and localized allocations cannot be located in the same permutation zone. Furthermore, in the legacy standard, a slot requires both a time and a subchannel dimension for completeness, and is the minimum possible data allocation unit. The definition of an OFDMA slot depends on the OFDM symbol structure, which varies for UL and DL, for Full Usage of Subchannels (FUSC) and Partial Usage of Subchannels (PUSC), and for the distributed sub-carrier permutations and the adjacent sub-carrier permutations. As shown in Figure 9-20, a data region in the legacy standard is defined as a two-dimensional resource unit consisting of a group of contiguous subchannels, in a group of contiguous OFDM symbols. A two-dimensional allocation may be visualized as a rectangle such as 4×3, 6×1 etc. Therefore, to address a particular rectangle in the DL or UL MAP, two parameters are required, i.e., time and frequency offsets (or alternatively symbol offset and subchannel offset). In order to reduce the overhead of signaling allocations in two-dimensional subchannelization schemes, as well as to allow both localized and distributed allocations in the same time-zone, IEEE 802.16m introduced new one-dimensional subchannelization schemes and their associated permutation methods to achieve frequency diversity or frequency scheduling gain depending on the user mobility and channel conditions. The comparison between the new and legacy subchannelization schemes is shown in Figure 9-20. The one-dimensional resource blocks in IEEE 802.16m simplify the addressing of user allocations in the assignment MAPs, and allow efficient addressing and indexing schemes such as binary tree-structured resource indexing. Furthermore, localized and distributed resource units can be simultaneously assigned to a user in the same time zone. Regardless of the type of legacy subchannelization and permutation scheme in the DL or UL, there

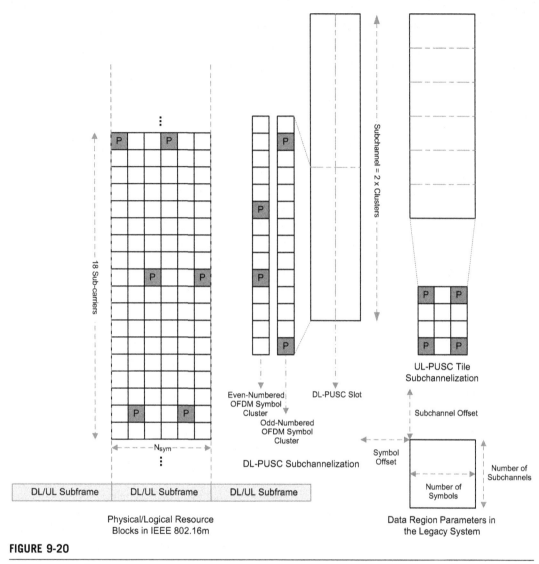

FIGURE 9-20

A comparison of the subchannelization schemes in IEEE 802.16m and the legacy system [1,2]

are 48 data sub-carriers and a number of pilot sub-carriers in each subchannel or slot (e.g., the DL-PUSC slot contains 4 clusters, 48 data sub-carriers, and 8 pilot sub-carriers; the UL-PUSC slot contains 6 tiles, 48 data sub-carriers, and 24 pilot sub-carriers, etc.). From the latter example, it can be noted that the pilot density (i.e., the ratio of pilot sub-carriers over the total number of sub-carriers in a resource block/subchannel) is excessively high, resulting in additional Layer 1 overhead and lowering the throughput.

9.8.1 Downlink Subchannelization and Permutation

The available frequency resources over a downlink subframe are divided into a number of frequency partitions, where each partition consists of a set of physical resource units over the available number of OFDM symbols in the subframe. Each frequency partition can include localized and/or distributed resource units. This is different from the legacy system, where each permutation zone could only accommodate localized or distributed sub-channels. The frequency partitions can be used for different purposes, such as Fractional Frequency Reuse (FFR). The downlink/uplink resource petitioning and mapping procedure is illustrated in Figure 9-22). In the example shown, three frequency partitions are defined, where some of them include both localized and distributed resource units.

A Physical Resource Unit (PRU) is the basic physical unit for resource allocation that comprises P_{sc} consecutive sub-carriers by N_{sym} consecutive OFDM symbols. The default value for P_{sc} is 18 sub-carriers, and N_{sym} can be 6, 7, or 5 OFDM symbols for type-1, type-2, and type-3 subframes, respectively. A Logical Resource Unit (LRU) is the basic logical unit for localized or distributed resource allocations. An LRU comprises $P_{sc} \times N_{sym}$ sub-carriers, inclusive of the pilot sub-carriers that are embedded in a PRU. A Distributed Resource Unit (DRU) is defined to achieve frequency diversity gain in multipath fading channels. The DRU contains a group of sub-carriers which are physically spread across the distributed resources within a frequency partition. The size of the DRU is the same size as the PRU. The minimum unit for forming a DRU is equal to one sub-carrier or a pair of sub-carriers, also known as a tone-pair.

A localized resource unit, alternatively known as Contiguous Resource Unit (CRU), is used to achieve frequency-selective scheduling gain. The CRU contains a group of sub-carriers which are physically contiguous across the localized resource allocations within a frequency partition. The size of the CRU is the same as the size of the PRU. The physical sub-carriers associated with an OFDM symbol are grouped into $N_{guard-left}$ left guard sub-carriers, $N_{guard-right}$ right guard sub-carriers, and N_{used} used sub-carriers. The DC sub-carrier is not used. The N_{used} sub-carriers are divided into an integer number of PRUs. Each PRU contains pilot and data sub-carriers. The number of used pilot and data sub-carriers depends on the MIMO mode, number of streams, and number of users, as well as the type of subframe.

To better understand the difference between the localized and distributed allocations, consider the example shown in Figure 9-21. The frequency response of a multipath fading channel varies with time and frequency due to path loss, shadowing, and user mobility effects. Let's assume that user A and user B are two users with different channel conditions. The BS receives channel quality reports (typically in the form of CINR or SINR measurements) from the mobile stations in the cell. The BS scheduler may adopt either of the following resource allocation schemes depending on the channel condition reports and other considerations that will be discussed later in this chapter. One allocation strategy is to allocate user A and user B in the subchannels where the corresponding SINR is the best for that user. This is when a frequency-selective scheduling gain can be achieved through allocation of a group of physically adjacent sub-carriers in the subchannels with the relatively best SINR for this user. Another strategy is to permute the sub-carriers over the entire channel and form a logical group of distributed sub-carriers, and to assign them to a user. In this case, frequency diversity gain can be achieved through the use of distributed resource units for this user.

The procedure for forming logical resource units from physical resource units can be described as follows. The PRUs are first subdivided into sub-bands (SB) and mini-bands (MB) where a sub-band

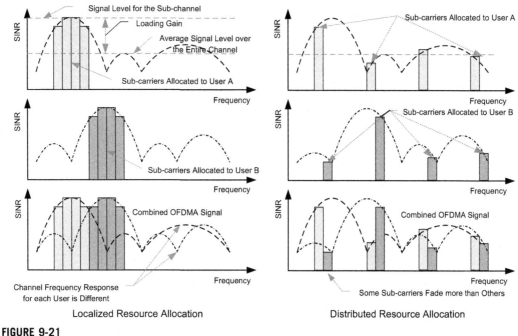

FIGURE 9-21

A comparison of localized and distributed allocation schemes (example)

comprises N_1 adjacent PRUs and a mini-band consists of N_2 adjacent PRUs. The sub-bands are suitable for frequency-selective allocations as they provide a contiguous allocation of PRUs in frequency. The mini-bands are suitable for frequency diversity allocation and are permuted in frequency. The downlink sub-carrier to resource unit mapping process is defined as follows and is illustrated in Figures 9-23 to 9-25:

1. An outer permutation is applied to the PRUs in the units of N_1 and N_2 PRUs where $N_1 = 4$ and $N_2 = 1$.
2. The reordered PRUs are distributed into frequency partitions.
3. The frequency partition is divided into localized and/or distributed resource allocations. The sizes of the distributed/localized groups are flexibly configured per sector. Adjacent sectors do not need to have same configuration for the localized and distributed groups.
4. The localized and distributed resource units are further mapped into LRUs by direct mapping for CRUs and by the use of sub-carrier permutation for DRUs.

The sub-carrier permutation defined for the downlink distributed resource allocations spreads the sub-carriers of the DRU across all the distributed resource allocations within a frequency partition. After mapping all pilots, the remaining usable sub-carriers are used to form the DRUs. To allocate the LRUs, the remaining sub-carriers are paired into contiguous sub-carrier pairs. Each LRU consists of a group of sub-carrier pairs. The number of sub-bands depends on the system bandwidth and is broadcast through the Downlink Sub-band Allocation Count (DSAC) parameter in the secondary superframe header [2].

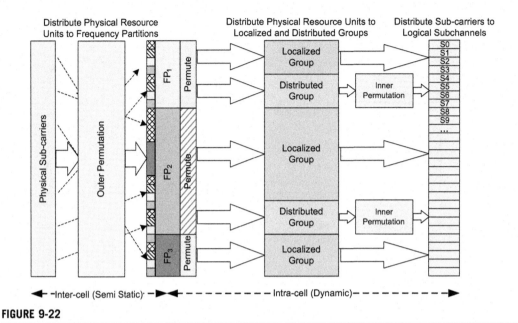

FIGURE 9-22

An illustration of the downlink/uplink physical to logical resource mapping concept [2]

The PRUs are divided and reordered into two groups: (1) sub-band PRUs; and (2) mini-band PRUs, denoted by PRU_{SB} and PRU_{MB}, respectively. The set of PRU_{SB} is numbered from 0 to $(L_{SB} - 1)$, and the set of PRU_{MB} is numbered from 0 to $(L_{MB} - 1)$, where L_{SB} and L_{MB} denote the number of PRUs in sub-bands and mini-bands, respectively (see Figure 9-23). The sub-band and mini-band permutation formulae are defined in reference [2]. The set of PRU_{SB} and Permuted PRU_{MB} ($PPRU_{MB}$) are allocated to one or more frequency partitions. By default, only one partition is present. The maximum number of frequency partitions within the transmission bandwidth is four. The mini-band permutation maps the PRU_{MB} to the $PPRU_{MB}$ to ensure frequency diversity among the PRUs that are assigned to each frequency partition. The frequency partition configuration is transmitted in the secondary superframe header using the parameter Downlink Frequency Partition Configuration (DFPC) whose value depends on the system bandwidth. The Frequency Partition Count (FPCT) parameter defines the number of frequency partitions. The Frequency Partition Size (FPS_i) parameter defines the number of PRUs allocated to the ith frequency partition FP_i. The FPCT and FPSi parameters are determined from the DFPC. The Downlink Frequency Partition Sub-band Count (DFPSC) parameter defines the number of sub-bands allocated to FP_i. The $PPRU_{FPi}$ resources are mapped to LRUs. Additional PRU and sub-carrier permutations are constrained to the PRUs within a frequency partition.

The distinction between CRUs and DRUs is done on a sector-specific basis. Let $L_{SB-CRU-FPi}$ and $L_{MB-CRU-FPi}$ denote the number of allocated sub-band CRUs and mini-band CRUs to the ith frequency partition FP_i, the total number of allocated sub-band and mini-band CRUs, in units of a sub-band (i.e., N_1 PRUs), in FP_i is given by the Downlink CRU Allocation Size ($DCAS_i$) parameter. For example, the number of sub-band and mini-band CRUs in FP_0 is given by $DCAS_{SB0}$ and $DCAS_{MB0}$, in units of a sub-band and mini-band, respectively. When DFPC = 0, $DCAS_i$ must be equal to 0. For FP_0, the

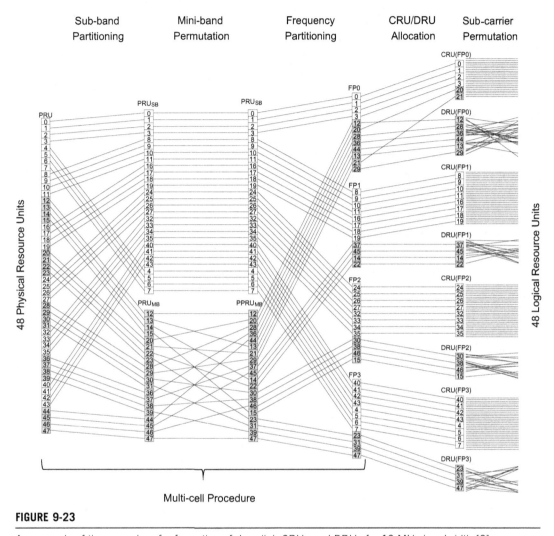

FIGURE 9-23

An example of the procedure for formation of downlink CRUs and DRUs for 10 MHz bandwidth [2]

value of $DCAS_{SB0}$ is explicitly signaled in the secondary superframe header to indicate the number of sub-bands in unsigned binary format. The Downlink mini-band CRU Allocation Size ($DCAS_{MB0}$) is sent in the secondary superframe header only for partition FP_0 whose value depends on the system bandwidth. As an example, the number of sub-band CRUs for FP_0 is given by $L_{SB-CRU-FP0} = N_1DCAS_{SB0}$. The mapping between $DCAS_{MB0}$ and the number of mini-band CRUs for FP_0 for system bandwidths of 5, 10, and 20 MHz are given in reference [2] and are summarized in Table 9-3.

The downlink DRUs are used to form 2-stream distributed LRUs via sub-carrier permutation. The sub-carrier permutation defined for downlink distributed resource allocations within a frequency partition spreads the sub-carriers of the DRU across the entire distributed resource allocation. The

Table 9-3 Summary of Downlink Subchannelization and Permutation Parameters [2]

Procedure		Signaling Field Bandwidth (20/10/5 MHz)	Signaling Mechanism	Parameters Calculated from Signaled Fields	Definition	Unit
Sector Common	Sub-band and Mini-band Partitioning	DSAC (5/4/3) bits	S-SFH Sub-packet 2	K_{SB} $L_{SB} = N_1 * K_{SB}$	Number of sub-bands Number of PRUs Assigned to sub-bands	Sub-band PRU
				L_{MB}	Number of PRUs Assigned to mMini-bands	PRU
	Frequency Partitioning	DFPC (4/3/3) bits		FPCT	Number of frequency partitions	Frequency Partition
				FPS_i	Number of PRUs in FP_i	PRU
		DFPSC (3/2/1) bits		K_{SB-FPi}	Number of sub-bands assigned to FP_i	Sub-band
				K_{MB-FPi}	Number of mini-bands assigned to FP_i	Sub-band (groups of N_1 PRUs)
				$L_{SB-FPi} = N_1 * K_{SB-FPi}$	Number of mini-bands assigned to FP_i	PRU
				$L_{MB-FA} = N_2 * K_{MB-FPi}$	Number of PRUs assigned to sub-bands in FP_i	PRU
Sector Specific	CRU/DRU Allocation	$DCAS_{SB0}$ (5/4/3) bits	S-SFH Sub-packet 1	$L_{SB-CRU-FPi}$	Number of sub-band CRUs in FP_i	CRU
		$DCAS_{MB0}$ (5/4/3) bits		$L_{MB-CRU-FPi}$	Number of mini-band CRUs in FP_i	CRU
		$DCAS_i$ (3/2/1) bits		$L_{CRU-FPi} = L_{SB-CRU-FPi} + L_{MB-CRU-FPi}$	Number of CRUs in FP_i	CRU
				$L_{DRU-FPi} = FPS_i - L_{CRU-FPi}$	Number of DRUs in FP_i	DRU
	Tone Permutation		Obtained from SA-Preamble			

granularity of the sub-carrier permutation is equal to a pair of sub-carriers. After mapping all pilots, the available sub-carriers are used to define the distributed LRUs. To allocate the LRUs, the remaining sub-carriers are paired into adjacent tone-pairs. Each LRU consists of a group of tone-pairs. Let's assume that there are N_{RU} DRUs. A permutation sequence $PermSeq(.)$ for the distributed group is defined and the subchannelization for downlink distributed resource allocations is performed using the following procedure. The permutation sequence is of length $L_{DRU\text{-}FPi}$ and is determined by $SEED = \{343 IDcell\} \bmod 2^{10}$, where the $IDcell$ parameter is the cell identifier associated with the serving cell. The permutation sequence is generated by the random sequence generation algorithm specified in reference [2]. For the kth OFDM symbol in the subframe:

1. Let n_k denote the number of pilot sub-carriers in the kth OFDM symbol within a PRU, allocate n_k pilots in the kth OFDM symbol within each PRU;
2. Let N_{RU} denote the number of DRUs within the Frequency Partition (FP), renumber the remaining $N_{RU}(P_{sc} - n_k)$ data sub-carriers of the DRUs and reorder from 0 to $N_{RU}(P_{sc} - n_k) - 1$ sub-carriers;
3. Group these contiguous and logically renumbered sub-carriers into $N_{RU}(P_{sc} - n_k)/2$ pairs and renumber them from 0 to $N_{RU}(P_{sc} - n_k)/2 - 1$;
4. Apply the sub-carrier permutation formula with the permutation sequence $PermSeq(.)$ or data sub-carrier pairs;
5. Map each set of logically contiguous $(P_{sc} - n_k)$ sub-carriers into distributed LRUs and form a total of N_{RU} distributed LRUs.

There is no sub-carrier permutation defined for the downlink localized resource allocations. The CRUs are directly mapped to the sub-band and mini-band LRUs within each frequency partition. Example procedures for formation of downlink CRUs and DRUs for 10 MHz bandwidth are illustrated in Figures 9-23 to Figure 9-25.

The guard sub-carriers between contiguous frequency channels can be utilized for data transmission, if the sub-carriers from adjacent frequency channels are precisely aligned. In mixed mode operation, the legacy channel raster of 250 kHz is maintained [6]. The channel raster is the frequency steps/increments that a mobile device will use when scanning for the desired downlink RF carrier. The frequency separation of contiguous RF carriers used for carrier aggregation must be an integer multiple of the OFDM sub-carrier spacing. In order to align the sub-carriers associated with adjacent frequency channels, a frequency offset can be applied to the center frequency. The guard sub-carriers can be utilized for data transmission, if the information of the available guard sub-carriers suitable for data transmission is communicated to the mobile station. This information includes the number of available sub-carriers in the upper and lower band edges.

As shown in Figure 9-26, the guard sub-carriers between adjacent RF carriers are grouped to form an integer multiple of PRUs. The structure of a guard PRU is identical to downlink subchannel structures. The guard PRU is used as a mini-band CRU at partition FP_0 for data transmission only. The number of useable guard sub-carriers is predefined, and is known to both MS and BS based on the RF carrier bandwidth. The number of guard PRUs in the left and right edges of each RF carrier are shown in Table 9-2. The number of guard PRUs in the left and right edges of the RF carrier are denoted by $N_{LG\text{-}PRU}$ and $N_{RG\text{-}PRU}$, respectively. The total number of guard PRUs are $N_{G\text{-}PRU} = N_{LG\text{-}PRU} + N_{RG\text{-}PRU}$. It must be noted that when an RF carrier occupies the left-most segment of the spectrum among a number of contiguous RF carriers, the number of guard PRUs in the left edge of the carrier is zero. Furthermore, when an RF carrier occupies the right-most of segment of the

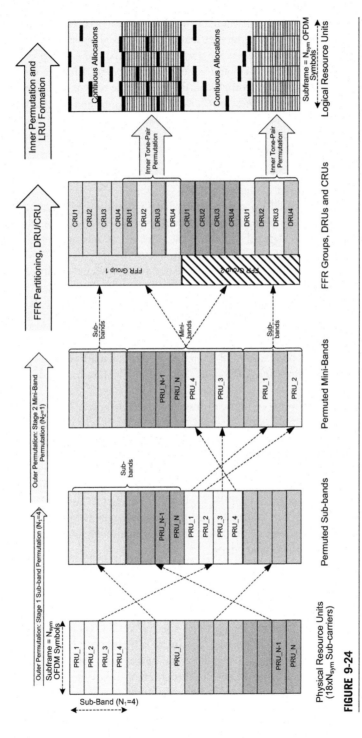

FIGURE 9-24

An illustration of physical to logical resource mapping

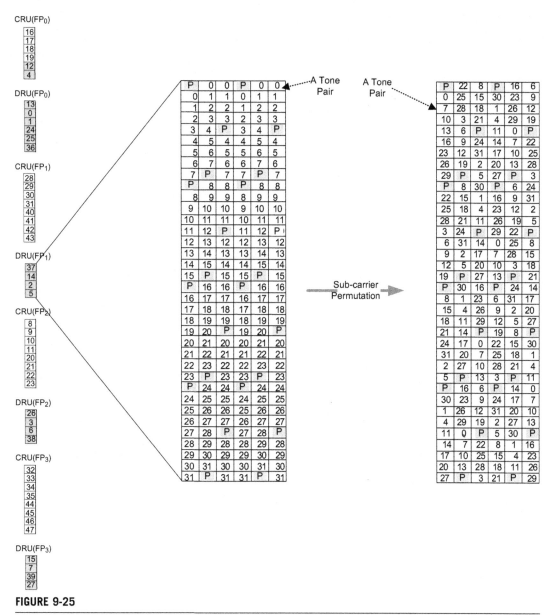

FIGURE 9-25

An example of downlink sub-carrier permutation ($K_{SB} = 7$, $FPCT = 4$, $FPS_O = FPS_i = 12$, $DFPSC = 2$, $DCAS_{SBO} = 1$, $DCAS_{MBO} = 1$, $DCAS_i = 2$, $Idcell = 2$, $BW = 10$ MHz, and 2 spatial streams) [2]

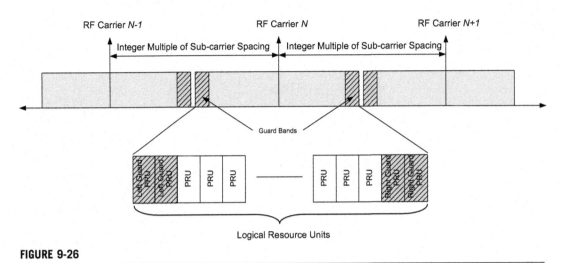

FIGURE 9-26

An example of data transmission using guard sub-carriers [2]

spectrum among a number of contiguous RF carriers, the number of guard PRUs in the right edge of the carrier is zero [2].

The left and right guard PRUs are denoted as G-PRU$_L$[0], G-PRU$_L$[2], ..., G-PRU$_L$[$N_{LG\text{-}PRU}$ − 1] and G-PRU$_R$[0], G-PRU$_R$[0], ..., G-PRU$_R$[$N_{RG\text{-}PRU}$ − 1], respectively, from the lowest frequency index. The guard PRUs are indexed by interleaving G-PRU$_L$ and G-PRU$_R$, i.e., G-PRU[i] = G-PRU$_L$[$i/2$] for even-valued i and G-PRU[i] = G-PRU$_R$[(i − 1)/2] for odd-valued i, where i is an integer. If N$_{LGPRU}$ = 0, then G-PRU[i] = G-PRU$_R$[i]. If N$_{RG\text{-}PRU}$ = 0, then G-PRU[i] = G-PRU$_L$[i]. The N$_{G\text{-}PRU}$ guard PRUs are used as mini-band LRUs, i.e., the mini-band LRUs at frequency partition FP$_0$ with no permutation for data transmission. The ith guard mini-band LRU is always allocated along with the last ith mini-band LRU in partition FP$_0$. In other words, when an MS with multicarrier support is provided, an allocation including the last ith mini-band LRU in partition FP$_0$ is allocated together with the ith guard mini-band LRU. When the adjacent RF carrier is not an active carrier for the MS, the guard sub-carriers in between active and non-active carriers are not utilized for data transmission. When the overlapped guard sub-carriers are not aligned in the frequency domain, they are not used for data transmission.

9.8.2 Uplink Subchannelization and Permutation

The subchannelization schemes in the uplink are similar to the downlink with the exception of the difference in permutation granularity. Similar to downlink, each uplink subframe is divided into a number of frequency partitions (i.e., FP$_i$, i = 0, 1, 2, 3), where each partition comprises a set of physical resource units across the number of OFDM symbols available in the subframe. Each frequency partition can include localized and/or distributed physical resource units. Each frequency partition can be used for different purposes, such as fractional frequency reuse.

A physical resource unit is the basic physical unit for resource allocation in the uplink which contains P_{sc} sub-carriers by $Nsym$ OFDM symbols. The Psc parameter is equal to 18 and the number of OFDM symbols $Nsym$ is equal to 6, 7, 5, or 9, depending on the type of subframe (i.e., type-1, type-2, type-3, or type-4). Similarly, a logical resource unit is defined as the basic unit for distributed and localized resource allocations which have the same size as a physical resource unit.

The size of logical resource units for uplink control channel transmission is the same as for the traffic channel. Multiple users are allowed to share a control LRU. The effective number of data sub-carriers in an LRU depends on the number of allocated pilots and control channel information fields. The distributed resource units contain a group of sub-carriers that are spread across distributed resource allocations within a frequency partition. The size of the DRU is equal to that of the PRU. The minimum unit for forming the DRU in the uplink is a tile. The uplink tile size is $6 \times N_{sym}$, where N_{sym} is the number of OFDM symbols in the uplink subframe. The UL multi-cell resource mapping consists of sub-band partitioning, mini-band permutation, and frequency partitioning, similar to the procedure described for the downlink. The PRUs are first divided into sub-bands and mini-bands as illustrated in Figure 9-23. Similar to downlink subchannelization, a sub-band comprises N_1 adjacent PRUs and a mini-band consists of N_2 adjacent PRUs where $N_1 = 4$ and $N_2 = 1$. The sub-bands are suitable for frequency-selective allocations as they can provide a continuous allocation of PRUs in the frequency-domain. The mini-bands are suitable for frequency-diversity allocations and are permuted in the frequency-domain.

The number of sub-bands is denoted by K_{SB}. The number of PRUs allocated to sub-bands is $L_{SB} = N_1 \times K_{SB}$. The Uplink Sub-band Allocation Count (*USAC*) parameter is used to determine the value of K_{SB}, depending on the system bandwidth. The *USAC* parameter is broadcast via the secondary superframe header. The remaining PRUs are allocated to mini-bands. The number of mini-bands in an allocation is denoted by K_{MB}. The number of PRUs allocated to mini-bands is $L_{MB} = N_2 \times K_{MB}$. The total number of PRUs is $N_{PRU} = L_{SB} + L_{MB}$. The maximum number of sub-bands is given by the N_{sub} where $N_{sub} = \lfloor N_{PRU}/N_1 \rfloor$. The relationship between *USAC* and K_{SB} values for system bandwidths of 5, 10, and 20 MHz are given in reference [2].

The PRUs in the uplink are partitioned and reordered into two groups of sub-band PRUs (PRU$_{SB}$) and mini-band PRUs (PRU$_{MB}$). The set of PRU$_{SB}$ and PRU$_{MB}$ are numbered from 0 to ($L_{SB} - 1$) and 0 to ($L_{MB} - 1$), respectively. The uplink mini-band permutation maps the set of PRU$_{MB}$ to the set of Permuted PRU$_{MBS}$ (PPRU$_{MB}$) to ensure frequency diversity is sufficiently utilized in each frequency partition. The PRU$_{SBS}$ and PPRU$_{MBS}$ are allocated to one or more frequency partitions. By default, only one frequency partition is present. The maximum number of frequency partitions is four and the uplink frequency partition configuration information is transmitted via the S-SFH sub-packet 1 using Uplink Frequency Partition Configuration (*UFPC*), whose value depends on the system bandwidth. The Frequency Partition Count (*FPCT*) parameter defines the number of frequency partitions. The Frequency Partition Size (*FPS$_i$*) defines the number of PRUs allocated to the *i*th frequency partition (i.e., FP$_i$). The *FPCT* and *FPS$_i$* parameters are determined from the *UFPC*, based on predefined mappings given in reference [2]. The Uplink Frequency Partition Sub-band Count (*UFPSC*) parameter defines the number of sub-bands allocated to FP$_i$ (for $i > 0$). If *UFPC* = 0, then *UFPSC* is set to 0. The PRU$_{FPi}$ is mapped to a logical resource unit. Additional PRU and tile permutations are limited to the PRUs within a frequency partition.

The mapping and permutation of CRUs and DRUs in each partition are performed on a sector-specific basis. Let $L_{SB\text{-}CRU\text{-}FPi}$ and $L_{MB\text{-}CRU\text{-}FPi}$ denote the number of allocated sub-band CRUs and mini-band CRUs for FP$_i$ ($i > 0$). The total number of allocated CRUs in a unit of sub-bands within FP$_i$ (for $i > 0$) is given by uplink CRU allocation size $UCAS_i$. The number of sub-band and mini-band CRUs in FP$_0$ is given by $UCAS_{SB0}$ and $UCAS_{MB0}$ in integer multiples of sub-band and mini-band, respectively. The value of $UCAS_{SB0}$ is broadcast in the S-SFH for FP$_0$. An uplink mini-band CRU allocation size $UCAS_{MB0}$ parameter is sent in the S-SFH only for partition FP_0 depending on system bandwidth. The number of sub-band-based CRUs for FP_0 is given as $L_{SB-CRU-FP0} = N_1 \times UCAS_{SB0}$. A summary of uplink subchannelization and permutation parameters are given in Table 9-4.

Figure 9-27 illustrates an example of the sub-band partitioning, mini-band permutation, frequency partitioning, and cell-specific resource mapping (CRU/DRU allocation) procedures for a 10 MHz system bandwidth. In this example, $K_{SB} = USAC = 7$, $FPCT = 4$, $FPS_i = 12$ (for $i > 0$), $UFPSC = 2$, $UCAS_{SB0} = 1$, $UCAS_{MB0} = 1$, $UCAS_i = 2$, and $IDcell = 2$. An example uplink tile permutation is further shown in Figure 9-28.

Each of the DRUs of a UL frequency partition is divided into three tiles of six adjacent sub-carriers over N_{sym} OFDM symbols. The tiles within a frequency partition are collectively tile-permuted to obtain frequency-diversity across the allocated resources. The tile permutation allocates physical tiles of the DRUs to logical tiles of subchannels.

The distributed LRUs in each uplink frequency partition may be further divided into data, bandwidth request, and feedback regions. A feedback region consists of feedback channels that can be used for both HARQ ACK/NACK and fast-feedback. In a multicarrier system with active downlink-only RF carriers, the primary UL RF carrier may contain multiple feedback regions for the primary downlink RF carrier and the active downlink-only RF carriers. The primary uplink RF carrier should contain one feedback region corresponding to each DL RF carrier with predefined mapping. The allocation order of data channels and UL control channels are UL HARQ feedback channels, UL fast-feedback channels, UL bandwidth request channels, and UL data channels. If fractional frequency reuse is used in a UL subframe, the UL control channels can be allocated in the reuse-1 partition or the highest-power reuse-3 partition.

The number of bandwidth request channels in frequency partition FP_i in a UL subframe is N_{BWR} where the value of N_{BWR} is 1 in MZone and 2 in LZone with UL-PUSC. In MZone, the bandwidth request channels are of the same size as LRUs, i.e., three 6×6 UL tiles. In the LZone with a UL-PUSC subchannelization scheme, the bandwidth request channels consist of three 4×6 UL tiles. Note that the UL tile granularity is different for the new and the legacy zones. The bandwidth request channels use LRUs constructed after the tile permutation. Let $UL\_FEEDBACK\_SIZE$ denote the number of distributed LRUs in frequency partition FP_i that are reserved for feedback channels, the number of feedback channels in frequency partition FP_i is then $L_{FB-FPi} = N_{FB} \times UL\_FEEDBACK\_SIZE$, where N_{FB} is 3 in MZone and 4 in the LZone with UL-PUSC.

The primary and secondary fast-feedback channels are formed by three permuted 2×6 Feedback Mini-Tiles (FMT) where a feedback mini-tile is a subdivision of the IEEE 802.16m UL tile (i.e., 2 sub-carriers by 6 OFDM symbols). The feedback mini-tile reordering procedure is as follows:

1. The uplink tiles in the distributed LRUs reserved for feedback channels are divided into 2×6 feedback mini-tiles. The FMTs are numbered from 0 to $3 \times L_{FB-FPi} - 1$;
2. A mini-tile reordering is applied to the available 2×6 FMTs to obtain the reordered FMTs;
3. Each group of three consecutive reordered FMTs form a feedback channel.

The HARQ feedback channel is used to transmit six HARQ feedback channels. The number of HARQ feedback channels is denoted by the $L_{HFB-FPi}$ parameter. A pair of HARQ feedback channels is formed by three 2×2 reordered HARQ Mini-Tiles (HMT), where a HARQ mini-tile is a subdivision of the IEEE 802.16m UL tile (i.e., 2 sub-carriers by 2 OFDM symbols). The HMTs reordering procedure and the construction of the HARQ feedback channel are described below and illustrated in Figure 9-29.

Table 9-4 Summary of Uplink Subchannelization and Permutation Parameters [2]

	Procedure	Signaling Field Bandwidth (20/10/5 MHz)	Signaling Mechanism	Parameters Calculated from Signaled Fields	Definition	Unit
Sector Common	Sub-band and Mini-band Partitioning	USAC (5/4/3) bits	S-SFH Sub-packet 2	K_{SB}	Number of sub-bands	Sub-band
				$L_{SB} = N_1 * K_{SB}$	Number of PRUs Assigned to sub-bands	PRU
				L_{MB}	Number of PRUs Assigned to mini-bands	PRU
	Frequency Partitioning	UFPC (4/3/3) bits		$FPCT$	Number of frequency partitions	Frequency Partition
		UFPSC (3/2/1) bits		FPS_i	Number of PRUs in FP_i	PRU
				K_{SB-FPi}	Number of sub-bands sssigned to FP_i	Sub-band
				K_{MB-FPi}	Number of mini-bands assigned to FP_i	Sub-band (groups of N_1 PRUs)
				$L_{SB-FPi} = N_1 * K_{SB-FPi}$	Number of mini-bands assigned to FP_i	PRU
				$L_{MB-FPi} = N_2 * K_{MB-FPi}$	Number of PRUs assigned to sub-bands in FP_i	PRU
Sector Specific	CRU/DRU Allocation	UCAS$_{SB0}$ (5/4/3) bits	S-SFH Sub-packet 1	$L_{SB-CRU-FPi}$	Number of sub-band CRUs in FP_i	CRU
		UCAS$_{MB0}$ (5/4/3) bits		$L_{MB-CRU-FPi}$	Number of mini-band CRUs in FP_i	CRU
		UCAS$_i$ (3/2/1) bits		$L_{CRU-FPi} = L_{SB-CRU-FPi} + L_{MB-CRU-FPi}$	Number of CRUs in FP_i	CRU
	Tile Permutation	IDcell (10 bits)	Obtained from SA-Preamble	$L_{DRU-FPi} = FPS_i - L_{CRU-FPi}$	Number of DRUs in FP_i	DRU

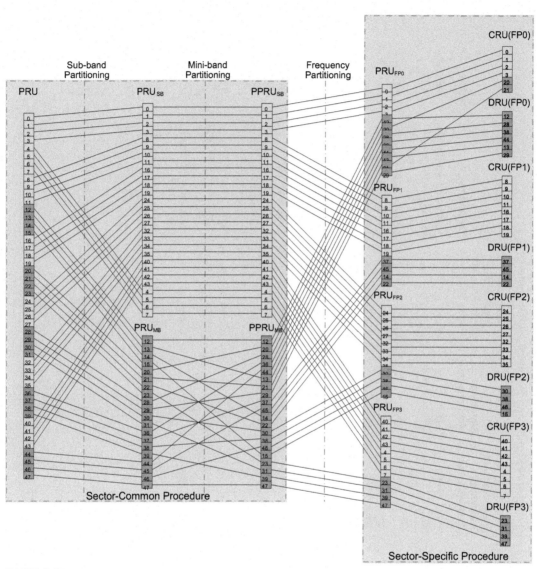

FIGURE 9-27

An example procedure for the formation of uplink CRUs and DRUs for 10 MHz bandwidth [2]

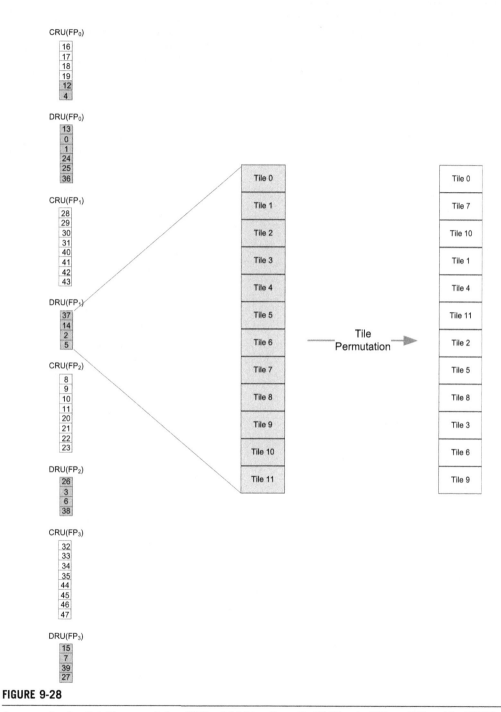

FIGURE 9-28

An example of uplink tile permutation (BW = 10 MHz, $K_{SB} = 7$, $FPCT = 4$, $FPS_0 = FPS_i = 12$, $UFPSC = 2$, $UCAS_{SB0} = 1$, $UCAS_{MB0} = 1$, $UCAS_i = 2$, $IDcell = 2$, and $Subframe\ Index = 0$) [2]

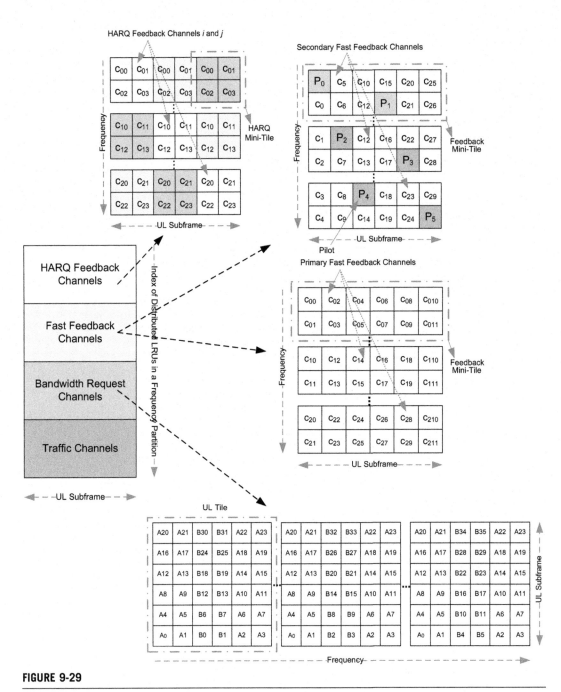

FIGURE 9-29

The location and structure of various uplink subchannels in IEEE 802.16m [2]

1. Each 2×6 reordered FMT is divided into three serially-indexed 2×2 HMTs. The HMTs are numbered from 0 to $3 \times L_{HFB\text{-}FPi} - 1$;
2. A reordering is applied to the HMTs to obtain the reordered HMTs;
3. Each group of three consecutive reordered HMTs forms a pair of HARQ feedback channels.

In FDD mode, there is one HARQ feedback region in each UL subframe. In TDD mode, the HARQ feedback channel in the UL subframe L may be associated with the DL bursts in several DL subframes whose indices are denoted by $M = \{m_0, m_1, \ldots, m_{K-1}\}$, where $m_0 < m_1 < \ldots < m_{K-1}$. The number of HARQ feedback regions is equal to the size of set M. For DL bursts starting at subframe m_k, the index of the associated HARQ feedback region is in the order of m_k in set M, with index zero corresponding to the first HARQ feedback region. Within each HARQ feedback region, the index for the HARQ feedback channel is calculated as follows. For persistent allocation, the index k is specified in the DL Persistent Allocation A-MAP IE. A primary or secondary fast-feedback channel consists of one feedback channel, which is allocated after the HARQ feedback regions.

In TDD systems, when the legacy mobile stations are supported in the uplink using a legacy UL-PUSC subchannelization scheme, the new and legacy uplink regions are frequency-division multiplexed and a new tile structure is used for the new resource allocations. The uplink tile consists of four consecutive sub-carriers over six OFDM symbols. The partitioning of the transmission bandwidth between the legacy and new systems is dynamic; however, the size of the frequency regions is not frequently changed, since the variation in bandwidth of the legacy uplink control channels requires signaling through the legacy uplink channel descriptor which is sent every second. The sub-carriers of an OFDM symbol are grouped into $N_{g\text{-}left}$ left-guard sub-carriers, $N_{g\text{-}right}$ right-guard sub-carriers, and N_{used} data/pilot sub-carriers. The DC sub-carrier is not used. The N_{used} sub-carriers are divided into multiple UL-PUSC tiles (see Figure 9-30 for the structure of the UL-PUSC tile). The uplink subchannelization procedure can be described as follows, and has been depicted in Figure 9-31:

1. All usable sub-carriers are divided into UL-PUSC tiles over the legacy system uplink bandwidth. As an example, shown in Figure 9-31, in the 10 MHz bandwidth, there are 840 sub-carriers and thereby there are 210 UL-PUSC tiles.
2. The UL-PUSC subchannelization procedure (i.e., physical to logical resource mapping) as defined in reference [1] is performed. Given the previous example, there are 35 UL-PUSC subchannels where each UL-PUSC subchannel consists of six UL-PUSC tiles.
3. The available subchannels in the new frequency zone are identified through subchannel bitmap information broadcast by the AAI_SCD MAC management message.
4. All UL-PUSC tiles of specified subchannels from step 3 are extended in the time-domain from three OFDM symbols to N_{sym} OFDM symbols where $N_{sym} = 6$ or 9, depending on the subframe type. The time extension is done by concatenating two or three contiguous UL-PUSC tiles to form a new UL-PUSC tile.
5. Based on the number of subchannels obtained in step 3 with the symbol-extended tiles of step 4, the DRUs for the new subchannels are formed.
6. Steps 4 and 5 are repeated for the remaining OFDM symbols of each uplink subframe. Note that the tile permutation is performed over the extent of an uplink subframe.
7. Renumber the distributed LRU index in reverse order of UL-PUSC subchannel index. The tile indexing is the same as the legacy indexing method.

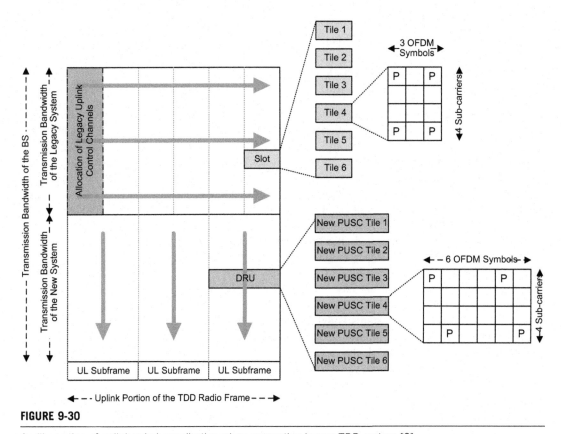

FIGURE 9-30

An illustration of uplink subchannelization when supporting legacy TDD system [2]

9.9 PILOT STRUCTURE AND CHANNEL ESTIMATION

The transmission of pilot sub-carriers in the downlink and uplink is required to allow channel estimation, measurements of channel quality indicators such as the SINR, frequency offset estimation, etc. To optimize the system performance in different propagation environments and applications, IEEE 802.16m supports both common (to all or group of users) and dedicated (to a single user) pilot structures. The common pilot structures can be used by all users and the pilots can be precoded (using a non-adaptive fixed precoding) similar to the data sub-carriers within the same resource block. The dedicated pilots can be used with both localized and distributed allocation, and are associated with a specific resource allocation which can be used by the users that are assigned to the specific resource block and therefore are precoded or beam-formed (adaptive precoding) in the same way as the data sub-carriers of the resource block. The pilot structure is defined for up to eight streams in the downlink and up to four streams in the uplink, using a unified structure for common and dedicated pilots. The pilot density scales with the number of transmit streams; however, there is not necessarily equal pilot density per OFDM symbol on the downlink/

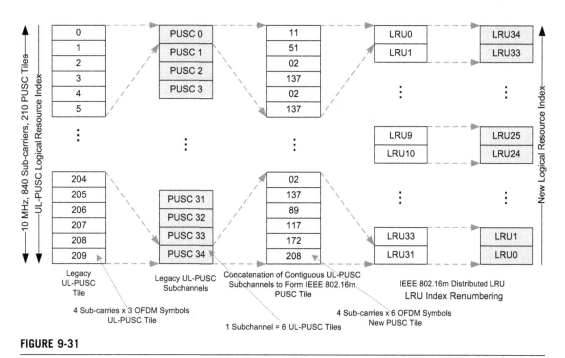

FIGURE 9-31

An illustration of an FDM-based PUSC subchannelization procedure [2].

uplink subframes. Furthermore, within the same subframe, there is an equal number of pilots for each PRU of a data burst assigned to an MS.

9.9.1 Pilot Structure Design Criteria

In this section we briefly discuss the theoretical criteria for efficient pilot pattern design for OFDM systems. Pilot-based channel estimation is to use the channel samples estimated at the pilot tones to reconstruct channel samples at the remaining data tones. As a result, the pilot pattern design is essentially a conventional sampling rate selection problem in two-dimensional signal processing. In order to avoid aliasing during reconstruction of the channel time-frequency function, the pilot tone selection should follow the two-dimensional sampling theorem. When multiple antennas are used, the receiver must estimate the channel impulse response (or the transfer function) from each of the transmit antennas in order to correctly detect the signal. This is achieved through distributing reference signals (or pilot tones) or MIMO midamble among the transmit antennas.

Let $1/T_u$ and T_s denote the sub-carrier spacing and the OFDM symbol duration (inclusive of the guard interval), respectively. Let's further assume that the pilot sub-carriers are transmitted at integer multiples of sub-carrier spacing and OFDM symbol duration in frequency and time directions, respectively (i.e., $f_p = m/T_u$ and $T_p = nT_s$ where m and n are integers). The (m, n) pair represent the pilots' separation in terms of sub-carrier spacing and OFDM symbol duration. From the sampling theorem point of view, the channel's two-dimensional delay-Doppler response $h(\tau, \nu)$ can be fully reconstructed, if the two-dimensional transform function $H(t, f)$ is sampled at greater than or equal to the Nyquist rate across time and frequency dimensions. Hence, the time-domain sampling rate must be greater than or equal to the channel's maximum Doppler spread, i.e., $T_p \leq 1/\nu_{max}$ (the sampling rate in time must be less than the coherence time) and the frequency-domain sampling rate must be greater than or equal to the channel's maximum delay spread, i.e., $f_p \leq 1/\tau_{max}$ (the sampling rate in frequency must be less than the coherence bandwidth). Assuming a wide-sense stationary uncorrelated scattering channel model and assuming the channel to be constant over one OFDM symbol, the frequency response $H(t, f)$ of the L-path channel is given as [46–48]:

$$H(t,f) = \frac{1}{\sqrt{L}} \sum_{l=1}^{L} e^{j(\psi_l + 2\pi\nu_l t - 2\pi f \tau_l)} \qquad (9\text{-}18)$$

Where ψ_1, ν_1, and τ_1 denote the phase, Doppler frequency, and delay of the lth path, respectively. All of these parameters are independent random variables. In general, the pilot signals are over-sampled in order to ensure a good trade-off between performance and overhead. Therefore, the choice of (m, n) depends on the channel's maximum delay spread and maximum Doppler spread and must satisfy the following equation according to the two-dimensional sampling theorem [25]:

$$n \leq \frac{1}{2T_s\nu_{max}}$$
$$m \leq \frac{T_u}{2\tau_{max}} \qquad (9\text{-}19)$$

It should be noted that the pilot density for a regular pattern can be calculated using Equation (9-19). From Equation (9-19), it can be seen that for large values of ν_{max} (i.e., large Doppler spread or alternatively small channel coherence time means that channel is time-varying), n should be small in order to appropriately track channel time variations. On the other hand, for large values of τ_{max} (i.e., large delay spread or alternatively small coherence bandwidth means that channel is frequency-selective), m should be small in order to follow channel frequency variation closely. In a regularly-spaced pilot pattern, the pilot symbols are evenly spaced in frequency and in time.

Using the default OFDM symbol duration and the sub-carrier spacing given in Table 9-2 (OFDMA parameters $T_s = 102.86$ μs and $1/T_u = 10.94$ kHz), and assuming that the channel maximum delay spread $\tau_{max} = 5$ μs for the radio channels under consideration in the IEEE 802.16m evaluation methodology document, the most appropriate range for the values of (m,n) can be determined. Since the IEEE 802.16m system is required to support vehicle speeds up to 350 km/h, the maximum Doppler spread is estimated as $\nu_{max} \approx 810$ Hz at 2.5 GHz RF carrier frequency. From Equation (9-19) and the above design parameters, the following pilot spacing constraints are obtained $n \leq 6$, $m \leq 9$. Figure 9-32 illustrates a regularly-spaced two-dimensional pilot pattern for one spatial stream in the time frequency-domain where $n = 4$ and $m = 5$. The above criteria can be generalized to the case of multiple transmit antennas where each transmit antenna (alternatively known as antenna port) is associated with a group of pilot sub-carriers.

In the case of irregular pilot patterns, the pilot symbols can be irregularly placed in time, in frequency, or in both. The irregular patterns can be chosen according to certain criteria, each yielding

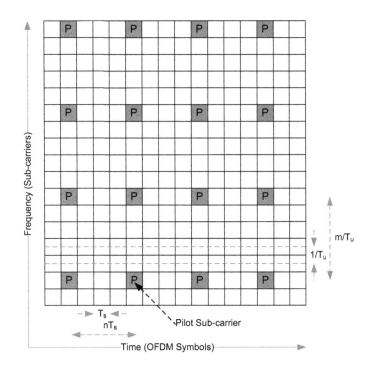

FIGURE 9-32

An example of a regular pilot pattern for one spatial stream with $n = 4$ and $m = 5$

a specific pattern. Other important considerations in the design of pilot structures for support of multi-antenna OFDMA systems include the following:

- Identical pilot pattern for each physical resource block: pilot spacing in time and frequency must be in proportion to the resource block size. If the location of the pilot tones within a resource block is not maintained the same across all resource blocks within the system bandwidth, or alternatively the pilot sub-carriers have different positions within each resource block, the filtering and interpolation operations during channel estimation become excessively complex.
- Pilot density: proper pilot overhead must be considered as a trade-off between accurate channel estimation under various mobility conditions and higher throughput.
- Types of pilots: pilots are typically classified as common and dedicated. Each MS in the cell can estimate the channel over the entire bandwidth using the common pilots, while dedicated pilots in the MS-assigned resource block can be used for channel estimation over a fraction of the total bandwidth. In conventional MIMO schemes, the dedicated pilots are typically adaptively precoded. The dedicated pilots can be used in conjunction with common pilots or replace the common pilots.
- Pilot power boosting: in general, the power of pilot sub-carriers is boosted relative to data sub-carriers in order to enhance the channel estimation with pilot hopping or shifting. To avoid cross-symbol power fluctuation when using pilot boosting, it is desirable to place the pilot sub-carriers regularly on every symbol or to employ power adjustment for data sub-carriers over a symbol.

- Per antenna power balance: when using multiple transmit antennas, it is important to maintain balanced power distribution across antennas. For this purpose, if there are two transmit antennas, each OFDM symbol should contain the same number of pilot tones associated with different antenna ports.

As an example and by taking the above additional criteria into consideration, the structure and link-level performance (i.e., bit error rate versus SNR curves) of several hypothetical common pilot patterns for two transmit antennas and two different multi-antenna transmission schemes are shown in Figure 9-33. In this example, the pilot overhead of type A and type B patterns is 11.11% (i.e., 5.5% per antenna port) and the transmit power per antenna is balanced. The type C pilot pattern has unbalanced power across antennas and symbols. The pilot tone powers have been boosted by 3 dB relative to data sub-carriers.

The accuracy of channel estimation using common pilots affects the performance of downlink/uplink data and control channel decoding. Figure 9-33 shows the link-level performance of spatial multiplexing with the example pilot patterns, where type B provides the best performance under low and high mobility conditions. The ideal channel estimation is used as a reference. Also shown in Figure 9-33 is the link-level performance of space-frequency block codes with the example pilot patterns. In this case, type A outperforms the other structures. It can be concluded that type A is the best among the three example patterns. Note that the example pilot patterns satisfy the sampling theorem constraint of Equation (9-19). This example underlines the criteria for the design of robust and high performance pilot structures for OFDMA cellular system operation in a wide range of mobility and channel conditions. The time/frequency characteristics of a number of typical radio channel models, mobility classes, and the associated time-frequency sampling rate requirements are summarized in Table 9-5.

9.9.2 Downlink Pilot Structure

The IEEE 802.16m downlink pilot patterns have been designed based on the principles that were described in the previous section. The pilot patterns are specified within a physical resource unit. The downlink pilot patterns used for data transmission for one and two spatial streams (dedicated and common types) are shown in Figure 9-34 with the sub-carrier index increasing from top to bottom and the OFDM symbol index increasing from left to right. In Figure 9-34, the pilot tones are marked with Pi where index i denotes the antenna port number. Note that there is at least one pilot tone on every OFDM symbol. The pilot density for one and two data streams is 5.5% and 11.11%, respectively. It must be further noted that the pilot structures in Figure 9-34 are distinguished based on the number of spatial streams and the number of transmit antenna ports. When a single data stream is transmitted from two antennas, the pilot tones for each antenna port are located in non-overlapping positions in the physical resource unit and the tones corresponding to pilots of the other antenna port are used for data transmission; therefore, the pilot density per antenna port remains the same and equal to 5.5%. However, if dual data streams are transmitted from two antennas, the tones corresponding to pilots of the other antenna port are blank and not used for data transmission, therefore reducing the interference of data sub-carriers with pilot tones of the other antenna port. The null sub-carriers are marked with "X" in Figure 9-34. The pilot patterns for four and eight data streams are shown in Figure 9-35. The pilot pattern of the type-3 subframes (five OFDM symbols) is obtained by deleting the last OFDM symbol of the type-1 subframe (six OFDM symbols). The pilot pattern of the type-2 subframes (seven OFDM symbols) is obtained by repeating the first OFDM symbol of the type-1 subframe at the end of the subframe.

A Collision-Free Interlaced Pilot (CoFIP) pattern is specifically designed for use in the type-1 open-loop MIMO region (see Section 10.5) to reduce inter-cell interference. Figure 9-36 shows the CoFIP

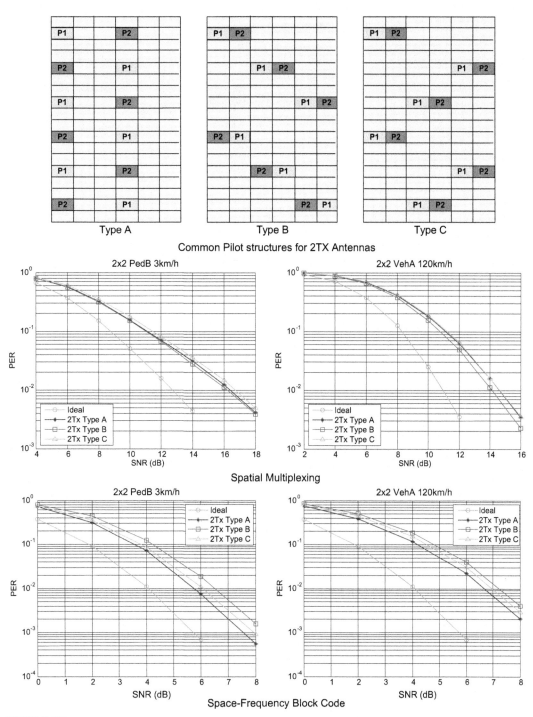

FIGURE 9-33

Channel estimation accuracy of various pilot patterns for two transmit antennas and QPSK 1/2 in different mobility and channel conditions [123]

Table 9-5 Characteristics of Typical Radio Channel Models

Channel Model	RMS Delay Spread (ns) ν_{RMS}	Maximum Delay Spread (µs) τ_{max}	Maximum Sampling Rate in Frequency (Number of Sub-carriers) $\dfrac{T_u}{2\tau_{max}}$
ITU-T Pedestrian A (PedA)	45	0.41	112
ITU-T Pedestrian B (PedB)	750	3.7	13
ITU-T Vehicular A (VehA)	370.4	2.51	19
ITU-T Vehicular B (VehB)	4000	20	3
Typical Urban 6-Ray Model (TU)	1000	5	10

Vehicular Speed Range (km/h)	Maximum Doppler Spread (Hz) ν_{max}	Coherence Time (ms) T_c	Maximum Sampling Rate in Time (Number of OFDM Symbols) $\dfrac{1}{2T_s\nu_{max}}$
0–15	~ 27.78	15.23	175
15–120	~ 222.22	1.90	22
120–350	~ 648.15	0.65	8

pilot pattern for subframes consisting of six OFDM symbols. The index of the pilot pattern set used by a particular BS with $IDcell = k$ is denoted by $P_k = (k \bmod 3)$. The index of the pilot pattern set is determined by the $IDcell$. For subframes consisting of seven OFDM symbols, the first OFDM symbol which contains pilot sub-carriers and null sub-carriers in each pilot pattern set is inserted as the seventh symbol. Note that in order to reduce the inter-cell interference due to overlapping data and pilot sub-carriers, the pilot sub-carriers used by another base station/sector are made null and not used. The difference of this pilot pattern with the cyclically-shifted interlaced pilot pattern is that in the latter the overlapping locations of pilot sub-carriers used by other base stations/cells can be used for data transmission.

To overcome the effects of pilot interference among the neighboring sectors or base stations, an interlaced pilot structure is utilized in the downlink by cyclically shifting the base pilot pattern such that the pilots of neighboring cells do not overlap. The base pilot patterns used for two downlink data streams are shown in Figure 9-37 with the sub-carrier index increasing from top to bottom and the OFDM symbol index increasing from left to right. The index of each pilot sub-carrier indicates the corresponding pilot stream (or the antenna port). The interlaced pilot patterns are used by different BSs for one and two data streams. Each BS selects one of the three pilot patterns. The index of the pilot pattern used by a particular BS with cell identifier $IDcell$ is denoted by $P_k = \lfloor IDcell/256 \rfloor$. For a single data stream, each BS has the choice of selecting one of the two possible locations for the pilot tone, depending whether the value of its cell identifier is even or odd (i.e., $IDcell \bmod 2$).

For three-stream MIMO transmissions, the first three of the four pilot streams will be used and the unused pilot tones are allocated for data transmission. Once again, the pilot pattern of the type-3 subframe is obtained by deleting the third OFDM symbol of the type-1 subframe. The pilot pattern of the type-2 subframe is obtained by adding the third OFDM symbol of the type-1 subframe to the end of the type-1

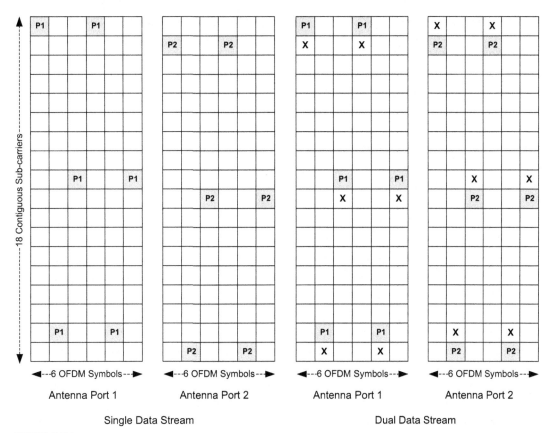

FIGURE 9-34

Pilot structures for single and dual spatial streams [2]

subframe. The pilot patterns for eight downlink data streams are shown in Figure 9-35 with the sub-carrier index increasing from top to bottom and the OFDM symbol index increasing from left to right.

The demodulation pilots or reference signals in a downlink PRU for a given antenna port are precoded in the same way that data tones on the same stream in that PRU are precoded. In a distributed LRU, the data transmitted in a PRU on a given stream may be sent to several mobile stations, but in different tones using the same precoder. Two pilot streams are transmitted in the distributed LRUs inside or outside the open-loop region[iii] regardless of what data is transmitted by the BS in all distributed LRUs. The precoder may be adaptive (user-specific) or non-adaptive (non-user-specific) depending on

---

[iii]An open-loop region with $k \leq \min(N_t, N_r)$ MIMO streams is defined as a time-frequency resource using the k stream's pilot pattern and a given open-loop MIMO mode without rank adaptation. The open-loop region allows base stations to coordinate their open-loop MIMO transmissions in order to create a stable interference environment where the precoders and number of MIMO streams are not time-varying. The LRUs used for the open-loop region are identified in AAI_SCD MAC control message. These LRUs are aligned across cells. A limited set of open-loop MIMO modes are allowed for transmission in the open-loop region.

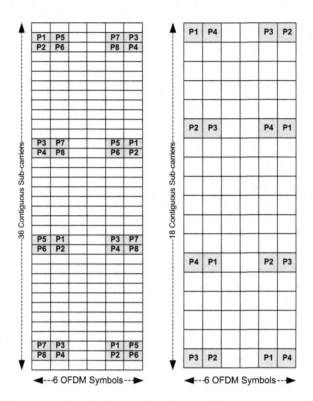

FIGURE 9-35

Pilot structure for 4 and 8 spatial streams [2]

the downlink MIMO mode. Non-adaptive precoders are determined according to the MIMO mode, the number of data streams, the type of LRU, operation inside or outside the open-loop region, and the physical index of the sub-band or mini-band where the precoder is applied. In multi-user MIMO transmissions using a contiguous LRU, each pilot stream is dedicated to one MS. The MS uses its dedicated pilot stream for channel estimation within the allocated resources. Other pilot streams may be used for inter-stream interference estimation. The total number of streams in the transmission and the index of the dedicated pilot stream are indicated in the DL Basic Assignment A-MAP IE, DL Persistent Allocation A-MAP IE, or DL Sub-band Assignment A-MAP IE. The channel estimation for demodulation of data burst at an MS is performed by taking the following rules into consideration:

- The dual stream non-adaptively precoded common pilots across the distributed LRU should be used for channel estimation by all mobile stations allocated a burst in the distributed LRU. Within each frequency partition, all pilots are shared by all mobile stations for demodulation in the distributed LRU. Only the pilots located within a physical sub-band should be used for channel estimation within that sub-band.
- The MS should use its dedicated pilot streams for channel estimation in the contiguous allocation. The pilots are not shared by mobile stations for demodulation in contiguous LRU whether they are non-adaptively or adaptively precoded.

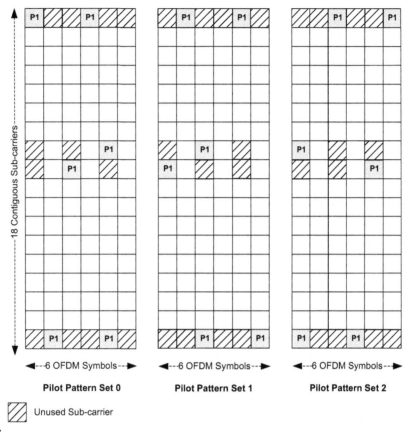

Pilot Pattern Set 0   Pilot Pattern Set 1   Pilot Pattern Set 2

Unused Sub-carrier

FIGURE 9-36

Collision-free interlaced pilot pattern for subframe type 1 [2]

MIMO feedback measurements make use of non-precoded and/or precoded pilots at the MS. For MIMO feedback reports requested with a MIMO feedback mode for operation in an open-loop region, measurements are conducted on the $k \leq \min(N_t, N_r)$ streams non-adaptively precoded pilots in that open-loop region. All pilots are shared by all mobile stations for MIMO feedback measurements in each open-loop region. The wideband Channel Quality Indicator (CQI)[iv] reports

---

[iv]Channel Quality Indicator (CQI) is a measure of quality of wireless channels as experienced by a mobile station. The CQI metric can be expressed as a value (or values) representing a measure of channel quality for a given channel. Typically, a high value CQI is indicative of a channel with high quality. The CQI for a communication link can be computed by making use of performance metric such as a signal-to-noise ratio, signal-to-interference plus noise ratio, etc. These values and others can be measured for a given channel and then used to compute a CQI for the channel. The CQI for a given channel is dependent on the physical transmission scheme used by the communication system. In communication systems which make use of multiple-input multiple-output and space-time coding schemes, the CQI can also be dependent on receiver type, MIMO mode, etc. Other factors that may be taken into account in CQI are performance impairments, such as Doppler shift, channel estimation error, interference, etc.

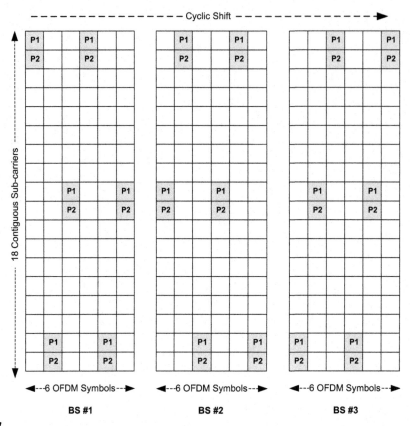

FIGURE 9-37

Interlaced pilots structures for two data streams [2]

inside an open-loop region are averaged over the open-loop region pilots. For MIMO feedback reports requested with a MIMO feedback mode for operation outside the open-loop region, measurements are conducted on the downlink MIMO midamble. The wideband CQI reports outside the open-loop region (i.e., measurements on MIMO midamble) are averaged over the frequency partition indicated by the Frequency Partition Indicator (*FPI*) parameter in Feedback Allocation A-MAP IE. For reports requested with a MIMO feedback mode for open-loop MIMO operation, the MS should adjust the non-precoded MIMO channel estimated from the midamble by applying it with the non-adaptive precoder according to the MIMO, the sub-band index, and the MIMO rate. For reports requested with a MIMO feedback mode for closed-loop operation, the MS should adjust the non-precoded MIMO channel estimated from the midamble with an estimated adaptive precoder. The sub-band CQI reports inside and outside the open-loop region should be reported for sub-bands in sub-band LRUs indicated by the secondary superframe header.

9.9.3 **Uplink Pilot Structure**

Uplink pilots within a logical resource unit are dedicated to each user and can be precoded or beam-formed similar to data sub-carriers of that logical resource unit. The uplink pilot structure is defined for up to four data streams. The pilot pattern may support variable pilot boosting factors. When pilots are boosted, each data sub-carrier has the same transmission power across all OFDM symbols in a resource block. Figure 9-38 shows the pilot structure for distributed logical resource units for single and dual data streams. Note that the pilot patterns for UL contiguous LRUs are the same as those in the downlink. For three data stream MIMO transmissions, the first three of the four pilot streams will be used and the unused pilot stream is allocated for data transmission. The pilot pattern of type-4 subframes (i.e., uplink subframes with 9 OFDM symbols that are used when supporting legacy 8.75 MHz bandwidth) is derived from the type-2 subframe pattern. The first seven symbols of the type-4 subframe pilot patterns are identical to the type-2 subframe patterns. The last two symbols of the type-4 subframe pilot patterns are generated by appending the first two symbols of the type-2 subframe pilot patterns.

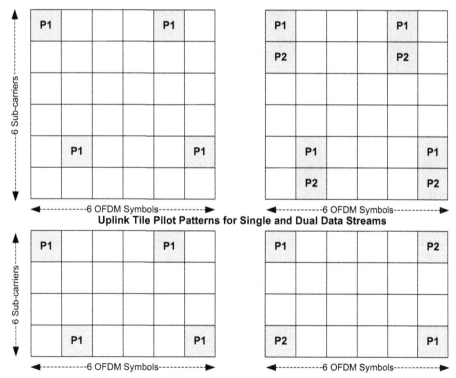

Uplink Tile Pilot Patterns for Single and Dual Data Streams

Uplink PUSC Tile Pilot Patterns for Single and Dual Data Streams

FIGURE 9-38

Uplink Pilot Patterns for Basic Resource Units [2]

9.10 MIMO MIDAMBLE

The pilot structures that have been discussed thus far were all narrowband type, i.e., the reference tones only provide channel information for a fraction of the transmission bandwidth due to the fact that the pilot sub-carriers are contained in physical or logical resource blocks. Some closed-loop MIMO modes require channel information over the entire transmission bandwidth. This information can only be provided through non-precoded periodic wideband signals. The MIMO midamble is used for pre-coding matrix index selection in closed-loop MIMO. For open-loop MIMO, the midamble can be used to calculate the channel quality indicator. The MIMO midamble is transmitted in every radio frame in the second downlink subframe. The midamble signal occupies the first OFDM symbol in a DL type-1 or type-2 subframe. For the type-1 subframe, the remaining five consecutive OFDM symbols form a type-3 subframe. For a type-2 subframe, the remaining six consecutive OFDM symbols form a type-1 subframe. The MIMO midamble signal transmitted by the BS antenna is described as follows:

$$s(t) = \text{Re}\left[e^{j2\pi f_c t} \sum_{k=0}^{N_{used}-1} \alpha_k e^{j2\pi\left(k - \frac{N_{used}-1}{2}\right)\left(\frac{t-T_g}{T_u}\right)} \right] \quad k \neq \frac{N_{used}-1}{2} \tag{9-20}$$

Where the α_k coefficients modulate sub-carriers in the midamble symbol and are defined as follows:

$$\alpha_k = \begin{cases} 2.18\{1 - 2G([k + u + D(N_{FFT})]\bmod N_{FFT})\} & k \neq \frac{N_{used}-1}{2}, \ (k-s)\bmod 3N_t \\ & = 3g + \left(\left\lfloor \frac{IDcell}{256} \right\rfloor + \left\lfloor \frac{k-s}{N_1 - N_{sc}} \right\rfloor\right)\bmod 3) \\ 0 & \text{otherwise} \end{cases} \tag{9-21}$$

where k is the sub-carrier index $0 \leq k \leq N_{used} - 1$, N_{used} is the number of used sub-carriers, N_t is a number of transmit antennas, $G(x)$ is the Golay sequence[v] ($0 \leq x \leq 2047$), N_{FFT} is the FFT size, u is a shift value ($0 \leq u \leq 127$) $u = IDcell \bmod 256$, $D(N_{FFT})$ is an FFT size-specific offset, g is the BS transmit antenna index ranging from 0 to $N_t - 1$, N_t is the number of BS transmit antennas, parameter $s = 0$ for $k \leq (N_{used} - 1)/2$ and $s = 1$ for $k > (N_{used} - 1)/2$, and n is the frame number [2]. The *IDcell*

---

[v]Golay sequences are pair of sequences with complementary autocorrelation functions. They were introduced as a pair of complementary binary sequences, which have the property that the sum of their autocorrelation functions equals zero for all time shifts, except zero. Mathematically complementary sequences $a(n)$ and $b(n)$ of length N can be defined by $R_a(k) + R_b(k) = 0 \forall k \neq 0$ and $R_a(k) + R_b(k) = 2N, k = 0$ where $R_a(k)$ and $R_b(k)$ are the autocorrelation functions of a (n) and $b(n)$ sequences. Binary complementary sequences are typically of length 2^N although some even lengths that can be expressed as a sum of two squares were introduced. Several recursive and non-recursive methods for generating complementary sequences have been proposed. The recursive methods are based on the following algorithm: $a_n = [a_{n-1}, b_{n-1}]$ and $b_n = [a_{n-1}, -b_{n-1}]$ where the operator [ ] denotes concatenation of sequences, and a_n and b_n represent a complementary pair of length 2^n. The number of different Golay sequences is large and increases rapidly with N, the sequence length. The number of different sequences of length 2^N is $N! \ 2^N/2$. Golay sequences have very good aperiodic autocorrelation properties. Golay sequences have the property that their spectral peaks are no more than 3 dB larger than the average of the spectrum [42].

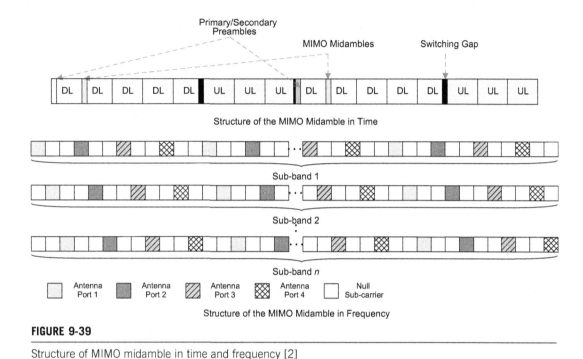

FIGURE 9-39

Structure of MIMO midamble in time and frequency [2]

is derived from the secondary preamble. The physical structure of MIMO midamble is shown in Figure 9-39 for four transmit antennas and *IDcell* = 0.

9.11 PILOT-BASED CHANNEL ESTIMATION

While perfect knowledge of the radio channel can be used to find an upper bound for system performance, such knowledge is not available in practice, and the channel needs to be frequently estimated. Channel estimation can be performed in various ways using frequency and/or time correlation properties of the wireless channel, blind or pilot (reference signal) based, adaptive or non-adaptive, etc. Nonparametric methods attempt to estimate the frequency response without relying on a specific channel model. In contrast, the parametric estimation methods assume a certain channel model and determine the parameters of this model. Spaced-time and spaced-frequency correlation functions, discussed earlier, are specific properties of the channel that can be incorporated in the estimation method, improving the quality of estimations. Pilot-based estimation methods are the most commonly used methods in OFDM systems, and are applicable in systems where the sender transmits some known signal. Blind estimation, on the other hand, relies on some properties of the signal and is rarely used in practical OFDM systems. Adaptive channel estimation methods are typically used for a rapidly time-varying channel.

In this section, we describe a generic pilot-based non-adaptive channel estimation method. In a linear time-invariant AWGN channel, the relationship between transmitted signal X_k, $k = 0, 1, \ldots, N_{FFT} - 1$ and received signal Y_k can be expressed as $Y_k = H_k X_k + Z_k$ where Z_k is the frequency-domain noise sampled at the kth sub-carrier frequency and H_k is the channel transfer-function sampled at the kth sub-carrier frequency. To estimate the channel, pilot symbols are used. Let's assume that every pth sub-carrier contains known pilot symbols X_{pk}, $kp = 0, p, 2p, \ldots, N_p - 1$ where N_p is the total number of pilot sub-carriers across frequency. Using the known pilot symbols X_{pk} and the received symbols Y_{pk} sampled at pilot sub-carriers, the channel transfer function $\widehat{H}_k$ sampled at pilot sub-carriers is:

$$\widehat{H}_{pk} = \frac{Y_{pk}}{X_{pk}} + \frac{Z_{pk}}{X_{pk}} = H_{pk} + W_{pk} \tag{9-22}$$

where W_{pk} is a scaled-noise component at the pkth sub-carrier. Different methods, such as one- or two-dimensional linear interpolation can be used to estimate the channel samples at other sub-carrier frequencies.

Channel estimation in OFDM systems is a two-dimensional (2-D) problem, i.e., the channel transfer function or channel impulse response is a function of time and frequency. Due to the computational complexity of 2-D estimators, one-dimensional (1-D) channel estimation methods are more practical. The idea behind 1-D estimators is to estimate the channel in one dimension (e.g., frequency) and then estimate the channel in another dimension (e.g., time), thus obtaining a 2-D channel estimate. Linear interpolation is a simple method using channel sample estimates at pilot sub-carriers. This is done by linearly interpolating the channel samples at the two nearest pilot sub-carriers. Although linear interpolation provides some limited noise reduction of the channel estimates at data locations (due to averaging function), it is the simplicity of the solution that is more attractive. It must be noted that averaging window length (i.e., the number of pilots contained in the averaging window) is inversely proportional to the coherence bandwidth of the channel. The one-dimensional linear interpolation method estimates the channel by interpolating the channel transfer function between $\widehat{H}_{m,k}$ and $\widehat{H}_{m,l}$ (interpolation across frequency) or between $\widehat{H}_{m,k}$ and $\widehat{H}_{n,k}$ (interpolation across time).

The channel transfer function $H_{m,k}$ at time frequency index (m,k) can be modeled as a linear weighed sum of 2-D basic functions evaluated at the kth sub-carrier (frequency index) and at the mth OFDM symbol (time index).

$$H_{m,k} = \sum_{n=0}^{N-1} \alpha_n \phi_n(m, k) \tag{9-23}$$

Where $\phi_n(m, k)$ denotes the nth basis function sampled at mth OFDM symbol and at the kth sub-carrier, α_n is the coefficient of the nth basis function, and N is number of basis functions used in the linear model. By taking a one-dimensional approach, the above two-dimensional problem is reduced to one-dimensional if one of the time or frequency indices is fixed.

$$H_k = \sum_{n=0}^{N-1} \alpha_n \phi_n(k) \tag{9-24}$$

The channel samples at pilot sub-carriers shown in Equation (9-22) can be written as:

$$\widehat{H}_{pk} = \sum_{n=0}^{N-1} \alpha_n \phi_n(pk) + W_{pk} \tag{9-25}$$

The above equation can be represented in matrix form as:

$$
\begin{bmatrix}
\widehat{H}_0 \\
\widehat{H}_p \\
\widehat{H}_{2p} \\
\vdots \\
\widehat{H}_{N_p-1}
\end{bmatrix}
=
\begin{bmatrix}
\phi_0(0) & \phi_1(0) & \phi_2(0) & \cdots & \phi_{N-1}(0) \\
\phi_0(p) & \phi_1(p) & \phi_2(p) & \cdots & \phi_{N-1}(p) \\
\phi_0(2p) & \phi_1(2p) & \phi_2(2p) & \cdots & \phi_{N-1}(2p) \\
\vdots & \vdots & \vdots & \ddots & \vdots \\
\phi_0(N_p-1) & \phi_1(N_p-1) & \phi_2(N_{p-1}) & \cdots & \phi_{N-1}(N_{p-1})
\end{bmatrix}
\begin{bmatrix}
\alpha_0 \\
\alpha_1 \\
\alpha 2 \\
\vdots \\
\alpha_{N-1}
\end{bmatrix}
+
\begin{bmatrix}
z_0 \\
z_p \\
z_{2p} \\
\vdots \\
z_{N_p-1}
\end{bmatrix}
$$

$$(9\text{-}26)$$

Alternatively using matrix notations the above equation can be rewritten as:

$$\widehat{\mathbf{H}} = \mathbf{\Phi}\boldsymbol{\alpha} + \mathbf{z} \tag{9-26a}$$

The Least Square (LS) estimate of the coefficients are calculated by minimizing the squared distance between the actual channel vector H and the estimated channel vector $\widehat{\mathbf{H}}$ as:

$$\widehat{\alpha}_{LS} = (\mathbf{\Phi}^H\mathbf{\Phi})^{-1}\mathbf{\Phi}^H\widehat{\mathbf{H}} = \mathbf{\Phi}^{-1}\widehat{\mathbf{H}} \tag{9-27}$$

Where superscript ^{H} denotes conjugate transpose operation, hence:

$$\widehat{H}_k = \sum_{n=0}^{N-1} \widehat{\alpha}_n\phi_n(k) \tag{9-28}$$

A number of different options exist while selecting the basic functions such as orthogonal polynomials, Fourier series, discrete cosine transform function, discrete sine transform series, etc. There are a number of reasons for using orthogonal polynomials, including ease of calculation of $\mathbf{\Phi}^H\mathbf{\Phi}$ since it becomes a diagonal matrix if matrix $\mathbf{\Phi}$ is orthogonal, resulting in ease of calculation of $\widehat{\alpha}_n's$ and the inverse matrix $(\mathbf{\Phi}^H\mathbf{\Phi})^{-1}$ with minimal computational complexity. The degree of orthogonal polynomials can be increased without changes in recursive calculation of $\widehat{\alpha}_n's$. When applying the polynomial-based channel estimation, a sliding window is typically used to estimate the channel. This is done to avoid the use of high-order polynomials to estimate highly frequency-selective channels. The sliding window approach allows for local approximations to the channel transfer function using a window length larger than the polynomial order, providing a better estimate.

In a special case where the OFDM channel estimation symbols are transmitted periodically and all sub-carriers over an OFDM symbol are used as pilots, the channel estimation algorithm is more simplified. In this case, the receiver uses the estimated channel transfer function to decode the received data within the block until the next pilot symbol arrives. If $Y_k = H_kX_k + Z_k$, $k = 0, 1, 2, ..., N_{FFT} - 1$ denotes the received signal at kth sub-carrier (pilot sub-carrier), the LS estimate of the channel transfer function in this case is given as

$$\widehat{\mathbf{H}}_{LS} = (Y_1/X_1, Y_2/X_2, ..., Y_{N_{FFT}-1}/X_{N_{FFT}-1})^H = (Y_1, Y_2, ..., Y_{N_{FFT}-1})^H; X_k = 1, \forall k \tag{9-29}$$

The channel transfer function at non-pilot sub-carriers can be estimated using interpolation in time. For convenience we assume that the pilot sub-carriers have unity magnitude.

An alternative approach to pilot-based channel estimation is Minimum Mean Squared Error (MMSE) method. The MMSE channel estimator exploits the second-order statistics of the channel

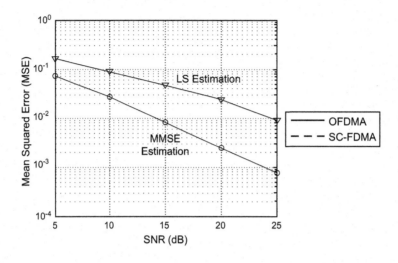

FIGURE 9-40

An example comparison of the LS and MMSE channel estimation algorithms

transfer function to minimize the mean-square error. Let $\mathbf{R_{HH}} = E(\mathbf{HH}^H)$, and σ_Z^2 denote the auto-correlation matrix of the channel vector $\mathbf{H}$ and the AWGN noise variance, respectively. The MMSE estimate of the channel transfer function is given as:

$$\widehat{\mathbf{H}}_{MMSE} = \mathbf{R_{HH}}[\mathbf{R_{HH}} + \sigma_Z^2(\mathbf{XX}^H)^{-1}]^{-1}\ \widehat{\mathbf{H}}_{LS} = \mathbf{R_{HH}}[\mathbf{R_{HH}} + \sigma_Z^2\mathbf{I}]^{-1}\ \widehat{\mathbf{H}}_{LS}; \qquad \mathbf{XX}^H = \mathbf{I} \tag{9-30}$$

The MMSE estimator performs much better than LS estimators, especially under low SNR conditions. A major drawback of the MMSE estimator is its high computational complexity, particularly if matrix inversion is necessary every time the input data changes. Figure 9-40 shows an example where the performance of the LS and MMSE channel estimation algorithms has been compared. In this example $N_{FFT} = 64$, and the channel is modeled as a two-tap delay line filter. The superiority of MMSE over LS estimation in low SNR conditions can be seen in Figure 9-40.

9.12 CHANNEL CODING AND MODULATION

Forward-Error-Correction (FEC) channel coding schemes are commonly used to improve the efficiency and robustness of wireless digital communication systems. On the transmitter side, an FEC encoder adds redundancy to the input data in the form of parity bits, while at the receiver, the redundancy is utilized by the FEC decoder to correct a number of channel errors. The use of FEC schemes would allow tolerance of more channel errors that otherwise can be tolerated without forward error correction. The use of FEC schemes enables wireless communication systems to operate with a lower transmit-power, to transmit over longer distances, to tolerate more interference, and to transmit at a higher data rate. A binary FEC encoder receives k bits at a time and produces a codeword of n bits, where $n > k$. While there are 2^n possible sequences of n bits, only 2^k codewords are used. The ratio k/n

is the code rate and is denoted by R. Lower code rates corresponding to small values of R can generally correct more channel errors than higher code rates, and therefore are more robust. However, higher code rates are more bandwidth efficient because of lower overhead due to parity bits. Thus, the selection of the code rate is a trade-off between energy efficiency and bandwidth efficiency. For every combination of code rate R, code word length n, baseband modulation scheme, channel type, and receiver noise power, there is a theoretical lower limit on the amount of energy that must be used to convey one bit of information. This limit is called the channel capacity of the communication channel [88]. Since the dawn of information theory, engineers and mathematicians have tried to construct codes that achieve performance close to Shannon capacity [89]. The introduction of Convolutional Turbo Codes (CTC) in 1993 followed by the invention of Block Turbo Codes (BTC) in 1994 closed much of the performance gap with maximum channel capacity. Turbo codes are essentially parallel concatenated convolutional codes with an internal interleaving mechanism combined with an iterative soft-decision decoding algorithm. In this section, we first review the principles of turbo coding and then we describe the channel coding and modulation schemes that are used in the IEEE 802.16m standard.

9.12.1 **Principles of Turbo Coding**

In information theory, one can ideally approach the Shannon limit as closely as desired using soft decision decoding of a long block code or a convolutional code with a large constraint length. In practice; however, decoding of these codes becomes very computationally intensive. Similar to any conventional error correcting codes, the turbo codes work by imposing a structure on the transmitted bit sequence. If the received bit sequence does not match this known structure, the receiver declares an error has occurred. If the number of errors is sufficiently small and the structure is robust, the receiver can detect the erroneous bits and reconstruct the correct bit sequence. A strong error correcting code has two key characteristics: (1) the encoder imposes a structure that maximizes the difference (or the distance) between any two valid bit sequences; (2) the decoder utilizes all the information available at the receiver's end, including the redundancy and previously unsuccessful transmissions. Thus, the decoder can determine which valid bit sequence is most likely the one that has been transmitted. The first significant difference between a Turbo code and a conventional code is the use of a recursive systematic encoder. A typical convolutional coder uses a non-recursive structure. By feeding one of the outputs back to the input, a recursive encoding structure is obtained. The recursive structure is systematic, i.e., the input bits appear directly as part of the encoded bit-stream. Therefore, a systematic structure enables the combination of two codes in order to construct a stronger composite code. The input bit-stream only needs to be transmitted once. The computed parity bits from each of the two constituent encoders are transmitted separately. The recursive structure interacts with the interleaver to give the composite code some unique performance characteristics. The optimum decoder for any code is a Maximum-Likelihood (ML) decoder. In ML decoding, the receiver produces a probability that each received bit is either 0 or 1. The ML decoder then uses these received bit probabilities, along with its knowledge of the structure imposed by the encoder, to compute the probability of every possible transmitted bit sequence. The most probable transmitted bit sequence is then chosen as the decoded bit sequence. The ML decoding is theoretically optimal and can achieve the lowest probability of decoder failure or probability of error. However, there is no practically efficient way to compute these probabilities when the block size is large. This leads to the use of suboptimal, but realizable, decoding methods. The second unique feature of turbo codes is their decoder, which is based on the iterative application of ML decoding. The ML decoding of a convolutional code would be easy, if the constraint length (register size) is short; however, short codes are relatively weak. Combining two such codes produces a much stronger code, but now ML decoding becomes intractable. Each shorter code can be ML decoded if we assume the other code has already been decoded. This leads to an iterative decoding strategy, which is to decode the first code and then use the resulting updated probabilities to decode the second code. Once this is done, the updated results can more accurately decode the first code. This decoding strategy shares the limitations of all iterative solutions, including the possibility of entrapment in a local minimum. The decoder output is not necessarily the true global ML solution. But the study of the corresponding link-level BER curves suggests that it tends to be a very good solution. In a convolutional code, the input bit sequence usually has tail bits added. This initializes the encoder state to all zeros at the end of the bit sequence. In a turbo coder, this implies that a valid bit sequence cannot contain a single one. It must either be all zeros or it must contain two or more ones. Because encoding is a linear process, this implies that any valid input sequence must differ from any other valid input sequence in at least two bits. If we consider only the first decoder's output, it can be seen that it differs from the correct output in two positions. However, the turbo encoder contains an interleaver. Thus, the two constituent encoders are operating on the same set of bits, but in a different order. Therefore, although the error bits are close together for the first encoder, they are widely separated in the second encoder's input. As a result, the second encoder's output differs from the correct output in many bit positions. It will see the incorrect bit sequence as highly implausible. This is the basic principle of operation of the turbo codes and the key to the high performance of turbo codes at low SNRs. Erroneous information sequences that look reasonable to one decoder are likely to be rejected by the other decoder. This is also the rationale for the BER floor at higher SNRs. Since the interleaver is random, there are a few error patterns that will look plausible to both decoders. As the SNR increases, these weak patterns come to dominate the BER curve.

One of the most interesting characteristics of a turbo code is that it is not just a single code. It is, in fact, a combination of two codes that work together to achieve a synergy that would not be possible by merely using one code by itself. As shown in Figure 9-41, a turbo code is formed from the parallel concatenation of two constituent encoders separated by an interleaver. Each constituent encoder can be any form of FEC coder used for conventional data communication. However, in practice, the constituent encoders are identical convolutional encoders. As can be seen in Figure 9-41, the turbo code consists of two identical constituent encoders. The input data stream and the parity outputs of the two parallel encoders are then serialized into a single turbo code word. The interleaver is a critical component of the turbo code. It is a simple functional block that rearranges the order of the data bits in a prescribed, but irregular, manner. Although the same set of data bits is present at the output of the interleaver, the order of these bits has been changed. It must be noted that the interleaver used by a turbo coder is quite different from the rectangular interleavers that are commonly used in wireless systems to help mitigate the effect of deep fading. While a rectangular channel interleaver tries to space the data out according to a regular pattern, a turbo interleaver randomizes the ordering of the data in an irregular manner.

The operation of a turbo encoder and decoder can be described as follows. Let the binary input bits $\{0,1\}$ be represented by bipolar levels $\{+1,-1\}$ and assigned to the variable d; which may take on the values $d = +1$ and $d = -1$. For an AWGN channel, the conditional probability density functions $f(x \mid d = -1)$ and $f(x \mid d = +1)$ are referred to as likelihood functions. The common hard decision criterion, known as maximum likelihood, selects the symbol $d_k = +1$ or $d_k = -1$ depending on the intercept point of received signal value x_k and the above conditional probability density functions using a fixed

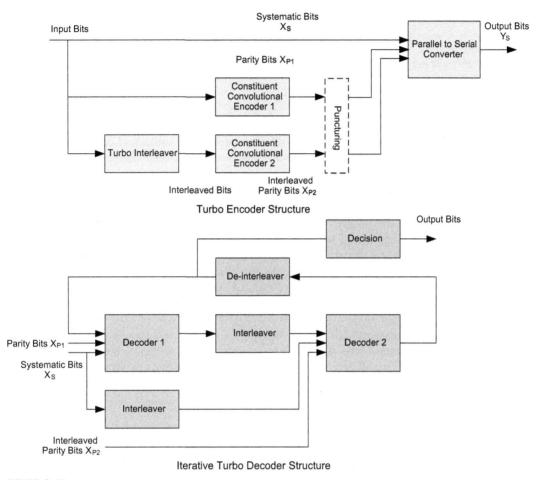

FIGURE 9-41

Generic structure of turbo encoder and iterative decoder

threshold of λ (decision point); thus, $d_k = +1$ if $x_k > \lambda$; otherwise, $d_k = -1$. Another decision rule, known as Maximum a Posteriori (MAP) criterion takes into account the *a posteriori* probabilities $f(d = +1|x)$ and $f(d = -1|x)$ to construct the hypotheses H_1 and H_2 as follows:

$$f(d = +1|x) \underset{H_2}{\overset{H_1}{\gtrless}} f(d = -1|x) \tag{9-31}$$

That is, one selects hypothesis $H_1(d = +1)$, if the $f(d = +1|x) > f(d = -1|x)$; otherwise, $H_2(d = -1)$ is selected. Using the Bayes' theorem, [66] the above *a posteriori* probabilities can be replaced by their equivalent expressions, yielding:

$$f(x|d = +1)p(d = +1) \underset{H_2}{\overset{H_1}{\gtrless}} f(x|d = -1)p(d = -1) \tag{9-32}$$

The likelihood ratio test is constructed as follows:

$$\frac{f(x|d = +1)p(d = +1)}{f(x|d = -1)p(d = -1)} \overset{H_1}{\underset{H_2}{\gtrless}} 1 \tag{9-33}$$

If the input bits are independent and identically distributed random variables, the above equation is simplified as follows:

$$\frac{f(x|d = +1)}{f(x|d = -1)} \overset{H_1}{\underset{H_2}{\gtrless}} 1 \tag{9-34}$$

By taking the logarithm of the above likelihood ratio, a useful metric known as Log-Likelihood Ratio (LLR) is obtained, which is a real-valued number representing a soft decision output of a detector.

$$L(d|x) = \log\left(\frac{f(x|d = +1)}{f(x|d = -1)}\right) + \log\left(\frac{p(d = +1)}{p(d = -1)}\right) = L(x|d) + L(d) \tag{9-35}$$

where $L(x|d)$ is the LLR of test statistics x obtained by measurement of the channel output x under the condition that either $d = +1$ or $d = -1$ may have been transmitted and $L(d)$ denotes the *a priori* LLR of the data bit d. Introduction of a decoder would improve the reliability of the above decision making process. For a systematic code, it can be shown that the LLR (soft output) out of the decoder can be written as $L_{output}(\widehat{d}) = L_{input}(\widehat{d}) + L_e(\widehat{d})$ where $L_{input}(\widehat{d})$ is the LLR of the data bit output of the detector (input to the decoder) and $L_e(\widehat{d})$ represents additional information that is obtained from the decoding process. The output sequence of a systematic decoder consists of values representing data and parity bits. Thus, the decoder LLR can be decomposed into a data component that is associated with the detector measurement and the extrinsic component that is represented by the decoder contribution due to parity. The soft decision $L_{output}(\widehat{d})$ is a real number that provides a hard decision, as well as the reliability of that decision. The sign of $L_{output}(\widehat{d})$ denotes the hard decision, i.e., for positive values of $L_{output}(\widehat{d})$ decide $\widehat{d} = +1$ and for negative values of $L_{output}(\widehat{d})$ decide $\widehat{d} = -1$. The magnitude of $L_{output}(\widehat{d})$ denotes the reliability of the decision [12–14].

Turbo decoding relies on the exchange of probabilistic information between the two soft-input soft-output decoders shown in Figure 9-41. The extrinsic information is the result of the decoder's estimation of bit d, but not taking its own input into account. It is precisely this extrinsic information that is exchanged iteratively between the two soft-decision decoders during the decoding process. Subtracting the decoder's input from its output prevents the decoder from acting as a positive feedback amplifier and introduces stability in the feedback process. Usually, after a given number of iterations, one observes that the two decoders converge to a stable final decision for d. In practice, depending on the nature of the soft-decision decoder, scaling or clipping operations may be applied to the extrinsic information in order to ensure convergence within a small number of iterations. It is shown in the literature that the Max-Log-MAP decoding algorithm can make a good trade-off between performance and complexity with the added advantage of not requiring any knowledge of the noise level (see Figure 9-43). A stop criterion facilitates the convergence of the iterative decoding process and helps reduce the average power consumption of the decoder by reducing the average number of iterations required to decode a block, without compromising performance [85,86,91].

For practical applications the turbo code parameters are very often required to be adapted to those of the application. The systematic bits are always present at the output of the turbo encoder; however, depending on the desired code rate, the parity bits from the two constituent encoders may be punctured prior to transmission (see Figure 9-41). The puncturing mechanism involves deleting some of the

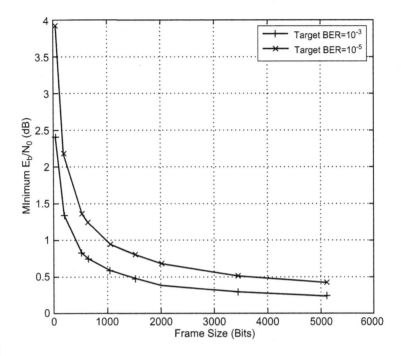

FIGURE 9-42

Minimum E_b/N_0 to achieve a target BER using the UMTS turbo coder ($R = 1/3$) with BPSK modulation over an AWGN channel [89]

parity bits prior to transmission. As an example, if the minimum code rate of a turbo coder is 1/5, a code rate of 1/3 can be generated by deleting the second parity output of each encoder. Note that the puncturing pattern for each code rate must be known by the decoder for correct interpretation of the received bits.

It is shown that for the smaller input frame sizes fewer iterations are required and the performance of the turbo coder improves with increasing the frame size. This is due to an increase in interleaver gain as the input frame size becomes larger. Figure 9-42 shows the minimum E_b/N_0 required to achieve a BER of 10^{-3} and 10^{-5} for various frame sizes using UMTS turbo encoder with minimum code rate of $R = 1/3$. In general, turbo codes are more suited for large packet size applications and their performance degrades for small packet sizes. Furthermore, the delay due to turbo interleaver might become prohibitive for delay-sensitive applications. Figure 9-43 shows the BER performance of a BTC coder as a function of E_b/N_0 for various code rates, frame sizes, and detector types. In Figure 9-43, notation (n, k) denotes the number of input (k) and output (n) bytes.

The BER performance of the IEEE 802.16 CTC coder with a minimum rate of 1/3 in the AWGN channel at different code rates and various modulation schemes is shown in Figure 9-44. Once again, the effect of code rate and frame size (in number of bytes) on performance can be seen in the figure.

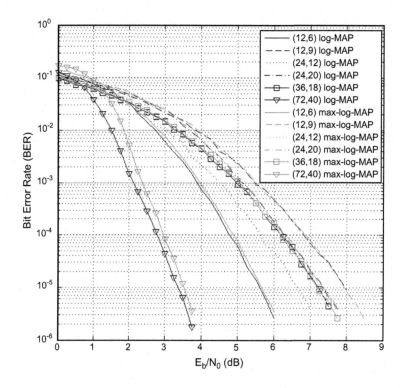

FIGURE 9-43

A comparison of the BER performance of a BTC coder in AWGN channel with different code rates, frame sizes, and detector types [124]

9.12.2 Coding and Modulation of Traffic Channels

In the IEEE 802.16-2009 standard, the BS selects the most suitable modulation and coding scheme from a finite set of supported modulation and coding schemes based on the channel quality reports from the MS, user traffic QoS requirements, and several other factors. This means that there are a limited number of combinations of code rate R and modulation order M that can be selected. If the allocation size, in terms of number of QAM symbols N, is known, the burst size can calculated as NRM for a given code rate and modulation order. In the latter approach, the burst size is derived based on the knowledge of the allocation size. A limited link adaptation is possible, if one assumes that the performance depends only on modulation order and coding rate. The main problem with MCS signaling, as utilized in the IEEE 802.16-2009 standard, [1] is that the set of possible burst sizes depend on allocation size. As an example, in the IEEE 802.16-2009 standard, the burst size of 45 bytes is only possible, if 5 slots are allocated. Therefore, it would be difficult or even impossible to avoid padding. For large burst sizes, padding is avoided by fragmentation or concatenation of the MAC PDUs. For small burst sizes, or delay-sensitive applications such as VoIP or interactive gaming, the PDU size is set by the higher layers and padding is inevitable. Furthermore, in adaptive HARQ operation, it might be impossible to signal the same burst size for different allocation sizes, assuming

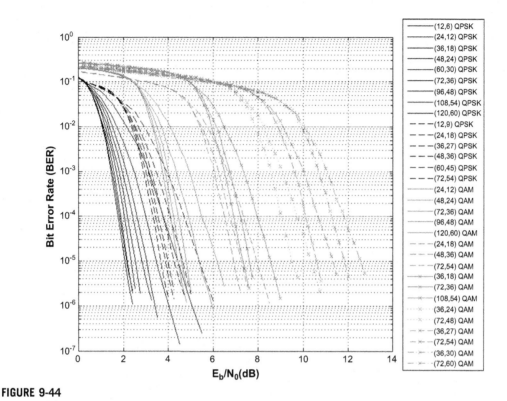

FIGURE 9-44

The performance of an IEEE 802.16 CTC coder in AWGN channel with various code rates and modulation schemes supported in the standard [124]

burst size is signaled in the re-transmission. A finite number of burst sizes independent of allocation size would allow a more efficient design of the channel coding functions.

As an example, considering that each DL-PUSC slot, in the IEEE 802.16-2009 standard, comprises 24 data sub-carriers, and assuming a burst size of 24 bytes and that the SNR supports the use of QPSK 1/2 and 3/4, then the data burst must be transmitted in four slots. If there are five slots available, then the data burst can be sent using QPSK 1/2 with 6 bytes of padding. If there are only three slots available, the data burst must be sent using QPSK 3/4 with 3 bytes of padding. Thus, there is a strong correlation between padding and allocation size. However, if 24 bytes of data could be transmitted without padding on three, four, or five slots, it would simplify the scheduling by allowing more flexibility in the resource allocation. This problem would have been more critical in the IEEE 802.16m standard due to the use of rate matching and variable slot sizes, i.e., with fixed MCS, the burst sizes supported in different MIMO modes or over unequal subframe sizes would have been significantly different. The use of conventional MCS signaling and rate-matching together would have limited the set of FEC block sizes and MCS granularity. Therefore, the IEEE 802.16m standard adopted a different approach from the legacy standard for data burst modulation and coding that does not rely on MCS selection.

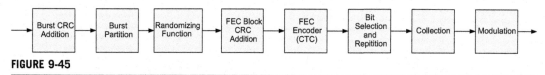

FIGURE 9-45

Coding and modulation procedures for traffic channels [2]

Figure 9-45 shows the IEEE 802.16m channel coding and modulation procedures. A cyclic redundancy check is appended to a burst (i.e., a physical layer data unit) prior to partitioning. The 16-bit CRC is calculated over the entire bits in the burst. If the burst size including burst CRC exceeds the maximum FEC block size, the burst is partitioned into K_{FB} FEC blocks, each of which is encoded separately. If a burst is partitioned into more than one FEC block, a FEC block CRC is appended to each FEC block before FEC encoding. The CRC of an FEC block is calculated based on the entire bits in that block. Each partitioned FEC block, including 16-bit FEC block CRC, has the same length. The maximum FEC block size is 4800 bits. Concatenation rules are based on the number of information bits and do not depend on the structure of the resource allocation (number of logical resource units and their size). The IEEE 802.16m utilizes convolutional turbo code with a minimum code rate of 1/3, as defined in reference [1]. The CTC scheme has been extended to support additional FEC block sizes. Furthermore, FEC block sizes can be regularly increased with predetermined block size resolutions.

The performance of adaptive modulation generally suffers from the power inefficiencies of multilevel modulation formats. This is due to the variations in bit reliabilities caused by the bitmapping onto the signal constellation. To overcome this issue, a constellation-rearrangement scheme is utilized where signal constellation of QAM signals between re-transmissions is rearranged, i.e., the mapping of the bits into the complex-valued symbols between successive HARQ re-transmissions is changed, resulting in averaging bit reliabilities over several re-transmissions and lower packet error rates. The mapping of bits to the constellation point depends on the constellation-rearrangement type used for HARQ re-transmissions and may also depend on the MIMO scheme. The complex-valued modulated symbols are mapped to the input of the MIMO encoder. Incremental redundancy HARQ is used in determining the starting position of the bit selection for HARQ re-transmissions.

The supported physical layer burst sizes N_{DB} are listed in Table 9-6. The burst size is inclusive of the burst CRC. Other burst sizes require padding to the next burst size. When the burst size including 16-bit burst CRC bits exceeds the maximum FEC block size, the burst is partitioned into K_{FB} FEC blocks. The burst size index is calculated as $index = I_{minimum-size} + I_{size-offset}$, where $I_{size-offset} \in \{0, 1, \ldots, 31\}$ a 5-bit index is signaled through A-MAP IE and $I_{minimum-size}$ is calculated based on the allocation size as shown in Table 9-7. The allocation size is defined as the number of logical resource units multiplied by the MIMO rank, which are allocated to the burst. In the case of long-TTI (i.e., concatenation of an integer number of subframes), the number of logical resource units is calculated as the number of subframes in TTI multiplied by the number of LRUs per subframe. The modulation order M (i.e., 2 for QPSK, 4 for 16 QAM, and 6 for 64 QAM) depends on the parameter $I_{size-offset}$ as shown in Table 9-8. The allocation size and the value of $I_{size-offset}$ are set by the serving BS scheduler, which takes into account the resulting modulation order and effective code rate and is subject to the link adaptation.

If a burst is partitioned into more than one FEC block, each partitioned FEC block has the same size. The size of the FEC encoder input is denoted by N_{FB}. The set of supported FEC encoder input sizes, including FEC block CRC when applicable, is the subset of the burst size, i.e., N_{DB} of $index$ from 1 to 39.

Table 9-6 Supported Burst Sizes in IEEE 802.16m [2]

Index	N_{DB} (Byte)	K_{FB}	Index	N_{DB} (Byte)	K_{FB}	Index	N_{DB} (Byte)	K_{FB}
1	6	1	23	90	1	45	1200	2
2	8	1	24	100	1	46	1416	3
3	9	1	25	114	1	47	1584	3
4	10	1	26	128	1	48	1800	3
5	11	1	27	145	1	49	1888	4
6	12	1	28	164	1	50	2112	4
7	13	1	29	181	1	51	2400	4
8	15	1	30	205	1	52	2640	5
9	17	1	31	233	1	53	3000	5
10	19	1	32	262	1	54	3600	6
11	22	1	33	291	1	55	4200	7
12	25	1	34	328	1	56	4800	8
13	27	1	35	368	1	57	5400	9
14	31	1	36	416	1	58	6000	10
15	36	1	37	472	1	59	6600	11
16	40	1	38	528	1	60	7200	12
17	44	1	39	600	1	61	7800	13
18	50	1	40	656	2	62	8400	14
19	57	1	41	736	2	63	9600	16
20	64	1	42	832	2	64	10800	18
21	71	1	43	944	2	65	12000	20
22	80	1	44	1056	2	66	14400	24

Table 9-7 Minimum-Size Index as a Function of the Allocation Size [2]

Allocation Size	$I_{minimum-size}$	Allocation Size	$I_{minimum-size}$	Allocation Size	$I_{minimum-size}$
1–3	1	16–18	15	58–64	26
4	2	19–20	16	65–72	27
5	4	21–22	17	73–82	28
6	6	23–25	18	83–90	29
7	8	26–28	19	91–102	30
8	9	29–32	20	103–116	31
9	10	33–35	21	117–131	32
10–11	11	36–40	22	132–145	33
12	12	41–45	23	146–164	34
13	13	46–50	24	165–184	35
14–15	14	51–57	25	192	36

Table 9-8 Modulation Order Determination

$I_{\text{size-offset}}$	M (Allocation Size > 2)	M (Allocation Size = 2)	M (Allocation Size = 1)
0–9	2	2	2
10–15	2	2	4
16–18	2	4	6
19–21	4	4	6
22–23	4	6	6
24–31	6	6	6

The burst size N_{DB} including burst CRC and FEC block CRC is defined as $N_{DB} = K_{FB}N_{FB}$. The payload size excluding burst CRC and FEC block CRC is given by $N_{PL} = N_{DB} - N_{DB-CRC} - I_{MFB} \cdot K_{FB} \cdot N_{FB-CRC}$ where $I_{MFB} = 0$ ($K_{FB} = 1$) or 1 ($K_{FB} > 1$), $N_{FB-CRC} = 16$, which is the size of FEC block CRC, $N_{DB-CRC} = 16$, which is the size of burst CRC. The data octets then sequentially enter into the randomization block. The data bits are XOR with the output of a pseudo-random binary sequence generator. The randomization function is initialized with the initial vector for each FEC block. The burst partition module generates K_{FB} FEC blocks and each FEC block is processed by the FEC block CRC encoding function. If $K_{FB} > 1$, the FEC block CRC generator appends a 16-bit FEC block CRC to each FEC block. The cyclic generator for FEC block CRC encoding is given by $G_{FB-CRC}(D) = D^{16} + D^{15} + D^2 + 1$. The step-by-step procedure for calculation of the modulation and coding schemes in IEEE 802.16m is shown in Figure 9-46.

The FEC encoder input bits are denoted by $d_1, d_2, d_3, \ldots, d_{N_{FB}}$ with d_1 representing the MSB and N_{FB} being the size of the FEC encoder input inclusive of 16-bit FEC block CRC and the parity bits produced by the FEC block CRC generator being denoted by $p_1, p_2, p_3, \ldots, p_{16}$. The FEC block CRC encoding is performed in a systematic form, which means that in GF(2),[vi] the polynomial

[vi]A finite field is a field with a finite number of elements, which is also called a Galois field. The order or the number of elements of a finite field is always a prime or a power of a prime. For each prime power, there exists exactly one finite field $GF(p^n)$, which is often expressed as F_{p^n} in the literature, where $GF(p)$ is called the prime field of order p, and is the field of residue classes modulo p, where the p elements are denoted 0, 1, 2,..., $p - 1$. In $GF(p)$ the two expressions $a \equiv b$ and $a \equiv b \bmod p$ are equivalent. The finite field $GF(2)$ consists of elements 0 and 1 which satisfy the following addition and multiplication properties: [67,68]

+	0	1
0	0	1
1	1	0

×	0	1
0	0	0
1	0	1

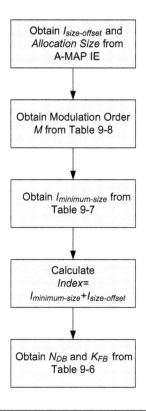

FIGURE 9-46

The procedure for calculation of coding and modulation parameters

$d_1 D^{N_{FB}-1} + d_2 D^{N_{FB}-2} + \ldots + d_{N_{FB}-16} D^{16} + p_1 D^{15} + p_2 D^{14} + \ldots + p_{15} D + p_{16}$ yields a remainder of zero when divided by $G_{\text{FB-CRC}}(D)$. Each FEC block is encoded using the convolutional turbo coder.

The IEEE 802.16m CTC encoder uses a double binary Circular Recursive Systematic Convolutional (CRSC) code. Duo-binary turbo codes are built from parallel concatenation of two RSC component codes, each with two inputs. There are several advantages for this construction compared to classical turbo codes, including better convergence of the iterative decoding, large minimum distances (i.e., large asymptotic gains), less sensitivity to puncturing patterns, reduced latency, and robustness towards the flaws of the component decoding algorithm, in particular when the MAP algorithm is simplified to the Max-Log-MAP scheme. The input bits are alternatively fed to the constituent encoders, starting with the MSB of the first byte that is fed to constituent encoder 1 followed by the next bit being fed to constituent encoder 2. The encoder is fed by blocks of N_{FB} bits. The order in which the encoded bits are fed to the bit separation block is as follows:

$$A, B, Y_1, Y_2, W_1, W_2 = A_0, A_1, A_2, \ldots, A_{N-1}, B_0, B_1, B_2, B_{N-1}, Y_{1,0}, Y_{1,1}, Y_{1,2}, \ldots Y_{1,N-1},$$
$$Y_{2,0}, Y_{2,1}, Y_{2,2}, \ldots Y_{2,N-1}, W_{1,0}, W_{1,1}, W_{1,2}, \ldots W_{1,N-1}, W_{2,0}, W_{2,1}, W_{2,2}, \ldots W_{2,N-1}$$
$$(9\text{-}36)$$

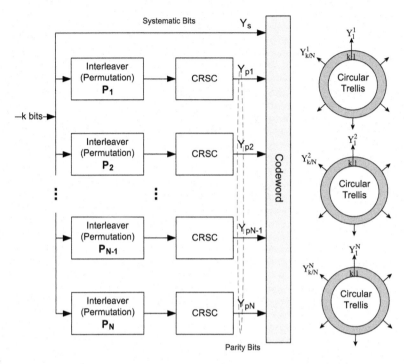

FIGURE 9-47

Parallel concatenation of N CRSC codes [85,86,91]

where A, B are the systematic bits, Y_1, W_1 are the parity bits from constituent encoder 1 and Y_2, W_2 are the parity bits from constituent encoder2.

It was mentioned earlier that the IEEE 802.16m CTC encoder relies on a parallel concatenation of N CRSC encoders. They are called circular recursive systematic because they use circular or tail-biting termination. Figure 9-47 illustrates the principle of the encoding for blocks of k bits with 1/2 global rate. Due to the circular termination mechanism, which involves making the initial and the final state of the encoder equal, no additional bit is needed to transform each RSC code into a block code. Moreover, if the permuted sequence at the input of a particular component encoder is not a return to zero sequence, the entire redundancy set of the corresponding circle is affected by this sequence, not only the part after it as would be the case with a "0" initialized encoder.

The CTC coder output bits are grouped into six sub-blocks and interleaved to break special patterns in the encoder output. As shown in Figure 9-48, the interleaved sub-blocks are multiplexed into four blocks where those four blocks consist of an interleaved A sub-block, an interleaved B sub-block, a bit-by-bit multiplexed sequence of the interleaved Y_1 and Y_2 sub-block sequences, which is referred to Y, and a bit-by-bit multiplexed sequence of the interleaved W_2 and W_1 sub-block sequences, which is referred to W. Information sub-blocks A and B are bypassed while parity sub-blocks are multiplexed bit by bit. The bit-by-bit multiplexed sequence of interleaved Y_1 and Y_2 sub-block sequences consists of the first output bit from the Y_1 sub-block interleaver, the first output bit from the Y_2 sub-block

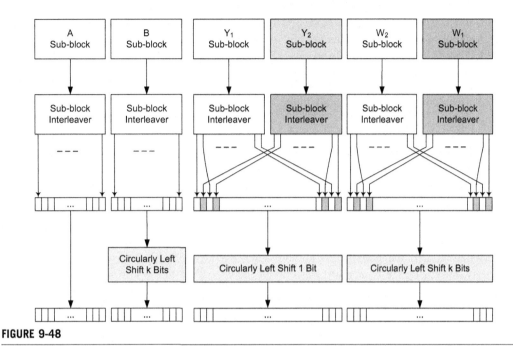

FIGURE 9-48

An illustration of the block interleaving procedure [2]

interleaver, the second output bit from the Y_1 sub-block interleaver, the second output bit from the Y_2 sub-block interleaver, etc. The bit-by-bit multiplexed sequence of interleaved W2 and W1 sub-block sequences comprises the first output bit from the W_2 sub-block interleaver, the first output bit from the W_1 sub-block interleaver, the second output bit from the W_2 sub-block interleaver, the second output bit from the W_1 sub-block interleaver, etc. After multiplexing sub-blocks into four blocks, sub-block B and sub-block W are circularly left-shifted by k bits, and sub-block Y is circularly left-shifted by 1 bit. When the FEC block size N_{FB} is equal to a multiple of the modulation order, k is set to one.

If $K_{FB} > 1$, the N_{RE} data sub-carriers allocated for a sub-packet are segmented into K_{FB} blocks, one for each FEC block. The number of data sub-carriers in the kth FEC block is given as follows:

$$N_{RE_k} = N_{RS}\left\lfloor \frac{N_{RE}/N_{RS}+(K_{FB}-k-1)}{K_{FB}} \right\rfloor \quad 0 \le k < K_{FB} \qquad (9\text{-}37)$$

where $N_{RS} = 1$ for a single spatial stream and $N_{RS} = 2$ for multiple spatial streams. Bit selection and repetition is performed to generate the sub-packets. Let N_{CTC_k} be the number of coded bits that are transmitted in the kth FEC block. The value of N_{CTC_k} is calculated as $N_{CTC_k} = N_{RE_k}N_{SM}M$ where N_{SM} is equal to the product of MIMO rank for the burst and the number of subframes when long TTI is used for the burst. The index in the HARQ buffer for the jth bit transmitted for the kth FEC block u_{ijk} is given as follows:

$$N_{shift_i} = iM$$
$$\beta_{ijk} = (N_{CTC_k} - N_{shift_i} + j) \bmod N_{crc_k} \qquad (9\text{-}38)$$
$$u_{ijk} = (P_{ik} + \beta_{ijk}) \bmod N_{FB\_Buffer_k}$$

where $k = 0, 1, ..., K_{FB} - 1, j = 0, 1, ..., N_{CTC_k} - 1$, i is the sub-packet identifier ($i = $ SPID), P_{ik} is the starting position for sub-packet i of the kth FEC block, and $N_{FB\_Buffer_k} = 3N_{FB_k}$ is the buffer size for the kth FEC block. The selected bits from each FEC block are collected in the order of FEC block for the HARQ transmission. After the repetition function, the data bits are serially fed to the constellation mapping block. The Gray-mapped QPSK, 16 QAM, and 64 QAM (as shown in Figure 9-49) are supported. The magnitude of the constellation points are normalized by a factor of γ to ensure equal average power.

9.12.3 Coding and Modulation of the Control Channels

Tail-Biting Convolutional Code (TBCC) is used for encoding downlink and uplink control channels. Terminating the trellis of a convolutional code is a key parameter in the code's performance for packet-based communications. Tail-biting convolutional coding is a mechanism for trellis termination which avoids the rate loss incurred by zero-tail termination at the expense of a more complex decoder. Tail-biting encoding ensures that the starting state of the encoder is the same as its ending state, and that this state is not necessarily the same as the all-zero state. For a rate $1/n$ feed-forward encoder, this is achieved by initializing the m memory elements of the encoder with the last m information bits of a block of data of length L and ignoring the output. All of the L bits are then fed to the encoder and the resulting Ln output bits are used as the codeword. In contrast, the zero-tail termination method appends m zeros to a block of data to ensure the feed-forward encoder starts from and ends in the all-zero state for each block. This incurs a rate loss due to the extra tail bits (i.e., non-informational bits) that are transmitted. The maximum likelihood tail-biting decoder of the TBCC codes involves determining the best path in the trellis under the constraint that it starts and ends in the same state. One way to implement this is to run $M_{trellis}$ parallel Viterbi algorithms where $M_{trellis}$ is the number of states in the trellis, and select the decoded bits based on the Viterbi algorithm that gives the best metric. However this makes the decoding $M_{trellis}$ times more complex than the decoding algorithm for zero-tailed encoding.

The structure of TBCC encoder is depicted in Figure 9-50. The output encoded bits of TBCC encoder with minimum rate of 1/5 are separated to five sub-blocks and each sub-block is individually interleaved. In the bit selection block, a rate-matching mechanism is performed by puncturing or repeating the bits. The minimum code rate of the TBCC encoder is 1/5, and the constraint length of the convolutional encoder is $K = 7$. The TBCC encoder structure illustrated in Figure 9-50 utilizes the following generator polynomials:

$$
\begin{aligned}
G_1 &= 171_{OCT} \rightarrow A \\
G_2 &= 133_{OCT} \rightarrow B \\
G_3 &= 165_{OCT} \rightarrow C \\
G_4 &= 117_{OCT} \rightarrow D \\
G_5 &= 127_{OCT} \rightarrow E
\end{aligned}
\tag{9-39}
$$

The encoded bits are de-multiplexed into five sub-blocks denoted as A, B, C, D, and E. Let L denotes the number of the information bits of the encoder. The encoded bits $u[i]$, $i = 0,...,5L - 1$ are distributed into five sub-blocks with $y^A[k] = u[5k]$ $y^B[k] = u[5k + 1]$ $y^C[k] = u[5k + 2]$ $y^D[k] = u[5k + 3]$ and $y^E[k] = u[5k + 4]$ where $k = 0, ..., L - 1$. The five sub-blocks are interleaved separately. The

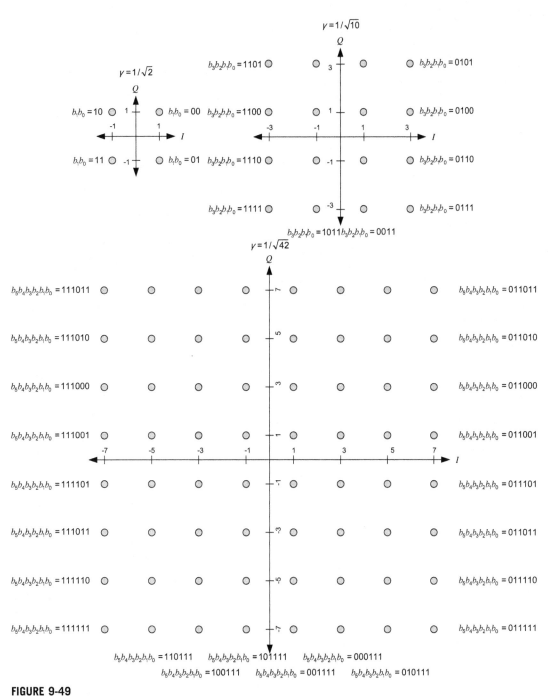

FIGURE 9-49

QPSK, 16 QAM, and 64 QAM constellations [1]

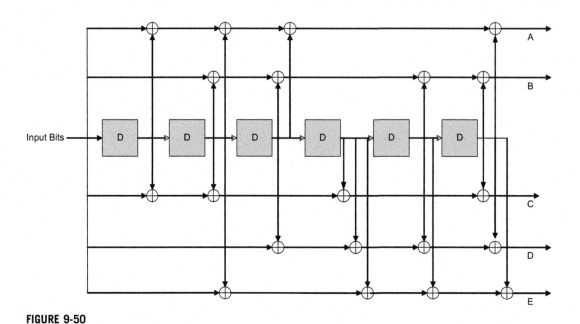

FIGURE 9-50

The structure of a TBCC encoder with minimum code rate of 1/5 [2]

interleaving is performed in the unit of bits. The output of the sub-block interleaver $z^j[k]$ with $j = A,B,C,D,E$, and $k = 0, ..., L - 1$ corresponds to the input sequence $y^j[k]$ and can be expressed as $z^j[k] = y^j [\Pi_k]$ where Π_k is the interleaver index. The maximum value of parameter L is 128. The output sequence of sub-block interleaver consists of the interleaved A, B, C, D, and E sub-blocks. Let $w[i]$ denote the bit grouping output with $i = 0, ..., 5L - 1$. It can be shown that $w[k] = z^A[k]$, $w[k + L] = z^B[k]$, $w[k + 2L] = z^C[k]$, $w[k + 3L] = z^D[k]$, and $w[k + 4L] = z^E[k]$ with $k = 0, ..., L - 1$. Suppose the desired number of the output bits is M. The output sequence $c[n]$ can be expressed as $c[n] = w[n \text{ mod } K_{buffer-size}]$, $n = 0, 1,..., M - 1$ where $K_{buffer-size} \leq 5L$ is the size of buffer that is used for repetition.

The performance of an example tail-biting convolution coder with a minimum rate of 1/2 in the AWGN channel with different modulation orders, coding rates, and data frame sizes using Viterbi decoder has been evaluated and is shown in Figure 9-51, where the wrap depth is a predetermined number of trellis stages used as a parameter in the figure. One way to estimate the required wrap depth for MAP decoding of a tail-biting convolutional code is to conduct hardware or software experimentations, requiring a circular MAP decoder with a variable wrap depth to be implemented and to measure the decoded bit error rate versus E_b/N_0 for successively increasing wrap depths. The minimum decoder wrap depth that provides the minimum probability of decoded bit error for a specified E_b/N_0 is found when further increases in wrap depth do not decrease the error probability, as shown in Figure 9-51. It can be seen that the link-level BER performance improves when increasing the depth and the performance does not significantly change after a certain limit. The effect of different data frame sizes (6, 9, 18, 27, and 36 bytes) on the BER performance has been investigated and is shown in the figure.

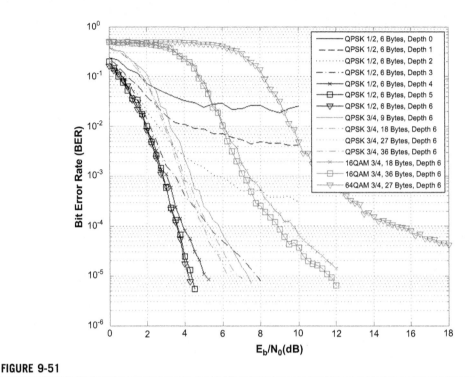

FIGURE 9-51

Performance of an example tail-biting convolutional coder

9.12.4 HARQ-IR Physical Layer Procedures

The incremental redundancy HARQ is performed by changing the starting position P_{ik} of the bit selection for HARQ re-transmissions. For downlink HARQ, the starting point for the bit selection algorithm is determined as a function of the HARQ sub-packet ID as shown in Table 9-9.

For the uplink HARQ, the starting position for the bit selection algorithm is determined as a function of HARQ sub-packet ID $P_{ik} = (iN_{CTC_k}) \bmod N_{FB\_Buffer_k}$. For the uplink HARQ, the sub-packets are transmitted in sequential order.

Table 9-9 Starting Position Determination for Downlink HARQ [2]

HARQ Sub-packet ID	Starting Position P_{ik}
0	0
1	$(-N_{CTC_k}) \bmod N_{FB\_Buffer_k}$
2	
$(N_{FB\_Buffer_k}/2 - N_{CTC_k}/2) \bmod N_{FB\_Buffer_k}$	3
$(N_{FB\_Buffer_k} - N_{CTC_k}/2) \bmod N_{FB\_Buffer}$	

9.12.5 Constellation Rearrangement

It was mentioned earlier that the performance of adaptive modulation generally suffers from the power inefficiencies of multi-level modulation formats. This is due to the variations in bit reliabilities caused by the bitmapping onto the signal constellation. To overcome this issue, a Constellation Rearrangement (CoRe) scheme is utilized where the signal constellation of QAM signals between re-transmissions is rearranged, i.e., the mapping of the bits into the complex-valued symbols between successive HARQ re-transmissions is changed, resulting in averaging bit reliabilities over several re-transmissions and lower packet error rates. The mapping of bits to the constellation point depends on the constellation-rearrangement type used for the HARQ re-transmissions and may also depend on the MIMO scheme. Two constellation rearrangement versions have been specified as shown in Table 9-10. The constellation rearrangement exclusively applies to 16 QAM and 64 QAM modulation schemes. Note that the two bits of higher quality in constellation rearrangement version 0 are of lower quality in the constellation rearrangement version 1, while the two bits of lower quality in constellation rearrangement version 0 are of higher quality constellation rearrangement version 1.

9.12.6 Performance of Channel Coding and HARQ-IR

The overall performance of the channel coding and HARQ transmission chains in the downlink and uplink for the new (as shown in Figure 9-45) and the legacy systems have been evaluated and are shown in Figure 9-52 and Figure 9-53. In these link-level simulations, AWGN channel, SISO antenna configuration, data frame size of 912 bits inclusive of the 16-bit CRC, 64 QAM modulation in the downlink and 16 QAM modulation in the uplink, a maximum of four HARQ re-transmissions, three LRUs in the downlink and five LRUs in the uplink, and max-log MAP detection have been assumed.

The figures show that the use of new schemes and parameters, as described earlier, significantly improve the overall performance compared to that of the legacy transmission chain. It is further shown that the use of HARQ re-transmissions significantly and progressively decreases the block error rate and consequently improves the performance. It must be noted that a CTC encoder rate of 1/3 has been used for both new and the legacy transmission chains.

In addition, the performance of the channel coding and HARQ transmission chains in the downlink and uplink for the new (as shown in Figure 9-45) and the legacy systems have been evaluated under a different channel model and is shown in Figure 9-54. In these link-level simulations, ITU PedB 3 km/h channel model (Rayleigh fading), SISO antenna configuration, data frame size of 1864 bits inclusive of the 16-bit CRC, 64 QAM modulation, 6 LRUs and 96 sub-carriers per LRU, a maximum of four HARQ re-transmissions, and max-log MAP detection have been assumed. Furthermore, a CTC encoder rate of 1/3 has been used for both the new and the legacy transmission chains. The figure shows that the use of new schemes and parameters as described earlier significantly improves the overall performance compared to that of the legacy transmission chain. It is further shown that the use of HARQ re-transmissions significantly and progressively decreases the block error rate and consequently improves performance.

In Figure 9-55, the performance of the channel coding and HARQ transmission chain of the IEEE 802.16m in the downlink direction has been compared with that of the legacy system, assuming ITU PedB 3 km/h channel model (Rayleigh fading), SISO antenna configuration, data frame size of 2328 bits inclusive of the 16-bit CRC, adaptation of modulation order (16 QAM and 64 QAM), and different number of LRUs per MCS.

Table 9-10 Constellation Rearrangement Version [2]

Modulation Scheme	Modulation Order	Constellation Rearrangement Version	Rank = 1 Mapping Rule	Rank > 1 Mapping Rule	
				Even Symbol	Odd Symbol
16 QAM	4	0	b_0 b_1 b_2 b_3 – –	b_0 b_1 b_2 b_3 – –	b_4 b_5 b_6 b_7 – –
16 QAM	4	1	b_3 b_2 b_1 b_0 – –	b_1 b_4 b_3 b_6 – –	b_5 b_0 b_7 b_2 – –
64 QAM	6	0	b_0 b_1 b_2 b_3 b_4 b_5	b_0 b_1 b_2 b_3 b_4 b_5	b_6 b_7 b_8 b_9 b_{10} b_{11}
64 QAM	6	1	b_5 b_4 b_3 b_2 b_1 b_0	b_2 b_7 b_0 b_5 b_{10} b_3	b_8 b_1 b_6 b_{11} b_4 b_9

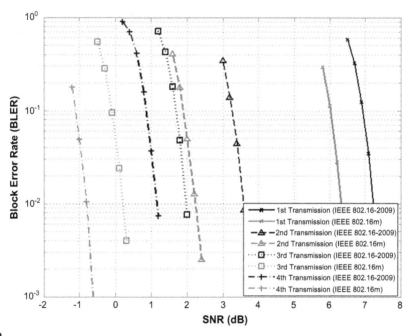

FIGURE 9-52

A comparison of the uplink channel coding and HARQ performance in IEEE 802.16m and IEEE 802.16-2009 [87]

One of the techniques that has been utilized in the IEEE 802.16m channel coding and HARQ-IR transmission is constellation rearrangement. The effect of constellation rearrangement on the coding and HARQ performance at link-level has been evaluated and is shown in Figures 9-56 and 9-57, for downlink and uplink, respectively. We can see from the figures that the constellation rearrangement method can slightly improve the HARQ-IR performance. In these link-level simulations, ITU PedB 3 km/h channel model (Rayleigh fading), SISO antenna configuration, data frame size of 512 bits

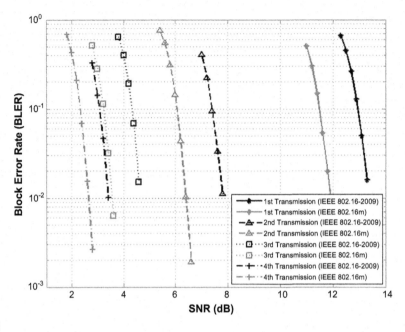

FIGURE 9-53

A comparison of the downlink channel coding and HARQ performance in IEEE 802.16m and IEEE 802.16-2009 [87]

inclusive of the 16-bit CRC in the downlink and 912 bits with 16-bit CRC in the uplink, 64 QAM modulation in the downlink and 16 QAM modulation in the uplink, one LRU in the downlink and five LRUs in the uplink (96 sub-carriers per LRU), a maximum of four HARQ re-transmissions, and max-log MAP detection have been assumed. Furthermore, a CTC encoder rate of 1/3 has been used for both the new and the legacy transmission chains. The effect of constellation rearrangement is more pronounced in the HARQ re-transmissions.

9.13 SYNCHRONIZATION CHANNEL

The Advanced Preamble (A-Preamble) or alternatively the synchronization channel is a downlink physical channel which provides a periodic reference signal for system discovery (IEEE 802.16m system beacon), timing and frequency acquisition, frame synchronization, RSSI and path loss measurements, channel estimation, and base station identification. The IEEE 802.16m supports a hierarchical synchronization scheme with two stages. There are two downlink synchronization signals; the Primary Advanced Preamble (PA-Preamble); and the Secondary Advanced Preamble (SA-Preamble). The PA-Preamble is a narrowband synchronization signal and is common to a group of cells that is used for initial acquisition, superframe synchronization, and broadcast of essential system information. The SA-Preamble is a wideband and cell/sector-specific synchronization signal that is used for fine time/frequency synchronization and cell/sector identification. The PA-Preamble and

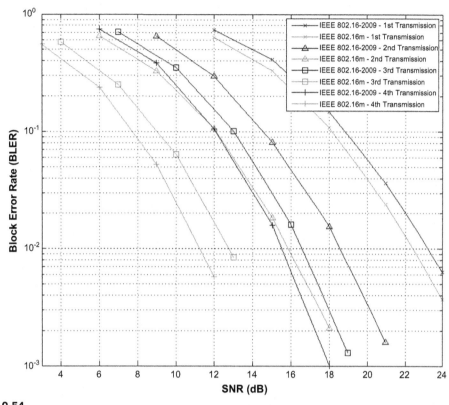

FIGURE 9-54

A comparison of the downlink/uplink channel coding and HARQ performance in IEEE 802.16m and IEEE 802.16-2009 [87]

SA-Preamble are time-division multiplexed across time. One PA-Preamble OFDM symbol and two SA-Preamble OFDM symbols present in every superframe period. The PA-Preamble is located at the first OFDM symbol of the second frame in a superframe, while the SA-Preamble is located at the first OFDM symbol of the first and third frames in the superframe. When the new and the legacy systems are co-deployed, the legacy preamble is located at the first OFDM symbol of the legacy frame. The locations of the A-Preamble OFDM symbols are fixed within the superframe. Figure 9-58 shows the structure of the IEEE 802.16m preambles in time and frequency.

The timing and Cell-ID acquisition in IEEE 802.16m is substantially different from that of the legacy system. In the IEEE 802.16-2009 standard, the preamble is the first OFDM symbol of the frame. For each FFT size, three different preamble sub-carrier sets are defined (i.e., frequency reuse-3), differing in the allocation of sub-carriers. These sub-carriers are modulated using a boosted BPSK modulation with a specific Pseudo-Noise (PN) code. The timing and frequency acquisition in the legacy system is a non-hierarchical procedure, i.e., both timing and Cell ID information are acquired in a single step. The legacy preamble sequences carry 114 distinct Cell IDs. One of the problems with the legacy preamble design is that when an MS is turned on at a location with a similar distance to

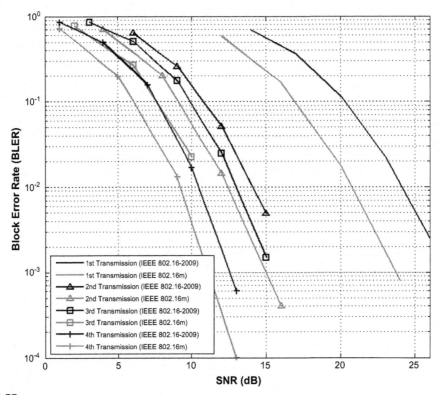

FIGURE 9-55

A comparison of the downlink channel coding and HARQ performance with MCS and resource adaptation in IEEE 802.16m and IEEE 802.16-2009 [87]

three base stations using three different segment preambles, the received preamble would lose its quasi-repetition property because all sub-carriers would be non-zero (except the DC carrier and the guard-band sub-carriers). Without this repetition property, the initial frame synchronization algorithms would only rely on the cyclic prefix in every observed OFDM symbol in a radio frame to decide the symbol boundary and to detect the PN sequence with unknown channel impulse response and frequency offset over 114 possible sequences. This would consume the power of the MS and increase the scanning time. The latter would result in poor cell selection/re-selection at the cell edge. Other problems with the legacy preamble design include the timing accuracy and an insufficient number of preamble sequences for cell planning and deployment (at least 256 distinct sequences are required based on information obtained from early deployments of mobile WiMAX Release 1.0).

The IEEE 802.16m timing and Cell ID acquisition, on the other hand, is based on a hierarchical scheme, where the timing and Cell ID and other information are obtained in two steps. Also the number of distinct Cell IDs has been increased to 768 sequences. The length of sequence for the PA-Preamble is 216, regardless of the FFT size. The PA-Preamble carries the information related to system bandwidth and RF carrier configuration, i.e., whether the carrier is fully- (all downlink and uplink

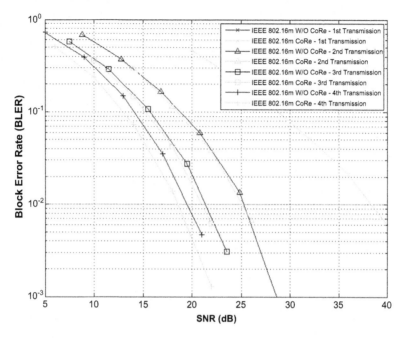

FIGURE 9-56

Performance with and without constellation rearrangement in the downlink [87]

control channels are contained) or partially-configured (some of the downlink and uplink control channels are included). The sub-carrier index 256 is reserved for the DC sub-carrier. The sequences of PA-Preamble indices from 3 to 9 in Table 9-11 are reserved for irregular channel bandwidths which support tone dropping. The sub-carriers associated with PA-Preamble are assigned as $m = 2k + 41$, $k = 0, 1, 2, ..., 215$.

The IEEE 802.16m PA-Preamble is designed in a manner so as not to degrade the performance of legacy system acquisition. The IEEE 802.16m PA-Preamble enables mobile stations to synchronize in frequency and time without requiring the IEEE 802.16-2009 standard preamble. The IEEE 802.16m PA-Preamble supports timing synchronization by autocorrelation with a repeated waveform. The structure of the PA-Preamble is not identical to that of the legacy preamble in the time-domain, as shown in Figure 9-59. The magnitude boosting levels in single carrier mode for different FFT sizes of 512, 1024, and 2048 are 2.3999, 3.4143, and 5.1320, respectively. As an example, for 512-point FFT, the boosted PA-Preamble at kth sub-carrier is $\alpha k = 2.3999 \, \beta k$ where βk denotes the kth sub-carrier of the PA-Preamble before boosting. The sub-carrier levels before boosting are $+1$ or -1.

The SA-Preamble sequences are partitioned and each partition is dedicated to a specific base station type, such as macro BS, macro Hot-zone BS,[vii] femto BS, etc. The partition information is

---

[vii]Macro hot-zone BS is a base station/access point with lower transmission power relative to macro BS (higher transmit power than a femto BS) and is possibly deployed overlaid with another macro BS. The macro hot-zone BS may be deployed by a service provider or, unlike a femto BS, it may be deployed in an outdoor environment.

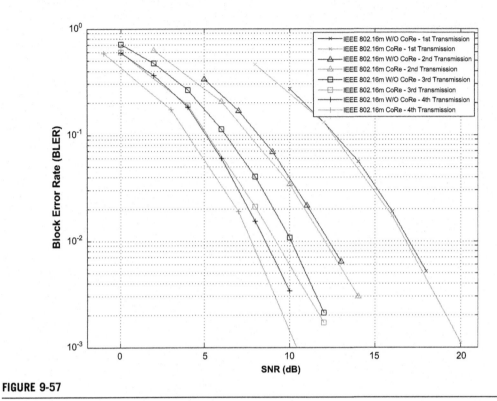

FIGURE 9-57

Performance with and without constellation rearrangement in the uplink [87]

broadcast in the secondary superframe header. For support of femto-cell deployments, a femto BS is expected to configure the segment or sub-carrier set for SA-Preamble transmission based on the segment information of the overlay macro-cell BS for minimized interference to macro-cell, if the femto BS is synchronized to the macro-cell base station. The segment information of the overlay macro-cell BS may be obtained through communication with the macro-cell BS via a backhaul network or active scanning of the SA-Preamble transmitted by neighbor macro-cell base stations. The sectors/cells are distinguished by the SA-Preamble. Each segment uses an SA-Preamble composed of a carrier-set out of the three available carrier-sets. For 512-point FFT, corresponding to the 5 MHz transmission bandwidth, the 288-bit SA-Preamble sequence is divided into eight main blocks, namely A, B, C, D, E, F, G, and H. The length of each block is 36 bits. Each segment has different sequence blocks. For 512-point FFT size, A, B, C, D, E, F, G, and H are modulated and mapped sequentially in ascending order onto the SA-Preamble sub-carrier set corresponding to the segment identifier. For larger FFT sizes, the basic blocks A, B, C, D, E, F, G, and H are repeated in the same order. For instance, in 1024-point FFT size, corresponding to a 10 MHz transmission bandwidth, E, F, G, H, A, B, C, D, E, F, G, H, A, B, C, D are modulated and mapped sequentially in ascending order onto the SA-Preamble sub-carrier-set corresponding to a certain segment.

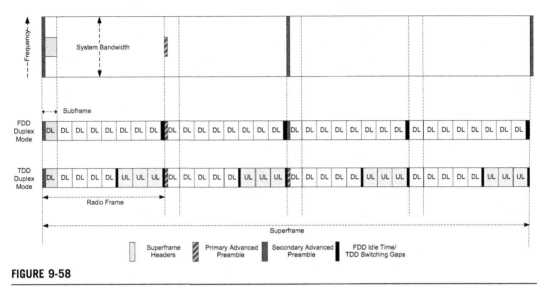

FIGURE 9-58

Structure of the IEEE 802.16m advanced preambles in time and frequency [2]

Table 9-11 PA-Preamble Information Content [2]

PA-Preamble Index	Carrier Configuration	Bandwidth (MHz)
0	Fully configured	5
1		7, 8.75, and 10
2		20
3		Reserved
4		Reserved
5		Reserved
6		Reserved
7		Reserved
8		Reserved
9		Reserved
10	Partially configured	N/A

A circular shift is applied over three consecutive sub-carriers after applying sub-carrier mapping. Each sub-block has a common offset. The circular shift pattern for each sub-block is defined as [2,1,0... .....,2,1,0... .....,2,1,0, 2,1,0, DC, 1,0,2, 1,0,2, ...,1,0,2, ...1,0,2] where the shift is right-circular. As an example, for the 512-point FFT size, the blocks (A, B, C, D, E, F, G, H) are circularly shifted as [0, 2, 1, 0, 1, 0, 2, 1,2], respectively. Figures 9-60 and 9-61 illustrate the structure of the SA-Preamble in the frequency-domain for 512-point FFT.

When multiple antennas are supported at the base station, the SA-Preamble blocks are interleaved on the number of transmit-antennas (1, 2, 4 or 8). The multi-antenna transmission is supported using

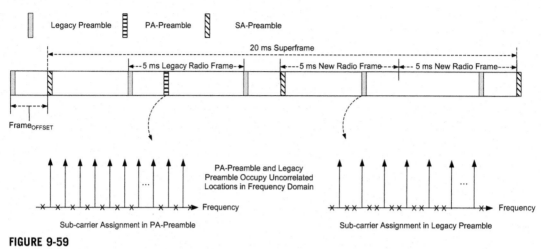

FIGURE 9-59

Structure of PA-preamble compared to the legacy preamble [2]

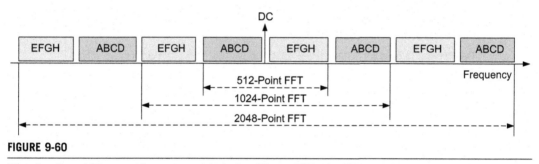

FIGURE 9-60

The allocation of SA-preamble sequence blocks for different FFT sizes [2]

Cyclic Delay Diversity (CDD) with antenna-specific delay values defined for the PA-Preamble and interleaving within a symbol (multiple antennas can transmit within a single symbol but on distinct sub-carriers) for the SA-Preamble (see Figure 9-62).

Each SA-Preamble segment uses a carrier-set from the three available carrier-sets in the following manner: segment 0 uses SA-Preamble carrier-set$_0$, segment 1 uses SA-Preamble carrier-set$_1$, and segment 2 uses SA-Preamble carrier-set$_2$. Each Cell ID has an integer value *IDcell* from 0 to 767. The *IDcell* parameter is defined by segment index and an index per segment as $IDcell = 256n + 2(m \bmod 128) + [m/128]$ where n and m are the index of the SA-Preamble carrier-set and a running index from 0 to 255, respectively. The SA-Preamble sequences are partitioned and each partition is dedicated to a specific base station type, such as a macro-cell BS, macro hot-zone BS, or femto BS. The base stations are categorized into macro BS and non-macro BS through hard partition with 258 sequences (86 sequences per segment × 3 segments) dedicated for the macro BS. The non-macro BS information is broadcast in a hierarchical structure which consists of S-SFH SP3 and an AAI_SCD MAC control

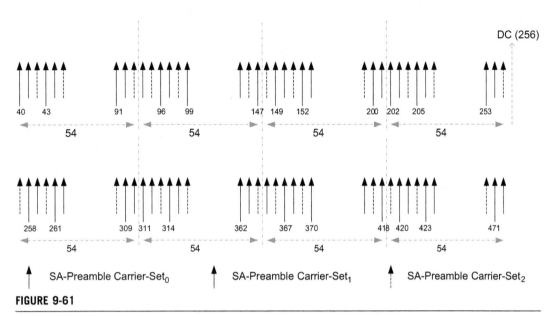

FIGURE 9-61

SA-preamble structure for 512-point FFT [2]

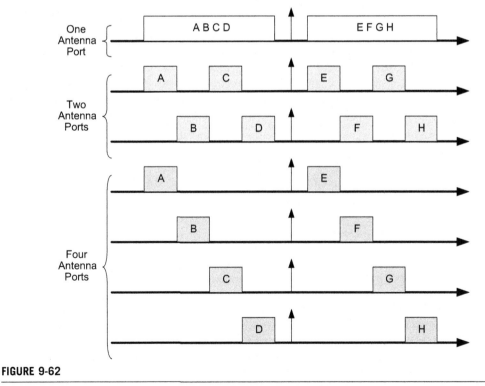

FIGURE 9-62

Multi-antenna transmission of SA-preamble at 5 MHz bandwidth [2]

message. In S-SFH SP3, the non-macro BS cell type is partitioned as open and closed subscriber group femto base stations. A total of 16 cases of *IDcell* partitioning for open and closed subscriber group femto base stations are specified in reference [2], consisting of Cell ID partitions based on 30 sequences, i.e., 10 sequences per segment, in each group.

The magnitude of sub-carriers corresponding to the SA-Preamble is amplified. The value of the boosting factor depends on the FFT size and the number of antennas [2]. For a single transmit antenna, the SA-Preamble is transmitted with a magnitude boosting factor of 1.87. The boosted SA-Preamble at kth sub-carrier is denoted as $\alpha_k = 1.87\,\beta_k$ where β_k denotes the magnitude of the kth sub-carrier of the SA-Preamble prior to boosting (i.e., $+1$, -1, $+j$ or $-j$). In order to reduce the PAPR of the SA-Preamble sequences, a block cover sequence that is transparent to the receiver is defined and multiplied by each sub-block. Each bit of block cover sequence is mapped to a real number $\{+1, -1\}$, and then multiplied by all the sub-carriers in the corresponding sub-block. The SA-Preamble sequences for tone dropping support are obtained by dropping the farthest sub-blocks of the reference bandwidth from the DC sub-carrier on both sides.

9.14 SUPERFRAME HEADERS (BROADCAST CHANNEL)

Mobile WiMAX networks may be configured differently according to specific deployment scenarios by various operators (e.g., urban micro-cell, urban macro-cell, and suburban). The system configuration information is periodically transmitted by a mobile WiMAX-compliant base station which provides details of operational parameters of the air interface. In the IEEE 802.16-2009 standard, the system configuration information is transmitted by a BS using Downlink Channel Descriptor (DCD) and Uplink Channel Descriptor (UCD) broadcast MAC management messages [1]. The DCD and UCD messages contain system configuration information and physical layer characteristics related to the downlink and uplink channels, respectively. The DCD and UCD messages are transmitted by every BS at regular intervals. The information contained in the DCD and UCD messages is received by the mobile stations within the coverage area of the BS. The mobile stations use the information contained in the DCD and UCD messages to learn about the downlink and uplink channel parameters, respectively. The DCD message is always transmitted on a DL burst described by a DL MAP IE with Downlink Interval Usage Code (DIUC) = 0; DIUC = 0 has burst profile parameters that are the same as those used for the transmission of the DL MAP message. Note that HARQ operation is optional in the IEEE 802.16-2009 standard. For FDD/H-FDD operation, if two parameters for a particular type occur in the UCD, the first parameter corresponds to H-FDD Group 1 and the second parameter corresponds to H-FDD Group 2 [1].

The contents of the DCD and UCD messages for the OFDMA physical layer are shown in Table 9-12 and Table 9-13, respectively. The different fields of these two messages are used for different purposes. While some of these fields are present in the DCD/UCD messages in all types of system configurations, other fields are present when certain system configurations are used. This is shown in the second columns of Table 9-12 and Table 9-13 for DCD and UCD messages, respectively. The information fields of the DCD message that are used for all types of system configurations are referred to as mandatory DCD information fields. On the other hand, the information fields of the DCD message that are used only for some system configurations are referred to as configuration-dependent DCD information fields. For example, BS Equivalent Isotropic Radiated Power (EIRP), Transmit/Receive Transition Gap (TTG),

Table 9-12 Contents of a DCD MAC Management Message [1]

Information Field in DCD Message	Usage	Description
Downlink_Burst_Profile	Always	The Downlink_Burst_Profile is a compound TLV encoding that defines and associates a particular DIUC and the physical characteristics that are used with that DIUC which includes a list of physical layer attributes, encoded as TLV values. A Downlink_Burst_Profile is included for each DIUC to be used in the DL-MAP.
BS EIRP	Always	BS Effective Isotropic Radiated Power in units of 1 dBm
Channel Number	Always	The channel center frequency is defined as $Channel\_Center\_Frequency$ (MHz) $= 5000 + 5\,n_{ch}$ where $n_{ch} = 0, 1, ..., 199$ is the Channel Number. This provides an 8-bit unique numbering system for all channels in 5 MHz Steps.
TTG	TDD Mode	Transmit/Receive Transition Gap for TDD and H-FDD.
RTG	TDD Mode	Receive/Transmit Transition Gap for TDD and H-FDD.
EIRxP$_{IR-max}$	Always	Initial ranging maximum equivalent isotropic received power at BS. Signed in units of 1 dBm.
Channel Switch Frame Number	Always	A BS informs its associated terminals of the new channel using the Channel Number in the DCD message. The new channel will be used starting from the frame with the number given by the Channel Switch Frame Number in the DCD message.
Frequency	Always	DL Center Frequency (kHz).
BS-ID	Always	48-bit Base Station Identifier.
HARQ ACK Delay for UL Burst	When HARQ is enabled	1, 2, or 3 frame offset.
Permutation Type of Broadcast Region In HARQ Zone	When HARQ is enabled	PUSC, FUSC, Optional FUSC, AMC.
Maximum Number of HARQ Retransmission	When HARQ is enabled	Maximum number of re-transmissions in DL HARQ and the default value is four re-transmissions.
Default RSSI And CINR Averaging Parameters	Always	Default averaging parameter for physical CINR measurements in multiples of 1/16 and the default averaging parameter for RSSI measurements in multiples of 1/16.

(Continued)

Table 9-12 Contents of a DCD MAC Management Message [1] *Continued*

Information Field in DCD Message	Usage	Description
DL AMC Allocated Physical Bands Bitmap	When AMC is Used	A bitmap describing the physical bands allocated to the segment in the DL when allocating AMC subchannels through the HARQ MAP, or through the normal MAP, or for band AMC CINR reports, or using the optional AMC permutation.
Available DL Radio Resources	Always	Indicates the average ratio of non-assigned DL radio resources to the total usable DL radio resources. The average ratio is calculated over a time interval defined by the *DL_RADIO_RESOURCES_WINDOW_SIZE* parameter. The reported average ratio will serve as a relative load indicator. This value can be adjusted by the operator provided that it reflects a consistent indication of the average loading condition of BSs across the operator network.
FDD DL Gap	FDD Mode	Indicates the location of the residual frame time (this is known as Idle Time in IEEE 802.16m).
FDD Frame Partition Change Timer	FDD Mode	Minimum number of frames (excluding current frame) before next possible change is given by *FDD Frame Partition Change Timer* parameter.
DL Region Definition	Always	OFDMA symbol offset (8 bits), Subchannel offset (6 bits), Number of OFDMA symbols (8 bits), Number of subchannels (6 bits).
Handover Type Support Bitmap	Always	Hard HO, MDHO, FBSS HO, BS_Controlled_HO.
H_Add_Threshold	Always	Threshold used by the MS to add a neighbor BS to the diversity set. When the CINR of a neighbor BS is higher than H_Add, the MS should send MOB_MSHO-REQ message to request adding this neighbor BS to the diversity set. This threshold is used for the MS that performs MDHO/FBSS HO. It is in the unit of decibels. If the BS does not support FBSS HO/MDHO, this value is not set.
H_Delete_Threshold	Always	Threshold used by the MS to drop a BS from the diversity set. When the CINR of a BS is lower than H_Delete, the MS should send MOB_MSHO-REQ to request dropping this BS from the diversity set. This threshold is used for the MS that performs MDHO/FBSS HO. It is in the unit of decibels. If the BS does not support FBSS HO/MDHO, this value is not set.

Parameter	Condition	Description
Anchor Switch Report Slot Length and Switching Period	Always	For values see reference [1].
Paging Group ID	Always	Paging Group Identifier; i.e., One or more logical affiliation grouping of BS.
TUSC1 Permutation Active Subchannels Bitmap	When TUSC1 Permutation is Used	This is a bitmap describing the subchannels allocated to the segment in the DL, when using the TUSC1 permutation.
TUSC2 Permutation Active Subchannels Bitmap	When TUSC2 Permutation is Used	This is a bitmap describing the subchannels allocated to the segment in the DL, when using the TUSC2 permutation.
Hysteresis Margin	Always	Hysteresis margin is used by the MS to include a neighbor BS to a list of possible target BSs. When the CINR of a neighbor BS is larger than the sum of the CINR of the current serving BS and the hysteresis margin for the time-to-trigger duration, then the neighbor BS is included in the list of possible target BSs in MOB_MSHO-REQ. It is the unit of dB and applicable only for handover.
Time-To-Trigger Duration	Always	Time-to-trigger duration is the time, measured in number of frames, used by the MS to decide to select a neighbor BS as a possible target BS. It is applicable only for handover.
Trigger	Always	The Trigger is a compound TLV value that indicates trigger metrics. The trigger in this encoding is defined for serving BS or commonly applied to neighbor BSs.
Noise + Interference	Always	The operator will define the $N + I$ (Noise + Interference) based on the related RF system design calculations.
Downlink Burst Profiles for Multiple FEC Types	When Multiple FEC Types are Used	The DL burst profile is encoded with a type of 1, an 8-bit length, and a 4-bit DIUC. The DIUC field is associated with the DL burst profile and thresholds. The DIUC value is used in the DL-MAP message to specify the burst profile to be used for a specific DL burst.
MBS Zone Identifier List	When MBS is Used	This parameter includes all MBS zone identifiers (i.e., $n \times$ MBS zone identifier) with which BS is associated. An MBS zone identifier is 1 byte long.
BS Restart Count	Always	The value is incremented by one whenever BS restarts.
CDD STC Descriptor	When CDD is Used	This parameter may be transmitted to specify CDD parameters to the MS. It applies to zones with 2 logical antennas and dedicated pilots in STC DL zone IE.

(Continued)

Table 9-12 Contents of a DCD MAC Management Message [1] *Continued*

Information Field in DCD Message	Usage	Description
CDD SISO/SIMO Descriptor	When CDD is Used	This parameter may be transmitted to specify CDD parameters to the MS. It applies to the first PUSC zone as well as zones with 1 logical antenna and dedicated pilots in STC DL zone IE.
Connection Identifier Descriptor	Always	For values see reference [1]
MAC Version	Always	This parameter specifies the version of IEEE 802.16 to which the message originator conforms. If the MAC version values exchanged between a BS and MS during network entry differ such that the BS/MS version is greater than the MS/BS version, the MS may attempt to perform normal operations. The BS may attempt to communicate with the MS per the version specified by the MS, or may decline to interoperate with the MS.
Default HO RSSI and CINR Averaging Parameter	Always	For values see reference [1].
Emergency Service	Always	The Emergency Service is a compound TLV that defines the parameters required for Emergency Service.
DL Coordinated Zone Indication	Always	For values see reference [1].
Cell Type	Always	Cell type parameter may be used by the MS in the network for cell selection and reselection.
NSP Change Count	When BS transmits NSP List Parameter	The value of *NSP Change Count* is programmable. *NSP Change Count* is an incrementing value. A change in *NSP Change Count* signals to an MS that *NSP List* and/or *Verbose NSP Name List* has changed. Inclusion of the *NSP Change Count* is only required if the base station transmits NSP List parameter in MAC management message.
MIH Capability Support	Always	Indicates the IEEE 802.21 Media Independent Handover Services capability of the BS.

Table 9-13 Content of a UCD MAC Management Message [1]

Information Field in UCD Message	Usage	Description
Uplink_Burst_Profile	Always	The Uplink_Burst_Profile is a compound TLV encoding that defines and associates a particular Uplink Interval Usage Code (UIUC) and the physical characteristics that are used with that UIUC which includes a list of physical layer attributes, encoded as TLV values. An Uplink_Burst_Profile is included for each UIUC to be used in the UL-MAP.
Contention-Based Reservation Timeout	Always	Number of UL-MAPs to receive before contention-based reservation is attempted again for the same connection.
Frequency	Always	UL center frequency (kHz).
HO_Ranging_Start	Always	Initial backoff window size for MS performing initial ranging during handover process, expressed as a power of 2.
HO_Ranging_End	Always	Final back-off window size for MS performing initial ranging during handover process, expressed as a power of 2.
Available UL Radio Resources	Always	Indicates the average ratio of non-assigned UL radio resources to the total usable UL radio resources. The average ratio is calculated over a time interval defined by the *UL_RADIO_RESOURCES_WINDOW_SIZE* parameter. The reported average ratio will serve as a relative load indicator. This value can be adjusted by the operator provided it reflects a consistent representation of the average loading condition of BSs across the operator network.
Initial Ranging Codes	Always	Number of initial ranging CDMA codes. Possible values are from 0 to 255. The total number of codes is less than 256.
Periodic Ranging Codes	Always	Number of periodic ranging CDMA codes. Possible values are from 0 to 255. The total number of codes is less than 256.
Bandwidth Request Codes	Always	Number of bandwidth request codes. Possible values are from 0 to 255. The total number of codes is less than 256.
Periodic Ranging Back-off Start	Always	Initial back-off window size for periodic ranging contention, expressed as a power of 2.
Periodic Ranging Back-off End	Always	Final back-off window size for periodic ranging contention, expressed as a power of 2.

(Continued)

Table 9-13 Content of a UCD MAC Management Message [1] *Continued*

Information Field in UCD Message	Usage	Description
Start of Ranging Codes Group	Always	Indicates the starting number of the group of codes used for this UL. If not specified, the default value is set to zero.
Permutation Base	Always	Determines the *UL_PermBase* parameter for the sub-carrier permutation to be used on this UL channel.
UL Allocated Subchannels Bitmap	When PUSC is used in the UL	This is a bitmap describing the physical subchannels allocated to the segment in the UL, when using the UL PUSC permutation.
Optional Permutation UL Allocated Subchannels Bitmap	When optional PUSC is used in the UL	This is a bitmap describing the physical subchannels allocated to the segment in the UL, when using the UL optional PUSC permutation.
Band AMC Allocation Threshold	When band AMC is used in the UL	Threshold of the maximum of the standard deviations of the individual bands CINR measurements over time to trigger mode transition from normal subchannel to band AMC expressed in decibel unit which ranges from -128 to $+127$ dB.
Band AMC Release Threshold	When band AMC is used in the UL	Threshold of the maximum of the standard deviations of the individual bands CINR measurements over time to trigger mode transition from band AMC to normal subchannel expressed in decibel unit which ranges from -128 to $+127$ dB.
Band AMC Allocation Timer	When band AMC is used in the UL	Minimum required number of frames to measure the average and standard deviation for the event of band AMC triggering, which ranges from 0 to 255 frames.
Band AMC Release Timer	When band AMC is used in the UL	Minimum required number of frames to measure the average and standard deviation for the event triggering from band AMC to normal subchannel, which ranges from 0 to 255 frames.
Band Status Reporting Maximum Period	When band AMC is used in the UL	Maximum period between refreshing the band CINR measurement by the unsolicited REP-RSP, which ranges from 0 to 255 frames.
Band AMC Retry Timer	When band AMC is used in the UL	Back-off timer between consecutive mode transitions from normal subchannel to band AMC when the previous request is failed, which ranges from 0 to 255 frames.
Safety Channel Allocation Threshold	Always	In decibels.
Safety Channel Release Threshold	Always	In decibels.
Safety Channel Allocation Timer	Always	In the number of frames.
Safety Channel Release Timer	Always	In the number of frames.
Bin Status Reporting Maximum Period	Always	In the number of frames.
Safety Channel Retry Timer	Always	In the number of frames.

H-ARQ ACK Delay for DL burst	When HARQ is enabled	1, 2, or 3 frame offset.
UL AMC Allocated Physical Bands Bitmap	When AMC is used in the UL	A bitmap describing the physical bands allocated to the segment in the UL. When using the optional AMC permutation with regular MAPs.
CQICH Band AMC-Transition Delay	When band AMC is used in the UL	In the number of frames from 0 to 255.
Maximum Number of HARQ Retransmission	When HARQ is enabled	Maximum number of re-transmissions in UL HARQ, where the default value is 4 re-transmissions.
Normalized C/N Override	Always	This is a list of numbers, where each number is encoded by one nibble (i.e., half an octet or 4 bits) and interpreted as a signed integer. The number encoded by each nibble represents the difference in normalized C/N relative to the previous one.
Size of CQICH_ID Field	Always	0 (default), 3, 4, 5, 6, 7, 8, or 9 bits.
Normalized C/N Override 2	Always	Bits 0–7 are interpreted as a signed integer in dB corresponding to the normalized C/N value, and Bits 8–63 are list of numbers, where each number is encoded by one nibble, and interpreted as a signed integer. The number encoded by each nibble represents the difference in normalized C/N relative to the previous one.
Band AMC Entry Average CINR	When band AMC is used in the UL	Threshold of the average CINR over the entire bandwidth to trigger mode transition from normal subchannel to AMC which ranges from -128 to $+127$ dB.
UpperBound$_{AAS\text{-}Preamble}$	When AAS is used in the UL	Signed in units of 0.25 dB.
LowerBound$_{AAS\text{-}Preamble}$	When AAS is used in the UL	Signed in units of 0.25 dB.
Allow AAS Beam Select Messages	When AAS is used in the UL	A binary flag to indicate whether unsolicited Advanced Antenna System (AAS) Beam Select messages may be sent by the MS.
Use CQICH Indicator Flag	Always	The N MSB values of this field represent the N-bit payload value on the fast-feedback channel reserved as indication flag for MS to initiate feedback on the feedback header, where N is the number of payload bits used for S/N measurement feedback on the fast-feedback channel.
MS-Specific Up Power Offset Adjustment Step	MS-specific and not always present	Unsigned value in units of 0.01 dB.
MS-Specific Down Power Offset Adjustment Step	MS-specific and not always present	Unsigned value in units of 0.01 dB.

(Continued)

Table 9-13 Content of a UCD MAC Management Message [1] *Continued*

Information Field in UCD Message	Usage	Description
Minimum Level of Power Offset Adjustment	Always	Signed value in units of 0.1 dB.
Maximum Level of Power Offset Adjustment	Always	Signed value in units of 0.1 dB.
Handover Ranging Codes	Always	Number of handover ranging CDMA codes. Possible values are 0–255.
Initial Ranging Interval	Always	Number of frames between initial ranging interval allocation.
Transmit Power Report	Always	—
Normalized C/N for Channel Sounding	Always	Signed integer for the required C/N (dB) for Channel Sounding which overrides C/N for the channel sounding.
Initial_Ranging_ Back-off_Start	Always	Initial back-off window size for initial ranging contention expressed as a power of 2.
Initial_Ranging_ Back-off_End	Always	Final back-off window size for initial ranging contention expressed as a power of 2.
Bandwidth_Request_Back-off_Start	Always	Initial back-off window size for contention-based bandwidth request expressed as a power of 2.
Bandwidth_Request_Back-off_End	Always	Final back-off window size for contention-based bandwidth request expressed as a power of 2.
Uplink_Burst_Profile for Multiple FEC Types	When multiple FEC types are used	It defines the format of the Uplink Burst Profile parameter, which is used in the UCD message. The UL burst profile is encoded with a type of 1, an 8-bit length, and a 4-bit UIUC. The UIUC field is associated with the UL burst profile and thresholds. The UIUC value is used in the UL-MAP message to specify the burst profile to be used for a specific UL burst.
Fast Feedback Region	Always	It contains same fields as in the FAST FEEDBACK Allocation IE, as well as the number of subchannels, number of OFDM symbols, subchannel offset, OFDM symbol offset, etc. Note that up to two parameters may be used for FDD/H-FDD, to indicate two fast-feedback regions in two groups.
HARQ ACK Region	Always	It contains the number of subchannels, number of OFDM symbols, subchannel offset, OFDM symbol offset, etc. Note that up to two parameters may be used for FDD/H-FDD.

Parameter	Condition	Description
Ranging Region	Always	The value of the parameter consists of up to four concatenated sections (one section per ranging method), each describing dedicated ranging indicator, ranging method, number of subchannels, number of OFDM symbols, subchannel offset, OFDM symbol offset, etc.
Sounding Region	Always	Five bytes per each sounding region, PAPR Reduction/Safety zone, number of subchannels, number of OFDM symbols, subchannel offset, OFDM symbol offset, etc.
MS Transmit Power Limitation Level	Always	An unsigned 8-bit integer which specifies the maximum allowed MS transmit power whose values indicate power levels in 1 dB steps starting from 0 dBm.
H-FDD Group Switch Delay	FDD/H-FDD Mode	The delay (in number of frames) of H-FDD Group Switching transition. If this parameter is not present, H-FDD Group Switching Delay is set to H-ARQ ACK Delay. For H-FDD, either H-FDD Group Switch Delay or H-ARQ ACK Delay should be included in the UCD.
Frame offset	Always	The offset between the frame of the corresponding CQI channel and the current frame. The offset between the frame of the corresponding UL burst and the current frame.
Country Code	Always	Country code according to List of Mobile Country or Geographical Area Code.
Number of Power Control Command Bits	Always	The value in the Power Control Bitmap is the change that MS applies to its transmit power by changing the offset value.

Receive/Transmit Transition Gap (RTG), and Base Station Identifier (BS-ID) are mandatory DCD information fields, whereas DL Adaptive Modulation and Coding (AMC) allocated physical bands bitmap is configuration-dependent in DCD information fields where it is present when AMC permutation is used. Similarly, Tile Usage of Subchannels Type 1 (TUSC1) permutation active subchannels bitmap is also a configuration-dependent DCD information field when TUSC1 permutation is used. In a similar note, the information fields of the UCD message that are used for all types of system configurations are referred to as mandatory UCD information fields. On the other hand, the information fields of the UCD message that are used only for some system configurations are referred to as configuration-dependent UCD information fields. For example, frequency, and periodic ranging codes are mandatory UCD information fields, whereas Band AMC Allocation Threshold is a configuration-dependent DCD information field when Band AMC permutations are used. For mandatory-DCD/UCD information fields, the third columns of Table 9-12 and Table 9-13 show "Always," whereas configuration-dependent DCD information fields the third column of these tables show the configuration for which configuration-dependent information is included in the DCD/UCD messages. It is noted that the information fields of the DCD/UCD messages are not categorized and transmitted based on necessity and time-sensitivity of the parameters for different system operations such as handover, network entry, etc.

A drawback of the above legacy mechanism to the broadcast system configuration is that the location and content of the DCD/UCD messages would not be known to the MS until the DL MAP (i.e., downlink control channel) is successfully decoded. If the location of the system configuration information (i.e., the broadcast channel) is fixed so that on successful DL synchronization and preamble detection, the broadcast channel containing the system configuration information can be acquired, this would enable the MS to make a decision for attachment to the BS without acquiring and decoding the legacy Frame Control Header (FCH) and DL MAP and waiting for arrival of the DCD/UCD messages (DCD/UCD messages are transmitted every few hundred milliseconds). This modification would further result in power saving in the MS due to shortening and simplification of the initialization procedure, thereby facilitating cell selection or re-selection.

In the IEEE 802.16m, the Superframe Header (SFH) carries essential system parameters and configuration information. The SFH consists of two components, i.e., Primary Superframe Header (P-SFH); and Secondary Superframe Header (S-SFH). The primary superframe header is transmitted every superframe, whereas the secondary superframe header is transmitted every few superframes. The primary and secondary superframe headers are located in the first subframe within a superframe and are time-division-multiplexed with the secondary advanced preamble. The superframe header occupies a narrower bandwidth relative to the system bandwidth (i.e., 5 MHz). The primary superframe header is transmitted using a predetermined modulation and coding scheme. The secondary superframe header is also transmitted using a predetermined modulation scheme, while its repetition factor is signaled via primary superframe header. The primary and secondary superframe headers are transmitted using two spatial streams and space-frequency block coding to improve coverage and reliability. The MS is not required to know the antenna configuration prior to decoding the primary superframe header. The information transmitted in the secondary superframe header is divided into three sub-packets. The secondary superframe header Sub-Packet 1 (SP1) includes information required for network re-entry. The secondary superframe header Sub-Packet 2 (SP2) contains information for initial network entry. The secondary superframe header Sub-Packet 3 (SP3) contains the remaining system information for maintaining communication with the BS. The contents of the superframe header components are shown in Table 9-14.

Table 9-14 Contents of Superframe Header [2]

Superframe Header Component	Periodicity (ms)	Parameters	Size (Bits)	Description
P-SFH	20	LSB of Superframe Number	4	Four least significant bits of superframe number.
		S-SFH Change Count	4	The value of S-SFH change count associated with the S-SFH SP1, SP2, or SP3 IE transmitted in this superframe.
		S-SFH Size Extension	2	
		S-SFH Number of Repetitions	2	The transmission format (repetition factor) used for S-SFH (1, 3, 6).
		S-SFH Scheduling Information	2	0b00: S-SFH includes SP1 IE; 0b01: S-SFH includes SP2 IE; 0b10: S-SFH includes SP3 IE; 0b11: No S-SFH.
		S-SFH SP Change Bitmap	3	The change of S-SFH SP1, SP2, or SP3 IE associated with the S-SFH change count: 0b000: No change in SP1-SP3; 0b001: SP1 has changed; 0b010: SP2 has changed; 0b100: SP3 has changed.
		S-SFH Applying Offset	1	This one-bit flag informs the MS of the S-SFH change count that is considered for applying the system parameters in the S-SFH SPx IEs. If the S-SFH applying offset is set to "0," the MS uses the system parameters of S-SFH SPx IEs associated with the current S-SFH change count. Otherwise, if the S-SFH applying offset is "1," the MS uses the parameters associated with previous S-SFH

(Continued)

Table 9-14 Contents of Superframe Header [2] *Continued*

Superframe Header Component	Periodicity (ms)	Parameters	Size (Bits)	Description
				change count. If the S-SFH contents associated with previous S-SFH change count is not available, the MS will acquire the relevant S-SFH contents by decoding the S-SFH sub-packets at their corresponding scheduled timing.
S-SFH SP1	40	MSB of superframe number	8	Remaining bits of superframe number except those already in P-SFH.
		LSB of 48-bit BS MAC ID (BS-ID)	12	Twelve least significant bits of BS-ID.
		Number of UL ACK/NACK Channels per HARQ Feedback Region	2	Channel numbers represented by the two bits (0, 1, 2, 3) are as follows: For 5 MHz band, 6, 12, 18, 24; For 10 MHz band, 6, 12, 24, 30; For 20 MHz band, 12, 24, 48, 60.
		Number of DL ACK/NACK Channels per HARQ Feedback A-MAP Region	2	Channel numbers represented by the two bits (0, 1, 2, 3) are as follows: For 5 MHz band, 4, 8, 12, 16; For 10 MHz band, 8, 16, 24, 32; For 20 MHz band, 16, 32, 48, 64.
		Power Control Channel Resource Size Indicator	2	Size of power control channel.
		Primary Frequency Partition Location	1	0b0: Reuse-1 partition; 0b1: Power-boosted reuse-3 partition.
		Assignment A-MAP MCS Selection	1	0b0: QPSK 1/2 and QPSK 1/4 can be used for Assignment A-MAP in reuse-1 partition. QPSK 1/2 is used for Assignment A-MAP in the

Parameter	Size (bits)	Description
		power-boosted reuse-3 partition of FFR. 0b1: QPSK 1/2 and QPSK 1/8 can be used for Assignment A-MAP in reuse-1 partition and QPSK 1/4 is used for Assignment A-MAP in the power-boosted reuse 3 partition of FFR.
$DCAS_{SB0}$	5/4/3	For 2048-point FFT size, 5 bits; For 1024-point FFT size, 4 bits; For 512-point FFT size, 3 bits.
$DCAS_{MB0}$	5/4/3	For 2048-point FFT size, 5 bits; For 1024-point FFT size, 4 bits; For 512-point FFT size, 3 bits.
$DCAS_i$	3/2/1	For 2048-point FFT size, 3 bits; For 1024-point FFT size, 2 bits; For 512-point FFT size, 1 bit.
Frame Configuration Index	6	The mapping between value of this index and frame configuration is listed in the reference [2] specification.
Legacy Support	1	Indicates whether frame configuration supports legacy system: 0b0: No legacy support; 0b1: Legacy is supported.
If Legacy System is Supported		Note: The following parameters are transmitted only if legacy systems are supported.
Allocation Periodicity of Ranging Channel	2	The periodicity of ranging channel allocation.
Subframe Offset of Ranging Channel	2	The value of the subframe offset of ranging channel allocation $0 \leq O_{SF} \leq 3$.
Start code Information of Ranging Channel	4	The value of the parameter for the beginning of code group (S) $0 \leq k_s \leq 15$.
Ranging Preamble Code Partition Information	4	The number of initial, handover and periodic codes (N, O and M).
UL_PermBase	7	The UL_PermBase parameter used in the legacy system.

(Continued)

Table 9-14 Contents of Superframe Header [2] *Continued*

Superframe Header Component	Periodicity (ms)	Parameters	Size (Bits)	Description
		If Legacy System is not Supported		Note: The following parameters are transmitted only if legacy systems are supported.
		If S-SFH is Transmitted by an IEEE 802.16m Femto-cell		Note: The following parameters are transmitted by femto BS.
		Allocation Periodicity of Ranging Channel for Synchronized Mobile Stations	2	The periodicity of ranging channel allocation.
		Subframe Offset of Ranging Channel	2	The value of the subframe offset of ranging channel allocation $0 \leq O_{SF} \leq 3$.
		Start code Information of Ranging Channel for Synchronized Mobile Stations	4	The value of the parameter controlling the start root index of ranging preamble codes (r_{nso}) $0 \leq k_{ns} \leq 15$.
		Ranging Preamble Code Partition Information	4	The number of initial, handover, and periodic codes (N, O, and M).
		If S-SFH is not transmitted by an IEEE 802.16m femtocell		Note: The following parameters are transmitted by non-femto BS.
		Allocation periodicity of ranging channel for non-synchronized mobile stations	2	The periodicity of ranging channel allocation.
		Subframe Offset of Ranging Channel	2	The value of subframe offset of ranging channel allocation $0 \leq O_{SF} \leq 3$.
		Start Code Information of Ranging Channel for Non-synchronized Mobile Stations	4	The value of the parameter controlling the start root index of ranging preamble codes (r_{nso}) $0 \leq k_{ns} \leq 15$.
		Ranging preamble code partition information for non-synchronized mobile stations	4	The number of initial ranging and handover ranging preamble codes (N_{IN} and N_{HO}).
		Number of cyclic shifted ranging preamble codes per root index for non-synchronized mobile stations	2	The number of cyclic shifted codes per root index (M_{ns}) for ranging preamble codes.

Ranging Channel Formats for Non-Synchronized Mobile Stations	1	The ranging channel formats.
UCAS_{SB0}	5/4/3	UL CRU/DRU allocation: For 2048-point FFT size, 5 bits; For 1024-point FFT size, 4 bits; For 512-point FFT size, 3 bits.
UCAS_{MB0}	5/4/3	UL CRU/DRU allocation: For 2048-point FFT size, 5 bits; For 1024-point FFT size, 4 bits; For 512-point FFT size, 3 bits.
UCAS_i	3/2/1	UL CRU/DRU allocation: For 2048-point FFT size, 3 bits; For 1024-point FFT size, 2 bits; For 512-point FFT size, 1 bits.
Uplink Subframes for Sounding	3	The number of uplink subframes with sounding symbols: 0b000 – No sounding symbols; 0b001 – 1 subframe; 0b010 – 2 subframes; 0b011 – 3 subframes; 0b100 – 4 subframes. The sounding symbols are located in subframes based on their type. Sounding symbols are allocated in uplink subframes with 6 OFDM symbols starting from the first in time subframe. If the number of uplink subframe of type 2 is less than the number of subframes for sounding, sounding symbols are allocated in the subframes of other types in the following order. The uplink subframes with 5 OFDM symbols are not used for sounding.
BS EIRP	7	Effective Isotropic Radiated Power of the BS in units of 1 dBm.
Cell Bar Information	1	A network load indicator, if Cell Bar bit = 1, the cell does not allowed for network entry or re-entry due to excessive load.

(Continued)

Table 9-14 Contents of Superframe Header [2] *Continued*

Superframe Header Component	Periodicity (ms)	Parameters	Size (Bits)	Description
		$UL\_N\_MAX\_ReTx$	1	The maximum number of re-transmissions for UL HARQ: 0b0: 4 (default); 0b1: 8.
		$DL\_N\_MAX\_ReTx$	1	The maximum number of re-transmissions for DL HARQ: 0b0: 4 (default); 0b1: 8.
		$T_{UL\_RX\_Processing}$	1	Specifies the base station's receive processing time for UL HARQ: 0b0: 3 AAI subframes; 0b1: 4 AAI subframes.
S-SFH SP2	80	If Duplex Mode is FDD		The duplex mode information is obtained from the frame configuration index. Note: The following parameters are transmitted only if operating in FDD mode.
		UL Carrier Frequency	6	The frequency spacing for UL channel is the same as DL channel: 0b000: 512-point FFT; 0b001: 1024-point FFT; 0b010: 2048-point FFT.
		UL Bandwidth	3	
		MSB of 48-bit BS MAC ID	36	36 most significant bits of BS-ID.
		MAC Protocol Revision	4	Version number of IEEE 802.16m MAC supported on this channel.
		DSAC	5/4/3	DL frequency partitioning: For 2048-point FFT size, 5 bits; For 1024-point FFT size, 4 bits; For 512-point FFT size, 3 bits.
		DFPC	4/3/3	DL frequency partitioning: For 2048-point FFT size, 4 bits; For 1024-point FFT size, 3 bits; For 512-point FFT size, 3 bits.
		DFPSC	3/2/1	DL frequency partitioning: For 2048-point FFT size, 3 bits; For 1024-point FFT size, 2 bits; For 512-point FFT size, 1 bit.

Field	Size (bits)	Description
USAC	5/4/3	UL frequency partitioning: For 2048-point FFT size, 5 bits; For 1024-point FFT size, 4 bits; For 512-point FFT size, 3 bits.
UFPC	4/3/3	UL frequency partitioning: For 2048-point FFT size, 4 bits; For 1024-point FFT size, 3 bits; For 512-point FFT size, 3 bits.
UFPSC	3/2/1	UL frequency partitioning: For 2048-point FFT size, 3 bits; For 1024-point FFT size, 2 bits; For 512-point FFT size, 1 bits.
MS Transmit Power Limitation Level	5	The maximum allowed MS transmit power whose values indicate power levels in 1 dB steps starting from 0 dBm.
$EIRxP_{IR,min}$	5	This is an initial ranging channel power control parameter indicating the minimum receive power. The BS_EIRP is the transmission power of the BS
S-SFH SP3	160/320	
Rate of Change of S-SFH Information	4	Minimum time interval in which the contents of the S-SFH do not change (S-SFH SP change bitmap determines which S-SFH sub-packet): 0b0000: 16 × the periodicity of S-SFH SPx superframes; 0b0001: 32 × the periodicity of S-SFH SPx superframes; 0b0010: 64 × the periodicity of S-SFH SPx superframes.
SA-Preamble Sequence Soft Partitioning Information	4	Information on SA-Preamble sequence partitioning for non-macro BS as open and closed subscriber group femto BS.
FFR Partition Resource Metrics	0/4/8	When an MS enters a network with Frequency Partition Count (FPCT) >1, it initially uses the frequency partition indicated by the BS. Once

(Continued)

Table 9-14 Contents of Superframe Header [2] *Continued*

Superframe Header Component	Periodicity (ms)	Parameters	Size (Bits)	Description
				the MS has receives the first superframe with resource metric information, it will start to use the resource metric to recommend its preferred partition to the BS in case of mini-band CRU/DRU, or select the preferred sub-bands in case of sub-band CRU subchannelization. Resource metric of frequency partition FP_0 (reuse-1 partition) has fixed value equal to 1. Resource metric of frequency partition FP_1 with power boosting is calculated as $Resource\_Metric\_FP_1 = 3 - Resouce\_Metric\_FP_2 - Resouce\_Metric\_FP_3$. The number of bits depends on the value of FPCT and the frequency partition size parameters.
		IoT Correction Value for UL Power Control	10	10-bit IoT value is used to support the correction of 5 IoT values ($IoT\_Sounding$, IoT_FP0, IoT_FP1, IoT_FP2, IoT_FP3) defined in AAI_ULPC-NI message (UL Noise and Interference Level Broadcast message), each 2 bits are expressed as the correction value: 0b00: +1 dB; 0b01: +0.5 dB; 0b10: 0 dB; 0b11: −0.5 dB. The correction value is accumulated on IoT values from the latest AAI_ULPC-NI message until the new AAI_ULPC-NI message is received and processed.
		Number of Distributed LRUs for UL Feedback Channel per a UL AAI Subframe	4	The number of UL feedback channels according to UL_FEEDBACK_SIZE.

Name	Size	Description
Number of BS Transmit Antennas	2	Number of BS transmit antennas: 0b00: 2 antennas; 0b01: 4 antennas; 0b10: 8 antennas; 0b11: reserved.
SP Scheduling Periodicity Information	4	The S-SFH sub-packet IEs (i.e., S-SFH SP1 IE, SP 2 IE and SP3 IE) are transmitted by the BS at the scheduled times with different periodicity (i.e., $T_{SP1} < T_{SP2} < T_{SP3}$).
HO Ranging Back-off Start	4	Initial back-off window size for MS performing initial ranging during HO process expressed as a power of 2 ($0 \leq n \leq 15$).
HO Ranging Backoff End	4	Final back-off window size for MS performing initial ranging during HO process expressed as a power of 2 ($0 \leq n \leq 15$).
Initial Ranging Backoff Start	4	Initial back-off window size for initial ranging contention expressed as a power of 2 ($0 \leq n \leq 15$).
Initial Ranging Backoff End	4	Final back-off window size for initial ranging contention expressed as a power of 2 ($0 \leq n \leq 15$).
UL BW-REQ Channel Information	3	0b000: First UL subframe in every superframe; 0b001: First UL subframe in every frame; 0b010: Every UL subframes in every frame.
Bandwidth Request S-SFH SP3 Backoff Start	4	Initial back-off window size for contention bandwidth request expressed as a power of 2 ($0 \leq n \leq 15$).
Bandwidth Request Backoff End	4	Final back-off window size for contention bandwidth request expressed as a power of 2 ($0 \leq n \leq 15$).
fpPowerConfig	4	Power level of each frequency partition in fractional frequency reuse scheme (power boosting/de-boosting values).

The last five OFDM symbols of the first subframe of a superframe are used to transmit the superframe header. All PRUs in the first subframe of a superframe have five OFDM symbols with a two-stream pilot pattern, as defined earlier. Note that the first OFDM symbol of this subframe is allocated to the secondary advanced preamble. The subframe, where the SFH is located, always contains one frequency partition, FP_0. All N_{PRU} PRUs in the subframe where the SFH is located are permuted to generate the distributed LRUs. The permutation and frequency partition of the SFH subframe is described by the parameters $DSAC = 0$ (all mini-bands with no sub-band), $DFPC = 0$ (frequency reuse-1), $DCAS_{SB0} = 0$ (no sub-band CRU allocated), and $DCAS_{MB0} = 0$ (no mini-band CRU allocated). The following parameters are further defined:

- N_{SFH} is the number of distributed LRUs which are occupied by the SFH where $N_{SFH} = N_{P\text{-}SFH} + N_{S\text{-}SFH}$;
- $N_{P\text{-}SFH}$ (fixed) is the number of distributed LRUs which are occupied by the P-SFH ($N_{P\text{-}SFH} = 4$);
- $N_{S\text{-}SFH}$ (variable) is the number of distributed LRUs which are occupied by the S-SFH.

The SFH occupies the first N_{SFH} distributed LRUs in the first subframe of the superframe where $N_{SFH} \leq 24$. The remaining distributed LRUs in that subframe are used for control and data transmission. The first $N_{P\text{-}SFH}$ distributed LRUs of the first subframe are allocated for P-SFH transmission, where $N_{P\text{-}SFH}$ has a fixed value. The S-SFH is mapped to the $N_{S\text{-}SFH}$ distributed LRUs following the $N_{P\text{-}SFH}$ distributed LRUs. The value of parameter $N_{S\text{-}SFH}$ is determined by the S-SFH repetition factor and the total number of information bits in S-SFH IE (variable depending on the type of S-SFH subpacket being transmitted) plus the 16-bit S-SFH CRC. The S-SFH can be repeated over two consecutive superframes to achieve time diversity. The information transmitted in S-SFH is divided into three sub-packets. The sub-packets of S-SFH are transmitted periodically where each sub-packet has a different transmission periodicity, as illustrated in Figure 9-63. The *SP Scheduling Periodicity Information* field of S-SFH SP3 is used to indicate the transmission periodicity of the S-SFH SP1, SP2, and SP3. When no S-SFH sub-packet is contained in the SFH, the S-SFH resources may be used for transmitting other control information or the A-MAPs.

The P-SFH and S-SFH are transmitted using predetermined modulation and coding schemes. The modulation for the P-SFH and the S-SFH is QPSK. The effective coding rate for the P-SFH is 1/24 and the effective coding rate for the S-SFH is configurable and signaled by the P-SFH. The physical processing of the P-SFH IE is shown in Figure 9-64. A 5-bit CRC is appended to P-SFH IE whose generating polynomial is described as $G(x) = x^5 + x^4 + x^2 + 1$. The CRC length for the S-SFH is 16 bits.

A Tail-Biting Convolutional Coder with a rate of 1/4 is used as the minimum code rate for the P-SFH and S-SFH. The modulated symbols are mapped to two transmission streams using an SFBC transmit diversity scheme. The two streams using SFBC are precoded and mapped to the transmit antennas. The MS is not required to know the antenna configuration prior to decoding the P-SFH. When using more than two transmit-antennas, the P-SFH and S-SFH are transmitted using two-stream SFBC with precoding, which is decoded by the MS without any information on the precoding and antenna configuration.

The S-SFH sub-packet IEs (i.e., S-SFH SP1 IE, SP 2 IE, and SP3 IE) are transmitted by the BS with certain period for each sub-packet where $T_{SP1} < T_{SP2} < T_{SP3}$. The periodicity of S-SFH sub-packet scheduling is transmitted in the S-SFH SP3 IE, as shown in Table 9-14. The BS transmits P-SFH IE containing the S-SFH scheduling information bitmap, S-SFH change count, S-SFH SP

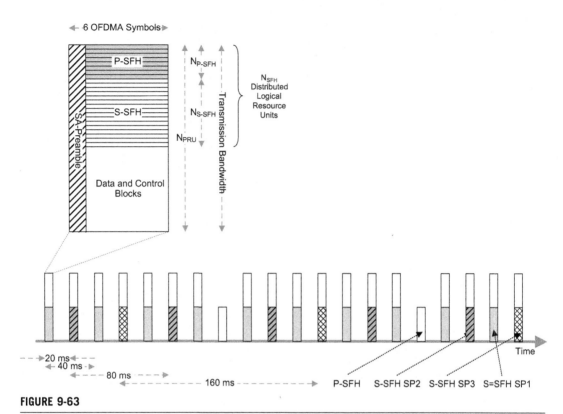

FIGURE 9-63

Structure and timing of superframe header components [2]

change bitmap, and start superframe offset where new S-SFH information is used in every super-frame. The S-SFH change count remains unchanged as long as all of the values (except MSB of superframe number in S-SFH SP1 IE) of the S-SFH SP IEs remain unchanged. The S-SFH change count is incremented by one (modulo 16) whenever any of the values (except MSB of superframe number in S-SFH SP1 IE) of the S-SFH IEs change. The changed S-SFH SP IEs are transmitted at the scheduled superframes intervals corresponding to each S-SFH SP IE. If the MS determines that the *S-SFH Change Count* field in P-SFH has not changed, then it concludes that its copy of the system information is up-to-date.

Each bit of the S-SFH sub-packet change bitmap is an indication of the change of the corresponding S-SFH sub-packet IE where the least and most significant bits are mapped to the S-SFH SP1 IE and S-SFH SP3 IE, respectively. The *Start Superframe Offset* parameter indicates the time relevance of the S-SFH change where the new system parameters are applied. Using the *S-SFH Change Count*, *S-SFH SP Change Bitmap*, and *Start Superframe Offset* parameters, an MS can determine whether it needs to decode S-SFH IE in the current superframe and to update its system parameters. Therefore, the MS does not have to decode the entire superframe headers at every superframe instance, if there is no change in system information.

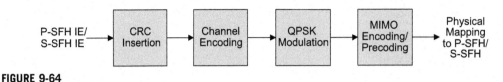

FIGURE 9-64

Physical processing of the P-SFH/S-SFH information element [2]

9.15 3GPP LTE PHYSICAL LAYER PROTOCOLS

This section describes the physical layer procedures in 3GPP LTE Releases 8 through 10, in order to provide the reader with sufficient technical background to understand and contrast the similarities and differences between the IEEE 802.16m and 3GPP LTE. The main functional elements of the physical layer processing and control signaling are described in the following sections [95–117].

9.15.1 Multiple Access Schemes

The 3GPP LTE uses asymmetric multiple access schemes in the downlink and uplink. The multiple-access scheme in the downlink is based on OFDMA with a cyclic prefix and a single-carrier frequency division multiple-access with cyclic prefix is used in the uplink. The OFDMA scheme is particularly suited for frequency-selective channels and high data rates. It transforms a wideband frequency-selective channel into a set of parallel flat-fading narrowband channels. This ideally, allows the receiver to perform a less complex equalization process in the frequency-domain, i.e., single-tap frequency-domain equalization. The 10 ms radio frame is divided into 10 equally sized subframes. Each subframe is further sub-divided into two slots of 0.5 ms length. The basic transmission parameters in the downlink are shown in Table 9-15.

The CP length is chosen to be longer than twice the maximum delay spread and propagation delays in the radio channel. For 3GPP LTE, the normal CP length is set to 4.69 μs, enabling the system to tolerate delay variations due to propagation over cells up to 0.7 km. Note that the insertion of CP increases the Layer 1 overhead and hence reduces the overall throughput. An extended CP length of 16.67 μs would allow support of cell radii as large as 2.5 km. To provide enhanced multicast and broadcast services, LTE has the capability to transmit Multicast and Broadcast content over a Single Frequency Network (MBSFN), where a time-synchronized common waveform is transmitted from multiple eNBs, allowing macro-diversity combining of multi-cell transmissions at the UE. The cyclic prefix is utilized to cover the difference in the propagation delays, which makes the MBSFN transmission appear to the UE as a transmission from a single large cell. Transmission on a dedicated carrier for MBSFN with the possibility of using a longer CP with a sub-carrier bandwidth of 7.5 kHz is supported, as well as transmission of MBSFN on a carrier with both MBMS and unicast transmissions using time division multiplexing [116].

E-UTRA is designed to operate in the frequency bands defined in Table 9-16. The requirements are defined for 1.4, 3, 5, 10, 15, and 20 MHz bandwidth with a specific configuration in terms of number of resource blocks. Figure 9-65 illustrates the relationship between the total channel bandwidth and the transmission bandwidth, i.e., the number of resource blocks. The channel raster is 100 kHz, which means the center frequency must be a multiple of 100 kHz. To support transmission in paired and

Table 9-15 3GPP LTE OFDMA Parameters [30,97]

Parameter		Value					
Channel Bandwidth (MHz)		1.4	3	5	10	15	20
Number of Resource Blocks		6	15	25	50	75	100
Number of Occupied Sub-Carriers		72	180	300	600	900	1200
FFT Size		128	256	512	1024	1536	2048
Sub-Carrier Spacing Δf (kHz)		15 (7.5)					
Sampling Rate (MHz)		1.92	3.84	7.68	15.36	23.04	30.72
Samples per Slot		960	1920	3840	7680	11520	15360
CP Size (μs)	Normal CP ($\Delta f = 15$ kHz)	5.21 (first symbol of the slot) 4.69 (other symbols of the slot) 7 symbols/slot					
	Extended CP ($\Delta f = 15$ kHz)	16.67 6 symbols/slot					
	Extended CP ($\Delta f = 7.5$ kHz)	33.33 3 symbols/slot					

unpaired spectrums, two duplexing schemes are supported: frequency division duplex (allowing both full and half duplex terminal operation); as well as time division duplex.

The basic transmission scheme in the uplink is single-carrier transmission (SC-FDMA) with a cyclic prefix to achieve uplink inter-user orthogonality and to enable efficient frequency-domain equalization at the receiver side. The frequency-domain generation of the SC-FDMA signal, also known as DFT-spread OFDM, is similar to OFDMA and is illustrated in Figure 9-10. This allows for a relatively high degree of commonality with the downlink OFDMA baseband processing using the same parameters, e.g., clock frequency, sub-carrier spacing, FFT/IFFT size, etc. The use of SC-FDMA in the uplink is mainly due to the relatively inferior peak-to-average power ratio properties of OFDMA that result in worse uplink coverage compared to SC-FDMA. The PAPR characteristics are important for cost-effective design of UE power amplifiers.

The sub-carrier mapping determines which part of the spectrum is used for transmission by inserting a suitable number of zeros at the upper and/or lower end, as shown in Figure 9-10. Between each DFT output sample $L - 1$ zeros are inserted. A mapping with $L = 1$ (as shown in Figure 9-10) corresponds to localized transmissions, i.e., transmissions where the DFT outputs are mapped to consecutive sub-carriers. There are two sub-carrier mapping schemes (localized and distributed) that could be used in the uplink. However, 3GPP LTE only specifies localized sub-carrier mapping in the uplink.

9.15.2 Frame Structure

Downlink and uplink transmissions are organized into radio frames with 10 ms duration. 3GPP LTE supports two radio frame structures, Type 1, applicable to an FDD duplex scheme, and Type 2, applicable to a TDD duplex scheme. Frame structure Type 1 is illustrated in the upper portion of Figure 9-66. Each 10 ms radio frame is divided into 10 equally sized subframes. Each subframe consists of two equally sized slots. For FDD, 10 subframes are available for downlink transmission and

Table 9-16 3GPP Band Classes [111,112]

Band Class	Uplink Operating Band eNB Receive/UE Transmit (MHz) F_{UL_low}–F_{UL_high}			Downlink Operating Band eNB Transmit/UE Receive (MHz) F_{DL_low}–F_{DL_high}			Duplex Mode
1	1920	–	1980	2110	–	2170	FDD
2	1850	–	1910	1930	–	1990	FDD
3	1710	–	1785	1805	–	1880	FDD
4	1710	–	1755	2110	–	2155	FDD
5	824	–	849	869	–	894	FDD
6	830	–	840	865	–	875	FDD
7	2500	–	2570	2620	–	2690	FDD
8	880	–	915	925	–	960	FDD
9	1749.9	–	1784.9	1844.9	–	1879.9	FDD
10	1710	–	1770	2110	–	2170	FDD
11	1427.9	–	1447.9	1475.9	–	1495.9	FDD
12	698	–	716	728	–	746	FDD
13	777	–	787	746	–	756	FDD
14	788	–	798	758	–	768	FDD
15	Reserved			–			-
16	Reserved			–			-
17	704	–	716	734	–	746	FDD
18	815	–	830	860	–	875	FDD
19	830	–	845	875	–	890	FDD
20	832	–	862	791	–	821	FDD
21	1447.9	–	1462.9	1495.9	–	1510.9	FDD
22	3410	–	3500	3510	–	3600	FDD
...		–			–		
33	1900	–	1920	1900	–	1920	TDD
34	2010	–	2025	2010	–	2025	TDD
35	1850	–	1910	1850	–	1910	TDD
36	1930	–	1990	1930	–	1990	TDD
37	1910	–	1930	1910	–	1930	TDD
38	2570	–	2620	2570	–	2620	TDD
39	1880	–	1920	1880	–	1920	TDD
40	2300	–	2400	2300	–	2400	TDD
41	3400	–	3600	3400	–	3600	TDD

10 subframes are available for uplink transmissions in each radio frame. Uplink and downlink transmissions are separated in the frequency-domain. The transmission time interval in the downlink and uplink is 1 ms [107].

Frame structure Type 2 is illustrated in the lower portion of Figure 9-66. Each 10 ms radio frame consists of two half-frames of 5 ms each. Each half-frame consists of eight slots 0.5 ms long and three

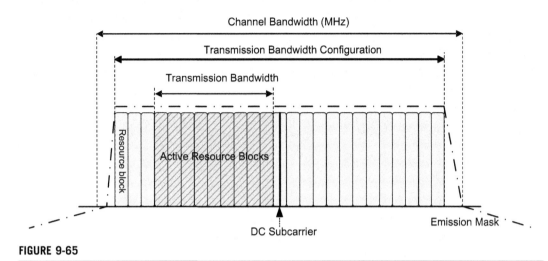

FIGURE 9-65

Illustration of the relationship between channel bandwidth and transmission bandwidth [111,112]

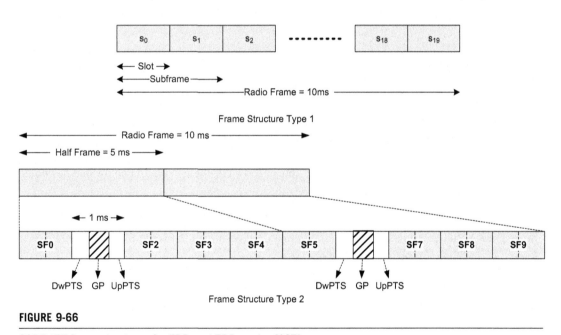

FIGURE 9-66

3GPP LTE frame structures for FDD and TDD modes [107]

special fields: Downlink Pilot Time Slot (DwPTS); Guard Period (GP); and Uplink Pilot Time Slot (UpPTS). The length of DwPTS and UpPTS is configurable subject to the total length of DwPTS, GP, and UpPTS being equal to 1 ms. Different configurations of the special subframe, as specified in 3GPP LTE, have been shown in Table 9-17. In this table T_s denotes the sampling time which is equal to the

Table 9-17 Configuration of Special Subframe (DwPTS/GP/UpPTS) [107]

Special Subframe Configuration	Normal Cyclic Prefix in Downlink			Extended Cyclic Prefix in Downlink		
		UpPTS			UpPTS	
	DwPTS	Normal Cyclic Prefix in Uplink	Extended Cyclic Prefix in Uplink	DwPTS	Normal Cyclic Prefix in Uplink	Extended Cyclic Prefix in Uplink
0	$6592 \cdot T_s$	$2192 \cdot T_s$	$2560 \cdot T_s$	$7680 \cdot T_s$	$2192 \cdot T_s$	$2560 \cdot T_s$
1	$19760 \cdot T_s$			$20480 \cdot T_s$		
2	$21952 \cdot T_s$			$23040 \cdot T_s$		
3	$24144 \cdot T_s$			$25600 \cdot T_s$		
4	$26336 \cdot T_s$			$7680 \cdot T_s$	$4384 \cdot T_s$	$5120 \cdot T_s$
5	$6592 \cdot T_s$	$4384 \cdot T_s$	$5120 \cdot T_s$	$20480 \cdot T_s$		
6	$19760 \cdot T_s$			$23040 \cdot T_s$		
7	$21952 \cdot T_s$			–	–	–
8	$24144 \cdot T_s$			–	–	–

inverse of sampling frequency, i.e., $T_s = 1/(15000 \times 2048)$ [107]. The length of the radio frame, subframe, and slot can alternatively be expressed in the number of time samples which are integer values. Both 5 ms and 10 ms switching-point periodicity are supported. The first subframe in all configurations and the sixth subframe in configuration with 5 ms switching-point periodicity consist of DwPTS, GP, and UpPTS. The sixth subframe in configuration with 10 ms switching-point periodicity consists of DwPTS only. All other subframes consist of two equally sized slots.

For TDD systems, the GP is reserved for downlink to uplink transition. Other subframes/fields are assigned for either downlink or uplink transmission, as shown in Table 9-18. The uplink and downlink transmissions are separated in the time domain.

Table 9-18 Various Permissible Uplink/Downlink Configurations in Frame Structure Type 2 where "S" Denotes the Special Subframe [107]

Configuration	Switching-Point Periodicity	Subframe Number									
		0	1	2	3	4	5	6	7	8	9
0	5 ms	DL	S	UL	UL	UL	DL	S	UL	UL	UL
1	5 ms	DL	S	UL	UL	DL	DL	S	UL	UL	DL
2	5 ms	DL	S	UL	DL	DL	DL	S	UL	DL	DL
3	10 ms	DL	S	UL	UL	UL	DL	DL	DL	DL	DL
4	10 ms	DL	S	UL	UL	DL	DL	DL	DL	DL	DL
5	10 ms	DL	S	UL	DL	DL	DL	DL	DL	DL	DL
6	10 ms	DL	S	UL	UL	UL	DL	S	UL	UL	DL

9.15.3 **Physical Resource Blocks**

The smallest time-frequency resource unit used for downlink/uplink transmission is called a resource element, defined as one sub-carrier over one OFDM/SC-FDMA symbol [107,116]. For both TDD and FDD duplex schemes, as well as in both downlink and uplink, a group of 12 sub-carriers contiguous in frequency over one slot in time form a Resource Block (RB), as shown in Figure 9-67 (corresponding to one slot in the time-domain and 180 kHz in the frequency-domain). Transmissions are allocated in units of resource blocks. One downlink/uplink slot using the normal CP length contains seven symbols. There are 6, 15, 25, 50, 75, and 100 resource blocks corresponding to 1.4, 3, 5, 10, 15, and 20 MHz channel bandwidths, respectively [107]. Note that the

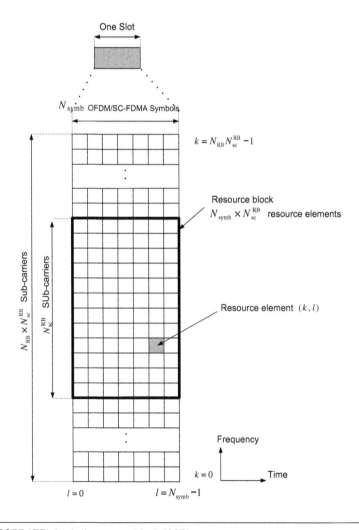

FIGURE 9-67

Structure of the 3GPP LTE physical resource block [107]

resource block size is the same for all bandwidths. The resource blocks are allocated in pairs in the time-domain by the scheduler.

In the uplink, the transmitted signal in each slot is described by a resource grid of $N_{RB}^{UL} N_{sc}^{RB}$ subcarriers and N_{symb}^{UL} SC-FDMA symbols. The resource grid is illustrated in Figure 9-67. The number of resource blocks N_{RB}^{UL} depends on the uplink transmission bandwidth configured in the cell and is bounded as $N_{RB}^{min,UL} \leq N_{RB}^{UL} \leq N_{RB}^{max,UL}$ where $N_{RB}^{min,UL} = 6$ and $N_{RB}^{max,UL} = 110$ are the minimum and maximum number of resource blocks in the uplink which depend on the transmission bandwidth (see Figure 9-65). A physical resource block in the uplink thus consists of $N_{symb}^{UL} N_{sc}^{RB}$ resource elements, corresponding to one slot in the time domain and 180 kHz in the frequency domain.

In the downlink, similarly, the transmitted signal in each slot is described by a resource grid of $N_{RB}^{DL} N_{sc}^{RB}$ sub-carriers and N_{symb}^{DL} OFDM symbols. The resource grid structure is illustrated in Figure 9-67. The quantity N_{RB}^{DL} depends on the downlink transmission bandwidth configured in the cell and satisfies $N_{RB}^{min,DL} \leq N_{RB}^{DL} \leq N_{RB}^{max,DL}$, where $N_{RB}^{min,DL} = 6$ and $N_{RB}^{max,DL} = 110$ are the lower and upper bounds, respectively. The set of allowed values for N_{RB}^{DL} is given in reference [107]. The number of OFDM symbols in a slot depends on the cyclic prefix length and the sub-carrier spacing. A physical resource block is defined as N_{symb}^{DL} consecutive OFDM symbols in the time-domain and N_{sc}^{RB} consecutive sub-carriers in the frequency-domain, where N_{symb}^{DL} and N_{sc}^{RB} are the number of symbols per slot and the number of sub-carriers per physical resource block, respectively. A physical resource block thus consists of $N_{symb}^{DL} N_{sc}^{RB}$ resource elements, corresponding to one slot in the time-domain and 180 kHz in the frequency-domain. Physical resource blocks are numbered from 0 to $N_{RB}^{DL} - 1$ in the frequency-domain.

For multi-antenna transmission, there is one resource grid defined per antenna port. An antenna port is defined by its associated reference signal. The antenna ports are logical entities and there is no one-to-one mapping to the physical antennas. The physical antennas are dynamically mapped to the antenna ports per slot in the time-domain. The set of antenna ports supported depends on the reference signal configuration in the cell and are as follows:

1. Cell-specific reference signals support 1, 2, or 4 antenna ports where the antenna port number p, $p = 0$ for one transmit antenna, $p = 0,1$ for two transmit antennas, and $p = 0,1,2,3$ for four transmit antennas;
2. MBSFN reference signals are transmitted on antenna port $p = 4$;
3. UE-specific reference signals are transmitted on antenna ports $p = 5$, $p = 7$, $p = 8$, or $p = 7,8$;
4. Positioning reference signals are transmitted on antenna port $p = 6$.

The cell-specific reference signals and UE-specific reference signals cannot be simultaneously used; rather, they are time-multiplexed across subframes.

A virtual resource block is defined as the same size as a physical resource block. There are two types of virtual resource blocks: virtual resource blocks of localized type; and virtual resource blocks of distributed type. For each type of virtual resource block, a pair of virtual resource blocks over two slots in a subframe is assigned together by a single virtual resource block number n_{VRB}. Virtual resource blocks of the localized type are mapped directly to physical resource blocks such that virtual resource block n_{VRB} corresponds to physical resource block $n_{PRB} = n_{VRB}$. Virtual resource blocks are numbered from 0 to $N_{VRB}^{DL} - 1$, where $N_{VRB}^{DL} = N_{RB}^{DL}$. In the localized allocations, there is a one-to-one mapping between virtual and physical resource blocks, whereas in the distributed case of resource allocation, the virtual resource block numbers are mapped to physical resource block numbers according to the rule specified in reference [107], and inter-slot hopping is utilized. In the latter, the first part of a virtual resource block pair is mapped to one physical resource block and the other part of the virtual resource block pair is mapped to a physical resource block which is at a predefined gap away (which causes the inter-slot hopping) in order to achieve frequency diversity. This mechanism is especially important for small resource block allocations, due to inherently lower frequency diversity. Note that the localized and distributed allocations in 3GPP LTE are conceptually similar to contiguous and distributed resource units in IEEE 802.16m.

9.15.4 Modulation and Coding

The baseband modulation schemes supported in the downlink and uplink of 3GPP LTE are QPSK, 16 QAM, and 64 QAM. While all three modulation schemes are supported in the downlink in all UE categories, the uplink 64 QAM is only supported in the highest UE category and lower UE categories only support QPSK and 16 QAM. The channel coding scheme for transport blocks in LTE is turbo coding similar to UTRA, with a minimum coding rate of R = 1/3, two 8-state constituent encoders, and a contention-free Quadratic Permutation Polynomial (QPP) turbo interleaver [108]. Trellis termination is performed by taking the tail bits from the shift register feedback after all information bits are encoded. Tail bits are padded after the encoding of information bits. Before the turbo coding, transport blocks are segmented into octet-aligned segments with a maximum information block size of 6144 bits. Error detection is supported by the use of a 24-bit CRC. Link adaptation (adaptive modulation and coding) with various modulation schemes and channel coding rates is applied to the shared data channel. The same coding and modulation is applied to all groups of resource blocks belonging to the same MAC PDU scheduled to one user within one TTI and within a single stream.

The downlink procedures for transport block processing are shown in Figure 9-68. The coding steps for each transport block include addition of CRC to the transport block, code block segmentation and code block CRC attachment, channel coding, rate matching, and code block concatenation. The input bits to the CRC calculation and appending unit are denoted as $a_0, a_1, \ldots, a_{A-1}$ and the parity bits by $p_0, p_1, \ldots, p_{L-1}$ where A is the size of the input sequence and L is the number of parity bits. The output bits after appending CRC bits are denoted by $b_0, b_1, \ldots, b_{B-1}$ where $B = A + L$. The input bit sequence to the code block segmentation is denoted by $b_0, b_1, \ldots, b_{B-1}$ where $B > 0$. If B is larger than the maximum block size of 6144, the input bit sequence is segmented an additional CRC sequence of length $L = 24$ bits is attached to each code block. The output bits from block segmentation unit are denoted by $c_{r0}, c_{r1}, \ldots, c_{r(k-1)}$ where r is the block number and K_r is the number of bits in block number r.

The input bit sequence for a given input block to the channel coding unit is denoted by $c_0, c_1, c_2, c_3, \ldots, c_{k-1}$, where K is the number of bits to be encoded. After encoding the bits are denoted by $d_0^{(i)}, d_1^{(i)}, d_2^{(i)}, d_3^{(i)}, \ldots, d_{D-1}^{(i)}$ where D is the number of encoded bits per output stream and the superscript i identifies the encoder output stream. The relationship between c_k and $d_k^{(i)}$ and further between K and D is determined by the channel coding scheme that is used for the transport block. Tail-biting convolutional coding and turbo coding schemes are used to encode control and traffic channels, respectively. The use of a coding scheme and coding rate for different types of control and traffic information is shown in Table 9-19. The value of D in a tail-biting convolutional coding with a rate of 1/3 is given as $D = K$, and for the turbo coding with a rate of 1/3 is given as $D = K + 4$. The value of the output stream index i is 0, 1, and 2 for both coding schemes.

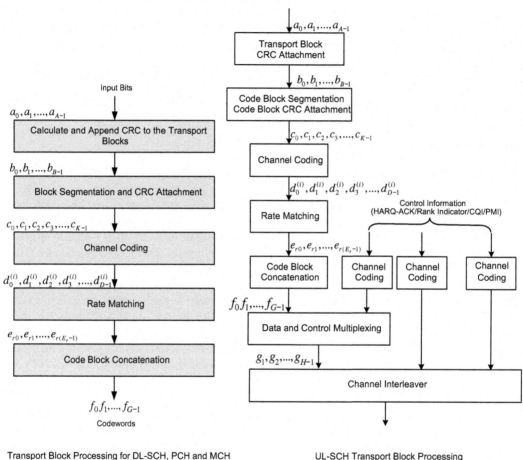

Transport Block Processing for DL-SCH, PCH and MCH

UL-SCH Transport Block Processing

FIGURE 9-68

Downlink/uplink transport block processing [107]

A tail-biting convolutional code with constraint length 7 and coding rate 1/3 is used in 3GPP LTE. The turbo encoder used in 3GPP LTE is a Parallel Concatenated Convolutional Code (PCCC) with two 8-state constituent encoders and one turbo code internal interleaver. The minimum coding rate of the turbo encoder is 1/3. Trellis termination is performed by taking the tail bits from the shift register feedback after all information bits are encoded. Tail bits are padded after the encoding of information bits.

The rate matching for turbo coded transport channels is defined per coded block and consists of interleaving of the three information bit streams $d_k^{(0)}$, $d_k^{(1)}$ and $d_k^{(2)}$ followed by the collection of bits and the generation of a circular buffer, as depicted in Figure 9-69. The output bits for each code block are transmitted as described in the next section. The rate matching output bit sequence is denoted as $e_k, k = 0, \dots, E-1$ where E denotes the rate matching output sequence length. The input bit sequence

Table 9-19 Coding Schemes for Various Channel Types [107]

Channel Types	Channel Types	Coding Scheme	Coding Rate
Traffic Channels	Uplink Shared Channel (UL-SCH)	Turbo Coding	1/3
	Downlink Shared Channel (DL-SCH)		
	Paging Channel (PCH)		
	Multicast Channel (MCH)		
	Broadcast Channel (BCH)	Tail-Biting Convolutional Coding	1/3
Control Channels	Downlink Control Information (DCI)	Tail-Biting Convolutional Coding	1/3
	Control Format Indicator (CFI)	Block Code	1/16
	HARQ Indicator (HI)	Repetition Code	1/3
	Uplink Control Information (UCI)	Block Code	Variable
		Tail-Biting Convolutional Coding	1/3

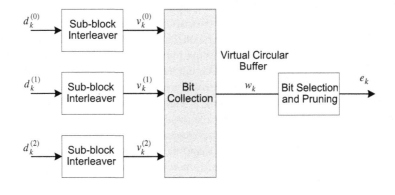

FIGURE 9-69

Rate matching for turbo coded transport channels [107]

to the code block concatenation module is the sequence $e_{rk}, r = 0, ..., C - 1$ and $k = 0, ..., E_r - 1$. The output bit sequence from the code block concatenation block is the sequence $f_k, k = 0, ..., G - 1$. The code block concatenation consists of sequentially concatenating the rate matching outputs for the different code blocks.

The downlink control information transports downlink or uplink scheduling information, requests for aperiodic CQI reports, notifications of MCCH change, or uplink power control commands associated with one RNTI (user identifier). The RNTI is implicitly encoded in the CRC; that is CRC is masked with RNTI. The physical layer processing for the downlink control information includes information element multiplexing, CRC attachment, channel coding, and rate matching.

Figure 9-69 shows the physical layer processing of the UL-SCH transport channel. The data bits $a_0, a_1, ..., a_{A-1}$ enter the coding module in the form of one transport block every transmission time interval. Error detection is provided for UL-SCH transport blocks through a cyclic redundancy check.

The entire transport block is used to calculate the CRC parity bits $p_0, p_1, \ldots, p_{L-1}$ and the output bits $b_0, b_1, \ldots, b_{B-1}$ are generated. Code blocks are then delivered to the channel coding module. The bits in a code block are denoted by $c_{r0}, c_{r1}, \ldots, c_{r(K-1)}$, where r is the code block number, and K_r is the number of bits in code block number r. The total number of code blocks is denoted by C and each code block is individually turbo encoded. After encoding the bits are denoted by $d_{r0}^{(i)}, d_{r1}^{(i)}, \ldots, d_{r(D_r-1)}^{(i)}$, with $i = 0, 1$, and 2 and where D_r is the number of bits in the ith coded stream for code block number r, thus $D_r = K_r + 4$. After rate matching, the bits are denoted by $e_{r0}, e_{r1}, \ldots, e_{r(E_r-1)}$, where r is the coded block number, and where E_r is the number of rate matched bits for code block number r. The bits after code block concatenation are denoted by $f_0, f_1, \ldots, f_{G-1}$, where G is the total number of coded bits for transmission excluding the bits used for control transmission, when control information is multiplexed with the UL-SCH transmission. Uplink control data is fed to the coding unit in the form of channel quality information (CQI and/or PMI), HARQ-ACK, and rank indication. Different coding rates for the control information are achieved by allocating a different number of coded symbols for the transmission. When uplink control data are transmitted in the PUSCH, the channel coding for HARQ-ACK, rank indication, and CQI information $o_0, o_1, \ldots, o_{(O-1)}$, where O is the number of HARQ-ACK bits or rank indicator bits, is individually performed. Two HARQ-ACK feedback modes are supported by higher layer configuration in TDD: HARQ-ACK bundling; and HARQ-ACK multiplexing. For TDD HARQ-ACK bundling, HARQ-ACK consists of one or two bits of information, whereas for the HARQ-ACK multiplexing the HARQ-ACK consists of one to four bits of information.

The control and data multiplexing is performed such that HARQ-ACK information is present on both slots and is mapped to resources around the demodulation reference signals. The multiplexing ensures that control and data information are mapped to different modulation symbols. The inputs to the data and control multiplexing are the coded bits of the control information denoted by $q_0, q_1, \ldots, q_{Q_{CQI}-1}$ and the coded bits of the UL-SCH denoted by $f_0, f_1, \ldots, f_{G-1}$. The output of the data and control multiplexing operation is denoted by $g_0, g_1, \ldots, g_{H-1}$ where H is the total number of coded bits allocated for UL-SCH data and CQI/PMI information. The channel interleaver, in conjunction with the resource element mapping for PUSCH, implements a time-first mapping of modulation symbols onto the transmit waveform while ensuring that the HARQ-ACK information is present on both slots in the subframe and is mapped to resources around the uplink demodulation reference signals.

When uplink control information is transmitted via PUSCH without UL-SCH data, the coding procedure includes channel coding of control information, control information mapping, and channel interleaver. In the physical layer processing of the BCH transport channel, the information arrives at the coding module in the form of one transport block every 40 ms and the coding steps include addition of CRC to the transport block, channel coding, and rate matching. The 16-bit parity is calculated and attached to the BCH transport block. After the attachment, the CRC bits are scrambled according to the eNB transmit antenna configuration, i.e., the CRC bits of the PBCH are masked with the cell identifier. The permissible modulation schemes for various physical channels in the downlink and uplink of LTE are shown in Table 9-20 [116].

9.15.5 Physical Channel Processing

Figure 9-70 illustrates different stages of 3GPP LTE physical channel processing in the downlink and uplink. In the downlink, the coded bits in each of the codewords are scrambled for transmission on a physical channel. The scrambled bits are modulated to generate complex-valued modulation symbols

Table 9-20 Physical Channels, Signals, and their Corresponding Modulation Schemes for 3GPP LTE Downlink and Uplink [107,108]

Downlink	Physical Channels	Physical Broadcast Channel (PBCH)	QPSK
		Physical Downlink Control Channel (PDCCH)	QPSK
		Physical Downlink Shared Channel (PDSCH)	QPSK, 16 QAM, 64 QAM
		Physical Multicast Channel (PMCH)	QPSK, 16 QAM, 64 QAM
		Physical Control Format Indicator Channel (PCFICH)	QPSK
		Physical HARQ Indicator Channel (PHICH)	BPSK modulated on I and Q with the spreading factor 2 or 4 orthogonal codes
	Physical Signals	Reference Signals (RS)	Complex I+jQ pseudo-random sequence of length 31 Gold sequence derived from cell ID
		Primary Synchronization Signal	One of 3 Zadoff-Chu sequences
		Secondary Synchronization Signal	Two 31-bit BPSK M-sequence
Uplink	Physical Channels	Physical Uplink Shared Channel (PUSCH)	QPSK, 16QAM, 64QAM
		Physical Uplink Control Channel (PUCCH)	BPSK, QPSK
		Physical Random Access Channel (PRACH)	u root Zadoff-Chu sequence
	Physical Signals	Demodulation Reference Signal (DRS) (narrowband)	Zadoff-Chu
		Sounding Reference Signal (SRS) (wideband)	Based on Zadoff-Chu

that are later mapped to one or several transmission layers. The complex-valued modulation symbols on each layer are precoded for transmission and are further mapped to resource elements for each antenna port. The complex-valued time-domain OFDMA signal for each antenna port is then generated following these stages [107]. The continuous-time signal $s_l^{(p)}(t)$ on antenna port p in OFDM symbol l in a downlink slot can be expressed as follows:

$$s_l^{(p)}(t) = \sum_{k=-\lfloor N_{RB}^{DL}N_{sc}^{RB}/2\rfloor}^{-1} a_{k'l}^{(p)} e^{j2\pi k\Delta f(t-N_{CP,l}T_s)} + \sum_{k=1}^{\lceil N_{RB}^{DL}N_{sc}^{RB}/2\rceil} a_{k''l}^{(p)} e^{j2\pi k\Delta f(t-N_{CP,l}T_s)}; \quad 0 \le t < (N_{CP,l}+N)T_s$$

(9-40)

where $k' = k + \lfloor N_{RB}^{DL}N_{sc}^{RB}/2\rfloor$ and $k'' = k + \lfloor N_{RB}^{DL}N_{sc}^{RB}/2\rfloor - 1$. The variable N is equal to 2048 for $\Delta f = 15$ kHz sub-carrier spacing and 4096 for $\Delta f = 7.5$ kHz sub-carrier spacing.

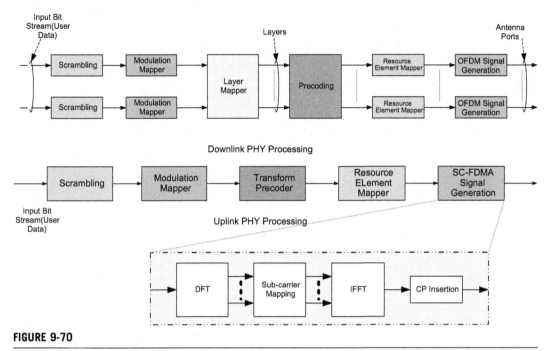

FIGURE 9-70

Overview of physical channel processing [107]

In the uplink, the baseband signal is processed by scrambling the input coded bits and then by modulation of the scrambled bits to generate complex-valued symbols. The complex-valued modulation symbols are transform-precoded (DFT-based precoding) to generate complex-valued symbols that are later mapped to resource elements. The complex-valued time-domain SC-FDMA signal for each antenna port is then generated. The continuous-time signal $s_l(t)$ in SC-FDMA symbol l in an uplink slot is defined as follows:

$$S_1(t) = \sum_{k=-\lfloor N_{RB}^{UL}N_{sc}^{RB}/2\rfloor}^{\lceil N_{RB}^{UL}N_{sc}^{RB}/2\rceil-1} a_{k'l}e^{j2\pi(k+1/2)\Delta f(t-N_{CP,l}T_s)}; \quad 0 \le t < (N_{CP,l}+N)T_s \tag{9-41}$$

where $k' = k + \lfloor N_{RB}^{UL}N_{sc}^{RB}/2 \rfloor$, $N = 2048$, $\Delta f = 15$ kHz, and a_{kl} is the content of resource element (k,l).

The detailed physical channel processing in the downlink, as shown in Figure 9-70, can be described as follows. For each codeword $N_c = 0, 1$, the block of bits $b^{(N_c)}(0), b^{(N_c)}(1), ..., b^{(N_c)}(M_{bit}^{(N_c)} - 1)$, where $M_{bit}^{(N_c)}$ is the number of bits in codeword N_c transmitted on the physical channel in one subframe, are scrambled prior to modulation, resulting in a block of scrambled bits $\tilde{b}^{(N_c)}(0), \tilde{b}^{(N_c)}(1), ..., \tilde{b}^{(N_c)}(M_{bit}^{(N_c)} - 1)$. The scrambling sequence generator is initialized at the start of each subframe, where the initialization value depends on the transport channel type. There are up to two codewords that can be transmitted in one subframe. In the case of single codeword transmission, N_c is equal to zero. For each codeword N_c, the block of scrambled bits $\tilde{b}^{(N_c)}(0), \tilde{b}^{(N_c)}(1), ..., \tilde{b}^{(N_c)}(M_{bit}^{(N_c)} - 1)$ are modulated using one of the permissible modulation schemes (QPSK, 16 QAM, and 64 QAM modulation schemes are used for PDSCH and PMCH), resulting in a block of complex-valued modulation symbols $d^{(N_c)}(0), d^{(N_c)}(1), ..., d^{(N_c)}(M_{symb}^{(N_c)} - 1)$. The complex-valued modulation symbols for each of the codewords to be transmitted are mapped onto one or several layers. The complex-valued modulation symbols $d^{(N_c)}(0), d^{(N_c)}(1), ..., d^{(N_c)}(M_{symb}^{(N_c)} - 1)$ for codeword N_c are mapped onto the layer $x(i) = (x^{(0)}(i), x^{(1)}(i), ..., x^{(N_l-1)}(i))^t, i = 0, 1, ..., M_{symb}^{layer} - 1$ where N_l is the number of layers and M_{symb}^{layer} is the number of modulation symbols per layer.

For transmission on a single antenna port, a single layer is used, i.e., $N_l = 1$ and the mapping is defined as $x^{(0)}(i) = d^{(0)}(i)$ with $M_{symb}^{layer} = M_{symb}^{(0)}$. For spatial multiplexing, the number of layers N_l is less than or equal to the number of antenna ports N_p that are used for transmission of the physical channel. A single codeword can only be mapped to two layers, when the number of antenna ports $N_l = 4$. For transmit diversity, there is only one codeword and the number of layers N_l is equal to the number of antenna ports N_p.

The precoder input is a block of vectors $x(i) = (x^{(0)}(i), x^{(1)}(i), ..., x^{(N_l-1)}(i))^t$, where $i = 0, 1, ..., M_{symb}^{layer} - 1$ from the layer mapping module, and the output of the precoder module is a block of vectors $y(i) = (y^{(0)}(i), y^{(1)}(i), ..., y^{(P-1)}(i))^t, i = 0, 1, ..., M_{symb}^{P} - 1$ that is mapped to resources on each of the antenna ports, where $y^{(p)}(i)$ represents the signal for antenna port p. For transmission on a single antenna port, the precoding function is performed as $y^{(p)}(i) = x^{(0)}(i)$ where $p = 0, 4, 5, 7, 8$ is the number of the single antenna port used for transmission of the physical channel, $i = 0, 1, ..., M_{symb}^{P} - 1$, and $M_{symb}^{P} = M_{symb}^{layer}$. The precoding function for the spatial multiplexing using antenna ports with cell-specific reference signals is only used in conjunction with layer mapping for spatial multiplexing. The spatial multiplexing supports two or four antenna ports and the set of antenna ports used is $p = 0, 1$ or $p = 0, 1, 2, 3$, respectively. If cyclic delay diversity is not used, then the precoding operation for spatial multiplexing can be expressed as follows:

$$
\begin{bmatrix} y^{(0)}(i) \\ \vdots \\ y^{(N_p-1)}(i) \end{bmatrix} = W(i) \begin{bmatrix} x^{(0)}(i) \\ \vdots \\ x^{(N_l-1)}(i) \end{bmatrix} \tag{9-42}
$$

where the precoding matrix $W(i)$ is an $N_p \times N_l$ matrix, $i = 0, 1, ..., M_{symb}^{P} - 1$, and $M_{symb}^{P} = M_{symb}^{layer}$. For spatial multiplexing, the precoding matrix $W(i)$ is selected from subsets of the base codebook configured in the eNB and the UE. The eNB may further limit the precoding matrix selection in the UE to a subset of the elements in the base codebook using codebook subset restrictions. For a large-delay CDD scheme, the precoding for spatial multiplexing is defined by:

$$
\begin{bmatrix} y^{(0)}(i) \\ \vdots \\ y^{(N_p-1)}(i) \end{bmatrix} = W(i)\Lambda(i)U \begin{bmatrix} x^{(0)}(i) \\ \vdots \\ x^{(N_l-1)}(i) \end{bmatrix} \tag{9-43}
$$

where the precoding matrix $W(i)$ is of size $N_p \times N_l$, $i = 0, 1, ..., M_{symb}^{P} - 1$, $M_{symb}^{P} = M_{symb}^{layer}$, $\Lambda(i)$ is a diagonal $N_l \times N_l$ matrix, supporting cyclic delay diversity, and U is a $N_l \times N_l$ matrix. The precoding matrix $W(i)$ is selected among the precoder elements in the codebook configured in the eNB and the UE. The eNB can further confine the precoder selection in the UE to a subset of the elements in the codebook using a codebook subset restriction. As an example, for transmission on two antenna ports $p = 0, 1$, the precoding matrix $W(i)$ is selected from Table 9-21 [107].

Table 9-21 Codebook for Transmission on Antenna Ports $p = 0,1$ [107]

Codebook Index	Number of Layers N_l	
	1	2
0	$\frac{1}{\sqrt{2}}\begin{bmatrix} 1 \\ 1 \end{bmatrix}$	$\frac{1}{\sqrt{2}}\begin{bmatrix} 1 & 0 \\ 0 & 1 \end{bmatrix}$
1	$\frac{1}{\sqrt{2}}\begin{bmatrix} 1 \\ -1 \end{bmatrix}$	$\frac{1}{2}\begin{bmatrix} 1 & 1 \\ 1 & -1 \end{bmatrix}$
2	$\frac{1}{\sqrt{2}}\begin{bmatrix} 1 \\ j \end{bmatrix}$	$\frac{1}{2}\begin{bmatrix} 1 & 1 \\ j & -j \end{bmatrix}$
3	$\frac{1}{\sqrt{2}}\begin{bmatrix} 1 \\ -j \end{bmatrix}$	–

Precoding for transmit diversity is only used in conjunction with layer mapping for transmit diversity. The precoding operation for transmit diversity is defined for two and four antenna ports. For transmission on two antenna ports, $p = 0,1$, the output $\mathbf{y}(i) = (y^{(0)}(i), y^{(1)}(i))^t, i = 0, 1, ..., M^p_{symb} - 1$ of the precoding operation is defined by:

$$\begin{bmatrix} y^{(0)}(2i) \\ y^{(1)}(2i) \\ y^{(0)}(2i+1) \\ y^{(1)}(2i+1) \end{bmatrix} = \frac{1}{\sqrt{2}} \begin{bmatrix} 1 & 0 & j & 0 \\ 0 & -1 & 0 & j \\ 0 & 1 & 0 & j \\ 1 & 0 & -j & 0 \end{bmatrix} \begin{bmatrix} \mathrm{Re}(x^{(0)}(i)) \\ \mathrm{Re}(x^{(1)}(i)) \\ \mathrm{Im}(x^{(0)}(i)) \\ \mathrm{Im}(x^{(1)}(i)) \end{bmatrix} \tag{9-44}$$

where $i = 0, 1, ..., M^{layer}_{symb} - 1$ with $M^p_{symb} = 2M^{layer}_{symb}$. The precoding for spatial multiplexing using antenna ports with UE-specific reference signals is only used in conjunction with layer mapping for spatial multiplexing. Spatial multiplexing using antenna ports with UE-specific reference signals supports two antenna ports and the set of antenna ports used is $p = 7,8$. As an example, for transmission over two antenna ports $p = 7,8$, the precoding operation is defined as:

$$\begin{bmatrix} y^{(7)}(i) \\ y^{(8)}(i) \end{bmatrix} = \begin{bmatrix} x^{(0)}(i) \\ x^{(1)}(i) \end{bmatrix} \tag{9-45}$$

where $i = 0, 1, ..., M^P_{symb} - 1$ and $M^P_{symb} = M^{layer}_{symb}$. For each of the antenna ports used for transmission of the physical channels, the block of complex-valued symbols $y^{(p)}(0), ..., y^{(p)}(M_{symb} - 1)$ are then mapped in sequence starting with $y^{(p)}(0)$ to resource element (k,l) such that they are in the resource blocks corresponding to the virtual resource blocks assigned for transmission, and that those resource elements are not used for transmission of PBCH, synchronization signals, cell-specific reference signals, MBSFN reference signals, or UE-specific reference signals, and further those resource elements are not in an OFDM symbol used for PDCCH. The mapping to resource element (k, l) on antenna port p that is not reserved for other purposes is in increasing order of index k and then index l, starting with the first slot in a subframe.

The detailed physical channel processing in the uplink, as shown in Figure 9-70, can be described as follows. The block of bits $b_0, b_1, ..., b_{M_{bit}-1}$ where M_{bit} denotes the number of bits transmitted on the physical uplink shared channel in one subframe, are scrambled with a UE-specific scrambling sequence prior to modulation, resulting in a block of scrambled bits $\tilde{b}_0, \tilde{b}_1, ..., \tilde{b}_{M_{bit}-1}$ according to a pseudo code. The block of scrambled bits $\tilde{b}_0, \tilde{b}_1, ..., \tilde{b}_{M_{bit}-1}$ are modulated with QPSK, 16 QAM, or 64 QAM, resulting in a block of complex-valued symbols $d_0, d_1, ..., d_{M_{symb}-1}$. Note that the 64 QAM modulation scheme is not supported by all UE categories. The block of complex-valued symbols $d_0, d_1, ..., d_{M_{symb}-1}$ is then divided into M_{symb}/M_{sc}^{PUSCH} sets, each corresponding to one SC-FDMA symbol. Transform precoding is further applied according to the following expression, resulting in a block of complex-valued symbols $z_0, z_1, ..., z_{M_{symb}-1}$:

$$z(lM_{sc}^{PUSCH} + k) = \frac{1}{\sqrt{M_{sc}^{PUSCH}}} \sum_{i=0}^{M_{sc}^{PUSCH}-1} d(lM_{sc}^{PUSCH} + i)e^{-j\frac{2\pi ik}{M_{sc}^{PUSCH}}}; \quad (9\text{-}46)$$

$$k = 0, ..., M_{sc}^{PUSCH} - 1; \quad l = 0, ..., M_{symb}/M_{sc}^{PUSCH} - 1$$

The parameter $M_{sc}^{PUSCH} = M_{RB}^{PUSCH} N_{sc}^{RB}$, where M_{RB}^{PUSCH} denotes the bandwidth of the PUSCH in terms of resource blocks, and $M_{RB}^{PUSCH} = 2^{\alpha_2} 3^{\alpha_3} 5^{\alpha_5} \leq N_{RB}^{UL}$, where α_2, α_3, and α_5 is a set of non-negative integers.

The block of complex-valued symbols $z_0, z_1, ..., z_{M_{symb}-1}$ are multiplied by the amplitude scaling factor β_{PUSCH} in order to conform to the transmit power P_{PUSCH} for the physical uplink shared channel, and sequentially mapped starting with z_0 to physical resource blocks that are allocated for transmission of PUSCH. The mapping to resource elements (k, l), which are assigned for data transmission and not used for transmission of reference signals and not reserved for possible synchronization reference signal transmission, is in increasing order of first the index k, then the index l, starting with the first slot in the subframe. If uplink frequency-hopping is disabled, the set of physical resource blocks to be used for transmission is given by $n_{PRB} = n_{VRB}$ where n_{VRB} is obtained from the uplink scheduling grant [107]. The hopping mode is decided by higher layers and determines whether the hopping is inter-subframe or intra- and inter-subframe. The set of physical resource blocks to be used for transmission are determined based on the type of uplink frequency-hopping. The set of physical resource blocks to be used for transmission in slot n_s is given by the scheduling grant together with a predefined pattern.

9.15.6 Reference Signals

Four types of downlink reference signals are defined: cell-specific reference signals associated with non-MBSFN transmission; MBSFN reference signals associated with MBSFN transmission; UE-specific reference signals (for single-layer and dual-layer transmission); and positioning reference signals. There is one reference signal transmitted per downlink antenna port [107]. The cell-specific downlink reference signals consist of predetermined reference symbols that are inserted in the first and the third before last OFDM symbols of each slot and are used for downlink channel estimation [116]. The exact sequence is derived from cell identifiers. The number of downlink antenna ports with common reference signals equals 1, 2, or 4 (this number has been extended to 8 in 3GPP LTE-Advanced). The reference signal sequence is derived from a pseudo-random sequence and results in a QPSK type constellation. Cell-specific frequency shifts are applied when mapping the reference signal sequence to the sub-carriers. The two-dimensional reference signal sequence is generated as the symbol-by-symbol product of a two-dimensional orthogonal sequence and a two-dimensional pseudo-random sequence. There are three different two-dimensional orthogonal sequences and 168 different two-dimensional pseudo-random sequences. Each cell identity corresponds to a unique combination of one orthogonal sequence and one pseudo-random sequence, thus allowing for 504 unique cell identities (i.e., 168 cell identity groups with three cell identities in each group).

Cell-specific reference signals are transmitted in all downlink subframes in a cell supporting PDSCH transmission. As mentioned earlier, cell-specific reference signals are transmitted on antenna ports 0 to 3 and are defined for sub-carrier spacing of $\Delta f = 15$ kHz. In an MBSFN subframe, the cell-specific reference signals are only transmitted in the non-MBSFN region of the MBSFN subframe. The MBSFN reference signals are transmitted only when the PMCH is transmitted and are defined for an extended cyclic prefix. In that case, the MBSFN reference signals are transmitted on antenna port 4. The UE-specific reference signals (alternatively known as precoded reference signals) are supported for single-antenna-port transmission of PDSCH and are transmitted on antenna ports 5, 7, or 8. The UE-specific reference signals are also supported for spatial multiplexing on antenna ports 7 and 8. These reference signals are utilized for PDSCH demodulation, if the PDSCH transmission is associated with the corresponding antenna port. They are transmitted only on the resource blocks upon which the corresponding PDSCH is mapped. The UE-specific reference signals are not transmitted in resource elements (k, l) on which one of the physical channels or physical signals other than UE-specific reference signal are transmitted using resource elements with the same index pair (k, l), regardless of their antenna port p.

Positioning reference signals are only transmitted in downlink subframes configured for positioning reference signal transmission. If both normal and MBSFN subframes are configured as positioning subframes within a cell, the OFDM symbols in an MBSFN subframe configured for positioning reference signal transmission use the same cyclic prefix as that used for the first subframe. Otherwise, if only MBSFN subframes are configured as positioning subframes within a cell, the OFDM symbols configured for positioning reference signals in these subframes use the extended cyclic prefix. The positioning reference signals are transmitted on antenna port 6 and are not mapped to resource elements (k, l) allocated to physical broadcast, primary, and secondary synchronization channels, regardless of their antenna port p. Moreover, these reference signals are only defined for sub-carrier spacing $\Delta f = 15$ kHz. Figures 9-71 and 9-72 illustrate the mapping of downlink reference signals to antenna ports 0, 1, 2, 3, 4, 5, 7, and 8 [107]. The downlink MBSFN reference signals consist of known reference symbols inserted every other sub-carrier in the 3rd, 7th, and 11th OFDM symbols of subframe in the case of 15 kHz sub-carrier spacing and extended cyclic prefix.

Two types of reference signals are used in the uplink: demodulation reference signals (embedded in each PUCCH and PUSCH transmission and use the same bandwidth as control/data transmission) that are associated with transmission of PUSCH or PUCCH; and sounding reference signals (located in last symbol of a subframe and which can be configured by the network to support uplink frequency-domain scheduling and channel sounding for downlink transmissions, especially for TDD mode; it uses interleaving in frequency-domain to provide additional support for multiple users transmitting sounding signals in the same bandwidth), which is not associated with transmission of PUSCH or PUCCH. The same set of base sequences is used for demodulation and sounding reference signals.

Uplink demodulation reference signals are used for channel estimation in coherent demodulation and are transmitted in the 4th SC-FDMA symbol of the slot assuming normal CP length. The uplink

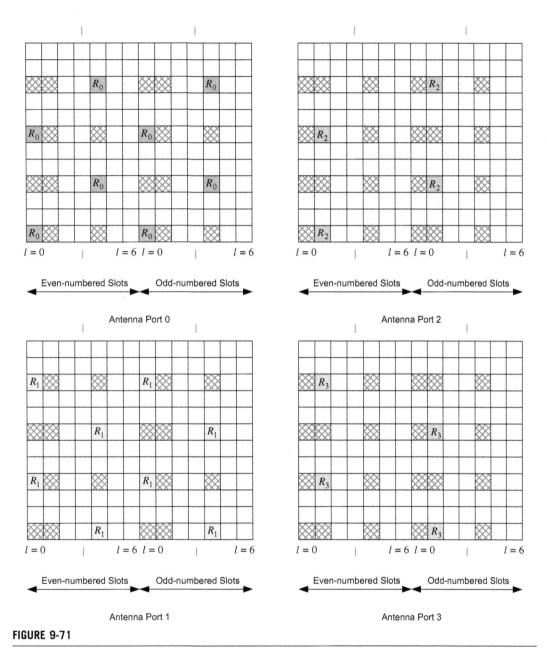

FIGURE 9-71

Mapping of downlink reference signals, antenna ports 0, 1, 2, and 3 with 4 transmit antennas [107]

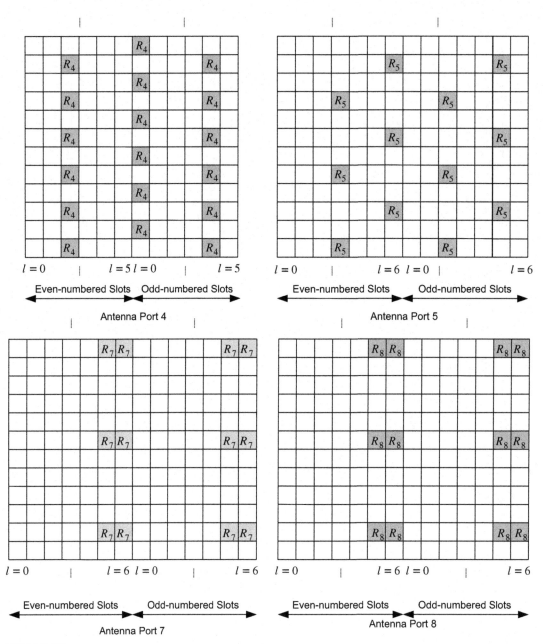

FIGURE 9-72

Mapping of downlink reference signals, antenna ports 4 (MBSFN), 5, 7, and 8 (UE-specific single-layer and dual-layer transmission) [107] reference signals' sequence length equals the size (number of sub-carriers) of the assigned resource. The uplink reference signals are based on prime-length Zadoff-Chu sequences[viii] that are cyclically extended to the desired length. Multiple reference signals can be created either based on different Zadoff-Chu sequences from the same set of Zadoff-Chu sequences or different shifts of the same sequence. The demodulation and sounding reference signals are shown in Figure 9-73.

9.15.7 Physical Control Channels

The Physical Downlink Control Channel (PDCCH) is primarily used to carry scheduling information to individual UEs, i.e., resource assignments for uplink and downlink data and control information. The PDCCH is located in the first few OFDM symbols of a subframe. For frame structure Type 2, PDCCH can also be mapped to the first two OFDM symbols of the DwPTS field. An additional Physical Control Format Indicator Channel (PCFICH), located on specific resource elements in the first OFDM symbol of the subframe, is used to indicate the number of OFDM symbols occupied by the PDCCH (1, 2, 3, or 4 OFDM symbols may be consumed where 4-OFDM-symbol PDCCH is only used when operating in the minimum supported system bandwidth). Depending on the number of users in a cell and the signaling formats conveyed on PDCCH, the size of PDCCH may vary. The information carried in PDCCH is referred to as Downlink Control Information (DCI), where depending on the purpose of the control message, different DCI formats are defined [108]. Note that DCI formats in 3GPP LTE are analogous to IEEE 802.16m A-MAP information elements. As an example, the information content of DCI format 1 is shown in Table 9-22. DCI format 1 is used for the assignment of downlink shared channel resources when no spatial multiplexing is used (i.e., the scheduling information is only provided for one codeword). This

---

[viii]A Zadoff–Chu sequence is a complex-valued sequence with constant amplitude property whose cyclically-shifted versions exhibit low cross-correlation. Thus, under certain conditions, the cyclically-shifted versions of each sequence remains orthogonal to one another. A Zadoff–Chu sequence that has not been shifted is referred to as a *Root Sequence*. The uth root Zadoff–Chu sequence of prime length N is defined as follows [125]:

$$x_u(n) \triangleq \begin{cases} e^{-j[\pi un(n+1)]/N} & 0 \leq n < N - 1 \quad (N \text{ is an odd integer}) \\ e^{-j\pi un^2/N} & 0 \leq n < N - 1 \quad (N \text{ is an even integer}) \end{cases}$$

where N is an integer, which denotes the length of the Zadoff-Chu sequence. It is easily verified that $x_u(n)$ is periodic with period N; i.e., $x_u(n) = x_u(n + N)$, $\forall n$. In other word, the sequence index u is prime relative to N. For a fixed value of u, the Zadoff-Chu sequence has an ideal periodic auto-correlation property (i.e., the periodic auto-correlation is zero for all time shifts other than zero). For different values of index u, the Zadoff-Chu sequences are not orthogonal, rather exhibit low cross-correlation. If the sequence length N is selected as a prime number, there are different sequences with periodic cross-correlation of $1/\sqrt{N}$ between any two sequences regardless of time shift. The Zadoff–Chu sequences are a subset of Constant Amplitude Zero Autocorrelation (CAZAC) sequences. The properties of Zadoff-Chu sequences can be summarized as follows [125]:

1. They are periodic with period N if N is a prime number; i.e., $x_u(n + N) = x_u(n)$
2. Given N is a prime number, the DFT of a Zadoff–Chu sequence is another Zadoff–Chu sequence conjugated and time-scaled multiplied by a constant factor; i.e., $X_u[k] = x_u^*(vk)X_u[0]$ where v is the multiplicative inverse of u modulo N. It can be shown that $x_u^*(vk) = x_v^*(k)e^{j\pi(1-v)k/N}$.
3. The autocorrelation of a prime-length Zadoff–Chu sequence with a cyclically shifted version of itself also yields zero auto-correlation sequence; i.e., it is non-zero only at one instant which corresponds to the cyclic shift 0.
4. The cross correlation between two prime-length Zadoff–Chu sequences; i.e., different u, is constant and equal to $1/\sqrt{N}$.
5. The Zadoff-Chu sequences have low PAPR.

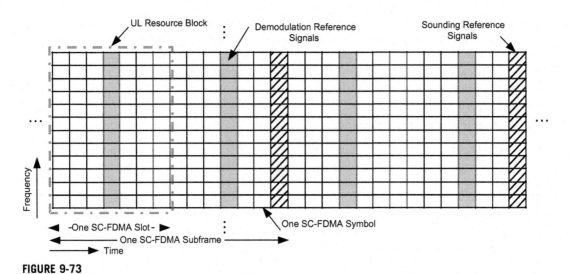

FIGURE 9-73

Uplink physical resource blocks and reference signals

Table 9-22 Information Content of DCI Format 1 [108]		
Information Type	**Number of Bits in PDCCH**	**Usage**
Resource allocation header	1	Indicates whether resource allocation type 0 or 1 is used.
Resource block assignment	Depending on resource allocation type	Indicates resource blocks to be assigned to the UE.
Modulation and coding scheme	5	Indicates modulation scheme and, together with the number of allocated physical resource blocks, the transport block size.
HARQ process number	3 (TDD), 4 (FDD)	Identifies the HARQ process to which the packet is associated.
New data indicator	1	Indicates whether the packet is a new transmission or a re-transmission.
Redundancy version	2	Identifies the redundancy version used for coding the packet.
Transmit Power Control (TPC) command for PUCCH	2	Transmit power control command for adapting the transmit power on the physical uplink control channel.
Downlink assignment index (TDD only)	2	Number of downlink subframes for uplink ACK/NACK bundling.

information enables the UE to identify the location and size of the resources in that subframe, as well as provide the UE with information on the modulation and coding scheme, and HARQ operation.

As mentioned earlier, the information fields in the downlink scheduling grant are used to convey the information needed to demodulate the downlink shared channel. They include resource allocation information such as resource block size and duration of assignment, transmission format such as multi-antenna mode, modulation scheme, payload size, and HARQ operational parameters such as process number, redundancy version, and new data indicator. Similar information is also included in the uplink scheduling grants. There are 11 DCI formats specified as follows [108]:

1. Format 0: uplink allocation (scheduling of PUSCH);
2. Format 1: downlink allocation for one PDSCH codeword;
3. Format 1A: compact downlink allocation for one PDSCH codeword and random access procedure initiated by a PDCCH order;
4. Format 1B: compact downlink allocation for one PDSCH codeword with precoding information;
5. Format 1C: very compact downlink allocation for one PDSCH codeword and notifying MCCH change;
6. Format 1D: compact downlink allocation for one PDSCH codeword with precoding and power offset information;
7. Format 2: downlink allocation for UEs configured in closed-loop spatial multiplexing mode;
8. Format 2A: downlink allocation for UEs configured in open-loop spatial multiplexing mode;
9. Format 2B: downlink allocation with dual-layer transmission (corresponding to antenna ports 7 and 8) with spatial multiplexing;
10. Format 3: transmission of Transmit Power Control (TPC) commands for PUCCH and PUSCH with 2-bit power adjustments;
11. Format 3A: transmission of TPC commands for PUCCH and PUSCH with 1-bit power adjustments.

The CRC of the DCI is scrambled with the UE identity that is used to find the allocation by the UE. In order to efficiently use the resources in PDCCH, other DCI formats are defined which are optimized for specific use cases and transmission modes, including scheduling of paging channel, random access response, and system information blocks. DCI formats 2 and 2A provide downlink shared channel assignments in case of closed-loop or open-loop spatial multiplexing, respectively. In these cases, scheduling information is provided for two codewords within one control message. In addition, DCI format 0 conveys uplink scheduling grants, and DCI formats 3 and 3A carry transmit power control commands for the uplink.

There are different ways to signal the resource allocation within the DCI, in order to trade-off between signaling overhead and flexibility. For example, DCI format 1 may use resource allocation types 0 or 1 as described in the following. An additional resource allocation type 2 is specified for other DCI formats. In resource allocation type 0, a bitmap indicates the resource block groups that are allocated to a UE. A Resource Block Group (RGB) is defined which consists of a set of consecutive physical resource blocks (1 to 4 depending on system bandwidth). The allocated resource block groups do not have to be adjacent in frequency-domain. Figure 9-74 illustrates example resource block groups for a 20 MHz bandwidth. In resource allocation type 1, a bitmap indicates physical resource blocks inside a selected resource block group subset. The information field for the resource block assignment on PDCCH is divided into three fields: one field indicates the selected resource block group subset; a 1-bit field indicates whether an offset shall be applied when the bitmap is mapped to the resource

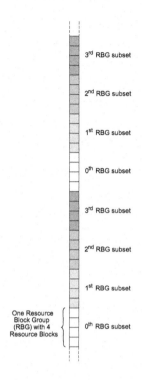

FIGURE 9-74

Resource block groups for resource allocation type 0/1 (example: 20 MHz bandwidth and each resource block group contains 4 resource blocks)

blocks; and another field contains the bitmap that indicates specific physical resource blocks inside the resource block group subset.

In resource allocation type 2, physical resource blocks are not directly allocated. Instead, virtual resource blocks are allocated which are then mapped to physical resource blocks. The information field for the resource block assignment carried on PDCCH contains a Resource Indication Value (RIV) from which the starting virtual resource block and a length in terms of contiguously allocated virtual resource blocks can be derived. Both localized and distributed virtual resource block assignment is possible; they are differentiated by a one-bit flag within the DCI. In the localized case, there is a one-to-one mapping between virtual and physical resource blocks. In the distributed case of resource allocation type 2, the virtual resource block numbers are mapped to physical resource block numbers according to a certain rule, [30] and inter-slot hopping is applied. Frequency diversity is achieved if the first part of a virtual resource block pair is mapped to one physical resource block and the other part of the virtual resource block pair is mapped to a physical resource block which is at a predefined distance in frequency from the first one, resulting in inter-slot hopping. This mechanism is especially useful for small resource block allocations. The UE decodes the resource allocation field depending on the PDCCH DCI format. A resource allocation field in each PDCCH includes two parts: a resource allocation header field; and information consisting of the actual resource block assignment. The PDCCH DCI formats 1, 2, 2A, and 2B with type 0, and PDCCH DCI formats 1, 2, 2A, and 2B with type 1 resource allocation have the same format and are distinguished from each other via the single bit resource allocation header field which exists depending on the downlink system bandwidth. The PDCCH with DCI format 1A, 1B, 1C, and 1D have a type 2 resource allocation while PDCCH with DCI format 1, 2, 2A, and 2B have a type 0 or type 1 resource allocation. The PDCCH DCI formats with a type 2 resource allocation do not have a resource allocation header field. The UE would discard the PDSCH resource allocation in the corresponding PDCCH, if consistent control information is not detected (see Figures 9-75 and 9-76).

In 3GPP LTE, several downlink control channels are scheduled in one subframe and a UE is expected to monitor a number of those control channels in two search spaces (blind detection): (1) common search space; and (2) UE-specific search space. Each channel carries information associated with an RNTI (UE identifier). The minimum code rate R = 1/3 with constraint length K = 7 tail-biting convolution code is used with QPSK modulation for the control channel. Higher and lower code rates are generated through rate matching. The control and data regions are time-division multiplexed. Each scheduling grant is defined based on fixed-size Control Channel Elements (CCEs), which are combined in a predetermined manner to achieve different coding rates. Each CCE consists of multiple mini-Ces, also known as Resource Element Groups (REGs), that are distributed across time and frequency control resource. Interleaving of the REGs is performed using a sub-block interleaver that is configured on a cell-specific basis. Note that the number of control channel elements or the number of control channel symbols in the subframe is transmitted by the eNB in every subframe. Since multiple control channel elements can be combined in order to reduce the effective coding rate, a UE control channel assignment would then be based on channel quality information reported. The UE monitors a set of candidate control channels in the common and/or UE-specific search spaces, as shown in Table 9-23, which may be configured by higher layer signaling. The size of the control channel elements varies with different bandwidth allocation and is a multiple of 6. It may be noted that 1, 2, 4, and 8 control channel elements can be aggregated to obtain approximate code rates of 2/3, 1/3, 1/6, and 1/12 (i.e., code repetition).

The control region consists of a set of CCEs numbered from 0 to $N_{CCE_k} - 1$, where N_{CCE_k} is the total number of CCEs in the control region of subframe k. The UE monitors a set of PDCCH candidates for control information in every non-DRX subframe, where monitoring implies attempting to decode each of the PDCCHs in the set according to all the monitored DCI formats. The set of PDCCH candidates to monitor are defined in terms of search spaces, where a search space at an aggregation level $L = 1, 2, 4, 8$ is defined by a set of PDCCH candidates [109].

The UE is required to monitor one common search space at each of the aggregation levels 4 and 8, and one UE-specific search space at each of the aggregation levels 1, 2, 4, 8. The common and UE-specific search spaces may overlap. The aggregation levels defining the search spaces are listed in Table 9-23. The DCI formats that the UE monitors depend on the configured transmission mode.

As mentioned earlier, a physical control channel is transmitted on an aggregation of one or several consecutive CCEs, where a control channel element corresponds to nine resource element groups. The number of resource-element groups not assigned to PCFICH or PHICH is N_{REG}. The CCEs available in the system are numbered from 0 to $N_{CCE} - 1$, where $N_{CCE} = floor(N_{REG}/9)$. The PDCCH supports multiple formats as listed in Table 9-24. The PDCCH format should not be confused with the DCI format. A PDCCH consisting of n consecutive CCEs may only start on a CCE satisfying $i \bmod n = 0$, where i is the CCE number. Multiple PDCCHs can be transmitted in a subframe.

The Physical HARQ Indicator Channel (PHICH) is used to acknowledge uplink data transmission. The resources used for the HARQ acknowledgment channel are configured on a semi-static basis and are defined independently of the grant channel (i.e., a set of resource elements are semi-statically allocated for

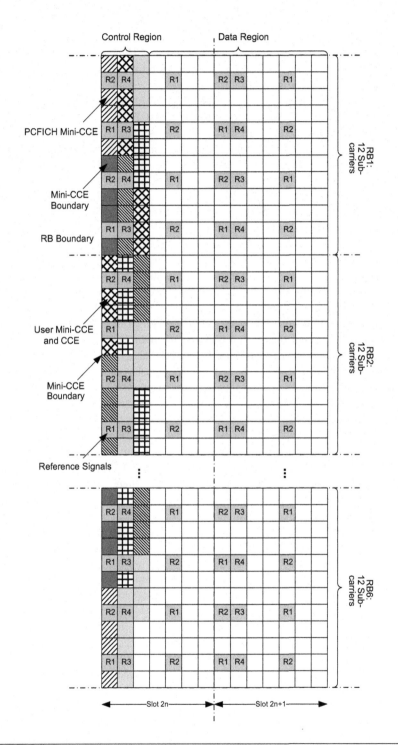

FIGURE 9-75

Structure of 3GPP LTE downlink control channels [92]

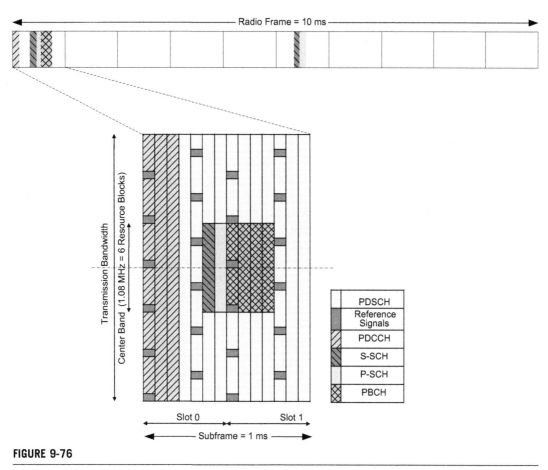

FIGURE 9-76

Structure of P-SCH, S-SCH, and PBCH in time and frequency

Table 9-23 PDCCH Candidates Monitored by a UE [109]

Type	Search Space — Aggregation Level	Search Space — Size in Number of CCEs	Number of PDCCH candidates	DCI Format
UE-specific search space	1	6	6	0, 1, 1A,1B, 2
	2	12	6	
	4	8	2	
	8	16	2	
Common search space	4	16	4	0, 1A, 1C, 3/3A
	8	16	2	

Table 9-24 PDCCH Formats [107]

PDCCH Format	Number of CCEs	Number of Resource-Element Groups	Number of PDCCH Bits
0	1	9	72
1	2	18	144
2	4	36	288
3	8	72	576

this purpose). Since only one information bit is to be transmitted, a combination of CDM/FDM multiplexing among acknowledgments is used. Hybrid CDM/FDM allows for power control between acknowledgments for different users and provides good interference averaging. In addition, it can provide frequency diversity for different users. The ACK/NACK resource assignment is based on an implicit relationship based on the resource block assignment. With BPSK modulation and I/Q multiplexing, each PHICH channel can carry up to eight acknowledgments for a normal cyclic prefix [107].

The Physical Control Format Indicator Channel (PCFICH) is used to dynamically indicate the number of OFDM symbols used for the control region in a subframe. The PCFICH is transmitted in the first OFDM symbol of the subframe and the three values are indicated by three sequences 16 QPSK symbols in length. Predefined codewords based on (3, 2) simplex coding with repetition and systematic bits with $d_{min} = 21$ is used. To provide maximum frequency diversity, the PCFICH is transmitted over the entire system bandwidth. Transmit diversity is also supported using the same diversity scheme as the PDCCH. In addition, cell specific scrambling, tied to the cell ID, is utilized.

The PBCH has a fixed and predefined transmission format and is broadcast over the entire coverage area of the cell. In 3GPP LTE, the broadcast channel is used to transmit the System Information which is essential for network access. Due to the large size of the System Information field, it is divided into two segments: Master Information Block (MIB) which is transmitted in PBCH; and System Information Blocks (SIB) that are transmitted in PDSCH. The PBCH contains the basic system configuration parameters necessary to demodulate the PDSCH which contains the remaining *System Information Blocks*. The PBCH is characterized by a single fixed-size transport block per TTI, QPSK modulation scheme, it is transmitted on 72 active sub-carriers that are centered around the DC sub-carrier, and no HARQ is used for PBCH (see Figure 9-76). The *Master Information Block* is transmitted on the PBCH over 40 ms as shown in Figure 9-77. The CRC masking is used to implicitly provide the UE the number of transmit antennas at the eNB (1, 2 or 4). Tail-biting convolutional coding with minimum rate of R = 1/3 and repetition is used, and the coded bits are rate-matched to 1920 bits for normal CP and 1728 bits for extended cyclic prefix. The coded BCH transport block is mapped to four subframes within a 40 ms interval and the 40 ms timing is blindly detected, i.e., there is no explicit signaling indicating 40 ms timing, and each subframe is assumed to be self-decodable, i.e., the BCH can be decoded from a single reception, assuming sufficiently good channel conditions.

The Physical Uplink Control Channel (PUCCH) carries Uplink Control Information (UCI) including ACK/NACK information related to data packets received in the downlink, CQI reports, precoding matrix index, Rank Indication (RI) for MIMO, and Scheduling Requests (SR). The PUCCH is transmitted on a reserved frequency region in the uplink which is configured by higher layers. Figure 9-78 shows an example for a PUCCH resource allocation. One resource block is reserved at the

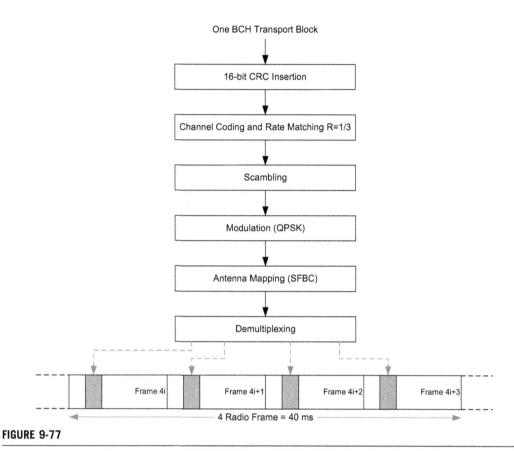

FIGURE 9-77

Master information block transmission over PBCH

edge of the bandwidth and inter-slot hopping is applied. In TDD mode, the PUCCH is not transmitted in special subframes. Note that a UE only uses PUCCH when it does not have any data to transmit on PUSCH. If a UE has data to transmit on PUSCH, it will multiplex the control information with data on PUSCH. Depending on the type of information that PUCCH can carry, different PUCCH formats are specified, see Table 9-25. When a UE sends a HARQ ACK/NACK in response to a downlink PDSCH transmission, it will derive the exact PUCCH resource to use from the PDCCH transmission (i.e., the number of the first control channel element used for the transmission of the corresponding downlink resource assignment). When a UE has a scheduling request or CQI to send, higher layers will configure the exact PUCCH resource. The PUCCH formats 1, 1a, and 1b are based on cyclic shifts from a Zadoff-Chu type of sequence [102,107]; i.e., the modulated data symbol is multiplied by the cyclically shifted sequence. The cyclic shift varies between symbols and slots. The higher layers may impose a limitation that disallows some of the cyclic shifts available in a cell. In addition, a spreading with an orthogonal sequence is applied. The PUCCH formats 1, 1a, and 1b carry three reference symbols per slot in case of a normal cyclic prefix (located on SC-FDMA symbols 2, 3, and 4 as shown in Figure 9-78). For PUCCH formats 1a and 1b, when both ACK/NACK and SR are transmitted in the

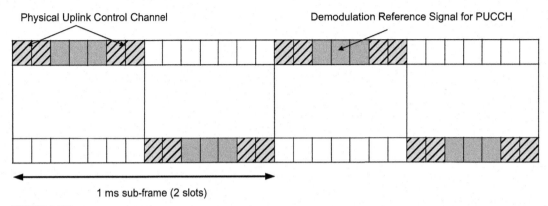

FIGURE 9-78

Structure of physical uplink control channel for frame structure type 1 and PUCCH format 1, 1a, and 1b

same subframe, the UE transmits ACK/NACK on its assigned ACK/NACK resource for negative SR transmission, and transmits ACK/NACK on its assigned SR resource for positive SR transmission. In PUCCH formats 2, 2a, and 2b, the bits for transmission are first scrambled and QPSK modulated. The resulting symbols are then multiplied with a cyclically shifted Zadoff-Chu type of sequence, where again the cyclic shift varies between symbols and slots [30]. PUCCH formats 2, 2a, and 2b carry two reference symbols per slot in case of normal cyclic prefix (located on SC-FDMA symbol numbers 1, 5). A resource block can either be configured to support a mix of PUCCH formats 2/2a/2b and 1/1a/1b, or to support formats 2/2a/2b exclusively [102].

As mentioned earlier, 3GPP LTE supports intra-subframe and inter-subframe frequency hopping, which is configured per cell by higher layers. Either both intra-subframe and inter-subframe hopping or only inter-subframe hopping is supported. In intra-subframe hopping or inter-slot hopping, the UE hops to another frequency allocation from one slot to another within one subframe. In inter-subframe

Table 9-25 PUCCH Formats and Contents

PUCCH Format	Contents	Modulation Scheme	Number of Bits per Subframe
1	Scheduling Request (SR)	N/A	N/A (*Information is carried by presence or absence of transmission*)
1a	ACK/NACK, ACK/NACK + SR	BPSK	1
1b	ACK/NACK, ACK/NACK + SR	QPSK	2
2	CQI/PMI or RI (any CP length), (CQI/PMI or RI) + ACK/NACK (extended CP)	QPSK	20
2a	(CQI/PMI or RI)+ACK/NACK (normal CP)	QPSK + BPSK	21
2b	(CQI/PMI or RI)+ACK/NACK (normal CP)	QPSK + QPSK	22

hopping, the frequency resource allocation changes from one subframe to another. The uplink scheduling grant in DCI format 0 contains a single-bit flag for switching hopping on or off. Also, the UE is instructed to use type 1 or type 2 frequency-hopping and it receives the index of the first resource block of the uplink allocation. Type 1 hopping corresponds to the use of an explicit offset in the second slot resource allocation. The offset between the consecutive slots can be different and adjustable. The offset is indicated to the UE within the resource block assignment or hopping resource allocation field in DCI format 0. Type 2 hopping refers to the use of a predefined hopping pattern [102]. The bandwidth available for PUSCH is divided into sub-bands (e.g., 4 sub-bands with 5 resource blocks each in 5 MHz bandwidth), and the hopping is performed between sub-bands (from one slot or subframe to another, depending on whether intra- or inter-subframe is configured). Furthermore, mirroring can be applied according to a mirroring function, which means that the resource block allocation starts from the other direction of the sub-band they are located in. Note that in case of type 2 hopping, the resource allocation for the UE cannot be larger than the sub-band configured. The UE will first determine the allocated resource blocks after applying all frequency hopping rules. Then the data is mapped into these resources, first in sub-carrier order and then in symbol order.

9.15.8 Downlink and Uplink HARQ

An asynchronous HARQ protocol is supported in the downlink. The UE can request re-transmission of data packets that were incorrectly received on PDSCH. ACK/NACK information is transmitted in uplink, either on PUCCH or multiplexed within uplink data transmission on PUSCH. Up to eight HARQ concurrent processes are supported in the downlink. The ACK/NACK transmission in FDD mode refers to the downlink packet that was received four subframes earlier. In TDD mode, the uplink ACK/NACK timing depends on the uplink/downlink configuration. For TDD, the use of a single ACK/NACK response for multiple PDSCH transmissions is possible; this is referred to as ACK/NACK bundling.

A synchronous HARQ re-transmission protocol is supported in the uplink. The eNB can request re-transmissions of incorrectly received data packets on PUSCH. The ACK/NACK feedback in the downlink is sent on PHICH. After a PUSCH transmission, the UE will monitor the corresponding PHICH resource four subframes later in FDD mode. In TDD mode, the PHICH subframe to monitor is derived from the uplink/downlink configuration and from the PUSCH subframe number. The PHICH resource is determined from the lowest index physical resource block of the uplink resource allocation and the uplink demodulation reference symbol cyclic shift associated with the PUSCH transmission, both indicated in the PDCCH with DCI format 0 granting the PUSCH transmission. A PHICH group consists of multiple PHICHs that are mapped to the same set of resource elements and are separated through different orthogonal sequences. The UE derives the PHICH group number and a specific PHICH inside that group from the information on the lowest resource block number in the PUSCH allocation and the cyclic shift of the demodulation reference signal. The UE can derive the redundancy version to use on PUSCH from the uplink scheduling grant in DCI format 0. The 3GPP LTE supports up to eight HARQ concurrent processes in the uplink for FDD, while for TDD the number of HARQ processes depends on the uplink-downlink configuration.

9.15.9 Physical Random Access Channel

The physical layer random access preamble, illustrated in Figure 9-79 consists of a cyclic prefix of length T_{CP} and a sequence part of length T_{SEQ}. There are five random access preamble formats specified by the standard, as shown in Table 9-26, where the parameter values depend on the frame structure and the random access configuration. Higher layers control the preamble format. Each random access preamble occupies a bandwidth corresponding to six consecutive resource blocks for both frame structures. The transmission of a random access preamble, if triggered by the MAC layer, is restricted to certain time and frequency resources. These resources are enumerated in increasing order of the subframe number within the radio frame and the physical resource blocks in the frequency domain, such that index 0 correspond to the lowest numbered physical resource block and subframe within the radio frame.

The random access preambles are generated from Zadoff-Chu sequences with a zero correlation zone. The network configures the set of preamble sequences the UE is allowed to use. There are 64 preambles available in each cell. The set of 64 preamble sequences in a cell is found by including first, in the order of increasing cyclic shift, all the available cyclic shifts of a root Zadoff-Chu sequence with the logical index RACH_ROOT_SEQUENCE, where RACH_ROOT_SEQUENCE is broadcast as part of the system information [107]. The random access procedure in 3GPP LTE consists of four steps. In step 1, the preamble is sent by UE. The time/frequency resource where the preamble is sent is associated with a Random Access Radio Network Temporary Identifier (RA-RNTI). In step 2, a random access response is generated by the eNB and sent on the downlink shared channel. It is addressed to the UE using the temporary identifier and contains a timing advance value, an uplink grant, and a temporary identifier C-RNTI. Note that the eNB may generate multiple random access responses for different UEs which can be concatenated inside one MAC protocol data unit. The preamble identifier is contained in the MAC sub-header of each random access response so that the UE can detect whether a random access response for the used preamble exists. In step 3, the UE will send an RRC CONNECTION REQUEST message on the uplink common control channel,

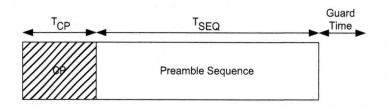

FIGURE 9-79

Random access preamble structure

Table 9-26 Random Access Parameters [107]		
Preamble Format	T_{CP}	T_{SEQ}
0	($\sim$100 μs) $3168T_s$	($\sim$800 μs) $24576T_s$
1	($\sim$680 μs) $21024T_s$	($\sim$800 μs) $24576T_s$
2	($\sim$200 μs) $6240T_s$	($\sim$1600 μs) $245762T_s$
3	($\sim$680 μs) $21024T_s$	($\sim$1600 μs) $245762T_s$
4 (frame structure type 2)	($\sim$15 μs) $448T_s$	($\sim$130 μs) $4096T_s$

based on the uplink grant received in the previous step. In step 4, the eNB sends back a MAC PDU containing the uplink CCCH service data unit that was received in step 3. The message is sent on the downlink shared channel and addressed to the UE via the temporary identifier C-RNTI. When the received message matches the one sent in step 3, the contention resolution is considered successful; [30,109] otherwise the procedure is restarted at step 1. Paging is used for setting up network-initiated connection. A power-saving and efficient paging procedure should allow the UE to sleep without having to process any information and only to briefly wake up at predefined intervals to monitor paging information (as part of control signaling in the beginning of each subframe) from the network [30,109].

9.15.10 Cell Search

Cell search is a procedure performed by a UE to acquire timing and frequency synchronization with a cell and to detect the cell ID of that cell. 3GPP LTE uses a hierarchical cell search scheme similar to WCDMA. E-UTRA cell search is based on successful acquisition of the following downlink signals: (1) the primary and secondary synchronization signals that are transmitted twice per radio frame; and (2) the downlink reference signals based on the broadcast channel that carries system information such as system bandwidth, number of transmit antennas, and system frame number. The 504 available physical layer cell identities are grouped into 168 physical layer cell identity groups, each group containing three unique identities. The secondary synchronization signal carries the physical layer cell identity group, and the primary synchronization signal carries the physical layer identity 0, 1, or 2 [95,102,116].

As shown in Figure 9-76, the primary synchronization signal is transmitted on the sixth symbol of slots 0 and 10 of each Type 1 radio frame and occupies 62 sub-carriers, centered on the DC sub-carrier. The primary synchronization signal is generated from a frequency-domain Zadoff-Chu sequence. The secondary synchronization signal is transmitted on the fifth symbol of slots 0 and 10 of each radio frame. It occupies 62 sub-carriers centered on the DC sub-carrier. The secondary synchronization signal is an interleaved concatenation of two length-31 M-sequences. The concatenated sequence is scrambled with a scrambling sequence given by the primary synchronization signal [30,95,102,116]. As shown in Figure 9-72, PBCH is transmitted on symbols 0 to 3 of slot 1 and occupies 72 sub-carriers centered on the DC sub-carrier. The coded BCH transport block is mapped to four subframes within a 40 ms interval (see Figure 9-77). The 40 ms timing is blindly detected, i.e., there is no explicit signaling to indicate 40 ms timing [30]. These channels are contained within the central 1.08 MHz (corresponding to six resource blocks) frequency band so that the system operation can be independent of the channel bandwidth.

As shown in Figure 9-80, during cell search different types of information need to be identified by the UE. This information includes symbol and radio frame timing, frequency, cell identification, overall transmission bandwidth, antenna configuration, and cyclic prefix length. The first step of cell search in 3GPP LTE is based on specific synchronization signals, a primary synchronization signal, and a secondary synchronization signal. The synchronization signals are transmitted twice per 10 ms on predefined slots. In addition, the Master Information Block carried on PBCH which contains the basic physical layer information such as system bandwidth, number of transmit antennas, and system frame number must be detected.

9.15.11 PDSCH Transmission Modes

The eNB determines the downlink transmit energy per resource element. A UE may assume downlink cell-specific RS Energy per Resource Element (EPRE) is constant across the downlink system bandwidth and constant across all subframes until a different cell-specific RS power information is received. The downlink reference-signal EPRE can be derived from the downlink reference-signal transmit-power given by the parameter *reference-signal-power* provided by higher layers. The downlink reference-signal transmit power is defined as the linear average over the power contributions of all resource elements that carry cell-specific reference signals within the operating system bandwidth. The UE is semi-statically configured via higher layer signaling to receive PDSCH data transmissions signaled via PDCCH according to one of eight transmission modes denoted by mode 1 to mode 8.

In FDD mode, the UE is not expected to receive PDSCH resource blocks transmitted on antenna port 5 in any subframe in which the number of OFDM symbols for PDCCH with normal CP is equal to four. Furthermore, the UE is not expected to receive PDSCH resource blocks transmitted on antenna port 5, 7, or 8 in the two PRBs to which a pair of VRBs is mapped if either one of the two PRBs

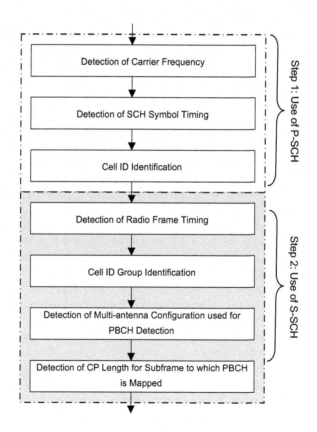

FIGURE 9-80

Cell search procedure

overlaps in frequency with a transmission of either PBCH or primary or secondary synchronization signals in the same subframe. The UE is also not expected to receive PDSCH resource blocks transmitted on antenna port 7, for which distributed VRB resource allocation is assigned. The UE may skip decoding the transport block(s) if it does not receive all assigned PDSCH resource blocks. If the UE skips decoding, the physical layer indicates to higher layer that the transport block(s) are not successfully decoded.

Table 9-27 PDCCH and PDSCH Configured by C-RNTI [109]

Transmission Mode	DCI Format	Search Space	Transmission Scheme of PDSCH Corresponding to PDCCH
Mode 1	DCI format 1A	Common and UE specific by C-RNTI	Single-Antenna Port, Port 0
	DCI format 1	UE specific by C-RNTI	Single-Antenna Port, Port 0
Mode 2	DCI format 1A	Common and UE specific by C-RNTI	Transmit Diversity
	DCI format 1	UE specific by C-RNTI	Transmit Diversity
Mode 3	DCI format 1A	Common and UE specific by C-RNTI	Transmit Diversity
	DCI format 2A	UE specific by C-RNTI	Large Delay CDD or Transmit Diversity
Mode 4	DCI format 1A	Common and UE specific by C-RNTI	Transmit Diversity
	DCI format 2	UE specific by C-RNTI	Closed-loop Spatial Multiplexing or Transmit Diversity
Mode 5	DCI format 1A	Common and UE specific by C-RNTI	Transmit Diversity
	DCI format 1D	UE specific by C-RNTI	Multi-User MIMO
Mode 6	DCI format 1A	Common and UE specific by C-RNTI	Transmit Diversity
	DCI format 1B	UE specific by C-RNTI	Closed-loop Spatial Multiplexing using a Single Transmission Layer
Mode 7	DCI format 1A	Common and UE specific by C-RNTI	If the Number of PBCH Antenna Ports is One, Single-Antenna Port, Port 0 is used; Otherwise, Transmit Diversity
	DCI format 1	UE specific by C-RNTI	Single-Antenna Port; Port 5
Mode 8	DCI format 1A	Common and UE specific by C-RNTI	If the Number of PBCH Antenna Ports is One, Single-Antenna Port, Port 0 is used, Otherwise, Transmit Diversity
	DCI format 2B	UE specific by C-RNTI	Dual-Layer Transmission; Port 7 And 8 or Single-Antenna Port; Port 7 or 8

In TDD mode, the UE is not expected to receive PDSCH resource blocks transmitted on antenna port 5 in any subframe in which the number of OFDM symbols for PDCCH with normal CP is equal to four. In addition, the UE is not expected to receive PDSCH resource blocks transmitted on antenna port 5 in the two PRBs to which a pair of VRBs is mapped if either one of the two PRBs overlaps in frequency with a transmission of PBCH in the same subframe. The UE is not expected to receive PDSCH resource blocks transmitted on antenna port 7 or 8 in the two PRBs to which a pair of VRBs is mapped if either one of the two PRBs overlaps in frequency with a transmission of primary or secondary synchronization signals in the same subframe. With normal CP configuration, the UE is not expected to receive PDSCH on antenna port 5 for which distributed VRB resource allocation is assigned in the special subframe with configuration 1 or 6. The UE is not expected to receive PDSCH on antenna port 7 for which distributed VRB resource allocation is assigned. The UE may skip decoding the transport blocks, if it does not receive all assigned PDSCH resource blocks. If the UE skips decoding, the physical layer indicates to higher layer that the transport block(s) are not successfully decoded. If a UE is configured by higher layers to decode PDCCH with CRC scrambled by the C-RNTI, the UE decodes the PDCCH and any corresponding PDSCH according to the respective combinations defined in Table 9-27.

References

[1]   IEEE 802.16-2009, IEEE Standard for Local and Metropolitan Area Networks, Part 16: Air Interface for Broadband Wireless Access Systems, May 2009.
[2]   P802.16m/D6, IEEE Standard for Local and metropolitan area networks – Part 16: Air Interface for Broadband Wireless Access Systems, Advanced Air Interface, May 2010.
[3]   IEEE 802.16m-07/002r10, IEEE 802.16m System Requirements, January 2010 <http://ieee802.org/16/tgm/index.html>.
[4]   IEEE 802.16m–08/004r5, IEEE 802.16m Evaluation Methodology Document, January 2009 <http://ieee802.org/16/tgm/index.html>.
[5]   IEEE 802.16m–08/0034r3, IEEE 802.16m System Description Document, May 2010 <http://ieee802.org/16/tgm/index.html>.
[6]   WiMAX Forum Mobile System Profile, Release 1.0 Approved Specification (Revision 1.7.1: 2008-11-07), <http://www.wimaxforum.org/technology/documents>.
[7]   WiMAX Forum Network Architecture Release 1.5 Version 1, Stage 2: Architecture Tenets, Reference Model and Reference Points, September 2009 <http://www.wimaxforum.org/resources/documents/technical/release>.
[8]   Report ITU-R M.2135-1, Guidelines for Evaluation of Radio Interface Technologies for IMT-Advanced, December 2009.
[9]   Report ITU-R M.2134, Requirements Related to Technical System Performance for IMT-Advanced Radio Interface(s), November 2008.
[10] IMT-Advanced submission and evaluation process <http://www.itu.int/ITU-R/>.
[11] WiMAX System Evaluation Methodology, July 2008 <http://www.wimaxforum.org/technology/documents>.
[12] Bernard Sklar, "Rayleigh Fading Channels in Mobile Digital Communication Systems. I. Characterization," IEEE Communications Magazine Volume 35 (Issue 7), July 1997.
[13] Bernard Sklar, "Rayleigh Fading Channels in Mobile Digital Communication Systems. II. Mitigation," IEEE Communications Magazine Volume 35 (Issue 7), July 1997.

[14] Bernard Sklar, Digital Communications: Fundamentals and Applications, second ed., Prentice Hall, January 2001.

[15] Arogyaswami Paulraj, et al., Introduction to Space-Time Wireless Communications Cambridge University Press, June 2008.

[16] Y. Okumura, E. Ohmori, K. Fukuda, "Field Strength and its Variability in VHF and UHF Land Mobile Radio Service," Review of the Electrical Communication Laboratory No. 16, September-October, 1968.

[17] M. Hata, "Empirical Formulae for Propagation Loss in Land Mobile Radio Services," IEEE Transactions on Vehicular Technology vol. VT-29 (No. 3), 1980.

[18] William C. Jakes, Microwave Mobile Communications John Wiley & Sons, 1974.

[19] 3GPP TR 25.892, Feasibility Study for Orthogonal Frequency Division Multiplexing (OFDM) for UTRAN Enhancement, March 2004.

[20] Yiyan Wu, William Y. Zou, "Orthogonal Frequency Division Multiplexing: A Multi-Carrier Modulation Scheme," IEEE Transactions on Consumer Electronics vol. 41 (No. 3), August 1995.

[21] Hu Su, et al., "Analysis of Tone Reservation Method for WiMAX System," International Symposium on Communications and Information Technologies, ISCIT '06, October 2006.

[22] Savo G. Glisic, Advanced Wireless Communications: 4G Cognitive and Cooperative Broadband Technology, second ed., Wiley Inter-science, September 2007.

[23] A. Osseiran, Jiann-Ching Guey, "Hopping Pilot Pattern for Interference Mitigation in OFDM," IEEE 19th International Symposium on Personal, Indoor and Mobile Radio Communications, September 2008.

[24] Richard Nilsson, Ove Edforst, Magnus Sandellt, Per Ola Borjesson, "An Analysis of Two-Dimensional Pilot-Symbol Assisted Modulation for OFDM," 1997 IEEE International Conference on Personal Wireless Communications Publication, December 1997.

[25] D.E. Dudgeon, R.M. Mersereau, Multidimensional Digital Signal Processing Prentice Hall, 1984.

[26] IEEE 802.16m–07/244r1, Yang Song, Liyu Cai, Keying Wu, Hongwei Yang, Collaborative MIMO, November 2007.

[27] IEEE C802.16m–10/0042, Zheng Yan-Xiu, et al., Modification to Resource Allocation in Sub-band Assignment A-MAP IE, March 2010.

[28] IEEE C802.16m–10/0320r1, Sudhir Ramakrishna, et al., Simplifications to the Specifications of the Sub-band Assignment A-MAP IE Proposed Amendment Text, March 2010.

[29] David Tse, Pramod Viswanath, Fundamentals of Wireless Communication Cambridge University Press, June 2005.

[30] Stefania Sesia, Issam Toufik, Matthew Baker, LTE, The UMTS Long Term Evolution: From Theory to Practice John Wiley & Sons, 2009.

[31] Ezio Biglieri, et al., MIMO Wireless Communications Cambridge University Press, 2010.

[32] Andrea Goldsmith, Wireless Communications Cambridge University Press, 2005.

[33] Hemanth Sampath, Petre Stoica, Arogyaswami Paulraj, "Generalized Linear Precoder and Decoder Design for MIMO Channels Using the Weighted MMSE Criterion," IEEE Transactions on Communications vol. 49 (No. 12), December 2001.

[34] Yang-Seok Choi, J. Peter Voltz, A. Frank Cassara, "On Channel Estimation and Detection for Multicarrier Signals in Fast and Selective Rayleigh Fading Channels," IEEE Transactions on Communications vol. 49 (No. 8), August 2001.

[35] M. Siavash Alamouti, "A Simple Transmit Diversity Technique for Wireless Communications," IEEE Journal on Selected Areas in Communications vol. 16 (No. 8), October 1998.

[36] Aria Nosratinia, E. Todd Hunter, Ahmadreza Hedayat, "Cooperative Communication in Wireless Networks," IEEE Communications Magazine, October 2004.

[37] G. Jeffrey Andrews, "Interference Cancellation for Cellular Systems: A Contemporary Overview," IEEE Wireless Communications, April 2005.

[38] David Gesbert, et al., "From Theory to Practice: An Overview of MIMO Space–Time Coded Wireless Systems," IEEE Journal on Selected Areas in Communications vol. 21 (No. 3), April 2003.

[39] D. RossMurch, Khaled Ben Letaief, "Antenna Systems for Broadband Wireless Access," IEEE Communications Magazine, April 2002.

[40] Severine Catreux, et al., "Adaptive Modulation and MIMO Coding for Broadband Wireless Data Networks," IEEE Communications Magazine, June 2002.

[41] J. Arogyaswami Paulraj, A. Dhananjay Gore, U. Rohit Nabar, Helmut Bölcskei, "An Overview of MIMO Communications – A Key to Gigabit Wireless," Proceedings of the IEEE vol. 92 (No. 2), February 2004.

[42] G. Matthew Parker, G. Kenneth Paterson, Chintha Tellambura, Golay Complementary Sequences, January 2004.

[43] J. David Love, W. Robert Heath Jr., "Grassmannian Beamforming for Multiple-Input Multiple-Output Wireless Systems," IEEE Transactions on Information Theory vol. 49 (No. 10), October 2003.

[44] Vahid Tarokh, Hamid Jafarkhani, "On the Computation and Reduction of the Peak-to-Average Power Ratio in Multicarrier Communications," IEEE Transactions on Communications vol. 48 (No. 1), January 2000.

[45] M. Timothy Schmid, C. Donald Cox, "Robust Frequency and Timing Synchronization for OFDM," IEEE Transactions on Communications vol. 45 (No. 12), December 1997.

[46] Markku Pukkila, Channel Estimation Modeling, Postgraduate Course in Radio-communications, Fall 2000.

[47] Muhammad Saad Akram, Pilot-based Channel Estimation in OFDM Systems Master Thesis, June 2007.

[48] Freescale Semiconductor Application Note, Channel Estimation in OFDM Systems AN3059 Rev. 0, January 2006.

[49] Vineet Srivastava, et al., "Robust MMSE Channel Estimation in OFDM Systems with Practical Timing Synchronization," IEEE Wireless Communications and Networking Conference, 2004.

[50] Lizhong Zheng, N.C. David Tse, "Diversity and Multiplexing: A Fundamental Tradeoff in Multiple-Antenna Channels," IEEE Transactions on Information Theory vol. 49 (No. 5), May 2003.

[51] Hanan Weingarten, Yossef Steinberg, Shlomo Shamai, "The Capacity Region of the Gaussian Multiple-Input Multiple-Output Broadcast Channel," IEEE Transactions on Information Theory vol. 52 (No. 9), September 2006.

[52] Florian Kaltenberger et al., "Capacity of Linear Multi-User MIMO Precoding Schemes with Measured Channel Data," EURECOM, Sophia-Antipolis, France.

[53] V. Stankovic and M. Haardt, "Multi-User MIMO Downlink Precoding for Users With Multiple Antennas," Proceedings of the 12th meeting of the Wireless World Research Forum (WWRF), Toronto, Canada, November 2004.

[54] Q.H. Spencer, A.L. Swindlehurst, M. Haardt, "Zero-Forcing Methods for Downlink Spatial Multiplexing in Multiuser MIMO Channels," IEEE Transactions on Signal Processing vol. 52 (No. 2), February 2004.

[55] Nihar Jindal, "MIMO Broadcast Channels with Finite-Rate Feedback," IEEE Transactions on Information Theory vol. 52 (No. 11), November 2006.

[56] H. Quentin Spencer, B. Christian Peel, A. Lee Swindlehurst, Martin Haardt, "An Introduction to the Multi-User MIMO Downlink," IEEE Communications Magazine, October 2004.

[57] David Gesbert, et al., "Shifting the MIMO Paradigm," IEEE Signal Processing Magazine, September 2007.

[58] Lu Wei, "Capacity of Hybrid Open-loop and Closed-loop MIMO with Channel Uncertainty at Transmitter," Helsinki University of Technology, ESPOO, March 2008.

[59] J. David Love, W. Robert Heath Jr., "Grassmannian Precoding for Spatial Multiplexing Systems," Proc. of the Allerton Conference on Communication Control and Computing, Monticello, October 2003.

[60] Tetsushi Abe, Gerhard Bauch, "Differential Codebook MIMO Precoding Technique," 2007 IEEE Global Telecommunications Conference, 2007.

[61] Bernd Bandemer, Martin Haardt, Samuli Visuri, "Linear MMSE Multi-User MIMO Downlink Precoding for Users with Multiple Antennas," 17th Annual IEEE International Symposium on Personal, Indoor and Mobile Radio Communications (PIMRC '06), 2006.

[62] Gavin Mitchell, Frank R. Kschischang, "An Augmented Orthogonal Code Design for the Non-coherent MIMO Channel," 24th Biennial Symposium on Communications, June 2008.

[63] Yong Fan, Petteri Lundén, Markku Kuusela, Mikko Valkama, "Efficient Semi-Persistent Scheduling for VoIP on EUTRA Downlink," IEEE 68th Vehicular Technology Conference, VTC, 2008. Fall 2008.

[64] S. Lawrence Marple, Digital Spectral Analysis: With Applications Prentice Hall, 1987.

[65] H. Gene Golub, F. Charles Van Loan, Matrix Computations, third ed., Johns Hopkins University Press, 1996.

[66] Athanasios Papoulis, Probability, Random Variables and Stochastic Processes, fourth ed., McGraw Hill Higher Education, 2002.

[67] Lennart Rade, Mathematics Handbook for Science and Engineering Springer, Berlin Heidelberg, 2010.

[68] A. Granino Korn, Theresa M. Korn, Mathematical Handbook for Scientists and Engineers: Definitions, Theorems, and Formulas for Reference and Review (Revised edition) Dover Publications, 2000.

[69] Goldsmith, et al., "Capacity Limits of MIMO Channels," IEEE Journal on Select Areas in Communications vol. 21, June 2003.

[70] H. Weingarten, Y. Steinberg and S. Shamai, "The Capacity Region of the Gaussian MIMO Broadcast Channel," Proceedings of Conference Information Sciences and Systems (CISS), Princeton, NJ, March 2004.

[71] M. Costa, "Writing on Dirty Paper," IEEE Transactions on Information Theory vol. 29, May 1983.

[72] T. Svantesson and A.L. Swindlehurst, "A Performance Bound for Prediction of a Multipath MIMO Channel," Proc. 37th Asilomar Conference on Signals, Systems, and Computers, Session: Array Processing for Wireless Communications, Pacific Grove, California, November 2003.

[73] C.B. Peel, B.M. Hochwald, A.L. Swindlehurst, "A Vector-Perturbation Technique for Near-Capacity Multi-Antenna Multi-User Communication," IEEE Transactions on Communications, June 2003.

[74] M. Bengtsson, B. Ottersten, Optimal and Suboptimal Beamforming," in: L.C. Godara (Ed.), Handbook of Antennas in Wireless Communications CRC Press, 2001.

[75] M. Schubert, H. Boche, "Solution of the Multiuser Downlink Beamforming Problem with Individual SINR Constraints," IEEE Transactions on Vehicular Technology vol. 53, January 2004.

[76] Q.H. Spencer, A.L. Swindlehurst, M. Haardt, "Zero-Forcing Methods for Downlink Spatial Multiplexing in Multi-User MIMO Channels," IEEE Transactions on Signal Processing vol. 52, February 2004.

[77] G. Caire, S. Shamai, "On the Achievable Throughput of a Multi-Antenna Gaussian Broadcast Channel," IEEE Transactions Information Theory vol. 49, July 2003.

[78] U. Erez, S. Shamai, R. Zamir, "Capacity and Lattice Strategies for Cancelling Known Interference," Proceedings International Symposium Information Theory and its Applications, November 2000.

[79] C. Windpassinger, R.F.H. Fischer, J.B. Huber, Lattice-Reduction-Aided Broadcast Precoding," Proceedings of 5th ITG Conference Source and Channel Coding, January 2004.

[80] Glavieux Berrou, Thitimajshima, Near Shannon Limit Error-Correcting Coding and Decoding: Turbo-codes, Proceedings of 1993 International Communication Conference, May 1993.

[81] Q.H. Spencer, A.L. Swindlehurst, "Channel Allocation in Multi-user MIMO Wireless Communications Systems," Proceedings of 2004 International Communication Conference, June 2004.

[82] Seghers Perez, Costello, "A Distance Spectrum Interpretation of Turbo Codes," IEEE Transactions on Information Theory vol. 42 (No. 6), November 1996.

[83] Shu Lin, Daniel J. Costello, Error Control Coding, second ed., Prentice Hall, 2004.

[84] J. Rothweller, "Turbo Codes," IEEE Potentials vol. 18 (Issue: 1), February–March 1999.

[85] C. Berrou, A. Glavieux, P. Thitimajshima, Near Shannon Limit Error-Correcting Coding and Decoding: Turbo Codes, Proceedings of 1993 International Communication Conference, May 1993.

[86] C. Berrou, A. Glavieux, "Near Optimum Error Correcting Coding and Decoding: Turbo-Codes," IEEE Transactions on Communications vol. 44 (No. 10), October 1996.

[87] IEEE C802.16m–09/1259, Jin Xu, Bo Sun, Evaluation Simulation for Channel Coding and HARQ, July 2009.

[88] E. Claude Shannon, Warren Weaver, The Mathematical Theory of Communication University of Illinois Press, 1998.

[89] Farid Dowla, Handbook of RF and Wireless Technologies, first ed., Newnes, 2003.

[90] Chris Heegard, B. Stephen Wicker, Turbo Coding Springer, 2010.

[91] C. Berrou, R. Pyndiah, P. Adde, C. Douillard, R. Le Bidan, An Overview of Turbo Codes and Their Applications, The European Conference on Digital Wireless Technology, 2005.

[92] H. Ma, J. Wolf, "On Tail Biting Convolutional Codes," IEEE Transactions on Communications vol. COM-34 (No. 2), February 1986.

[93] C. Weiss, C. Bettstetter, S. Riedel, "Code Construction and Decoding of Parallel Concatenated Tail-Biting Codes," IEEE Transactions on Information Theory vol. 47 (No. 1), January 2001.

[94] Y.E. Wang, R. Ramesh, "To Bite or not to Bite – A study of Tail Bits versus Tail-Biting," 7th IEEE International Symposium Personal, Indoor and Mobile Radio Communications, PIMRC'96 vol. 2, October 1996.

[95] 3GPP Long Term Evolution, System Overview, Product Development, and Test Challenges, Agilent Technologies, June 2009.<http://www.agilent.com>.

[96] G. Hyung Myung, "Technical Overview of 3GPP LTE," <http://hgmyung.googlepages.com/scfdma>, May 2008.

[97] E. Dahlman, et al., 3G Evolution: HSPA and LTE for Mobile Broadband, second ed., Academic Press, 2008.

[98] E. Basuki Priyanto, Humbert Codina, "Initial Performance Evaluation of DFT-Spread OFDM Based SC-FDMA for UTRA LTE Uplink," IEEE 65th Vehicular Technology Conference VTC2007, 2007.

[99] 3G Americas, MIMO and Smart Antennas for 3G and 4G Wireless Systems, May 2010.<http://www.3gamericas.org/>.

[100] Hyung G. Myung, "Single Carrier FDMA," <http://hgmyung.googlepages.com/scfdma>, May 2008.

[101] Juho Lee, Jin-Kyu Han, Jianzhong Zhang, "MIMO Technologies in 3GPP LTE and LTE-Advanced," EURASIP Journal on Wireless Communications and Networking, 2009.

[102] Rohde & Schwarz Application Notes, LTE-Advanced Technology Introduction, March 2010.

[103] 3G Americas, 3GPP Mobile Broadband Innovation Path to 4G: Release 9, Release 10 and Beyond: HSPA+, SAE/LTE and LTE-Advanced, February 2010.

[104] 3G Americas, HSPA to LTE-Advanced: 3GPP Broadband Evolution to IMT-Advanced (4G), September 2009.

[105] 3G Americas, The Mobile Broadband Evolution: 3GPP Release 8 and Beyond HSPA+, SAE/LTE and LTE-Advanced, February 2009.

[106] 3GPP TS 36.201, Evolved Universal Terrestrial Radio Access (E-UTRA) Physical Layer – General Description, March 2010.

[107] 3GPP TS 36.211, Evolved Universal Terrestrial Radio Access (E-UTRA), Physical Channels and Modulation, March 2010.

[108] 3GPP TS 36.212, Evolved Universal Terrestrial Radio Access (E-UTRA) Multiplexing and Channel Coding, March 2010.

[109] 3GPP TS 36.213, Evolved Universal Terrestrial Radio Access (E-UTRA); Physical Layer Procedures, March 2010.

[110] 3GPP TS 36.214, Evolved Universal Terrestrial Radio Access (E-UTRA); Physical Layer – Measurements, March 2010.

[111] 3GPP TS 36.104, Evolved Universal Terrestrial Radio Access (E-UTRA); Base Station (BS) Radio Transmission and Reception, March 2010.

[112] 3GPP TS 36.101, Evolved Universal Terrestrial Radio Access (E-UTRA); User Equipment (UE) Radio Transmission and Reception, March 2010.

[113] 3GPP TS36.321, Evolved Universal Terrestrial Radio Access (E-UTRA); Medium Access Control (MAC) Protocol Specification, March 2010.

[114] 3GPP TR 36.912, Feasibility Study for Further Advancements for E-UTRA (LTE-Advanced), March 2010.

[115] 3GPP TR 36.913, Requirements for Evolved UTRA (E-UTRA) and Evolved UTRAN (E UTRAN), March 2010.

[116] 3GPP TS 36.300, Evolved Universal Terrestrial Radio Access (E-UTRA) and Evolved Universal Terrestrial Radio Access Network (E-UTRAN) Overall Description, Stage 2, March 2010.

[117] 3GPP TR 36.814, Feasibility Study for Further Advancements for E-UTRA (LTE-Advanced), March 2010.

[118] IEEE C802.16m–09/1940r1, Ping Wang, et al., Assignment A-MAP Group Size Indication in Non-User Specific A-MAP IE, August 2009.

[119] IEEE C802.16m–10/0445r1, Roshni Srinivasan, Alexei Davydov, et al., UL MIMO Transmission Format in the 802.16m UL Basic Assignment A-MAP Information Element, March 2010.

[120] MATLAB CENTRAL, An Open Exchange for the MATLAB and SIMULINK User Community <http://www.mathworks.com/matlabcentral>.

[121] 3GPP TS 25.101, User Equipment (UE) radio transmission and reception (FDD), March 2009.

[122] 3GPP R1-060385, Cubic Metric in 3GPP-LTE Motorola, February 2006.

[123] IEEE C802.16m–08/153, Bin-Chul Ihm, Jinsoo Choi, Wookbong Lee, Pilot Related to DL MIMO, March 2008.

[124] The Coded Modulation Library (The Iterative Solutions Coded Modulation Library (ISCML) is an open source toolbox for simulating capacity approaching codes in Matlab) <http://www.iterativesolutions.com>.

[125] David Chu, "Polyphase codes with good periodic correlation properties," IEEE Transaction on Information Theory Vol. 18 (No. 4), July 1972.

The IEEE 802.16m Physical Layer (Part II)

10

INTRODUCTION

This chapter describes the control and signaling mechanisms, as well as the multi-antenna techniques used in IEEE 802.16m and 3GPP LTE/LTE-Advanced. The theoretical aspects and basic concepts of practical multi-antenna techniques including single-user and multi-user schemes, transmit and receive diversity techniques, beamforming, feedback requirements, and collaborative multi-antenna communication are discussed so that the multi-antenna operation at the physical layers of IEEE 802.16m and 3GPP LTE/LTE-Advanced can be better understood. The control signaling mechanisms of the legacy system have undergone a substantial change in IEEE 802.16m to improve reliability and coverage, and to reduce the signaling overhead and control-plane and user-plane latencies. The IEEE 802.16m system requirements called for improved link-budget in the downlink and uplink for both traffic and control channels. As a result, the use of dual transmit-antenna at the base station as a minimum configuration has been mandated. Unlike the legacy system, all downlink control channels in IEEE 802.16m use frequency-domain transmit diversity as the default multi-antenna mode of operation. The reduction in Layer 2 signaling overhead, use of more robust multi-antenna techniques along with user-dedicated precoded reference signals, and improved feedback schemes have resulted in a significant increase of VoIP and data capacity compared to the legacy system.

We start with the description of the control channels and signaling mechanisms and compare them to the corresponding techniques in the legacy system, and then will proceed to detailed discussion of multi-input multi-output techniques. Similar to the previous topics, the corresponding methods in both IEEE 802.16m and 3GPP LTE/LTE-Advanced are discussed to allow the reader to draw parallels and to better understand the design philosophies in the 4th generation cellular systems.

10.1 CONTROL CHANNELS

In the IEEE 802.16-2009 standard, the control signaling for downlink or uplink resource allocations and other operational aspects of the system is performed using downlink or uplink information elements included in the Medium Access Protocol (MAP) MAC management messages. The information elements of all active users in the cell, as well as broadcast messages, are jointly encoded and transmitted using a robust transmission scheme. The information elements contained in the DL or UL MAPs are distinguished using the connection identifiers that are associated with user connections.

Mobile WiMAX. DOI: 10.1016/B978-0-12-374964-2.10010-4

There are different types of MAP structure that are specified in the standards specification; nevertheless, due to inefficiency and lack of use cases for some of the variants, only normal MAPs, compressed and HARQ MAPs, and sub-MAPs have been utilized in Release 1.5 of the mobile WiMAX profile [6]. The sub-MAPs are inherently multicast messages that can be encoded with different modulation and coding schemes, depending on the channel conditions of the users that are assigned to each sub-MAP. While the main MAP is a broadcast message which is sent with the most robust modulation and coding scheme and can be received across the base station's coverage area, the sub-MAPs are multicast to a group of users with similar geometries (i.e., statistical distribution of SINR) and long-term channel conditions using optimized modulation and coding schemes. This can reduce the overhead due to common control and traffic assignment messaging, which is the major part of the MAP, while maintaining the reliability of broadcast and common control signaling.

Figure 10-1 shows various legacy MAP structures. The size of the DL/UL MAPs, DL/UL MAP IEs, DL/UL subframes, and DL/UL data bursts are not drawn to scale and the actual sizes may differ depending on the type of the MAP. Among the legacy MAP structures, the HARQ MAP structure is of more practical interest. The HARQ MAP management message includes a compressed/compact DL/UL-MAP IE and defines the access information for the DL and UL burst of HARQ-enabled mobile stations. This message is sent without a generic MAC header. The BS may broadcast multiple HARQ MAP messages using multiple bursts after the MAP message. Each HARQ MAP message has a different modulation and coding rate. The DL-MAP IEs in the MAP message describe the location and coding and modulation schemes of the bursts. The order of DL-MAP IEs in the MAP message and the bursts for HARQ MAP messages is determined by the coding and modulation scheme of the burst. The burst for a HARQ MAP message with a lower rate coding and modulation is placed before other bursts for the HARQ MAP message. The presence of the HARQ MAP message format is indicated by the contents of the of the first data byte of a burst [1]. The trade-off between the efficiency and reliability of the control channels resulted in the design of new individually coded MAPs for the IEEE 802.16m systems. While the jointly coded MAPs can be more efficient, since the overhead is shared among the users, they cannot be equally reliable for geographically-dispersed users in the cell with various channel conditions. The individually-coded MAPs are conceptually similar to legacy sub-MAPs when containing allocations of a single user. The main difference between the legacy jointly-coded MAPs and the new individually-coded MAPs is that the legacy MAPs are time-division multiplexed with data, whereas the new MAPs are frequency-division multiplexed with data. The FDM and TDM schemes have advantages and disadvantages in terms of MS power-saving, efficiency of user resource allocations, and transmission reliability of the control messages. As shown in Figure 10-1, the control signaling in the legacy uplink subframe consists of three parts: HARQ Acknowledgement Channel (ACKCH); Ranging and Fast-Feedback; and the Channel Quality Indication Channel (CQICH). The uplink control channels of the legacy system occupy the first three OFDM symbols of the uplink subframe immediately following the downlink-to-uplink switching gap in the TDD mode. The UL-PUSC subchannelization scheme is used in the uplink control region.

10.2 DOWNLINK CONTROL CHANNELS

In order to minimize the control signaling overhead, and to improve the reliability and coverage of the control channels particularly at the cell edge, the control signaling schemes have been redesigned in

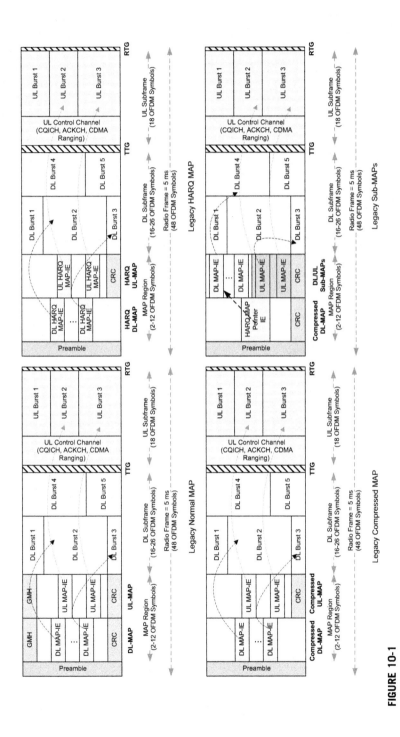

FIGURE 10-1

Structure of various legacy MAPs (TDD mode) [1]

IEEE 802.16m. The jointly-coded MAPs in the legacy system were replaced by individually-coded Advanced MAPs (A-MAPs). As shown in Figure 10-2, the control signaling in IEEE 802.16m comprises a superframe header for the broadcast of the essential system configuration information, downlink control channels to carry HARQ feedback, power control signals, the downlink/uplink assignment and allocation information for each user or a group of users, uplink control channels for transmission of channel quality reports, channel state information, bandwidth request, synchronized and non-synchronized ranging, and sounding.

Comparison of Figure 10-2 with Figure 10-1 reveals several structural differences between IEEE 802.16m control signaling and that of the legacy. Aside from migration from the concept of joint-coding to individual-coding of the user control channels, the time-relevance of the downlink and uplink assignments have also changed, control channels in the downlink and uplink are frequency-division-multiplexed with traffic channels as opposed to the time-division-multiplex used in the legacy system, and there has been the addition of new control channels such as the superframe header.

In the following sections, the physical structure and transmission characteristics of the A-MAPs and the corresponding information elements will be discussed. The detailed description of the uplink control channels will follow in later sections.

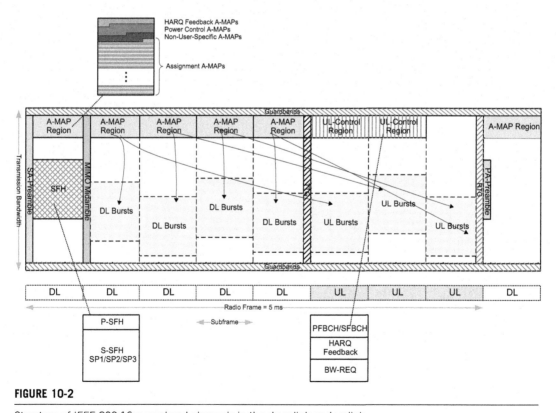

FIGURE 10-2

Structure of IEEE 802.16m overhead channels in the downlink and uplink

10.2.1 **Physical Structure of Advanced MAPs**

The Advanced MAP carries unicast control information for the active users in the cell. The unicast control information consists of both user-specific and non-user-specific control information. The user-specific control is further divided into assignment, HARQ feedback and power control fields, and is transmitted in the assignment A-MAP, HARQ feedback A-MAP, and power control A-MAP, respectively. These A-MAPs are contained in a physical region called the A-MAP region that is frequency-division multiplexed with data. The A-MAP regions are located in all DL unicast subframes. When the default TTI of one subframe is used, the DL data allocations corresponding to an A-MAP region occupy resources in the subframe where the A-MAP region is located. If Fractional Frequency Reuse (FFR) is used in a DL subframe, either reuse-1 or a power-boosted reuse-3 partition may contain the A-MAP region. In a DL subframe, the non-user specific, HARQ feedback, and power control A-MAPs are located in the primary frequency partition. The primary frequency partition can be either reuse-1 or a power-boosted reuse-3 partition, which is identified by the S-SFH SP1 broadcast channel, whereas the assignment A-MAP can be either in a reuse-1 or a power-boosted reuse-3 partition, or both. The number of assignment A-MAPs in each frequency partition is signaled through a non-user specific A-MAP. The structure of an A-MAP region is illustrated in Figure 10-3. The resources occupied by each A-MAP physical channel may vary depending on the system configuration and scheduler operation. The A-MAP region consists of a number of distributed LRUs. In all DL subframes except the first subframe of a superframe, an A-MAP region consists of the first $N_{A\text{-}MAP}$ distributed LRUs in the primary frequency partition and the LRUs with N_{sym} symbols. In the first DL subframe of a superframe, the A-MAP region is formed using the $N_{A\text{-}MAP}$ distributed LRUs after N_{SFH} distributed LRUs occupied by the superframe headers.

The non-user-specific control information consists of information that is not dedicated to a specific user or a specific group of users. It includes information required to decode the user-specific control blocks.

The Minimum A-MAP Logical Resource Unit (MLRU) is used for the assignment A-MAPs. The minimum logical resource unit in the assignment A-MAP consists of $N_{MLRU} = 56$ sub-carriers. The assignment A-MAP IE is transmitted with one MLRU or multiple concatenated MLRUs in the A-MAP region. The number of logically contiguous MLRUs is determined based on the assignment A-MAP IE size and channel coding rate where the channel coding rate is selected based on the mobile station's channel quality reports. The assignment A-MAP IEs are grouped together based on the channel coding rate. Assignment A-MAP IEs in the same group are transmitted in the same frequency partition with the same channel coding rate. Each assignment A-MAP group contains several logically contiguous MLRUs. The number of assignment A-MAP IEs in each assignment A-MAP group is signaled through a non-user specific A-MAP in the same subframe. If two assignment A-MAP groups using two channel coding rates exist in an A-MAP region, the assignment A-MAP group using the lower channel coding rate is allocated first, followed by the assignment A-MAP group using higher channel coding rate.

If a broadcast assignment A-MAP IE (i.e., an assignment A-MAP IE sent to all mobile stations) exists in a DL subframe, it is located at the beginning of either assignment A-MAP Group 1 or assignment A-MAP Group 2. All the multicast assignment A-MAP Ies (i.e., assignment A-MAP IEs that are sent to specific groups of users present in any assignment A-MAP group) occupy contiguous set of MLRUs from the beginning of the assignment A-MAP group. If the broadcast assignment

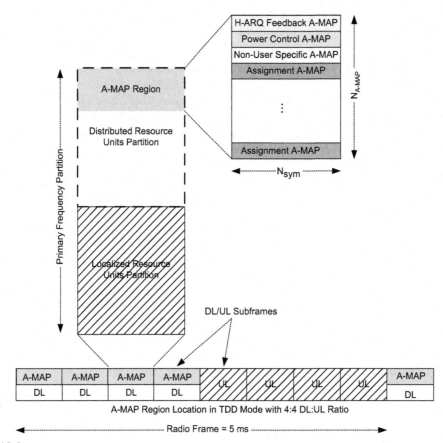

FIGURE 10-3

The structure of the A-MAP region

A-MAP IE is present in the assignment A-MAP group, the multicast assignment A-MAP IEs are located immediately after the broadcast assignment A-MAP IE. The Group Resource Allocation A-MAP IE is an example of a multicast assignment A-MAP IEs [2].

The unicast assignment A-MAP IE sent to a particular mobile station is transmitted in the same assignment A-MAP group. The DL/UL Basic Assignment A-MAP IE is an example of unicast assignment A-MAP IE. The BS may allocate a maximum of eight assignment A-MAP IEs to a particular MS in any subframe. This number includes all of the assignment A-MAP IEs that are decoded by the MS. For an assignment A-MAP IE that can be segmented (i.e., an assignment A-MAP IE that occupies more than one MLRU when modulated with QPSK ½), each segment is counted as one assignment A-MAP IE. The physical layer procedures for various A-MAP IE types are shown in Figure 10-4.

The non-user specific A-MAP IE carries 12 bits of information and is encoded using Tail Biting Convolutional Code (TBCC) with an effective code rate of 1/12. More specifically, when using

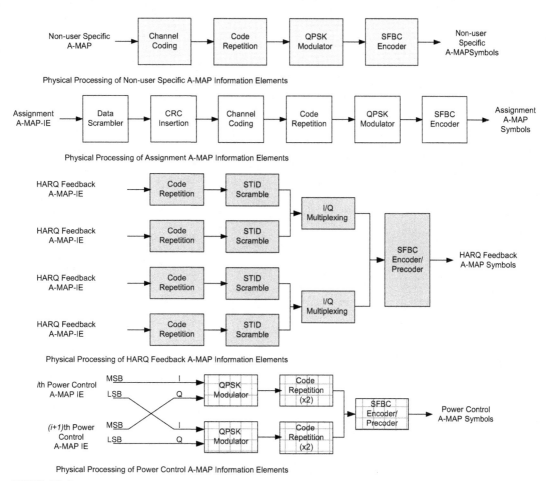

Physical Processing of Non-user Specific A-MAP Information Elements

Physical Processing of Assignment A-MAP Information Elements

Physical Processing of HARQ Feedback A-MAP Information Elements

Physical Processing of Power Control A-MAP Information Elements

FIGURE 10-4

Physical layer procedures for various A-MAPs [2]

Fractional Frequency Reuse in frequency reuse-1 partition, the non-user-specific A-MAP is encoded with an effective code rate of 1/12; otherwise, when the non-user-specific A-MAP is located in the power-boosted frequency reuse-3 partition, it is encoded with code rate of 1/4. The encoded bit sequence is modulated using QPSK modulation. As shown in Figure 10-5, for each transmit antenna, the complex-valued symbols at the output of the SFBC encoder denoted by $s_{N\text{-}US}[0]$ to $s_{N\text{-}US}[L_{N\text{-}US} - 1]$ are mapped to renumbered tone-pairs $A[(L_{HF} + L_{PC})/2]$ to $A[(L_{HF} + L_{PC} + L_{N\text{-}US})/2 - 1]$, where A refers to the array of renumbered A-MAP tone-pairs obtained via renumbering all tone pairs within the distributed LRUs in the A-MAP region in a time first manner (i.e., tone-pairs of distributed LRUs in the A-MAP region are rearranged into a one-dimensional array in a time first manner and subsequently an A-MAP channel is formed by tone-pairs in a particular segment of that array), L_{HF} is the number of tones required to transmit the entire HARQ feedback A-MAP, L_{PC} is the number of tones required to

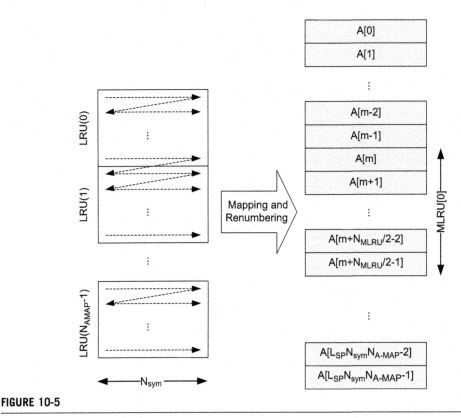

FIGURE 10-5

An illustration of the formation of MLRUs for A-MAPs

transmit the entire power control A-MAP; $L_{N\text{-}US}$ is the number of tones required to transmit the non-user specific A-MAP. As shown in Figure 10-5, it is assumed that each distributed LRU consists of L_{SP} tone-pairs per symbol, the renumbered A-MAP tone-pairs are denoted by $A[m]$, $m =$ $0,1,\ldots,L_{SP}N_{sym}N_{A\text{-}MAP} - 1$ where $N_{A\text{-}MAP}$ is the number of distributed LRUs allocated to the A-MAP and N_{sym} is the number of OFDM symbols per subframe. The MS initially decodes the non-user-specific A-MAP in the primary frequency partition to obtain the information for decoding the assignment A-MAPs and HARQ feedback A-MAPs.

The non-user-specific A-MAP IE comprises: assignment A-MAP size (8 bits), i.e., the number of assignment A-MAPs in each assignment A-MAP group; HARQ feedback A-MAP index (1 bit), i.e., a parameter that is used to calculate HARQ feedback A-MAP index; HARQ feedback channel index (1 bit), i.e., a parameter that is used to calculate the HARQ feedback channel index; and the non-user-specific A-MAP extension flag (1 bit), i.e., whether the non-user specific A-MAP is extended. The extended non-user-specific part uses the same PHY structure as the non-user-specific A-MAP. The provision for the potential extension of the non-user-specific A-MAP has been reflected in the location of the non-user-specific A-MAP, as shown in Figure 10-5. A one-bit extension flag is also included to allow future extensions of the non-user-specific A-MAP fields. If the non-user-specific A-MAP extension flag is set, it indicates that the non-user-specific A-MAP is extended. The extended non-user-specific part has 12 bits and uses the same physical structure, i.e., channel coding, modulation, and MIMO encoding, as the non-user-specific A-MAP. The number of assignment A-MAPs in each assignment A-MAP group can be derived from the Assignment A-MAP Size indication field in the non-user-specific A-MAP IE using a lookup table. The actual number of assignment A-MAP IEs in each assignment A-MAP group is less than the value obtained from the lookup tables. The lookup tables, each with 256 entries (see Table 10-1), can be generated by the following equations for a particular assignment A-MAP modulation and coding set and fractional frequency reuse configuration.

It must be noted that only one assignment A-MAP IE size of 56 bits inclusive of CRC is supported. The maximum number of MLRUs for the assignment A-MAP is 48 per subframe and the maximum number of assignment A-MAP information elements per subframe is limited to 32, whereas the maximum number of assignment A-MAP information elements per subframe that the BS may allocate to an MS is 8. Four assignment A-MAP group-sizes are defined: (1) when QPSK (1/2, 1/4) is used in the non-FFR case, there are two assignment A-MAP groups with the minimum resource units of one and two MLRUs in each group; (2) when QPSK (1/2, 1/8) is used in the non-FFR case, there are two assignment A-MAP groups with the minimum resource units of one and four MLRUs in each group; (3) when QPSK 1/2 is used in the FFR reuse-3 region assignment A-MAP and QPSK (1/2, 1/4) is used for the reuse-1 region assignment A-MAP, there are three assignment A-MAP groups with the minimum resource units of one, one, and two MLRUs in each group; and (4) when QPSK 1/2 is used for the FFR reuse-3 assignment A-MAP and QPSK (1/2, 1/8) is used for the reuse-1 assignment A-MAP, there are three assignment A-MAP groups with the minimum resource units of one, one, and four MLRUs in each group. There is an 8-bit index that is used for assignment A-MAP group size indication. The assignment A-MAP group with more robust MCS is located before the assignment A-MAP group with less robust MCS. The exact size of the group with more robust MCS is signaled when possible, to minimize resource consumption.

The following describes the methodology for allocation of MLRUs for non-FFR region assignment A-MAPs using QPSK (1/2, 1/4). The total number of MLRUs for assignment A-MAP N_{total} is given by $N_{total} = [2G_1(I) + G_2(I)]$ where $0 \leq N_{total} \leq 48$, $0 \leq G_1(I) \leq 24$ is the number of assignment A-MAPs in Group 1 using QPSK 1/4 and $0 \leq G_2(I) \leq 48$ is the number of assignment A-MAPs in Group 2 using QPSK 1/2. For non-FFR configuration with Group 1 using QPSK 1/4 and Group 2 using QPSK 1/2, the lookup tables shown in Table 10-1 can be generated iteratively using the following equations for all values of N_{total} and k [2]:

$$G_1(I) = \begin{cases} k & , & k = 0, 1, ..., \lfloor N_{total}/2 \rfloor, & 0 \leq N_{total} \leq 24 \\ N_{total} - 24 + 3k, & k = 0, 1, 2, 3, & 25 \leq N_{total} \leq 29 \\ N_{total} - 29 + 4k, & k = 0, 1, 2, 3, & 30 \leq N_{total} \leq 32 \\ N_{total} - 30 + 4k, & k = 0, 1, 2, 3, & 33 \leq N_{total} \leq 34 \\ N_{total} - 32 + 3k, & k = 0, 1, 2, 3, & 35 \leq N_{total} \leq 39 \\ N_{total} - 32 + 4k, & k = 0, 1, 2, & 40 \leq N_{total} \leq 42 \\ N_{total} - 32 + 3k, & k = 0, 1, 2, & 43 \leq N_{total} \leq 48 \end{cases} \quad (10\text{-}1)$$

$$G_2(I) = N_{total} - 2G_1(I)$$

Table 10-1 The Number of Assignment A-MAPs in each Assignment A-MAP Group for Non-FFR Configuration with Group 1 using QPSK 1/4 and Group 2 using QPSK 1/2 [118]

Index (l)	Assignment A-MAP Group 1 using QPSK 1/4	Assignment A-MAP Group 2 using QPSK 1/2	Index (l)	Assignment A-MAP Group 1 using QPSK 1/4	Assignment A-MAP Group 2 using QPSK 1/2	Index (l)	Assignment A-MAP Group 1 using QPSK 1/4	Assignment A-MAP Group 2 using QPSK 1/2	Index (l)	Assignment A-MAP Group 1 using QPSK 1/4	Assignment A-MAP Group 2 using QPSK 1/2
0	0	0	64	0	15	128	7	7	192	0	31
1	0	1	65	1	13	129	8	5	193	3	25
2	0	2	66	2	11	130	9	3	194	7	17
3	1	0	67	3	9	131	10	1	195	11	9
4	0	3	68	4	7	132	0	22	196	14	3
5	1	1	69	5	5	133	1	20	197	2	28
6	0	4	70	6	3	134	2	18	198	6	20
7	1	2	71	7	1	135	3	16	199	10	12
8	2	0	72	0	16	136	4	14	200	13	6
9	0	5	73	1	14	137	5	12	201	1	31
10	1	3	74	2	12	138	6	10	202	5	23
11	2	1	75	3	10	139	7	8	203	8	17
12	0	6	76	4	8	140	8	6	204	12	9
13	1	4	77	5	6	141	9	4	205	16	1
14	2	2	78	6	4	142	10	2	206	4	26
15	3	0	79	7	2	143	11	0	207	8	18
16	0	7	80	8	0	144	0	23	208	12	10
17	1	5	81	0	17	145	1	21	209	15	4
18	2	3	82	1	15	146	2	19	210	4	27
19	3	1	83	2	13	147	3	17	211	8	19
20	0	8	84	3	11	148	4	15	212	11	13
21	1	6	85	4	9	149	5	13	213	15	5
22	2	4	86	5	7	150	6	11	214	5	26

Index			Index			Index			Index		
23	3	2	87	6	5	151	7	9	215	8	20
24	4	0	88	7	3	152	8	7	216	12	12
25	0	9	89	8	1	153	9	5	217	16	4
26	1	7	90	0	18	154	10	3	218	5	27
27	2	5	91	1	16	155	11	1	219	9	19
28	3	3	92	2	14	156	0	24	220	13	11
29	4	1	93	3	12	157	1	22	221	16	5
30	0	10	94	4	10	158	2	20	222	7	24
31	1	8	95	5	8	159	3	18	223	11	16
32	2	6	96	6	6	160	4	16	224	14	10
33	3	4	97	7	4	161	5	14	225	18	2
34	4	2	98	8	2	162	6	12	226	9	21
35	5	0	99	9	0	163	7	10	227	13	13
36	0	11	100	0	19	164	8	8	228	16	7
37	1	9	101	1	17	165	9	6	229	8	24
38	2	7	102	2	15	166	10	4	230	12	16
39	3	5	103	3	13	167	11	2	231	15	10
40	4	3	104	4	11	168	12	0	232	19	2
41	5	1	105	5	9	169	2	21	233	11	19
42	0	12	106	6	7	170	6	13	234	14	13
43	1	10	107	7	5	171	10	5	235	18	5
44	2	8	108	8	3	172	0	26	236	11	20
45	3	6	109	9	1	173	4	18	237	14	14
46	4	4	110	0	20	174	8	10	238	18	6
47	5	2	111	1	18	175	11	4	239	11	21
48	6	0	112	2	16	176	1	25	240	14	15
49	0	13	113	3	14	177	5	17	241	18	7
50	1	11	114	4	12	178	8	11	242	12	20
51	2	9	115	5	10	179	12	3	243	15	14
52	3	7	116	6	8	180	2	24	244	19	6
53	4	5	117	7	6	181	5	18	245	13	19
54	5	3	118	8	4	182	9	10	246	16	13

(Continued)

Table 10-1 The Number of Assignment A-MAPs in each Assignment A-MAP Group for Non-FFR Configuration with Group 1 using QPSK 1/4 and Group 2 using QPSK 1/2 [118] *Continued*

Index (I)	Assignment A-MAP Group 1 using QPSK 1/4	Assignment A-MAP Group 2 using QPSK 1/2	Index (I)	Assignment A-MAP Group 1 using QPSK 1/4	Assignment A-MAP Group 2 using QPSK 1/2	Index (I)	Assignment A-MAP Group 1 using QPSK 1/4	Assignment A-MAP Group 2 using QPSK 1/2	Index (I)	Assignment A-MAP Group 1 using QPSK 1/4	Assignment A-MAP Group 2 using QPSK 1/2
55	6	1	119	9	2	183	13	2	247	20	5
56	0	14	120	10	0	184	1	27	248	15	16
57	1	12	121	0	21	185	5	19	249	18	10
58	2	10	122	1	19	186	9	11	250	22	2
59	3	8	123	2	17	187	12	5	251	17	13
60	4	6	124	3	15	188	1	28	252	20	7
61	5	4	125	4	13	189	5	20	253	16	16
62	6	2	126	5	11	190	8	14	254	20	8
63	7	0	127	6	9	191	12	6	255	24	0

$$I = I_b(N_{total}) + k$$

$$
I_b(N_{total}) =
\begin{cases}
0 & , & N_{total} = 0 \\
I_b(N_{total} - 1) + \lfloor (N_{total} - 1)/2 \rfloor + 1, & 1 \leq N_{total} \leq 25 \\
I_b(N_{total} - 1) + 4 & , & 26 \leq N_{total} \leq 40 \\
I_b(N_{total} - 1) + 3 & , & 41 \leq N_{total} \leq 48 \\
0 & , & otherwise
\end{cases}
\tag{10-2}
$$

where $G_1(I)$ is the number of assignment A-MAPs for Group 1 using QPSK 1/4, $G_2(I)$ is the number of assignment A-MAPs for Group 2 using QPSK 1/2, I is the 8-bit index included in the non-user-specific A-MAP IE, which ranges from 0 to 255, and $I_b(N_{total})$ is the base index for a given N_{total} value. Table 10-1 shows the lookup tables based on which the number of assignment A-MAPs can be determined when the 8-bit assignment A-MAP group size index is given.

In the case of non-FFR region assignment A-MAPs using QPSK (1/2, 1/8), the following method can be used to generate the lookup tables and thereby to find the number of assignment A-MAPs in each group. The total number of MLRUs for assignment A-MAP N_{total} is given by $N_{total} = [4G_1(I) + G_2(I)] \leq 48$ where $0 \leq G_1(I) \leq 12$ is the number of assignment A-MAP in Group 1 using QPSK 1/8 and $0 \leq G_2(I) \leq 48$ is the number of assignment A-MAPs in Group 2 using QPSK 1/2. The assignment A-MAP Group 1 is not adjacent to data transmission, whereas Group 2 is adjacent to data transmission. For non-FFR configuration with Group 1 using QPSK 1/8 and Group 2 using QPSK 1/2, the lookup tables shown in Table 10-2 can be generated iteratively using the following equations for all values of N_{total} and k [2]:

$$
G_1(I) =
\begin{cases}
k & , & k = 0, 1, \ldots, \lfloor N_{total}/4 \rfloor, & 0 \leq N_{total} \leq 24 \\
k + 1 & , & k = 0, 1, 2, 3, 4, 5 & , & 25 \leq N_{total} \leq 27 \\
k + 1 & , & k = 0, 1, 2, 3, 4, 5, 6 & , & 28 \leq N_{total} \leq 32 \\
k + \lfloor N_{total}/3 \rfloor - 10, & k = 0, 1, 2, 3, 4, 5, 6 & , & 33 \leq N_{total} \leq 48
\end{cases}
\tag{10-3}
$$

$$G_2(I) = N_{total} - 4G_1(I)$$

$$I = I_b(N_{total}) + k$$

$$
I_b(N_{total}) =
\begin{cases}
0 & , & N_{total} = 0 \\
I_b(N_{total} - 1) + \lfloor (N_{total} - 1)/4 \rfloor + 1, & 1 \leq N_{total} \leq 25 \\
I_b(N_{total} - 1) + 6 & , & 26 \leq N_{total} \leq 28 \\
I_b(N_{total} - 1) + 7 & , & 29 \leq N_{total} \leq 48 \\
0 & , & otherwise
\end{cases}
\tag{10-4}
$$

The HARQ feedback A-MAP consists of HARQ feedback IEs to acknowledge success or failure of uplink data transmission. Each HARQ feedback A-MAP IE carries one bit of information which is then repeated eight times. The resulting bits are scrambled by the eight least significant bits of the Station Identifier (STID) of the MS assigned during network entry. Depending on the channel conditions, a certain amount of power boosting may be applied to each scrambled bit sequence. Prior to SFBC encoding, each

Table 10-2 The Number of Assignment A-MAPs in each Assignment A-MAP Group for Non-FFR Configuration with Group 1 using QPSK 1/8 and Group 2 using QPSK 1/2 [118]

Index (I)	Assignment A-MAP Group 1 using QPSK 1/8	Assignment A-MAP Group 2 using QPSK 1/2	Index (I)	Assignment A-MAP Group 1 using QPSK 1/8	Assignment A-MAP Group 2 using QPSK 1/2	Index (I)	Assignment A-MAP Group 1 using QPSK 1/8	Assignment A-MAP Group 2 using QPSK 1/2	Index (I)	Assignment A-MAP Group 1 using QPSK 1/8	Assignment A-MAP Group 2 using QPSK 1/2
0	0	0	64	4	4	128	4	14	192	5	19
1	0	1	65	5	0	129	5	10	193	6	15
2	0	2	66	0	21	130	6	6	194	7	11
3	0	3	67	1	17	131	7	2	195	8	7
4	0	4	68	2	13	132	0	31	196	9	3
5	1	0	69	3	9	133	1	27	197	3	28
6	0	5	70	4	5	134	2	23	198	4	24
7	1	1	71	5	1	135	3	19	199	5	20
8	0	6	72	0	22	136	5	11	200	7	12
9	1	2	73	1	18	137	6	7	201	8	8
10	0	7	74	2	14	138	7	3	202	9	4
11	1	3	75	3	10	139	0	32	203	10	0
12	0	8	76	4	6	140	1	28	204	3	29
13	1	4	77	5	2	141	2	24	205	4	25
14	2	0	78	0	23	142	3	20	206	5	21
15	0	9	79	1	19	143	4	16	207	6	17
16	1	5	80	2	15	144	5	12	208	7	13
17	2	1	81	3	11	145	6	8	209	8	9
18	0	10	82	4	7	146	8	0	210	10	1
19	1	6	83	5	3	147	1	29	211	4	26
20	2	2	84	0	24	148	2	25	212	5	22
21	0	11	85	1	20	149	3	21	213	6	18
22	1	7	86	2	16	150	4	17	214	7	14
23	2	3	87	3	12	151	5	13	215	8	10

24	0	12	88	4	8	152	6	9	216	9	6
25	1	8	89	5	4	153	7	5	217	10	2
26	2	4	90	6	0	154	8	1	218	4	27
27	3	0	91	0	25	155	2	26	219	6	19
28	0	13	92	1	21	156	3	22	220	7	15
29	1	9	93	2	17	157	4	18	221	8	11
30	2	5	94	3	13	158	5	14	222	9	7
31	3	1	95	4	9	159	6	10	223	10	3
32	0	14	96	5	5	160	7	6	224	4	28
33	1	10	97	6	1	161	8	2	225	5	24
34	2	6	98	0	26	162	1	31	226	6	20
35	3	2	99	1	22	163	2	27	227	7	16
36	0	15	100	3	14	164	4	19	228	9	8
37	1	11	101	4	10	165	5	15	229	10	4
38	2	7	102	5	6	166	6	11	230	11	0
39	3	3	103	6	2	167	7	7	231	5	25
40	0	16	104	0	27	168	8	3	232	6	21
41	1	12	105	1	23	169	2	28	233	7	17
42	2	8	106	2	19	170	3	24	234	8	13
43	3	4	107	3	15	171	4	20	235	9	9
44	4	0	108	4	11	172	5	16	236	10	5
45	0	17	109	6	3	173	7	8	237	5	26
46	1	13	110	0	28	174	8	4	238	6	22
47	2	9	111	1	24	175	9	0	239	7	18
48	3	5	112	2	20	176	2	29	240	8	14
49	4	1	113	3	16	177	3	25	241	9	10
50	0	18	114	4	12	178	4	21	242	10	6
51	1	14	115	5	8	179	5	17	243	11	2
52	2	10	116	6	4	180	6	13	244	5	27
53	3	6	117	7	0	181	7	9	245	6	23
54	4	2	118	1	25	182	9	1	246	8	15
55	0	19	119	2	21	183	2	30	247	9	11

(Continued)

Table 10-2 The Number of Assignment A-MAPs in each Assignment A-MAP Group for Non-FFR Configuration with Group 1 using QPSK 1/8 and Group 2 using QPSK 1/2 [118] *Continued*

Index (I)	Assignment A-MAP Group 1 using QPSK 1/8	Assignment A-MAP Group 2 using QPSK 1/2	Index (I)	Assignment A-MAP Group 1 using QPSK 1/8	Assignment A-MAP Group 2 using QPSK 1/2	Index (I)	Assignment A-MAP Group 1 using QPSK 1/8	Assignment A-MAP Group 2 using QPSK 1/2	Index (I)	Assignment A-MAP Group 1 using QPSK 1/8	Assignment A-MAP Group 2 using QPSK 1/2
56	1	15	120	3	17	184	3	26	248	10	7
57	2	11	121	4	13	185	4	22	249	11	3
58	3	7	122	5	9	186	5	18	250	6	24
59	4	3	123	6	5	187	6	14	251	7	20
60	0	20	124	7	1	188	7	10	252	8	16
61	1	16	125	0	30	189	8	6	253	9	12
62	2	12	126	1	26	190	9	2	254	10	8
63	3	8	127	3	18	191	4	23	255	12	0

scrambled bit sequence is mapped to real or imaginary parts of the signal constellation and multiplexed with other scrambled sequences. Figure 10-5 illustrates the physical layer procedures for the HARQ feedback A-MAP, which consists of four HARQ feedback A-MAP channels. For each transmit antenna, the complex-valued symbols at the output of SFBC encoder, denoted as $s_{HF}[0]$ to $s_{HF}[L_{HF} - 1]$, are mapped to tone-pairs from $A[0]$ to $A[L_{HF}/2 - 1]$, where A is the array of renumbered A-MAP tone-pairs and L_{HF} is the number of tones required to transmit the entire HARQ feedback A-MAP. The groups of HARQ feedback A-MAP bits are indexed sequentially from index 0 within a HARQ feedback A-MAP region in the mapping process [2]. There is one HARQ feedback A-MAP region in each DL subframe. The HARQ feedback A-MAP IE consists of one bit corresponding to HARQ ACK/NACK information.

The power control A-MAP contains closed-loop power control commands for uplink transmission which are transmitted by the BS to every MS operating in closed-loop power control mode. Figure 10-5 illustrates the physical layer procedures for processing the power control A-MAP-IE. The ith power control A-MAP-IE comprises two bits. The ith and $(i + 1)$th power control A-MAP IEs are mapped to two QPSK symbols which are repeated twice, as depicted in Figure 10-5. The ith power control A-MAP may also be mapped to two QPSK symbols for transmitting to the corresponding MS with poor channel conditions. An appropriate power boosting factor is applied to the ith power control A-MAP IE; $0 \leq i \leq N_{PC\text{-}A\text{-}MAP}$ where $N_{PC\text{-}A\text{-}MAP}$ is the number of power control A-MAP IEs. The power boosting factors are determined by the management entity to satisfy the link performance. For each transmit antenna, symbols at the output of SFBC encoder, denoted by $s_{PC}[0]$ to $s_{PC}[L_{PC} - 1]$, are mapped to tone-pairs from $A[L_{HF}/2]$ to $A[(L_{HF} + L_{PC})/2 - 1]$, where A refers to the array of the renumbered A-MAP tone-pairs and L_{PC} is the number of tones required to transmit the entire power control A-MAP. The power control A-MAP IE comprises two bits corresponding to values $(-0.5, 0, 0.5, 1$ dB$)$ for transmit power adjustment.

The Assignment A-MAP includes one or multiple Assignment A-MAP IEs where each assignment A-MAP IE is encoded separately. Figure 10-4 describes the procedure for constructing Assignment A-MAP symbols. The Assignment A-MAP IE is masked by a sequence generated by a pseudo-random sequence generator. The mask randomizes the Assignment A-MAP IE data, making it practically impossible for unauthorized users to decode the content of the Assignment A-MAPs not belonging to them. A seed is used to initialize the pseudo-random sequence generator which is provided to the MS by the BS during network-entry or re-entry. A 16-bit CRC is generated based on the contents of the randomized Assignment A-MAP IE. If the randomized Assignment A-MAP IE bits are represented by polynomial $m(x) = \alpha_{N-1}x^{N-1} + \alpha_{N-2}x^{N-2} + \dots + \alpha_1 x + \alpha_0$, where α_{N-1} is the MSB of the randomized Assignment A-MAP IE and α_0 is the LSB of the randomized Assignment A-MAP IE. The 16-bit CRC is calculated as the remainder of dividing $m(x)x^{16}$ by the 16-bit CRC generator polynomial $g(x) = x^{16} + x^{12} + x^5 + 1$. The resulting CRC is denoted by $c(x) = c_{15}x^{15} + c_{14}x^{14} + \dots + c_1 x + c_0$ where c_{15} is the MSB of the CRC and c_0 is the LSB of the CRC. A 16-bit CRC is generated based on the contents of assignment A-MAP IE and then is masked by a 16-bit CRC mask using the bitwise XOR operation ($\oplus$) to allow the mobile stations to uniquely identify their corresponding A-MAP IEs. The CRC can be masked using one of the following methods depending on the value of the masking prefix:

- The 16-bit mask comprises 1-bit masking prefix, 3-bit message type indicator, and 12-bit masking code. The 3-bit type indicator determines whether a 12-bit masking code based on the MS STID is used, or alternatively a 12-bit masking code is the same as that used to mask the Broadcast Assignment A-MAP IE, BR-ACK A-MAP IE, and Group Resource Allocation A-MAP IE.
- The 16-bit mask comprises 1-bit masking prefix and 15-bit Random Access Identifier (RAID).

The masked CRC is then appended to the Assignment A-MAP IE, resulting in a bit sequence of $m(x)$ $x_{16}c(x) \oplus u(x)$ where $u(x)$ is the masking polynomial. The resulting sequence of bits is encoded by the TBCC encoder with minimum code rate of 1/4 and the encoded bits are repeated to improve the robustness of Assignment A-MAP information transmission. The Assignment A-MAP IEs may be encoded with two different effective code rates, i.e., code rate sets (1/2, 1/4) or (1/2, 1/8), where the indication is signaled via the superframe header. Note that each code rate set consists of two code rates where the first code rate is applied to the first group of Assignment A-MAP IEs within the control region and the second code rate is applied to the second group of Assignment A-MAP IEs within the control region. The number of Assignment A-MAPs in each group is indicated through parameter *Assignment A-MAP Size* in non-user-specific A-MAP. This would allow grouping of the users depending on their channel conditions and geographical location (e.g., near or far to the BS), and ensure robustness of the Assignment A-MAP IE transmission to each user group. If FFR is used, the above code rates can be used in reuse-1 partition. The 1/2 or 1/4 code rate is used in the power-boosted reuse-3 partition. The encoded bit sequences are modulated using QPSK baseband modulation scheme.

In most cases, the MS needs to decode all MLRUs in an A-MAP region in order to determine whether there is any relevant assignment A-MAP. The MS does not need to use different MCS to decode the same MLRU. The non-user-specific A-MAP carries information about the MCS used by each MLRU. The MS does not need to decode the MLRU using different rate de-matching for different IE sizes. All assignment A-MAP IEs or segmented IEs have fixed sizes, i.e., 56 bits. The MS determines if an assignment A-MAP is relevant by performing a CRC test using its STID (for unicast assignments), group ID (for group scheduling), or RAID (CDMA allocation) to unmask the CRC. If the CRC test is successful, the MS may continue parsing the content of the decoded Assignment A-MAP. As mentioned earlier in this section, there are four lookup tables, each with 256 entries, which are used to signal the assignment A-MAP group sizes with the following configurations:

- Non FFR (single A-MAP region): QPSK ¼, QPSK ½;
- Non FFR (single A-MAP region): QPSK $\frac{1}{8}$, QPSK ½;
- FFR (two A-MAP regions): reuse-1 QPSK ¼, reuse 1 QPSK ½, reuse-3 QPSK ½;
- FFR (two A-MAP regions): reuse-1 QPSK $\frac{1}{8}$, reuse 1 QPSK ½, reuse-3 QPSK ½.

In all cases, the maximum number of Assignment A-MAPs is 32 per subframe, i.e., the maximum number of blind detection hypothesis testing is 32 per subframe and the total utilized physical resource is no more than 48 MLRUs. In the example shown in Figure 10-6, a total of 13 Assignment A-MAPs

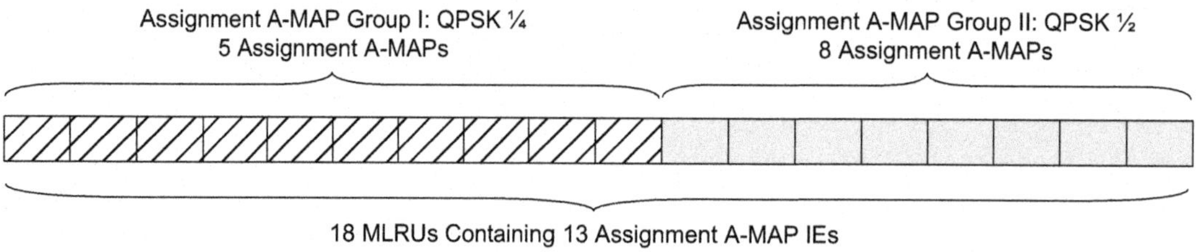

Assignment A-MAP Group I: QPSK ¼
5 Assignment A-MAPs

Assignment A-MAP Group II: QPSK ½
8 Assignment A-MAPs

18 MLRUs Containing 13 Assignment A-MAP IEs

FIGURE 10-6

An illustration of the Assignment A-MAP groups

are divided into groups with each containing a different number of Assignment A-MAPs. The (QPSK 1/4, QPSK 1/2) modulation and coding scheme has been used. Therefore, there are 13 hypothesis-testing trials for the purpose blind detection of the Assignment A-MAPs.

10.2.2 Advanced MAP Information Elements

The A-MAPs carry control and signaling information and scheduling parameters. The unicast or broadcast control and signaling parameters are used for configuration of the mobile stations and base stations during normal operation. The parameters and control signal are contained in various fields of the Information Elements (IEs) that constitute the A-MAPs. Each A-MAP IE consists of type information (16 possible types) and a number of parameter fields. The length of the A-MAP IEs is limited to 40 bits unless fragmented into additional information elements. Table 10-3 shows the various A-MAP IE types that are defined in IEEE 802.16m. In the following, we describe each information element type's main parameters and usage. The reserved fields are not shown. It must be noted that the exact format of all information elements are defined in the IEEE 802.16m standard specification [2]. Furthermore, the existence of some parameters of the information elements are conditioned to use of certain system configurations; those conditions are not shown in the following tables in order to avoid distraction and confusion due to complexity.

Table 10-3 Various A-MAP IE Types and their Usage [2]

A-MAP IE Type	Usage	Property
DL Basic Assignment	Allocation information for an MS to decode DL bursts using continuous logical resources	Unicast
UL Basic Assignment	Allocation information for an MS to transmit UL bursts using continuous logical resources	Unicast
DL Sub-band Assignment	Allocation information for an MS to decode DL bursts using sub-band based resources	Unicast
UL Sub-band Assignment	Allocation information for an MS to transmit UL bursts using sub-band based resources	Unicast
Feedback Allocation	Allocation or de-allocation of UL fast-feedback control channels assigned to an MS	Unicast
UL Sounding Command	Control information for an MS to start UL sounding transmission	Unicast
CDMA Allocation	Allocation for an MS requesting bandwidth using a ranging or bandwidth request codes	Unicast
DL Persistent	DL persistent resource allocation	Unicast
UL Persistent	UL persistent resource allocation	Unicast
Feedback Polling	Allocation for an MS to send MIMO feedback using MAC messages or extended headers	Unicast
Group Resource Allocation	Group scheduling and resource allocation	Multicast
BW-REQ ACK	Indication of decoding status of bandwidth request opportunities and resource allocation of bandwidth request header	Multicast
Broadcast	Broadcast burst allocation and other broadcast information	Broadcast

The **DL Basic Assignment A-MAP IE** is used for resource assignment in the downlink. The DL Basic Assignment A-MAP IE consists of several information fields as summarized in Table 10-4.

The *Resource Index* in the DL/UL Basic Assignment A-MAP IE identifies a resource block comprising a set of contiguous LRUs in a DL/UL subframe. The index determines the size of the allocated resource in a number of contiguous LRUs in the frequency-domain over the subframe and the index of the LRU from which the scheduled allocation is measured. The contiguous LRUs can be constructed from distributed LRUs, mini-band LRUs, or sub-band LRUs. The mapping between LRU and physical PRU indices is derived as follows. For each frequency partition i, let TC_{LRUi}, denote the total number of contiguous LRUs and distributed LRUs up to and including that partition i. Let's define $TC_{LRU,i} = \sum_{m=0}^{i} FPS_m; 0 \leq i \leq 3$ then:

$LRU[k]$

$$
= \begin{cases}
DLRU_{FP_i}[k - TC_{LRU,i-1}] & TC_{LRU,i-1} \leq k < (LRU_{DRU,FP_i} + TC_{LRU,i-1}) \\
NLRU_{FP_i}[k - (LRU_{DRU,FP_i} + TC_{LRU,i-1})] & (LRU_{DRU,FP_i} + TC_{LRU,i-1}) \leq k < (LRU_{DRU,FP_i} + \\
& \quad LRU_{MB-CRU,FP_i} + TC_{LRU,i-1}) \\
SLRU_{FP_i}[k - (LRU_{DRU,FP_i} + \\
LRU_{MB-CRU,FP_i} + TC_{LRU,i-1})] & (LRU_{DRU,FP_i} + LRU_{MB-CRU,FP_i} + TC_{LRU,i-1}) \leq k < TC_{LRU,i}
\end{cases}
$$

$$(10-5)$$

where $0 \leq k < \sum_{m=0}^{3} FPS_m; 0 \leq i \leq 3$ and $TC_{LRU,-1} = 0$. In the above equation, the CRUs are directly mapped into contiguous LRUs including sub-band LRUs (SLRU) and mini-band LRUs (NLRU). The mapping between CRU_{FPi} and $SLRU_{FPi}$ or $NLRU_{FPi}$ is defined as follows:

$$
\begin{aligned}
SLRU_{FPi}[j] &= CRU_{FPi}[j] & 0 \leq j \leq L_{SB-CRU,FPi}[j]; 0 \leq i \leq 3 \\
NLRU_{FPi}[j] &= CRU_{FPi}[j + L_{SB-CRU,FPi}] & 0 \leq j \leq L_{MB-CRU,FPi}[j]; 0 \leq i \leq 3
\end{aligned}
$$

$$(10-6)$$

Multiple non-contiguous sub-bands are indexed using the DL/UL Sub-band Assignment A-MAP IE. For a given system bandwidth with N_{max} LRUs, the size of allocations comprising contiguous LRUs can be chosen from 1 to N_{max}. For each allocation size, resources can be allocated starting from different LRUs in the subframe. Any allocation size $1 \leq s \leq N_{max}$ can be assigned in any location starting at LRU index l where $s + l \leq 24$ for 5 MHz and $s + l \leq 48$ for a 10 MHz system bandwidth.

For a 20 MHz system bandwidth, the number of assignable resource sizes is reduced; however, there is no constraint on where those allocations may be made in the subframe as long as $s + l \leq 96$. The assignable resource sizes are specified such that all resource sizes in increments of one LRU are assignable in the range of 1 to 12 LRUs, only even resource sizes are assignable in increments of two LRUs in the range of 12 to 24 LRUs, resource sizes are assignable in increments of four LRUs in the range of 24 to 48 LRUs, resource sizes are assignable in increments of eight LRUs in the range of 48 to 88 LRUs, and all resource sizes in increments of one LRU are assignable in the range of 92 to 96 LRUs.

As shown in Figure 10-7, in order to determine the resource index RI for an allocation, the BS requires the size s and index l of the LRU from which the scheduled allocation begins. The BS also maintains a vector I_a of length N_{max} in which the non-zero entries contain the starting index for each of

Table 10-4 DL Basic Assignment A-MAP IE Parameters [2]

Information Field	Size (Bits)	Description
A-MAP IE Type	4	It defines what parameters are included in this A-MAP IE. The A-MAP IE Type distinguishes between assignment A-MAP IEs used for the UL/DL, and assignment A-MAP IEs used for resource allocation as well as control signaling. Some additional A-MAP IE types are reserved for future use.
$I_{SizeOffset}$	5	Offset used to calculate the burst size index.
MIMO Encoder Format	2	This parameter specifies the MIMO encoder type that is used with this resource assignment (i.e., Space-Frequency Block Coding, Vertical Encoding, Multi-layer Encoding, or Conjugate Data Repetition).
Number of Streams Used in the Transmission	3	In the case of vertical encoding, this parameter identifies is the number of streams that ranges from 1 to 8. Note that the number of streams must be smaller than the number of base station transmitting antennas.
S_i	5	In the case of multi-layer encoding, an index is used to identify the combination of the number of streams and the allocated pilot stream index in a transmission with MU-MIMO and the modulation constellation of paired user in the case of dual-stream transmission.
Resource Index	11	This parameter identifies the starting LRU index and size of a single allocation spanning contiguous LRUs. The resource index includes location and allocation size.
Long TTI Indicator	1	Indicates number of subframes spanned by the allocated resource: 0b0: 1 subframe (default); 0b1: 4 DL subframes for FDD or all DL subframes for TDD.
HARQ Feedback Allocation	3	In the case of de-allocation of persistent allocations in the DL/UL, the base station transmits a HARQ Feedback Allocation in the DL/UL Persistent Allocation A-MAP IE.
HARQ Identifier Sequence Number (AI_SN)	1	In order to specify the start of a new transmission, one-bit HARQ identifier sequence number is toggled on every new HARQ transmission with the same ACID. If the AI_SN changes, the receiver assumes that the corresponding HARQ transmission belongings to a new encoder packet and discards the previous HARQ transmission with the same ACID.
HARQ Channel Identifier (ACID)	4	HARQ channel identifier.
HARQ Sub-packet Identifier (SPID)	2	HARQ sub-packet identifier for HARQ-IR (0, 1, 2, or 3).
Constellation Rearrangement Version (CRV)	1	–

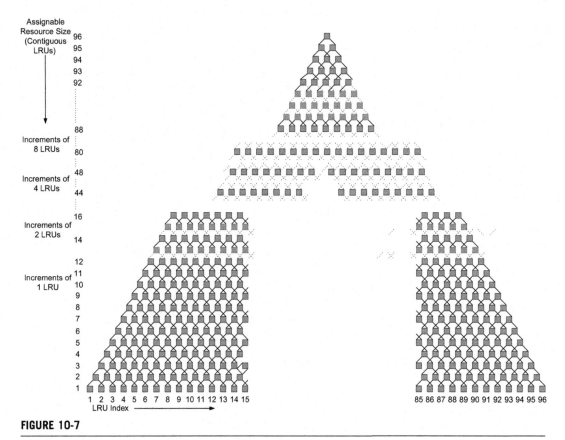

FIGURE 10-7

An illustration of resource assignment concept (11-bit resource indexing for 20MHz bandwidth)

the assignable resource sizes. The ith element of I_a ($1 \le i \le 96$) for an 11-bit resource indexing using 20 MHz bandwidth is defined as follows:

$$
I_a(i) = \begin{cases}
0 & i = 1 \\
I_a(i-1) + [96 - (i-1) + 1] & 2 \le i \le 12 \\
I_a(i-2) + [96 - (i-2) + 1] & i = 2k, 7 \le k \le 12 \\
I_a(i-4) + [96 - (i-4) + 1] & i = 4k, 7 \le k \le 12 \\
I_a(i-8) + [96 - (i-8) + 1] & i = 8k, 7 \le k \le 11 \\
2048 - \sum_{k=i}^{96} 96 - k + 1 & 92 \le i \le 96 \\
0 & \textit{Otherwise}
\end{cases}
\tag{10-7}
$$

The resource index RI for an allocation of size s LRUs beginning at LRU l is calculated as follows:

$$RI = \begin{cases} I_a(i) + \ell & \text{if } I_a(i) > 0 \\ not - assignable & \text{if } I_a(i) = 0 \end{cases} \tag{10-8}$$

where $0 \leq l \leq 96$. The BS first determines if the required resource size is assignable by checking whether the ith element in I_a has a non-zero value. If the size s is assignable, then the 11-bit resource index is determined by adding l to the value of the ith element in I_a. If the required resource is not assignable, the next higher or lower non-zero element in I_a is selected based on the link adaptation scheme. Note that the same scheme can be applied to the 5 and 10 MHz system bandwidths where the length of I_a is 24 and 48, respectively, and all values in I_a correspond to non-zero indices for every allocation size.

At the receiver, the 11-bit resource index RI contained in the DL/UL Basic Assignment A-MAP IE is used to determine the assigned resource size s. The ith entry in the vector I_a with the maximum possible value is found such that $I_a(i) \leq RI$; the value i that satisfies the latter equation is the assigned resource size, i.e., $s = \max\limits_{i \in \{1,2,...,N_{max}\}} \{I_a(i) | I_a(i) \leq RI\}$. The starting LRU index of the allocation is determined by subtracting the value of I_a corresponding to the assigned resource size s from the index RI contained in the DL/UL Basic Assignment A-MAP IE, i.e., $l = RI - I_a(s)$.

The **UL Basic Assignment A-MAP IE** is used for resource assignment in the uplink. Regardless of the duration of the TTI of the UL burst, the number of LRUs that are allocated to an MS within a specific subframe must be smaller than the number of LRUs in that subframe over all used sub-carriers. Table 10-5 shows the parameters of this information element.

The procedure depicted in Figure 10-8 describes the parsing of information fields in the UL Basic Assignment A-MAP IE to determine the MIMO transmission format and the associated parameters used for transmission on the uplink. The PMI Indicator is used for explicit signaling of PMI. The CSM flag is used to distinguish between SU-MIMO and collaborative spatial multiplexing operation in the uplink. The non-adaptive precoding is used for SFBC. The precoding flag PF is not always signaled, and is only used when PMI is not signaled in order to identify the type of precoding. For CSM and SU-MIMO with vertical encoding (single codeword), when PF is set to 0b0, the MS uses non-adaptive precoding, and when the PF is set to 0b1, the MS uses adaptive precoding. This is performed using the precoder of rank N_s based on the mobile station's choice. The MIMO encoder format (SFBC or vertical encoding) is signaled only when neither CSM nor PMI is signaled.

The **DL Sub-band Assignment A-MAP IE** is relatively more efficient method to allocate larger amounts of radio resources to a user. It has a similar format compared to the DL Basic Assignment A-MAP IE except the *A-MAP IE Type* and the *Resource Index* fields are defined differently. The *A-MAP IE Type* field is set to 0b0010 and the *Resource Index* field is 11-bits, regardless of the system bandwidth. In all cases, the BS or the MS perform the following pre-processing steps to derive the parameters that are used in the indexing and in interpretation of the RA field. The mapping between the sub-band logical resource unit (SLRU) index and the PRU index depends on the total number of sub-bands over all partitions. The MEF field is one bit whose value (0b0 or 0b1) indicates horizontal or vertical coding. There is a one-bit flag following the MEF field known as the Mode-Indicator field which is interpreted differently according to the number of sub-bands in all partitions. The total number of sub-bands over all partitions is found by

Table 10-5 UL Basic Assignment A-MAP IE Parameters [2]

Information Field	Size (Bits)	Description
A-MAP IE Type	4	UL Basic Assignment A-MAP IE.
$I_{SizeOffset}$	5	The offset that is used to calculate burst size index.
Number of Streams in Uplink Transmission	2	There can be up to 4 uplink streams per MS provided that the number of streams are less than or equal the number of MS transmit antennas.
SU-MIMO/Precoding Matrix Index (PMI) Indicator	1	This is a flag to indicate whether both SU-MIMO and PMI are signaled (SU-MIMO without PMI indication or Collaborative Spatial Multiplexing without PMI indication, SU-MIMO with PMI indication).
Collaborative Spatial Multiplexing (CSM)	1	This is a flag to indicate use of CSM or SU-MIMO mode when no PMI indication.
MIMO Encoder Format	1	This is an indicator for the MIMO encoding mode that is used in the uplink (SFBC or vertical encoding). Non-adaptive precoding is used at the MS with SFBC mode.
Precoding Flag	1	This is the precoding flag for SU-MIMO when PMI is not signaled (Non-adaptive precoding, Adaptive precoding using the precoder of rank equal to the number of uplink streams based on the MS choice.
Total Number of Uplink Streams	2	This parameter specifies the total number of streams in the LRU for CSM where the values can be 2, 3, or 4 streams.
S_i	2	This is the first pilot index for CSM which depends on the total number of streams.
PMI	6	This parameter signals the precoding matrix index where 4-bit PMI is used for 2 MS transmit antennas and 6-bit PMI is used for 4 MS transmit antennas.
Resource Index	11	The Resource Index field in the UL Basic Assignment A-MAP IE is interpreted the same as in the DL Basic Assignment A-MAP IE. The resource index specifies the resource location and the allocation size. If all bits are set to 1, the MS will not transmit the corresponding HARQ sub-packet for the ACID in this IE.
Long TTI Indicator	1	Indicates number of uplink subframes occupied by the allocated resource where transmission time interval can be one uplink subframe or 4 uplink subframes in FDD mode or all uplink subframes in TDD mode. If in a particular frame configuration, the number of DL subframes is less than the number of UL subframes, the Long TTI Indicator will be set to one.
HARQ Feedback Allocation	3	–
HARQ Identifier Sequence Number	1	–
HARQ Channel Identifier	4	–

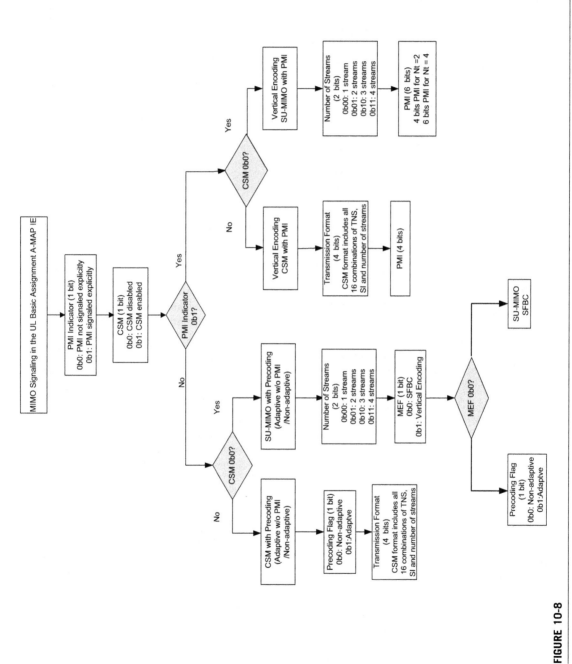

FIGURE 10-8

MIMO signaling in the UL basic assignment A-MAP IE [119]

dividing the length of the sub-band CRUs in each partition by $N_1 = 4$ and summing over the total number of partitions, hence:

$$N_{SB} = \sum_{i=0}^{3} \frac{L_{SB-CRU,FPi}}{N_1} \tag{10-9}$$

For each frequency partition i, the total number of sub-band CRUs is given as follows:

$$X_i = \sum_{m=0}^{i} L_{SB-CRU-FPm}; 0 \leq i \leq 3 \tag{10-10}$$

Thus, the SLRUs can be indexed as follows:

$$SLRU[k] = CLRU_{FP_i}[k - X_{i-1}]; \qquad X_{i-1} \leq k < X_i; \ \ 0 \leq i \leq 3; \ 0 \leq k < N_1 N_{SB} \tag{10-11}$$

The sub-bands are indexed as:

$$SB[m] = \forall (SLRU[k], k) \ni \left\lfloor \frac{k}{N_1} \right\rfloor = m; \ 0 \leq m < N_{SB} \tag{10-12}$$

The interpretation of the RA field in DL Sub-band Assignment A-MAP IE is different from that of DL Basic Assignment A-MAP IE, depending on the value of N_{SB}.

For $N_{SB} \leq 4$, SLRUs are allocated in a single instance using one Sub-band Assignment A-MAP IE. Each j-bit position in the RA field $0 \leq j < 3N_{SB}$ indicates the allocation of particular SLRU indices using a bit-map method as shown in Figure 10-9. The j-bit positions in the RA field, for $3N_{SB} \leq j < 12$ are set to 0.

For $5 \leq N_{SB} \leq 11$, SLRUs are allocated in a single instance using one Sub-band Assignment A-MAP IE. The MSB of the RA field is denoted as the IE Indication Field (IEF) which is identical to the Mode-Indicator. If IEF $= 0$, the 11 least significant bits of the RA field are denoted as the Resource Indexing Field (RIF) and each j-bit position in the RIF ($0 \leq j < N_{SB}$) indicates the allocation or non-allocation of all $N_1 = 4$ SLRUs within a particular sub-band, as shown in Figure 10-9. The j-bit positions in the RIF ($N_{SB} \leq j < 11$) are set to 0. Otherwise, if IEF $= 1$, the second MSB of the RA field is denoted by the Type Indication Field (TIF) followed by a 7-bit Resource Indication Field (RIF) and 3-bit Pattern Indication Field (PIF) as illustrated in Figure 10-9. The 5, 6, 7, 9, 10, or 11 SLRUs are allocated using the following method:

- If TIF $= 0$, three sub-bands $\{SB[u], SB[v], SB[w]\}, 0 \leq u < v < w < 10$ are defined and the 8 SLRUs are allocated in sub-bands $SB[u]$ and $SB[v]$, and 1, 2, or 3 SLRUs are allocated in $SB[w]$. The 1, 2, or 3 SLRUs in $SB[w]$ are identified by the PIF bits where u, v, and w are indexed using Cover's indexing method, i.e., $RIF = \binom{u}{1} + \binom{v}{2} + \binom{w}{3}$ where $\binom{s}{i}$ is the extended binomial coefficient.
- If TIF $= 1$ and the MSB of RIF $= 0$, other 6 bits in RIF indicate two sub-bands $\{SB[u], SB[v]\}, 0 \leq u < v \leq 10$ and 8 SLRUs are allocated in sub-bands $SB[u]$ and $SB[v]$. The other 1, 2, or 3 SLRUs in $SB[w]$ are identified by PIF bits where u and v are indexed as $RIF = \binom{u}{1} + \binom{v}{2}$.
- If TIF $= 1$ and MSB of RIF $= 1$, other 6 bits in RIF indicate two sub-bands $\{SB[u], SB[v]\}, 0 \leq u < v \leq 10$ and the 4 SLRUs are allocated in sub-band $SB[u]$. The other 1, 2, or 3 SLRUs in $SB[v]$ are identified by PIF bits where u and v are indexed as $RIF = \binom{u}{1} + \binom{v}{2}$.

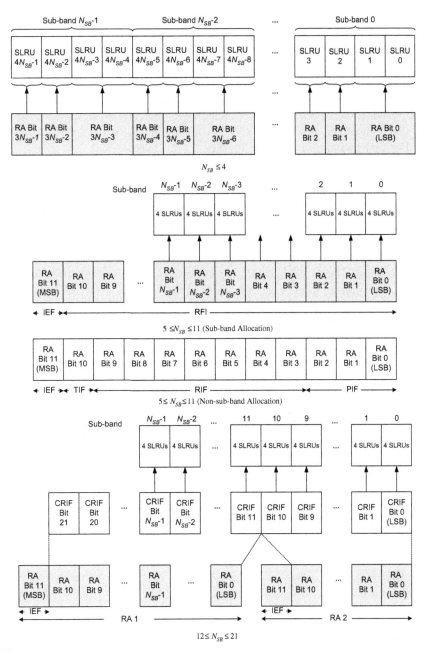

FIGURE 10-9

Interpretation of the RA field in downlink sub-band assignment A-MAP IE [2]

For $12 \leq N_{SB} \leq 21$, SLRUs are allocated in a single instance using one Sub-band Assignment A-MAP IE or two concatenated Sub-band Assignment A-MAP IEs. The MSB of the RA field is denoted as the IEF. If IEF $= 1$, a single instance of resource allocation is made using two IEs. The IEF in both IEs are 1. The RA fields of the two IEs with IEF $= 1$ is concatenated to form a 22-bit field referred to as the Concatenated Resource Indication Field (CRIF).

The LSB of the RA field of the Sub-band Assignment A-MAP IE occurring last in the AMAP region is interpreted as the LSB of the CRIF. Each of the j-bit position in the CRIF ($0 \leq j < N_{SB}$) indicates the allocation or non-allocation of all 4 SLRUs within a particular sub-band, as illustrated in Figure 10-9. The j-bit positions in the CRIF for $N_{SB} \leq j < 22$ are set to 0.

The structure of a **UL Sub-band Assignment A-MAP IE** is similar to the UL Basic Assignment A-MAP IE except the Resource Index fields are different and the IE Type field is set to 0b0011. The Resource Allocation field contains 11 bits for all operating bandwidths. The pre-processing steps are the same as those defined for the DL Sub-band Assignment A-MAP IE, except all DL terms are replaced by their UL equivalents. For the UL Sub-band Assignment A-MAP IE, a single IE is used to indicate a single instance of an allocation in all cases. The RA field interpretation is different from the downlink counterpart. If $0 \leq N_{SB} \leq 3$ each of the j-bit positions in the RA field $0 \leq j < 11$ indicates the allocation or non-allocation of particular SLRU indices using a bitmap as shown in Table 10-6. When $4 \leq N_{SB} \leq 10$, the RA field is interpreted in the same manner as in the $5 \leq N_{SB} \leq 10$ case in the DL Sub-band Assignment A-MAP IE by the MS and the BS. Furthermore, if $11 \leq N_{SB} \leq 21$, the MS and the BS's understanding of the RA field is the same as that in the $11 \leq N_{SB} \leq 21$ with Mode-Indicator $= 0b0$.

The **Feedback Allocation A-MAP IE** is used for dynamic allocation or de-allocation of uplink fast-feedback channels assigned to a mobile station. If an MS (with an existing fast-feedback channel corresponding to an active downlink carrier) receives a new allocation for a feedback channel on the same active carrier, the original fast-feedback channel is automatically de-allocated. The parameters of the Feedback Allocation A-MAP IE are listed in Table 10-7.

In Table 10-7 if the *Measurement Method Indication* flag is 0b0, then using a suitable MIMO feedback mode which can support MU-MIMO, (the MS is informed that) the maximum number of users that are going to be scheduled in the next MU-MIMO allocations is equal to N_s. Based on this information, the MS calculates and includes the multi-user interference in the CQI report. If N_s is set to one, then the MS assumes it will not be paired with any other mobile stations when it calculates the CQI. If MFM $= 5$ is included in the Feedback Allocation A-MAP IE, N_s indicates the rank of the open-loop codebook subset which belongs to the unitary matrix from which the MS should feedback its preferred stream index on a given physical sub-band. If a MIMO feedback mode is used which supports SU-MIMO, N_s would indicate the maximum spatial rate allowed in the feedback. The BS use of N_s takes into consideration the fact that the spatial rate is less than $\min(N_t, N_r)$ for the MS. If the *Measurement Method Indication* flag is 0b1 then N_s indicates the rank of the open-loop region for which the feedback is provided. The open-loop region type is derived from the N_s and MIMO feedback

Table 10-6 RA Bit to SLRU Index Mapping ($N_{SB} \leq 3$) [2]

RA[*j*]	0 (LSB)	1	2	3	4	5	6	7	8	9, 10 (Not Used)
SLRU[*k*] Indexed by RA[*j*]	0, 1		2	3	4, 5	6	7	8, 9	10	11	–

Table 10-7 Feedback Allocation A-MAP IE Parameters [2]

Parameter	Size (Bits)	Description
Channel Index	6	A unique identifier of a fast feedback channel on which an MS can transmit fast feedback information. A one-to-one relationship is established between channel index and the MS using this allocation.
Short-Term Feedback Period (p)	3	The short-term feedback is transmitted on the PFBCH/SFBCH every 2^p frames.
Long-Term Feedback Period (q)	2	A long-term feedback is transmitted on the FBCH every 2^q short-term feedback opportunity (i.e., every $2^p 2^q$ frames). The long-term feedback is not transmitted, if $q = 0b00$.
Frame Number	2	The MS starts reporting at the frame which number in the superframe is equal to $Frame\_Number$. If the current frame is specified, the MS starts reporting in four frames. Frames are numbered from 0 to 3 in the superframe.
Subframe Index	3	Indicate the UL subframe index in the UL portion of the frame.
Allocation Duration (d)	3	A parameter that indicates transmission of a feedback on the fast feedback channel indexed by $Channel\ Index$ during the next $2^{(3+d)}$ frames. If $d = 0b000$, the fast feedback channel is de-allocated. If $d = 0b111$, the MS continues the reporting until terminated by the BS.
ACK Allocation Flag	1	Indicate if one ACK channel is allocated to acknowledge the successful detection of this IE. $ACK\ Allocation\ Flag$ is set by the BS to 0b1, if $Allocation\ Duration$ equals 0b000, or is set to 0b1, if $Allocation\ Duration$ is not equal to 0b000 and the $Channel\ Index$ of the newly allocated fast feedback channel is the same as that of the de-allocated channel.
HARQ Feedback Allocation	6	Index of HARQ Feedback Allocation to acknowledge receipt of this A-MAP IE.
MIMO Feedback Mode	3	MIMO Feedback Mode (see Table 10-25).
Maximum Number of Streams	*Variable 1–2*	Variable number of bits whose value depends on number of transmit antenna N_t (2,4,8).
Feedback Format	2	This parameter specifies the feedback format index when reporting fast feedback information.
FP$_i$	1	Frequency partition indication. The BS asks the MS to send wideband CQI and spatial rate of the frequency partition and reuse factor in the future: 0b0: Frequency partition index 1; 0b1: Frequency partition index 2; BS sets FPI to a value that $FPS_{FPi} > 0$.
Frequency Partition Indication	2	Frequency partition indication. BS indicates MS to send wideband CQI and STC rate of the frequency partition and reuse factor in the future: 0b00: Frequency partition index 0; 0b01: Frequency partition index 1; 0b10: Frequency partition index 2;

(Continued)

Table 10-7 Feedback Allocation A-MAP IE Parameters [2] *Continued*

Parameter	Size (Bits)	Description
		0b11: Frequency partition index 3; BS sets FP_i to a value such that $FPS_{FPI} > 0$.
Long-Short FP_i Switch Flag	1	Used to inform the MS to switch short and long-term reporting based on the FP_i of the latest data allocation: 0b0: FP_i for long and short-term period report remains constant for the allocation duration; 0b1: FP_i for long and short-term period changes after every update of FP_i of last data allocation at the subsequent long-term feedback opportunity.
Long-Term FP_i	2	Frequency partition indication. The BS asks the MS to send wideband CQI and spatial rate for the second frequency partition using long term feedback: 0b00: Frequency partition index 0; 0b01: Frequency partition index 1; 0b10: Frequency partition index 2; 0b11: Frequency partition index 3; BS sets long-term FP_i to a different value than FP_i and $FPS_{long\ term\ FPI} > 0$.
Codebook Mode	2	Codebook Feedback Mode and Codebook Coordination: 0b00: Base mode with codebook coordination disabled; 0b01: Transformation mode with codebook coordination disabled; 0b10: Differential mode with codebook coordination disabled; 0b11: Base mode with codebook coordination enabled.
Codebook Subset	1	0b0: report PMI of the base codebook or transformed base codebook; 0b1: report PMI of the codebook subset or transformed codebook subset.
Codebook_Mode	2	Codebook Feedback Mode and Codebook Coordination: 0b01: transformation mode with codebook coordination disabled; 0b10: differential mode with codebook coordination disabled; 0b11: base mode with codebook coordination enabled.
Codebook_Subset	1	0b0: report PMI from the base codebook or transformed base codebook; 0b1: report PMI from the codebook subset or transformed codebook subset.
Codebook_Coordination	1	Codebook Feedback Mode and Codebook Coordination: 0b0: base mode with codebook coordination disabled; 0b1: base mode with codebook coordination enabled.
Codebook_Subset	1	0b0: report PMI from the base codebook; 0b1: report PMI from the codebook subset.
Measurement Method Indication	1	0b0: Use the midamble for CQI measurements; 0b1: Use pilots in open-loop region with N_s streams for CQI measurements.

Table 10-8 UL Sounding Command A-MAP IE Parameters [2]

Parameter	Size (Bits)	Description
A-MAP IE type	4	UL Sounding Command A-MAP IE
Sounding Subframe	2	Indentifies the sounding subframe. The subframes with sounding symbol are renumbered in time starting from 0
Sounding Sub-band Bitmap	Variable	FFT size dependant (maximum 24)
Decimation Offset	5	Unique decimation offset
Cyclic Time Shift	5	Unique cyclic shift
Periodicity (p)	3	If p = 0b000 no periodicity or terminate the periodicity; otherwise, repeat sounding every $2^{(p-1)}$ frames
Antenna Switching	1	0b0: Antenna switching; 0b1: No antenna switching.
Power Boosting	1	0b0: No power boosting; 0b1: 3 dB power boosting.

mode. The N_s indicates the pilot pattern used in the open-loop region where CQI measurements are conducted by the MS.

The **UL Sounding Command A-MAP IE** is used to schedule and configure the uplink sounding channel. The details of the physical structure and operation of sounding will be described in the next section.

Table 10-8 describes the parameters of the UL Sounding Command A-MAP IE. The number of UL sounding transmissions per frame that the BS allocates to an MS does not exceed one instance. The sounding sub-band bitmap is used to identify the sounding sub-bands used in the sounding allocation. For this purpose, the N_{used} contiguous sub-carriers are divided into sounding sub-bands, where each sounding sub-band consists of $N_1 P_{SC}$ adjacent sub-carriers with $N_1 = 4$ and $P_{SC} = 18$ for $N_{FFT} = 512,1024,2048$. The MSB of the sounding sub-band bitmap corresponds to the sounding sub-band with lowest sub-carrier index. The three periodicity bits are used to instruct the MS to periodically repeat the sounding transmission. If the antenna switching flag is set to zero, the MS performs sounding using antenna switching; otherwise, the MS performs sounding using all transmit antennas. The MS performs sounding using one transmit antenna at the each sounding opportunity starting from antenna 1 to antenna N_t. If power boosting is enabled, 3 dB power-boosting per sub-carrier is applied to each MS transmit antenna.

The **CDMA Allocation A-MAP IE** is used by the BS for uplink bandwidth allocation to a user that requested bandwidth using asynchronous ranging code or bandwidth request preamble. Furthermore, in a contention-based random access ranging procedure, the BS transmits the AAI_RNG-RSP message through a downlink burst assigned by CDMA Allocation A-MAP IE to an MS with no STID or TSTID assignment. The maximum number of the HARQ re-transmission is set to the default value and HARQ re-transmission control information cannot be changed during the re-transmission process. The parameters of the CDMA Allocation A-MAP information element and their corresponding values are shown in Table 10-9.

In wireless networks, persistent scheduling takes advantage of the traffic characteristics of the application (e.g., VoIP) to increase the number of users running that application in the system. The

Table 10-9 CDMA Allocation A-MAP IE Parameters [2]

Parameter	Size (bits)	Description
A-MAP IE Type	4	CDMA Allocation A-MAP IE.
CDMA Allocation Indication	1	0b0: Bandwidth allocation for bandwidth request; 0b1: Bandwidth allocation for ranging.
Resource Index	11	Resource index includes location and allocation size. The interpretation of bits is as follows: 5 MHz: 0 in 2 most significant bits + 9 bits for resource index; 10 MHz: 11 bits for resource index; 20 MHz: 11 bits for resource index.
Long TTI Indicator	1	Indicates number of subframes spanned by the allocated resource: 0b0: 1 subframe (default); 0b1: 4 uplink subframes for FDD or all uplink subframes for TDD. If number of downlink subframes is less than number of uplink subframes, then Long TTI Indicator is set to one.
HARQ Feedback Allocation	3	–
Uplink/Downlink Indicator	1	Indication of resource assignment in uplink or downlink: 0b0: Uplink; 0b1: Downlink.
$I_{SizeOffset}$	5	Offset used to compute burst size index.
IoT of the Frequency Partition	7	The IoT of the frequency partition used for MS resource assignment that is quantized in 0.5 dB steps as IoT level varies from 0 dB to 63.5 dB.
Offset Control	6	A parameter used for transmit power adjustment which represents a value from −15.5 dB to 16 dB in 0.5 dB steps
HARQ Channel Identifier	4	–

periodic nature of the packet arrivals from a VoIP source allows the required resource to be allocated persistently for the duration of an active voice session. Consequently, base stations can send the allocation information once at the beginning of a talk spurt, and avoid sending different allocation information for each subsequent packet. Significant resources that would otherwise be occupied by unnecessary control signals can be used to accommodate VoIP packets from other users. Because VoIP packets are usually small in size, the savings from reduced overhead can significantly increase the overall system capacity. The capacity gain from persistent scheduling; however, can be compromised significantly due to dynamic link adaptation, which is a very common technique in wireless systems to adapt to dynamic variation in wireless channel quality. The baseline concept of persistent and dynamic scheduling is illustrated in Figure 10-10.

The **DL Persistent Allocation A-MAP IE** is used to persistently allocate resources to a user in the downlink. It can also be used to de-allocate persistently allocated resources. Table 10-10 shows the parameters and their corresponding values in the DL persistent allocation information element.

The **UL Persistent Allocation A-MAP IE** is used to persistently allocate resources to a user in the uplink. It can further be used to de-allocate persistently allocated resources. Table 10-11 shows the parameters and their corresponding values in the DL persistent allocation information element.

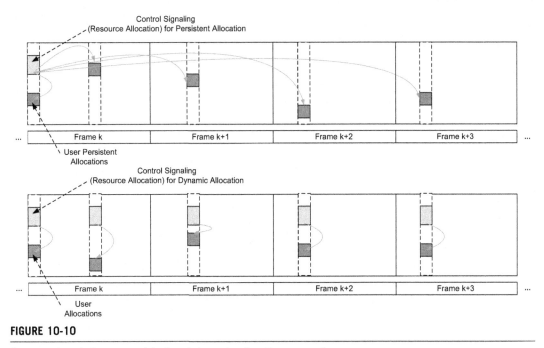

FIGURE 10-10

A comparison between persistent and dynamic allocation concepts

The **Group Resource Allocation A-MAP IE** is used to allocate resources to one or multiple mobile stations within a user group. The Group Resource Allocation A-MAP IE is used for signaling group resource allocation in the downlink or uplink. Group scheduling requires assignment of a service flow of an MS to a group. In order to add a flow of an MS to a group in the DL or UL, the BS transmits a Group Configuration MAC control message. It further requires allocation of resources to mobile stations within a group. In order to assign resources to one or more mobile stations in a group, the BS sends the Group Resource Allocation A-MAP IE which is included in the user-specific resource assignment in an A-MAP region. This information element contains bitmaps to identify scheduled mobile stations, and to signal MIMO mode HARQ burst size and resource size. Table 10-12 shows the parameters and their corresponding values in the group allocation information element.

The **Feedback Polling A-MAP IE** is used by the BS to schedule MIMO feedback transmission by the MS. The MS sends MIMO feedback using a MAC control message or a MAC signaling header, depending on the requested feedback content. If a dedicated UL resource is allocated, it will be used by the MS to transmit feedback at the designated feedback transmission frame; otherwise, in the designated transmission frame, the MS composes the feedback and the BS either includes an uplink allocation for the transmission using UL Basic Assignment A-MAP IE or UL Sub-band Assignment A-MAP IE, or the MS transmits in a dedicated uplink allocation assigned by a previous Feedback Polling A-MAP IE in the same subframe with feedback periods designating the same transmission frames. The MS sends the feedback in a MAC signaling header when wideband information for any combinations of MIMO feedback modes 0, 4, and 7 that are requested.

Table 10-10 DL Persistent Allocation A-MAP IE Parameters [2]

Parameter	Size (Bits)	Description
A-MAP IE Type	4	DL Persistent Allocation A-MAP IE.
Allocation Period	2	Period of persistent allocation: 0b00 de-allocation of a persistently allocated resource; 0b01 2 frames; 0b10 4 frames; 0b11 8 frames.
Resource Index	11	Confirmation of the resource index for a previously assigned persistent resource which has been de-allocated (Resource index includes location and allocation size): 5 MHz 0 in 2 most significant bits + 9 bits for resource index; 10 MHz 11 bits for resource index; 20 MHz 11 bits for resource index.
Long TTI Indicator	1	The number of subframes spanned by the allocated resource: 0b0 1 subframe (default); 0b1 4 DL subframes for FDD or all DL subframes for TDD.
HARQ Feedback Allocation	6	Index for HARQ Feedback Allocation to acknowledge receipt of de-allocation A-MAP IE.
$I_{SizeOffset}$	5	Offset used to compute burst size index.
MIMO Encoder Format	2	MIMO encoder format: 0b00 SFBC; 0b01 Vertical encoding; 0b10 Multi-layer encoding; 0b11 CDR.
Ns	3	Number of streams in transmission less than or equal to the number of transmit antennas at the BS (1 to 8 streams).
Si	4	Index to identify the combination of the number of streams and the allocated pilot stream index in a transmission with MU-MIMO and the modulation constellation of paired user in the case of dual-stream transmission.
Number of ACIDs	2	Number of ACIDs for implicit cycling of HARQ channel identifier (2,3,4,8).
Initial_ACID	4	Initial value of HARQ channel identifier for implicit cycling of HARQ channel identifiers.

The MS also sends feedback for other MIMO feedback modes using MAC control messages. The coefficients of the quantized transmit correlation matrix are fed back in the Correlation Matrix Feedback Header (CMFH) when no AAI_SingleBS_MIMO_FBK message is sent in the same packet and the BS has two or four transmit antennas; otherwise, the coefficients of the quantized transmit correlation matrix are sent in an AAI_SingleBS_MIMO_FBK message. The coefficients of the quantized transmit correlation matrix are fed back in an AAI_SingleBS_MIMO_FBK message, when the BS has eight transmit antennas. If the transmission of the transmit correlation matrix is previously scheduled and the MS has long-term feedback allocations, then the previous feedback allocation for the transmit correlation matrix is de-allocated with Feedback Polling A-MAP IE. In the case of

Table 10-11 DL Persistent Allocation A-MAP IE Parameters [2]

Parameter	Size (Bits)	Description
A-MAP IE Type	4	UL Persistent Allocation A-MAP IE.
Allocation Period	2	Period of persistent allocation: 0b00 de-allocation of persistent resource; 0b01 2 frames; 0b10 4 frames; 0b11 8 frames.
Resource Index	11	Confirmation of the resource index for a previously assigned persistent resource which has been de-allocated (Resource index includes location and allocation size): 5 MHz 0 in 2 most significant bits + 9 bits for resource index; 10 MHz 11 bits for resource index; 20 MHz 11 bits for resource index.
Long TTI Indicator	1	Indicates number of subframes spanned by the allocated resource: 0b0 1 subframe (default); 0b1 4 DL subframes for FDD or all DL subframes for TDD.
HARQ Feedback Allocation	6	Index for HARQ Feedback Allocation to acknowledge receipt of de-allocation A-MAP IE.
$I_{SizeOffset}$	5	Offset used to compute burst size index.
Ns	1	Number of streams in transmission where up to 2 streams per MS supported: 0b0 1 stream; 0b1 2 streams.
Total Number of Streams	2	Total number of streams in the LRU for CSM (2,3,4 streams).
Si	2	First pilot index for CSM with TNS = 2 streams: 0b00; 0b01. First pilot index for CSM with TNS = 3,4 streams: 0b00; 0b01; 0b10; 0b11.
MIMO Encoder Format	1	MIMO encoder format: 0b0 SFBC; 0b1 Vertical encoding.
Precoding Flag	1	Precoding flag: 0b0 Non-adaptive precoding; 0b1 Adaptive precoding using the precoder of rank Ns based on user's choice.
N_ACID	2	Number of ACIDs for implicit cycling of HARQ channel identifier (2,3,4,8).
Initial_ACID	4	Initial value of HARQ channel identifier for implicit cycling of HARQ channel identifiers.
Allocation Relevance	1	0b0 Allocation in the first UL subframe relevant to an A-MAP region; 0b1 Allocation in the second UL subframe relevant to an A-MAP region.

Table 10-12 Group Resource Allocation A-MAP IE Parameters [2]

Parameter	Size (Bits)	Description
A-MAP IE Type	4	Group Resource Allocation A-MAP IE.
Allocation Relevance	1	0b0 Allocation in the first UL subframe relevant to an A-MAP region; 0b1 Allocation in the second UL subframe relevant to an A-MAP region.
User Bitmap	Variable	Bitmap to identify scheduled mobile stations in a group. The size of the bitmap is equal to the User Bitmap Size signaled to each MS in the Group configuration MAC control message: 0b0 MS not allocated in this subframe; 0b1 MS allocated in this subframe.
Resource Offset	7	Indicates starting LRU for resource assignment to this group.
HARQ Feedback Allocation Offset	6	Indicates the start of the HARQ feedback channel index used for scheduled allocations.
MIMO Bitmap	Variable	Bitmap to indicate MIMO mode for the scheduled mobile stations: 0b0 Mode 0; 0b1 Mode 1.
Resource Assignment Bitmap	Variable	Bitmap which indicates burst size/resource size for each scheduled MS. The resource allocation bitmap uses 5 bits per flow to signal the HARQ burst size and the resource size for the user allocation in that subframe. The first 2 bits signal the HARQ burst size and the next 3 bits signal the resource size. The 2-bit and 3-bit codes for burst sizes and resource sizes respectively for the group are determined by each MS based on the information in the Group Configuration MAC control message.

feedback for MIMO feedback mode 0 with *Measurement Method Indication* = 0b0, and MIMO feedback modes 4 or 7, the MS feeds back the CQI for FP_0, if $FPS_0 > 0$ or for FP_k if $FPS_0 = 0$, where FP_k is determined by frequency partition index $k = 0,1,2,3$. The MIMO feedback reported by an MS in frame *N* pertains to measurements performed at least up to frame *N-1*. The first MIMO feedback report following the Feedback Polling A-MAP IE may contain invalid MIMO feedback information, if the MIMO feedback is sent in the frame immediately following the frame in which the Feedback Polling A-MAP IE was received [2]. Table 10-13 shows the parameters and their corresponding values in the feedback polling information element.

The **Bandwidth Request (BR) Acknowledgement A-MAP IE** indicates the decoding status of the bandwidth request opportunities in the uplink frame that is identified by the BR ACK offset. The successfully received preamble sequences, if any, are included in ascending order. The bandwidth request opportunities are encoded in ascending order, based on the number of the uplink subframe in which they are contained in single or multiple BR-ACK A-MAP IEs. The BR-ACK A-MAP IE cannot be segmented. In addition, the BR-ACK A-MAP IE also includes the allocation information for the fixed-sized bandwidth request header. The uplink resource and HARQ feedback channel will be allocated to the preamble sequences whose grant indicator is set to one. The allocations are in ascending order based on the index of preamble sequences. If uplink resource for the BR header is not allocated through the BR-ACK A-MAP IE, CDMA Allocation A-MAP IE is used for the uplink

Table 10-13 Feedback Polling A-MAP IE Parameters [2]

Parameter	Size (Bits)	Description
A-MAP IE Type	4	Feedback polling A-MAP IE.
Polling Sub-type	1	0b0 uplink resource allocation or de-allocation; 0b1 feedback mode allocation or de-allocation.
Allocation Duration (*d*)	3	The allocation is valid for $2^{(d-1)}$ superframes starting from the superframe defined by allocation relevance. If $d = 0b000$, the pre-scheduled feedback header transmission is released. If $d = 0b111$, the pre-scheduled feedback header transmission is valid until further notification from the BS.
Resource Index	11	Confirmation of the resource index for a previously assigned persistent resource that has been de-allocated (resource index includes location and allocation size): 5 MHz 0 in 2 most significant bits + 9 bits for resource index; 10 MHz 11 bits for resource index; 20 MHz 11 bits for resource index.
Polling De-allocation Bitmap	3	The MS is required to support up to 3 distinct and concurrent feedback allocations including one or several MIMO feedback modes, the transmit correlation matrix, and multi-BS MIMO feedback. The ordering of concurrent allocations in this bitmap is MFM i, MFM j ($j > i$), MFM k ($k > j$), transmit correlation matrix, and multi-BS MIMO feedback.
HARQ Feedback Allocation	6	HARQ feedback channel allocation to acknowledge the successful detection of this information element.
I$_{SizeOffset}$	5	Offset used to compute burst size index.
MIMO Encoder Format	1	MIMO encoder format for uplink feedback transmission (non-adaptive precoding is used at the MS): 0b0 SFBC; 0b1 Vertical encoding with two streams or one stream, if the MS has only one transmit antenna.
Long TTI Indicator	1	Indicates number of subframes spanned by the allocated resource: 0b0 1 subframe (default); 0b1 4 UL subframes for FDD or all UL subframes for TDD.
ACID	4	HARQ channel identifier.
MFM Allocation Index	2	0b00 MFM 0 with Measurement Method Indication = 0; 0b01 MFM 3 with all subbands; 0b10 MFM 6 with all subbands; 0b11 MFM is defined in Feedback Polling A-MAP IE with Polling Sub-type = 0b1.
Maximum Number of Transmit Streams	1	1 or 2 depending on the MIMO feedback mode.
Short-term and Long-term Periods	4	The resources are allocated in frames with short or long period. The short feedback period is *p* frames. The long feedback period is *q* superframes. The first allocation starts two frames later. The frame index is given by *i*+2, where *i* is the index of the frame where the Feedback Polling A-MAP IE is transmitted.

(Continued)

Table 10-13 Feedback Polling A-MAP IE Parameters [2] *Continued*

Parameter	Size (Bits)	Description
		The transmission of MIMO feedback modes in MFM Allocation Index is allocated with a short period, whereas the transmit correlation matrix is scheduled with a long period. Short and long period reports start at the first allocation. When short and long period feedback reports coincide in the same frame, both short period feedback content and long period feedback content are sent in the same burst.
ACK Allocation Flag	1	
MIMO Feedback IE Type	1	0b0 feedback allocation for single-BS MIMO operation; 0b1 feedback allocation for multi-BS MIMO operation.
Measurement Method Indication	1	0b0 Use the midamble for CQI measurements; 0b1 Use pilots in open-loop region with N_s streams for CQI measurements.
Number of Best Sub-bands ($N_{sub\text{-}bands}$)	2	0b00 report all sub-bands; 0b01 1 best sub-band; 0b10 min(6,N_{SB}) best sub-bands; 0b11 min(6,N_{SB}) best sub-bands. $1 \leq N_{sub\text{-}bands} \leq N_{SB}$.
Codebook Coordination	1	0b0 base mode with codebook coordination disabled; 0b1 base mode with codebook coordination enabled.
Codebook Subset	1	0b0 report PMI from the base codebook or transformed base codebook; 0b1 report PMI from codebook subset or transformed codebook subset.
Period (p)	3	Transmit feedback header every $4p$ frames. The first report is to start at the next frame.
Target Resource Unit	2	This parameter indicates which resource units or which type of resource unit to use for feedback: 0b00 Latest best sub-bands reported for single-BS MIMO; 0b01 The entire transmission bandwidth; 0b10 FFR partition 0; 0b11 boosted FFR partition.
Interference Coordination Type	2	0b00 PMI restriction for single-BS precoding; 0b01 PMI recommendation for single-BS precoding; 0b10 CL-MD for multi-BS precoding; 0b11 Co-MIMO for multi-BS precoding.
Number of Multi-BS Reports	3	This parameter indicates the number of reports.
Maximum Number of Users	2	Maximum number of users supported in Co-MIMO on the same resource (2, 3, 4).

resource allocation. For the uplink HARQ burst allocated through the BR-ACK A-MAP IE, the maximum number of the HARQ re-transmissions is set to the default value (i.e., 4). Non-adaptive HARQ re-transmission is used. Table 10-14 shows the parameters and their corresponding values in the bandwidth request acknowledgement information element.

Table 10-14 BR-ACK A-MAP IE Parameters [2]

Parameter	Size (Bits)	Description
A-MAP IE Type	4	BR-ACK A-MAP IE.
BR-ACK Bitmap	<5	Number of BR opportunities where each bit indicates the decoding status of BR preamble sequence in the corresponding BR opportunity. The length of bitmap determines based on the number of BR opportunities in a frame: 0b0 No BR preamble sequence is detected; 0b1 At least one preamble sequence is detected.
MSB of Resource Start Offset	2	These are the most significant bits of the start offset of the logical resource units for BR header.
LSB of Resource Start Offset	5	These are the least significant bits of the start offset of the logical resource units for BR header.
HARQ Feedback Allocation Start Offset	6	This field is start offset of HARQ feedback allocation.
Allocation Size	1	Resource size for each BR header (1 or 2 logical resource units).
Long TTI Indicator	1	Indicates number of subframes spanned by the allocated resource for BR header: 0b0 1 subframe (default); 0b1 4 UL subframes for FDD or all UL subframes for TDD.
Number of Received Preamble Sequences	2	The number of BR preamble sequence indices included in this ACK A-MAP IE (1, 2, 3, 4).
Preamble Sequence Index	5	Preamble sequence index received in the BR opportunity.
Message Decoding Indicator	1	Indicate the decoding status of quick access message: 0b0 Message not decoded; 0b1 Message decoded corresponding to preamble sequence index.
Grant Indicator	1	To indicate whether grant of BR header for the BR preamble sequence index is included. If this bit is set, the UL resource is allocated.

The **Broadcast Assignment A-MAP IE** is used to allocate resources for a broadcast burst. A broadcast burst contains one or more broadcast MAC control messages. Table 10-15 describes the parameters of the broadcast assignment information element. An example of usage of the Broadcast Assignment A-MAP IE for indications of traffic, paging, or neighbor advertisement messages is shown in Figure 10-11. Note that if this information element exists in a control region, it will be the first assignment A-MAP immediately following the non-user-specific A-MAP.

The broadcast burst signaled by the Broadcast Assignment A-MAP IE is always transmitted using an SFBC transmit diversity scheme and QPSK as the modulation scheme. The periodicity of the time-domain repetition is one frame. During the time interval where a broadcast burst is repeated, other broadcast bursts can only be transmitted without time-domain repetition. For each broadcast burst transmission with time-domain repetition, the BS sends this information element by decrementing the remaining repetition number by one in each transmission, i.e., if the BS sends the first transmission at the kth frame while signaling the remaining repetition number as n ($n = 0, 1, 2, 3$), the transmission

Table 10-15 Broadcast Assignment A-MAP IE Parameters [2]

Parameter	Size (Bits)	Description
A-MAP IE Type	4	Broadcast A-MAP IE.
Function Index	1	0 This IE carries broadcast assignment information; 1 This IE carriers ranging channel allocation information.
Burst Size	6	Burst size as indicated in the first 39 entries of Table 9-6.
Resource Index	11	Resource index includes location and allocation size: 5 MHz 0 in the 2 most significant bits + 9 bits for resource index; 10 MHz 11 bits for resource index; 20 MHz 11 bits for resource index.
Long TTI Indicator	1	Indicates number of subframes spanned by the allocated resource: 0b0 1 subframe (default); 0b1 4 DL subframes for FDD or all DL subframes for TDD.
Transmission Format	1	0b0 no time-domain repetition; 0b1 with time-domain repetition.
Repetition	2	0b00 no more repetition of the same burst; 0b01 the same burst is transmitted one time; 0b10 the same burst is transmitted two times; 0b11 the same burst is transmitted three times.
Number of Ranging Opportunities	2	–
Subframe Index	3	–

will be completed at the $(k + n)$th frame. The value of the repetition parameter is decremented in each transmission. For other information element's fields, only the resource index may change during the period when the time-domain repetition is performed. The HARQ sub-packet identifier is always set to zero for broadcast burst transmission.

10.3 UPLINK CONTROL CHANNELS

Multiple types of control and signaling information are carried via uplink control channels in order to support various air interface procedures. The uplink control information is categorized into the following:

- Primary Fast-Feedback Channel;
- Secondary Fast-Feedback Channel;
- HARQ Feedback (ACK/NACK) Channel;
- Bandwidth Request (BW-REQ) Channel;
- Initial, Periodic, and Handover Ranging Channel;
- Sounding Channel.

Channel quality feedback provides information about channel conditions as seen by the mobile station. This information is used by the BS for link adaptation, resource allocation, power control, etc. The channel quality measurement includes narrowband and wideband measurements. The channel quality

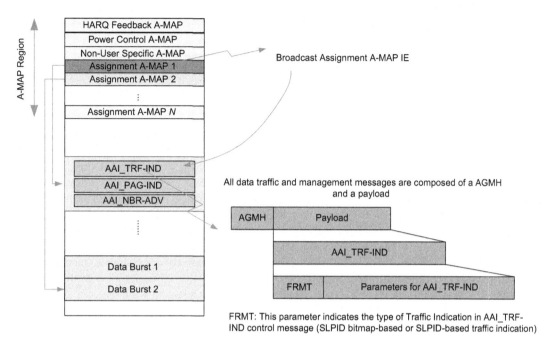

FIGURE 10-11

An example of usage of broadcast assignment A-MAP IE

feedback overhead can be reduced through differential feedback or other compression techniques. Examples of channel quality feedback include Physical CINR, Effective CINR, band selection, etc. Channel sounding can also be used to measure uplink channel quality. MIMO feedback provides wideband and/or narrowband spatial characteristics of the channel that are required for MIMO operation. The MIMO mode, precoding matrix index, rank adaptation information, channel covariance matrix entries, power loading factor, eigenvectors, and channel sounding are examples of MIMO feedback information. HARQ feedback (ACK/NACK) is used to acknowledge downlink burst transmissions. Multiple codewords in MIMO transmission can be acknowledged in a single ACK/NACK transmission. Uplink synchronization signals are needed to acquire uplink synchronization during initial access or handover, and also to periodically maintain synchronization. Reference signals for measuring and adjusting the uplink timing offset are used for these purposes. Bandwidth requests are sent by the mobile station to signal the required uplink bandwidth to the BS. A bandwidth request indicator notifies the BS of an uplink grant request by the mobile station. The bandwidth request messages can include information about the status of queued traffic at the MS, such as buffer size and quality of service, including QoS parameters.

The E-MBS feedback (if enabled) would provide information about downlink multicast and broadcast service to one or multiple cells. Note that multicast and broadcast service is typically a downlink-only transmission and there is no uplink mechanism to send feedback to the base station; however, if a bi-directional RF carrier is aggregated with a downlink-only RF carrier, the feedback may be sent on the bi-directional RF carrier. In that case, the E-MBS may utilize a common uplink channel which is used by mobile stations to transmit feedback. If a predefined feedback condition is met, an ACK or NACK is transmitted through a common E-MBS feedback channel. The uplink subframe size for transmission of control information requires at least six OFDM symbols. In the following sections, more details on the physical structure and operation of the uplink control channels are provided.

10.3.1 Fast-Feedback Channels

The uplink fast-feedback channel carries channel quality indicators and MIMO feedback. There are two types of fast-feedback uplink control channels: Primary Fast-Feedback Channel (PFBCH); and Secondary Fast-Feedback Channel (SFBCH). The PFBCH carries 6 bits of information, providing wideband and narrowband channel quality and MIMO feedback. It is used to support robust feedback reports. The SFBCH carries narrowband CQI and MIMO feedback information. The number of information bits carried in the SFBCH may vary from 7 to 24. A set of predefined numbers of bits in this range is supported. The SFBCH can be used to support CQI reporting at a higher code rate and thus more CQI information bits. The SFBCH can be allocated in a non-periodic manner based on traffic or channel conditions. The number of bits carried in the secondary fast-feedback channel can be adaptive. The fast-feedback channels are frequency-division multiplexed with other uplink control and data channels. The fast-feedback channel starts at a predetermined location with the size defined in a downlink broadcast control message. The fast-feedback channel allocations to an MS can be periodic and the allocations are configurable. For periodic allocations, the specific type of feedback information carried on each fast-feedback opportunity can be different. The fast-feedback channel carries one or more types of fast-feedback information. The physical layer processing of PFBCH and SFBCH is illustrated in Figure 10-12. For PFBCH transmission, four encoding types are specified. The encoding types correspond to MIMO Feedback Mode (MFM) and feedback format in Feedback Allocation A-MAP IE. The primary fast-feedback channel parameters are shown in Table 10-16. As mentioned earlier, the primary fast-feedback channel carries 6 bits of information whose 64 combinations specify the value of Space Time Coding (STC) Rate, Modulation, and Coding Scheme (MCS) of the feedback channel, Various Event-Driven Indicators (EDI), and Sub-band Index for the best sub-band depending on the encoding type. Encoding Type 2 in PFBCH is used for PMI reporting where the ith codebook entry $C(N_t, M_t, N_B, i)$ is mapped to sequence index i in PFBCH.

The SFBCH carries narrowband CQI and MIMO feedback information and the number of information bits that can be adaptively changed is between 7 and 24. Feedback formats define the information content carried by the fast-feedback channels. The format of the content of the PFBCH is determined by the PFBCH encoding type. The SFBCH payload information bits may carry sub-band indices (i.e., best sub-band index), STC rate, wideband STC rate, PFBCH indicator, sub-band CQI, sub-band average CQI, differential CQI, stream index, wideband PMI, sub-band PMI, base PMI (i.e., base PMI is a PMI generated from the base codebook or the base codebook subset), and differential PMI. The timing and content of the fast-feedback channels are mainly controlled by the MIMO feedback mode, as well as the short-term and long-term feedback period parameters carried via Feedback Allocation A-MAP IE. The secondary fast-feedback channel parameters are shown in Table 10-17.

The MS reports the CQI values by selecting a modulation and coding index from Table 10-18. The MCS index is selected assuming allocation size of 4 LRUs in a six OFDM symbol subframe, and 10% target error rate for the first HARQ transmission. It must be noted that the channel conditions may vary from the time that the CQI measurement is conducted on the reference signal until the CQI is reported

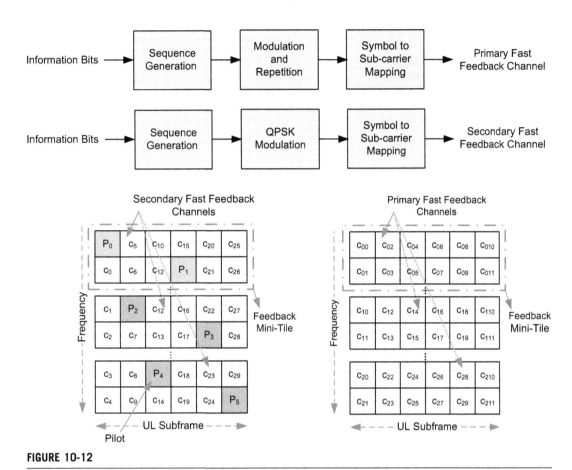

FIGURE 10-12

Physical processing and structure of primary and secondary fast-feedback channels [2]

(i.e., CQI report delay). Therefore, the reported CQI in frame n which might have been measured on a reference signal at frame $n - 1$ or earlier is associated with an appropriate MCS index for frame n. In order to allocate an appropriate MCS level and rank to an MS based on its actual conditions, the BS would make further adjustments in the MS reports by taking into consideration the CQI delay and the mobility conditions. For MU-MIMO feedback modes with codebook-based feedback, the CQI is calculated assuming that the interfering users are scheduled by the serving BS using rank-1 precoders, which are mutually orthogonal and orthogonal to the rank-1 precoder associated with the reported PMI.

The parameters contained in the PFBCH and SFBCH are encoded as follows:

- The STC rate is encoded with 1, 2, or 3 bits when the BS utilizes 2, 4, or 8 transmit antennas, respectively.
- The sub-band CQI and sub-band average CQI are encoded with 4 bits corresponding to the nominal MCS of Table 10-18. The differential CQI is encoded with 2 bits indicating MCS index increment/

Table 10-16 Primary Fast-Feedback Channel Parameters [2]

Parameter	MIMO Feedback Mode (MFM)	Usage
Channel Quality Indicator (CQI)	0,1,2,3,4,5,6,7	Wideband CQI or sub-band CQI for the best sub-band
Space Time Coding (STC) Rate Indicator	0,1,2,3	–
Sub-band Index	2,3,5,6	Sub-band selection for the best sub-band
Precoding Matrix Index (PMI)	3,4,6,7	Wideband PMI or sub-band PMI for the best sub-band
Event-driven Indicator to Request MFM Switching	N/A	The MS requests switching MIMO feedback mode between distributed and localized allocations.
Event-driven Indicator for Bandwidth Request Indicator	N/A	The MS requests uplink bandwidth (resources for uplink traffic).
Event-driven Indicator for Frequency Partition Selection (FPS)	0,1,4,7	The MS informs the serving BS about the frequency partition index.
Event-driven Indicator for Buffer Management	N/A	This parameter is an indication of the mobile station's available HARQ soft buffer (*The MS reports its buffering capability by stating the steady amount of information bits in 4800 byte units it can receive*). Buffer overflow is a condition where the receiving buffer is full and the soft bits of a failed (re) transmission can no longer be stored. Upon buffer overflow condition, the MS informs the BS using this EDI. This EDI is transmitted in the mobile station's first fast feedback opportunity. Even upon buffer overflow condition, the MS will still attempt to receive DL HARQ bursts and to send HARQ feedback to the BS. Upon reception of overflow EDI from the MS, the BS may assume that all soft bits of bursts in the last transmission that the MS failed to receive were not buffered. Consequently, to ensure that the systematic bits are used for decoding by the MS, the BS may retransmit SPID 0 for the failed bursts that were transmitted once. The BS may take into consideration the HARQ feedbacks from the MS for different retransmissions when selecting the amount of information bits, the SPID, or the MCS of the bursts.

decrement values of -1, 0, $+1$, or $+2$. The CQI of sub-band m in the first short-term report following a long-term report is calculated as *Sub-band m CQI Index = Sub-band Average CQI Index + Sub-band m Differential CQI* where *Sub-band Average CQI Index* is an average measure of the CQI over the reported sub-bands. The reported differential CQI from the MS must generate a value of *Sub-band m CQI Index* in the range of 0 to 15.

- The wideband PMI, sub-band PMI, and base PMI are encoded with 3, 4, or 6 bits according to the codebook size.

Table 10-17 Secondary Fast Feedback Channel Parameters [2]

Parameter	Related MIMO Feedback Mode	Usage
Sub-band CQI	2, 3, 5, 6	This parameter is used to report the average and differential CQI of selected sub-bands.
Sub-band Index	2, 3, 5, 6	This parameter is used to identify the selected sub-bands
Sub-band PMI	3, 6	*Precoding Matrix Indicator* of one sub-band for closed-loop MIMO
Stream Indicator	5	This parameter is used in open-loop MU-MIMO to indentify the spatial stream for CQI estimation
STC Rate Indicator	2, 3, 5, 6	–
PFBCH Indicator	2, 3, 5, 6	One bit indicator is used to indicate transmission of PFBCH in the next SFBCH opportunity. In the transmission of PFBCH, *Encoding Type* 0 is used.

- The differential PMI is encoded with 2, 4, or 4 bits when the BS uses 2, 4, or 8 transmit antennas, respectively. When the MS estimates PMI using the midamble, it should consider channel variations and report the PMI with reference to the frame the PMI has reported, i.e., if the reported PMI in frame n is based on the measurements conducted on the midamble in frame $n - 1$ or earlier, the appropriate value of PMI for frame n should be estimated and reported.

Figure 10-13 shows two example timing diagrams for uplink fast-feedback channels to visualize the operation assuming certain values for the parameters obtained from Feedback Allocation A-MAP IE. The physical structure of the uplink fast-feedback channels can be described as follows. An uplink

Table 10-18 The Modulation and Coding of the CQI Channel [2]

MCS Index	Modulation	Code Rate
0000	QPSK	31/256
0001	QPSK	48/256
0010	QPSK	71/256
0011	QPSK	101/256
0100	QPSK	135/256
0101	QPSK	171/256
0110	16 QAM	102/256
0111	16 QAM	128/256
1000	16 QAM	155/256
1001	16 QAM	184/256
1010	64 QAM	135/256
1011	64 QAM	157/256
1100	64 QAM	181/256
1101	64 QAM	205/256
1110	64 QAM	225/256
1111	64 QAM	237/256

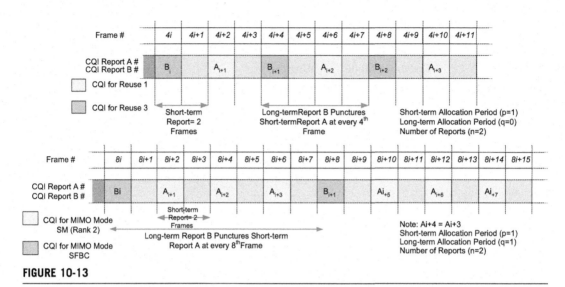

FIGURE 10-13

An example of primary and secondary uplink fast-feedback channel timing

feedback mini-tile is formed by two contiguous sub-carriers over six OFDM symbols. The uplink feedback control channels are formed by applying a UL mini-tile permutation to the LRUs allocated to the control resource. The primary and secondary fast-feedback channels each consists of three reordered FMTs. The process of forming a PFBCH is illustrated in Figure 10-12. The PFBCH information bits are used to generate a PFBCH sequence according to certain mapping [2]. The resulting sequence of bits is modulated, repeated, and mapped to uplink PFBCH symbol $s[k]$, where 0 is mapped to $+1$ and 1 is mapped to -1. The mapping of the primary fast-feedback channel symbol $s[k]$ to the UL FMTs is given by $c_{ij} = s[K_{ij}]$, $i = 0,1,2$ and $0 \leq j \leq 11$, where c_{ij} and K_{ij} are the FMT elements and jth element of the ith mapping sequence. This set of sequences can carry up to six information bits. The SFBCH comprised three distributed FMTs with two pilots allocated in each FMT. The pilot sequence $(p_0, p_1, p_2, p_3, p_4, p_5)$ is modulated as [1, 1, 1, 1, 1, 1] with pilot boosting. The SFBCH symbol generation procedure can be described as follows. The SFBCH information bits $a_0 a_1 \ldots a_{L-1}$ are encoded with M bits $b_0 b_1 \ldots b_{M-1}$ using the TBCC encoder with minimum code rate of 1/5. The value of parameter M is set to 60. The value of parameter $7 \leq L \leq 24$ depends on the size of the buffer [2]. The coded sequence $b_0 b_1 \ldots b_{M-1}$ is then modulated with $M/2$ symbols using QPSK modulation. The modulated symbols $c_0 c_1 \ldots c_{M/2-1}$ are mapped to the data sub-carriers of the SFBCH FMTs, as shown in Figure 10-12. When the length of feedback information bits is less than 7, a number of ones are padded to the end of feedback message to adjust the length.

10.3.2 HARQ Feedback Channel

The HARQ feedback channels are used to carry ACK/NACK information corresponding to downlink transmissions. The HARQ feedback channels start at predetermined time offsets relative to the corresponding DL transmissions and are frequency-division multiplexed with other control and data channels over an uplink subframe. A combined TDM/FDM and TDM/CDM scheme is used to multiplex multiple HARQ feedback channels. It was mentioned earlier that each HMT is identified by two indices: m is the HMT index in a HARQ feedback channel where m takes values 0, 1 or 2; and k is the HARQ feedback channel index where k takes an integer value in the range 0 to $L_{HFB\text{-}FPi} - 1$. The HARQ feedback channel comprises three distributed uplink feedback mini-tiles where the uplink FMT, as defined earlier, consists of two sub-carriers over six OFDM symbols. A total of three distributed 2×6 UL FMTs can support six UL HARQ feedback channels. The 2×6 UL FMTs are further divided into UL HARQ mini-tiles. An uplink HARQ mini-tile is constructed by two sub-carriers over two OFDM symbols, as illustrated in Figure 10-14.

The uplink HARQ feedback channels are physically processed as follows. Each pair of HARQ feedback channels (i.e., a pair of one-bit uplink HARQ feedback channels with even- and odd-valued indices) are mapped to a pair of orthogonal 4-bit sequences (c_{i0} c_{i1} c_{i2} c_{i3}, $i = 0,1,2,3$ shown in Table 10-19), depending on whether their corresponding index is odd or even. The resulting orthogonal sequences occupy one HMT which is repeated three times over three HMTs, as shown in Figure 10-14. In other words, each pair of uplink HARQ feedback channels is code-division multiplexed over one HMT and then repeated three times over three pairs of OFDM symbols to achieve time and frequency diversity. It can be seen that each group of three reordered FMTs can support up to six single-bit HARQ feedback channels. At the receiver, even- and odd-indexed HARQ feedback channels are decoded by the inner product of the orthogonal sequences shown in Table 10-19, and the received content of the uplink HARQ feedback channel. Since the HARQ feedback channels are sequentially indexed, there will be one even and one odd index in each pair whose corresponding sequences are orthogonal.

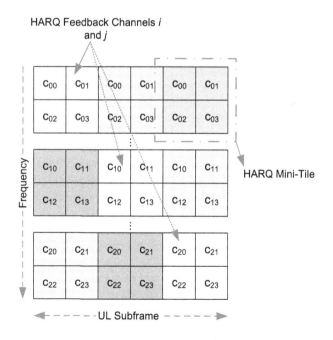

FIGURE 10-14

Physical structure of HARQ feedback channel [2]

Table 10-19 Orthogonal Sequences for the Uplink HARQ Feedback [2]

Sequence Index i	Orthogonal Sequence $c_{i0}c_{i1}c_{i2}c_{i3}$	HARQ Feedback Channel Information and Index
0	[+1 +1 +1 +1]	Even-numbered channel ACK
1	[+1 −1 +1 −1]	Even-numbered channel NACK
2	[+1 +1 −1 −1]	Odd-numbered channel ACK
3	[+1 −1 −1 +1]	Odd-numbered channel NACK

10.3.3 Sounding Channel

The sounding channel is used by a mobile station to send sounding signals (alternatively known as a sounding reference signal in 3GPP LTE) for MIMO feedback, channel quality feedback, and uplink channel measurement at the base station. Furthermore, uplink sounding enables sounding-based downlink MIMO in TDD mode, and uplink closed-loop MIMO in TDD and FDD modes. The sounding channel occupies specific uplink sub-bands (narrowband sounding signal) or the entire bandwidth (wideband sounding signal) over one OFDM symbol. It is obvious that the use of narrowband or wideband sounding signals depends on the uplink MIMO mode, which will be discussed in the next section. The sounding signal is transmitted over predefined sub-carriers within the sub-bands. The sounding channel parameters are transmitted in *System Configuration Descriptor* MAC control message and SFH SP1 broadcast channel.

The periodicity of the sounding signal for each MS is configurable. The sounding channel is frequency-division multiplexed (narrowband sounding) or time-division multiplexed (wideband sounding) with other control and data channels. The BS can configure a group of mobile stations to transmit sounding signals on the corresponding sounding channels. The sounding channels from multiple users or multiple antennas per user are orthogonally multiplexed. Power control for the UL sounding channel is supported to improve sounding quality. Each mobile station's transmit power for the sounding channel is individually controlled according to its sounding channel target CINR value. Transmission of the sounding signal from one or multiple antennas and multiple users are supported to provide MIMO channel information for DL and UL transmission.

The sounding signal occupies the first OFDM symbol in the uplink subframe. Each subframe can contain only one sounding symbol. In uplink subframes with six OFDM symbols, the sounding signal is not transmitted in the LRUs containing other control channels. The sounding signals can be transmitted in any resource unit when using uplink subframes with seven OFDM symbols. The remaining OFDM symbols in subframes containing sounding signals are used for data transmission, as shown in Figure 10-15. Multiple uplink subframes in a radio frame can be used for sounding.

The uplink sounding channels of multiple users and multiple antennas per mobile station can be multiplexed, and thereby simultaneously transmitted through decimation or cyclic shift in each sounding channel allocation. When multiple uplink subframes are used for sounding, time-division separation can be applied by assigning different mobile stations to different uplink subframes. When the sounding channels of multiple mobile stations are multiplexed using cyclic shifts, each mobile station occupies all the sub-carriers that are allocated to the sounding channel using different sounding

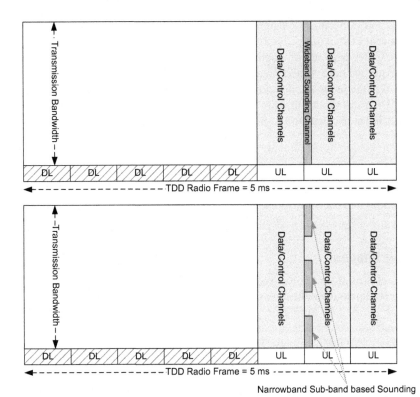

FIGURE 10-15

Structure of narrowband/wideband uplink sounding channel in TDD mode

waveforms. In a frequency-decimation separation scheme, each mobile station utilizes decimated sub-carrier sets from the sounding allocation with different frequency offsets. The sounding signal transmitted by the MS in the uplink is described as follows:

$$s(t) = \mathrm{Re}\left[e^{j2\pi f_c t} \sum_{k=0}^{N_{used}-1} \beta_k e^{j2\pi \left(k-\frac{N_{used}-1}{2}\right)\left(\frac{t-T_g}{T_u}\right)} \right] \quad k \neq \frac{N_{used}-1}{2} \tag{10-13}$$

When separating multiple users based on decimation, the occupied sub-carriers are decimated starting with offset g relative to the first used sub-carrier ($k = 0$). The occupied sub-carriers for each user or antenna are modulated using BPSK modulation and the coefficients are extracted from the Golay sequence of length 2048 as follows [2]:

$$\beta_k = \begin{cases} 2\sqrt{D}\left\{ \frac{1}{2} - G([k + u + \mathit{offset}_D(N_{FFT})]\mathrm{mod}2048) \right\} & k \in B, k \neq \frac{N_{used}-1}{2}, k \bmod D = g \\ 0 & \text{Otherwise} \end{cases}$$

$$\tag{10-14}$$

where D is the sub-carrier decimation parameter transmitted in the AAI_SCD MAC control message, k is the sub-carrier index $0 \leq k \leq N_{used} - 1$, N_{used} is the number of used sub-carriers over the sounding symbol, $G(x)$ is the Golay sequence of length 2048 bits ($0 \leq x \leq 2047$), N_{FFT} is the FFT size, u is a shift whose value is transmitted in the superframe header, $offset_D(N_{FFT})$ is an offset whose value depends on the FFT size, and B is the group of sub-carrier indices allocated to the sounding channel. If sounding signals transmitted by users or antennas are separated using cyclic shifts, the sequence associated with the nth cyclic shift index is calculated as follows [2]:

$$\beta_k = \begin{cases} 2\left\{\dfrac{1}{2} - G([k + u + offset_D(N_{FFT})]\mod 2048)\right\}e^{-j2\pi\frac{k}{P}n} & k \in B, k \neq \dfrac{N_{used} - 1}{2} \\ 0 & \text{Otherwise} \end{cases} \quad (10\text{-}15)$$

where P is the maximum cyclic-shift whose value is conveyed in the AAI_SCD MAC control message and $0 \leq n \leq P - 1$ denotes the cyclic shift index.

10.3.4 Ranging Channel

The ranging channel is used for uplink synchronization. The BS processes the received signal on the ranging channel to estimate the radio-link parameters, such as channel impulse response, SINR, and time of arrival, that are used for timing/frequency/power adjustment of the uplink transmissions from the MS. The ranging process consists of transmission of a deterministic sequence known as a ranging code over a number of OFDM symbols. The phase continuity of the received signal at the OFDM symbol boundaries is crucial to the ranging process. The ranging process is further classified into ranging for non-synchronized mobile stations (i.e., mobile stations that are not uplink-synchronized) and synchronized mobile stations (i.e., mobile stations that are uplink-synchronized with the BS). Note that in an OFDMA system, a mobile station cannot perform any uplink transmission prior to successful completion of the ranging procedures. A contention-based or non-contention based random access procedure is used for ranging. Contention-based random access is used for initial ranging and hand-over. Non-contention based random access is used for periodic ranging.

The physical ranging channel for non-synchronized mobile stations consists of a Ranging Cyclic Prefix (RCP) and a Ranging Preamble (RP). The value of the ranging cyclic prefix must be greater than the sum of the maximum channel delay spread and the round trip delay corresponding to the maximum cell size. The duration of the ranging preamble must be greater than RCP. To support large cell sizes, the ranging channel for non-synchronized mobile stations may be extended to multiple concatenated subframes (as shown in Figure 10-16). The physical ranging channel for non-synchronized mobile stations consists of a Ranging Preamble (RP) of length T_{RP} whose value depends on the ranging channel sub-carrier spacing Δf_{RP} (which is a fraction of the OFDMA system sub-carrier spacing Δf given in Table 9-2) and a Ranging Cyclic Prefix (RCP) of length T_{RCP} in time-domain. The ranging channel occupies a number of contiguous sub-carriers corresponding to one sub-band. Power control can be applied to the ranging channel to ensure robust signal transmission. The parameters of the initial ranging channel for the two supported formats are provided in Table 10-20, assuming a six symbol OFDM subframe. The OFDMA parameters T_g, T_u, and Δf are given in Table 9-2. The maximum coverage of ranging channels of the legacy and new systems are compared in Table 10-20 where it is shown that the legacy ranging channel may not meet the rigorous system requirements of IEEE 802.16m [3].

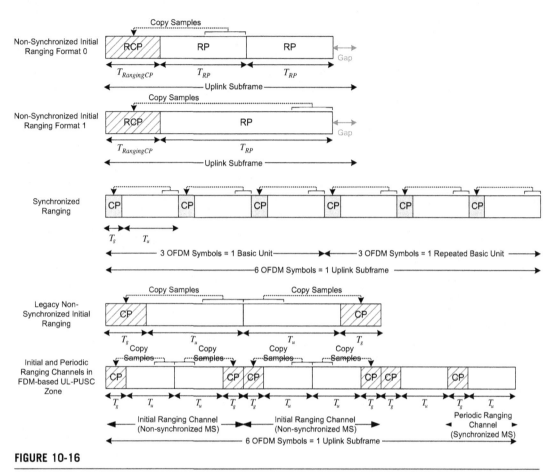

FIGURE 10-16

Ranging channel structure in the time-domain [2]

In the IEEE 802.16-2009 standard, the initial ranging channel is located in the first three OFDM symbols of the uplink subframe (TDD mode), and occupies six adjacent UL-PUSC subchannels, as shown in Figure 10-1 and Figure 9-20. The first OFDM symbol of the ranging sub-channels is formed in the same way as other OFDM symbols, i.e., performing an IFFT operation on the initial ranging code

Table 10-20 Initial Ranging Channel Parameters and Coverage [2]

Format	T_{RCP}	T_{RP}	Δf_{RP}	Physical Resource	Maximum Coverage
IEEE Std 802.16-2009	–	–	Δf	144 sub-carriers × 3 OFDM symbols	12 km
IEEE 802.16m Format 0	$3.5T_g + T_u$	$2T_u$	$\Delta f/2$	1 sub-band × 1 subframe	18 km
IEEE 802.16m Format 1	$3.5T_g + 7T_u$	$8T_u$	$\Delta f/8$	1 sub-band × 3 subframe	100 km

and then appending the cyclic prefix to the beginning of the symbol. The cyclic prefix for the second OFDM symbol is a copy of the first few samples of the symbol and is appended to the end of the OFDM symbol (as shown in the bottom of Figure 10-16) to ensure phase continuity at the symbol boundary. The virtual extension of the ranging opportunity would increase the probability of success for the mobile station making simultaneous random access. In order to perform the initial ranging, a ranging code is repeated twice and transmitted over two consecutive OFDM symbols, as described earlier. The ranging codes in the IEEE 802.16-2009 standard are PN sequences of length 144, and are chosen from a set of 256 codes. Note that the set of pseudo-random codes are divided into four subsets, where one subset is dedicated to initial ranging, periodic ranging, handover ranging, and bandwidth request [1]. The MS randomly selects one of the initial ranging codes, modulates it using BPSK, and sends it to the BS during the initial ranging opportunity over the allocated sub-channels and OFDM symbols for the ranging channel. The BS can separate colliding codes and extract timing information and power. In the process of user code detection, the BS obtains the Channel Impulse Response (CIR) of the code, thus acquiring information about the user's channel condition. The timing and power measurements allow the system to compensate for the near/far problem, as well as the propagation delay caused by transmission over large cells.

The ranging channel for non-synchronized users in IEEE 802.16m is formed using one or three uplink subframes known as *Initial Ranging Channel Format 0* or *Format 1*, respectively. The ranging preamble *RP* is repeated twice in non-synchronized initial ranging *Format 0*, as shown in Figure 10-16, to increase ranging opportunity and to avoid phase discontinuity at the OFDM symbol boundaries. The *RCP* is a copy of the last portion of the ranging preamble. The beginning of the transmission time in the ranging channel is aligned with that of the uplink subframe at the MS. There is a time gap at the end of uplink subframes containing the ranging channel to prevent interference between the adjacent subframes. A non-synchronized MS does not transmit any other uplink burst or uplink control channel signal in the subframe where a ranging channel is transmitted.

The ranging channel for synchronized mobile stations is used for periodic ranging. The mobile stations that are already synchronized with the target BS are allowed to transmit a periodic ranging signal. The synchronized ranging channel occupies 72 sub-carriers over six OFDM symbols, starting from the first OFDM symbol within the uplink subframe. There are two repeated signal waveforms, a basic unit which is generated by the ranging preamble code over 72 sub-carriers, three OFDM symbols, and a repeated basic unit. In the *Initial Ranging Channel Format 0* (i.e., the default format), the ranging preamble is repeated to increase ranging opportunity. The non-synchronized mobile stations can only use one instance of the ranging preamble with an RCP. When the preamble is repeated as an extended single opportunity, the second RCP is omitted for coverage extension. The guard sub-carriers are reserved at the edge of the non-synchronized physical ranging channel. The use of code-division multiplexing would allow a number of mobile stations to share the same ranging channel. The ranging channel for synchronized mobile stations is used for periodic ranging. The synchronized ranging channel starts at a configurable location (as defined in a downlink broadcast control message) and is frequency-division multiplexed with other control and data channels.

In an FDM-based UL-PUSC zone, non-contiguous resource blocks are used for the ranging channel. The ranging channel for FDM-based UL-PUSC zone consists of six distributed LRUs. The transmission of the initial ranging signal by a non-synchronized mobile station is performed within two consecutive OFDM symbols, as shown at the bottom of Figure 10-16. The same ranging code is transmitted on the ranging channel in each OFDM symbol to avoid phase discontinuity between the two symbols. The transmission of the ranging signal by the synchronized mobile stations is performed within an OFDM symbol, as shown in Figure 10-16. The physical layer procedures for generation and detection of the initial ranging sequences are illustrated in Figure 10-17.

The ranging preamble codes are grouped into initial ranging and handover ranging preamble codes. The initial ranging preamble codes are used for initial network entry, whereas handover ranging preamble codes are used for ranging with a target base station during handover. For a ranging code opportunity, each MS randomly chooses one of the ranging preamble codes from the available ranging preamble code set in a cell, except in the handover ranging case where a dedicated ranging code is assigned, the MS uses the assigned dedicated preamble code.

Zadoff-Chu sequences with cyclic shifts are used to generate the ranging preamble codes for non-synchronized access. The pth ranging preamble code denoted as $x_p(k)$ is defined as follows:

$$x_p(k) = e^{\left[-j\pi\frac{\alpha p k(k+1)+2k\beta p N_{CS}}{N_{RP}}\right]}, k = 0, 1, ..., N_{RP} - 1 \qquad (10\text{-}16)$$

where N_{CS} is the unit of cyclic shift according to the cell size, N_{RP} is the length of ranging preamble codes ($N_{RP} = 139$ for ranging channel Format 0 and $N_{RP} = 557$ for ranging channel Format 1), and P denotes the index of the Pth ranging preamble code, which is constructed by the β_P cyclic-shifted sequence derived from the root index α_P of a Zadoff-Chu sequence [2]. The number of cyclic-shifted codes per root index, the start root index of the Zadoff-Chu code, and the ranging preamble code partitioning information are broadcast via a secondary superframe header. The ranging preamble code

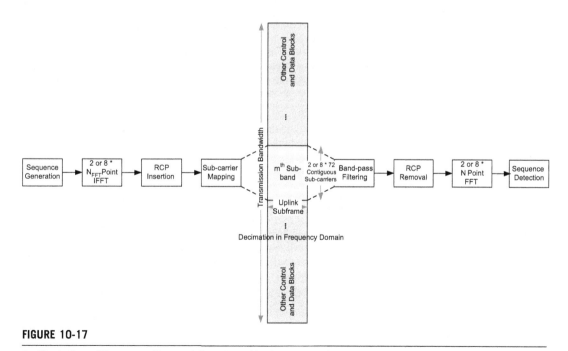

FIGURE 10-17

An illustration of the generation and detection of initial ranging sequences partitioning information indicates the number of initial and handover ranging preamble codes. The transmitted ranging signal waveform as function of time can be expressed as follows:

$$s(t) = \text{Re}\left[e^{j2\pi f_c t} \sum_{k=-(N_{RP}-1)/2}^{(N_{RP}-1)/2} x_P[k + (N_{RP} - 1)/2]e^{j2\pi(k+N_{offset})\Delta f_{RP}(t-t_{RCP})}\right]$$

In Format $0 : 0 \leq t \leq T_{RCP} + 2T_{RP}$

In Format $1 : 0 \leq t \leq T_{RCP} + T_{RP}$

(10-17)

where Noffset is a frequency position adjustment parameter due to the fact that the sub-carrier spacing for a ranging channel is a fraction of the normal sub-carrier spacing. The information regarding ranging channel configuration and allocation is carried through the secondary superframe header. This information includes a subframe offset (O_{SF}) for the ranging channel in the time-domain and the sub-band index for the ranging channel allocation in the frequency-domain, where the latter is determined once the *Cell-ID* and the number of sub-bands is given. The ranging channel can be scheduled in every frame, in the first frame of every superframe, in the first frame of every even-numbered superframes, and in the first frame of every fourth superframe. Therefore, in IEEE 802.16m, the initial ranging channel can be scheduled as frequently as 5 ms. The handover ranging channel can also be allocated though A-MAP in any subframe, except the subframe that has already been used for a regular resource allocation.

Padded Zadoff-Chu sequences with cyclic shifts are used for the periodic ranging preamble codes as follows:

$$x_p(n, k) = e^{\left[-j\pi\frac{\alpha p(71n+k)(71n+k+1)}{211} + \frac{2k\beta p N_{TCS}}{N_{FFT}}\right]}, k = 0, 1, ..., N_{RP} - 1; n = 0, 1, 2$$

(10-18)

where N_{TCS} denotes the time-domain cyclic shift per OFDM symbol. The start root index of the Zadoff-Chu sequence and the periodic ranging preamble code information are broadcast via the AAI_SCD MAC control message for non-femto-cell deployment scenarios. The number of periodic ranging preamble codes can be 8, 16, 24, or 32, depending on the system configuration. The information regarding periodic ranging channel configuration includes subframe offset (O_{SF}) for ranging channel allocation in the time-domain, where O_{SF} is the same as that for the initial ranging, as well as the sub-band index for ranging channel allocation in the frequency domain. For FDM-based UL-PUSC legacy system support, the same pseudo noise codes specified in the IEEE 802.16-2009 standard are used by the new mobile stations for the initial, periodic, and handover ranging.

10.3.5 Bandwidth Request Channel

A contention-based random access mechanism is used by the mobile stations to transmit bandwidth request information to the base station. Prioritized bandwidth requests are supported over the bandwidth request channel. In the legacy uplink zone with PUSC subchannelization, a BW-REQ tile is formed by four contiguous sub-carriers over six OFDM symbols. There are three BW-REQ tiles per BW-REQ channel, where each BW-REQ tile carries one BW-REQ preamble. In the new uplink zones, a BW-REQ tile is formed by six contiguous sub-carriers over six OFDM symbols, where each BW-REQ channel consists of three distributed BW-REQ tiles, and each BW-REQ tile carries one BW-REQ preamble and one quick access message. The MS may only transmit the BW-REQ sequence and leave the resources for the quick access message unused. As shown in Figure 10-18, the BW-REQ tile

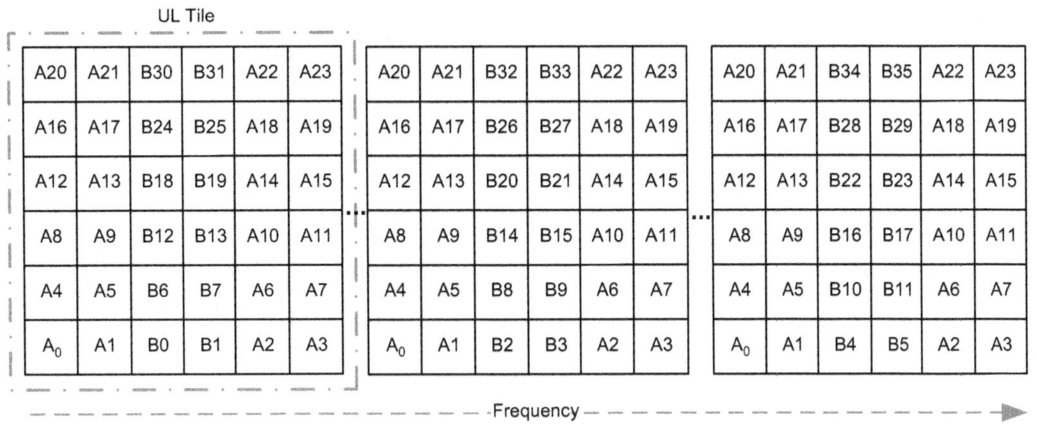

UL Tile

A20	A21	B30	B31	A22	A23
A16	A17	B24	B25	A18	A19
A12	A13	B18	B19	A14	A15
A8	A9	B12	B13	A10	A11
A4	A5	B6	B7	A6	A7
A_0	A1	B0	B1	A2	A3

A20	A21	B32	B33	A22	A23
A16	A17	B26	B27	A18	A19
A12	A13	B20	B21	A14	A15
A8	A9	B14	B15	A10	A11
A4	A5	B8	B9	A6	A7
A_0	A1	B2	B3	A2	A3

A20	A21	B34	B35	A22	A23
A16	A17	B28	B29	A18	A19
A12	A13	B22	B23	A14	A15
A8	A9	B16	B17	A10	A11
A4	A5	B10	B11	A6	A7
A_0	A1	B4	B5	A2	A3

UL Subframe

Frequency →

FIGURE 10-18

Bandwidth request channel physical structure [2]

consists of a preamble (A_{ij}) and data (B_{ij}) sub-carriers. The BW-REQ preamble is transmitted over four sub-carriers and six OFDM symbols. The data portion of the BW-REQ tile occupies two contiguous sub-carriers over six OFDM symbols and is used for transmission of the quick access message for the three-step bandwidth request procedure.

In a three-step bandwidth request procedure, 16-bit BW-REQ or BR information is constructed from a 12-bit STID and a 4-bit predefined BR index (i.e., the mapping between *Predefined BR Index*, *BR Size*, and *QoS Level* is performed during DSx procedure. The *BR Size* and *QoS Level* are determined based on the QoS parameters of the service flow in the DSx messages. The *BR Size* is described in number of bytes and is mapped to *Maximum Traffic Burst* parameter, and *QoS Level* is mapped to *UL Grant Scheduling Type* parameter). Let $s_0s_1s_2s_3s_4s_5s_6s_7s_8s_9s_{10}s_{11}$ and $s_{12}s_{13}s_{14}s_{15}$ denote the STID and predefined BR index, respectively.

The 16-bit BW-REQ information is formed by reordering the STID and predefined BR index bits as $s_0s_1s_2s_3s_4s_5s_6s_7s_8s_9s_{10}s_{11}s_{12}b_0b_1b_2$ where $b_i = (s_i + s_{i+3} + s_{i+6} + s_{i+9} + s_{i+12}\text{mod } 2)$, $i = 0,1,2$. The 3-bit $b_0b_1b_2$ index is carried in the BR preamble using the preamble index. The remaining physical resources in the BW-REQ channel are used to transmit the other 13 bits of information. The frame number and 16 bits of the bandwidth request message are used to select a 24-bit preamble sequence which is transmitted in the preamble portion of the BW-REQ tiles. The BR preamble sequences are defined as $S_P(k,u), 0 \leq k < 24$ where k is symbol index and u denotes the sequence index [2]. The selected preamble sequence $S_P(k,u), 0 \leq k < 24$ is BPSK modulated (0 mapped to $+1$ and 1 mapped to -1) and mapped to A_{ij} in the BW-REQ channel (see Figure 10-18).

The 16-bit information in the quick access message transmitted in the BW-REQ channel is used to generate a 5-bit CRC using generating polynomial $G(x) = x^5 + x^4 + x^2 + 1$. The 13 bits of information together with the 5 bits of CRC are encoded into 72 bits using the TBCC encoder. The 72 coded bits are QPSK modulated to generate 36 data symbols. The physical resources in the data portion of the three distributed BR tiles are used to transmit these data symbols.

In order to support mixed-mode operation, a BR tile is defined as four contiguous sub-carriers over six OFDM symbols (to be consistent with UL-PUSC subchannelization). As shown in Figure 10-19,

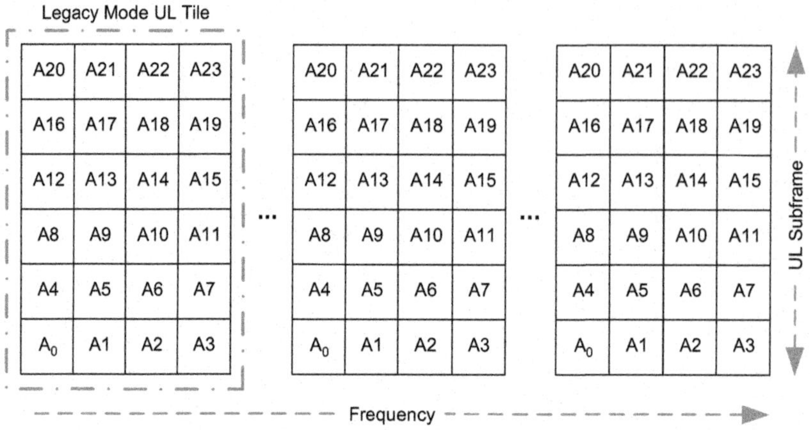

Legacy Mode UL Tile

A20	A21	A22	A23
A16	A17	A18	A19
A12	A13	A14	A15
A8	A9	A10	A11
A4	A5	A6	A7
A_0	A1	A2	A3

...

A20	A21	A22	A23
A16	A17	A18	A19
A12	A13	A14	A15
A8	A9	A10	A11
A4	A5	A6	A7
A_0	A1	A2	A3

...

A20	A21	A22	A23
A16	A17	A18	A19
A12	A13	A14	A15
A8	A9	A10	A11
A4	A5	A6	A7
A_0	A1	A2	A3

UL Subframe

Frequency

FIGURE 10-19

Physical structure of 4 × 6 bandwidth request tile [2]

the BR preamble is only transmitted over the 24 available sub-carriers. In this case, the preamble index u is randomly selected in the range of 0 to 23.

10.3.6 Power Control

Power control is a mechanism where the transmit power of the downlink or uplink control or traffic channels are adjusted at the base station (downlink power control) or at mobile stations (uplink power control) based on instructions from the base station such that with minimal impact on the reliability of the downlink/uplink transmissions and throughput, the interference among users and base stations are reduced. Therefore, power control can be considered as a link adaptation mechanism that is utilized for interference mitigation in cellular systems. While increasing the transmit power over a communication link has certain advantages, such as higher signal-to-noise ratio at the receiver, which reduces the bit error rate and allows a higher data rate resulting in greater spectral efficiency, as well as more protection against signal attenuation over fading channels. A higher transmit power; however, has several drawbacks including increased power consumption for the transmitting device, reduced MS battery life and increased interference to other users in the same frequency band. The following sections describe the power control algorithms that are incorporated in IEEE 802.16m.

Downlink Power Control

The IEEE 802.16m base station is capable of controlling the transmit power per subframe and per user. With downlink power control, each user-specific information or control information would be received with appropriately adjusted power level. As an example, the downlink A-MAP is power-controlled based on the MS uplink channel quality reports. The pilot and data tones' power can be jointly adjusted for adaptive downlink power control. In the case of dedicated pilots, power control is done on a per user basis, and in the case of common pilots this is done jointly for the users sharing the pilots. The power control in the downlink further supports SU-MIMO and MU-MIMO modes.

Uplink Power Control

IEEE 802.16m provides uplink power control mechanisms to compensate for the effects of path loss, shadowing, fast-fading, and implementation loss. The uplink power control is further used to mitigate the inter-cell and intra-cell interference level, thereby enhancing the overall throughput and reducing power consumption. The uplink power control includes open-loop and closed-loop power control. The base station transmits necessary power control information through the transmission of power control channels or MAC control messages. The parameters of the power control algorithm are optimized on a system-wide basis by the BS, and are broadcast periodically or trigged by certain events. The MS provides the necessary information through the control channels or MAC control messages to the serving BS, in order to enable uplink power control. The BS can exchange necessary information with neighboring base stations through backhaul to support uplink power control to facilitate the handover process.

The power control scheme may not be effective in high mobility scenarios for compensating for the effects of a fast-fading channel due to variation in the channel impulse response. As a result, the power control is used to mitigate the distance-dependent path loss, shadowing, and implementation loss. The uplink power control takes into consideration the MIMO transmission mode, and whether a single user or multiple users are supported on the same resource at the same time. The open-loop power control compensates for the channel variations and implementation loss without requiring frequent interactions with the serving BS. The MS can determine the transmit power based on the transmission parameters sent by the BS, uplink channel quality, downlink channel state information, or the interference knowledge obtained from downlink transmissions. The open-loop power control provides a coarse initial transmit power setting for the mobile station before establishing connection with the base station.

A power control mechanism takes into consideration the serving BS link target SINR and/or interference level to other cells/sectors for mitigating inter-cell interference. In order to achieve the target SINR, the serving BS path-loss can be fully or partially compensated based on a trade-off between overall system throughput and cell edge performance. The mobile station's transmit power is adjusted in order to ensure the level of interference is less than the permissible interference level. The compensation factor and interference targets for each frequency partition are determined and broadcast by the serving BS considering the FFR pattern, cell loading, etc. The closed-loop power control, on the other hand, compensates channel variations through periodic power-control commands from the serving BS. The base station measures the uplink channel state information and interference level using uplink data and/or control channel transmissions, and sends power control commands to the mobile stations. On receiving the power control command from the serving BS, the MS adjusts its uplink transmit power. The closed-loop power control is active during data and control channel transmissions.

An MS is expected to maintain the transmit power density (i.e., total transmit power normalized by transmission bandwidth) for each data and control subchannel below a certain level that is determined by the maximum permissible power level for the MS, emission mask, and other regulatory constraints. In other words, when the number of active logical resource units assigned to a particular user is reduced, the total transmitted power must be reduced proportionally by the MS in the absence of any additional change of power control parameters. When the number of resource blocks is increased, the total transmitted power must be proportionally increased such that the transmitted power level does not exceed the permissible power levels specified by the regulatory specifications. For interference level control, the information about the current interference level of each BS may be shared among the base stations via backhaul. In an OFDMA system, the power per sub-carrier and per stream is calculated as follows:

$$P_T = P_L + SINR_{target} + P_{NI} + P_{offset} \quad (dBm) \quad (10\text{-}19)$$

where $SINR_{target}$ is the target uplink SINR, P_T is the uplink transmit power level in dBm per stream and per sub-carrier for the current transmission, P_L denotes the average downlink path loss calculated by the MS based on the total power received on the active subcarriers of the frame preamble which is inclusive of mobile station's antenna gain, P_{NI} is the estimated average power level in dBm of the noise plus interference per sub-carrier at the BS as indicated in the AAI_ULPC-NI MAC control message, and P_{offset} is an MS-specific power correction factor which is controlled by the BS through power control signals. The power offset values for data and control are different and thereby are individually specified. The $SINR_{target}$ for the traffic channels is calculated as follows:

$$SINR_{target} = 10\log[\max(\text{SINR}_{\min}, \gamma SIR_{DL} - \alpha)] - \beta 10\log(n_{stream}) \quad (dB) \quad (10\text{-}20)$$

where $SINR_{min}$ is the required SINR for the minimum data rate expected by the BS (typically 0 dB), γ denotes the Interference over Thermal (IoT) control factor, SIR_{DL} is the linear ratio of the downlink signal to interference power measured by the MS, α is an adjustment factor proportional to the number of receive antennas at the BS, β is a masking parameter which can be set to zero or one to exclude or include n_{stream} effect on $SINR_{target}$, and n_{stream} is the number of streams in the logical resource unit that is signaled by the uplink Basic Assignment A-MAP IE. When SU-MIMO mode is used, n_{stream} is set to the number of streams for the MS. When a collaborative spatial multiplexing mode is utilized, n_{stream} will denote the aggregated number of streams. If the calculated $SINR_{target}$ for the traffic channel is higher than the maximum SINR defined in the AAI_SCD MAC control message, the $SINR_{target}$ is set to the maximum value. For a traffic channel, the P_{offset} is set to the value that is signaled in the AAI_UL-POWER-ADJUST MAC control message. Figure 10-20 illustrates the statistical distribution of the IoT with γ as a parameter which is obtained from exhaustive system-level simulations using IEEE 802.16m evaluation methodology general assumptions. Based on the same assumptions, the effect of open-loop power control on the cell edge and cell spectral efficiencies as a function of control command rate and γ is shown in Figure 10-21. It can be seen that slightly higher cell spectral efficiency and slightly lower cell edge spectral efficiency are obtained with slower power control rate.

For the uplink control channels, with the exception of initial ranging and sounding, the $SINR_{target}$ in Equation (10-20) is set separately for each control channel type, e.g., $SINR_{target\_HARQ}$ for the uplink HARQ feedback channel. The target SINR for each control channel is signaled through the AAI_SCD MAC control message.

The BS may change the mobile station's transmit power by signaling through the Power Control A-MAP. When the MS receives these commands from the serving BS, it updates the control channel P_{offset} value, based on the power correction value transmitted by the BS. The control channel power offset is initially set to zero until the MS receives the first AAI-UL-POWER-ADJUST MAC control message. If the MS receives an AAI-UL-POWER-ADJUST message at the ith frame, the local value is updated according to the BS instruction applies from the $(i+1)$th frame.

For initial ranging, the MS sends an initial ranging code at a randomly selected ranging opportunity. The initial transmission power is decided according to the Received Signal Strength (RSS) measured on the downlink reference signals such as the secondary advanced preamble. If the MS does not receive a response, it may increase the transmit power level by $\Delta P_{IR} = 2$ dB or may send a new

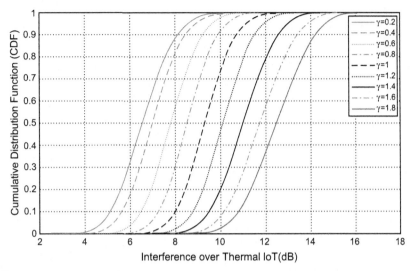

FIGURE 10-20

Distribution of IoT for different values of γ

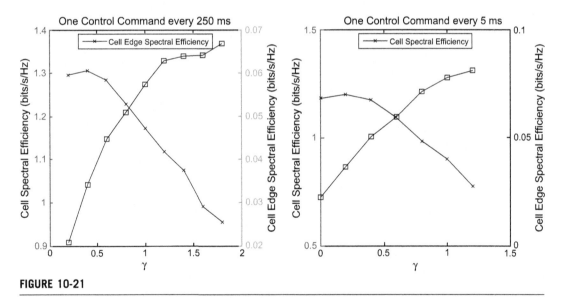

FIGURE 10-21

Effect of open-loop power control on cell edge and cell spectral efficiencies

initial ranging code, where ΔP_{IR} is the power step size. The MS can repeatedly increase the power on failure until the maximum transmit power is reached. The initial transmit power of the MS in the initial ranging channel is given as $P_{TX-IR_{\min}} = P_{RX-IR_{\min}} + EIRP_{BS} - RSS + (G_{MS-RX} - G_{MS-TX})$ where $P_{RX-IR_{\min}}$, $EIRP_{BS}$, G_{MS-TX}, and G_{MS-RX} denote the minimum BS detectable power for receiving the initial ranging preamble, Effective Isotropic Radiation Power of the BS, the MS receiver and transmitter antenna gains, respectively. If the MS receiving and transmitting antenna gains are equal, the last term of latter equation vanishes. The other terms in the latter equation are obtained from S-SFH SP2 and SP1.

Power control for the sounding channel is supported to ensure the quality of channel measurements. The mobile station's transmit-power for the sounding channel is controlled separately according to the target SINR value for the corresponding sounding channel. The power per sub-carrier is adjusted according to Equation (10-19) for uplink sounding signal transmission. In Equation (10-20), $SINR_{target}$ is replaced with the sounding channel target SINR. The quality of sounding channel measurements is maintained by assigning different target SINR values according to the downlink SIR of each mobile station, i.e., a relatively higher target SINR is set for the MS with a higher downlink SIR. The $SINR_{target}$ for the sounding channel is calculated based on the values of $SINR_{min}$ (the minimum SINR requirement for the sounding channel), γ (the IoT control factor for sounding channel), and SIR_{DL} (the ratio of the downlink signal to interference power) with α and β set to zero. Some of these parameters are signaled through the AAI_SCD MAC control message. In the case of simultaneous transmission of several uplink control channels, the transmission power of each channel is appropriately set in a certain order (i.e., HARQ feedback, PFBCH/SFBCH, synchronized ranging, sounding, traffic channels, bandwidth request) provided that the total transmit power does not exceed the maximum power limit of the MS (typically 24 dBm).

The base uplink transmission Power Spectral Density (PSD) and the SIR_{DL} are two parameters that are measured and reported by the MS to the serving BS. The base uplink transmission PSD is derived from Equation (10-19) by setting $SINR_{target}$ and P_{offset} to zero. The PSD value is reported in 0.5 dBm steps ranging from -74 dBm to 53.5 dBm, and is encoded in 8 bits. The value of SIR_{DL} is encoded using 10 bits in 0.05 dB steps ranging from -10 dB to 41.15 dB.

Link Adaptation

Link adaptation schemes adaptively adjust over-the-air transmission parameters in response to a change of radio channel for both downlink and uplink. The IEEE 802.16m supports an Adaptive Modulation and Coding (AMC) scheme for downlink and uplink transmissions. The serving BS can adapt the MCS level based on the DL channel quality reports and HARQ feedback from the MS in the downlink. The transmit power of downlink control channels is adapted based on channel quality reports from the MS. The MIMO mode is adapted according to CQI reports from the MS, while considering system parameters such as number of users, ACK/NACK, CQI variation, preferred MIMO feedback mode, etc. In the uplink direction, the serving BS may adapt the MCS level based on the uplink channel quality estimation, allocated resource size, and the maximum transmission power of the MS. The transmit power of the uplink control channels (excluding initial ranging channel) is adapted according to power control commands.

10.4 MULTI-ANTENNA TRANSMISSION SCHEMES

Design of mobile broadband wireless communication systems capable of data transmission rates in the excess of 1 Gbps has been of practical interest in the past decade to enable the 4th generation of cellular systems and IMT-Advanced. The use of multiple antennas at transmitter and/or receiver, commonly known as the Multiple-Input Multiple-Output (MIMO) system has become a pragmatic and cost-effective approach that offers substantial gain in making 1 Gbps wireless links a reality. This section provides an overview of MIMO wireless technology, including channel models, performance limits, coding, and transceiver design. MIMO techniques can increase system throughput and transmission reliability without increasing the required bandwidth. While most communication systems suffer from multi-path channels, a MIMO system benefits from the propagation over different paths through which the signals arrive at the receiver. A MIMO system typically performs better in an indoor environment, while it may not properly perform in Line-of-Sight (LoS) environments.

The performance improvements resulting from the use of multi-antenna systems are mainly due to array gain, diversity gain, spatial multiplexing gain, and interference reduction that may be achieved through the appropriate MIMO configurations. The main advantages of multi-antenna techniques can be summarized as follows [15,41]:

- Array gain can be achieved through signal processing at the transmitter and/or the receiver, and results in an increase in average SINR at the receiver due to a coherent combining effect. The increased SINR further results in improved coverage and user throughput. Transmit or receive array gain requires channel knowledge in the transmitter and receiver, respectively, and further depends on the number of transmit and receive antennas. The channel knowledge in the receiver is typically available (through channel estimation based on downlink reference signals), whereas channel state information in the transmitter is in general more difficult to attain.
- The signal power is faded randomly in a wireless communication channel. Diversity is an effective technique to mitigate fading in a wireless channel. The diversity schemes rely on transmitting the signal over multiple and ideally independent fading channels in time, frequency, and/or space. Spatial diversity is preferred over time or frequency diversity, as it does not consume additional transmission time or bandwidth. If the $N_r \times N_t$ links comprising the MIMO channel fade independently and the transmitted signal is properly constructed, the receiver can combine the arriving signals such that the resulting signal exhibits considerably reduced amplitude variability relative to a SISO link and a diversity gain on the order of $N_r \times N_t$ can be achieved. Spatial diversity gain in the absence of channel knowledge at the transmitter can be achieved using suitably designed transmit signals, which are known as space–time coding.
- MIMO channels provide a linear increase in capacity as a function of $\min(N_r, N_t)$ without requiring additional power or transmission bandwidth. This gain, referred to as spatial multiplexing gain, is realized by transmitting independent data streams from individual transmit antennas. Note that the number of independent data streams is limited to $\min(N_r, N_t)$. Under good channel conditions and sufficiently high SINR values, the receiver can detect different data streams, yielding a linear increase in capacity.
- Co-channel interference is generated due to frequency reuse in wireless channels. When multiple antennas are used, the differentiation between the spatial signatures of the desired signal and co-channel signals can be utilized to reduce interference. While interference cancellation requires proper knowledge of the desired signal's channel, exact knowledge of the interferers' channels may not be necessary. Interference avoidance schemes can also be implemented at the transmitter where the goal is to minimize the interference energy sent toward the co-channel users while delivering the signal to the desired user. Interference reduction allows aggressive

frequency reuse, and thereby increases multi-cell capacity. It must be noted that not all advantages of MIMO schemes can be simultaneously achieved due to conflicting demands on the spatial degrees of freedom or the number of transmit and/or receive antennas. The degree to which these conflicts can be resolved depends on the signaling scheme and transceiver design.

10.4.1 Capacity of MIMO Channels

As shown in Figure 10-22, a generic MIMO system consists of a MIMO transmitter with N_t transmit antennas, a MIMO receiver with N_r receive antennas, and $N_r \times N_t$ paths or channels between the transmit and receive antennas. Let $x_k(t)$ denote the transmitted signal from the kth transmit antenna at time t, then the received signal at the lth antenna can be expressed as follows:

$$y_l(t) = \sum_{k=0}^{N_t-1} h_{lk}(t) * x_k(t) + n_l(t) \tag{10-21}$$

where $h_{lk}(t)$ and $n_l(t)$ are the channel impulse response between the kth transmit and lth receive antenna and the additive noise at the lth receive antenna port, respectively. The above equation can be written in the frequency-domain as

$$Y_l(\omega) = H_{lk}(\omega)X_k(\omega) + N_l(\omega) \tag{10-22}$$

If $\mathbf{x}(\omega) = [X_1(\omega), X_2(\omega), ..., X_{N_t}(\omega)]^T$, $\mathbf{y}(\omega) = [Y_1(\omega), Y_2(\omega), ..., Y_{N_r}(\omega)]^T$, and $\mathbf{n}(\omega) = [N_1(\omega), N_2(\omega), ..., N_{N_r}(\omega)]^T$ denote the Fourier transform vectors of $x_k(t)$, $y_l(t)$, and $n_l(t)$, respectively, then $\mathbf{y}(\omega) = \mathbf{H}(\omega)\mathbf{x}(\omega) + \mathbf{n}(\omega)$ where $\mathbf{H}(\omega)$ is an $N_r \times N_t$ channel matrix with $H_{lk}(\omega)|_{k=1,2,...,N_t; l=1,2,...,N_r}$ entries.

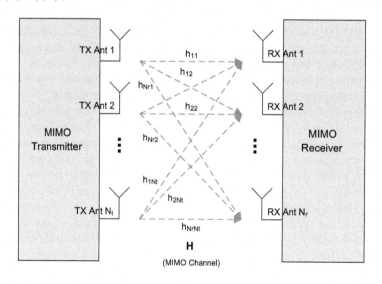

FIGURE 10-22

An illustration of the general principle of a MIMO system

Assuming a linear time-invariant MIMO channel, the channel input-output relationship can be further described in the discrete time-domain as follows:

$$y_l(nT) = \sum_{k=1}^{N_t} \sum_{m=0}^{M-1} h_{lk}(mT)x_k[(n-m)T] + n_l(nT) \qquad 0 \le n \le M-1; \ 1 \le l \le N_r \quad (10\text{-}23)$$

where $x_k(n)|_{k=1,2,\dots,N_t}$ and $y_l(n)|_{i=1,2,\dots,N_r}$ represent the channel input and output time-domain signals, respectively. In the case of time-varying channels, Equation (10-23) can be written as $y_l(t) = \sum_{k=1}^{N_t} h_{lk}(t,\tau)x_k(\tau) + n_l(t)$ where $h_{lk}(t,\tau)$ denotes the time-varying impulse response of the lkth channel. The matrix form of Equation (10-22) in a frequency-domain sampled at single frequency ω_m can be written as follows:

$$\mathbf{y}(\omega_m) = \mathbf{H}(\omega_m)x(\omega_m) + \mathbf{n}(\omega_m) \qquad (10\text{-}24)$$

In an OFDM system the signal processing is inherently performed in the frequency-domain. Furthermore, OFDM transforms a frequency-selective fading channel to a flat-fading channel when considering narrowband orthogonal sub-carriers. In such a system, the MIMO signal processing can be performed at each sub-carrier. This is the main reason for the suitability of the extension of MIMO to an OFDM system. When MIMO processing is done at each sub-carrier, the MIMO channel input-output relationship can be demonstrated as $\mathbf{y} = \mathbf{H}\mathbf{x} + \mathbf{n}$, where the channel between the transmitter and the receiver is typically modeled as a Finite Impulse Response (FIR) filter. In this case, each tap is typically a complex-valued Gaussian random variable with exponentially decaying magnitudes. The tap delays correspond to the RMS delay spread and the channel type (e.g., low delay spread or flat-fading, high delay spread or selective fading). There is a new realization of the channel at every transmitted packet, if the channel remains invariant for the duration of the packet; otherwise, the variation of the channel is explicitly modeled in the signal detection. As mentioned earlier, there are $N_r \times N_t$ paths between the transmitter and the receiver, where each channel is the sum of several FIR filters with different delay spreads. The channels may or may not be correlated. The MIMO schemes can be used with non-OFDM systems when the channel is modeled as flat-fading such that:

$$y_l(nT) = \sum_{k=1}^{N_t} h_{lk}x_k(nT) + n(nT) \qquad (10\text{-}25)$$

An important question is to what extent MIMO techniques can increase the throughput and improve the reliability of wireless communication systems. This question can be answered by calculating the information theoretic capacity of a Single-Input Single-Output (SISO) channel and comparing with that of Single-Input Multiple-Output (SIMO), Multiple-Input Single-Output (MISO), and MIMO channels. For a memoryless SISO channel (i.e., one transmit and one receive antenna), the channel capacity is given by

$$C_{SISO} = \log_2(1 + \gamma|h|^2) \qquad (10\text{-}26)$$

where h is the normalized complex-valued gain/attenuation of a fixed wireless channel or that of a particular realization of a random channel and γ denotes the SNR at any receive antenna port. As the number of receive antennas increases, the statistics of channel capacity improve. Using N_r receive antennas and one transmit antenna, a SIMO system is formed with a capacity given by:

$$C_{SIMO} = \log_2\left(1 + \gamma\sum_{i=1}^{N_r}|h_i|^2\right) \qquad (10\text{-}27)$$

where h_i is the gain of the ith channel corresponding to the ith receive-antenna. Note that increasing the value of N_r results in a logarithmic increase in average channel capacity. In the case of MISO or transmit diversity, where the transmitter usually does not have knowledge of the channel, the capacity is given by:

$$C_{MISO} = \log_2\left(1 + \frac{\gamma}{N_t}\sum_{i=1}^{N_t}|h_i|^2\right) \tag{10-28}$$

The power normalization factor N_t ensures that the total transmit power is uniformly distributed among the transmit antennas, and the absence of an array gain in this case (MISO) compared to a receive diversity scenario where the energy of the multipath channels can be coherently combined. In addition, the MISO capacity has a logarithmic relationship with N_t similar to that of a SIMO scheme. The use of diversity at both transmitter and receiver ends gives rise to a MIMO system. The capacity of a MIMO system with N_t transmit antennas and N_r receive antennas is expressed as follows:

$$C_{MIMO} = \log_2\left(\det\left[\mathbf{I} + \frac{\gamma}{N_t}\mathbf{HH}^H\right]\right) \tag{10-29}$$

where $\mathbf{I}$ is an $N_r \times N_r$ identity matrix and $\mathbf{H}$ is the $N_r \times N_t$ channel matrix. Note that both MISO and MIMO channel capacities are based on equal power uncorrelated sources. It is demonstrated in the literature that the capacity of the MIMO channel increases linearly with $\min(N_r, N_t)$ rather than logarithmically, as in case of MISO or SIMO channel capacity, since the determinant operator yields the product of non-zero eigenvalues of its channel-dependent matrix argument, each eigenvalue characterizing the SNR over a SISO eigen-channel. It will be shown later that the overall MIMO channel capacity is the sum of capacities of each of these SISO eigen-channels. The increase in capacity is dependent on the properties of the channel eigenvalues. If the channel eigenvalues decay rapidly, then linear growth in capacity will not occur. However, the eigenvalues have a known limiting distribution and tend to be spaced out along the range of this distribution. Hence, it is unlikely that most eigenvalues are very small and linear growth is indeed achieved.

Figures 10-23 and 10-24 show the impulse responses and the frequency responses of large delay-spread MIMO channels (two transmit and two receive antennas) measured over the extent of a subframe. It is noted that channel characteristics vary from one path (between one transmit and one receive antenna) to another and across time from one symbol to another.

The capacity of the MIMO channel can be calculated under various conditions and different assumptions. Depending on whether the receiver has perfect channel knowledge or whether the channel is flat-fading or frequency-selective fading, different expressions for the channel capacity can be obtained. The MIMO channel capacity in Equation (10-29) can be analyzed under two different assumptions: (1) the transmitter has no channel knowledge; and (2) the transmitter has perfect channel knowledge through feedback from the receiver or reciprocity of the downlink and uplink channels. Let's denote by Ξ the $N_t \times N_t$ covariance matrix of the channel input vector $\mathbf{x}$ and let us further assume that the channel is unknown to the transmitter; then it can be shown that the MIMO channel capacity can be written as:

$$C_{MIMO} = \log_2(\det[\mathbf{I} + \mathbf{H}\Xi\mathbf{H}^H]) \tag{10-30}$$

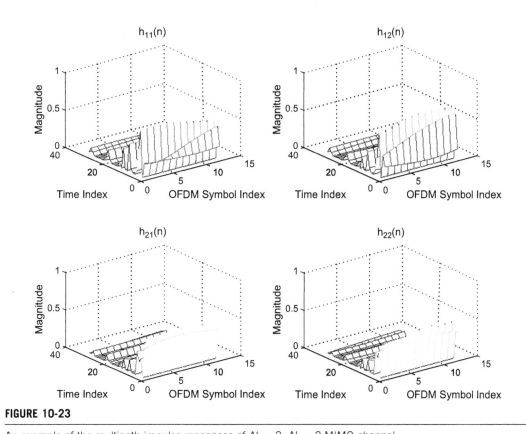

FIGURE 10-23

An example of the multipath impulse responses of $N_t = 2$, $N_r = 2$ MIMO channel

where $\text{tr}(\Xi) \leq \gamma$ ensures the total signal power does not exceed a certain limit. It can be shown that for equal transmit power uncorrelated sources $\Xi = (\gamma/N_t)\mathbf{I}$ and Equation (10-30) becomes identical to Equation (10-29). This is true when the channel matrix is unknown to the transmitter and the input signal is Gaussian-distributed, maximizing the mutual information. If the receiver measures and sends channel quality feedback or channel state information to the transmitter, the covariance matrix Ξ is not proportional to the identity matrix, rather it is constructed from a water-filling algorithm. If one compares the capacity achieved assuming equal transmit power and unknown channel with that of perfect channel estimation through feedback, then the capacity gain due to the use of feedback is obtained. For the independent identically-distributed Rayleigh fading scenario, the linear capacity growth discussed earlier will be observed. It is shown that Equation (10-29) can be written as follows:

$$C_{MIMO} = \sum_{i=1}^{\min(N_t,N_r)} \log_2\left(1 + \frac{\gamma\lambda_i^2}{N_t}\right) \tag{10-31}$$

where $\lambda_i = 1,2,\ldots,\min(N_t,N_r)$ are the non-zero eigenvalues of $\mathbf{HH}^H$. We can decompose the MIMO channel into $K \leq \min(N_t,N_r)$ equivalent parallel SISO channels using the Singular Value

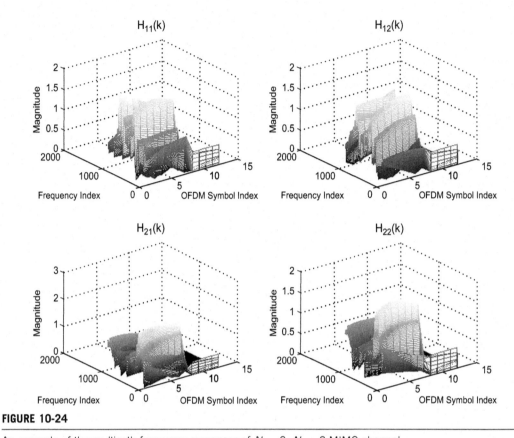

FIGURE 10-24

An example of the multipath frequency responses of $N_t = 2$, $N_r = 2$ MIMO channel

Decomposition (SVD) theorem.[15,31,32][i] Let $\mathbf{y} = \mathbf{Hx} + \mathbf{n}$ describe the input-output relationship of the MIMO channel where $\mathbf{y}$ is the output vector with N_r components, $\mathbf{x}$ is the input vector with N_t components, $\mathbf{n}$ is the additive noise vector with N_r components, and $\mathbf{H}$ is the $N_r \times N_t$ channel matrix. Using SVD, we can show $\mathbf{H} = \mathbf{U}\Sigma\mathbf{V}^H$. Let $\widehat{\mathbf{x}} = \mathbf{V}^H\mathbf{x}$, $\widehat{\mathbf{y}} = \mathbf{U}^H\mathbf{y}$, and $\widehat{\mathbf{n}} = \mathbf{U}^H\mathbf{n}$ denote the unitary transformation of the channel input and output and noise vectors, it can be shown that

---

[i]The concept of decomposition of an $N \times N$ Hermitian matrix in terms of quadratic product of the $N \times N$ unitary matrix composed of eigenvectors, an $N \times N$ diagonal matrix of eigenvalues can be generalized to $M \times N$ complex-valued matrices of rank K. If $\mathbf{A}$ is an $M \times N$ ($M > N$) complex-valued matrix of rank K, then $\mathbf{A} = \mathbf{U}\sum\mathbf{V}^H$ denotes the Singular Value Decomposition of $\mathbf{A}$ where the $M \times M$ unitary matrix $\mathbf{U}$ is composed of the eigenvectors of $\mathbf{AA}^H$; i.e., $\mathbf{U} = (\mathbf{u_1}, \mathbf{u_2}, \dots \mathbf{u_m})$ and the $N \times N$ unitary matrix $\mathbf{V}$ is composed of eigenvectors of $\mathbf{A}^H\mathbf{A}$ (i.e., $\mathbf{V} = \mathbf{v_1}, \mathbf{v_2}, \dots \mathbf{v_n}$), $\mathbf{A}^H\mathbf{A}\mathbf{v}_i = \sigma_i^2\mathbf{v}_i$). The elements of the $M \times N$ matrix $\Sigma = \begin{pmatrix} \Lambda & 0 \\ 0 & 0 \end{pmatrix}$, $\Lambda = \mathbf{diag}(\sigma_1, \sigma_2, \dots, \sigma_K)$ are the square roots of the eigenvalues of matrix $\mathbf{A}^H\mathbf{A}$ which are referred to as the singular values of matrix $\mathbf{A}$. Therefore, matrix $\mathbf{A}$ may be written as $\mathbf{A} = \sum_{i=1}^{K} \sigma_i\mathbf{u}_i\mathbf{v}_i$. The number of non-zero singular values of matrix $\mathbf{A}$ is the rank of $\mathbf{A}$. The singular values of matrix $\mathbf{A}$ are positive real numbers which satisfy $\sigma_1 \geq \sigma_2 \geq \dots \geq \sigma_K > 0$. [64]

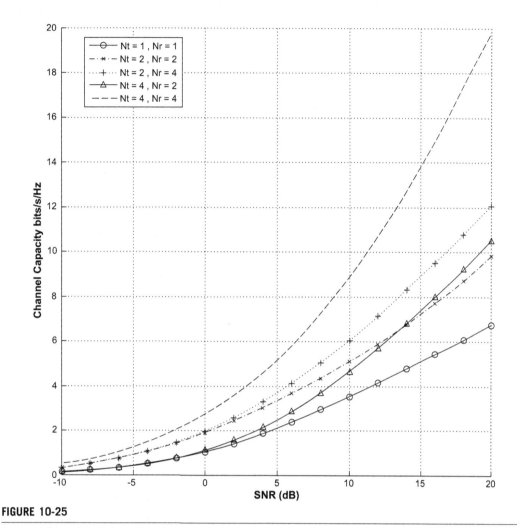

FIGURE 10-25

A comparison of the information theoretic capacity of some MIMO channels

$\widehat{\mathbf{y}} = \mathbf{\Sigma}\widehat{\mathbf{x}} + \widehat{\mathbf{n}}$. Since $\mathbf{U}$ and $\mathbf{V}$ are unitary matrices and $\mathbf{\Sigma} = \mathbf{diag}(\lambda_1, \lambda_2, ..., \lambda_{\min(N_r,N_t)}, 0, 0, ..., 0)$, it is clear that the capacity of this model is the same as the capacity of the model $\mathbf{y} = \mathbf{Hx} + \mathbf{n}$. However, $\mathbf{\Sigma}$ is a diagonal matrix with K non-zero elements on the main diagonal, thus $\widehat{y}_1 = \lambda_1 \widehat{x}_1 + \widehat{n}_1, ..., \widehat{y}_k = \lambda_k \widehat{x}_k + \widehat{n}_k, \widehat{y}_{k+1} = \widehat{n}_{k+1}$. The latter equations are conceptually equivalent to K parallel SISO eigen-channels, each with a signal power of $\lambda_i^2, i = 1, 2, ..., \min(N_t, N_r)$. Hence, the MIMO channel capacity can be rewritten in terms of the eigenvalues of the input signal covariance matrix $\mathbf{\Xi}$.

When the channel knowledge is available at the transmitter and receiver, then $\mathbf{H}$ is known and we can optimize the capacity over $\mathbf{\Xi}$, subject to the power constraint $\mathrm{tr}(\mathbf{\Xi}) \leq \gamma$. It is shown in the literature that the optimal Ξ in this case exists and is known as a water-filling solution. The channel capacity in this case is given by:

$$C = \sum_{k=1}^{K} \log_2(\eta\lambda_i^2)^+$$

$$\gamma = \sum_{k=1}^{K} (\eta - \lambda_i^{-2})^+$$

(10-32)

where $(x)^+ = x \forall x \geq 0, (x)^+ = 0 \forall x < 0$ and η is a non-linear function of eigenvalues of the channel input covariance matrix. The effect of various channel conditions on the channel capacity has been extensively studied in the literature. For example, increasing the LoS signal strength at fixed SNR reduces the capacity in Ricean channels [38]. This can be explained in terms of the channel matrix rank or through various eigenvalue properties. The issue of correlated fading is of considerable importance for implementations where the antennas are required to be closely spaced. The optimal water-filling allocation strategy is obtained when the power allocated to each spatial sub-channel is non-negative. Figure 10-26 illustrates an example where the water-filling algorithm is applied to an OFDM system assuming *Number of Sub-channels* = 16, *Total Power* = −20 dBm, N_0 = −80 dBm, and

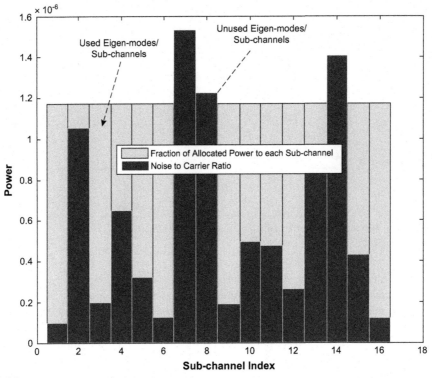

FIGURE 10-26

An example illustration of a water-filling algorithm in an OFDM system

Bandwidth = 1 MHz. It can be observed that depending on the inverse carrier-to-noise ratio of each sub-channel, the amount of power allocated to each sub-channel differs, provided that the sum of sub-channel powers does not exceed the total available power. The power allocated to each sub-channel corresponds to parameters λ_i^2.

10.4.2 Spatial Multiplexing and Diversity

The main objective of space–time diversity coding schemes is to exploit the inherent spatial (multi-path) diversity in MIMO channels through properly-designed space–time codes. In this section, two transmit diversity schemes, namely the Alamouti scheme and Cyclic Delay Diversity (CDD), are described where both realize full spatial diversity of the channel without requiring channel knowledge at the transmitter. In the Alamouti scheme, we consider a MIMO channel with two transmit antennas and any number of receive antennas. The Alamouti scheme can be described as follows. Two different complex-valued data symbols s_1 and s_2 are transmitted simultaneously from antenna ports 1 and 2, respectively, during the first symbol period. In the second symbol period, symbols $-s_2^*$ and s_1^* are and are sent from antennas 1 and 2, respectively. It must be noted that the rate equals one in the Alamouti scheme since two independent data symbols are transmitted over two symbol periods. Let's consider the following case with two transmit-antennas and one receive-antenna, thereby we have two channels H_1 and H_2. The received signal at the first and second symbol period can be expressed as follows:

$$
\begin{aligned}
y_1 &= H_1 s_1 + H_2 s_2 + n_1 \\
y_2 &= -H_1 s_2^* + H_2 s_1^* + n_2
\end{aligned}
\tag{10-33}
$$

Let $\mathbf{r} = [y_1, y_2^*]^H$, $\mathbf{s} = [s_1\ s_2]^H$, and $\mathbf{n} = [n_1\ n_2]^H$ and $\mathbf{H} = \begin{bmatrix} H_1 & H_2 \\ H^* & -H^* \end{bmatrix}$, the relationship between the output and the input of the channel can be written as follows $\mathbf{r} = \mathbf{Hs} + \mathbf{n}$ where multiplying both sides of the latter equation by $\mathbf{H}^H$ yields $\tilde{\mathbf{r}} = \mathbf{H}^H \mathbf{Hs} + \tilde{\mathbf{n}}$. Since $\mathbf{H}$ is orthogonal; i.e., $\mathbf{H}^H \mathbf{H} = \alpha \mathbf{I}$ where $\alpha = |H_1|^2 + |H_2|^2$; dividing both sides of the equation by α yields:

$$
\begin{aligned}
\frac{\tilde{r}_1}{|H_1|^2 + |H_2|^2} &= s_1 + \frac{H_1^* n_1}{|H_1|^2 + |H_2|^2} + \frac{H_2 n_2}{|H_1|^2 + |H_2|^2} \\
\frac{\tilde{r}_2}{|H_1|^2 + |H_2|^2} &= s_2 + \frac{H_2^* n_1}{|H_1|^2 + |H_2|^2} + \frac{-H_1 n_2}{|H_1|^2 + |H_2|^2}
\end{aligned}
\tag{10-34}
$$

It is assumed that the channels are independent and identically distributed with frequency flat-fading and remain constant over (at least) two consecutive symbol periods. Appropriate processing at the receiver, as shown in Equations (10-33) and (10-34), collapses the vector channel into a scalar channel for either of the transmitted data symbols where $\tilde{r}$ is the processed received signal corresponding to transmitted symbol $\mathbf{s}$ and $\tilde{n}$ is the processed noise. Even though the channel knowledge is not available to the transmitter, the Alamouti scheme achieves $2N_r$ order diversity. We note, however, that array gain is realized only at the receiver since the transmitter does not have channel state information. The Alamouti scheme may be extended to channels with more than two transmit antennas through orthogonal space–time block coding. The link-level performance of the Alamouti scheme is compared to the SISO and SIMO cases in Figure 10-27, assuming BPSK-modulated input symbols and Rayleigh fading channel.

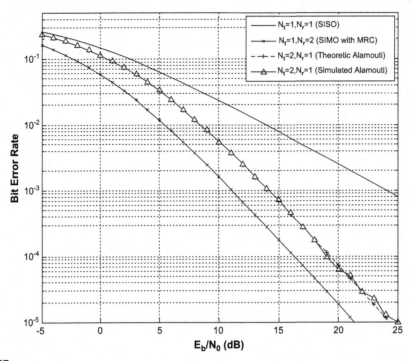

FIGURE 10-27

Performance of the Alamouti scheme

Delay diversity is an alternative method in which the same symbol is transmitted at different time instances, such that each symbol experiences different fading. The delay diversity converts spatial diversity into frequency diversity by transmitting the data signal from the first antenna port and a delayed replica of the signal from the second antenna port. Assuming two transmit antennas and one receive antenna and the delay induced by the second antenna equals one symbol period, the effective channel seen by the data signal is a frequency-selective fading SISO channel with impulse response $h(k) = h_1\delta(k) + h_2\delta(k-1)$ where h_1 and h_2 are complex-valued flat-fading channels. We note that the effective channel resembles a two-path SISO channel with independently fading paths and equal average path energy. A Maximum-Likelihood (ML) detector will realize full second-order diversity at the receiver. If the same signal is transmitted from two or more transmit antennas, a beamforming effect is created where in some directions the signal is attenuated while in other directions it may be amplified. The CDD scheme mitigates this problem by transmitting cyclically shifted signals from the transmit antennas. Let's consider an OFDM symbol with constellation d_k at the kth frequency index, the time-domain signal corresponding to this symbol which is transmitted from antenna port 1 can be expressed as follows:

$$s_1(t) = \mathrm{Re}\left\{ e^{j\omega_c t} \sum_{k=-(N_{FFT}-1)/2}^{(N_{FFT}-1)/2} d_k e^{j2\pi k(t-t_g)/T_u} \right\} \qquad mT_u \le t \le (m+1)T_u \qquad (10\text{-}35)$$

Using CDD scheme, the transmitted signal from the second antenna port is given by

$$s_2(t) = \mathrm{Re}\left\{ e^{j\omega_c t} \sum_{k=-(N_{FFT}-1)/2}^{(N_{FFT}-1)/2} d_k e^{j2\pi k(t-t_g-T_{CDD})/T_u} \right\} \qquad mT_u \le t \le (m+1)T_u \quad (10\text{-}36)$$

This is equivalent to a cyclic rotation of the signal in the time-domain. The implementation of the CDD scheme is simplified, if the cyclic shift T_{CDD} is an integer multiple of the sampling time. At the receiver, the effect of CDD at the kth index is of receiving through the channel $H_1(k) + \exp(j2\pi k T_{CDD}/T_u)H_2(k)$. The advantage over the case of $H_1(k) + H_2(k)$, which is the result of transmission of the same signal over the two channels, is the prevention of the beamforming effect, i.e., directed transmission in one direction and undesired emission in the other directions.

The main objective of spatial multiplexing, as opposed to space–time diversity coding, is to maximize transmission data rate. Therefore, $\min(N_t, N_r)$ independent data symbols are transmitted per symbol period, such that the spatial rate is $\min(N_t, N_r)$. There are several encoding options that can be used in conjunction with spatial multiplexing to achieve the increased data rate. Horizontal Encoding (HE) and Vertical Encoding (VE) are the prominent encoding options that are typically used in wireless communication systems. In an HE scheme, the bit stream to be transmitted is de-multiplexed into separate data streams where each stream is processed through the transmission chain, i.e., independent temporal encoding, symbol mapping, and interleaving, and is then transmitted from the corresponding antennas. The antenna-to-stream mapping remains static over time. The spatial rate is $\min(N_t, N_r)$ and the overall signaling rate is $n_{modulation\text{-}order} r_b \min(N_t, N_r)$, where r_b is the channel coder input bit rate and $n_{modulation\text{-}order}$ denotes the QAM modulation order. The HE scheme can at most achieve N_r order diversity, since any given information bit is transmitted from only one transmit antenna and received by N_r receive antennas. While this is a source of sub-optimality of the HE architecture, it does simplify receiver design. The coding gain achieved by HE depends on the coding gain of the temporal code. Furthermore, a maximum array gain of N_r can be realized using the HE scheme.

In the VE architecture, the bitstream undergoes temporal encoding, symbol mapping, and interleaving after which it is de-multiplexed into $\min(N_t, N_r)$ streams transmitted from the individual antennas. This form of encoding can achieve full $N_t \times N_r$ order diversity, provided that the temporal code is properly designed, since each information bit can be spread across all the transmit antennas. Nevertheless, the VE scheme requires joint decoding of the sub-streams, which increases receiver implementation complexity compared to an HE scheme where the individual data streams can be decoded separately. The spatial rate of VE is $\min(N_t, N_r)$ and the overall signaling rate is $n_{modulation\text{-}order} r_b \min(N_t, N_r)$. The coding gain achieved by VE will depend on the temporal channel coding characteristic and a maximum array gain of N_r can be achieved.

Space-Frequency Block Codes (SFBC) is the frequency-domain version of the Space-Time Block Codes (STBC). This family of codes is designed to achieve diversity gain by transmitting orthogonal streams which yield optimal SNR with a linear receiver. Since in the OFDMA systems, the number of OFDM symbols in a subframe may often be an odd number, while STBC (e.g., Alamouti codes) operates on pairs of adjacent symbols in the time-domain, the application of STBC would not be straightforward. For SFBC transmission, the adjacent sub-carriers on each OFDM symbol are paired, encoded, and transmitted from two BS antenna ports where the encoding is defined as follows:

$$\mathbf{W} = \begin{bmatrix} x_k & x_{k+1} \\ -x_{k+1}^* & x_k^* \end{bmatrix} \tag{10-37}$$

where $(x_k, -x_{k+1}^*)$ and (x_{k+1}, x_k^*) are transmitted from antenna port 1 and 2, respectively. Note that the spatial rate of the SFBC codes in this case is one.

10.4.3 MIMO Receivers

The orthogonal space-time block coding transforms a vector detection problem into a number of scalar detection problems. Similar extensions can be made to frequency-selective fading MIMO channels. Therefore, receiver design techniques such as Zero-Forcing (ZF), Minimum-Mean Square Error (MMSE) estimation, and (optimal) Maximum Likelihood (ML) sequence detection can be utilized [15,41]. The use of transmit diversity techniques, such as delay diversity or frequency-offset diversity, transforms a MISO channel into a number of SISO channels and allows the application of SISO receiver architectures. For a general space–time trellis code, a vector Viterbi decoder can be employed. The space–time trellis coding in general provides improved performance over orthogonal space-time block coding at the expense of receiver complexity. The problem encountered by a receiver for spatial multiplexing is the presence of multi-stream interference since the signals sent from different transmit antennas interferes with each other. Note that in spatial multiplexing the different data streams are transmitted in the same channel and occupy the same resources in time and frequency. For the sake of simplicity we restrict our attention to the case where $N_r \geq N_t$.

The optimal detection for spatially-multiplexed signals is to use maximum likelihood criterion. The ML receiver performs vector decoding and is optimal in the sense of minimizing the error probability. Assuming equally likely, temporally uncoded vector symbols, the ML receiver generates an estimate of the transmitted signal vector as follows:

$$\widehat{\mathbf{s}} = \underset{\mathbf{s}}{\operatorname{argmin}} \left\| \mathbf{y} - \sqrt{\frac{E_s}{N_t}} \mathbf{H}\mathbf{s} \right\|^2 \tag{10-38}$$

where the minimization is performed over all possible transmit vector symbols $\mathbf{s}$. Let Z denote the alphabet size of the scalar constellation transmitted from each antenna, a complete implementation requires an exhaustive search over a total of Z^{N_t} vector symbols or hypotheses, making the decoding complexity of this receiver grow exponentially with the number of transmit antennas. The recent development of fast ML detection algorithms reduces the computational complexity of the ML decoder. The ML receiver realizes N_r diversity order for the HE scheme and $N_r \times N_t$ diversity order for the VE scheme [15,41].

An alternative non-linear approach to detection of spatially multiplexed signals is Successive Interference Cancellation (SIC) by assuming that the spatially-multiplexed signals are separately coded (i.e., horizontal encoding or multi-codeword transmission). As shown in Figure 10-28, in an SIC receiver, one of the spatially multiplexed signals is initially demodulated and decoded. On successful decoding, the data is re-encoded and subtracted from the received signals. The second spatially multiplexed signal can then be demodulated and decoded in the absence of interference from the first detected signal. The subtraction would ideally improve the signal-to-interference ratio. The procedure

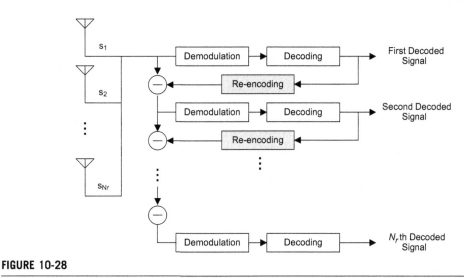

FIGURE 10-28

The structure of a successive interference cancellation MIMO receiver

is iterated until all spatially multiplexed signals are demodulated and decoded. It is noted that the first signals to be decoded using this approach are subject to higher interference levels compared to those that are decoded later in the process. Therefore, the ith signal to be decoded must be more robustly coded than the $(i + 1)$th signal. This can be achieved in multi-codeword transmission by applying different modulation and coding schemes to different spatially multiplexed signals. The latter scheme is often referred to as Per-Antenna Rate Control (PARC).

The decoding complexity of the ML receiver can be reduced using a linear filter to separate the transmitted data streams, and then independently decode each stream. The ZF and MMSE linear detectors are examples of such receivers. The ZF matrix filter that separates the received signal into its component streams detector is given by:

$$\mathbf{G}_{ZF} = \sqrt{\frac{N_t}{E_s}}\mathbf{H}^{\dagger} = \sqrt{\frac{N_t}{E_s}}(\mathbf{H}^H\mathbf{H})^{-1}\mathbf{H}^H \tag{10-39}$$

where $\mathbf{G}_{ZF}$ is an $N_t \times N_r$ matrix that inverts the channel matrix and $\mathbf{H}^{\dagger}$ denotes the Moore–Penrose inverse[ii] of the channel matrix $\mathbf{H}$. The output of the ZF receiver is obtained as follows:

$$\mathbf{y} = \mathbf{s} + \sqrt{\frac{N_t}{E_s}}\mathbf{H}^{\dagger}\mathbf{n} \tag{10-40}$$

[ii]Given a matrix $\mathbf{A}$ and vector $\mathbf{b}$ and equation $\mathbf{Ax} = \mathbf{b}$, we find the vector $\mathbf{x}$ that minimizes $\|\mathbf{Ax} - \mathbf{b}\|$. Using the orthogonality principle, the error is taken perpendicular to the estimation vector such that $\mathbf{x}^H\mathbf{A}^H(\mathbf{Ax} - \mathbf{b}) = 0$ which implies that $\mathbf{A}^H\mathbf{Ax} = \mathbf{A}^H\mathbf{b}$(since $\mathbf{x} \neq 0$), or $\mathbf{x} = (\mathbf{A}^H\mathbf{A})^{-1}\mathbf{A}^H\mathbf{b}$. One should note that $\mathbf{Ax} = \mathbf{A}(\mathbf{A}^H\mathbf{A})^{-1}\mathbf{A}^H\mathbf{b}$. The matrix $\mathbf{A}^{\dagger} = (\mathbf{A}^H\mathbf{A})^{-1}\mathbf{A}^H$ is defined as the Moore–Penrose inverse of the matrix $\mathbf{A}$. If $\mathbf{A}^H\mathbf{A}$ is not invertible, the SVD algorithm is used to calculate the matrix pseudo-inverse. In that case, we may write $\mathbf{A} = \mathbf{V}_K\mathbf{\Sigma}_K\mathbf{U}_K^H$ and seek x such that $\|\mathbf{V}_K\mathbf{\Sigma}_K\mathbf{U}_K^H\mathbf{x} - \mathbf{b}\|$ is minimized. Using the orthogonality principle, we have $\mathbf{x}^H\mathbf{U}_K\mathbf{\Sigma}_K\mathbf{V}_K^H(\mathbf{V}_K\mathbf{\Sigma}_K\mathbf{U}_K^H\mathbf{x} - \mathbf{b}) = 0$ or (assuming $\mathbf{x} \neq 0$) $\mathbf{U}_K\mathbf{\Sigma}_K^2\mathbf{U}_K^H\mathbf{x} - \mathbf{U}_K\mathbf{\Sigma}_K\mathbf{V}_K^H\mathbf{b} = 0$ such that $\mathbf{\Sigma}_K^2\mathbf{U}_K^H\mathbf{x} - \mathbf{\Sigma}_K\mathbf{V}_K^H\mathbf{b} = 0$, multiplying both sides by $\mathbf{\Sigma}_K^{-2}$ yields $\mathbf{U}_K^H\mathbf{x} - \mathbf{\Sigma}_K^{-1}\mathbf{V}_K^H\mathbf{b} = 0$. The latter is true when $\mathbf{x} = \sum_{i=1}^{K}(\frac{1}{\sigma_i}\mathbf{v}_i^H\mathbf{b})\mathbf{u}_i$; therefore $\mathbf{x} = \mathbf{U}_K\mathbf{\Sigma}_K^{-1}\mathbf{V}_K^H\mathbf{b}$.

which shows that the ZF receiver decomposes the channel matrix into parallel scalar channels with additive spatially-colored noise. Each scalar channel is then decoded independently, irrespective of noise correlation across the processed streams. The ZF receiver transforms the joint decoding problem into single stream decoding problems, thereby significantly reducing receiver complexity. This complexity reduction is achieved at the expense of noise enhancement, which in general results in a significant performance degradation (compared to the ML decoder). The diversity order achieved by each of the individual data streams equals $N_r - N_t + 1$. Figure 10-29 illustrates the link-level performance of a 2×2 MIMO system with a ZF receiver. It is shown that the performance is close to that of a SISO system since the diversity order is one.

The MMSE receiver balances multi-stream interference mitigation and noise enhancement effect and minimizes the total error, and is given by:

$$\mathbf{G}_{MMSE} = \underset{\mathbf{G}}{\mathrm{argmin}}\, E\left\{ \|\mathbf{G}y - \mathbf{s}\|^2 \right\} = \sqrt{\frac{N_t}{E_s}} \left(\mathbf{H}^H \mathbf{H} + \frac{N_t N_0}{E_s} \mathbf{I} \right)^{-1} \mathbf{H}^H \qquad (10\text{-}41)$$

where $\mathbf{I}$ is an $N_r \times N_r$ identity matrix. In low signal-to-interference plus noise ratios (i.e., SINR < 0 dB), the performance of the MMSE receiver approaches that of a matched-filter receiver, out-performing the ZF detector which tends to enhance noise. In high signal-to-interference plus noise

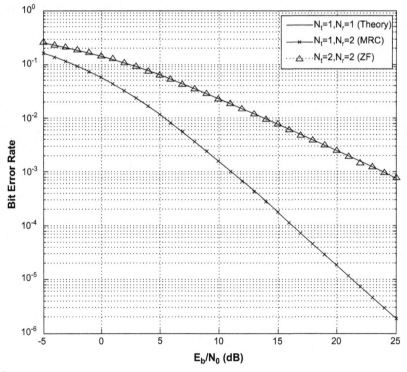

FIGURE 10-29

Link-level performance of 2 × 2 MIMO with ZF equalizer (Rayleigh channel and BPSK modulation)

ratios; however, the performance of the MMSE receiver approaches that of a ZF receiver, and therefore realizes $N_r - N_t + 1$ order diversity for each data stream.

In Maximal Ratio Combining (MRC), the output is a weighted sum of all receiver branches, thus the coefficients $\alpha_i, i = 1, 2, ..., N_r$ are all non-zero. The signals are co-phased and $\alpha_i = \beta_i e^{-j\theta_i}, i = 1, 2, ..., N_r$ where θ_i is the phase of the received signal on the ith branch (see Figure 10-30). The magnitude of the combined output can be written as $y = \sum_{i=1}^{N_r} \beta_i r_i$. Let's assume that the noise power spectral density is $N_0/2$ in each branch, which yields the combined noise at the output $n = \sum_{i=1}^{N_r} \beta_i^2 N_0/2$. The SNR at the output of the combiner is given as:

$$\gamma = \frac{1}{N_0} \frac{(\sum_{i=1}^{N_r} \beta_i r_i)^2}{\sum_{i=1}^{N_r} \beta_i^2} \tag{10-42}$$

The objective is to find $\alpha_i, i = 1, 2, ..., N_r$ which maximizes the output SNR. It can be understood that branches with higher SNR are weighted more than those with lower SNR; therefore, the coefficients $\alpha_i, i = 1, 2, ..., N_r$ are proportional to the branch SNRs. The coefficients are found by taking partial derivatives of the output SNR expression and solving the resulting equations for the optimal weights which yields $\alpha_i = r_i^2/N_0, i = 1, 2, ..., N_r$, and the output SNR will be the sum of the SNRs of the receiving branches. Therefore, the output SNR and the corresponding array gain increase linearly with the number of the diversity branches (i.e., the number of received antennas).

The trade-off among transmission rate, error rate, and SINR for the scenarios where the transmitter has no channel knowledge and the receiver has perfect knowledge of channel state information has been studied in the literature. If the MIMO channel is assumed to be block-fading (A Block-Fading Channel is a model which assumes the fading-gain stochastic process is piece-wise constant over the blocks of N input symbols. The channel is modeled as a sequence of independent random variables, each representing the fading gain in a block.) and that the length of the transmitted codewords is less than or equal to the channel block length, and if the channel were perfectly known to the transmitter, we could choose a signaling rate equal to or less than channel capacity and achieve error-free transmission. The coding scheme to achieve capacity consists of performing modal decomposition, which decouples the MIMO channel into parallel SISO channels, and then using ideal SISO channel coding. In practice, turbo codes can achieve near MIMO channel capacity performance. If the channel is

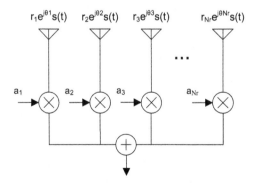

FIGURE 10-30

A generic linear combining receiver

unknown to the transmitter, modal decomposition is not possible. Since the channel is selected randomly according to a given fading distribution, there will always be a non-zero probability that a given transmission rate is not supported by the channel. We assume that the transmitted codeword is decoded successfully if the rate is at or below the mutual information, assuming a spatially white transmit covariance matrix, associated with the given channel realization. A decoding error is declared if the rate exceeds the mutual information. Therefore, if the transmitter does not have any knowledge of the channel, the PER would be equal to the outage probability associated with the transmission rate. We define the diversity order for a given transmission rate as:

$$d(R) = -\lim_{\gamma \to \infty} \frac{\log[P_e(R, \gamma)]}{\log(\gamma)} \tag{10-43}$$

where $P_e(R,\gamma)$ is the PER corresponding to transmission rate R and SNR γ. Thus, the diversity order is the magnitude of the slope of the PER plotted as a function of the SNR on a logarithmic scale.

10.4.4 Precoding and Beamforming

The multiple antennas at the transmitter and receiver can be used to achieve array and diversity gain instead of capacity gain. In this case, the same symbol weighted by a complex-valued scale factor is sent from each transmit antenna so that the input covariance matrix has unit rank. This scheme is referred to as beam-forming. It must be noted that there are two conceptually and practically different classes of beamforming: (1) direction-of-arrival beamforming (i.e., adjustment of transmit or receive antenna directivity); and (2) eigen-beamforming (i.e., a mathematical approach to maximize signal power at the receive antenna based on certain criterion). In this section, we only consider eigen-beamforming schemes.

A classic eigen-beamforming scheme usually performs linear, single-layer, complex-valued weighting on the transmitted symbols, such that the same signal is transmitted from each transmit antenna using appropriate weighting factors. In this scheme, the objective is to maximize the signal power at the receiver output. When the receiver has multiple antennas, the single-layer beamforming cannot simultaneously maximize the signal power at every receive antenna, hence, precoding is used for multi-layer beamforming in order to maximize the throughput of a multi-antenna system. Pre-coding is a generalized beamforming scheme to support multi-layer transmission in a MIMO system. Using precoding, multiple streams are transmitted from the transmit antennas with independent and appropriate weighting per antenna such that the throughput is maximized at the receiver output.

In a single-user MIMO system, identity matrix precoding (for open-loop) and SVD precoding (for closed-loop) are used to achieve link-level MIMO channel capacity. In addition, random unitary precoding can achieve the open-loop MIMO channel capacity with no signaling overhead in the uplink. The SVD precoding, on the other hand, has been shown to achieve the MIMO channel capacity when channel state information is signaled to the transmitter. In a precoded SU-MIMO system with N_t transmit antennas and N_r receive antennas, the input-output relationship can be described as $\mathbf{y} = \mathbf{HWs} + \mathbf{n}$ where $\mathbf{s} = [s_1, s_2,..., s_M]^t$ is an $M \times 1$ vector of normalized complex-valued modulated symbols, $\mathbf{y} = [y_1, y_2,..., y_{Nr}]^t$ and $\mathbf{n} = [n_1, n_2,...,n_{Nr}]^t$ are the $N_r \times 1$ vectors of received signal and noise, respectively, $\mathbf{H}$ is the $N_r \times N_t$ complex-valued channel matrix, and $\mathbf{W}$ is the $N_t \times M$ linear precoding matrix. The superscript "t" denotes the transpose operator.

In the receiver, a hard decoded symbol vector $\hat{\mathbf{s}}$ is obtained by decoding the received vector $\mathbf{y}$ by a vector decoder, assuming perfect knowledge of the channel and the precoding matrices. We assume the entries of $\mathbf{H}$ are independent and distributed according to $Z(0, 1)$ and the entries of noise vector $\mathbf{n}$ are independent and distributed according to $Z(0, N_0)$. The input vector $\mathbf{s}$ is assumed to be normalized, thus $E[\mathbf{ss}^H] = \mathbf{I}$ where $\mathbf{I}$ is an identity matrix. Let's further assume that precoding matrix $\mathbf{W}$ is unitary, thus $\mathbf{WW}^H = \mathbf{I}$. The receiver selects a precoding matrix $\mathbf{W}_{i}, i = 1, 2, \ldots,$ $N_{codebook}$ from a finite set of quantized precoding matrices $\Omega = \{\mathbf{W}_1, \mathbf{W}_2, \ldots, \mathbf{W}_{N codebook}\}$ and sends the index of the chosen precoding matrix back to the transmitter over a low-delay feedback channel. There are two questions that need to be answered: (1) the optimal selection criterion for choosing a precoding matrix from Ω; and (2) the design of codebook Ω. The matrix $\mathbf{W}_{i}, i = 1, 2, \ldots, N_{codebook}$ can be selected from Ω by using either of the following optimization criterion [57]: (1) minimizing the trace of the mean squared error (MMSE-trace selection); (2) minimizing the determinant of the mean squared error (MMSE-determinant selection); (3) maximizing the minimum singular value of $\mathbf{HW}$ (singular value selection); (4) maximizing the instantaneous capacity (capacity selection); or (5) maximizing the minimum received symbol vector distance (minimum distance selection). The above selection criteria may be evaluated at the receiver using a full search over all matrices in Ω. Using distortion functions based on the selection criteria, it can be shown that the codebook Ω is designed using Grassmannian subspace packing [57]. If MMSE-trace, singular value, or minimum distance selection is used, the codebook is designed such that $\varepsilon = \min_{W_i \neq W_j} \|\mathbf{W}_i \mathbf{W}_i^H - \mathbf{W}_j \mathbf{W}_j^H\|_2^{\text{iii}}$ is maximized. If the MMSE-determinant or capacity selection optimization method is used, the codebook is designed such that $\varepsilon = \min_{W_i \neq W_j} \arccos|\det(\mathbf{W}_i^H \mathbf{W}_j)|$ is maximized.

The MIMO codebooks are designed based on a trade-off between performance and complexity. The following are some desirable properties of the codebooks:

1. Low-complexity codebooks can be designed by choosing the elements of each constituent matrix or vector from a small binary set, e.g., a 4 alphabet $\{\pm1, \pm j\}$ set, which eliminates the need for matrix or vector multiplication. In addition, nested property of the codebooks can further reduce the complexity of CQI calculation when rank adaptation is performed.
2. The base station may perform rank overriding which results in significant CQI mismatch, if the codebook structure cannot adapt to it. A nested property with respect to rank overriding can be exploited to mitigate the mismatch effects.
3. Power amplifier balance is taken into consideration when designing codebooks with constant modulus property, which may eliminate an unnecessary increase in PAPR.
4. Good performance for a wide range of propagation scenarios, e.g., uncorrelated, correlated, and dual-polarized channels, is expected from the codebook design algorithms. A DFT-based codebook is optimal for linear array with small antenna spacing, since the vectors match with the structure of the transmit array response. Additionally, with an optimal selection of the matrices and entries of the codebook (rotated block diagonal structure), significant gains can be obtained in dual-polarized scenarios.
5. Low feedback and signaling overhead are desirable from operation and performance perspective. As such, a 3-bit codebook for two transmit-antennas and a 6-bit codebook for four transmit-antennas

---

[iii]The Euclidean norm of square matrix $\mathbf{A}$ is defined as $\|\mathbf{A}\|_2 = \sup_{\|\mathbf{x}\|=1} \|\mathbf{Ax}\|_2$. The spectral norm of matrix $\mathbf{A}$ is the largest singular value of $\mathbf{A}$ or the square root of the largest eigenvalue of the positive-semi-definite matrix $\mathbf{A}^H \mathbf{A}$, i.e., $\|\mathbf{A}\|_2 = \sqrt{\lambda_{\max}(\mathbf{A}^H \mathbf{A})}$ where $\mathbf{A}^H$ denotes the conjugate transpose of $\mathbf{A}$. [67,68]

have been considered for IEEE 802.16m. Additional gain can be obtained using a differential codebook.

6. Low memory requirement is another design consideration for the MIMO codebooks.

The IEEE 802.16-2009 standard codebooks were optimized exclusively for uncorrelated scenarios and the performance loss in other scenarios (correlated antennas, dual-polarized antennas) is significant. In order to improve the performance of the closed-loop MIMO schemes in various deployment scenarios, DFT-based transform codebooks and associated precoding matrices were investigated that were optimized for cross-polarized channels.

An alternative approach to codebook-based precoding is to use differential codebook precoding, in which the transmitter computes the precoding matrix by using a differential codebook precoding matrix specified by the receiver and the previous precoding matrix used in the transmitter, thus $\mathbf{W}_k = \mathbf{C}_{m_{opt}} \mathbf{W}_{k-1}$ where each differential codebook precoding matrix $\mathbf{C}_m, m = 1, 2, ..., N'_{codebook}$ is an $N_t \times N_t$ matrix. The receiver selects the optimum differential precoding matrix which maximizes the instantaneous capacity as follows:

$$Optimum - Codebook - Index : m_{opt} = \arg\max_m \log_2 \det\left(\mathbf{I} + \frac{\gamma}{N_t}\mathbf{H}_k\mathbf{W}\mathbf{C}'_m\mathbf{C}'^H_m\mathbf{H}^H_k\right) \quad (10\text{-}44)$$

where $\{\mathbf{C}'_1, \mathbf{C}'_2, ..., \mathbf{C}'_{N'_{codebook}}\} = \{\mathbf{C}_1\mathbf{F}_{k-1}, \mathbf{C}_2\mathbf{F}_{k-1}, ..., \mathbf{C}'_{N'_{codebook}}\mathbf{F}_{k-1}\}$ and γ is the signal-to-noise ratio. The initial precoding matrix could be set as the first M columns of the $N_t \times N_t$ identity matrix. One advantage of the differential precoding scheme is that the candidate precoding matrices $\{\mathbf{C}'_1, \mathbf{C}'_2, ..., \mathbf{C}'_{N'_{codebook}}\}$ are refined every precoding update instance, which virtually increases the codebook size to realize finer quantization of the channel [60]. Since the differential precoding schemes compute the precoding matrix by using the previous precoding matrix, different criteria should be used to generate precoding matrices. In a slow-fading environment where precoding is more effective, the channel variation during the precoding matrix update interval is relatively small. In this case, the ideal codebook matrix which is multiplied by the previous precoding matrix should be close to the identity matrix. Thus, the codebook design strategy is to optimize the codebook such that the codebook includes quasi-diagonal matrices. More specifically, codebook matrices are computed from two unitary matrices as $\mathbf{C}_m = \mathbf{G}^H(t)\mathbf{G}(t + \Delta t)$ where $\mathbf{G}(t) \in \mathbf{C}^{N_t \times N_t}$ is a random unitary matrix. Note that since both $\mathbf{G}(t)$ and $\mathbf{G}(t + \Delta t)$ are unitary matrices, each codebook matrix $\mathbf{C}_m \in \mathbf{C}^{N_t \times N_t}$ also has unitary properties and small perturbation of Δt produces quasi-diagonal matrices whose diagonal entries are more dominant than other entries. A set of codebooks $\{\mathbf{C}_1, \mathbf{C}_2, ..., \mathbf{C}_{N'_{codebook}}\}$ is generated by using various Δt values where the codebook is optimized by capacity maximization criterion.

In SVD precoding, assuming perfect knowledge of the channel state information, the channel matrix is $\mathbf{H} = \mathbf{U\Sigma V}^H$ where $\mathbf{U}$ and $\mathbf{V}$ are unitary matrices (i.e., $\mathbf{U}^H\mathbf{U} = \mathbf{I}, \mathbf{V}^H\mathbf{V} = \mathbf{I}$) and $\mathbf{\Sigma}$ is a diagonal matrix containing eigenvalues of the channel matrix. If the relationship between the output and input of the channel is described as $\mathbf{y} = \mathbf{Hs} + \mathbf{n}$, then by replacing the channel matrix with its SVD form and multiplying both sides by $\mathbf{U}^H$, one can write $\mathbf{U}^H\mathbf{y} = \mathbf{U}^H\mathbf{U\Sigma V}^H\mathbf{s} + \mathbf{U}^H\mathbf{n}$. Further simplification of the latter equation would yield $\tilde{\mathbf{y}} = \mathbf{\Sigma x} + \tilde{\mathbf{n}}$, if the input signal is precoded with $\mathbf{x} = \mathbf{V}^H\mathbf{s}$. Using SVD precoding, the channel matrix is diagonalized and the spatial interference is removed without any matrix inversion or non-linear processing. Since $\mathbf{U}$ is unitary, $\mathbf{U}^H\mathbf{n}$ still has the same variance as $\mathbf{n}$, thus, the SVD precoding does not result in noise enhancement. Due to substantial feedback and the complexity of singular value decomposition of the channel matrix, the SVD precoding is not considered a viable precoding scheme. This procedure is illustrated in Figure 10-31.

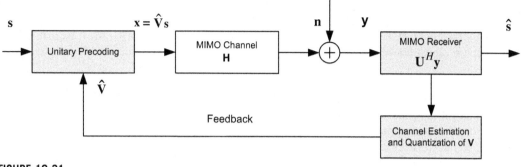

FIGURE 10-31

An illustration of unitary precoding procedures

The precoding algorithms for multi-user MIMO can be classified into linear and nonlinear classes. Linear precoding schemes can achieve reasonable performance with lower complexity compared to nonlinear precoding techniques. Linear precoding includes unitary and zero-forcing precoding schemes. Nonlinear precoding can achieve near optimal capacity at the expense of increased complexity. Nonlinear precoding is designed based on the concept of dirty paper coding, where any known interference at the transmitter can be canceled, if the optimal precoding scheme is applied to the transmit signal.

Unitary precoding includes unitary and semi-unitary precoding both of which are a simple extension of SVD precoding in single-user MIMO, with the addition of the SDMA-based user scheduling technique. The SDMA-based opportunistic user scheduling technique groups near orthogonal users to avoid intra-group interferences at the cost of minimal signaling overhead, which results in higher performance relative to the single-user MIMO. For example, it can increase diversity order to approximately the number of transmit antennas, even with simple linear decoding at the receiver.

The zero-forcing precoding consists of zero-forcing and regularized zero-forcing precoding. If the transmitter has perfect knowledge of the downlink channel state information, ZF-based precoding can achieve near channel capacity performance when the number of users is large. With limited channel state information at the transmitter, ZF-precoding requires increased feedback to achieve the full multiplexing gain. Hence, inaccurate channel state information at the transmitter may result in significant loss of performance due to residual interference among transmit streams.

Dirty paper coding is a coding technique that cancels known interference without power penalty, provided that the transmitter has perfect knowledge of the interfering signals regardless of whether the receiver has knowledge of channel state information. This category includes Costa precoding [71], Tomlinson-Harashima precoding [15,31], and the vector perturbation technique [73]. Vector perturbation uses modulo operation at the transmitter to perturb the transmitted signal vector to avoid the transmit power enhancement incurred by zero-forcing methods. The optimal perturbation method is found by solving a minimum distance problem and thus can be implemented using sphere encoding or full search-based algorithms [57].

In the downlink direction of a precoded MU-MIMO system (alternatively known as broadcast channel in the literature) with N_t transmitter antennas at the BS and one receive antenna for the kth user, the input-output relationship can be written as $y_k = \mathbf{h}_k^H \mathbf{x} + n_k, k = 1, 2, ..., N_u$, where $\mathbf{x} = \sum_{i=1}^{N_u} s_i \mathbf{w}_i$ is the $N_t \times 1$ vector of weighted transmitted symbols s_i, y_k and n_k are the received signal and noise, respectively, $\mathbf{h}_k$ is the kth $N_t \times 1$ channel vector, where matrix $\mathbf{H} = [\mathbf{h}_1, \mathbf{h}_2, ..., \mathbf{h}_{Nu}]^t$ is the $N_u \times N_t$ complex-valued downlink channel matrix and $\mathbf{w}_k$ is the kth $N_t \times 1$ normalized linear precoding vector.

The mathematical relationship for the input and output of a precoded MU-MIMO system in the uplink (alternatively known as multiple access channel in the literature) with N_r receive antennas at BS and one transmit antenna for each user can be written as $\mathbf{y} = \sum_{k=1}^{N_u} s_k v_k \mathbf{h}_k + \mathbf{n}$ where $s_k v_k$ is the weighted complex-valued modulated symbol from user k, $\mathbf{y} = [y_1, y_2, ..., y_{Nr}]^t$ and $\mathbf{n} = [n_1, n_2, ..., n_{Nr}]^t$ are the $N_r \times 1$ vectors of received signal and noise, respectively, $\mathbf{h}_k$ is the kth $N_r \times 1$ channel vector, where matrix $\mathbf{H} = [\mathbf{h}_1, \mathbf{h}_2, ..., \mathbf{h}_{Nu}]$ is the $N_r \times N_u$ complex-valued uplink channel matrix. As mentioned earlier, perfect knowledge of the channel state information is necessary at the transmitter in order to achieve the capacity of a multi-user MIMO channel. However, in practical systems, the receiver only provides partial channel state information through uplink feedback channels to the transmitter, i.e., the multi-user MIMO precoding with limited feedback.

The received signal in the downlink of an MU-MIMO system with limited feedback precoding is mathematically expressed as $y_k = \mathbf{h}_k^H \sum_{i=1}^{N_u} s_i \widehat{\mathbf{w}}_i + n_k, k = 1, 2, ..., N_u$. Since the transmit vector for limited feedback precoding is $\widehat{\mathbf{w}}_i = \mathbf{w}_i + \boldsymbol{\varepsilon}_i$, where $\boldsymbol{\varepsilon}_i$ is the error vector generated as a result of the limited feedback and vector quantization, the received signal can be rewritten as:

$$y_k = \mathbf{h}_k^H \sum_{i=1}^{N_u} s_i \mathbf{w}_i + \mathbf{h}_k^H \sum_{i=1}^{N_u} s_i \boldsymbol{\varepsilon}_i + n_k, k = 1, 2, ..., N_u \tag{10-45}$$

where $\mathbf{h}_k^H \sum_{i=1}^{N_u} s_i \boldsymbol{\varepsilon}_i$ is the residual interference due to the limited feedback precoding. To reduce the residual interference term, one should use more accurate channel state information feedback which results in the use of more uplink resources for the feedback. It is shown in reference [55] that the number of feedback bits per mobile B must be increased linearly with the SNR ρ_{dB} (in decibels) at the rate of $B = (N_t - 1)\log_2 \rho = \rho_{dB}(N_t - 1)/3$ in order to achieve the full multiplexing gain of N_t. In addition, the scaling of B guarantees that the throughput loss relative to zero-forcing precoding with perfect channel state information knowledge at the transmitter is upper-bounded by N_t bps/Hz, which corresponds to a 3 dB power offset. The throughput of a feedback-based zero-forcing system is bounded, if the SNR approaches infinity and the number of feedback bits per mobile is fixed. Reducing the number of feedback bits according to $B = \alpha \log_2 \rho$ for any $\alpha < N_t - 1$ results in a strictly inferior multiplexing gain of $N_t[\alpha/(N_t - 1)]$ where N_t is the number of transmit antennas and ρ is the SNR of the downlink channel. In order to calculate the amount of the feedback required to maintain certain throughput, the difference between the feedback rates of zero-forcing precoding with perfect feedback and with limited feedback is required to satisfy the following constraint $\Delta R(\rho) = R_{PF-ZF}(\rho) - R_{LF-ZF}(\rho) \leq \log_2 b$. In order to maintain a rate offset less than $\log_2 b$ (per user) between zero-forcing with perfect CSI and with finite-rate feedback (i.e., $\Delta R(\rho) \leq \log_2 b, \forall \rho$), it is sufficient to scale the number of feedback bits per mobile according to $B = \rho_{dB}(N_t - 1)/3 - (N_t - 1)\log_2(b - 1)$. The rate offset of $\log_2 b$ (per user) is translated into a power offset, which is a more useful metric from the design perspective. Since a multiplexing gain of N_t is achieved with zero-forcing, the zero-forcing curve has a slope of N_t bps/Hz/3 dB at asymptotically high SNR. Therefore, a rate offset of $\log_2 b$ bps/Hz per user corresponds to a power offset of $3\log_2 b$ decibels. To feedback B bits through uplink channel, the throughput of the uplink feedback channel should be larger than or equal to B, i.e., $w_{FB}\log_2(1 + \rho_{FB}) \geq B$

where ρ_{FB} denotes the SNR of the feedback channel. Thus, the required feedback resource to satisfy the constraint $\Delta R(\rho) \leq \log_2 b$ can be shown to be given as follows $w_{FB} \geq [\rho_{dB}(N_t - 1)/3 - (N_t - 1) \log_2(b - 1)]/\log_2(1 + \rho_{FB})$, i.e., the required feedback resource is a function of both downlink and uplink channel conditions.

We defined precoding as adaptive or non-adaptive weighting of the spatial streams prior to transmission from each antenna port (in a multi-antenna configuration) using a precoding matrix to improve reception or separation of the spatial streams at the receiver. Both feed-back and feed-forward precoder matrix selection schemes can be used in order to select the optimal weights. Feed-back precoding matrix selection techniques do not rely on channel reciprocity, rather they use feedback channels, provided that the feedback latency is less than the channel coherence time. In feed-forward approaches, the necessary CSI can be theoretically obtained through direct channel feedback where the CSI is explicitly signaled to the transmitter by the receiver or channel sounding reference signals. The direct channel feedback methods preclude the channel reciprocity requirement, whereas channel sounding methods rely on channel reciprocity. Therefore, explicit control signaling is required for PMI-based (feedback) schemes. However, in reciprocity-based schemes, the sounding signals in the uplink and precoded pilots in the downlink are used to assist the transmitter and receiver to appropriately select the precoding matrix. Reciprocity-based schemes have the additional advantage of not being constrained to a finite set of codebooks. Beamforming relies on long-term statistics of the radio channel and, unlike reciprocity-based techniques, does not require short-term correlation between the uplink and downlink in order to function properly.

10.4.5 Single-User and Multi-User MIMO

Single-user MIMO (SU-MIMO) techniques are point-to-point schemes that improve channel capacity and reliability through the use of space-time/space-frequency codes (transmit/receive diversity) in conjunction with spatial multiplexing transmission. In a SU-MIMO transmission, the advantage of MIMO processing is obtained from the coordination of processing among all the transmitters or receivers. In the multi-user channel, on the other hand, it is usually assumed that there is no coordination among the users. As a result of the lack of coordination among users, uplink and downlink multi-user MIMO channels are different. In the uplink scenario, users transmit to the base station over the same channel. The challenge for the base station is to separate the signals transmitted by the users, using array processing or multi-user detection methods. Since the users are not able to coordinate with each other, there is not much that can be done to optimize the transmitted signals with respect to each other. If some channel feedback is allowed from the transmitter back to the users, some coordination may be possible, but it may require that each user knows all the other users' channels, rather than only its own. Otherwise, the challenge in the uplink is mainly in the processing done by the base station to separate the users. In the downlink channel, where the base station simultaneously transmits to a group of users over the same channel, there is some inter-user interference for each user which is generated by the signals transmitted to other users. Using multi-user detection techniques, it may be possible for a given user to overcome multiple access interference, but such techniques are often extremely complicated for use at the receivers. Ideally, one would like to mitigate the interference at the transmitter by carefully designing the transmitted signal. If CSI is available at the transmitter, it is aware of what interference is being created for user i by the signal it is transmitting to user j and *vice versa*. The inter-user interference can be mitigated by beamforming or the use of dirty paper codes. In general, the MU-MIMO and SU-MIMO comparison can be summarized as follows [15]:

- SU-MIMO is a point-to-point link with predictable link capacity, whereas MU-MIMO channel is a broadcast channel (BC) in the downlink direction and a multiple access channel (MAC) in the uplink direction whose link-level data rates are characterized in terms of capacity regions.
- Multi-stream SU-MIMO schemes offer stream diversity in the sense that if one stream has a poor SNR, the system will not necessarily experience an outage, whereas in the same situation, a MU-MIMO system be in outage. This is due to the fact that in MU-MIMO schemes, users typically have an equal target data rate and symbol error rate on their respective links, while in SU-MIMO systems, only the sum rate of the overall link is considered since all streams are delivered to the same user.
- The MU-MIMO schemes suffer from a near–far problem due to the significant difference between the path losses experienced by each user, resulting in large deviations in the SINR of the corresponding user links. This would benefit the users with better channel conditions, while there is no near–far problem in SU-MIMO systems. The near–far problem in MU-MIMO systems may be alleviated via appropriate grouping of the users with similar channel conditions.
- The use of cooperative collocated transmit antennas in SU-MIMO schemes can facilitate encoding at the transmitter and decoding at the receiver. In contract, the users in a MU-MIMO scheme can cooperate in encoding at the BS in the downlink and decoding in the uplink; however, the users cannot cooperate in decoding in the downlink or encoding in the uplink directions.
- The capacity of the downlink and uplink are theoretically identical in the downlink and uplink for SU-MIMO systems (given the same transmit power and perfect channel knowledge in the transmitter and the receiver); however, the capacity of the MU-MIMO broadcast channel and multiple access channel are not identical.
- The capacity of SU-MIMO schemes is less impacted by lack of channel state information at the transmitter, whereas the capacity of the MU-MIMO broadcast channel suffers significantly from lack of channel state information at the transmitter.

An important metric for measuring the performance of any communication channel is the information theoretic capacity. In a single-user MIMO channel, the capacity is the maximum amount of information that can be transmitted as a function of available bandwidth given a constraint on transmitted power. In single-user MIMO channels, it is common to assume that the total power distributed among all transmit antennas is limited. For the multi-user MIMO channel, the problem is somewhat more complex. Given a constraint on the total transmit power, it is possible to allocate varying fractions of that power to different users in the network, thus for any value of total power, different information rates are obtained. The result is a capacity region shown in Figure 10-33 for a two-user MU-MIMO channel. The maximum capacity for user 1 is achieved when 100% of the power is allocated to user 1, and for user 2, the maximum capacity is also obtained when it is allocated the full power. For every possible power distribution there is an achievable information rate, which results in the capacity regions depicted in the figure. Two regions are shown in Figure 10-33, the larger one for the case where both users have roughly the same maximum capacity (similar channel conditions), and the other region for a case where one of the users has much better channel condition than the other. For N_u users, the capacity region is characterized by a N_u-dimensional hyper-region.

Figure 10-25 illustrates a generic multi-user MIMO scenario where N_u users, each having N_{tNu} transmit antennas, transmitting simultaneously to the BS in the uplink. The maximum achievable throughput of the system is determined by the point on the N_u-dimensional hyper-region that maximizes the sum of all the users' information rates and is referred to as the sum capacity of the channel (see Figure 10-33). In the case of a near–far problem where one user has a more strongly attenuated channel than other users or some users are closer to the BS than others, obtaining the sum capacity would cause the user with the worse channel to achieve a lower rate.

As shown in Figure 10-32, in the uplink of a multi-user MIMO system, the received signal at the BS can be written as $\mathbf{y} = \sum_{k=1}^{N_u} \mathbf{H}_k^H \mathbf{x}_k + \mathbf{n}$ where $\mathbf{x}_k$ is the $N_{tk} \times 1$ of the kth user complex-valued

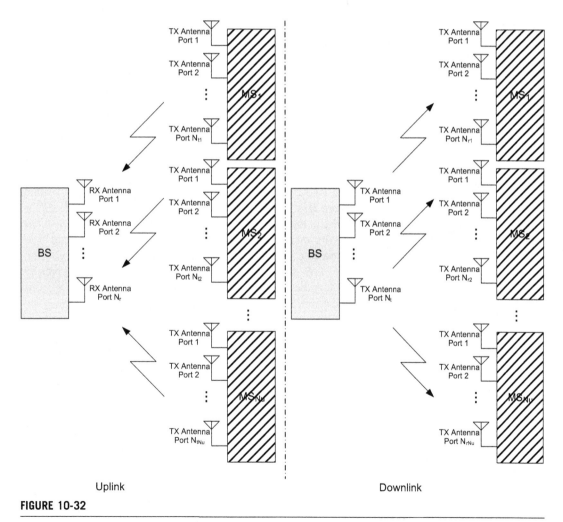

Uplink   Downlink

FIGURE 10-32

A generic multi-user MIMO transmission scheme

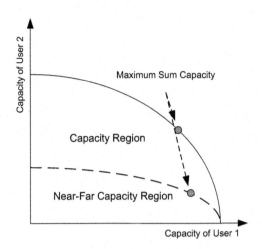

FIGURE 10-33

An example illustration of capacity region

signal vector, $\mathbf{H}_k \in C^{N_{rk} \times N_r}$ denotes the flat-fading channel matrix and $\mathbf{n}$ is the independent and identically distributed additive white Gaussian noise $\mathbf{n} = (n_1, n_2, ..., n_{Nr}), n_k \sim N(0, 1)$ vector at the BS. We assume that the receiver k has perfect and instantaneous knowledge of the channel matrix $\mathbf{H}_k$.

In the downlink as illustrated in Figure 10-32, the received signal at the kth receiver can be written as $\mathbf{y}_k = \mathbf{H}_k \mathbf{x} + \mathbf{n}_k \, \forall k = 1, 2, ..., N_u$ where $\mathbf{H}_k \in C^{N_{rk} \times N_t}$ is the downlink channel and $\mathbf{n}_k \in C^{N_{rk} \times N_t}$ is the complex-valued additive Gaussian noise at the kth receiver. We assume that each receiver also has perfect and instantaneous knowledge of its own channel matrix $\mathbf{H}_k$. The transmitted signal $\mathbf{x}$ is a function of the multiple users' information data, i.e., $\mathbf{x} = \sum_{k=1}^{N_u} \mathbf{x}_k$ where $\mathbf{x}_k$ is the signal carrying kth user's message with covariance matrix $\mathbf{\Omega}_k = E\{\mathbf{x}_k \mathbf{x}_k^H\}$. The power allocated to the kth user is given by $\rho_k = tr\{\mathbf{\Omega}_k\}$. Under a sum power constraint at the BS, the power allocation needs to maintain $\sum_{k=1}^{N_u} \rho_k \leq P_{total}$. Assuming a unit variance for the noise, it is now known that the capacity region for a given matrix channel realization can be written as:

$$C_{DL} = \bigcup_{(\rho_1, \rho_2, ..., \rho_{Nu} | \sum \rho_k \leq P_{total})} \left\{ (R_1, R_2, ..., R_{Nu}) \in \Re^{+N_u}, R_l \leq \log_2 \frac{\det\left[\mathbf{I} + \mathbf{H}_i(\sum_{j \geq i} \mathbf{\Omega}_j)\mathbf{H}_i^H\right]}{\det\left[\mathbf{I} + \mathbf{H}(\sum_{j > i} \mathbf{\Omega}_j)\mathbf{H}_i^H\right]} \right\}$$

(10-46)

where $\Re^{+N_u}$ is the N_u-dimensional set of positive real numbers. The above equation may be optimized over each possible user ordering. Although difficult to realize in practice, the computation of the capacity region can be simplified using the assumption that the downlink capacity region can be calculated through the union of regions of the dual multiple access channel with all uplink power allocation vectors meeting the sum power constraint. The fundamental effect of the use of multiple antennas at either the BS or the user terminals in increasing the channel capacity is best understood by examining how the sum capacity, i.e., the point obtained by the maximum $\sum_{k=1}^{N_u} R_k$ in the capacity region, scales with the number of active users.

The capacity region in the uplink of an MU-MIMO system can be calculated for two different cases: (1) joint decoding, i.e., the signals from a group of users are decoded in a cooperative manner; and (2) independent decoding, where the signals from users are independently decoded in parallel.

The capacity of the MU-MIMO channels can be calculated using dirty paper coding techniques. The method was inspired by interference cancellation schemes. It is shown that the capacity of a channel where the transmitter has the knowledge of interference is the same as if there were no interference. The dirty paper coding concept can be better comprehended by comparing the scheme to writing on dirty paper with a properly-selected ink color so that the scripts have high contrast and are clearly readable. It can be shown that the capacity of the MU-MIMO multiple-access channel in the above cases is given as follows:

$$R_i \le \log_2 \left[\frac{\det(\mathbf{R_y})}{\det(\mathbf{R_y} - E_{si}\mathbf{h}_i\mathbf{h}_i^H)} \right], i = 1, 2, ..., N_u \qquad \text{(Independent Decoding)}$$

$$\sum_{k \in \Re} R_k \le \log_2 \det \left(\mathbf{I}_{Nr} + \frac{1}{N_0} \mathbf{H}_\Re \mathbf{R}_{s\Re} \mathbf{H}_\Re^H \right) \qquad \text{(Joint Decoding)}$$

(10-47)

In the above equations, N_r denotes the number of receive antennas at the base station, N_u is the number of users each with only one transmit antenna, $\Re$ is a subset of the set of users whose signals are jointly decoded, $\mathbf{H}_\Re$ is the $N_r \times M$ complex-valued channel matrix where M is the cardinality of the set $\Re$, $\mathbf{R}_{s\Re}$ denotes the covariance matrix of the signals transmitted by the users contained in the set $\Re$, $\mathbf{I}_{Nr}$ is an $N_r \times N_r$ identity matrix, R_k is the rate that can be reliably achieved by user k, s_i is the complex-valued symbol transmitted by user $i, i = 1, 2, ..., N_u$, E_{si} is the average energy of the signal from user i which differs from one user to another, $\mathbf{R_y}$ is the covariance matrix of the received signal at the base station, and $\mathbf{H} = [\mathbf{h}_1, \mathbf{h}_2, ..., \mathbf{h}_{Nu}]$ is the normalized $N_r \times N_u$ complex-valued channel matrix composed of individual user complex-valued channel vectors $\mathbf{h}_i$.

If the conventional AWGN channel were modified to include an additive interference term that is known at the transmitter, and the received signal were expressed as the sum of the transmitted signal plus interference plus noise, one could set the transmitted signal equal to the desired data minus the interference term; however, such an approach requires increased power and perfect knowledge of interference. It is shown in the literature that the capacity in such scenario is the same as if the interference were not present; no additional power is required to cancel the interference than is used in a nominal additive white Gaussian noise channel. Therefore, writing on dirty paper, from an information theory perspective, is equivalent to writing on clean paper when one knows in advance where the dirt is located. This concept has been used to characterize the sum-capacity and capacity region of multi-antenna multi-user channels. A dirty paper coding technique for the downlink of a MU-MIMO system uses the QR decomposition[iv] of the channel matrix, as the product of a lower triangular matrix $\mathbf{R}$ with a unitary matrix $\mathbf{Q}$ such that $\mathbf{H} = \mathbf{QR}$. The signal to be transmitted is precoded with the $\mathbf{Q}^H$ matrix, resulting in the effective channel $\mathbf{R}$. The first user of this equivalent system sees no interference from other users, i.e., the signal may be chosen without regard for the other users. The second user sees

---

[iv]In linear algebra, the QR decomposition of a matrix is defined as decomposition of a matrix into an orthogonal and an upper triangular matrix. The QR decomposition is often used to solve the linear least squares problem. The QR decomposition of a real square matrix $\mathbf{A}$ is $\mathbf{A} = \mathbf{QR}$ where $\mathbf{Q}$ is an orthogonal matrix, i.e., $\mathbf{Q}^H\mathbf{Q} = \mathbf{I}$ and $\mathbf{R}$ is an upper triangular matrix. This can be generalized to a complex-valued square matrix $\mathbf{A}$ and a unitary matrix $\mathbf{Q}$. If $\mathbf{A}$ is non-singular, then this factorization is unique if one requires the diagonal elements of $\mathbf{R}$ to be positive-valued.

interference only from the first user; this interference is known and thus may be cancelled using dirty paper coding. The other users are dealt with in a similar manner. Alternative approaches apply dirty paper techniques directly, rather than for individual users. An important difference between the multi-user MIMO channel and the interference channels for which dirty paper techniques are designed is that the interference depends on the signal being designed.

Block Diagonalization (BD) is a linear precoding technique for the downlink of MU-MIMO systems. It decomposes a MU-MIMO downlink channel into multiple parallel orthogonal single-user MIMO channels. The signal of each user is pre-processed at the transmitter using a modulation matrix that lies in the null space of all other users' channel matrices; thereby the multi-user interference in the system is effectively zero. Block diagonalization is restricted to channels where the number of transmit antennas N_t is greater or equal to the total number of receive antennas in the network N_r. Let's define the precoder matrices as $\mathbf{F} = [\mathbf{F}_1, \mathbf{F}_2, ..., \mathbf{F}_{Nu}] \in C^{N_r \times N_s}$ where $\mathbf{F}_i \in C^{N_r \times N_{si}}$ is the ith user precoder matrix. We assume that $N_s \leq N_r$ is the total number of the transmitted data streams, while $N_{si} \leq N_{ri}$ is the number of data streams transmitted to the ith user. We can find the optimal precoding matrix $\mathbf{F}$ such that multi-user interference is zero by choosing a precoding matrix $\mathbf{F}_i$ that lies in the null space of the other users' channel matrices. Thereby, a MU-MIMO downlink channel is decomposed into multiple parallel independent SU-MIMO channels.

Minimum mean square error beamforming improves the system performance by allowing a certain amount of interference, especially for users equipped with a single antenna. However, it suffers from performance loss when it attempts to mitigate the interference between two closely-spaced antennas, a situation that usually occurs when the user terminal is equipped with more than one receive antenna. An improved version of this algorithm known as Successive MMSE (SMMSE) was developed to mitigate this problem by successively calculating the columns of the precoding matrix for each of the receive antennas separately. Linear equalization suffers from noise enhancement and hence has poor power efficiency in some cases. The same drawback is experienced by linear precoding, which alleviates noise by boosting the transmit power.

The Tomlinson-Harashima Precoding (THP) algorithm is a non-linear precoding technique originally developed for multipath SISO channels. The THP algorithm is performed by moving the feedback part of the decision-feedback equalizer to the transmitter. It has also been applied for the pre-equalization of multi-user interference in MU-MIMO systems; hence, no error propagation occurs and the precoding can be performed for the interference-free channel. The MMSE precoding in combination with THP is another approach to balance the multi-user interference in order to reduce the performance loss that occurs with zero interference techniques, while THP is used to reduce the multi-user interference and to improve the diversity.

Space-Division Multiple Access (SDMA) is a special case of MU-MIMO where several users share the same time-frequency physical resource block due to their spatial separation. SDMA uses antenna beam direction or angle (using directional antennas) as another dimension in signal space which can be channelized and assigned to different users (see Figure 10-34). Orthogonal channels can be assigned only if the angular separation between users exceeds the angular resolution of the directional antenna. In practice, SDMA is often implemented using sectorized antenna arrays. The 360° angular space is divided into a number of sectors and in each sector a highly-directional antenna array is used to minimize the inter-user interference. The users within each sector can be served using multiple access schemes such as OFDMA, CDMA, etc. The SDMA system adapts to the mobility of the users by beam-steering or assigning the users to a new channel when the user changes its geographical location.

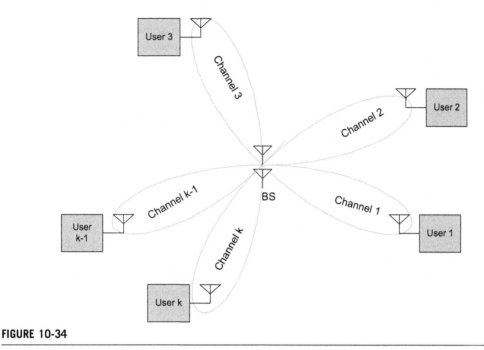

FIGURE 10-34

An illustration of the SDMA concept

10.4.6 Collaborative MIMO and Collaborative Spatial Multiplexing

Collaborative spatial multiplexing is a virtual MIMO technique where users transmit over the same time-frequency resource unit in the uplink. This type of spatial multiplexing improves the sector throughput without requiring multiple transmit antennas at the mobile station. The received signals from the simultaneous uplink transmission of the users are processed in the BS using maximum likelihood detection techniques. The base station's scheduler groups the users with similar channel conditions and assigns their uplink transmissions to the same time-frequency region. The operation of a collaborative spatial multiplexing scheme is conceptually similar to the regular spatial multiplexing scheme using transmit antennas that are located at different locations (user terminals are geographically apart). The throughput of collaborative spatial multiplexing theoretically increases by the number of users which are scheduled to transmit simultaneously over the same physical resource units.

The Collaborative MIMO (Co-MIMO) is an extension of conventional single-BS based MIMO techniques. It allows multiple base stations to serve one or multiple mobile stations simultaneously over the same radio resource through coordination among the participating base stations. The Co-MIMO is characterized by the following three features:

1. In Co-MIMO, each MS is jointly served by multiple base stations, which is different from conventional MIMO techniques where each MS is only served by a single BS. Through BS coordination, the inter-cell interference among these coordinated base stations can be significantly reduced when using low frequency reuse factors.

2. In Co-MIMO, each BS can also simultaneously serve multiple mobile stations using the same radio resource in order to increase the system throughput. The co-channel interference among these mobile stations can be minimized by use of spatial division multiple access techniques such as beamforming or multi-user MIMO schemes.
3. In Co-MIMO, SDMA processing is implemented independently within individual coordinated base stations, based on the channel state information between the BS and the mobile stations that are served by the BS. Therefore, the Co-MIMO limits the amount of information exchange between base stations, which is different from other BS coordination approaches requiring considerable information exchange.

A central scheduler, a logical functional module in the network, is responsible for coordinating a group of base stations, collecting information from the participating base stations, and determining the serving relationship between base stations and mobile stations. The overall complexity of the network operation increases with increasing numbers of coordinated base stations in Co-MIMO schemes. The Co-MIMO has the advantages of inter-cell interference reduction and spectral efficiency improvement. The inter-cell interference mitigation is realized by turning interference from neighboring cells into useful signals and separating signals for different users via multi-user MIMO or SDMA techniques. The spectral efficiency improvement is achieved via interference mitigation and the number of users supported over the same time-frequency resource region. Moreover, the Co-MIMO provides a simple approach to implementation of BS coordination and a reasonable trade-off between the performance gain, and the stringent requirement on the core network traffic and implementation complexity. In practice, the user data delivery to multiple coordinated base stations in a Co-MIMO scheme can be realized in the same manner as the multicast over backhaul is done in multicast and broadcast services. There is no additional computational complexity in base stations supporting Co-MIMO compared to the base stations which support multi-user MIMO techniques.

In order to demonstrate the advantages of Co-MIMO, an example system-level simulation is conducted where the collaborative zone is defined over three neighboring sectors belonging to different BSs, as shown in Figure 10-35. As the inter-cell interference is most severe in the cell edge, the evaluation of Co-MIMO performance is performed in the cell-edge area. The cell-edge area is characterized by the cell-edge length in the simulation, as shown in Figure 10-35. We assume a 19-cell collaborative-zone configuration in the simulation. The users are randomly dropped with a uniform distribution in the cell-edge area, one user per sector. The system performance with different sizes of cell-edge area is evaluated by setting the cell-edge length to 150 m, 100 m, and 50 m. In each collaborative region, three coordinated base stations communicate with three mobile stations located in their cell-edge areas, allocating the same time and frequency resource. Each MS in the cell-edge area will be served by two coordinated base stations, and each BS serves two mobile stations simultaneously. The precoding matrices for the two served mobile stations are calculated independently within each single sector of a coordinated BS. Each coordinated BS transmits with full power and equal power allocation to the two mobile stations. A single narrowband sub-carrier is assumed in the system without loss of generality. A conventional SU-MIMO scheme is used as the benchmark for comparison where three mobile stations in the cell-edge area are served independently by different base stations without any coordination in frequency reuse 1 deployment. As shown in Figure 10-36, even though there remains a certain amount of residual inter-cell interference, the interference power from neighboring cells within the collaborative zone is considerably reduced.

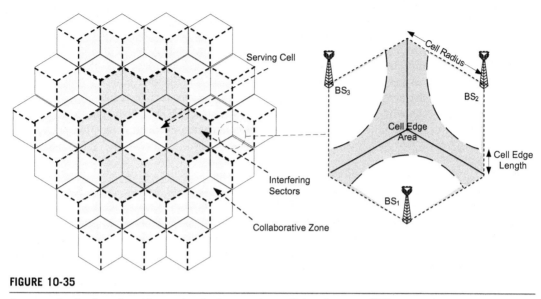

FIGURE 10-35

An example of cell configuration and cell edge area in a collaborative zone [26]

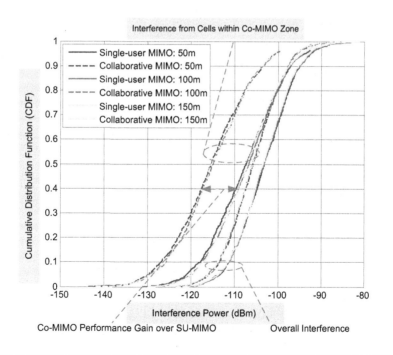

FIGURE 10-36

CDF of interference power of the cell-edge users as a function of cell-edge length [26]

If the interference from cells outside the collaborative zone is taken into consideration, the reduction in overall interference level is less noticeable, and can be further improved by extending the collaborative zone to contain more coordinated base stations. The cumulative distribution function of the cell-edge users within the cell-edge area as a function of cell-edge length for SU-MIMO and Co-MIMO are compared and shown in Figure 10-36. It is shown that the interference power increases with increasing cell-edge length (see Figure 10-35). There is a considerable reduction in the interference power from cells within the Co-MIMO zone when using Co-MIMO techniques compared to the conventional SU-MIMO schemes. The performance gain decreases when comparing the overall interference [26].

10.5 IEEE 802.16M DOWNLINK MIMO SCHEMES

The IEEE 802.16m standard supports several advanced multi-antenna techniques including single and multi-user MIMO (spatial multiplexing and beamforming), as well as a number of transmit diversity schemes. The classification of the IEEE 802.16m downlink-MIMO schemes is depicted in Figure 10-37. In this section, a detailed description of these schemes and their advantages over the legacy schemes will be provided.

As mentioned in earlier sections, in a single-user MIMO scheme only one user can be scheduled over one (time, frequency, space) resource unit. In a multi-user MIMO, on the other hand, multiple users can be scheduled in one resource unit. Vertical encoding (or single codeword) utilizes one encoder block (or layer), whereas horizontal encoding (or multi-codeword) uses multiple encoders (or multiple layers). For SU-MIMO vertical encoding is utilized, whereas for MU-MIMO horizontal or multi-layer encoding is employed at the BS, and up to two streams can be transmitted to each MS.

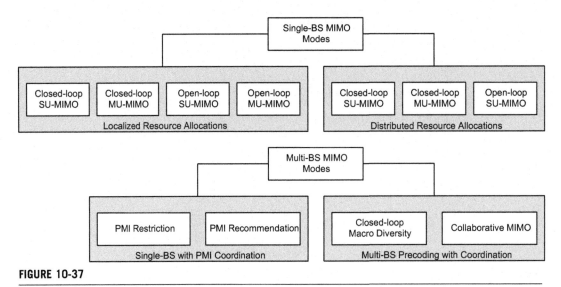

FIGURE 10-37

Downlink MIMO mode classification

A layer is defined as an encoding and modulation input path to the MIMO encoder. A stream is defined as the output of the MIMO encoder that is further processed through the beamforming or the precoder block. For spatial multiplexing, the rank is defined as the number of streams to be used for the user. Each of various SU-MIMO or MU-MIMO open-loop or closed-loop schemes is defined as a MIMO mode. The MIMO encoder block maps $L \geq 1$ layers into $N_s \geq L$ streams, which are fed to the precoder block. The physical layer procedures in the downlink MIMO on the transmitter side are shown in Figure 10-38.

The encoder block contains the channel encoder, interleaving, rate-matching, and modulation blocks per layer. The resource mapping block maps the complex-valued modulation symbols to the corresponding time-frequency resources. The MIMO encoder block maps the layers onto the streams, which are further processed through the beamforming or the precoder block. The beamforming/pre-coding block maps the streams to antennas by generating the antenna-specific data symbols according to the selected MIMO mode. The sub-carrier mapping function is responsible for mapping the logical sub-carriers to physical sub-carriers before applying OFDM modulation. The OFDM modulation block maps antenna-specific data to the OFDM symbols. The feedback block contains feedback information, such as CQI or channel state information from the MS.

The scheduler block schedules users to resource units and decides their MCS level, MIMO mode, MIMO rank, etc. This block is responsible for making a number of decisions with regard to each resource allocation, including allocation type (distributed or contiguous allocation), single-user versus multi-user MIMO (the resource allocation supports a single user or more than one user), MIMO mode (open-loop or closed-loop transmission scheme), user grouping for MU-MIMO (users that are assigned to the same resource unit), MIMO rank (the number of streams to be used for the user assigned to the resource unit when spatial multiplexing modes in SU-MIMO are used), MCS level per layer (modulation and coding rate used on each layer), the power boosting values used on the data and pilot sub-carriers, and band selection (location and the number of the contiguous resources). In the IEEE 802.16m systems, the BS is required to have a minimum of two transmit antennas, nonetheless,

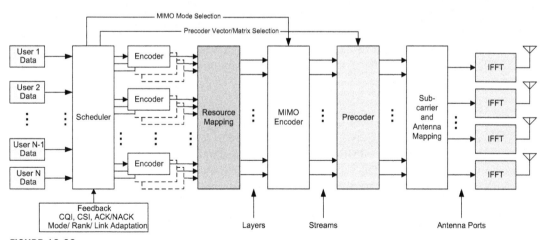

FIGURE 10-38

IEEE 802.16m downlink MIMO architecture

2, 4, and 8 antenna configurations are also supported. The MS is required to have two receive antennas as a minimum.

The layer-to-stream mapping is performed by the MIMO encoder. The MIMO encoder is a block processor that operates on M input symbols at a time. The input to the MIMO encoder is an $M \times 1$ vector $\mathbf{s} = [s_1, s_2, \ldots, s_M]^t$ where s_i is the ith symbol within the input block. The layer to stream mapping of the input symbols is done starting with the spatial dimension. The output of the MIMO encoder is an $N_s \times N_f$ matrix given by $\mathbf{x} = \mathbf{S}(\mathbf{s})$ which serves as the input to the precoder block where N_s denotes the number of streams, N_f is the number of sub-carriers occupied by one MIMO resource block, $\mathbf{x}$ is the output matrix of the MIMO encoder, $\mathbf{s}$ is the input vector, and $\mathbf{S}(\mathbf{s})$ is the MIMO encoder transformation matrix. The output of the MIMO encoder can be described as:

$$
\mathbf{x} = \begin{bmatrix} x_{11} & x_{12} & \cdots & x_{1N_f} \\ x_{21} & x_{22} & \cdots & x_{2N_f} \\ \vdots & \vdots & \ddots & \vdots \\ x_{N_s1} & x_{N_s2} & \cdots & x_{N_sN_f} \end{bmatrix} \tag{10-48}
$$

For SU-MIMO transmissions, the spatial rate is defined as in $R = M/N_f$. For MU-MIMO transmissions, the spatial rate per layer R_i is equal to 1 or 2. There are four MIMO Encoder Formats (MEF) specified in the IEEE 802.16m as follows:

- Space-frequency block coding;
- Vertical encoding;
- Horizontal or multi-layer encoding;
- Conjugate Data Repetition (CDR).

In SU-MIMO, MIMO encoding would allow the use of spatial multiplexing and transmit diversity transmission schemes. Spatial multiplexing MIMO employs vertical encoding within a single layer (codeword). Transmit diversity utilizes either vertical encoding with a single stream or space-frequency block coding. In MU-MIMO, horizontal or multi-layer encoding is used at the base station, while up to two streams can be transmitted to each mobile station. For open-loop transmit diversity with SFBC encoding, the input to the MIMO encoder is represented by 2×1 vector $\mathbf{s} = [s_1 \ s_2]^t$. The MIMO encoder generates the 2×2 SFBC matrix as follows:

$$
\mathbf{x} = \begin{bmatrix} s_1 & -s_2^* \\ s_2 & s_1^* \end{bmatrix} \tag{10-49}
$$

The SFBC output matrix occupies two consecutive sub-carriers. For open-loop transmit diversity with CDR encoding, the input to the MIMO encoder is represented by a 1×1 vector $\mathbf{s} = s_1$ and the MIMO encoder generates the 1×2 CDR vector $\mathbf{x} = [s_1 \ s_1^*]$. The CDR matrix occupies two consecutive sub-carriers in the frequency domain. For horizontal encoding and vertical encoding, the input and the output of the MIMO encoder is represented by an $M \times 1$ vector $\mathbf{x} = \mathbf{s} = [s_1, s_2, \ldots, s_M]^t$ where s_i is the ith modulated symbol in the input block. In the case of vertical encoding $s_1, s_2, \ldots, s_M$ belong to the same layer and the MIMO encoder is an identity operation, whereas for horizontal encoding $s_1, s_2, \ldots, s_M$ belong to different layers. Note that multi-layer encoding is only used in conjunction with MU-MIMO schemes. The number of streams depends on the MIMO encoder. For open-loop and closed-loop spatial multiplexing SU-MIMO, the number of streams is $N_s \leq \min(N_t, N_r)$, $N_t \leq 8$. For open-loop transmit diversity, N_t depends on the space-time coding scheme that is used by the MIMO encoder. The IEEE 802.16m MU-MIMO supports up to two spatial streams with two transmit-antennas and up to four spatial streams for four and eight transmit antennas.

Stream to antenna mapping is performed by the precoder block. The output of the MIMO encoder is multiplied by $N_t \times N_s$ precoding matrix $\mathbf{W}$. The output of the precoder block is denoted by the $N_t \times N_f$ matrix $\mathbf{Z}$ as follows:

$$\mathbf{Z} = \mathbf{Wx} = \begin{bmatrix} z_{11} & z_{12} & \cdots & z_{1N_f} \\ z_{21} & z_{22} & \cdots & z_{2N_f} \\ \vdots & \vdots & \ddots & \vdots \\ z_{N_t1} & z_{N_t2} & \cdots & z_{N_tN_f} \end{bmatrix} \tag{10-50}$$

Where N_f denotes the number of sub-carriers occupied by one MIMO block and z_{jk} is the output complex-valued symbol to be transmitted by the jth physical antenna on the kth sub-carrier. The pilot sub-carriers within a physical resource unit are precoded in the same way as the data sub-carriers. The successive symbols at the output of the precoder block are mapped to the corresponding antenna ports. The sub-carrier mapping within a logical resource unit starts from the data sub-carrier with the lowest OFDM symbol index and lowest sub-carrier index, and continues with increasing sub-carrier index. If the edge of the resource unit is reached, the mapping is continued on the next OFDM symbol.

Non-adaptive precoding is used with open-loop SU-MIMO and open-loop MU-MIMO modes. For non-adaptive precoding on a given sub-carrier k, the matrix $\mathbf{W}_k$ is selected from a predefined base codebook or from a subset of the base codebook for a given rank. There are three types of codebooks and their associated feedback modes for adaptive precoding, defined in the IEEE 802.16m, as follows:

1. The base codebook mode where the PMI feedback from an MS is indicative of an entry in the base codebook that is used by the serving BS to determine a new precoder.
2. The transformation codebook mode where the PMI feedback from an MS represents an entry in the transformed base codebook according to long-term channel information.
3. The differential codebook mode where the PMI feedback from an MS provides an entry in the differential codebook or an entry in the base codebook on PMI reset time. The feedback from an MS provides a differential knowledge of the short-term channel information. This feedback represents information that is used along with other feedback information known at the BS for determining a new precoder.

The base codebook for two transmit antennas comprises either a 3-bit codebook for adaptive precoding or a codebook subset for non-adaptive precoding. For four transmit antennas, the base codebook consists of either a 6-bit codebook or a 4-bit codebook which is a subset of the 6-bit codebook when using adaptive precoding, and a codebook subset when non-adaptive precoding is used. The base codebook for eight transmit antennas is formed from either a 4-bit codebook for adaptive precoding or a codebook subset for non-adaptive precoding. The mobile stations are required to support the base and transformation modes and may optionally support the differential mode. The transformation and differential feedback modes are applied to the base codebook or to a subset of the base codebook. In order to provide additional spatial diversity, the precoding coefficients $\mathbf{W}_k$ change every uP_{sc} sub-carriers and every v sub-frames where the values of u and v depend on the MIMO scheme and type of resource unit. The subsets of the base codebooks for open-loop SU-MIMO are optimized for performance in diversity allocations for both spatially correlated and uncorrelated channels, with the additional constraints of small-sized sets with nested property among the codebooks of different ranks, and containment of matrices with elements of constant modulus, in order to reduce the implementation complexity at the MS and BS. The subsets of the base codebooks for open-loop MU-MIMO are selected to provide sub-band selection gain and are composed of matrices with elements of constant modulus. The choice of the precoder from subsets of the base codebook does impact on the performance, especially in correlated channels with distributed allocations and single-stream transmissions, thus optimization of the open-loop precoders is necessary.

For adaptive precoding, the form and derivation of the precoding matrix $\mathbf{W}_k = [\mathbf{w}_{1k}, \mathbf{w}_{2k}, \ldots, \mathbf{w}_{Mk}]$ is specific to implementation and is not specified in the standard. The precoding vector $\mathbf{w}_{jk}$ at the kth sub-carrier and the jth stream is derived at the BS using the feedback received from the MS. Beamforming is enabled with this precoding mechanism. If the columns of the composite precoding matrix are orthogonal to each other, it is defined as unitary precoding. Otherwise, it is defined as non-unitary precoding. Non-unitary precoding is only allowed with closed-loop MU-MIMO.

In the downlink closed-loop SU-MIMO and MU-MIMO, all demodulation pilots are precoded similar to the data sub-carriers, regardless of the number of transmit antennas, allocation type, and MIMO transmission mode. The precoding matrix is signaled to the MS via precoding of the demodulation pilots. The IEEE 802.16m specifies six downlink MIMO transmission modes as listed in Table 10-21. The associated MIMO mode parameters and usage are given in Table 10-22 and Table 10-23, respectively. All downlink MIMO modes support either distributed or localized resource permutation.

An open-loop MIMO region is defined as a time-frequency region in which the base stations coordinate their open-loop MIMO transmissions in order to stabilize the interference environment where the precoders and numbers of streams are not time-varying. The open-loop MIMO region with N_s streams uses the N_s-stream pilot pattern and a given open-loop MIMO mode with N_s streams without rank adaptation. The resource units used for the open-loop region are indicated in a downlink broadcast message. These resource units are aligned across cells. A limited set of open-loop MIMO modes are allowed for transmission in the open-loop region (see Table 10-23 and Table 10-24). There is no limitation on the use of any open-loop MIMO mode outside the open-loop region. The logical resource units used for the open-loop region are designated through the AAI_SCD MAC management message. An open-loop region is associated with a specific set of parameters such as the type of open-loop MIMO region (number of streams, MIMO mode, MIMO feedback mode, type of supported permutation). All base stations that are coordinated over the same open-loop MIMO region use the same number of streams in order to minimize interference fluctuation and to improve the CQI prediction at the MS. All pilots are precoded by non-adaptive precoding with N_s streams in the open-loop MIMO region. The CQI measurements are conducted by the MS on the precoded demodulation pilots, rather than on the downlink reference signals. The precoded pilots are transmitted in all the resource units in the open-loop MIMO region even if data is not transmitted by the BS on some or all of those resource units.

In order to configure any MIMO mode, one or multiple MIMO feedback modes are specified to indicate the parameters that the mobile station needs to measure and send to the BS through the uplink feedback channels. When allocating a feedback channel, the MIMO feedback mode is signaled to the MS and the MS sends feedback according to the feedback mode. The feedback corresponding to the quantized wideband correlation matrix can be requested by the BS for operation with transform codebook-based feedback mode using the Feedback Polling A-MAP IE. The BS may request the

Table 10-21 Downlink MIMO Modes outside Open-loop MIMO Region

Mode Index	Description	MIMO Encoding Format	MIMO Precoding	DRU Permutation Support	Mini-Band based CRU Permutation Support (Diversity Allocation)	Sub-Band based CRU Permutation Support (Localized Allocation)
Mode 0	Open-Loop SU-MIMO	SFBC	Non-adaptive	Yes	Yes	Yes
Mode 1	Open-Loop SU-MIMO (spatial multiplexing)	Vertical encoding	Non-adaptive	Yes ($N_s = 2$))	Yes ($2 \leq N_s \leq 4$)	Yes
Mode 2	Closed-Loop SU-MIMO (spatial multiplexing)	Vertical encoding	Adaptive	No	Yes ($N_s \leq 4$)	Yes
Mode 3	Open-Loop MU-MIMO (spatial multiplexing)	Horizontal encoding	Non-adaptive	No	No	Yes
Mode 4	Closed-Loop MU-MIMO (spatial multiplexing)	Horizontal encoding	Adaptive	No	Yes	Yes
Mode 5	Open-Loop SU-MIMO (transmit diversity)	Conjugate data repetition	Non-adaptive	No	No	No

Table 10-22 Downlink MIMO Parameters

Mode Index	Number of Transmit Antennas (N_t)	Spatial Rate per Layer (R)	Number of Streams (N_s)	Number of Sub-carriers	Number of Layers (L)
MIMO Mode 0	2	1	2	2	1
	4	1	2	2	1
	8	1	2	2	1
MIMO Mode 1 and MIMO Mode 2	2	1	1	1	1
	2	2	2	1	1
	4	1	1	1	1
	4	2	2	1	1
	4	3	3	1	1
	4	4	4	1	1
	8	1	1	1	1
	8	2	2	1	1
	8	3	3	1	1
	8	4	4	1	1
	8	5	5	1	1
	8	6	6	1	1
	8	7	7	1	1
	8	8	8	1	1
MIMO Mode 3 and MIMO Mode 4	2	1	2	1	2
	4	1	2	1	2
	4	1	3	1	3
	4	1	4	1	4
	8	1	2	1	2
	8	1	3	1	3
	8	1	4	1	4
MIMO Mode 4	4	2 and 1[i]	3	1	2
	4	2 and 1[ii]	4	1	3
	4	2	4	1	2
	8	2 and 1[i]	3	1	2
	8	2 and 1[ii]	4	1	3
	8	2	4	1	2
MIMO Mode 5	2	1/2	1	2	1
	4	1/2	1	2	1
	7	1/2	1	2	1

[i]*Two streams to one MS and one stream to another MS, each with a single layer.*
[ii]*Two streams to one MS and one stream to each of other two mobile stations, each with a single layer.*

Table 10-23 Different Types of Open-loop Region and Associated Parameters [2]

Type of Open-loop MIMO Region	Number of Streams	MIMO Mode	Supported Permutation
Type 0	2	MIMO Mode 0 MIMO Mode 1 ($N_s = 2$)	DRU
Type 1	1	MIMO Mode 5 ($N_s = 1$)	Mini-band based CRU (diversity allocation) Sub-band based CRU (localized allocation)
Type 2	2	MIMO Mode 1 ($N_s = 2$) MIMO Mode 3 ($N_s = 2$)	Sub-band based CRU (localized allocation)

Table 10-24 Downlink MIMO Modes inside Open-loop MIMO Region [2]

Mode Index	DRU Permutation Support	Mini-Band based CRU Permutation Support (Diversity Allocation)	Sub-Band based CRU Permutation Support (Localized Allocation)
MIMO Mode 0	Yes ($N_s = 2$)	No	No
MIMO Mode 1	Yes ($N_s = 2$)	No	Yes ($N_s = 2$)
MIMO Mode 2	No	No	No
MIMO Mode 3	No	No	Yes ($N_s = 2$)
MIMO Mode 4	No	No	No
MIMO Mode 5	No	Yes ($N_s = 1$)	Yes ($N_s = 1$)

feedback on the quantized wideband correlation matrix independent of the MIMO feedback mode requested in the Feedback Polling A-MAP IE. The quantized wideband correlation matrix may be used for wideband beamforming. As shown in Table 10-25, MIMO feedback mode 0 is used for the open-loop single-user MIMO (SFBC) and spatial multiplexing. The MS measures wideband CQI for both SFBC and spatial multiplexing and reports the CQI and the spatial rate. It must be noted that spatial rate 1 indicates SFBC mode with precoding and spatial rate 2 denotes rank-2 spatial multiplexing with precoding. MIMO feedback mode 1 is associated with the CDR scheme with spatial rate 1/2 which is used with distributed logical resource units. MIMO feedback mode 2 is used with spatial multiplexing in localized permutation zones to facilitate frequency-selective scheduling, in which the spatial rate indicates the preferred number of MIMO streams. The sub-band CQI, in this case, corresponds to the selected rank. MIMO feedback mode 3 is used in conjunction with the closed-loop SU-MIMO mode and spatial multiplexing in localized permutation zones in order to achieve frequency selective scheduling. The spatial rate indicates the preferred number of MIMO streams and the sub-band CQI indicates the selected rank. MIMO feedback mode 4 is used with the closed-loop SU-MIMO and wideband beamforming. In this mode, the MS measures and sends the wideband CQI. The wideband CQI is estimated at the MS assuming short-term or long-term precoding at the BS according to the feedback period. The channel state information may be obtained at the BS based on the feedback of the correlation matrix or via wideband PMI. MIMO feedback mode 5 is used for an open-loop MU-MIMO

Table 10-25 MIMO Feedback Modes [2]

MIMO Feedback Mode	Type of Resource Unit	Feedback Components	MIMO Parameters outside Open-loop MIMO Region	MIMO Parameters inside Open-loop MIMO Region
Mode 0	Open-loop SU-MIMO SFBC/Spatial Multiplexing (DRU and mini-band LRU) Sounding-based closed-loop SU-MIMO and MU-MIMO	Spatial rate and wideband CQI	MIMO mode 0 and MIMO mode 1 and flexible adaptation between the two modes: $R = 1$ (SFBC) CQI $2 \leq R \leq 4$ (spatial multiplexing) CQI $N_s = 2$ (spatial multiplexing) when using DRU $N_s \geq 2$ (spatial multiplexing) when using mini-band LRU For sounding-based closed-loop SU-MIMO and MU-MIMO $(R = 1)$: SFBC CQI	MIMO mode 0 and MIMO mode 1 with flexible adaptation between the two modes: $R = 1$ (SFBC) CQI $R = 2$ (spatial multiplexing) CQI DRU
Mode 1	Open-loop SU-MIMO CDR (mini-band LRU)	Wideband CQI	N/A	MIMO mode 5 $R = 1/2$
Mode 2	Open-loop SU-MIMO spatial multiplexing (sub-band LRU)	Spatial rate, sub-band CQI, and sub-band selection	MIMO mode 1 $1 \leq R \leq 8$	MIMO mode 5 $R = 1/2$
Mode 3	Closed-loop SU-MIMO (sub-band LRU)	Spatial Rate, sub-band CQI, sub-band PMI, sub-band selection, and wideband correlation matrix	MIMO mode 2 $1 \leq R \leq 8$	N/A
Mode 4	Closed-loop SU-MIMO (mini-band LRU)	Spatial rate, wideband CQI, wideband PMI, and wideband correlation matrix	MIMO mode 2 $(N_s \leq 4)$	N/A
Mode 5	Open-loop MU-MIMO (sub-band LRU)	Sub-band CQI, sub-band selection, and MIMO stream indicator	MIMO mode 3	MIMO mode 3

Table 10-25 MIMO Feedback Modes [2] *Continued*

MIMO Feedback Mode	Type of Resource Unit	Feedback Components	MIMO Parameters outside Open-loop MIMO Region	MIMO Parameters inside Open-loop MIMO Region
Mode 6	Closed-loop MU MIMO (localized: sub-band LRU)	Sub-band CQI, sub-band PMI, sub-band selection, and wideband correlation matrix	MIMO mode 4	N/A
Mode 7	Closed-loop MU-MIMO (mini-band LRU)	Wideband CQI, wideband PMI, and wideband correlation matrix	MIMO mode 4	N/A

scheme in localized permutation zones to facilitate frequency selective scheduling. In this mode, the MS feeds back the sub-band selection, MIMO stream indicator, and the corresponding CQI. MIMO feedback mode 6 is used for closed-loop MU-MIMO in localized permutation zones to assist frequency selective scheduling. In this mode, the MS feedback includes the sub-band selection, corresponding CQI, and sub-band PMI. The sub-band CQI refers to the CQI of the best PMI in the sub-band. A rank-1 codebook or its subset is used to estimate the PMI in a sub-band. MIMO feedback mode 7 is used for closed-loop MU-MIMO in distributed permutation zones using wideband beam-forming. In this mode, the MS is required to feedback the wideband CQI. The wideband CQI is estimated at the MS assuming short-term or long-term precoding at the BS according to the feedback period. The channel state information may be obtained at the BS, the correlation matrix, or the wideband PMI that is sent by the MS. The downlink MIMO parameters are summarized in Table 10-26.

The serving BS provides common configuration parameters for downlink MIMO operation using broadcast messages. The broadcast information is carried through the S-SFH SP3 IE or in additional broadcast information such as the AAI_SCD or AAI_DL-IM MAC management messages. Individual downlink MIMO configuration parameters are transmitted through unicast messages. The unicast information is carried in the DL Basic Assignment A-MAP IE, DL Sub-band Assignment A-MAP IE, DL Persistent A-MAP IE, Feedback Polling A-MAP IE, and Feedback Allocation A-MAP IE.

There are three types of codebook feedback modes as follows: (1) base mode, where the PMI feedback from an MS indicates an entry of the base codebook that is used by the BS to determine a new precoder; (2) transformation mode, where the PMI feedback from an MS represents an entry in the transformed base codebook according to long-term channel information; and (3) differential mode where the PMI feedback from an MS refers to an entry in the differential codebook or an entry in the base codebook on PMI reset. In the latter case, the feedback from an MS provides differential knowledge of the short-term channel information. The mobile station is required to support both base and transformation modes, and it may optionally support a differential mode. The transformation and differential feedback modes are applied to the base codebook or to a subset of the base codebook.

Table 10-26 Downlink MIMO Parameters

Type of Information	Description of the Parameter	Value
Broadcast	Number of transmit antennas at the BS which is indicated in S-SFH SP3 IE	2, 4, 8
	Open-loop MIMO region, which is used to indicate the location of the open-loop MIMO region and the number of MIMO streams (1 or 2). This parameter is sent via an AAI_SCD MAC management message.	Refer to AAI_SCD message
	Rank-1 base codebook subset indication bitmap for interference mitigation with PMI coordination. Rank-1 codebook element restriction or recommendation information (This parameter is enabled by *Codebook_Mode* or *Codebook_Coordination* parameters).	8 bits if $N_t = 2$ 16 bits if $N_t = 4,8$
Unicast	MIMO encoder format which is indicated in DL basic assignment A-MAP IE, DL sub-band assignment A-MAP IE, or DL persistent A-MAP IE.	SFBC Vertical encoding Horizontal encoding
	Number of MIMO streams in transmission which is indicated in DL basic assignment A-MAP IE, DL sub-band assignment A-MAP IE, or DL persistent A-MAP IE.	1 to 8
	Index of allocated pilot stream which is indicated if MIMO encoder format is horizontal encoding. This parameter is signaled in DL basic assignment A-MAP IE, DL sub-band assignment A-MAP IE, or DL persistent A-MAP IE.	1 to 4
	MIMO feedback mode in order to decide the feedback content and the related MS measurements and reporting process. This parameter is indicated in feedback allocation A-MAP IE, or feedback polling A-MAP IE.	See Table 10-25
	Maximum number of spatial streams depending on the MIMO mode and the minimum number of transmit and receive antennas depending on MIMO feedback mode (this parameter is signaled through the feedback allocation A-MAP IE, or feedback polling A-MAP IE.	$N_s \leq \min(N_t,N_r)$
	Codebook subset type for closed-loop MIMO modes 2 and 4 Depending on the MIMO feedback mode and the value of *Codebook_Subset* parameter, the MS feeds back PMI from the SU-MIMO, MU-MIMO base codebook, or from a subset of the SU-MIMO or MU-MIMO base codebook (This parameter is conveyed via the feedback allocation A-MAP IE, or feedback polling A-MAP IE.	Base codebook or codebook subset
	Codebook feedback mode for closed-loop MIMO modes 2 and 4	Possible selections are as follows:

Table 10-26 Downlink MIMO Parameters *Continued*

Type of Information	Description of the Parameter	Value
	This parameter specifies the codebook feedback mode. If codebook coordination is enabled by setting *Codebook_Mode* parameter to 0b11 or by setting *Codebook_Coordination* parameter to 0b1, and if the MS reports spatial rate equal to 1, then the MS finds the rate-1 PMI from the codebook entries broadcasted in BC_SI filed (Rank-1 base codebook subset indication for interference mitigation with PMI coordination) in AAI_DL-IM MAC management message.	Base mode with codebook coordination disabled; Transformation mode with codebook coordination disabled; Differential mode with codebook coordination disabled; Base mode with codebook coordination enabled.

The base codebook is a unitary codebook. A codebook is defined as unitary if each of the constituent matrices is unitary. The MS selects its preferred matrix from the base codebook based on the channel measurements and sends the index of the preferred codeword. The BS calculates the precoder matrix $\mathbf{W}$ according to the index. Both BS and MS use the same codebook for proper operation. For the base mode, the PMI feedback from a mobile station refers to an entry in the base codebook where the base codebooks are defined for two, four, and eight transmit antennas at the BS. The notation $\mathbf{C}(N_t, N_s, N_B)$ denotes the codebook that consists of 2^{N_B} complex-valued $N_t \times N_s$ matrices and N_s denotes the number of streams. The notation $\mathbf{C}(N_t, N_s, N_B, i)$ denotes the ith codebook entry of $\mathbf{C}$ (N_t, N_s, N_B). The base codebook of SU-MIMO with two transmit antennas consists of rank-1 codebook $\mathbf{C}(2,1,3)$ and rank-2 codebook $\mathbf{C}(2,2,3)$. Example rank-1 and rank-2 base codebooks and their subset codebooks are shown in Table 10-27. The base codebook for MU-MIMO is the same as the rank-1 base codebook for SU-MIMO.

The base codebooks and their rank-1 subsets for SU-MIMO and MU-MIMO can be transformed as a function of the BS transmit correlation matrix. The $N_t \times N_t$ transmit correlation matrix $\mathbf{R} = E\{\mathbf{H}_{ij}^H \mathbf{H}_{ij}\}$ is quantized and fed back to the serving BS with longer period than PMI, where $\mathbf{H}_{ij}$ denotes the correlated channel matrix corresponding to the ith OFDM symbol and the jth sub-carrier. The transmit correlation matrix $\mathbf{R}$ contains the averaged directions for beamforming. The transform codebook improves the performance in highly correlated channels, with non-calibrated antenna arrays, and with cross-polarized antennas. For the transformation mode, the PMI feedback from a mobile station represents an entry of the transformed base codebook according to long-term channel information. In transformation mode, both the BS and MS transform the rank-1 base codebook to a rank-1 transformed codebook using the correlation matrix. The transformation for codewords of rank-1 codebook is defined as follows:

$$\mathbf{u}_i = \frac{\mathbf{R}\mathbf{v}_i}{\|\mathbf{R}\mathbf{v}_i\|} \tag{10-51}$$

Where $\mathbf{v}_i$ and $\mathbf{u}_i$ are the ith codeword of the base and the transformed codebook, respectively. Once the transformed codebook is calculated, both the MS and the serving BS use the transformed codebook for

Table 10-27 Example Base Code and Subset Codebooks [2]

Base Codebook	Codebook Index	m	C(2,1,3,m) = [c₁,c₂]	
			c_1	c_2
$\mathbf{C}(2,1,3)$	000	0	0.7071	−0.7071
	001	1	0.7071	−0.5000 −0.5000i
	010	2	0.7071	−0.7071i
	011	3	0.7071	0.5000 −0.5000i
	100	4	0.7071	0.7071
	101	5	0.7071	0.5000 +0.5000i
	110	6	0.7071	0.7071i
	111	7	0.7071	−0.5000 +0.5000i
$\mathbf{C}(2,2,3)$			$\mathbf{C}(2,2,3,m) = \begin{bmatrix} c_{11} & c_{12} \\ c_{21} & c_{22} \end{bmatrix}$	
			c_{11}	c_{12}
			c_{21}	c_{22}
	000	0	0.7071	−0.7071
			0.7071	0.7071
	001	1	0.7071	−0.5000 −0.5000i
			0.7071	0.5000 + 0.5000i
	010	2	0.7071	−0.7071i
			0.7071	0.7071i
	011	3	0.7071	0.5000 −0.5000i
			0.7071	−0.5000 + 0.5000i
	100~111	4~7	–	–

feedback and precoding of rank-1 closed-loop MIMO schemes. The codebooks with ranks greater than one are used without transformation when the MS is operating in transformation codebook-based feedback mode. The transmit correlation matrix $\mathbf{R}$ is fed back to support transformation mode of codebook-based precoding. $\mathbf{R}$ is fed back periodically and one correlation matrix is valid for the entire band. As an example, the correlation matrix for eight transmit antennas has the following format:

$$\mathbf{R} = \begin{bmatrix} r_{11} & r_{12} & r_{13} & r_{14} & r_{15} & r_{16} & r_{17} & r_{18} \\ r_{12}^* & r_{22} & r_{23} & r_{24} & r_{25} & r_{26} & r_{27} & r_{28} \\ r_{13}^* & r_{23}^* & r_{33} & r_{34} & r_{35} & r_{36} & r_{37} & r_{38} \\ r_{14}^* & r_{24}^* & r_{34}^* & r_{44} & r_{45} & r_{46} & r_{47} & r_{48} \\ r_{15}^* & r_{25}^* & r_{35}^* & r_{45}^* & r_{55} & r_{56} & r_{57} & r_{58} \\ r_{16}^* & r_{26}^* & r_{36}^* & r_{46}^* & r_{56}^* & r_{66} & r_{67} & r_{68} \\ r_{17}^* & r_{27}^* & r_{37}^* & r_{47}^* & r_{57}^* & r_{58}^* & r_{77} & r_{78} \\ r_{18}^* & r_{28}^* & r_{38}^* & r_{48}^* & r_{58}^* & r_{68}^* & r_{78}^* & r_{88} \end{bmatrix} \qquad (10\text{-}52)$$

where the "∗" denotes the conjugate operator. The diagonal entries of matrix **R** are positive and the non-diagonal entries are complex-valued. Due to the symmetry property of the correlation matrix, only the upper triangular elements need to be fed back following quantization. The **R** matrix is normalized by the maximum element amplitude, and then quantized to reduce the feedback overhead. The normalized diagonal elements are quantized by 1 bit and the normalized complex-valued elements are quantized by 4 bits. The total number of bits of feedback is 6 bits for two transmit antennas, 28 bits for four transmit antennas, and 120 bits for eight transmit antennas. The MS and the serving BS use the same transformation based on the correlation matrix fed back by the MS. Figure 10-39 illustrates the various distributions of the precoding vectors in multi-dimensional space.

The differential feedback is a different approach to the enhancement of feedback accuracy. The differential feedback exploits the correlation between precoding matrixes adjacent in time or frequencies. It only feeds back the difference between the current and the previous beamforming matrixes. If the channel variation between two feedbacks is small, quantization codewords are concentrated in a small region on the Grassmannian manifold and not uniformly distributed over the entire beamforming space. For low-mobility users, differential feedback may improve the feedback accuracy without increasing the number of feedback bits. The feedback starts initially and restarts periodically by sending a single feedback that fully describes the precoder. Differential feedbacks follow the start and restart feedbacks. The start and restart feedbacks employ a codebook that is self-contained; e.g., the base codebook. Figure 10-40 illustrates the operation of differential feedback across time.

An example differential feedback is transformation-based differential codebook. Let t, $D(t)$, and V (t) denote the feedback index, the corresponding feedback matrix, and the corresponding precoder, respectively. The sequential index is reset to 0 at $T_{max} + 1$. The index for the start and restart feedbacks are 0. Let U be a vector or a matrix and $\mathbf{Q_U}$ be a rotation matrix determined by U. The precoders of the

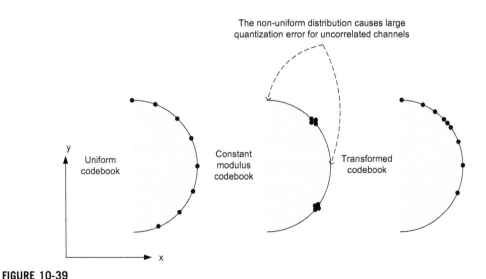

FIGURE 10-39

Distribution of the precoding vector coordinates in various transformation methods

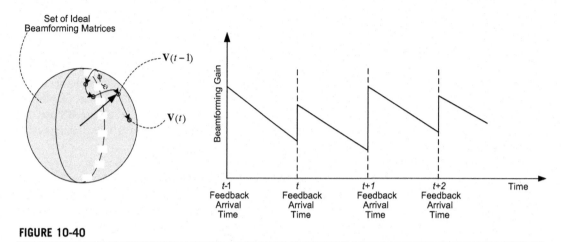

FIGURE 10-40

An example illustration of differential feedback and effect on beamforming gain

subsequent differential feedbacks are $\mathbf{D}(t)$ for $t = 1,2,..., T_{\max}$ and the corresponding precoders are given as follows:

$$\begin{aligned} \mathbf{V}(t) &= \mathbf{Q}_{\mathbf{V}(t-1)}\mathbf{D}(t) \\ \mathbf{D}(t) &= \mathbf{Q}^H_{\mathbf{V}(t-1)}\mathbf{V}(t) \end{aligned} \qquad t = 0, 1, 2, ..., T_{\max} \qquad (10\text{-}53)$$

where the rotation matrix $\mathbf{Q}_{\mathbf{v}(t-1)}$ is a unitary $N_t \times N_t$ matrix derived from the previous precoder matrix $\mathbf{v}$ $(t-1)$ and N_t denotes the number of transmit antennas. The dimension of the feedback matrix $\mathbf{D}(t)$ is $N_t \times N_s$, where N_s denotes the number of spatial streams. The feedback matrix $\mathbf{D}(t)$ can be viewed as a description of time/frequency correlation properties which does not change over the entire frequency band and long periods of time, and $\mathbf{Q}_{\mathbf{v}(t-1)}$ can be viewed as a description of narrowband and short-term channel properties which correspond to a specific sub-band and subframe. Due to correlation between beamforming matrices across time and frequency, feedback overhead can be reduced if the correlation property can be efficiently exploited. Let's take time-domain correlation as an example to explain the differential feedback scheme. The operations at the MS and at the BS can be described as follows:

1. Derivation of the differential codeword matrix at the MS, $\mathbf{D} = \mathbf{Q}^H_{\mathbf{V}_{t-1}} \mathbf{V}(t) = [\widehat{\mathbf{V}}(t-1), \quad \widehat{\mathbf{V}}^\perp(t-1)]^H \mathbf{V}(t)$
2. Quantization of the differential codeword matrix at the MS, $\widehat{\mathbf{D}} = \arg\max_{\mathbf{D}_i \in \mathbf{\Omega}} \|\mathbf{D}^H \mathbf{D}_i\|_F$ where $\|\mathbf{D}^H \mathbf{D}_i\|_F$ denotes the chordal distance[v] between the two codeword matrices $\mathbf{D}$ and $\mathbf{D}_i$. Therefore, the optimal codebook is selected based on the maximum between $\mathbf{D}$ and the codeword matrices contained in the codebook $\mathbf{\Omega}$.
3. Beamforming matrix reconstruction at the BS, $\widehat{\mathbf{V}}(t) = \widehat{\mathbf{Q}}_{\mathbf{V}(t-1)}\widehat{\mathbf{D}} = [\widehat{\mathbf{V}}(t-1) \quad, \widehat{\mathbf{V}}^\perp(t-1)]\widehat{\mathbf{D}}$
4. Beamforming at the BS, $\mathbf{y} = \mathbf{H}\,\widehat{\mathbf{V}}(t)\,\mathbf{s} + \mathbf{n}$

---

[v]The asymptotic performance of a coding scheme is dominated by the shortest distance between any pair of codewords. The relevant distance measure between two codewords, $\mathbf{X}_1$ and $\mathbf{X}_2$, of an orthogonal code for a non-coherent MIMO system is the chordal distance $d^2(\mathbf{X}_1, \mathbf{X}_2) = M - \|\mathbf{X}_1\mathbf{X}_2^H\|_F^2$. [62]

In the above equations, $N_t \times N_s$ matrices $\mathbf{V}(t)$ and $\widehat{\mathbf{V}}(t)$ denote the ideal and quantized beamforming matrices at time instant t, respectively, the columns of $\widehat{\mathbf{V}}^{\perp}(t-1)$ have unit norm and are orthogonal to $\widehat{\mathbf{V}}(t-1)$ where $\widehat{\mathbf{V}}^{\perp}(t-1)$ can be calculated from $\widehat{\mathbf{V}}(t-1)$ by a Householder transformation,[vi] $\mathbf{D}_i$ is the codeword matrix of the differential codebook denoted by $\mathbf{\Omega}$. The quantization criterion maximizes the received signal power. Other criteria such as channel capacity or mean squared error can also be used. If $N_s = 1$, $\mathbf{V}(t-1)$ would be a vector (with unit norm) and the rotation matrix is given by $\mathbf{Q}_{V(t-1)} = I - 2\phi\phi^H$, $\|\phi\| > 0$ wherein $\phi = \exp(-j\theta) V(t-1) - \varepsilon$ and θ is the phase of the first entry of $V(t-1)$ and $\varepsilon = [1,0,0,\ldots,0]^t$.

The advantage of the transformed codebook is demonstrated in the example given in Figure 10-39, where a 2×1 non-complex channel is assumed. The ideal beamforming vector is uniformly distributed over a semicircle for uncorrelated channels. The uniform codeword distribution of the IEEE 802.16-2009 standard base codebook adheres to this input distribution and thereby has a good performance. In contrast, the DFT codebook suffers from the constant modulus constraint that requires all vector entries to have the same magnitude. Although more than two codewords exist, the DFT codebook only has two valid codewords, in this case as shown by the two clusters at $45°$ and $-45°$ on the semicircle. This leaves large uncovered areas in the quantization space and causes large quantization errors for input vectors around $[1,0]^t$ and $[0,1]^t$. On the other hand, for highly correlated channels, the entry magnitudes of the input beamforming vector are close, due to the high correlation of the channel responses. The DFT codewords with constant modulus entries match to the input distribution that has close entry magnitudes and thus has a good performance. Note that there are two clusters of the DFT codewords and only one of them is used for each correlated scenario. Since the other cluster is too far away from the input vector, it is not used. This is the drawback of the DFT codebook because it only exploits rough information about the magnitude similarity. In contrast, the transformed codebook makes use of both the magnitude and phase information about the antenna correlation. The uniform codewords of the IEEE 802.16-2009 standard base codebook are dynamically transformed to only one cluster pointing to one direction, as shown on the right side of Figure 10-39. As the correlation diminishes, the codeword concentration of the transformed codebook decreases and the codebook becomes similar to the legacy base codebook for uncorrelated channels.

The link-level performance of some closed-loop SU-MIMO rate-1 codebooks with high-correlation (i.e., closely-spaced antennas and small angle spread) and low-correlation have been compared and are shown in Figures 10-41 and 10-42. The channel model is the modified ITU PedB 3 km/h. There are four transmit-antennas at the BS and two receive antennas at the MS. The antenna spacing of BS transmit antennas is assumed to be 4 λ (wavelength) and 0.5 λ for low and high correlation cases, respectively. One spatial stream is transmitted. The information is coded with $R = 0.5$ and modulated with 16 QAM. Three types of codebooks are compared in the figures; i.e. the conventional IEEE 802.16-2009 standard base codebook, the DFT codebook, and the transformed codebook. It is observed that under the above assumptions, the transformed codebook has better performance relative

---

[vi] In linear algebra, Householder transformation is defined as a linear transformation that describes a reflection about a plane or hyper-plane containing the origin. Householder transformations are widely used in linear algebra to perform QR decompositions. This is a linear transformation given by the Householder matrix $\mathbf{Q} = I - 2\mathbf{uu}^H$ where $\mathbf{I}$ denotes the identity matrix and u is given as a column unit vector. The Householder matrix has the following properties: it is hermitian, i.e., $\mathbf{Q} = \mathbf{Q}^H$; and it is unitary, i.e., $\mathbf{Q}^{-1} = \mathbf{Q}^H$

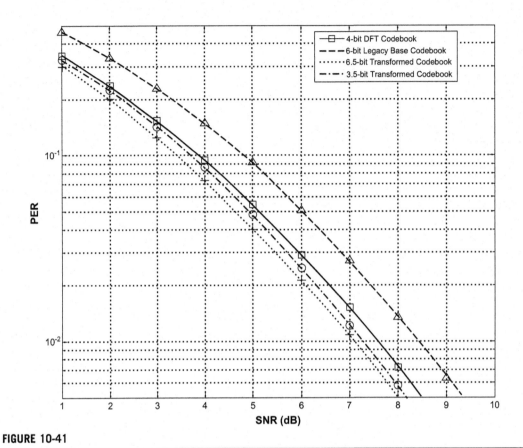

FIGURE 10-41

A comparison of the performance of closed-loop SU-MIMO rate-1 codebooks in highly-correlated channels (0.5 λ and angle spread 3°)

to the legacy base codebook and the DFT codebook (see Figure 10-43 for a comparison of the performance of transformed, DFT, and legacy base codebooks).

The user only knows the instantly average sub-band interference power for CQI calculation, and an identity matrix for the interference cell is used for average power calculation. After scheduling of all cells, the real interference precoding is used for detection because wrap-around simulation is used. All of these three cases are considered to have a CQI feedback delay of 5 ms.

To assist in the calculation of the precoding matrix for SU-MIMO or MU-MIMO, the BS may request the MS to transmit a sounding signal in a UL sounding channel. The BS may map the measured UL channel response to an estimated DL channel response. The transmitter and receiver hardware of the BS may be calibrated to assist the channel response mapping. The precoding matrix in this case is the same for all sub-carriers within a physical resource unit.

A SU-MIMO open-loop technique that provides diversity gain is used for the broadcast control channel. The 2-stream SFBC with two transmit antennas is used for P-SFH and S-SFH transmission.

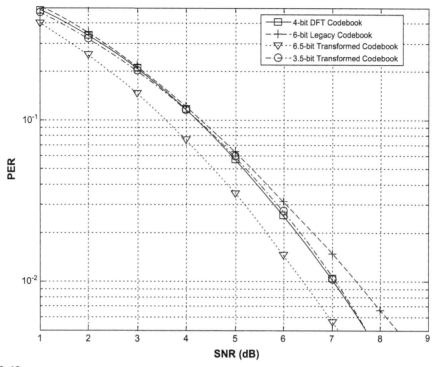

FIGURE 10-42

A comparison of the performance of closed-loop SU-MIMO rate-1 codebooks in low-correlated channels (4 λ and angle spread 3°)

For a more than two transmit-antenna configuration, P-SFH and S-SFH are transmitted by a 2-stream SFBC with precoding, which is decoded by the MS without any information on the precoding and antenna configuration. The 2-stream SFBC is used for the Downlink Unicast Control Channel.

In summary, the IEEE 802.16m supports the following MIMO schemes in the downlink:

- Single-BS and multi-BS MIMO operation;
- Single-user MIMO and multi-user MIMO schemes:
  - Vertical encoding for SU-MIMO;
  - Horizontal encoding for MU-MIMO;
- Adaptive-precoding (closed-loop) and non-adaptive (open-loop) precoding;
- Codebook and sounding-based precoding:
  - Short and long-term adaptive precoding;
  - Dedicated (precoded) pilot sub-carriers for MIMO operation;
- Enhanced codebook design;
  - Enhanced base codebook;
  - Transform codebook;
  - Differential codebook.

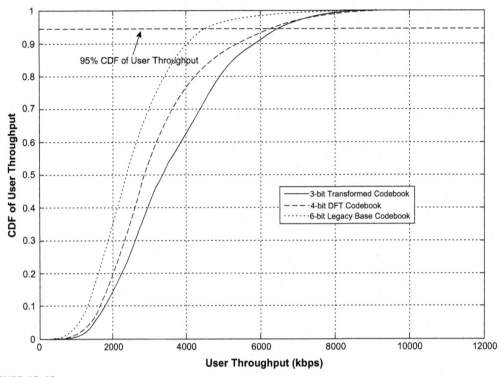

FIGURE 10-43

A comparison of CDF of user throughput for highly-correlated channel (MU-MIMO with zero-forcing precoding, four transmit antennas at BS and two receive antennas at MS)

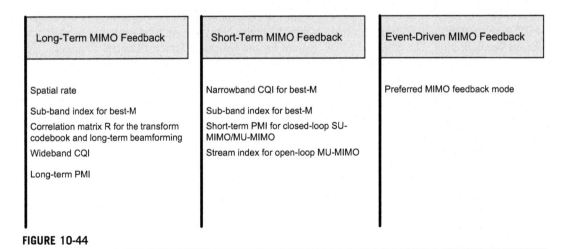

FIGURE 10-44

Summary of IEEE 802.16m downlink MIMO feedback

The feedback schemes to support various MIMO modes are classified into three classes: (1) long-term; (2) short-term; and (3) event-driven. The elements of MIMO feedback in each category are summarized in Figure 10-44.

Since multicast and broadcast services are typically a downlink-only type of transmission, only open-loop transmit diversity or spatial multiplexing schemes are used for E-MBS, and no closed-loop MIMO scheme is supported in E-MBS.

10.6 THE IEEE 802.16M UPLINK MIMO SCHEMES

The physical layer procedures in the uplink MIMO on the transmitter side are shown in Figure 10-45.

The MIMO encoder block maps a single $L = 1$ layer into $N_s \geq L$ streams, which are fed to the precoder block. A layer is defined as a coding and modulation processing path that is the input to the MIMO encoder. A stream is defined as the output of the MIMO encoder which is the input to the precoder block. For SU-MIMO and collaborative spatial multiplexing, there is only one FEC block per allocated resource unit, i.e., vertical MIMO encoding is only supported in the uplink at the transmitter side. The precoder block maps the streams to antennas by generating the antenna-specific data symbols according to the selected MIMO mode. The MIMO encoder and precoder blocks do not exist when the MS has only one transmit antenna. The decisions corresponding to resource allocations in the uplink take into consideration the allocation type (use of distributed or localized allocation), single-user/ multi-user operation mode, MIMO mode (use of open-loop or closed-loop scheme), user grouping (which users are assigned to the same resource unit in MU-MIMO), rank (for the spatial multiplexing modes in SU-MIMO modes is the number of streams transmitted to the user), modulation and coding rate to be used on each layer, power boosting (power boosting values to be used on the data and pilot sub-carriers), and band selection (location of the localized resource units across frequency). The uplink

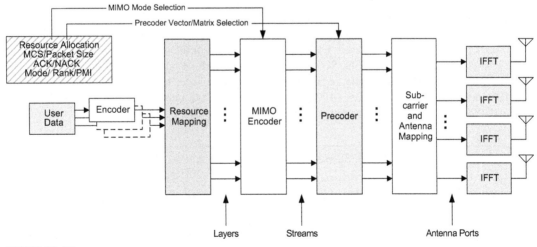

FIGURE 10-45

Uplink MIMO architecture [2]

antenna configuration is denoted by $N_t \times N_r$ where $N_t = 1,2,4$ denotes the number of MS transmit antennas and $N_r \geq 2$ denotes the number of BS receive antennas. There are two MIMO encoder formats in the uplink: (1) space-frequency block-coding; and (2) collaborative spatial multiplexing (vertical encoding). Uplink SU-MIMO physical processing is similar to that in the downlink. Uplink MU-MIMO is performed through collaborative spatial multiplexing with vertical encoding at each MS.

The number of spatial streams in the uplink depends on the MIMO encoder. For open-loop and closed-loop spatial multiplexing SU-MIMO, the number of streams is $N_s \leq \min(N_t, N_r)$, $N_s \leq 4$. For open-loop transmit diversity, N_s depends on the space-time coding scheme used by the MIMO encoder and the number of transmit antennas at the MS. The MU-MIMO schemes can support up to four spatial streams in the uplink. The number of spatial streams allocated to one user is limited to one, if there is only one transmit antenna at the MS. In that case ($N_t = 1$), SFBC encoding is not used at the MS. Vertical encoding with one spatial stream is allocated to an MS with one transmit antenna. Non-adaptive and adaptive precoding are supported on the uplink where non-adaptive precoding is used with open-loop SU-MIMO and open-loop MU-MIMO modes, and adaptive precoding is used with closed-loop SU-MIMO and closed-loop MU-MIMO modes.

For non-adaptive precoding at sub-carrier k, the $N_t \times N_s$ precoding matrix $\mathbf{W}_k$ is selected from a subset of size N_w precoders derived from the base codebook for a given rank and changes every uP_{SC} sub-carriers and every v subframes in order to provide additional spatial diversity. The values of u and v depend on the MIMO scheme and type of resource unit [2]. The values of N_w for non-adaptive pre-coding and different subchannelization schemes are given in Table 10-28. No precoding function is utilized at an MS with one transmit antenna.

For adaptive precoding, the precoding matrix $\mathbf{W}$ is derived at the BS or at the MS per instructions by the BS. Unitary codebook-based adaptive precoding is supported with two or four transmit antennas at the MS in FDD and TDD modes. In this mode, the MS transmits a sounding signal on the uplink to assist the precoder selection at the BS. Based on the measurements of the sounding channel at the BS, the BS indicates the uplink precoding matrix index to be used by the MS in a downlink control message. Adaptive precoding based on the measurement of downlink reference signals can be performed with two or four transmit antennas at the MS in TDD systems. The MS chooses the precoder based on the downlink measurements. The form and derivation of the precoding matrix does not need to be known at the BS. In the uplink SU-MIMO and MU-MIMO modes, all demodulation pilots are precoded in the same way as the data, regardless of the number of transmit antennas, allocation type, or MIMO mode. There are five uplink MIMO modes for traffic and control channels that are listed in Table 10-29. The uplink MIMO parameters are given in Table 10-30 and Table 10-32.

Table 10-28 Size of the Uplink Subset Codebooks (Nw)

Uplink Subchannelization Type	Type of Precoding	Rank	1	2	3	4
Uplink distributed LRU and mini-band LRU	Non-adaptive	$N_t = 2$	2	1	N/A	N/A
		$N_t = 4$	4	4	4	4
Uplink sub-band LRU		$N_t = 2$	8	4	N/A	N/A
		$N_t = 4$	64	64	64	64

Table 10-29 Uplink MIMO Modes [2]

Mode Index	Description	MIMO Encoding Format	MIMO Precoding
Mode 0	Open-loop SU-MIMO	SFBC	Non-adaptive
Mode 1	Open-loop SU-MIMO (spatial multiplexing)	Vertical encoding	Non-adaptive
Mode 2	Closed-loop SU-MIMO (spatial multiplexing)	Vertical encoding	Adaptive
Mode 3	Open-loop collaborative spatial multiplexing (MU-MIMO)	Vertical encoding	Non-adaptive
Mode 4	Closed-loop collaborative spatial multiplexing (MU-MIMO)	Vertical encoding	Adaptive

Table 10-30 Uplink MIMO Parameters [2]

	Number of Transmit Antennas (N_t)	Spatial Rate per Layer (R)	Number of Streams (N_s)	Number of Sub-carriers	Number of Layers (L)
MIMO Mode 0	2	1	2	2	1
	4	1	2	2	1
MIMO Mode 1 and MIMO Mode 2	2	1	1	1	1
	2	2	2	1	1
	4	1	1	1	1
	4	2	2	1	1
	4	3	3	1	1
	4	4	4	1	1
MIMO Mode 3 and MIMO Mode 4	2	1	1	1	1
	4	1	1	1	1
	4	2	2	1	1
	4	3	3	1	1

It must be noted that N_s denotes the number of streams that are transmitted by a single MS. In MIMO modes 3 and 4, N_t refers to the number of transmit antennas at each participating mobile station in collaborative spatial multiplexing. All MIMO modes are supported in either distributed or localized subchannelization schemes. Table 10-31 shows the permutations supported by each MIMO mode.

The successive symbols at the output of the precoder block are mapped to the corresponding antenna ports. The sub-carrier mapping within a logical resource unit starts from the data sub-carrier with the lowest OFDM symbol index and lowest sub-carrier index, and continues with increasing sub-carrier index. If the edge of the resource unit is reached, the mapping is continued on the next OFDM symbol. The serving BS provides the necessary parameters for uplink MIMO operation in a unicast message. The parameters may be transmitted depending on the type of operation, and are carried in the UL Basic Assignment A-MAP IE, UL Sub-band Assignment A-MAP IE, or UL Persistent Allocation A-MAP IE.

The base codebook is a unitary codebook. A codebook is defined as unitary, if each of the constituent matrices is unitary. The notation $\mathbf{C}(N_t, N_s, N_B)$ denotes the codebook that consists of 2^{N_B}

Table 10-31 Supported Subchannelization Schemes by Each Uplink MIMO Mode [2]

	Tile-based DRU	Mini-band CRU (Diversity Allocation)	Mini-band CRU Sub-band CRU (Localized Allocation)
MIMO mode 0	Yes	Yes	No
MIMO mode 1	Yes ($N_s \leq 2$)	Yes	Yes
MIMO mode 2	Yes ($N_s \leq 2$)	Yes	Yes
MIMO mode 3	Yes ($N_s = 1$)	Yes	Yes
MIMO mode 4	Yes ($N_s = 1$)	Yes	Yes

Table 10-32 Uplink MIMO Configuration/Control Parameters [2]

Parameter	Description	Value
MIMO Encoding Format (MEF)	MIMO encoding format.	SFBC Vertical encoding
Collaborative Spatial Multiplexing (CSM)	SU-MIMO is used if CSM is disabled; MU-MIMO is used if CSM is enabled.	Disabled or enabled
Number of MIMO Streams	Number of MIMO streams in the uplink transmission.	1 to 4
Total Number of Spatial Streams per LRU (TNS)	This parameter is enabled when CSM is enabled. This parameter is an indication of the total number of spatial streams per resource unit.	1 to 4
First Pilot Index (SI)	Enabled when CSM is enabled; 1 bit for two transmit antennas and 2 bits for four transmit antennas.	1 to 4
Precoding Flag (PF)	It does not apply when MS has only one transmit antenna.	Non-adaptive precoding or adaptive codebook precoding.
PMI Indicator	This field is relevant only when PF indicates adaptive codebook precoding. When PMI indication is set, the BS can select the precoder matrix to be used at the MS.	The MS uses the precoder of rank N_s of its choice if PMI indicator is zero; otherwise the indicated PMI of rank N_s will used by the MS for precoding.
Precoding Matrix Index (PMI) in the Uplink Base Codebook	This parameter is enabled when PF indicates use of the adaptive codebook precoding, and PMI indication is set.	0 to 9 when $N_t = 2$ 0 to 63 when $N_t = 4$

complex-valued $N_t \times N_s$ matrices and N_s denotes the number of streams. The notation $\mathbf{C}(N_t, N_s, N_B, i)$ denotes the ith codebook entry of $\mathbf{C}(N_t, N_s, N_B)$. The base codebooks of SU-MIMO with two transmit antennas consist of rank-1 codebook $\mathbf{C}(2,1,4)$ and rank-2 codebook $\mathbf{C}(2,2,3)$. The rank-2 base codebook $\mathbf{C}(2,2,3)$ for two transmit antennas in the uplink is the same as the downlink two transmit antennas rank-2 base codebook. The base codebook for uplink collaborative spatial multiplexing is identical to the base codebook for SU-MIMO.

Table 10-33 C(2,1,4) Base Codebook Content [2]

| Binary Index | m | C(2,1,4,m) = $[c_1, c_2]$ | |
		c_1	c_2
0000	0	0.7071	−0.7071
0001	1	0.7071	−0.5000 −0.5000i
0010	2	0.7071	−0.7071i
0011	3	0.7071	0.5000 −0.5000i
0100	4	0.7071	0.7071
0101	5	0.7071	0.5000 + 0.5000i
0110	6	0.7071	0.7071i
0111	7	0.7071	−0.5000 + 0.5000ι
1000	8	1	0
1001	9	0	1
1010–1111	10–15	–	–

The uplink base codebook for SU-MIMO with four transmit-antennas consists of rank-1 codebook **C**(4,1,6), rank-2 codebook **C**(4,2,6), rank-3 codebook **C**(4,3,6), and rank-4 codebook **C**(4,4,6). The rank-1 codebook entry **C**(4,1,6,*m*) consists of the first column of **C**(4,4,6,*m*), rank-2 codebook entry **C**(4,2,6,*m*) consists of the first two columns of **C**(4,4,6,*m*), and rank-3 codebook entry **C**(4,3,6,*m*) consists of the first three columns of **C**(4,4,6,*m*) base codebook. As an example, Table 10-33 provides the content of base codebook **C**(2,1,4).

The uplink open-loop MU-MIMO codebook subset is same as the uplink open-loop SU-MIMO codebook subset. The base codebook for uplink collaborative spatial multiplexing is a 4-bit subset of the base codebook for uplink SU-MIMO. The uplink MIMO features of the IEEE 802.16m can be summarized as follows:

- Support of single-user MIMO and collaborative spatial multiplexing;
- Vertical encoding for SU-MIMO and collaborative spatial multiplexing;
- Open-loop and closed-loop MIMO operation;
- Codebook-based and vendor-specific precoding:
  - Short-term and long-term precoding;
  - MS-specific precoded (dedicated) pilots for MIMO operation;
- Enhanced codebook design:
  - Enhanced base codebook for both correlated and uncorrelated channels;
  - Antenna selection codewords to reduce MS power consumption.

10.7 MULTI-BS MIMO

Multi-BS MIMO techniques are used for improving sector throughput and cell-edge throughput through multi-cell collaborative precoding, network coordinated beamforming, or inter-cell interference nulling. Both open-loop and closed-loop multi-BS MIMO techniques are supported in the IEEE

802.16m standard. For closed-loop multi-BS MIMO, CSI feedback via codebook-based feedback or sounding channel is used. The feedback information may be shared by neighboring base stations via a core network interface. Mode adaptation between single-BS MIMO and multi-BS MIMO can also be utilized.

The multi-BS MIMO principle can be described as follows. When a mobile station is at the cell edge, it may be able to receive signals from multiple cell sites, and the mobile station's transmission may also be received at multiple cell sites (see Figure 10-46). If the data transmission and signaling from multiple cell sites can be coordinated, the downlink performance can be significantly improved. This coordination can be similar to the interference avoidance techniques or the case where the same data is transmitted from multiple cell sites. In the uplink, since the signal can be received by multiple cell sites, the system can take advantage of coordinated multiple reception to significantly improve the link performance. The Coordinated MultiPoint (CoMP) transmission used in both the 3GPP LTE-Advanced and the IEEE 802.16m is a method of MIMO transmission for interference reduction which enables features such as network synchronization, cell- and user-specific pilots, feedback of multi-cell channel state information, and synchronous data exchange between the base stations that can be used for interference mitigation and to achieve macro diversity gain. The collaborative MIMO and the

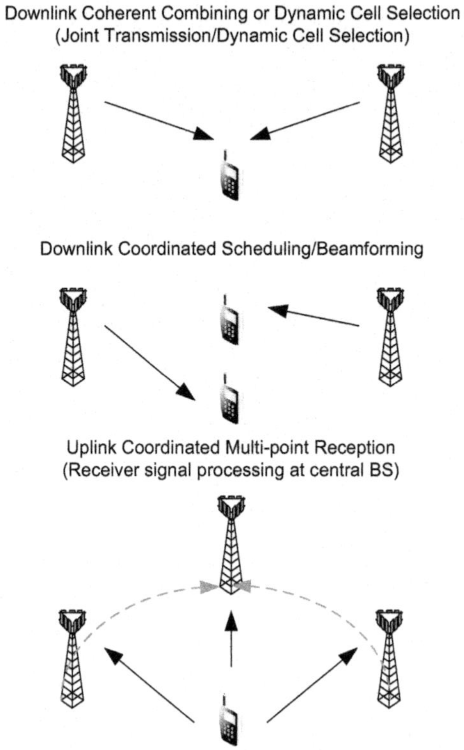

Downlink Coherent Combining or Dynamic Cell Selection
(Joint Transmission/Dynamic Cell Selection)

Downlink Coordinated Scheduling/Beamforming

Uplink Coordinated Multi-point Reception
(Receiver signal processing at central BS)

FIGURE 10-46

Downlink multi-BS transmission schemes

closed-loop macro diversity schemes are optional features supported in the IEEE 802.16m standard. For downlink collaborative MIMO, multiple base stations perform joint MIMO transmissions to multiple mobile stations located in different cells. Each BS performs multi-user precoding when transmitting to multiple mobile stations and each MS benefits from collaborative MIMO by receiving multiple streams from multiple base stations. When collaborative MIMO is enabled, several mobile stations are jointly served by multiple coordinated base stations through MU-MIMO scheduling and precoding, whereas when closed-loop macro-diversity is enabled, a single MS is served jointly by multiple coordinated base stations.

In terms of downlink CoMP, two different approaches are under consideration in 3GPP LTE-Advanced: coordinated scheduling and/or beamforming; and joint processing/transmission. In the first category, the downlink transmission to a single UE is sent from the serving cell, exactly as in the case of non-CoMP transmissions. However, the scheduling, including any beamforming functionality, is dynamically coordinated between the cells in order to control/reduce the interference between different transmissions. In principle, the best serving set of users will be selected so that the transmitter beams are constructed to reduce the interference to other neighboring users, while increasing the served users' signal strength. For joint processing/transmission, the transmission to a single UE is simultaneously transmitted from multiple transmission points, in practice cell sites. The multipoint transmissions will be coordinated as a single transmitter with antennas that are geographically separated. This scheme has the potential for higher performance, compared to coordination in scheduling only, but at the expense of more stringent requirement on backhaul communications [101]. Network and collaborative MIMO have been studied for the 4th generation of cellular systems. Their application depends on the geographical separation of the antennas, coordinated multipoint processing method, and the coordinated zone definition. Depending on whether the same data to a UE is shared at different cell sites, collaborative MIMO includes single-cell antenna processing with multi-cell coordination, or multi-cell antenna processing. The former technique can be implemented via precoding with interference nulling by exploiting the additional degrees of spatial freedom at a cell site. The latter technique includes collaborative precoding and closed-loop macro diversity. In collaborative precoding, each cell site performs multi-user precoding towards multiple UEs, and each UE receives multiple streams from multiple cell sites. In closed-loop macro diversity, each cell site performs precoding independently, and multiple cell sites jointly serve the same UE.

As shown in Figure 10-46, uplink coordinated multipoint reception implies reception of the transmitted signal at multiple geographically separated points. Scheduling decisions can be coordinated among cells to control interference. It should be noted that in different instances, the cooperating units can be separate base station's remote radio units, relays, etc. Moreover, since uplink CoMP mainly impacts the scheduler and receiver, it is primarily an implementation issue.

As shown in Figure 10-47, the single-BS precoding with multi-BS coordination is mainly an interference mitigation and coordination mechanism. These interference mitigation techniques are applicable to downlink MIMO modes 2 and 4 with codebook-based feedback mode with additional inter-BS coordination mechanisms and interference measurement support. The inter-BS coordination mechanisms do not require data forwarding between different base stations. Two types of single-BS precoding techniques with Multi-BS coordination are supported in IEEE 802.16m: (1) PMI coordination, supported by codebook-based feedback; and (2) interference nulling, supported by codebook-based feedback or uplink sounding. Single-BS precoding with multi-BS coordination may be enabled

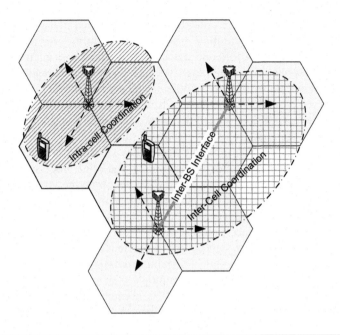

FIGURE 10-47

Illustration of inter-cell and intra-cell coordinated transmission

by the BS for one or several mobile stations when closed-loop MIMO precoding with downlink MIMO modes 2 or 4 are used in the serving and neighboring cells.

The inter-cell interference can be mitigated by coordinating the precoders applied in the neighboring cells using higher layer signaling and based on feedback from mobile stations to their respective serving base stations. The PMI coordination with codebook-based feedback can be used in the form of PMI recommendation or PMI restriction, depending on the instructions by the BS through the Feedback Polling A-MAP IE. If the *Interference Coordination Type* parameter is set to 0b00 in the Feedback Polling A-MAP IE, then the MS finds the PMI which demonstrates the strongest interference for the neighboring cell in the frequency resource unit indicated by the *Target Resource Unit* field signaled in Feedback Polling A-MAP IE; otherwise, the MS finds the PMI which demonstrates the weakest interference for the neighboring cell in the frequency resource unit. Restricting or recommending the usage of rank-1 codebook elements as a response to the neighboring cell's request is performed by the BS transmission of BC_SI field (Base Codebook Subset Indication is the rank-1 base codebook subset indication for interference mitigation with PMI coordination) in the AAI_DL_IM MAC control message. The procedure for a PMI coordination multi-BS MIMO scheme can be described as follows:

1. If the Feedback Polling A-MAP IE is received by an MS, it should periodically send the AAI_MultiBS_MIMO_FBK MAC control message containing the requested information (e.g., PMI or a set of PMIs, diversity set member identifier, or additional measurements) by the serving BS.

2. The BS should communicate with the neighboring base stations on receiving the feedback from multiple mobile stations, to coordinate their PMIs using higher layer signaling. The BS should then broadcast codebook subset information in BC_SI in an AAI_DL_IM MAC control message to all mobile stations within its coverage area.

3. The BS may send a Feedback Allocation A-MAP IE with the corresponding *Codebook Mode* field set to 0b11 to the selected mobile stations. Those mobile stations that received the message should feed back their desired PMI from the codebook subset that is signaled through the BC_SI field.

Inter-cell interference nulling can be performed using PMI which may be based on either the strongest interference from the neighboring cells or by detecting the neighboring cell's sounding signal. The interference mitigation techniques described in this section are based on joint MIMO transmission across multiple base stations. The base station and the mobile station may support one or both adaptive-precoding-based multi-BS joint processing techniques, e.g., Closed-loop Macro Diversity (CL-MD) or Collaborative MIMO (Co-MIMO)[vii] transmission. The CL-MD scheme is used with downlink MIMO mode 2, whereas Co-MIMO is used in conjunction with downlink MIMO mode 4. The multi-BS joint MIMO processing may be enabled by the BS for one or multiple mobile stations when adaptive precoding is utilized in the serving and the neighboring cells, and the user data is shared among multiple base stations. Using adaptive precoding, the precoding matrix $\mathbf{W}$ is derived from the feedback received from the mobile stations, with codebook-based feedback or sounding-based feedback.

In the CL-MD scheme, a single MS is served jointly by multiple coordinated base stations, whereas in the Co-MIMO scheme several mobile stations are served jointly by the multiple coordinated base stations through MU-MIMO joint scheduling and precoding. For codebook-based feedback, the mobile stations choose the PMI for the serving cell and the neighboring cells based on the channel state information. The serving BS may instruct the mobile stations to feed back 3-bit uniformly-quantized phase information for each neighboring cell, such that the BS can form a concatenated PMI based on the phase information for the neighboring cells. When a downlink multi-BS joint processing is utilized, radio resource allocation, data mapping, and pilot pattern allocation are coordinated among the participating base stations. The same data packet is transmitted by the coordinated base stations over the same radio resources. The same pilot patterns without interlacing are applied to the coordinated base stations. The operation of downlink multi-BS joint processing can be described as follows:

1. If an MS receives a Feedback Polling A-MAP IE, it will periodically send an AAI_MultiBS_MIMO_FBK MAC control message containing the codebook feedback or will send an uplink sounding per the UL Sounding Command A-MAP IE, depending on the designated mode of operation.

2. The BS will forward the PMI(s) and coordinated PMI(s) to the neighboring base stations on receiving feedback from the mobile stations, to coordinate the utilization of the PMI(s). In the case of sounding feedback, each participating base station can perform precoding based on the received sounding signal(s) from a single MS for the CL-MD mode or from multiple mobile stations for the Co-MIMO scheme. The default number of neighboring base stations to be coordinated in order to support collaborative MIMO is three.

[vii]The downlink collaborative MIMO scheme must be differentiated from the uplink collaborative spatial multiplexing scheme that has been inherited from the legacy systems.

A collaborative MIMO zone is defined in order to facilitate inter-BS coordination when supporting Co-MIMO transmission. The Co-MIMO zone is a time-frequency region comprising a number of LRUs over a number of subframes. The Co-MIMO zone utilized by the neighboring base stations is associated to the same radio resource region. The permutation scheme used in the Co-MIMO zone is identical across the participating base stations. On the other hand, for uplink sounding-based CL-MD and Co-MIMO schemes, the same sounding sequence can be used for DL/UL channel calibration. Therefore, the mobile stations will send the calibration sounding sequence on receiving the AAI_MULTI_BS_SOUNDING-CAL MAC management message. The phases of the received uplink calibration sounding sequences are used by the BS to calculate the mismatch between the uplink and downlink channels. Each participating BS compensates the downlink and uplink phase mismatch based the above-mentioned method.

The Co-MIMO transmission is managed by the core network, including the selection of base stations participating in the Co-MIMO transmission. If sounding-based Co-MIMO transmission is utilized, each Co-MIMO zone is allocated to one or more mobile stations and each BS can allocate multiple Co-MIMO zones to different mobile stations. The corresponding resource units and the permutation schemes in each Co-MIMO zone are negotiated and coordinated by the participating base stations through the backhaul prior to allocation. The BS broadcasts normalized interference power thresholds for multi-BS MIMO schemes and one common CINR threshold. The normalized interference power is defined as the ratio of average interference power (with or without transmitter pre-coder) from one dominant interfering BS to the total interference power plus noise received at the MS. The MS may accordingly request the preferred multi-BS MIMO scheme. The single-BS precoding with multi-BS coordination scheme is performed by combining two PMIs in order to mitigate the inter-cell interference when closed-loop MIMO precoding is used in the serving and the neighboring cells. One of the PMIs maximizes the transmission power of the serving cell, while the others minimize the interference to the neighboring cell.

The PMI combination procedure is performed as follows. The PMI combination may be triggered by the BS in an unsolicited manner. Using uplink sounding, the serving and the neighbor base stations can measure the channel conditions of the mobile stations. Following the channel measurements, the serving and neighbor base stations can determine the *PMI* which maximizes the transmission power and the PMI_{min} which minimizes the interference. The neighboring base stations may inform the serving BS of PMI_{min} through the backhaul. The above PMIs are calculated based on the following criteria [2]:

$$PMI = \arg\max_{i} \|\mathbf{H}_s\, \mathbf{v_i}\|^2$$

$$PMI_{min} = \arg\min_{i} \|\mathbf{H}_i\, \mathbf{v_i}\|^2$$

(10-54)

where $\mathbf{H}_s$ and $\mathbf{H}_i$ denote the channel matrices from the MS to its serving and the neighboring base stations, respectively, and $\mathbf{v}_i$ is the ith codeword of the base codebook. The serving BS informs the MS of the PMI, PMI_{min}, and *PMI Combination Ratio* γ through UL Basic Assignment A-MAP IE, UL Sub-band Assignment A-MAP-IE, UL CSM Beamforming A-MAP IE, or AAI_UL_MultiB-S_MIMO_SBP MAC management messages. The transmitted precoder $\mathbf{W}$ is generated by combining the two precoders $(\mathbf{W}_{PMI_{min}}, \mathbf{W}_{PMI})$ according to the following equation [2]:

$$\mathbf{W} = \frac{\gamma \mathbf{W}_{PMI} + (1-\gamma)\mathbf{W}_{PMI_{min}}}{\|\gamma\mathbf{W}_{PMI} + (1-\gamma)\mathbf{W}_{PMI_{min}}\|} \quad 0 \le \gamma \le 1$$

(10-55)

The IEEE 802.16m further supports uplink multi-BS MIMO to allow joint reception by multiple base stations, e.g., macro-diversity combining and cooperative beamforming. The collaborative zone initialization is defined as a common radio resource that is allocated as a collaborative zone among the participating base stations in the uplink multi-BS MIMO operation. The uplink sounding signals are assigned orthogonally among mobile stations in the collaborative zone. The inter-BS information exchange and joint processing, by enabling macro-diversity combining, soft decision information in the form of log-likelihood ratios are generated at the neighbor base stations, transmitted to anchor BS accompanied with scheduling information through M-SAP or C-SAP primitives over the core network, and combined at the anchor BS. Using cooperative beamforming, the quantized versions of received signals are generated at neighbor base stations, transmitted to the anchor BS accompanied with channel state information and scheduling information through M-SAP or C-SAP primitives over backhaul and jointly processed at the anchor BS.

10.8 INTERFERENCE MITIGATION

Fractional Frequency Reuse (FFR) is an interference mitigation scheme which allows the use of different frequency reuse factors over different frequency partitions in certain time intervals in downlink and uplink transmissions. The FFR is usually combined with other schemes such as power control or multi-BS MIMO technologies for adaptive control and joint optimization. The power control and multi-BS MIMO interference mitigation schemes were described in previous sections. In a generic FFR scheme as shown in Figure 10-48, the sub-carriers over the entire transmission bandwidth are grouped into frequency partitions with different reuse factors. In general, the received signal quality can be improved by assigning the mobile stations in the frequency partitions with a higher frequency reuse factor due to lower interference levels. This will benefit the mobile stations near the cell edge or the mobile stations that are suffering from severe inter-cell interference. The BS may apply a lower frequency reuse factor to some frequency partitions in order to serve the mobile stations which do not experience significant inter-cell interference, resulting in an improvement in spectral efficiency. The resource allocation when using the FFR scheme may take several factors into consideration, including reuse factor in partition, power in partition, and multi-antenna mode, as well as interference-based measurements conducted at the mobile station.

In the IEEE 802.16m standard, the size of the frequency partitions are selected as an integer multiple of physical resource units. The frequency partitions are indexed from the lowest LRU index to the highest index. The frequency partitions always start with a reuse-1 partition and continue with three reuse-3 or two reuse-2 partitions, depending on the value of the DFPC parameter and the transmission bandwidth. The frequency partitions are numbered as frequency partition 0 (FP_0), frequency partition 1 (FP_1), frequency partition 2 (FP_2), or frequency partition 3 (FP_3). In the downlink, the frequency partition configuration is signaled using the DFPC parameter in S-SFH SP0. The boosted frequency partition FP_1 is a partition which has the highest power level. Each partition may have different power level per cell. The transmission power level relative to the reference power level on different frequency partitions is adjusted by the BS and signaled using an AAI_DL-IM MAC management message or an S-SFH SP3. When FFR is used in the cell, i.e., FPCT > 1, different FFR power patterns are used by different base stations. For example, when FPCT = 4, each BS selects one of the three FFR patterns, as shown in Figure 10-8. When FPCT = 3 and $FPS_3 > 0$, the same

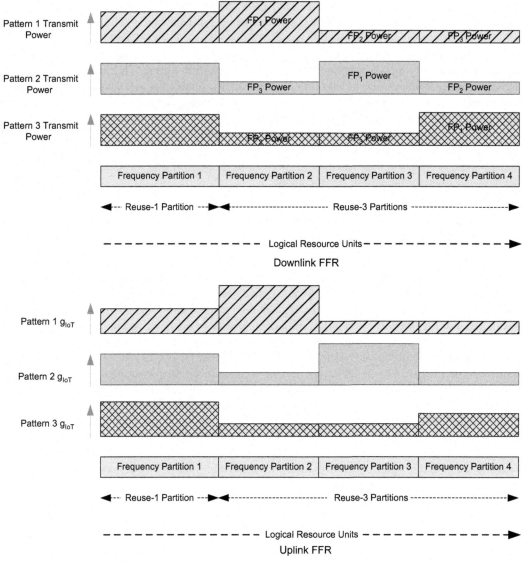

FIGURE 10-48

Illustration of fractional frequency reuse concept [2]

FFR patterns exist excluding FP_0. The index of the FFR power pattern is set by a particular BS using the frequency partition number k.

In downlink FFR, the mobile station reports the interference measurement to the serving base station. The serving base station will instruct the mobile station to perform interference measurement over the designated radio resource region in a solicited/unsolicited manner, or the mobile station may perform the independent interference measurement. Examples of interference measurements include SINR, SIR, interference power, RSSI, etc. The mobile station can also recommend its preferred frequency partition to the serving base station, based on interference measurements, the resource metric of each partition, etc. The measurement results can then be reported through messaging or feedback channels. The base station can transmit necessary information through signaling or message to facilitate the measurements conducted by the mobile station. The information includes the frequency reuse parameters of each frequency partition, the corresponding power levels, and the associated metric for each partition. The resource metric of each frequency partition is the measure of the overall system resource usage in the partition (e.g., effective bandwidth due to reuse, transmission power, multi-antenna mode, and interference to other cells). The serving BS may instruct the MS to feedback interference or SINR measurements for one or more frequency partitions through the AAI_FFR-CMD message.

The MS reports interference and/or SINR measurements corresponding to one or more frequency partitions through the AAI_FFR-REP message as a response to the AAI_FFR-CMD message. For MIMO feedback modes 0, 4, and 7, the BS instructs the MS to feedback wideband CQI and spatial rate for the active frequency partition using Feedback Allocation A-MAP IE. For MIMO feedback mode 0, the BS may request feedback on wideband CQI and spatial rate for alternative frequency partitions in addition to the active partition by puncturing the reporting period of the active frequency partition. The BS may use the interference statistics to schedule downlink data transmissions. A Preferred Frequency Partition Indicator (PFPI) is defined in the primary fast-feedback channel to indicate the change in the preferred active frequency partition.

The FFR partition information is broadcast in S-SFH SP2, and the resource metric and transmission power level are signaled in an AAI_DL-IM message and/or S-SFH SP3. When the MS attempts network entry to a system with FPCT > 1, it initially uses the frequency partition indicated by the BS. Once the MS detects the first superframe with resource metric information, it may start using the resource metric to recommend its preferred partition to the BS in distributed DRU regions or to select the preferred sub-bands in sub-band CRU allocations. In the latter case, the MS estimates the average SINR in each sub-band of the frequency partition and calculates the expected spectral efficiency for each sub-band. The expected SE can be calculated based on data rate, PER, and partition bandwidth as $SE_{expected} = (1 - PER)Data\_Rate/BW$ where $Data\_Rate$ is the uncoded data in the number of bits transmitted on four LRUs in type-1 subframe, PER is the MS estimated packet error rate in the same four LRUs, and BW is the amount of bandwidth occupied by four LRUs. The data rate is a function of average SINR per partition, and is determined according to the modulation and coding rate selected by link adaptation procedure. The MS calculates the normalized SE_i of the ith frequency partition as $Normalized\_SE_i = Expected\_SE_i/Resource\_Metric_i$. The MS compares the $Normalized\_SE_i$ of all partitions and selects the partition with maximum normalized SE as the preferred frequency partition, i.e., $i = \max(Normalized\_SE_i)$, $\forall i$. If the preferred partition is different from the previously selected one and has relatively stable normalized SE, the MS will send an indicator through the primary fast-feedback channel to notify the BS of its preferred frequency partition.

For uplink FFR, the base station estimates the interference statistics in each frequency partition and may send messages containing interference information to the mobile stations. The interference information includes the frequency reuse parameters of each partition and the corresponding uplink power control parameters, as well as the target IoT level. A mobile station assigned to a partition needs to properly perform power control considering the target IoT level of other cells for that partition. If the target IoT level of other cells for a partition is low, a mobile station assigned to that partition should transmit with lower power in order not to interfere with users in other cells. If the target IoT level of other cells for a partition is high, then a user assigned for that partition may transmit with higher power. To control system-wide interference, the base station can adjust the frequency partitions and the corresponding target IoT level in coordination with other base stations. The uplink FFR mechanism allows the BS to designate different uplink IoT control parameters γ_{IoT} in frequency partition. The uplink FFR configuration parameters, including the number of frequency partitions and size of each frequency partition, are broadcast through S-SFH SP2. The uplink IoT control parameter for each frequency partition is broadcast through the AAI_SCD MAC control message.

If uplink FFR is enabled in a cell, different FFR patterns with power-boosted partitions are used by different base stations. In the example shown in Figure 10-48, FPCT = 4 and each cell selects one of the three FFR patterns. The index of the FFR pattern is set by a particular cell with its frequency partition number k. The FFR partition configuration is semi-static and does not change frequently. The interval between two FFR partition changes is determined by the FFR *Partition Update Interval* parameters [2]. In each FFR partition update interval, the BS reports to the network controller the following information: BS-ID; the number of mobile stations in the cell; the mobile station location distribution; and the MS UL/DL SINR distributions, UL/DL traffic distribution, base station's transmission power in each frequency partition, uplink IoT control parameter γ_{IoT} corresponding to each FFR partition that is used to calculate the FFR partition size, power levels, relative load indicator, and reference uplink IoT control parameter $\gamma_{IoT_{reference}}$ for each partition that will be used for FFR partition configuration.

The base station may use various mechanisms to mitigate the inter-cell and/or intra-cell interference experienced by the mobile stations or to reduce interference to other cells. The interference mitigation techniques may include sub-channel scheduling, dynamic transmit power control, dynamic antenna pattern adjustment, and adaptive modulation and coding schemes. The base station may allocate different modulation and coding schemes to mobile stations through uplink scheduling which indirectly controls the MS transmit power and thereby interference to other cells. The base station can exchange power control information with neighbor base stations. The mobile station may use interference information and its downlink measurements to control the uplink interference it causes to adjacent cells. The base station can schedule mobile stations with high mutual interference on different sub-channels or frequency partitions. Cell/sector specific interleaving may be used to randomize the transmitted signal, in order to allow for interference suppression at the receiver.

10.9 MULTI-ANTENNA TECHNIQUES IN 3GPP LTE

Figure 10-49 illustrates different stages of 3GPP LTE physical channel processing in the downlink and uplink. In the downlink, the coded bits in each of the codewords are scrambled for transmission on a physical channel. The scrambled bits are modulated to generate complex-valued modulation symbols that are later mapped to one or several transmission layers. The complex-valued modulation symbols on each layer are pre-coded for transmission and are further mapped to resource elements for each antenna port. The complex-valued time-domain OFDMA signal for each antenna port is then generated following these stages [30,97]. In the uplink, the baseband signal is processed by scrambling the

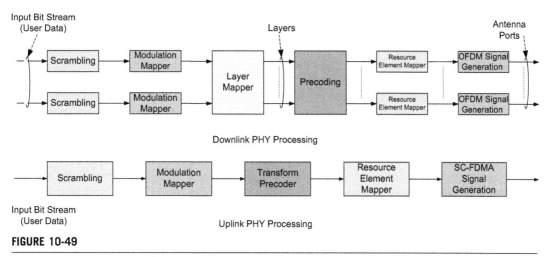

FIGURE 10-49

Overview of 3GPP LTE multi-antenna signal processing

input coded bits and then by modulation of scrambled bits to generate complex-valued symbols. The complex-valued modulation symbols are transform-precoded (DFT-based precoding) to generate complex-valued symbols that are later mapped to resource elements. The complex-valued time-domain SC-FDMA signal for each antenna port is then generated.

A 2 × 2 antenna configuration for MIMO is assumed as the baseline in the 3GPP LTE downlink; i.e., two transmit antennas at the base station and two receive antennas at the terminal side [107]. Configurations with four transmit or receive antennas are also supported in the specifications. Different downlink MIMO modes are supported in 3GPP LTE, which can be adapted based on channel conditions, traffic requirements, and UE capability. These include transmit diversity, open-loop spatial multiplexing (no UE feedback), closed-loop spatial multiplexing (with UE feedback), multi-user MIMO (more than one UE is assigned to the same resource block), and closed-loop rank-1 precoding. Multi-antenna transmission with two and four transmit antennas is supported. The maximum number of codewords is two, irrespective of the number of antennas with fixed mapping between codewords to layers. There is semi-static switching between SU-MIMO and MU-MIMO per UE. In addition, the following techniques are supported: codebook-based precoding with a single precoding feedback per full system bandwidth when the system bandwidth (or subset of resource blocks) is smaller or equal to 12 resource blocks and per 5 adjacent resource blocks or the full system bandwidth (or subset of resource blocks) when the system bandwidth is larger than 12 resource blocks, as well as rank adaptation with single rank feedback referring to full system bandwidth. The eNB can override rank report.

In 3GPP LTE spatial multiplexing, up to two codewords can be mapped onto different layers. One codeword represents an output from the channel coder. The number of layers available for transmission is equal to the rank of the channel matrix. Precoding in the transmitter side is used to support spatial multiplexing. This is achieved by multiplying the signal with a precoding matrix prior to transmission. The optimum precoding matrix is selected from a predefined codebook which is known to both the eNB and the UE. The optimum precoding matrix is the one which maximizes the capacity. The UE estimates the channel and selects the optimum precoding matrix. This feedback is provided to the eNB. Depending on the available bandwidth, this information is made available per resource block or group of resource blocks, since the optimum precoding matrix may vary between resource blocks. The network may configure a subset of the codebook that the UE is able to select from. In case of UEs with high velocity, the quality of the feedback may deteriorate. Thus, an open loop spatial multiplexing mode is also supported which is based on predefined settings for spatial multiplexing and precoding. In the case of four antenna ports, different precoders are assigned cyclically to the resource elements. The eNB will select the optimum MIMO mode and precoding configuration. The information is conveyed to the UE as part of the downlink control information on the PDCCH.

In order for MIMO schemes to work properly, each UE has to report information about the channel to the base station. Several measurement and reporting schemes are available which are selected according to MIMO mode of operation and network choice. The reporting may include wideband or narrowband CQI, which is an indication of the downlink radio channel quality as experienced by this UE, PMI which is an indication of the optimum precoding matrix to be used in the base station for a given radio condition, and Rank Indication (RI) which is the number of useful transmission layers when spatial multiplexing is used [30,107]. In the case of transmit diversity mode, only one codeword can be transmitted. Each antenna transmits the same information stream, but with different coding. The LTE employs space frequency block coding as a transmit diversity scheme. A special precoding matrix is applied at the transmitter side in the precoding stage in Figure 10-49. Cyclic delay diversity is an additional type of diversity which can be used in conjunction with spatial multiplexing in LTE. An antenna-specific delay is applied to the signals transmitted from each antenna port. This effectively introduces an artificial multipath to the signal as seen by the receiver. As a special method of delay diversity, cyclic delay diversity applies a cyclic shift to the signals transmitted from each antenna port [30,97].

For the 3GPP LTE uplink, multi-user MIMO can be used. Multiple user terminals may transmit simultaneously on the same resource block. The scheme requires only one transmit antenna at the UE side. The UEs sharing the same resource block have to apply mutually orthogonal pilot patterns. To take advantage of two or more transmit antennas, transmit antenna selection can be used. In this case, the UE has two transmit antennas but only one transmission chain. A switch will then choose the antenna that provides the best channel to the eNB [30,99,107]. To support MU-MIMO, the reference signals for users involved should be distinct and should have good cross-correlation properties. If two or more UEs in one sector are assigned to the same resource blocks, their reference signals are derived from the same sequence with cyclic-shift in the time-domain. In 3GPP LTE, the cyclic-shifts for a UE's reference signal can take eight different values. These values can ideally support uplink MU-MIMO with two to six UEs. However, in practice, only two users are paired to lower the receiver complexity. It may be noted that the length of the reference signals are based on the number of resource blocks allocated to a user. In general, reference signals of 36 or more in length are based on extended Zadoff-Chu sequences, while reference signals of length 12 and 24 are computer generated sequences. The interference seen by a user's signal comprises two parts: the intra-cell interference due to other users' involved in the uplink MU-MIMO; and inter-cell interference due to users in other sectors. As the UEs in a sector can operate at different MCS levels and multipath can create nulls and peaks in one transmission, the inter-cell interference due to these UEs seen at neighboring sectors can be quite irregular in the frequency-domain. One can choose to estimate the spatial pattern of the inter-cell interference over one or multiple resource blocks. There are a number of uplink MIMO receiver algorithms which can be used for 3GPP LTE uplink MIMO. Using MU-MIMO, two UE's transmit on the same frequency sub-carriers and at the same time. Thus, there is cross-user interference between these two UEs. Using a MMSE receiver at the BS, the interference between these two UEs can be significantly reduced.

The 3GPP LTE Release 8 supports downlink transmissions on one, two, or four cell-specific antenna ports, each corresponding to one, two, or four cell-specific reference signals, where each reference signal corresponds to one antenna port. An additional antenna port, associated with one UE-specific reference signal is also available. This antenna port can be used for conventional beamforming, especially in the case of TDD operation. All physical layer processing (i.e., up to and including the scrambling module) for the nth transport block in a certain subframe is denoted by codeword n. Up to two transport blocks can be transmitted simultaneously, while up to $Q = 4$ layers can be transmitted for the rank-4 case, so there is a need to map the codewords (transport blocks) to the appropriate layer. The layers form a sequence of $Q \times 1$ symbol vectors as $\mathbf{s}_n = (s_{n1}, s_{n2},\ldots,s_{nQ})_t$ which are input to a precoder that in general can be modeled on the form of a linear dispersion encoder. The precoder only exists if the PDSCH is configured to use cell-specific reference signals, which are then added after the precoding and thus do not undergo any precoding. If the PDSCH is configured to use the UE-specific reference signals, which would then also undergo the same precoding operation as the resource elements for data, then the precoder operation is transparent to the standard and therefore purely an eNB implementation issue. The precoder is block-based and outputs a block $\mathbf{X}_n = (\mathbf{x}_{nL}\,\mathbf{x}_{nL+1}\ldots\mathbf{x}_{nL+L-1})$ of precoded $N_t \times 1$ vectors for every symbol vector $\mathbf{s}_n$ where L is the block size and the parameter N_t denotes the number of antenna ports, if PDSCH is configured to use cell-specific reference signals. If a transmission mode using UE-specific reference signals is configured, then, similarly to above, N_t is entirely up to the eNB implementation, but it typically corresponds to the number of transmit antennas assumed in the baseband implementation.

The vectors $\mathbf{x}_k$ are distributed over the grid of data resource elements belonging to the resource block assignment for the PDSCH. Let k denote the resource element index, the corresponding received $N_r \times 1$ vector $\mathbf{y}_k$ on the UE side after DFT operation can then be written as $\mathbf{y}_k = \mathbf{H}_k\mathbf{x}_k + \mathbf{w}_k$ where $\mathbf{H}_k$ is an $N_r \times N_t$ matrix that represents the MIMO channel and $\mathbf{w}_k$ is an $N_r \times 1$ vector denoting noise and interference. Considering the resource elements associated with a certain block $\mathbf{x}_n$ output from the precoder and assuming that the channel is constant over the block, i.e., the block size L is small and the resource elements are localized in the resource element grid, the following block-based received data model is obtained $\mathbf{Y}_n = (\mathbf{y}_{nL}\,\mathbf{y}_{nL+1}\ldots\mathbf{y}_{nL+L-1}) = \mathbf{H}_n\mathbf{X}_n + \mathbf{W}_n$. The transmission rank is given by the average number of complex-valued symbols per resource element. Since Q symbols are transmitted over L resource elements, the transmission rank r is obtained as $r = Q/L$.

The 3GPP LTE supports rank-1 transmit-diversity through Alamouti-based space-time codes in frequency-domain (SFBC). The encoding operation is performed over space and frequency so the output block $\mathbf{X}_n$ from the precoder is limited to consecutive data resource elements over a single OFDM symbol. If single-codeword transmission is assumed, the modulated symbols of a single codeword are mapped to all layers. The transmit-diversity schemes for two and four cell-specific antenna ports are supported. Note that these transmit-diversity schemes are also used for the PBCH, PDCCH, and PCFICH. Furthermore, the number of cell-specific antenna ports used to encode the PBCH is the same as the total number of configured cell-specific antenna ports and these are also used for other control channels. Thus, all UEs must support up to four cell-specific antenna ports and the corresponding transmit-diversity schemes. For the case of two antenna ports, the output from the precoder is as follows:

$$\mathbf{X}_n = \begin{bmatrix} s_{n1} & s_{n2} \\ -s_{n2}^* & s_{n1}^* \end{bmatrix} \tag{10-56}$$

where the rows corresponds to the antenna ports and the columns to consecutive data resource elements over the same OFDM symbol. In the case where four antenna ports are utilized, a combination of SFBC and frequency-switched transmit diversity is employed to provide robustness against the correlation between channels from different transmit antennas and for less-complex UE receiver implementation. The output from the precoder is given as:

$$\mathbf{X}_n = \begin{bmatrix} s_{n0} & s_{n1} & 0 & 0 \\ 0 & 0 & s_{n2} & s_{n3} \\ -s_{n1}^* & s_{n0}^* & 0 & 0 \\ 0 & 0 & -s_{n3}^* & s_{n2}^* \end{bmatrix} \tag{10-57}$$

The above code is composed of two SFBC codes which are transmitted on antenna ports 0, 2, and 1, 3, respectively. The reason for distributing a single SFBC code in such an interlaced manner on every other antenna port instead of consecutive antenna ports is related to the fact that the first two cell-specific antenna ports have a higher reference signal density than the last two; and hence, provide better estimation of the channel. The above transmit-diversity scheme can be used for all downlink channels other than PHICH. For the latter, four different ACK/NACK bits are multiplexed using orthogonal codes with a spreading factor of four over a group of four sub-carriers, and the resulting group is repeated three times in the frequency-domain to achieve frequency-diversity gain. To maintain the orthogonality between different codes in each repetition of four sub-carriers, antenna switching is not applied within each repetition. Instead, the set of antennas changes across different repetitions. When there are multiple PHICHs transmitted, using type 1 or type 2 alternatively for different PHICHs would be helpful to keep uniform power distribution over the eNB transmit antennas.

Precoding for spatial multiplexing using antenna ports with cell-specific reference signals is only used in combination with layer mapping for spatial multiplexing. Spatial multiplexing supports two or four antenna ports and the set of antenna ports used is $p = 0,1$ or $p = 0,1,2,3$, respectively with p denoting the antenna port number. The precoding for spatial multiplexing without CDD is defined as:

$$\begin{bmatrix} y^{(0)}(i) \\ \vdots \\ y^{(N_p-1)}(i) \end{bmatrix} = \mathbf{W}(i) \begin{bmatrix} x^{(0)}(i) \\ \vdots \\ x^{(N_l-1)}(i) \end{bmatrix} \tag{10-58}$$

where $\mathbf{W}(i)$ is the $N_p \times N_l$ precoding matrix, N_p is the number of antenna ports, N_l denotes the number of layers, $i = 0, 1, ..., M_{symb}^p - 1$, and $M_{symb}^p = M_{symb}^{layer}$. For spatial multiplexing, the values of $\mathbf{W}(i)$ are selected among the precoder elements in the codebook configured in the eNB and the UE. The eNB can further confine the precoder selection in the UE to a subset of the elements in the codebook using

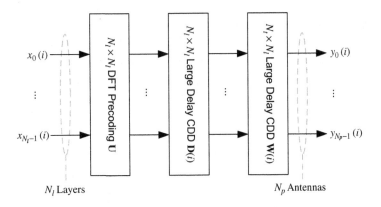

FIGURE 10-50

Open-loop spatial multiplexing with N_t antennas and N_l transmission layers [101]

codebook subset restrictions. For large-delay CDD, precoding for spatial multiplexing is defined by (see Figure 10-50):

$$\begin{bmatrix} y^{(0)}(i) \\ \vdots \\ y^{(N_p-1)}(i) \end{bmatrix} = \mathbf{W}(i)\mathbf{D}(i)\mathbf{U} \begin{bmatrix} x^{(0)}(i) \\ \vdots \\ x^{(N_l-1)}(i) \end{bmatrix} \tag{10-59}$$

where $\mathbf{D}(i)$ and $\mathbf{U}$ are $N_l \times N_l$ matrices supporting cyclic delay diversity, as shown in Table 10-34, for a different number of layers. For two antenna ports, the precoder is selected according to $\mathbf{W}(i) = \mathbf{C}_1$ where $\mathbf{C}_1$ denotes the precoding matrix corresponding to precoder index 0 in Table 10-35. For four antenna ports, the UE may assume that the eNB cyclically assigns different precoders to different vectors $(x^{(0)}(i), x^{(1)}(i), \ldots, x^{(N_l-1)}(i))^t$ on the physical downlink shared channel. A different precoder is used every N_l vectors, where N_l denotes the number of transmission layers in the case of spatial multiplexing. In particular, the precoder is selected according to $\mathbf{W}(i) = \mathbf{C}_k$, where k is the precoder index given by $k = (\lfloor i/N_l \rfloor \bmod 4) + 1 \in \{1, 2, 3, 4\}$ and $\mathbf{C}_1$, $\mathbf{C}_2$, $\mathbf{C}_3$, $\mathbf{C}_4$, denote precoder matrices. When two antenna ports are configured, the number of codewords is equal to the transmission rank and codeword n is mapped to layer n.

For transmission on two antenna ports ($p = 0,1$), the precoding matrix $\mathbf{W}(i)$ is selected from Table 10-35 or a subset of it. The open-loop spatial multiplexing may be used when reliable PMI feedback is not available at the eNB, for example, when the UE moves fast or when the feedback overhead on uplink is too high. The open-loop spatial multiplexing with N_l layers and N_p antenna ports ($N_p \geq N_l$) is illustrated in Figure 10-50. The feedback consists of the RI and the CQI in open-loop spatial multiplexing. In contrast to closed-loop spatial multiplexing, the eNB only determines the transmission rank, and a fixed set of precoding matrices are applied cyclically across all the scheduled sub-carriers in the frequency-domain. It is noted that in the open-loop spatial multiplexing mode, the transmit diversity scheme is applied when the transmission rank is set to one [101].

The 3GPP LTE supports MU-MIMO in the uplink and downlink. In the uplink, the eNB can schedule a number of UEs to transmit over the same time-frequency resource, which forms a MU-MIMO transmission configuration. However, in order for the eNB to correctly separate and

Table 10-34 Large-Delay Cyclic Delay Diversity [107]

Number of Layers N_l	U	D(i)
2	$\dfrac{1}{\sqrt{2}}\begin{bmatrix} 1 & 1 \\ 1 & e^{-j2\pi/2} \end{bmatrix}$	$\begin{bmatrix} 1 & 0 \\ 0 & e^{-j2\pi i/2} \end{bmatrix}$
3	$\dfrac{1}{\sqrt{3}}\begin{bmatrix} 1 & 1 & 1 \\ 1 & e^{-j2\pi/3} & e^{-j4\pi/3} \\ 1 & e^{-j4\pi/3} & e^{-j8\pi/3} \end{bmatrix}$	$\begin{bmatrix} 1 & 0 & 0 \\ 0 & e^{-j2\pi i/3} & 0 \\ 0 & 0 & e^{-j4\pi i/3} \end{bmatrix}$
4	$\dfrac{1}{2}\begin{bmatrix} 1 & 1 & 1 & 1 \\ 1 & e^{-j2\pi/4} & e^{-j4\pi/4} & e^{-j6\pi/4} \\ 1 & e^{-j4\pi/4} & e^{-j8\pi/4} & e^{-j12\pi/4} \\ 1 & e^{-j6\pi/4} & e^{-j12\pi/4} & e^{-j18\pi/4} \end{bmatrix}$	$\begin{bmatrix} 1 & 0 & 0 & 0 \\ 0 & e^{-j2\pi i/4} & 0 & 0 \\ 0 & 0 & e^{-j4\pi i/4} & 0 \\ 0 & 0 & 0 & e^{-j6\pi i/4} \end{bmatrix}$

Table 10-35 Codebook for Transmission on Antenna Ports 0 and 1 [107]

Codebook Index	Number of Layers N_l	
	1	2
0	$\dfrac{1}{\sqrt{2}}\begin{bmatrix} 1 \\ 1 \end{bmatrix}$	$\dfrac{1}{\sqrt{2}}\begin{bmatrix} 1 & 0 \\ 0 & 1 \end{bmatrix}$
1	$\dfrac{1}{\sqrt{2}}\begin{bmatrix} 1 \\ -1 \end{bmatrix}$	$\dfrac{1}{2}\begin{bmatrix} 1 & 1 \\ 1 & -1 \end{bmatrix}$
2	$\dfrac{1}{\sqrt{2}}\begin{bmatrix} 1 \\ j \end{bmatrix}$	$\dfrac{1}{2}\begin{bmatrix} 1 & 1 \\ j & -j \end{bmatrix}$
3	$\dfrac{1}{\sqrt{2}}\begin{bmatrix} 1 \\ -j \end{bmatrix}$	–

demodulate the UEs' signals, the eNB needs to assign orthogonal reference signals for these UEs scheduled for the MU-MIMO transmission. In the 3GPP LTE uplink slot structure, the reference signal is transmitted using the fourth SC-FDMA symbol and the data is transmitted using the other symbols. For a given slot and subframe in each cell, a Zadoff-Chu sequence is defined as the base sequence for uplink reference signals. The cyclically-shifted versions of a given Zadoff-Chu sequence form an orthogonal set of sequences. Each UE scheduled for MU-MIMO transmission is assigned a distinctive cyclic-shift value. The UE combines this cyclic-shift value with the knowledge of the base Zadoff-Chu sequence to form a reference signal sequence that is orthogonal to other UEs' reference signal sequences. It is noted that the cyclic shift value is always contained in the control signaling, which the UE has to receive for data transmission on uplink, regardless of whether the MU-MIMO is used.

In the downlink, if a UE is configured to use the MU-MIMO transmission mode, only rank-1 transmission can be scheduled for the UE. The eNB can schedule multiple UEs, which are configured to operate in the MU-MIMO transmission mode, in the same time-frequency resource using different rank-1 precoding matrices from Table 10-35 for two transmit antennas. Note that the UE receives only the information about its own precoding matrix. The scheduled UE then decodes the information utilizing the common reference signal in conjunction with the precoding information obtained from the control signaling. The UE generates the PMI/CQI feedback without any knowledge of other simultaneously scheduled UEs. Thus, there could be a mismatch between the UE's CQI report and the actual CQI experienced due to lack of knowledge of interference caused by other UEs scheduled simultaneously. In 3GPP LTE, in order to allow the use of higher-order modulation schemes such as 16 QAM or 64 QAM without adding more complexity in the UE, the transmit-power level for each UE is adjusted in a long-term manner. The per-UE pre-configured power level is hard to maintain in MU-MIMO transmission mode, since the eNB power amplifier has to support multiple UEs scheduled on the same time-frequency resource. Single-bit signaling is used to indicate whether a 3 dB power reduction with respect to the per-UE configured power-level is necessary, if a UE is operating in MU-MIMO transmission mode.

Codebook-based beamforming is supported through channel-dependent precoding with rank-1 transmission. Therefore, the UEs can be configured for the single-rank channel-dependent precoding scheme and report precoder vectors accordingly to the eNB. A critical functionality related to MU-MIMO is the need of the UE to derive the power ratio between the reference signals and the power per data resource element and antenna that is applied to its own transmission in order to assist demodulation. The derivation of the power ratio is important since multiple UEs share the same time-frequency resources and thus may share the finite power of the power amplifiers. This can result in power fluctuations during transmission to a particular UE. Since the UE is not mandated to blindly estimate this power ratio for all different modulations, the power fluctuations need to be signaled to the UE. For a QPSK scheme, the UE cannot rely on the knowledge of the power ratio, but for higher modulation (64 QAM and 16 QAM) it is assumed that the UE is informed about the power ratio. The UEs can only be simultaneously scheduled if their preferred beams are spatially separated. This is an additional constraint on the scheduler that needs to group the UEs that have data to send and have sufficiently high geometry (SINR distribution). Thus, in order for MU-MIMO to be useful, the system load should be high with many active UEs requesting data in each subframe. This enables the scheduler to find a group of UEs that can be concurrently scheduled on beams which will limit intra-cell interference [99].

Dedicated beamforming is supported for improving data coverage when the UE supports data demodulation using the UE-specific reference signal. The eNB generates a beam using the array of antenna elements (e.g., an array of eight antenna elements), and then applies the same precoding to both the data sub-carriers and the UE-specific reference signals. It is noted that the UE-specific reference signal is transmitted such that its time-frequency location does not overlap with the cell-specific reference signal [101].

10.10 MULTI-ANTENNA TECHNIQUES IN 3GPP LTE-ADVANCED

In 3GPP LTE-Advanced, the existing SU-MIMO technologies are extended to support configurations with up to eight transmit antennas in the downlink, and up to four transmit antennas in the uplink. In addition, multi-cell Coordinated Multipoint (CoMP) transmission is also under discussion and evaluation. In the case of uplink single-user spatial multiplexing, up to two transport blocks can be transmitted from a scheduled UE in a subframe per uplink component carrier. Each transport block has its own modulation and coding scheme. Depending on the number of transmission layers, the modulation symbols associated with each of the transport blocks are mapped onto one or two layers according to the same principle used in 3GPP LTE Rel-8 downlink spatial multiplexing. The transmission rank can be dynamically adapted. It is possible to configure the uplink single-user spatial-multiplexing transmission with or without the layer shifting. In the case of the layer shifting, shifting in time-domain is supported. If layer shifting is supported, the HARQ feedbacks for all transport blocks are bundled into a single HARQ feedback; thus, a one-bit ACK is transmitted to the UE, if all transport blocks are successfully decoded by the eNB; otherwise, a one-bit NACK is transmitted to the UE. If layer shifting is not configured, each transport block has its own HARQ feedback [114].

In FDD and TDD modes, precoding is performed according to a predefined codebook. If layer shifting is not used, precoding is applied after layer mapping. If layer shifting is configured, precoding is applied after the layer shifting operation. The 3GPP LTE-Advanced supports the application of a single precoding matrix per uplink component carrier. In the case of full-rank transmission, only an identity precoding matrix is supported. For uplink spatial multiplexing with two transmit-antennas, a 3-bit precoding codebook as shown in Table 10-36 is used. For uplink spatial multiplexing with four transmit-antennas, a 6-bit precoding codebook is used. A subset of the precoding codebook is used for single-layer and dual-layer transmission. For 3-layer transmission, the number of precoding matrices is 20; only BPSK or QPSK alphabets are used for non-zero elements in precoding matrices [114].

For UEs with multiple transmit antennas, an uplink single-antenna port mode is defined, where the UE's behavior is the same as having a single-antenna from the eNB's point of view. For a given UE, the uplink single antenna port mode can be independently configured for its PUCCH, PUSCH, and sounding reference signal transmissions. The uplink single antenna port mode is the default mode before the eNB is made aware of the UE's transmit antenna configuration.

Table 10-36 3-bit Precoding Codebook for Uplink Spatial Multiplexing with Two Transmit Antennas [114]

	Number of Layers	
Codebook Index	**1**	**2**
0	$\frac{1}{\sqrt{2}}\begin{bmatrix} 1 \\ 1 \end{bmatrix}$	$\frac{1}{\sqrt{2}}\begin{bmatrix} 1 & 0 \\ 0 & 1 \end{bmatrix}$
1	$\frac{1}{\sqrt{2}}\begin{bmatrix} 1 \\ -1 \end{bmatrix}$	–
2	$\frac{1}{\sqrt{2}}\begin{bmatrix} 1 \\ j \end{bmatrix}$	–
3	$\frac{1}{\sqrt{2}}\begin{bmatrix} 1 \\ -j \end{bmatrix}$	–
4	$\frac{1}{\sqrt{2}}\begin{bmatrix} 1 \\ 0 \end{bmatrix}$	–
5	$\frac{1}{\sqrt{2}}\begin{bmatrix} 0 \\ 1 \end{bmatrix}$	–

For uplink control channels with Rel-8 PUCCH format 1, 1a, and 1b, the Spatial Orthogonal-Resource Transmit Diversity (SORTD) scheme is supported for transmissions with two antenna ports. In this transmit diversity scheme, the same modulation symbol from the uplink channel is transmitted from two antenna ports, on two separate orthogonal resources. For the UEs with four transmit antennas, the two-transmit antenna transmit diversity scheme is applied [114].

The 3GPP LTE-Advanced extends LTE Rel-8 downlink spatial multiplexing with support for up to eight layers of spatial multiplexing. In the downlink, using single-user spatial multiplexing with eight transmit antennas, up to two transport blocks can be transmitted to a scheduled UE in a subframe per downlink component carrier. Each transport block is assigned its own modulation and coding scheme. For HARQ feedback on the uplink, one bit is used for each transport block. A transport block is associated with a codeword. For up to four layers, the codeword-to-layer mapping is the same as for 3GPP LTE Rel-8. For more than four layers, as well as the case of mapping one codeword to three or four layers, which is for re-transmission of one out of two codewords that were initially transmitted with more than four layers, layer mapping can be performed. Complex-valued modulation symbols for codewords q are mapped onto the layers, where v is the number of layers and is the number of modulation symbols per layer.

Downlink coordinated multipoint transmission (CoMP) is a relatively general term referring to different types of coordination in downlink transmission from multiple geographically separated transmission points (TP). This includes coordination in scheduling, including any beam-forming functionality, between geographically separated transmission points and joint transmission from geographically separated transmissions points. Uplink CoMP reception is a relatively general term referring to different types of coordination in the uplink reception at multiple, geographically separated points. This includes coordination in the scheduling, including any beam-forming functionality, between geographically separated reception points.

References

[1] IEEE 802.16-2009, IEEE Standard for Local and Metropolitan Area Networks, PART 16: Air Interface for Broadband Wireless Access Systems, May 2009.

[2] P802.16m/D6, IEEE Standard for Local and Metropolitan Area Networks – Part 16: Air Interface for Broadband Wireless Access Systems, Advanced Air Interface, May 2010.

[3] IEEE 802.16m-07/002r10, IEEE 802.16m System Requirements, January 2010 <http://ieee802.org/16/tgm/index.html>.

[4] IEEE 802.16m–08/004r5, IEEE 802.16m Evaluation Methodology Document, January 2009 <http://ieee802.org/16/tgm/index.html>.

[5] IEEE 802.16m–08/0034r3, IEEE 802.16m System Description Document, May 2010 <http://ieee802.org/16/tgm/index.html>.

[6] WiMAX Forum Mobile System Profile, Release 1.0 Approved Specification (Revision 1.7.1: 2008-11-07), <http://www.wimaxforum.org/technology/documents>.

[7] WiMAX Forum Network Architecture Release 1.5 Version 1 – Stage 2: Architecture Tenets, Reference Model and Reference Points, September 2009 <http://www.wimaxforum.org/resources/documents/technical/release>.

[8] Report ITU-R M.2135, Guidelines for Evaluation of Radio Interface Technologies for IMT-Advanced, November 2008.

[9] Report ITU-R M.2134, Requirements Related to Technical System Performance for IMT-Advanced Radio Interface(s), November 2008.

[10] IMT-Advanced submission and evaluation process <http://www.itu.int/ITU-R/>.

[11] WiMAX System Evaluation Methodology, July 2008 <http://www.wimaxforum.org/technology/documents>.

[12] Bernard Sklar, "Rayleigh Fading Channels in Mobile Digital Communication Systems. I. Characterization.", IEEE Communications Magazine volume 35 (Issue 7) (July 1997).

[13] Bernard Sklar, "Rayleigh Fading Channels in Mobile Digital Communication Systems. II. Mitigation.", IEEE Communications Magazine volume 35 (Issue 7) (July 1997).

[14] Bernard Sklar, Digital Communications: Fundamentals and Applications, second ed. Prentice Hall, 2001.

[15] Arogyaswami Paulraj, et al., Introduction to Space-Time Wireless Communications, Cambridge University Press, 2008.

[16] Y. Okumura, E. Ohmori, K. Fukuda, "Field Strength and its Variability in VHF and UHF Land Mobile Radio Service.", Review of the Electrical Communication Laboratory No. 16 (September-October, 1968).

[17] M. Hata, "Empirical Formulae for Propagation Loss in Land Mobile Radio Services." IEEE Transactions on Vehicular Technology vol. VT-29 (No. 3) (1980).

[18] William. C. Jakes, Microwave Mobile Communications, John Wiley & Sons, 1974.

[19] 3GPP TR 25.892, Feasibility Study for Orthogonal Frequency Division Multiplexing (OFDM) for UTRAN Enhancement, March 2004.

[20] Yiyan Wu, William Y. Zou, "Orthogonal Frequency Division Multiplexing: A Multi-Carrier Modulation Scheme", IEEE Transactions on Consumer Electronics vol. 41 (No. 3) (August 1995).

[21] Hu Su, et al., "Analysis of Tone Reservation Method for WiMAX System." International Symposium on Communications and Information Technologies, ISCIT'06, October 2006.

[22] Savo G. Glisic, Advanced Wireless Communications: 4G Cognitive and Cooperative Broadband Technology, second ed. Wiley Inter-Science, 2007.

[23] A. Osseiran, Jiann-Ching Guey, "Hopping pilot pattern for interference mitigation in OFDM." IEEE 19th International Symposium on Personal, Indoor and Mobile Radio Communications, September 2008.

[24] Richard Nilsson, Ove Edforst, Magnus Sandellt, Per Ola Borjesson, "An Analysis of Two-Dimensional Pilot-Symbol Assisted Modulation for OFDM.", 1997 IEEE International Conference on Personal Wireless Communications Publication, December 1997.

[25] D.E. Dudgeon, R.M. Mersereau, Multidimensional Digital Signal Processing, Prentice Hall, 1984.

[26] IEEE 802.16m–07/244r1, Yang Song, Liyu Cai, Keying Wu and Hongwei Yang, Collaborative MIMO, November 2007.

[27] IEEE C802.16m–10/0042, Zheng Yan-Xiu, et al., Modification to Resource Allocation in Sub-band Assignment A-MAP IE, March 2010.

[28] IEEE C802.16m–10/0320r1, Sudhir Ramakrishna, et al., Simplifications to the Specifications of the Sub-band Assignment A-MAP IE Proposed Amendment Text, March 2010.

[29] David Tse, Pramod Viswanath, Fundamentals of Wireless Communication, Cambridge University Press, 2005.

[30] Stefania Sesia, Issam Toufik, Matthew Baker, LTE, The UMTS Long Term Evolution: From Theory to Practice, John Wiley & Sons, 2009.

[31] Ezio Biglieri, et al., MIMO Wireless Communications, Cambridge University Press, 2010.

[32] Andrea Goldsmith, Wireless Communications, Cambridge University Press, 2005.

[33] Hemanth Sampath, Petre Stoica, Arogyaswami Paulraj, "Generalized Linear Precoder and Decoder Design for MIMO Channels Using the Weighted MMSE Criterion." IEEE Transactions on Communications vol. 49 (No. 12) (December 2001).

[34] Yang-Seok Choi, Peter J. Voltz, Frank A. Cassara, "On Channel Estimation and Detection for Multicarrier Signals in Fast and Selective Rayleigh Fading Channels." IEEE Transactions on Communications vol. 49 (No. 8) (August 2001).

[35] Siavash M. Alamouti, "A Simple Transmit Diversity Technique for Wireless Communications." IEEE Journal on Selected Areas in Communications vol. 16 (No. 8) (October 1998).

[36] Aria Nosratinia, Todd E. Hunter, Ahmadreza Hedayat, "Cooperative Communication in Wireless Networks.", IEEE Communications Magazine (October 2004).

[37] Jeffrey G. Andrews, "Interference Cancellation for Cellular Systems: A Contemporary Overview." IEEE Wireless Communications (April 2005).

[38] David Gesbert, et al., "From Theory to Practice: An Overview of MIMO Space–Time Coded Wireless Systems.", IEEE Journal on Selected Areas in Communications vol. 21 (No. 3) (April 2003).

[39] D. Ross Murch, Khaled Ben Letaief, "Antenna Systems for Broadband Wireless Access." IEEE Communications Magazine (April 2002).

[40] Severine Catreux, et al., "Adaptive Modulation and MIMO Coding for Broadband Wireless Data Networks.", IEEE Communications Magazine (June 2002).

[41] Arogyaswami J. Paulraj, Dhananjay A. Gore, Rohit U. Nabar, and Helmut Bölcskei, "An Overview of MIMO Communications—A Key to Gigabit Wireless." Proceedings of the IEEE, Vol. 92, (No. 2), (February 2004).

[42] Matthew G. Parker, Kenneth G. Patersonyand, Chintha Tellambura, Golay Complementary Sequences, January 2004.

[43] J. David Love, Robert W. Heath Jr., "Grassmannian Beamforming for Multiple-Input Multiple-Output Wireless Systems.", IEEE Transactions on Information Theory vol. 49 (No. 10) (October 2003).

[44] Vahid Tarokh, Hamid Jafarkhani, "On the Computation and Reduction of the Peak-to-Average Power Ratio in Multicarrier Communications." IEEE Transactions on Communications vol. 48 (No. 1) (January 2000).

[45] Timothy M. Schmid, Donald C. Cox, "Robust Frequency and Timing Synchronization for OFDM." IEEE Transactions on Communications vol. 45 (No. 12) (December 1997).

[46] Markku Pukkila, Channel Estimation Modeling, Postgraduate Course in Radio-communications, Fall 2000.

[47] Muhammad Saad Akram, Pilot-based Channel Estimation in OFDM Systems, Master Thesis, June 2007.

[48] Freescale Semiconductor Application Note, Channel Estimation in OFDM Systems, AN3059 Rev. 0, January 2006.

[49] Vineet Srivastava, et al., "Robust MMSE Channel Estimation in OFDM Systems with Practical Timing Synchronization." IEEE Wireless Communications and Networking Conference, 2004.

[50] Lizhong Zheng, David N.C. Tse, "Diversity and Multiplexing: A Fundamental Tradeoff in Multiple-Antenna Channels." IEEE Transactions on Information Theory vol. 49 (No. 5) (May 2003).

[51] Hanan Weingarten, Yossef Steinberg, Shlomo Shamai, "The Capacity Region of the Gaussian Multiple-Input Multiple-Output Broadcast Channel.", IEEE Transactions on Information Theory vol. 52 (No. 9) (September 2006).

[52] Florian Kaltenberger et al., "Capacity of Linear Multi-User MIMO Precoding Schemes with Measured Channel Data." EURECOM, Sophia-Antipolis, France.

[53] V. Stankovic and M. Haardt, "Multi-User MIMO Downlink Precoding for users with multiple antennas." Proceedings of the 12th meeting of the Wireless World Research Forum (WWRF), Toronto, Canada, (November 2004).

[54] Q.H. Spencer, A.L. Swindlehurst, M. Haardt, "Zero-Forcing Methods for Downlink Spatial Multiplexing in Multiuser MIMO Channels." IEEE Transactions on Signal Processing vol. 52 (No. 2) (February 2004).

[55] Nihar Jindal, "MIMO Broadcast Channels with Finite-Rate Feedback." IEEE Transactions on Information Theory vol. 52 (No. 11) (November 2006).

[56] Quentin H. Spencer, Christian B. Peel, A. Lee Swindlehurst, Martin Haardt, "An Introduction to the Multi-User MIMO Downlink." IEEE Communications Magazine, October 2004.

[57] David Gesbert, et al., "Shifting the MIMO Paradigm." IEEE Signal Processing Magazine (September 2007).

[58] Lu Wei, "Capacity of Hybrid Open-loop and Closed-loop MIMO with Channel Uncertainty at Transmitter." ESPOO, Helsinki University of Technology, March 2008.

[59] David J. Love and Robert W. Heath Jr., "Grassmannian Precoding for Spatial Multiplexing Systems." Proc. of the Allerton Conference on Communication Control and Computing, Monticello, October 2003.

[60] Tetsushi Abe, Gerhard Bauch, "Differential Codebook MIMO Precoding Technique." 2007 IEEE Global Telecommunications Conference, 2007.

[61] Bernd Bandemer, Martin Haardt, Samuli Visuri, "Linear MMSE Multi-User MIMO Downlink Precoding for Users with Multiple Antennas," 17th Annual IEEE International Symposium on Personal, Indoor and Mobile Radio Communications (PIMRC '06), 2006.

[62] Gavin Mitchell, Frank R. Kschischang, "An Augmented Orthogonal Code Design for the Non-coherent MIMO Channel," 24th Biennial Symposium on Communications, June 2008.

[63] Yong Fan, Petteri Lundén, Markku Kuusela and Mikko Valkama, "Efficient Semi-Persistent Scheduling for VoIP on EUTRA Downlink." IEEE 68th Vehicular Technology Conference, VTC 2008, Fall 2008.

[64] S. Lawrence Marple, Digital Spectral Analysis: With Applications, Prentice Hall, 1987.

[65] Gene H. Golub, Charles F. Van Loan, Matrix Computations, third ed. Johns Hopkins University Press, 1996.

[66] Athanasios Papoulis, Probability, Random Variables and Stochastic Processes, fourth ed. McGraw Hill Higher Education, 2002.

[67] Lennart Rade, Mathematics Handbook for Science and Engineering, Springer Berlin Heidelberg, 2010.

[68] Granino A. Korn, Theresa M. Korn, Mathematical Handbook for Scientists and Engineers: Definitions, Theorems, and Formulas for Reference and Review, Revised ed. Dover Publications, 2000.

[69] Goldsmith, et al., "Capacity Limits of MIMO Channels," IEEE Journal on Select Areas in Communications vol. 21 (June 2003).

[70] H. Weingarten, Y. Steinberg and S. Shamai, "The Capacity Region of the Gaussian MIMO Broadcast Channel." Proceedings of Conference Information Sciences and Systems (CISS), Princeton, NJ, Mar. 2004.

[71] M. Costa, "Writing on Dirty Paper." IEEE Transactions on Information Theory, vol. 29 (May 1983).

[72] T. Svantesson and A.L. Swindlehurst, "A Performance Bound for Prediction of a Multipath MIMO Channel." Proc. 37th Asilomar Conference on Signals, Systems, and Computers, Session: Array Processing for Wireless Communications, Pacific Grove, California, November 2003.

[73] C.B. Peel, B.M. Hochwald, A.L. Swindlehurst, "A Vector-Perturbation Technique for Near-Capacity Multi-Antenna Multi-User Communication.", IEEE Transactions on Communications (June 2003).

[74] M. Bengtsson, B. Ottersten, Optimal and Suboptimal Beamforming." in: L.C. Godara (Ed.), Handbook of Antennas in Wireless Communications, CRC Press, 2001.

[75] M. Schubert, H. Boche, "Solution of the Multiuser Downlink Beamforming Problem with Individual SINR Constraints.", IEEE Transactions on Vehicular Technology vol. 53 (January 2004).

[76] Q.H. Spencer, A.L. Swindlehurst, M. Haardt, "Zero-Forcing Methods for Downlink Spatial Multiplexing in Multi-User MIMO Channels.", IEEE Transactions on Signal Processing vol. 52 (February 2004).

[77] G. Caire, S. Shamai, "On the Achievable Throughput of a Multi-Antenna Gaussian Broadcast Channel." IEEE Transactions Information Theory vol. 49 (July 2003).

[78] U. Erez, S. Shamai, R. Zamir, "Capacity and Lattice Strategies for Cancelling Known Interference," Proceedings International Symposium Information Theory and its Applications, November 2000.

[79] C. Windpassinger, R.F.H. Fischer, J.B. Huber, "Lattice-Reduction-Aided Broadcast Precoding." Proceedings of 5th ITG Conference Source and Channel Coding, January 2004.

[80] Berrou, Glavieux, Thitimajshima, "Near Shannon Limit Error-Correcting Coding and Decoding: Turbo-codes." Proceedings of 1993 International Communication Conference, May 1993.

[81] Q.H. Spencer, A.L. Swindlehurst, "Channel Allocation in Multi-user MIMO Wireless Communications Systems," Proceedings of 2004 International Communication Conference, June 2004.

[82] Perez Seghers, Costello, "A Distance Spectrum Interpretation of Turbo Codes," IEEE Transactions on Information Theory vol. 42 (No. 6) (November 1996).

[83] Shu Lin, Daniel J. Costello, Error Control Coding, second ed. Prentice Hall, 2004.

[84] J. Rothweller, "Turbo Codes." IEEE Potentials vol. 18 (Issue 1) (February–March 1999).

[85] C. Berrou, A. Glavieux, P. Thitimajshima, "Near Shannon Limit Error-Correcting Coding and Decoding: Turbo Codes." Proceedings of 1993 International Communication Conference, May 1993.

[86] C. Berrou, A. Glavieux, "Near Optimum Error Correcting Coding and Decoding: Turbo-Codes." IEEE Transactions on Communications vol. 44 (No. 10) (October 1996).

[87] IEEE C802.16m–09/1259, Jin Xu, Bo Sun, Evaluation Simulation for Channel Coding and HARQ, July 2009.

[88] Claude E. Shannon, Warren Weaver, The Mathematical Theory of Communication, University of Illinois Press, 1998.

[89] Farid Dowla, Handbook of RF and Wireless Technologies, first ed. Newnes, 2003.

[90] Chris Heegard, Stephen B. Wicker, Turbo Coding. Springer, 2010.

[91] C. Berrou, R. Pyndiah, P. Adde, C. Douillard, R. Le Bidan, "An overview of turbo codes and their applications." The European Conference on Digital Wireless Technology, 2005.

[92] H. Ma, J. Wolf, "On Tail Biting Convolutional Codes." IEEE Transactions on Communications vol. COM-34 (No. 2) (February 1986).

[93] C. Weiss, C. Bettstetter, S. Riedel, "Code Construction and Decoding of Parallel Concatenated Tail-Biting Codes." IEEE Transactions on Information Theory vol. 47 (No. 1) (January 2001).

[94] Y.E. Wang, R. Ramesh, "To Bite or not to Bite – A study of Tail Bits versus Tail-Biting." 7th IEEE International Symposium Personal, Indoor and Mobile Radio Communications, PIMRC '96, Vol. 2, (October 1996).

[95] 3GPP Long Term Evolution: System Overview, Product Development, and Test Challenges, Agilent Technologies, June 2009 <http://www.agilent.com>.

[96] Hyung G. Myung, "Technical Overview of 3GPP LTE." May 2008 <http://hgmyung.googlepages.com/scfdma>.

[97] E. Dahlman, et al., 3G 4Evolution: HSPA and LTE for Mobile Broadband, second ed. Academic Press, 2008.

[98] Basuki E. Priyanto, Humbert Codina, "Initial Performance Evaluation of DFT-Spread OFDM Based SC-FDMA for UTRA LTE Uplink.", IEEE 65th Vehicular Technology Conference VTC2007, 2007.

[99] 3G Americas, MIMO and Smart Antennas for 3G and 4G Wireless Systems, May 2010 <http://www.3gamericas.org/>.

[100] Hyung G. Myung, "Single Carrier FDMA." May 2008 <http://hgmyung.googlepages.com/scfdma>.

[101] Juho Lee, Jin-Kyu Han, Jianzhong Zhang, "MIMO Technologies in 3GPP LTE and LTE-Advanced." EURASIP Journal on Wireless Communications and Networking, vol (2009).

[102] Rohde & Schwarz Application Notes, LTE-Advanced Technology Introduction, March 2010.

[103] 3G Americas, 3GPP Mobile Broadband Innovation Path to 4G: Release 9, Release 10 and Beyond: HSPA+, SAE/LTE and LTE-Advanced, February 2010.

[104] 3G Americas, HSPA to LTE-Advanced: 3GPP Broadband Evolution to IMT-Advanced (4G), September 2009.

[105] 3G Americas, The Mobile Broadband Evolution: 3GPP Release 8 and Beyond HSPA+, SAE/LTE and LTE-Advanced, February 2009.

[106] 3GPP TS 36.201 Evolved Universal Terrestrial Radio Access (E-UTRA) Physical Layer – General Description, March 2010.

[107] 3GPP TS 36.211 Evolved Universal Terrestrial Radio Access (E-UTRA), Physical Channels and Modulation, March 2010.

[108] 3GPP TS 36.212 Evolved Universal Terrestrial Radio Access (E-UTRA) Multiplexing and Channel Coding, March 2010.

[109] 3GPP TS 36.213 Evolved Universal Terrestrial Radio Access (E-UTRA); Physical layer procedures, March 2010.

[110] 3GPP TS 36.214 Evolved Universal Terrestrial Radio Access (E-UTRA); Physical layer – Measurements, March 2010.

[111] 3GPP TS 36.104 Evolved Universal Terrestrial Radio Access (E-UTRA); Base Station (BS) Radio Transmission and Reception, March 2010.

[112] 3GPP TS 36.101 Evolved Universal Terrestrial Radio Access (E-UTRA); User Equipment (UE) Radio Transmission and Reception, March 2010

[113] 3GPP TS36.321 Evolved Universal Terrestrial Radio Access (E-UTRA); Medium Access Control (MAC) Protocol Specification, March 2010.

[114] 3GPP TR 36.912 Feasibility Study for Further Advancements for E-UTRA (LTE-Advanced), March 2010.

[115] 3GPP TR 36.913 Requirements for Evolved UTRA (E-UTRA) and Evolved UTRAN (E UTRAN), March 2009.

[116] 3GPP TS 36.300 Evolved Universal Terrestrial Radio Access (E-UTRA) and Evolved Universal Terrestrial Radio Access Network (E-UTRAN) Overall Description, Stage 2, March 2010.

[117] 3GPP TR 36.814 Feasibility Study for Further Advancements for E-UTRA (LTE-Advanced), March 2010.

[118] IEEE C802.16m–09/1940r1, Ping Wang, et al., Assignment A-MAP Group Size Indication in Non-User Specific A-MAP IE, August 2009.

[119] IEEE C802.16m–10/0445r1, Roshni Srinivasan, Alexei Davydov, et al., UL MIMO Transmission Format in the 802.16m UL Basic Assignment A-MAP Information Element, March 2010.

[120] MATLAB CENTRAL, An Open Exchange for the MATLAB and SIMULINK User Community <http://www.mathworks.com/matlabcentral>.

[121] 3GPP TS 25.101 User Equipment (UE) Radio Transmission and Reception (FDD), March 2009.

[122] 3GPP R1-060385 Cubic Metric in 3GPP-LTE, Motorola, February 2006.

[123] IEEE C802.16m–08/153, Bin-Chul Ihm, Jinsoo Choi, Wookbong Lee, Pilot related to DL MIMO, March 2008.

[124] The Coded Modulation Library (The Iterative Solutions Coded Modulation Library (ISCML) is an open source toolbox for simulating capacity approaching codes in Matlab), <http://www.iterativesolutions.com>.

Multi-Carrier Operation

11

INTRODUCTION

The World Radio Conference (WRC) 2007 identified new frequency bands for International Mobile Telecommunications (IMT) systems (which consist of both IMT-2000 and IMT-Advanced), some of which are Region 1[i] specific

- 450–470 MHz globally;
- 698–806 MHz in Region 2 and nine countries in Region 3;
- 790–862 MHz in Region 3 and part of Region 1 countries;
- 2.3–2.4 GHz globally;
- 3.4–3.6 GHz in a large number of countries in Regions 1 and 3.

The intention was that the bands previously identified in the radio regulations for IMT-2000 are now applicable to the IMT systems. The work on the frequency arrangement is ongoing in ITU-R Working Party (WP) 5D and finalization of the band plans is expected in February 2011 [33]. Spectrum, and more recently carrier aggregation, have become particularly important for the successful deployment and adoption of IMT-Advanced systems. The concept of spectrum aggregation consists of exploiting multiple, small spectrum fragments simultaneously to deliver a wider band service (i.e., not otherwise achievable when using a single spectrum fragment). Spectrum aggregation can be useful when an operator's dedicated band is not continuous; rather it is split into two or more segments. In addition, spectrum aggregation can happen in scenarios in which an operator accesses both a dedicated band, and a spectrum sharing band which is separated in frequencies from the dedicated operator's band. Spectrum aggregation allows new high data rate wireless communication systems to coexist with their legacy systems when deployed in the same spectrum. This is also valid for the inter-operator scenario. In this context, it can be very beneficial to explore the scenarios for joint use of spectrum aggregation techniques and radio resource management in radio access networks [33].

Support of wider bandwidths up to 100 MHz is one of the distinctive features of IMT-Advanced systems [3,4]. The IMT-Advanced systems target a peak data rate up to 1 Gbps for low mobility and 100 Mbps for high mobility [4]. In order to support wider transmission bandwidths IEEE 802.16m and 3GPP LTE-Advanced systems introduced the carrier aggregation concept, where two or more Component Carriers (CC) belonging to a single frequency band or different frequency bands can be aggregated [2,21]. The support of system bandwidths up to 100 MHz will

---

[i]Region 1: Europe, the Middle East and Africa (EMEA); Region 2: North and South America (Americas); Region 3: Asia-Pacific (APAC) (see http://www.itu.int/ITU-R/).

Mobile WiMAX. DOI: 10.1016/B978-0-12-374964-2.10011-6
Copyright © 2011 Elsevier Inc.

allow an increase in both the peak data rate and the system capacity. In the scenarios of interest, the peak data rate increases almost linearly proportional to increasing bandwidth. Hence, for example, to obtain the required peak data rate of 1 Gbps in the downlink with 30 bps/Hz peak spectral efficiency (assuming 8-layer MIMO transmission), a bandwidth of only 40 MHz is sufficient. It is noted that in addition to an increase in the peak data rate, the bandwidth extension results in higher data rates for all mobile terminals in a cell, due to increased average and cell-edge throughputs.

Using the carrier aggregation scheme, it would be possible to simultaneously schedule a user on multiple component carriers for downlink or uplink data transmission, resulting in some challenges in resource scheduling and load balancing across the network. In a non-adjacent inter-band aggregation scenario, where the aggregated carriers belong to different frequency bands, the fading characteristics might be different between component carriers; consequently, the coverage may vary from one carrier to another. With different locations in the cell, some users can only be scheduled on certain carriers, while other users be scheduled on all aggregated carriers; this has a negative impact on the fairness of allocation among users. The original proportional fair algorithms can be used as a trade-off between the system throughput and fairness, but they cannot deal with the above-mentioned problem, since all of them assume that the users can be scheduled on the same number of carriers [5–33].

The base station performs admission control and carrier load balancing in order to allocate the users to different component carriers. Different methods for balancing the load across component carriers are possible and will impact on system performance [10]. Once the users are assigned to a certain component carrier(s), packet scheduling is performed where time-frequency resources are assigned to each of the allocated users on different component carriers. While independent physical layer transmission is assumed, packet scheduling can be performed either independently within each component carrier or jointly across multiple component carriers. The objective is to optimize the resource allocation process with the existence of multiple component carriers.

In this chapter, we review the physical layer and MAC layer aspects of bandwidth extension and carrier aggregation schemes that have been utilized in the IEEE 802.16m and the 3GPP LTE-Advanced [2,21–24]. It will be seen that the two technologies use very similar techniques in order to support larger bandwidths in a contiguous or non-contiguous spectrum.

11.1 PRINCIPLES OF MULTI-CARRIER OPERATION

One of the distinctive features of the 4th generation of cellular systems is the ability to operate at extremely large bandwidths. The RF spectrum utilized for such large bandwidth operation may comprise contiguous or non-contiguous bands (virtual wideband operation). The component carriers corresponding to each frequency band can be assigned to unicast and/or multicast and broadcast services. The multi-carrier operation enables control and operation of a number of contiguous or non-contiguous component carriers (several physical layers) using a single MAC instantiation. In FDD mode, each downlink or uplink frequency channel is individually referred to as a component carrier and is assigned a distinct physical carrier index, whereas in TDD mode each frequency channel with downlink and uplink partitions is designated as a component carrier and is

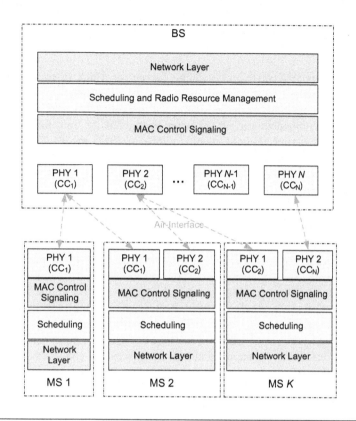

FIGURE 11-1

The concept of multi-carrier operation

assigned a distinct physical carrier index. The physical carrier index is an identifier for all available component carriers across the entire network. The mobile station is not required to support multi-carrier operation in certain device classes. However, if it does support multi-carrier operation, it may receive control and signaling, broadcast, and synchronization channels through a primary component carrier and traffic assignments (or services) may be made on the secondary component carriers. The generalization of the protocol structure to multi-carrier support using a single MAC instance is shown in Figure 11-1. The load balancing functions and component carrier mapping/assignment are performed via radio resource control and management functional class. The component carriers utilized in a multi-carrier system, from the perspective of a mobile station, can be divided into two categories:

- Primary component carrier: a standalone carrier through which the MS conducts initial network entry or network re-entry. When supporting multi-carrier operations, an MS only has one primary component carrier.
- Secondary component carrier: additional carrier(s) which may be assigned to the MS by the BS. All the unicast MAC control messages corresponding to multi-carrier operation are sent to the MS through its primary component carrier.

Depending on the use case, the component carriers may be configured differently in one of the following configurations:

- Fully configured component carrier: a standalone carrier for which all control channels including synchronization, broadcast, unicast, and multicast control channels are configured. A fully configured carrier is supported by all the mobile stations regardless of multi-carrier capability.
- Partially configured component carrier: a carrier configured for downlink-only transmission in TDD or a downlink carrier without paired uplink in FDD mode. The partially-configured carriers may be used only in conjunction with a primary carrier and cannot operate as a standalone to provide service to mobile stations.

If a partially-configured carrier is used for downlink unicast traffic, the uplink feedback channels are then provided by the primary carrier. The uplink control channels corresponding to the secondary partially-configured carriers are located in distinct non-overlapping control regions in the uplink of the primary carrier following the uplink control region of the primary carrier where their location is signaled through an AAI_SCD MAC control message. The MS uses these control channels to send HARQ ACK/ NACK and channel quality measurements corresponding to data transmission over a downlink-only secondary carrier. Only an FDD primary carrier may be used to provide uplink feedback channels for partially-configured carriers. A partially-configured carrier may be used for multicast and broadcast services. In this case, it would not need feedback channels on the primary carrier. A primary component carrier is fully-configured, while a secondary carrier may be fully or partially-configured depending on the deployment scenarios. The information about component carrier configuration (fully or partially-configured) is carried through the primary advanced preamble of the corresponding carrier.

A secondary carrier for an MS, if fully-configured, may be used as a primary carrier for other mobile stations; therefore, the designation of a carrier component as primary or secondary is relative rather than abstract. Multiple mobile stations with different primary carriers may also share the same physical carrier as their secondary carrier. The following multi-carrier operation modes may be supported in various deployment scenarios:

- Carrier Aggregation: a multi-carrier mode in which the MS maintains its physical layer connection with the primary component carrier while transmitting or receiving data on the secondary carrier(s). The resource allocation to an MS may span the primary and one or more secondary carriers. The link adaptation mechanisms will then rely on channel quality measurements on primary and secondary carriers. In this mode, the system may assign secondary carriers to an MS in the downlink and/or uplink asymmetrically based on MS capability, system load (i.e., for static/ dynamic load balancing), peak data rate, or QoS requirements.
- Carrier Switching: a multi-carrier mode in which the MS switches its physical layer connection from the primary to the (partially or fully-configured) secondary carrier based on the serving base station's instruction in order to receive E-MBS data on the secondary carrier. The MS will connect to the secondary carrier for the specified time interval before returning to the primary carrier. When the MS is connected to the secondary carrier, it is not required to maintain connection with the primary carrier.
- Basic Multi-carrier Mode: an operating mode in which the MS uses only one component carrier for normal operation, but supports the primary carrier switching as well as optimized scanning of carriers involved in multi-carrier operation.

The MS may support at least one of the above multi-carrier modes if it is multi-carrier capable. The following features are common to all multi-carrier modes:

- The system designates N standalone fully-configured carriers, each containing synchronization, broadcast, multicast, and unicast control channels required to support a single-carrier MS. Each MS in the serving cell is connected to a primary carrier.
- The system may also designate $M \geq 0$ partially-configured carriers, which can only be used as secondary carriers, along with a primary carrier for downlink-only data transmissions.
- The set of all component carriers used by a BS are called the available carriers. The available carriers may be in different parts of the contiguous spectrum block or in non-contiguous spectrum blocks.
- A sub-set of the available carriers are designated as assigned carriers, which can be activated for data transmission.
- A sub-set of assigned carriers are designated as active carriers, which are used for data transmission between the MS and the BS.
- In addition to the information about the primary carrier, the serving BS can also provide the MS with some configuration information about its available carriers through the primary carrier by the AAI_Global-Config and AAI_MC-ADV MAC control messages.

11.2 SUB-CARRIER ALIGNMENT AND USE OF GUARD SUB-CARRIERS

If the operating frequency bands are physically contiguous and the sub-carriers of the adjacent frequency channels are perfectly aligned, the guard sub-carriers between two adjacent frequency bands can be utilized for data transmission. In the current IEEE 802.16-2009 standard [1]. the center frequency of each component carrier is an integer multiple of 250 kHz, also known as a frequency raster, thus the separation of any adjacent carriers will be an integer multiple of 250 kHz (e.g., 10 MHz frequency separation for 10 MHz channel bandwidth). In this case, the sampling frequency is 11.2 MHz and the sub-carrier spacing is 10.94 kHz. Given that 10 MHz is not an integer multiple of the sub-carrier spacing, the use of the guard sub-carriers will cause significant Inter-Carrier Interference (ICI) to the adjacent bands. Figure 11-2 illustrates the ICI problem which could be caused by using sub-carriers in guard bands between the adjacent component carriers. On the other hand, if the frequency raster is modified to make the frequency separation an integer multiple of sub-carrier spacing, backward compatibility and interoperability with legacy system is compromised, but the guard-band sub-carriers can be used for data transmission.

The guard sub-carrier can then be utilized for data transmission, if the information of the available guard sub-carriers suitable for data transmission is communicated to multi-carrier-capable mobile stations. This information includes the number of available sub-carriers in the upper-side and the lower-side relative to the DC sub-carrier. As a result, when two physically contiguous carriers are used to deploy the IEEE 802.16m system in a multi-carrier operation, the overlapped sub-carriers should be aligned in the frequency-domain. In order to align the overlapped sub-carriers of the OFDM symbols transmitted over the adjacent carriers, a permanent frequency offset Δf_{offset} is applied to one of the carriers. The basic principle is illustrated in Figure 11-3 where the overlapping of guard sub-carriers before and after application of frequency offset Δf_{offset} is shown. A carrier

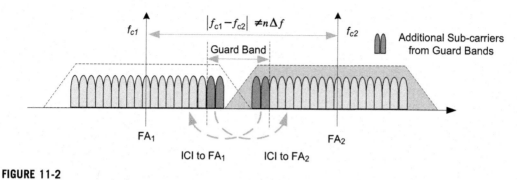

FIGURE 11-2

An illustration of adjacent channel ICI due to use of guard bands

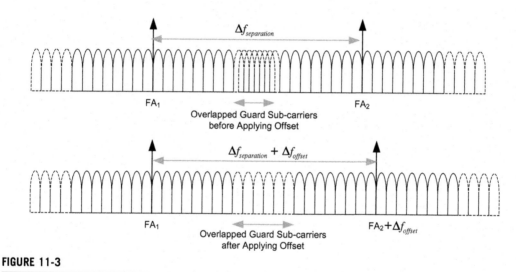

FIGURE 11-3

An illustration of a sub-carrier alignment scheme [2]

group is a group of contiguous carriers whose sub-carriers are aligned with a sub-carrier spacing Δf (see Table 9-2).

When one component carrier is used for co-deployment of the IEEE 802.16m and the legacy system, and the adjacent component carrier is exclusively dedicated to the IEEE 802.16m system, the overlapped sub-carriers should also be aligned. The center frequency of the mixed-mode component carrier will be located exactly on the channel raster grid. Different base stations may have different multi-carrier configurations according to the available spectral resources and the restriction due to the support of the legacy system. If the two adjacent component carriers are used to deploy a mixed-mode system, the overlapped sub-carriers may not be aligned. If the MS cannot support carrier aggregation due to hardware restriction under a sub-carrier misalignment configuration, it is required to inform the

Table 11-1 Multi-Carrier Configurations and Component Carrier Offset Values [2]		
Index	**Multi-Carrier Configuration (MHz)**	**Frequency Offset (kHz)**
1	(5)	(0)
2	(7)	(0)
3	(8.75)	(0)
4	(10)	(0)
5	(20)	(0)
6	(10, 10)	(0, −3.1248)
Other values are reserved		

serving BS about the component carriers that it can simultaneously process through an AAI_MC-REQ MAC control message.

During network entry, the serving BS will notify the MS of the frequency offset to be applied to each component carrier for the purpose of sub-carrier alignment through the AAI_MC-ADV message. According to the multi-carrier configuration index and the physical carrier index of the current component carrier that is broadcast by the BS, the MS can derive the center frequency of the available component carriers by the associated frequency offset Δf_{offset} using Table 11-1. The frequency offset Δf_{offset} specified in Table 11-1 is calculated as follows:

$$\Delta f_{offset} = \begin{cases} |f_{c2} - f_{c1}| \bmod \Delta f & \Delta f \geq 2|f_{c2} - f_{c1}| \bmod \Delta f \\ (|f_{c2} - f_{c1}| \bmod \Delta f) - \Delta f & \Delta f < 2|f_{c2} - f_{c1}| \bmod \Delta f \end{cases} \tag{11-1}$$

where f_{c2} is the center frequency of the component carrier before applying frequency offset Δf_{offset} and f_{c1} denotes the center frequency of the reference component carrier in each carrier group. Note that the frequency offset is calculated relative to the center frequency of the reference component carrier. The center frequency of the reference component carrier in each carrier group is always located on the channel raster grid. The center frequency of the offset component carrier is given as:

$$f_{c2} = \begin{cases} f_{c2} - \Delta f_{offset} & f_{c2} \geq f_{c1} \\ f_{c2} + \Delta f_{offset} & f_{c2} < f_{c1} \end{cases} \tag{11-2}$$

Note that the absolute value of the frequency offset Δf_{offset} is less than the OFDMA sub-carrier spacing Δf. Table 11-1 is used to identify the permissible configurations for a group of contiguous component carriers. If the network supports multiple groups of contiguous component carriers, each group of the contiguous carriers is identified in the AAI_MC-ADV MAC control message by an index in this table. As an example, the multi-carrier configuration {5, 10} represents two contiguous component carriers of 5 and 10 MHz. Based on the center frequency of the component carrier on which the MS receives the AAI_MC-ADV message, and the bandwidth of each component carrier, the center frequency of each component carrier before sub-carrier alignment can be calculated. Thus, the MS can obtain the frequency offset Δf_{offset} to be applied to each component carrier based on the multi-carrier configuration index, the physical carrier index of the current carrier, and the information in Table 11-1. If two adjacent component carriers are used to deploy the IEEE 802.16m and the legacy systems, they are considered as two non-contiguous carriers and identified with different carrier group indices in the

AAI_MC-ADV message. The BS and the (multi-carrier capable) MS are required to encode and decode all multi-carrier configurations in Table 11-1, and to apply the corresponding frequency offset before activating multi-carrier operation. Under the conditions discussed earlier, the guard sub-carriers between contiguous frequency channels may be utilized for data transmission. During the secondary carrier assignment procedure, the serving BS sends the information about available guard sub-carriers, which can be used for data transmission, to the MS.

The support for multiple component carriers is provided with the same frame structure as that used for single-carrier operation; however, some considerations in the design of protocol and channel structure may be needed to facilitate multi-carrier operation. In general, each MS compliant with the IEEE 802.16m standard is served by one component carrier, which is referred to as the primary component carrier. When multi-carrier operation is enabled, the system may define and utilize additional component carriers to improve the user experience and to meet QoS requirements for various services, or to provide services through additional component carriers which might have been configured or optimized for those services. Figures 11-4 and 11-5 depict the frame structure used for single-carrier and multi-carrier operation. A number of narrowband component carriers may be aggregated to support virtual wideband operation. Each component carrier may include synchronization and broadcast channels. A multi-carrier-enabled MS can utilize radio resources across multiple component carriers using a common MAC. Depending on the mobile station's capabilities, such utilization may include aggregation or switching of traffic across multiple component carriers controlled by a single MAC instance.

As mentioned earlier, the component carriers utilized in multi-carrier operation may be contiguous or non-contiguous. If component carriers are in the same frequency band and

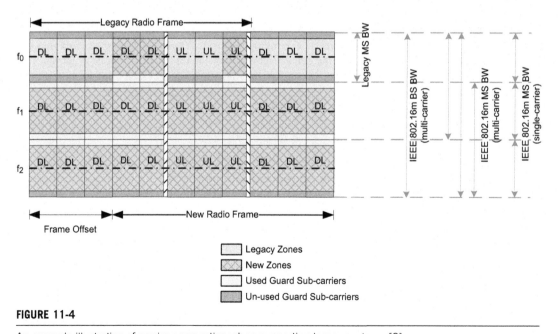

FIGURE 11-4

An example illustration of carrier aggregation when supporting legacy systems [2]

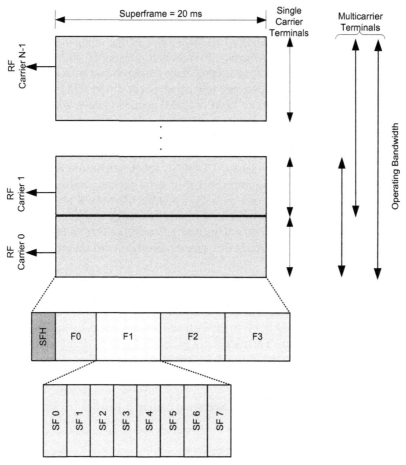

FIGURE 11-5

An example of a multi-carrier frame structure [2]

physically adjacent, and the frequency separation of two adjacent carriers is an integer multiple of OFDM sub-carrier spacing, no guard sub-carriers are necessary between adjacent component carriers (see Figure 11-4). When component carriers are in a non-contiguous spectrum, the number of uplink subframes is not necessarily the same for all the carriers in the TDD mode. The same frame structure is used for each component carrier in a multi-carrier operation. Each component carrier is required to carry a superframe header which may include part of a super-frame header. Figure 11-5 illustrates an example of frame structure to support multi-carrier operation. The preamble and superframe headers are replaced with OFDM symbols carrying user data and control in the uplink direction in FDD mode.

Figure 11-4 shows an example of an IEEE 802.16m frame structure supporting legacy systems. A number of narrowband component carriers supporting an IEEE 802.16m system can be aggregated to enable wideband operation of mobile stations. One or more component carriers can be designated as the legacy carrier(s), depending on the number of legacy terminals in the network. A different number of usable guard sub-carriers can be allocated on both sides of the component carrier. For uplink transmissions, either time-division or frequency-division multiplexing of the new and legacy mobile stations may be supported over the mixed-mode component carrier. If two adjacent component carriers include the IEEE 802.16m and the legacy zones (e.g., component carrier 0 in Figure 11-4), the frame offset and uplink multiplexing scheme (i.e., TDM or FDM) applied to these component carriers must be the same. The information about the frame configuration of other component carriers is provided through the S-SFH SP1 or via the AAI_MC-ADV and AAI_NBR-ADV messages. If two adjacent component carriers contain the IEEE 802.16m and legacy zones, the guard sub-carriers in the uplink are not used in order to allow use of the legacy UL-PUSC subchannelization scheme.

Each component carrier may utilize sub-carriers at its band edges as additional data sub-carriers. The guard sub-carriers are grouped to form an integer number of physical resource units. The PRU structure used for the guard resource is the same as the structure of the regular physical resource units described in Chapter 9. Contiguous resource units may be constructed from physical resource units, including those formed by guard sub-carriers. The guard sub-carriers are not used for the transmission of control channels.

As shown in Figure 11-6, the guard sub-carriers between adjacent component carriers are grouped to form an integer number of PRUs. The structure of a guard-PRU is identical to that of a downlink subchannel. The guard-PRUs are used as mini-band CRUs in partition FP_0 for data transmission only. The number of useable guard sub-carriers is predefined and is known to both the MS and the BS based on the component carrier bandwidth. The number of guard-PRUs in the left and right edges of each component carrier is shown in Table 9-2 (OFDMA parameters). The number of guard-PRUs in the left and right edges of the component carrier is denoted by N_{LG-PRU} and N_{RG-PRU}, respectively. The total number of guard-PRUs is $N_{G-PRU} = N_{LG-PRU} + N_{RG-PRU}$. It must be noted that when a component carrier occupies the left-most segment of the spectrum among a number of contiguous component

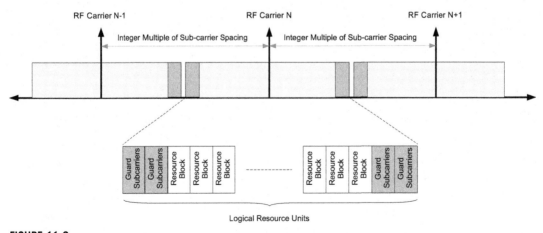

FIGURE 11-6

An example of data transmission using guard sub-carriers

Table 11-2 Number of Guard Resource Units in the Downlink and Uplink [2]

Component Carrier Bandwidth (MHz)	Number of Guard-PRUs in the Left Edge of Component Carrier	Number of Guard-PRUs in the Right Edge of Component Carrier
5	0	0
7	0	0
8.75	0	0
10	1	1
20	2	2

carriers, the number of guard-PRUs in the left edge of the carrier is zero. Furthermore, when a component carrier occupies the right-most segment of the spectrum among a number of contiguous component carriers, the number of guard-PRUs in the right edge of carrier is set to zero [2]. Note that the procedures for the formation of the guard-PRUs are the same in the downlink and uplink.

The left and right guard-PRUs are denoted as $\text{G-PRU}_L[0]$, $\text{G-PRU}_L[1]$, ..., $\text{G-PRU}_L[N_{LG\text{-}PRU}-1]$ and $\text{G-PRU}_R[0]$, $\text{G-PRU}_R[1]$, ..., $\text{G-PRU}_R[N_{RG\text{-}PRU}-1]$, respectively, from the lowest frequency index. The guard-PRUs are indexed by interleaving G-PRU_L and G-PRU_R, i.e., $\text{G-PRU}[i] = \text{G-PRU}_L[i/2]$ for even-valued i, and $\text{G-PRU}[i] = \text{G-PRU}_R[(i-1)/2]$ for odd-valued i, where i is an integer. If $N_{LG\text{-}PRU} = 0$, then $\text{G-PRU}[i] = \text{G-PRU}_R[i]$. If $N_{RG\text{-}PRU} = 0$, then $\text{G-PRU}[i] = \text{G-PRU}_L[i]$. The $N_{G\text{-}PRU}$ guard-PRUs are used as mini-band LRUs, i.e., the mini-band LRUs at frequency partition FP_0 with no permutation for data transmission. The ith guard mini-band LRU is always allocated along with the last ith mini-band LRU in partition FP_0. In other words, when an MS with multi-carrier capability is provided an allocation, it includes the last ith mini-band LRU in partition FP_0 together with the ith guard mini-band LRU. When an adjacent component carrier is not an active carrier for the MS, the guard sub-carriers between active and inactive carriers are not utilized for data transmission. When the overlapped guard sub-carriers are not aligned in the frequency domain, they are not used for data transmission.

The primary and secondary preambles, as well as the superframe headers, are present in fully-configured and partially-configured carriers. The location and transmission format of these overhead channels are the same as that of the single-carrier. The A-MAP is transmitted over a fully-configured component carrier. The location and transmission format of the A-MAP on the fully-configured carrier is the same as that for the single-carrier mode. Additional broadcast information related to multi-carrier operation is carried over the fully-configured carrier, except uplink information, additional broadcast information related to the operation of a partially-configured carrier can be carried on the partially-configured carrier. The uplink control channels corresponding to the single-carrier operation are supported for the fully-configured carrier. A partially-configured carrier may not include any uplink transmission capability, and is exclusively optimized for downlink-only transmission (e.g., multicast and broadcast services).

The BS configures a set of component carriers on which the MS conducts measurements and sends feedback information. The BS may only allocate resources to the MS over a subset of configured carriers. The required feedback for link adaptation and information for closed-loop MIMO operation can be sent through the primary carrier. The HARQ feedback corresponding to the PDUs transmitted over the primary and secondary carriers can be sent via the primary carrier. The HARQ feedback related to the PDUs transmitted on the secondary carrier is sent over the secondary carrier, if the secondary carrier configuration permits.

The initial ranging for a non-synchronized MS is conducted on a fully-configured carrier. The periodic ranging for a synchronized MS is performed on the primary carrier, but may also be performed on a secondary carrier, depending on the secondary carrier configuration. The serving BS transmits the ranging response on the same carrier that received the initial/periodic ranging message. The uplink sounding is conducted on the primary and secondary carriers. The bandwidth request is transmitted only on the primary carrier. Depending on the correlation between different component carriers, separate uplink power control for active carriers is necessary.

11.3 CARRIER AGGREGATION AND SPECTRAL MASK CONSIDERATIONS

Following the introduction of carrier aggregation and support of wider bandwidths, we should also consider the effects of bandwidth extension and the use of guard sub-carriers on the design of the spectral mask. The studies suggest that the spectral mask of the aggregated spectrum is different from that of the non-aggregated spectrum [32]. Figure 11-7 illustrates the difference between the two cases. Note that the spectrum emission mask is scaled in proportion with the channel bandwidth and the relationship between the single-carrier and aggregated-carrier spectral masks. An example spectral mask specification is shown in Figure 11-7.

The contiguous/non-contiguous carrier aggregation is a new concept for the BS and the MS RF specifications, and requires the appropriate extension of transmission (e.g., Adjacent Channel Leakage Power Ratio [ACLR][ii] and spurious emissions) and reception (e.g., Adjacent Channel Selectivity [ACS],[iii] blocking) RF requirements. The relative power difference between the adjacent component carriers can be an issue if the guard sub-carriers between the two component carriers are utilized for data transmission. The symmetrical property of the spectral mask may help in the design of transmit/receive filters.

The carrier aggregation will also change some radio resource management requirements, considering that a terminal will have to conduct measurements on several carriers at the same time. In practice, a mobile station is typically able to measure two carriers at the same time in active mode. The mobile station's battery consumption may be increased due to the activation of several carriers which need to be continuously monitored and measured. The MS power consumption increases with an increasing number of component carriers. The increased number of component carriers implies that the MS will have to monitor more downlink control channels.

While the cost and complexity of some hardware and software components may only depend on the total bandwidth, the cost/complexity, in general, would scale with the number of component carriers. As mentioned earlier, there are two types of carrier aggregation: contiguous; and non-contiguous. Non-contiguous carrier aggregation can be in the form of intra-band or inter-band. The different types of carrier aggregation will result in different deployment scenarios. In the case of contiguous carrier aggregation and intra-band carrier aggregation, if each component carrier has the same transmit power,

---

[ii]The Adjacent Channel Leakage power Ratio (ACLR) is the ratio of the filtered mean power centered on the assigned channel frequency to the filtered mean power centered on an adjacent channel frequency.
[iii]Adjacent Channel Selectivity (ACS) is a measurement of a receiver's ability to process a desired signal while rejecting a strong signal in an adjacent frequency channel. ACS is defined as the ratio of the receiver filter attenuation on the assigned channel frequency to the receiver filter attenuation on the adjacent channel frequency.

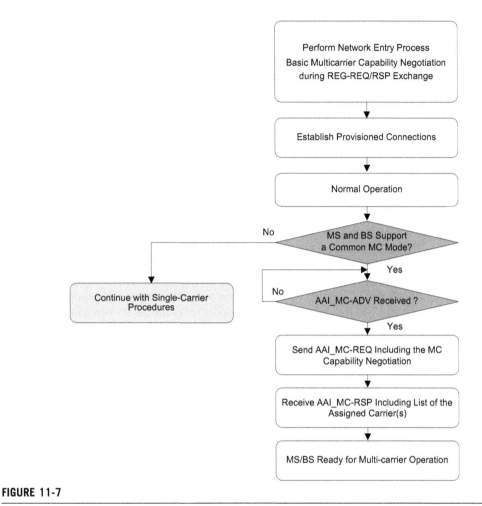

FIGURE 11-7

Carrier aggregation and spectral mask considerations [32]

then the coverage of each component carrier will be approximately the same, and it would be easier for mobile stations to use carrier aggregation in this case. However, for inter-band carrier aggregation, it is more difficult for all aggregated carriers to have the same coverage relative to the contiguous case. This is mainly due to the large path loss difference for the inter-band component carriers. The support of carrier aggregation when the component carriers have different coverage would require more efficient management and measurement procedures [33].

For non-contiguous carrier aggregation, the component carriers are usually separated by a sufficient frequency gap; therefore, the interference between aggregated bands is negligible. However, there are still frequency bands belonging to other systems adjacent to each component carrier that may cause interference. In a high-speed mobile environment, large Doppler frequency shift, nonlinear frequency response of a power amplifier and/or the asymmetric characteristic of a crystal oscillator and the effect of frequency aliasing may affect the orthogonality between adjacent frequency bands and may potentially cause inter-band interference. The aliasing effect may significantly degrade the BER performance, especially when high-order modulation schemes are used. Therefore, for both contiguous and non-contiguous carrier aggregation, the guard bands for a component carrier should be carefully set to suppress the intra-system and/or inter-system interference, while maintaining high spectral efficiency in data transmission.

11.4 MAC ASPECTS OF MULTI-CARRIER OPERATION

The MAC layer in multi-carrier mode operates in the same manner as in a single-carrier mode. There is no difference between single-carrier and multi-carrier operation from a MAC addressing point of view. The security procedures between the MS and BS are performed over the mobile station's primary carrier. The security context created and maintained through this process is managed by the BS using the primary carrier. The network entry procedures in multi-carrier and single-carrier modes are the same, where the MS and BS indicate their support for multi-carrier mode during registration. The MS performs initial ranging and network entry only with a fully-configured carrier. Thus, a multi-carrier-enabled MS will have to obtain information about a base station's fully-configured carriers. During the initial network entry, the MS will inform the BS of its support for multi-carrier transmission through the AAI_REG-REQ message, and the BS will notify the MS whether it can support multi-carrier operation through an AAI_REG-RSP message. The basic multi-carrier capability exchange uses a 3-bit field in the AAI_REG-REQ/RSP message [2]. The basic multi-carrier mode includes informing the MS of the multi-carrier capability of the BS, including support for primary carrier change, as well as scanning of available carriers. The support for both carrier aggregation and switching does not imply E-MBS support, which is negotiated separately. The multi-carrier initialization procedure following network entry for a multi-carrier-enabled MS is shown in Figure 11-8. This procedure includes obtaining multi-carrier configuration for available carriers at the BS and information about assigned carriers consisting of the following two steps: (1) the BS provides the MS with information on its supported component carriers and their configuration; and (2) the MS obtains information regarding a subset of available component carriers, i.e., the assigned carriers, which the BS may utilize in a multi-carrier operation concerning the MS. The MS does not perform any processing on an assigned component carrier until that carrier is activated by the BS.

The BS broadcasts the P-SFH/S-SFH on each component carrier with a format similar to that of a single-carrier. The BS also provides the MS with basic radio configuration for all available component carriers through the AAI_MC-ADV (multi-carrier advertisement message) MAC control message. This message contains the component carriers' configurations which are supported by the BS and is periodically broadcast. The multi-carrier configuration information is used by all mobile stations regardless of their multi-carrier or single-carrier capability. Following initial network entry and after obtaining the information about the base station's multi-carrier configuration, a multi-carrier-enabled MS (multi-carrier aggregation or switching) is required to send an AAI_MC-REQ message to the BS. The MS informs the BS of its multi-carrier capability via the parameters defined in the AAI_MC-REQ message. Depending on the mobile station's multi-carrier capability, the BS responds using an AAI_MC-RSP message and assigns one or more secondary component carriers from the set of available carriers to the MS.

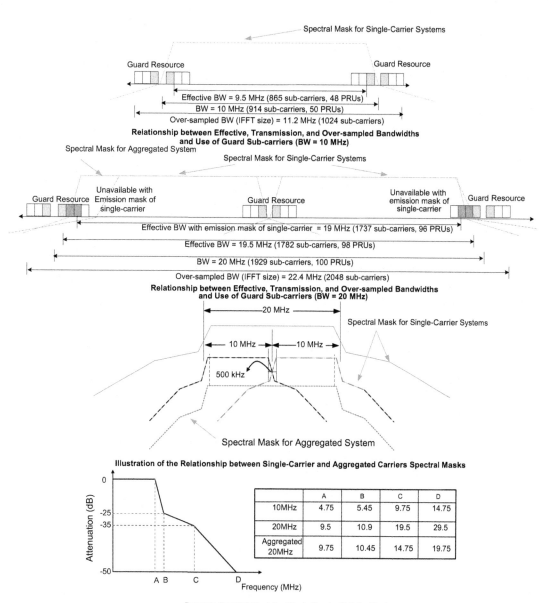

Spectral Mask for Single-Carrier Systems

Guard Resource

Guard Resource

Effective BW = 9.5 MHz (865 sub-carriers, 48 PRUs)
BW = 10 MHz (914 sub-carriers, 50 PRUs)
Over-sampled BW (IFFT size) = 11.2 MHz (1024 sub-carriers)

Relationship between Effective, Transmission, and Over-sampled Bandwidths and Use of Guard Sub-carriers (BW = 10 MHz)

Spectral Mask for Aggregated System

Spectral Mask for Single-Carrier Systems

Guard Resource Unavailable with Emission mask of single-carrier

Guard Resource

Unavailable with emission mask of single-carrier

Guard Resource

Effective BW with emission mask of single-carrier = 19 MHz (1737 sub-carriers, 96 PRUs)
Effective BW = 19.5 MHz (1782 sub-carriers, 98 PRUs)
BW = 20 MHz (1929 sub-carriers, 100 PRUs)
Over-sampled BW (IFFT size) = 22.4 MHz (2048 sub-carriers)

Relationship between Effective, Transmission, and Over-sampled Bandwidths and Use of Guard Sub-carriers (BW = 20 MHz)

20 MHz

10 MHz — 10 MHz

Spectral Mask for Single-Carrier Systems

500 kHz

Spectral Mask for Aggregated System

Illustration of the Relationship between Single-Carrier and Aggregated Carriers Spectral Masks

	A	B	C	D
10MHz	4.75	5.45	9.75	14.75
20MHz	9.5	10.9	19.5	29.5
Aggregated 20MHz	9.75	10.45	14.75	19.75

Attenuation (dB)

0
-25
-35
-50

A B   C   D
Frequency (MHz)

Example Spectral Mask for Single-Carrier Cellular Systems

FIGURE 11-8

Initialization of multi-carrier operation [2]

In some cases, the MS may not be able to communicate with the BS over the secondary carrier(s) without time/frequency synchronization and power adjustments. If the radio channels corresponding to the primary and the secondary carriers are highly correlated, the transmission parameters of the secondary carrier(s) are likely to be similar to those of the primary carrier. In this case, if the MS has already completed network entry with the BS using the primary component carrier, the MS is not required to perform initial ranging on the secondary carrier(s). Therefore, only periodic ranging may be performed on the secondary carrier(s). Once the secondary carriers are activated, the MS may perform periodic ranging on active secondary carrier(s), if instructed by the BS through the AAI_CM-CMD (carrier management command). If the MS skips initial ranging on the secondary carrier(s), it may use the same timing, frequency, and power adjustment parameters that are used in the primary carrier for initial transmission. The MS may perform the fine timing, frequency, and power adjustment on the secondary carrier(s) by measuring the primary preamble timing/frequency as well as the common pilots of the secondary carrier(s). The initial/periodic ranging procedures with a fully-configured carrier are the same as those for a single-carrier. Periodic ranging may only be performed on the activated secondary carrier(s) on instruction by the BS through the AAI_CM-CMD. The handover ranging is only performed with one of the fully-configured carriers of the target BS.

The construction and transmission of MAC PDUs are the same as in single-carrier operation. The ARQ protocol operates based on a common MAC instance. The MAC PDUs are processed in the physical layer similar to that of a single-carrier and are mapped to a data region in one of the primary or secondary carriers. The A-MAP IE corresponding to the allocation is transmitted on the component carrier where the data region is located. Therefore, each component carrier contains its own downlink control channels. In the uplink, the bandwidth requests are transmitted on the mobile station's primary carrier. The bandwidth request may also be transmitted in MAC PDUs over the secondary carrier(s) using the piggyback method [2]. The serving BS may allocate downlink or uplink resources on a specific active carrier or on a combination of multiple active carriers based on available resources, QoS requirements, load balancing, etc. The STID and the associated FIDs assigned to an MS are unique identifiers for the common MAC, and are used on all mobile station's component carriers. The service flow set-up/change messages are transmitted only through the mobile station's primary carrier. The service flow is defined for a common MAC.

The BS may assign uplink feedback channels to each fully-configured active carrier with uplink radio resources. When an uplink fast-feedback channel is allocated the MS reports to the CINR for an active carrier over the assigned fast-feedback channel of the corresponding carrier. If the radio resources of a fully-configured carrier are only allocated to downlink transmission, the uplink fast-feedback channel will be located in the uplink control region on the primary carrier. In this case, the BS may allocate one uplink fast-feedback channel per secondary carrier on the primary carrier. In multi-carrier aggregation with a downlink-only secondary partially-configured carrier, the MS is instructed to report the CINR measurements corresponding to downlink-only secondary carriers through the fast-feedback channel(s) on the primary carrier. In this case, the feedback region is allocated using the physical carrier index for the primary carrier. The starting logical index and the number of the distributed LRUs allocated to fast-feedback channels and the number of HARQ feedback channels are signaled in the AAI_SCD MAC control message that is transmitted on the active DL-only secondary carrier. The feedback region of the active DL-only secondary carriers are located after the feedback region of the primary carrier. Figure 11-9 illustrates an example of allocation of uplink control channels in the primary carrier uplink control region for downlink-only component carriers in a multi-carrier

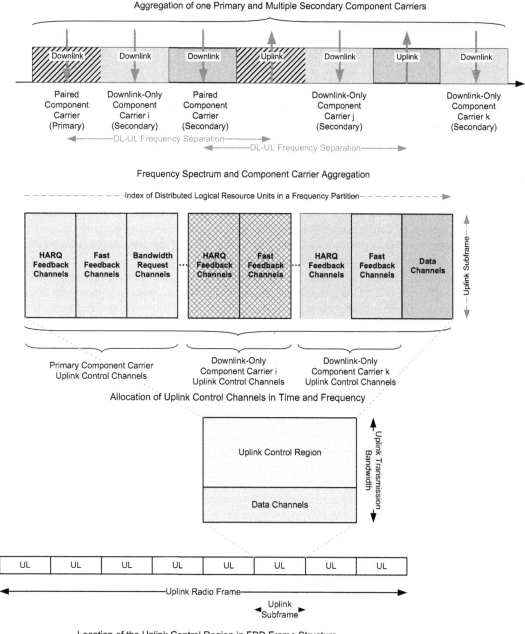

Aggregation of one Primary and Multiple Secondary Component Carriers

Frequency Spectrum and Component Carrier Aggregation

Allocation of Uplink Control Channels in Time and Frequency

Location of the Uplink Control Region in FDD Frame Structure

FIGURE 11-9

An example of allocation of uplink control channels corresponding to downlink-only component carriers aggregation scenario. When MAC control messages such as the AAI_SingleBS_MIMO_FBK, AAI_MultiBS_MIMO_FBK, MIMO feedback header, and correlation matrix feedback header are used for downlink CINR reporting, they are transmitted on the mobile station's primary carrier. The Feedback Polling A-MAP IE is used to notify an MS of an uplink burst region and is transmitted on each active downlink carrier.

11.4.1 Activation/Deactivation and Switching of Component Carriers

The activation or deactivation of secondary carriers is determined by the BS based on service flow QoS requirements, the loading status of the active carriers, channel quality measurements for active carriers, and a scan report for inactive carriers. The trigger condition associated with each carrier can be obtained via the AAI_SCD control message, which is broadcast on the primary carrier of the MS. The BS activates or deactivates secondary carriers using the AAI_CM-CMD MAC control message transmitted on the primary carrier. The latter control message includes activation/deactivation indicators, a list of secondary carriers referenced by their physical carrier indices, ranging parameters for the activated carriers, and the time for the MS to confirm activation of the secondary carrier by sending the AAI_CMD-IND message before expiration of the activation.

When an MS with a single radio transceiver goes through secondary carrier activation, the MS reconfigures its RF front-end and baseband parameters, based on the characteristics of the activated carrier. Once the hardware reconfiguration is complete and the MS is synchronized with the new carrier, the BS is notified of the MS preparedness by sending an AAI_CM-IND control message. After the BS receives the AAI_CM-IND control message, it starts data transmission on the newly activated secondary carrier. An MS operating in multi-carrier aggregation mode is assigned the same STID for the primary and secondary carriers. The MS in this mode of operation is required to monitor all active carriers and to follow resource allocations. The data communication in the carrier aggregation mode may be continued during the handover procedure.

As shown in Figure 11-10, the primary carrier of an MS can be changed without changing the MAC layer security and mobility contexts. A multi-carrier-enabled MS is required to support the primary carrier change. The BS may instruct the MS, through the AAI_CM-CMD message on the current primary carrier, to change its primary carrier to one of the assigned fully-configured carriers as a result of load balancing or varying channel conditions. For an MS supporting basic multi-carrier mode, the target primary carrier may be one of the available fully-configured carriers within the same cell. When an MS receives the AAI_CM-CMD MAC control message, it transmits the AAI_MSG-ACK message, and subsequently disconnects from the present carrier and switches to the target fully-configured carrier within the action time specified by the BS.

If the MS supports carrier aggregation mode and the target carrier is one of its active secondary carriers, it may receive data and control signals on the target carrier immediately after switching; otherwise, the MS first reconfigures its hardware setting and then switches to the target carrier. The MS may be required to perform periodic ranging with the target carrier. Given that a common MAC entity manages both serving and target primary carriers, network re-entry with the target primary carrier is not required. For a multi-carrier-enabled MS, the physical carrier indices of the current and target primary carriers are swapped after the primary carrier change process.

The MS may spontaneously or by instruction scan other assigned carriers which are not serving it. In that case, the MS reports the scanning results to the serving BS, which may be used by the BS to

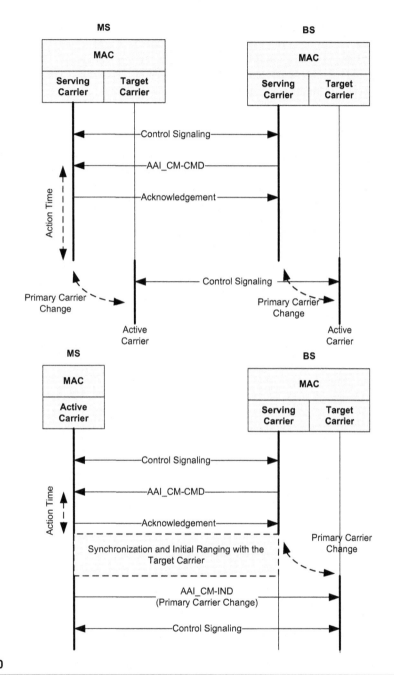

FIGURE 11-10

Primary carrier change scenarios [2]

determine the target carrier for the MS. If the target carrier is not currently used by the MS, synchronization with the target carrier may be necessary. The AAI_CM-CMD MAC control message for primary carrier change is transmitted on the primary carrier and includes the target primary carrier index, action time, and ranging indicator. The primary-to-secondary carrier switching in multi-carrier mode is only supported for multicast and broadcast services.

11.4.2 Multi-Carrier Handover

Multi-carrier handover is defined as the handover procedure between component carriers. A multi-carrier-enabled MS may use the single-carrier handover procedures alternatively perform the multi-carrier handover procedures. If an MS supports multi-carrier aggregation, it is required to use its primary carrier as the reference for serving BS scanning, and to follow the information broadcast in the AAI_SCD message for handover or scanning related operations. The AAI_NBR-ADV message contains neighbor base stations' multi-carrier configuration information to facilitate the mobile station's scanning of neighbor base stations' fully-configured carriers.

A multi-carrier capable MS may perform the multi-carrier scanning procedure, which is a scanning procedure extended to multiple RF carriers. In this case, the MS scans the component carriers of neighboring base stations whose information is included in an AAI_NBR-ADV message. The MS may further scan other fully-configured carriers of the serving BS which are not in use by the MS. Figure 11-11 illustrates an example of the procedure for neighbor BS advertisement and scanning of fully-configured carriers of serving and neighbor base stations. If the MS is capable of simultaneous processing of multiple RF carriers, it may perform scanning of the neighbor base stations while connected to one or more of its available component carriers without interruption to its normal communication with the serving BS. In this case, the MS and the BS may negotiate through AAI_SCN-REQ/RSP messages the component carriers be assigned for scanning operations to avoid resource allocation on those carriers (see Figure 11-12). Those carriers are identified by their carrier indices in AAI_SCN-REQ/RSP and AAI_SCN-REP messages. The handover from one component carrier to another associated with same BS for an MS supporting basic multi-carrier mode is the same as the primary carrier change procedure.

During handover preparation, the MS may request or be instructed by the serving BS to perform multi-carrier handover using AAI_HO-REQ/AAI_HO-CMD messages. The serving BS informs the MS of primary carrier indices of the target base stations through an AAI_HO-CMD message. The serving BS may communicate with the target base stations to assist the MS in order to obtain a pre-assigned secondary carrier prior to handover execution. The serving BS will forward the information received through the AAI_MC-REQ message to the target base stations for secondary carrier assignment.

When the *HO_Reentry_Mode* and *HO_Reentry_Interleaving_Interval* parameters are set to 1 and 0, respectively, the MS performs network re-entry with the target BS on one component carrier while communicating with the serving BS on another carrier. The MS may use the primary carrier to perform network re-entry with the target BS, as illustrated in Figure 11-13. It may also use another carrier other than its original primary carrier for network re-entry procedures, as shown in Figure 11-13. In this case, the disconnect time interval should be long enough so that the network re-entry procedure with the target BS can be completed prior to the expiration of this timer. When using an AAI_HO-CMD message with multiple target BSs and carriers, the physical index of each candidate carrier is provided

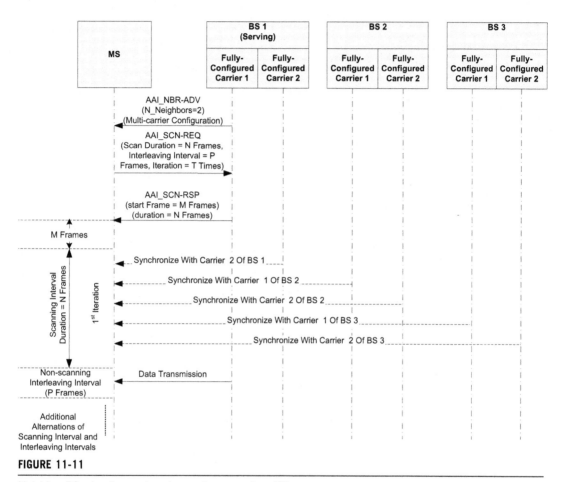

FIGURE 11-11

Neighbor-BS advertisement and scanning procedure [2]

by each target BS, and is included in the AAI_HO-CMD message. The candidate base stations identified in the AAI_HO-CMD message can instruct the MS to make a decision on handover, if the CINR/RSSI measurement on one of the base stations is sufficiently good to meet the QoS requirements of the MS. The MS may inform the serving BS of the component carrier that it is going use for network re-entry in order to avoid resource allocation on that carrier during the process. Once network entry is completed, the MS stops communication with the serving BS and may transmit data (if the user-plane is already established) or a bandwidth request message to the target BS.

The serving and target base station may negotiate secondary carrier pre-assignment as illustrated in Figure 11-14. If the *Carrier_Preassignment_Indication* parameter is set to 1 in the AAI_HO-CMD message, the pre-assignment information is forwarded by the target base stations to the serving BS via the core network. The serving BS will then send this information to the MS using the AAI_HO-CMD MAC control message. As a result, the pre-assigned secondary carriers (identified by the Carrier Status Bitmap) may be activated on completion of the network re-entry. The MS starts the activation

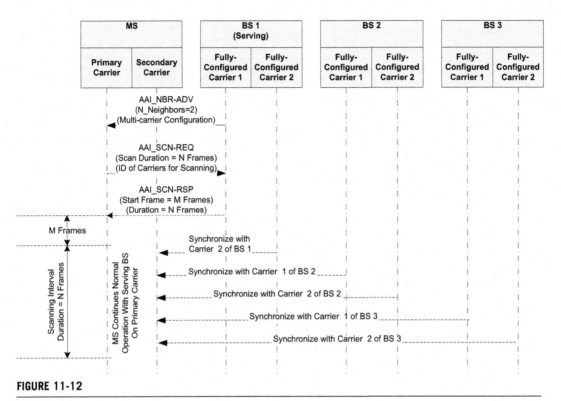

FIGURE 11-12

Scanning procedure in multi-radio mobile station [2]

procedure for these secondary carriers after receiving the AAI_RNG-RSP message from the target BS. The target BS may initiate data transmission on the activated secondary carriers on reception of the AAI_CM-IND message from the MS after network re-entry. If the AAI_CM-IND message is not received by the target BS within the *Activation Deadline* specified in the AAI_RNG-RSP, the target BS concludes that the secondary carrier activation for this mobile station has failed. In this case, the target BS may send an AAI_CM-CMD message to the MS to activate another secondary carrier. It must be noted that the serving BS is required to inform the target base stations of the multi-carrier capability of the MS. During this process, if the multi-carrier configuration of the target BS is different from the serving BS and unsuitable for the MS, the MS may negotiate its desirable multi-carrier configuration with the target BS through AAI_MC-REQ/RSP messages. If the *Carrier_Preassignment_Indication* flag in the AAI_HO-CMD message is set to zero, the MS follows the normal secondary carrier assignment procedure and carrier activation following completion of network re-entry.

11.4.3 Multi-Carrier Power Management

The power management in a multi-carrier operation is the same as that in the single-carrier mode. All MAC messaging, including idle mode procedures and state transitions, are performed using the primary carrier. The sleep mode parameters are negotiated with the serving BS before an MS

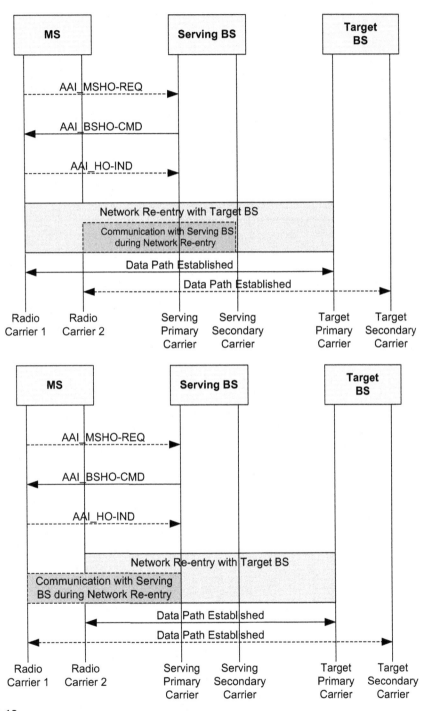

FIGURE 11-13

Multi-carrier handover scenarios [2]

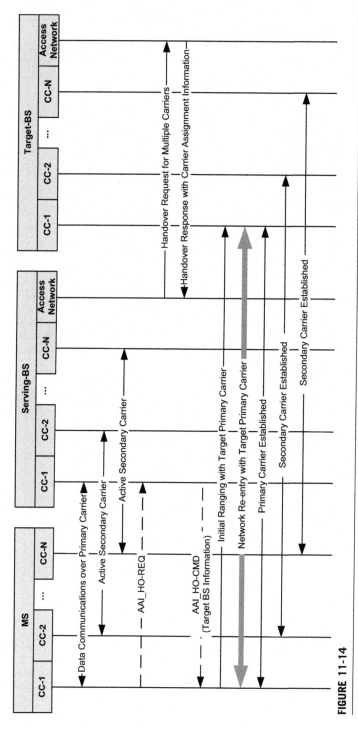

FIGURE 11-14

Multi-carrier handover with secondary carrier pre-assignment

transitions to sleep mode. The sleep mode parameters are applied to all mobile station's active carriers. Note that the serving BS may request the MS to change its primary carrier on entering the sleep mode or during the listening window using an AAI_CM-CMD message for load balancing or power conservation. In the beginning of the listening window, data transmission is permissible on all the active carriers assigned to the MS. The MS monitors the traffic indication message on its primary carrier. On receiving a negative traffic indication in the traffic indication message, the MS resumes the sleep cycle. If a positive traffic indication is received, the MS begins to receive data on one or more active carriers. If the traffic indication is disabled, data transmission and scheduling will be the same as in normal operation during the listening window. In this case, the MS monitors the active carriers during the listening window, and the BS may allocate downlink data on the primary carrier and the active secondary carriers. If the traffic indication is disabled and downlink data transmission on the active secondary carriers is complete, the BS may instruct the MS to stop data transmission on the active secondary carriers through the sleep control header, which is transmitted in the primary carrier during listening window. If the BS receives a bandwidth request from the MS on its primary carrier during the listening window, it considers the possibility of data transmission on all active carriers. During the sleep window, the MS may transmit a bandwidth request on the primary carrier on availability of uplink data. In that case, the serving BS may schedule an uplink transmission on all active carriers that have already been assigned to the MS. The MS resumes the normal sleep cycle after completion of the uplink transmission. Note that the sleep cycle parameters are applied to all active carriers of the MS, and downlink/uplink transmissions on all active carriers are suspended during the sleep window.

The idle mode procedures are similar in single-carrier and multi-carrier operation. In the multi-carrier mode, the PGID_Info message is transmitted on all fully-configured carriers. The AAI_PAG-ADV message is only transmitted on one of the fully-configured carriers. While in the idle mode, the MS determines the carrier index for monitoring the paging message within the paging listening interval as follows: *Paging Carrier Index* = (DID mod N), where N denotes the number of carriers per PGID that are used for conveying paging messages for idle-mode mobile stations. The paging carrier index corresponds to the physical carrier index of the carriers used for transmission of the paging message. A paging carrier indication flag is used to identify a paging carrier. The paging carrier indication flags of different carriers are included in the PGID_Info message and may be further included in an AAI_NBR-ADV or AAI_MC-ADV messages. The AAI_PAG-ADV message is transmitted in one or multiple frames starting from the second subframe in the superframe. For an MS with multicast and broadcast capability, the AAI_PAG-ADV message may be transmitted on the E-MBS dedicated carrier. An MS may perform a location update to acquire its preferred carrier during idle mode, if it cannot find the paging carrier. When carriers have different coverage areas, the most suitable active carrier is used for transmitting the paging message.

11.5 3GPP LTE-ADVANCED MULTI-CARRIER OPERATION

The 3GPP LTE-Advanced extends 3GPP LTE Rel-8 and 9 physical layer capabilities with the support of carrier aggregation, where two or more component carriers can be aggregated in order to support wider transmission bandwidths up to 100 MHz. It is possible to configure all component carriers which are required to be 3GPP LTE Rel-8 (backward) compatible when the number of component carriers in the uplink and downlink are the same. Some of the component carriers may be non-backward compatible with 3GPP LTE Rel-8. A user terminal may simultaneously receive or transmit one or multiple component carriers depending on its capabilities. An LTE-Advanced terminal with reception and/or transmission capabilities for carrier aggregation can simultaneously receive and/or transmit on multiple component carriers. A 3GPP LTE user terminal can only receive and transmit on a single component carrier which is Rel-8 compatible. As shown in Figure 11-15, the spectrum aggregation scenarios can be generally classified into three categories: (1) intra-band contiguous; (2) intra-band non-contiguous; and (3) inter-band.

Carrier aggregation is supported in both contiguous and non-contiguous scenarios with each component carrier limited to contain a maximum of 110 resource blocks in the frequency-domain, based on Rel-8 OFDMA/SC-FDMA numerology. It is possible to configure a UE to aggregate a different number of component carriers originating from the same eNB, and of possibly different bandwidths, in the downlink or uplink. Since OFDMA is utilized as the multiple access method in the downlink of 3GPP LTE, such carrier aggregation is a simple extension of the FFT sizes from the baseband perspective. In the uplink however, the 3GPP LTE Rel-8 single-carrier FDMA scheme does not allow such a simple extension of the bandwidth. Thus, a separate DFT module per component carrier is required prior to the IFFT module being utilized for transmission in the uplink [7]. This type of carrier aggregation slightly increases the uplink peak-to-average power ratio compared to that of Rel-8; however, the PAPR is still kept lower than that of OFDMA and therefore the increase in terminal power consumption as result of wideband transmissions is minimized. Note that if the sub-carriers of the contiguous bands are perfectly aligned, the baseband processing of the OFDMA signals can be performed with a single FFT/IFFT module (if the size is practically feasible from implementation point of view) or multiple FFT/IFFT modules (one per component carrier). As an example, if two 10 MHz contiguous bands are aggregated and the sub-carriers are perfectly aligned, we can use one 2048-point FFT/IFFT unit or two 1024-point FFT/IFFT modules for baseband processing.

In typical TDD deployments, the number of component carriers and the bandwidth of each component carrier in the uplink and downlink will be the same. The component carriers originating from the same eNB are not required to provide the same coverage, which can be the case when the center frequencies of the component carriers are far apart. The frequency separation between center frequencies of contiguously aggregated component carriers is a multiple of 300 kHz in order to be compatible with the 100 kHz frequency raster of 3GPP LTE Rel-8 and at the same time to preserve orthogonality of the sub-carriers with 15 kHz sub-carrier spacing. Depending on the aggregation scenario, the gap between two contiguous carriers can be filled by insertion of a small number of unused sub-carriers. In the 3GPP LTE technical specification, 10% of the total system bandwidth is allocated to the guard bands between adjacent component carriers.

The multi-carrier load balancing method is important when carrier aggregation is utilized. There are two carrier load balancing techniques that may be considered. Round Robin load balancing (also referred to as Combined Carrier Channel Assignment) is used to assign the newly arrived user to the carrier that has the least number of users. Thus, it tries to distribute the load evenly over all the component carriers. However, there might be small load variations on different component carriers, as the number of users per cell does not always divide equally into the number of component carriers, or because of the random departure of users. Mobile Hashing load balancing, also known as the Independent Carrier Channel Assignment, relies on the output from the

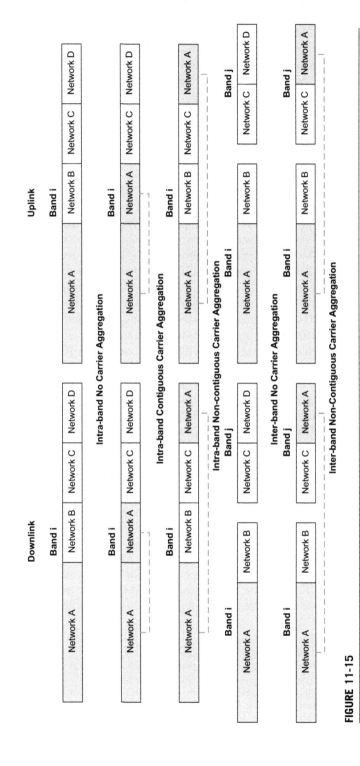

FIGURE 11-15

Carrier aggregation scenarios

terminal's hashing algorithm. The output hash values are uniformly distributed among a finite set, which maps directly onto the component carrier indices. Thereby, it provides a balanced load across the component carriers in the long-term. However, at each time-instant, the load across component carriers is not guaranteed to be balanced, and as a consequence the system will suffer from reduced trunking efficiency.[iv] [5,10,14,15].

In multi-carrier aggregation, each component carrier can have an independent control region due to the different number of users and allocation sizes on each carrier. In order to maintain backward compatibility, the 3GPP LTE Rel-8 PCFICH design (modulation, coding, and mapping to resource elements) is reused. Furthermore, the downlink (PDSCH) and uplink (PUSCH) resource assignments and data transmission support the following mechanisms: (1) the PDCCH on a component carrier assigns PDSCH resources on the same component carrier and PUSCH resources on a single-linked uplink component carrier, i.e., same coding, CCE-based resource mapping, and DCI formats, are used on each component carrier; (2) the PDCCH on a component carrier can assign PDSCH or PUSCH resources in one or multiple component carriers using the carrier indicator field, where 3GPP LTE Rel-8 DCI formats are extended with 1- to 3-bit carrier indicator field, and PDCCH structure, i.e., modulation and coding and CCE-based resource mapping, is reused. The presence of a carrier indicator field is semi-statically configured. For downlink HARQ feedback, PHICH physical transmission aspects such as orthogonal code design, modulation, scrambling sequence, and mapping to resource elements are maintained the same. The PHICH is transmitted only on the downlink component carrier that was used to transmit the uplink grant. If the number of downlink component carriers are more than or equal to the number of uplink component carriers and no carrier indicator field is used, the 3GPP LTE Rel-8 PHICH resource mapping principle is reused [21].

In the uplink, the HARQ ACK/NACK, scheduling request, and channel state information on PUCCH will function according to the following rules: (1) the Rel-10 PUCCH design can support up to five component carriers in the downlink; (2) the HARQ ACK/NACK on PUCCH is associated with downlink transmission on PDSCH; (3) the HARQ feedback for a UE can be transmitted on PUCCH in the absence of PUSCH transmission. In general, transmission of one HARQ ACK/NACK for each downlink component carrier transport block and limited transmission of ACK/NACK for downlink component carrier transport blocks (when power is limited) are supported. The design of the ACK/NACK resource allocation takes into consideration the performance and power control aspects without optimizing for a large number of UEs being simultaneously scheduled on multiple downlink component carriers. The scheduling request is transmitted on PUCCH and is semi-statically mapped to one UE-specific uplink component carrier. Periodic CSI reporting on PUCCH is supported for up to five downlink component carriers. The CSI is semi-statically mapped onto one UE-specific UL component carrier and the design follows the Rel-8 principles for the transmission of CQI/PMI/RI. This option can provide for maximum reuse of Rel-8 functionalities and better HARQ performance, due to a carrier component-based link adaptation, with the drawback of multiple HARQ feedback in

---

[iv]The capacity gain provided by cellular systems is offset to some extent by loss of trunking efficiency, which is the queuing efficiency resulting from a large number of users receiving service from a set of servers, rather than proportionally assigning each user to one of the servers. If a disproportionate number of mobile stations are simultaneously located in a single cell, a cellular system might practically end up supporting fewer users than a wide area radio system. Because relatively few of the users who are aggregated in the cell can receive service, due to the fact that only a subset of channels is available in the cell, the cellular system may appear ineffective. If the cell can only support m channels, the $(m + 1)$th simultaneous user could be blocked from receiving service.

each TTI. This also implies that the uplink transmission format will be a multi-carrier transmission consisting of aggregation of N DFT-S-OFDM carriers ($N \times$ DFT-S-OFDM). The alternative case of multiple transport blocks and HARQ entities where each transport block can be mapped into multiple component carriers was studied during the development of IEEE 802.16m, and was ruled out due to complexity.

- From the UE perspective, the Layer 2 aspects of HARQ operation are similar to that of 3GPP LTE Rel-8. There is one transport block without spatial multiplexing and up to two transport blocks, in the case of spatial multiplexing, and one independent HARQ entity per scheduled component carrier. Each transport block is mapped to a single component carrier. A UE may be scheduled over multiple component carriers simultaneously, but only one random access procedure can be performed at any time. In multi-carrier operation, the same DRX operation inherited from Rel-9 is applied to all configured component carriers (i.e., identical active time for PDCCH monitoring). The RLC and PDCP protocols of 3GPP LTE Rel-8 are also relevant to carrier aggregation.

In 3GPP LTE, a cell is identified by the E-UTRAN Cell Global Identifier, which corresponds to the transmission of system information on one component carrier. The 3GPP Rel-8 relevant system information is broadcast on backward-compatible component carriers. Each component carrier provides system information which is specific to that carrier. In multi-carrier mode, the UE only has one RRC connection with the network. A single cell also known as the special cell provides the security context, i.e., E-UTRAN Cell Global Identifier (ECGI), Physical Cell Identifier (PCI), Absolute Radio-Frequency Channel Number (ARFCN), and Non-Access Stratum (NAS) mobility information such as tracking area identity. There is only one such cell per UE in the RRC_Connected mode. After RRC connection is established, the reconfiguration, addition, and removal of component carriers can be performed using *RRCConnectionReconfiguration* including *mobilityControlInfo* (i.e., intra-cell handover) or *RRCConnectionReconfiguration* without *mobilityControlInfo* for addition of component carriers, and for the removal of component carriers with the exception of the component carrier corresponding to the special cell. In intra-LTE handover scenario, the *RRCConnectionReconfiguration* with *mobilityControlInfo* or the handover command can remove, reconfigure, or add component carriers for usage in the target cell. When adding a new component carrier, dedicated RRC signaling is used for sending the component carrier's system information which is necessary for component carrier transmission/reception similar to 3GPP LTE Rel-8 for handover. Detection of a component carrier failure by the UE does not necessarily trigger re-establishment of a connection. The RRC connection re-establishment triggers by the UE include the failure of all component carriers on which the UE is configured to receive PDCCH, loss of all uplink communications, or an indication from the RLC sub-layer that the maximum number of re-transmissions has been reached. The UE sees a component carrier as any other carrier frequency and a measurement object needs to be set up for a component carrier for the UE to measure it. Inter-frequency neighbor measurements, in which no serving cell is defined for measurement purposes, include all the carrier frequencies which are not configured as component carriers. The 3GPP LTE Rel-8 idle mode mobility procedures are applied to a network with carrier aggregation. It is possible for a network to configure only a subset of component carriers for idle mode operation. Table 11-3 shows four scenarios that are being investigated for the carrier aggregation in 3GPP RAN4 [21,24].

As mentioned earlier, the impacts of carrier aggregation on the RF requirements for the base station and mobile station need to be carefully considered. The areas which are impacted include transmitter aspects (base station output power, transmitted signal quality, unwanted emissions, transmitter spurious

Table 11-3 Carrier Aggregation Scenarios under Consideration in 3GPP LTE Release 10

Option	Deployment Scenarios
A	Single-band contiguous allocation for FDD (UL: 40 MHz; DL: 80 MHz)
B	Single-band contiguous allocation for TDD (100 MHz)
C	Multi-band non-contiguous allocation for FDD (UL: 40MHz; DL: 40 MHz)
D	Multi-band non-contiguous allocation for TDD (90 MHz)

Table 11-4 Transmit/Receive Characteristics for Carrier Aggregation Scenarios

		Transmit Characteristics		
		Intra-Band Aggregation		Inter-Band Aggregation
Option	Transmitter Architecture	Contiguous (CC)	Non-Contiguous (CC)	Non-Contiguous (CC)
A	Single (Baseband + IFFT + D/A Converter + Mixer + Power Amplifier)	Yes	–	–
B	Multiple (Baseband + IFFT + D/A Converter), Single (Stage-1 IF Mixer + IF Combiner + Stage-2 RF Mixer + Power Amplifier)	Yes	Yes	–
C	Multiple (Baseband + IFFT + D/A Converter + Mixer), Low-Power RF Combiner and Single Power Amplifier	Yes	Yes	–
D	Multiple (Baseband + IFFT + D/A Converter + Mixer + Power Amplifier), High-Power Combiner to Single Antenna or Dual Antennas	Yes	Yes	Yes (depending on the specific E-UTRA bands being aggregated)
		Receive Characteristics		
		Intra-Band Aggregation		Inter-Band Aggregation
Option	Receiver Architecture	Contiguous (CC)	Non-Contiguous (CC)	Non-Contiguous (CC)
A	Single (RF + FFT + Baseband) with BW > 20 MHz	Yes	–	–
B	Multiple (RF + FFT + Baseband) with BW ≤ 20 MHz	Yes	Yes	Yes

emissions) and receiver aspects (reference sensitivity level, adjacent channel selectivity, narrow-band blocking, receiver inter-modulation, and demodulation performance requirements). Some of the transmit/receive characteristics for carrier aggregation scenarios of interest are summarized in Table 11-4.

References

[1] IEEE 802.16-2009, IEEE Standard for Local and Metropolitan Area Networks, Part 16: Air Interface for Broadband Wireless Access Systems, May 2009.

[2] P802.16m/D6, IEEE Standard for Local and Metropolitan Area Networks – Part 16: Air Interface for Broadband Wireless Access Systems, Advanced Air Interface, May 2010.

[3] Report ITU-R M.2135–1, Guidelines for Evaluation of Radio Interface Technologies for IMT-Advanced, December 2009.

[4] Report ITU-R M.2134, Requirements Related to Technical System Performance for IMT-Advanced Radio Interface(s), November 2008.

[5] Y. Wang, K.I. Pedersen, T.B. Sorensen, P.E. Mogensen, "Resource allocation considerations for multi-carrier LTE-Advanced systems operating in backward compatible mode," IEEE 20th International Symposium on Personal, Indoor and Mobile Radio Communications, 2009.

[6] Shi Songsong, Feng Chunyan, Guo Caili, "A Resource Scheduling Algorithm Based on User Grouping for LTE-Advanced System with Carrier Aggregation," International Symposium on Computer Network and Multimedia Technology, 2009.

[7] Guangxiang Yuan, Xiang Zhang, Wenbo Wang, Yang Yang, "Carrier Aggregation for LTE-advanced Mobile Communication Systems," IEEE Communications Magazine Vol. 48 (Issue 2) (2010).

[8] Li Chen, Wenwen Chen, Xin Zhang and Dacheng Yang, "Analysis and Simulation for Spectrum Aggregation in LTE-Advanced System," IEEE 70th Vehicular Technology Conference, VTC-2009, Fall 2009.

[9] Vit Stencel, Andreas Muller and Philipp Frank, "LTE Advanced – A Further Evolutionary Step for Next Generation Mobile Networks," 20th International Conference Radioelektronika, 2010.

[10] Y. Wang, K.I. Pedersen, T.B. Sorensen, P.E. Mogensen, "Load Balancing and Packet Scheduling for Multi-Carrier Systems," IEEE Transactions on Wireless Communications Vol. 9 (Issue 5) (2010).

[11] Wang Xiaoyong, Xiao Dengkun and Jing Xiaojun, "A Novel Power Allocation Algorithm under CoMP with Carrier Aggregation," 2nd IEEE International Conference on Broadband Network & Multimedia Technology (IC-BNMT '09), 2009.

[12] P.E. Mogensen et al., "LTE-Advanced: The Path towards Gigabit/S in Wireless Mobile Communications," 1st International Conference on Wireless Communication, Vehicular Technology, Information Theory and Aerospace & Electronics Systems Technology, 2009. Wireless VITAE 2009.

[13] L. Garcia, et al., Comparison of Spectrum Sharing Techniques for IMT-Advanced Systems in Local Area Networks," IEEE 69th Vehicular Technology Conference, VTC-2009 Spring 2009.

[14] Y. Wang, K.I. Pedersen, M. Navarro, P.E. Mogensen, T.B. Sorensen, "Uplink Overhead Analysis And Outage Protection For Multi-Carrier LTE-Advanced Systems," IEEE 20th International Symposium on Personal, Indoor and Mobile Radio Communications, 2009.

[15] Y. Wang, K.I. Pedersen, P.E. Mogensen, T.B. Sorensen, "Carrier Load Balancing Methods With Bursty Traffic For LTE-Advanced Systems," IEEE 20th International Symposium on Personal, Indoor and Mobile Radio Communications, 2009.

[16] L.G.U. Garcia, K.I. Pedersen, P.E. Mogensen, "Autonomous Component Carrier Selection: Interference Management in Local Area Environments for LTE-Advanced," IEEE Communications Magazine Vol. 47 (Issue 9) (2009).

[17] Jian Li, Yang Liu, Jun Duan and Xuejun Liang, "Flexible Carrier Aggregation for Home Base Station in IMT-Advanced System," 5th International Conference on Wireless Communications, Networking and Mobile Computing (WiCom '09), 2009.

[18] T. Dean, P. Fleming, "Trunking Efficiency in Multi-carrier CDMA Systems," Proceedings of IEEE VTC Vol. 1 (September 2002).

[19] T. Wong, V. Prabhu, "Capacity Growth for CDMA System: Multiple Sectors and Multiple Carrier Deployment," Proceedings of IEEE VTC Vol. 2 (May 1998).

[20] B. Song, J. Kim, S. Oh, "Performance Analysis of Channel Assignment Methods for Multiple Carrier CDMA Cellular Systems," Proceedings IEEE VTC Vol. 1 (May 1999).

[21] 3GPP TR 36.912, Feasibility Study for Further Advancements for E-UTRA (LTE-Advanced), March 2010.

[22] 3GPP TR 36.913, Requirements for Evolved UTRA (E-UTRA) and Evolved UTRAN (E UTRAN), March 2009.

[23] 3GPP TS 36.300, Evolved Universal Terrestrial Radio Access (E-UTRA) and Evolved Universal Terrestrial Radio Access Network (E-UTRAN) Overall description, Stage 2, March 2010.

[24] 3GPP TR 36.814, Feasibility Study for Further Advancements for E-UTRA (LTE-Advanced), March 2010.

[25] 3GPP Long Term Evolution: System Overview, Product Development, and Test Challenges, Agilent Technologies, June 2009 <http://www.agilent.com>.

[26] Rohde & Schwarz Application Notes, LTE-Advanced Technology Introduction, March 2010.

[27] E. Dahlman, et al., 3G Evolution: HSPA and LTE for Mobile Broadband, second ed., Academic Press, 2008.

[28] 3G Americas, 3GPP Mobile Broadband Innovation Path to 4G: Release 9, Release 10 and Beyond: HSPA+, SAE/LTE and LTE-Advanced, February 2010.

[29] 3G Americas, HSPA to LTE-Advanced: 3GPP Broadband Evolution to IMT-Advanced (4G), September 2009.

[30] 3G Americas, The Mobile Broadband Evolution: 3GPP Release 8 and Beyond HSPA+, SAE/LTE and LTE-Advanced, February 2009.

[31] R. Attar, et al., "Evolution of cdma2000 Cellular Networks: Multicarrier EV-DO," IEEE Communications Magazine Vol. 44 (Issue 3) (March 2006).

[32] IEEE C802.16m–08/991, Youngsoo Yuk, Inuk Jung, Ronny Yongho Kim and Kiseon Ryu, Considerations for Carrier Aggregation, September 2008.

[33] Wireless World Initiative New Radio – WINNER+, Matthias Siebert, Albena Mihovska, Eiman Mohyeldin, Pierre Nguyen, Jean-Philippe Desbat, Miia Mustonen and Werner Mohr, Strategies and Technologies for Spectrum Utilization and Sharing Aspects of IMT, D3.3, March 2010.

Performance of IEEE 802.16m and 3GPP LTE-Advanced

INTRODUCTION

Link-level and system-level simulations are used to evaluate the performance of mobile radio access technologies under various operating conditions and deployment scenarios. While the simulations do not model the entire deployment parameters and propagation conditions that may be involved in a practical scenario due to increased computational complexity of the model, the statistical modeling of the parameters and estimation/measurement errors should be sufficiently accurate such that the simulation results are a faithful representative of the performance in an actual deployment [1].

The evaluation of the IMT-Advanced candidates was comprehensively performed in strict compliance with the technical parameters and the methodology that were specified by the ITU-R [2,3]. Each requirement is independently evaluated, except for the cell and cell edge user spectral efficiencies criteria that were jointly assessed using the same system-level simulation, consequently the candidates were required to simultaneously satisfy the corresponding minimum requirements. Furthermore, the system-level simulation set-up used in the assessment of the mobility requirement was the same as that used for the evaluation of cell spectral efficiency and cell edge user spectral efficiency. The evaluation of the IMT-Advanced candidate technologies (i.e., 3GPP LTE-Advanced and IEEE 802.16m) followed the following principles [6]:

1. Use of reproducible methods including computer simulation, analytical/theoretical approach, and inspection;
2. Technical evaluation against performance targets that were set for each test environment;
3. Consistency between self-evaluations and technical descriptions provided in technology description templates;
4. Use of unified methodology, software, and common parameter sets by the evaluation groups including channel models, link-level parameters, and link-to-system mapping;
5. Evaluation of multiple proposals using one simulation platform by each evaluation group to ensure comparability of the results;
6. Inclusion of L1/L2 overhead in the evaluation of cell spectral efficiency, cell edge user spectral efficiency, and VoIP capacity;
7. Independent drop of users with uniform distribution over the coverage area where each mobile user corresponds to an active session that runs for the duration of the drop;
8. Random assignment of LoS and NLoS path loss models to the users;

9. Cell assignment to a user based on the air interface cell selection schemes;
10. Fading signal and fading interference computed from each mobile station into each cell and from each cell into each mobile station (in both directions on an aggregated basis);
11. Consideration of the Interference over Thermal (IoT) constraint in the uplink design, such that the average IoT value remains less than or equal to 10 dB;
12. In the full-buffer traffic model, packets are not blocked upon their arrival, i.e., buffer sizes are assumed to be infinite;
13. Data and VoIP packets are scheduled with appropriate packet schedulers for full buffer and VoIP traffic models separately;
14. Modeling of channel quality feedback delay, feedback channel errors, packet errors, and real channel estimation (as opposed to ideal channel estimation) effects and re-transmission of erroneous packets;
15. Realistic modeling of the overhead channels, i.e., the overhead due to feedback and control channels.

For each drop, the system-level simulations were conducted and the process was repeated with the users dropped at new random geographical locations within the network. A large number of drops were simulated to ensure convergence in the system performance metrics and the width of confidence intervals of the user, and system performance metrics were calculated and reported. The confidence interval and the associated confidence level indicate the reliability of the estimated parameter value [7]. The confidence level is the probability that the true parameter value is within the confidence interval, such that the higher the confidence level, the larger the confidence interval. All cells in the network were simulated assuming dynamic channel models and the use of a 19-cell wrap-around scheme to model the inter-cell interference, except for the indoor environment.

This section describes the link-level and system-level evaluation of the IEEE 802.16m and 3GPP LTE-Advanced against the ITU-R requirements for IMT-Advanced systems using the methodology specified by the ITU-R [4,5,6]. The theoretical background, including the definition of performance metrics, channel models, physical layer abstraction schemes, and traffic models are provided to ensure in-depth understanding of the evaluation process and the results.

12.1 DEFINITION OF THE PERFORMANCE METRICS

The link budget is a commonly used metric to evaluate the coverage of a cellular system in the downlink or uplink. In order to calculate the link budget, one must account for all the gains and losses from the transmitter to the receiver over the air interface. It accounts for the attenuation of the transmitted signal due to propagation, as well as the antenna gains, cable and implementation loss, and other miscellaneous losses. The time-varying channel gains, such as fading, are taken into account by adding some margin depending on the anticipated severity of its effects. The amount of margin required can be reduced by the use of mitigating techniques such as antenna diversity or frequency hopping. An abstract description of the link budget in any direction (downlink/uplink) has the following form: [8]

$$Received\ Power\ (dBm)\ =\ Transmitted\ Power\ (dBm) + Gains\ (dB) - Losses\ (dB) \qquad (12\text{-}1)$$

Link budget evaluation is a well-known method for initial cell planning that needs to be carried out separately for downlink and uplink. Although the link budget can be calculated separately for each link, it is the combination of the links that determines the performance of the system. Using the margins in the link budget, the expected signal-to-noise ratio can be evaluated at given distances. Using these results, the noise limited range can be evaluated for the system. The typical link budget parameters are given in Table 12-1.

Although a user may be sufficiently covered within certain areas of a cell for a given service, when multiple users are in a cell the radio resources (time, frequency, space, power) are shared among the active users. It can be expected that a user's average data rate may be reduced by a factor of N when there are N active users, assuming resources are equally shared among active users, and there is no multi-user diversity gain, relative to that of a single user. Therefore, any coverage assessment for any particular service must be coupled with the number of active users in the cell and their associated quality of service requirements. If users with poor channel conditions are provided with more radio resources, it would adversely impact the total cell throughput. Thus, there is a trade-off between coverage and capacity. The performance metric that is typically used is the number of admissible users, parameterized by the service minimum bit rate, maximum tolerable latency, permissible outage probability, etc. It is assumed that simulation statistics are collected from sectors belonging to the test cells of a 19-cell deployment scenario in order to model the effects of inter-cell interference. In the

TABLE 12-1 Typical Link Budget Parameters [8]

Link Budget Parameters	Link Budget Parameters
Carrier Frequency/Channel Bandwidth	Number of Receive Antennas
BS/MS Antenna Heights (m)	Receive Antenna Gain (dBi)
Area Coverage (km)	Receiver Cable, Connector, or Body Losses (dB)
Type of Service (VoIP, Full Buffer Data, etc.)	Receiver Noise Figure (dB)
Modulation and Coding Scheme	Thermal Noise Density (dBm/Hz)
Multipath Channel Model	Receiver Interference Density (dBm/Hz)
Mobile Speed (km/h)	Required SNR (dB)
Number of Transmit Antennas	Receiver Implementation Margin (dB)
Maximum Transmitter Power per Antenna (dBm)	Fast-Fading Margin including Scheduler Gain (dB)
Transmitter Power Amplifier Back-off (dB)	HARQ Combining Gain (dB)
Transmit Antenna Gain (dBi)	Handover Gain (dB)
Transmit Array Gain (dB)	BS/MS Diversity Gain (dB)
Control Channel Power Boosting Gain (dB)	Lognormal Shadow Fading Standard Deviation (dB)
Data Carrier Power Loss due to Pilot/Control Boosting (dB)	Shadow Fading Margin (dB)
Transmitter Cable, Connector, Combiner, Body Losses (dB)	Penetration Loss (dB)

following, downlink and uplink throughputs are separately evaluated and the corresponding metrics are differentiated by superscripts "DL" and "UL," respectively.

The user throughput is defined as the ratio of the number of information bits that the user successfully received divided by the total simulation time. If user k has $p^{DL(UL)}$ downlink (uplink) packet transmissions during the simulation period (T_{sim}), and if there are $q_{ki}^{DL(UL)}$ packets in the ith transmission, and if b_k^{ij} denotes the number of correctly received bits in the jth packet; then the average user throughput for user k is:

$$R_k^{DL(UL)} = \frac{\sum_{i=1}^{p_k^{DL(UL)}} \sum_{j=1}^{q_{ik}^{DL(UL)}} b_k^{ij}}{T_{sim}} \tag{12-2}$$

The average user throughput is defined as the sum of the user throughput of each user in the cell divided by the total number of users in the cell. Let N_u denote the number of users in the sector/cell and assume user $k, k = 1, 2, \ldots, N_u$ has a throughput of $R_k^{DL(UL)}$, then DL (or UL) sector throughput is defined as:

$$R^{DL(UL)} = \sum_{k=1}^{N_u} R_k^{DL(UL)} \tag{12-3}$$

The user throughput per transmission is the number of correctly received bits per packet divided by the duration of the transmission. If user k has $p_k^{DL(UL)}$ downlink (uplink) transmissions, if there are $q_{ki}^{DL(UL)}$ packets in the ith transmission, and if there are b_k^{ij} correctly received bits in the jth packet, then the average transmission throughput can be defined as follows:

$$R_k^{DL(UL)} = \frac{1}{p_k^{DL(UL)}} \left(\sum_{i=1}^{p_k^{DL(UL)}} \frac{\sum_{j=1}^{q_{ki}^{DL(UL)}} b_k^{ij}}{\Delta T_{ki}^{DL(UL)}} \right) \tag{12-4}$$

where $\Delta T_{ki}^{DL(UL)}$ denotes the duration of the ith transmission to user k. The average user throughput per transmission is defined as the sum of the user throughputs per transmission divided by the number of users in the cell.

The outage throughput $\eta_{outage}(R_{min})$ is defined as the percentage of users with data rate R^{DL} less than a predefined minimum data rate R_{min}. The cell edge user throughput is defined as the 5th percentile point of the CDF of users' average throughput per transmission, assuming that 95% of the users are expected to achieve a certain throughput per transmission regardless of their geographical location in the cell.

If the jth packet of the ith transmission belongs to user k, and if the packet arrives at the BS (or MS) MAC at time instant $T_{arrival}^{kij}$ and is delivered to the MS (or BS) MAC at time instant $T_{departure}^{kij}$, then the packet delay can be calculated as follows:

$$D_{kij}^{DL(UL)} = -T_{arrival}^{kij} + T_{departure}^{kij} \tag{12-5}$$

The downlink and uplink delays are denoted by superscript "DL" or "UL," respectively. The packets that are dropped or erased may not be included in the analysis of packet delays, depending on the traffic model. For example, in the modeling of traffic for delay-sensitive applications, packets may be dropped if packet transmissions are not completed within a specified delay bound. The impact of the dropped packets is included in the packet loss rate. The CDF of the packet delay per user provides a basis in which maximum latency, 2nd percentile, average latency, as well as jitter can be derived. The 2nd percentile point of the CDF of packet delay denotes the packet delay value for which 98% of packets have a delay less than that value. As an example, the VoIP capacity is defined such that the percentage of users in outage is less than 2%, where a user is assumed to have experienced a service outage, if less than 98% of the VoIP packets have been delivered successfully to the user within a one-way radio access delay bound of 50 ms.

The average packet delay is defined as the average interval between packets originated at the source (either an MS or a BS) and received at the destination (either a BS or an MS) in a system for a given duration of transmission. The average packet delay for user k, $\overline{D}_k^{DL(UL)}$ is given by:

$$\overline{D}_k^{DL(UL)} = \frac{\sum_{i=1}^{p_k} \sum_{j=1}^{q_{ki}} (T_{departure}^{kij} - T_{arrival}^{kij})}{\sum_{i=1}^{p_k} q_{ki}} \tag{12-6}$$

The CDF of a users' average packet delay is the cumulative distribution of the average packet delay observed by all users in the cell. The packet loss ratio per user is defined as:

$$Packet\ Loss\ Ratio = 1 - \frac{Total\ Number\ of\ Successfully\ Delivered\ Packets}{Total\ Number\ of\ Packets} \tag{12-7}$$

where the total number of packets includes packets that were transmitted over the air interface and packets that were dropped prior to transmission. The data throughput of a BS is defined as the number of information bits per second that a cell can successfully deliver or receive via the air interface using appropriate scheduling algorithms.

We consider both physical layer spectral efficiency and MAC layer spectral efficiency as important performance indicators for a cellular system. The physical layer spectral efficiency is the system throughput measured at the interface between the physical layer and the MAC layer, thus including physical layer overhead but excluding MAC and upper layer protocol overheads. The MAC layer spectral efficiency represents the system throughput measured at the interface of the MAC layer and the upper layers, thus including both physical layer and MAC protocol overheads. The MAC efficiency of the system is evaluated by dividing the MAC layer spectral efficiency by the physical layer spectral efficiency. The average cell spectral efficiency is defined as:

$$SE_{cell} = \frac{R_{aggregate}}{BW_{effective}} \tag{12-8}$$

where $R_{aggregate}$ denotes the aggregate cell throughput and $BW_{effective}$ is the effective channel bandwidth. The effective channel bandwidth is defined as $BW_{effective} = \alpha_{DL-UL}BW_{transmission}$, where $BW_{transmission}$ denotes the channel bandwidth, and α_{DL-UL} is the downlink/uplink ratio in the TDD systems. Note that for FDD systems $\alpha_{DL-UL} = 1$, and for TDD systems with a DL:UL ratio of 2:1 $\alpha_{DL-UL} = 2/3$ for DL, and 1/3 for UL, respectively.

The CDF of SINR is defined as the cumulative distribution function for the signal-to-interference and noise ratio as measured by the BS (or the MS) for each MS (or the BS) on the uplink or downlink, respectively. This metric would allow comparison between different reuse scenarios, network loading conditions, smart antenna algorithms, resource allocation, power control schemes, etc.

12.2 CALCULATION OF STATIC AND DYNAMIC OVERHEAD

The calculation and analysis of the signaling and control channels' overhead as well as the physical layer overhead associated with synchronization and broadcast channels, guard time and guard bands, reference signals, and that of the cyclic prefix are important in determining the actual performance of a cellular system. In general, the overhead channels are essential for proper operation of a cellular system and statically (independent of the number of the users in the cell) or dynamically (depending on the number of the users in the cell) consume some radio resources and make them permanently or temporarily unavailable for data transmission. There are two types of overhead channels: static and dynamic. A static overhead channel requires fixed base station power, time slot, and/or bandwidth. On the other hand, a dynamic overhead channel requires base station power, time, and/or bandwidth which dynamically change over time as a function of the number of active users.

The Layer 1 (L1) and Layer 2 (L2) overhead are accounted for in time, frequency, or space for calculation of system performance metrics such as spectral efficiency, user throughput, VoIP capacity, etc. Examples of L1 overhead include synchronization channel (preambles), guard sub-carriers and DC sub-carrier, guard time (in TDD systems), reference signals, and cyclic prefix. Examples of L2 overheads include common and dedicated control channels, HARQ ACK/NACK feedback, channel quality feedback, random access channels, packet headers, and sub-headers, as well as CRC. The power allocation/boosting should also be accounted for in modeling resource allocation for control channels. Based on the example two-dimensional time-frequency model shown in Figure 12-1, the L1 and L2 overhead can be calculated as:

$$O_{L1+L2} = \lim_{T \to \infty} \frac{\sum\limits_{k \in S(L1) \cup S(L2)} B_k T_k}{BT}, \quad 0 \le O_{L1+L2} < 1 \tag{12-9}$$

where B denotes the system bandwidth and T is the radio frame size, B_k is the resource occupied by the kth L1 (or L2) overhead channel across the frequency dimension and T_k denotes the required time resource for transmission of the kth L1 (or L2) overhead channel in time.

Using this model, the total resources consumed by the L1 overhead over relatively long period of time is given as:

$$O_{L1} = \lim_{T \to \infty} \frac{1}{BT} \sum_{k \in S(L1)} B_k T_k, \quad 0 \le O_{L1} < 1 \tag{12-10}$$

where $S(L1)$ denotes the set of resources used for static L1 overhead channels. The total resources consumed by the L2 overhead over relatively long period of time is given as:

$$O_{L2} = \lim_{T \to \infty} \frac{1}{BT(1 - O_{L1})} \sum_{k \in S(L2)} B_k T_k, \quad 0 \le O_{L2} < 1 \tag{12-11}$$

Note that the amount of resources that have already been used by the L1 overhead must not be recounted in the calculation of O_{L2} and that $O_{L1+2} \ne O_{L1} + O_{L2}$. Also S(L2) denotes the set of resources used for L2 overhead channels The average spectral efficiency inclusive of the effect of the L1/L2 overhead can be obtained as $SE(1 - O_{L1+2})$, where SE denotes the spectral efficiency that has

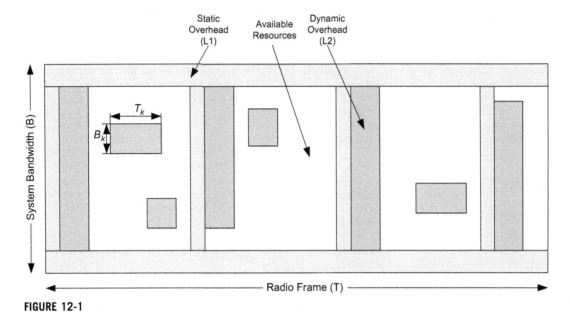

FIGURE 12-1

An illustration of the static and dynamic overhead

been obtained via system-level simulation excluding the L1/L2 overhead. Based on the above methodology, the minimum and maximum values of L1/L2 overhead of IEEE 802.16m were calculated and the results are shown in Table 12-2. The minimum and maximum values are obtained based on the consumption of radio resources by the A-MAP, which may vary from three to ten logical resource units per subframe. It must be noted that the L1/L2 overhead for the case of a 4×4 antenna configuration is the same as that for a 4×2 antenna configuration in the downlink, since the 4-stream pilot pattern is used in the overhead calculation for the 4×2 case. Unlike the system/user peak rate calculation, the number of receiver antennas does not affect the L1/L2 overhead. For the two downlink MU-MIMO cases, i.e., a BS with four transmit antennas and two mobile stations each with two receive antennas, and a BS with four transmit antennas and one MS with four receive antennas, the same pilot pattern is used.

TABLE 12-2 IEEE 802.16m L1/L2 Overhead Estimation [2]

Assumption	Overhead	Minimum Fraction	Maximum Fraction
10 MHz bandwidth CP = 1/8 DL 2 × 2 antenna configuration	L1 overhead Total overhead (L1/L2)	0.3104 0.3532	0.3104 0.4380
20 MHz bandwidth CP = 1/16 DL 4 × 2 antenna configuration	L1 overhead Total overhead (L1/L2)	0.2824 0.3029	0.2824 0.3441

TABLE 12-3 Error Sources and Impacts in the Air Interface [8]

Overhead Channel	Error Sources	Impact on System Operation
Downlink/Uplink HARQ ACK/NACK channel	Misinterpretation, misdetection, or false detection of the HARQ ACK/NACK signals	Transmission (frame or encoder packet) error or duplicate transmissions
Explicit rate indication/mode selection	Misinterpretation of rate/mode selection	One or more transmission errors due to decoding at a different rate (modulation and coding scheme) or selection of a different mode
User identification channel	A user tries to decode a transmissions belonging to another user or when a user fails to detect its own transmission	One or more transmission errors due to HARQ-IR combining of erroneous transmissions
Rate or C/I feedback channel	Misinterpretation of rate or C/I value	Potential transmission errors
Transmit sector indication, transfer of HARQ states, etc.	Misinterpretation of selected sector or misinterpretation of frames to be re-transmitted.	Transmission errors

The performance (i.e., detection error rate) of the downlink control channel can be estimated using the physical layer abstraction method that is used to perform system-level simulations (for the traffic channels) with proper modifications to reflect any difference in the transmission format of the control channels. For static overhead channels, the received SINR is calculated through system-level simulation in order to estimate the demodulation performance. For dynamic modeling of overhead channels with open-loop power control, the required downlink power or bandwidth for transmission of the overhead channel is taken into consideration during system-level simulation. The system-level simulations calculate the received SINR during the reception of overhead information.

The feedback errors, e.g., power control, HARQ feedback, MIMO rate indication, CQI channel, etc., and measurements errors (e.g., C/I measurement) are modeled in system-level simulations. Table 12-3 shows some typical signaling errors and their sources, and their potential impact on system performance.

12.3 TRAFFIC MODELS

12.3.1 Statistical Model for Conversational Speech

Voice-over-Internet Protocol (VoIP) refers to real-time delivery of coded voice packets across networks using Internet protocols. A VoIP session is defined as the entire user call time. Several robust voice codecs for encoding conversational speech have been developed, such as ITU-T G.729 (8 kbps) [14] and 3GPP Adaptive Multi-Rate (AMR) codec (4.75–12.2 kbps) [9,10]. A typical conversation is characterized by periods of active speech or talk spurts followed by silent periods. Figure 12-2 illustrates a two-state Markov model for conversational speech.

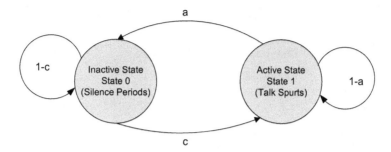

FIGURE 12-2

The two-state Markov model for speech [6]

The steady-state condition of the model requires that $P_0 = a/(a + c)$ and $P_1 = c/(a + c)$ where P_0 and P_1 are the probability of being in state 0 and state 1, respectively. As shown in Figure 12-2, the probability of a transition from state 1 (the active speech state) to state 0 (the inactive state) while in state 1 is equal to a, whereas the probability of a transition from state 0 to state 1 while in state 0 is c. The model is updated at the speech encoder frame rate $R = 1/T$, where T is the encoder frame duration, typically 20 ms for most speech encoders. Packets are generated at time intervals $iT + \tau$, where τ is the packet arrival delay jitter, and i denotes the encoder frame index. During the active state, packets of fixed size are generated at these time intervals, while the model is updated at regular frame intervals. The size of packet and the rate at which the packets are sent depends on the corresponding voice codecs and compression schemes. The voice activity factor λ is given as $\lambda = P_1 = c/(a + c)$. A talk-spurt is defined as the time period τ_{TS} between entering the active state and leaving the active state. The probability that a talk spurt has duration m speech frames is given by $P(\tau_{TS} = n) = a(1 - a)^{n-1}, n = 1,2,\dots$, whereas the probability that a silence period has a duration of n speech frames is given as $P(\tau_{SP} = m) = c(1 - c)^{m-1}, m = 1,2,\dots$. The average talk spurt duration μ_{TS} (in number of speech frames) is defined as $\mu_{TS} = E[\tau_{TS}] = 1/a$; the mean silence period duration μ_{SP} (in number of speech frames) is given by $\mu_{SP} = E[\tau_{SP}] = 1/c$. The distribution of the time period τ_{AE} (in number of speech frames) between successive active state entries is defined as the convolution of the distributions of τ_{SP} and τ_{TS}:

$$P(\tau_{AE} = n) = \frac{c}{c - a} a(1 - a)^{n-1} + \frac{a}{a - c} c(1 - c)^{n-1} \quad n = 1, 2, \cdots \quad (12\text{-}12)$$

It must be noted that τ_{AE} can be further considered as the time between MAC-layer resource reservations, provided that a single reservation is made per user per talk spurt. Note that in practice, very small values of τ_{AE} may not result in separate reservation requests. Since the transitions from state 1 to state 0 and *vice versa* are independent, the mean time μ_{AE} between active state entries is the sum of the mean time in each state; i.e., $\mu_{AE} = \mu_{TS} + \mu_{SP}$. Therefore, the mean rate of arrival $\overline{R}_{AE}$ of transitions into the active state is given as $\overline{R}_{AE} = 1/\mu_{AE}$.

The voice capacity assumes the use of a 12.2 kbps mode of 3GPP AMR codec with a 50% voice activity factor, provided that the percentage of users in outage is less than 2%, where a user is defined to have experienced outage if more than 2% of the VoIP packets are dropped, erased, or otherwise not delivered successfully to the user within the delay bound of 50 ms. The packet delay is defined based on the 98th percentile of the CDF of all individual users' 98th percentiles of packet delay (i.e., the 98th percentile of the packet delay CDF first determined for each user, and then the 98th percentile of the

TABLE 12-4 VoIP Traffic Model Parameters [8]

VoIP Model Attribute	Statistical Distribution	Parameters	Probabilistic Distribution
Active/inactive state duration	Exponential	$\mu = 1.25$ s	$f_x = \lambda e^{-\lambda x}, x \geq 0$ $\lambda = 1/\mu$
Probability of state transition	N/A	$c = 0.01, d = 0.99$	N/A
Packet arrival delay jitter (downlink only)	Laplacian	$\beta = 5.11$ ms	$f_x = \dfrac{1}{2\beta}e^{\frac{-|\tau|}{\beta}}$ $-80\text{ms} \leq \tau \leq 80\text{ms}$

TABLE 12-5 Assumptions for VoIP Capacity Calculation [6,8,9,10]

Parameter	Characterization
Codec	3GPP AMR12.2 kbps
Encoder Frame Length	20 ms
Voice Activity Factor (VAF)	50%
Payload	Active: 33 bytes (octet-aligned mode) Inactive: 7 bytes SID Packet is sent every 160 ms during silence intervals
Protocol Overhead with Compressed Header	10 bits + padding (RTP-pre-header) 4 bytes (RTP/UDP/IP) 2 bytes (RLC/security) 16 bits (CRC)

CDF that describes the 98th percentiles of the individual user's delay is obtained). VoIP capacity is measured as active users/MHz/cell. It is the minimum of the capacity calculated for either downlink or uplink divided by the effective bandwidth in the respective direction. In other words, the effective bandwidth is the operating bandwidth normalized appropriately considering the uplink/downlink ratio.

During each VoIP session, a user will be in the active or inactive state. The duration of time that the user stays in each state is exponentially distributed. In the active or inactive state, packets of fixed sizes will be generated at intervals of $iT + \tau$ seconds, where T is the voice frame size equal to 20 ms, τ is the network delay jitter in the downlink and i is the VoIP frame index. In the uplink direction, τ is equal to 0. As the range of the delay jitter is limited to 120 ms, the model may be implemented by generating packets at times $iT + \tau'$ seconds, where $\tau' = \tau + 80$ ms is a positive value. The air interface delay is the time elapsed from the packet arrival time $iT + \tau'$ to successful reception and decoding of the packet. Statistical distribution and parameters associated with the VoIP traffic model are summarized in Table 12-4. The assumptions that were used for VoIP capacity calculation are shown in Table 12-5.

12.3.2 Full Buffer Traffic Model

In the full buffer user traffic model, all the users in the system are assumed to have data to send or receive at any given time. In other words, there is always a constant amount of data that needs to be transferred, as opposed to bursts of data with a probabilistic arrival time distribution. This model allows the assessment of the spectral efficiency of the system independent of actual user traffic distribution type. A user is in outage if the residual packet error rate after HARQ re-transmissions exceeds 1%.

12.4 LINK-TO-SYSTEM MAPPING (PHY ABSTRACTION)

The objective of physical layer abstraction is to model link-level performance and to simplify the computations in system-level simulations. The motivation for abstraction of the physical layer is that simulating the physical layer between multiple base stations and mobile stations in a cellular network simulation platform can be computationally cumbersome and practically impossible [25]. The abstraction must be reasonably accurate, computationally simple, relatively independent of channel models, and extensible to interference models and multi-antenna processing. The system-level simulations are used to characterize the average system performance, which might have been useful in providing insights for deployment of the system in terms of cell and frequency planning. For such simulations, the average performance of a system is quantified using the topology and macro-channel characteristics to calculate a geometric SINR distribution across the cell (alternatively known as geometry). Each subscriber's SINR distribution is then mapped to the highest modulation and coding scheme that could be supported, based on link-level SINR tables which included the fast-fading characteristics of the channel. The link-level SINR-PER lookup tables served as the PHY abstraction for predicting average link-level performance. The instantaneous channel conditions are now used to improve the performance of the cellular systems. Channel dependent scheduling and adaptive coding and modulation are examples of channel-adaptive schemes employed to improve system performance. Therefore, new system-level simulation methodologies do explicitly model the dynamic behavior of the system [8,11].

In system-level simulations, an encoded data/control packet may be transmitted over a time-frequency selective channel. In OFDMA-based systems, the encoded block is transmitted over several sub-carriers; the post-processing SINR values of the pre-decoded streams are thus non-uniform and the channel gains of sub-carriers can be time-varying. As a result, in the transmission of a large encoded packet, the encoded symbols with unequal SINR ratios at the input of the decoder are considered due to the time or frequency selectivity of the channel impulse or frequency response over the duration of packet transmission.

The goal of PHY abstraction or link-to-system mapping is to predict the encoded Block Error Rate (BLER) for a given received channel realization across the OFDM sub-carriers that are used to transmit the channel-coded blocks. In order to predict the performance, the post-processing SINR values at the input to the FEC decoder are considered as input to the PHY abstraction. As the link-level curves are generated assuming a frequency flat channel response at a given SINR, an effective SINR is defined to accurately map the system-level SINR to the link-level curves to determine the resulting BLER. This mapping is known as Effective SINR Mapping (ESM) in the literature. The ESM PHY abstraction compresses the vector of received SINR values to a single effective SINR value, which can then be further mapped to a BLER value. Several ESM approaches to predict instantaneous link-level performance have been studied in the literature. Examples include mean instantaneous capacity, Exponential Effective SINR Mapping (EESM), and Mutual Information (MI) effective SINR mapping. Each of these PHY abstractions uses a different function to map the vector of SINR values to a single

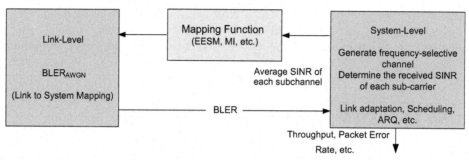

FIGURE 12-3

The link-to-system mapping process [8]

number. Given the instantaneous EESM SINR, mean capacity or mutual information effective SINR, the BLER for each MCS is calculated using a suitable mapping function. The general process of link-to-system mapping is illustrated in Figure 12-3.

In general, the ESM PHY abstraction methods can be described as follows:

$$SINR_{effective} = \Phi^{-1}\left(\frac{1}{N}\sum_{n=1}^{N_{FFT}}\Phi(SINR_n)\right) \qquad (12\text{-}13)$$

where $SINR_{effective}$ denotes the effective SINR, $SINR_n$ is the SINR measured at the nth sub-carrier, N_{FFT} is the number of symbols in a coded block, or the number of sub-carriers used in an OFDMA system and $\Phi(.)$ is an invertible function. In the case of the mutual information-based ESM, the function $\Phi(.)$ is derived from the constrained capacity, whereas in the case of EESM, the function $\Phi(.)$ is derived from the Chernoff bound[i] on the probability of error.

The accuracy of a mutual information-based metric depends on the equivalent channel over which this metric is defined. The capacity is defined as the mutual information based on a Gaussian channel with Gaussian distributed inputs. The modulation constrained capacity is the mutual information of a symbol channel, i.e., constrained by the input symbols drawn from a complex set. The computation of the mutual information per coded bit can be derived from the received symbol-level mutual information which is regarded as Received Bit mutual Information Rate (RBIR) link-to-system mapping. An alternative approach directly arrives at the bit-level mutual information and is known as the Mean Mutual Information per Bit (MMIB) PHY abstraction method. The procedure of the MI ESM approach is illustrated in Figure 12-4. Given a set of N received encoder symbol SINRs from the

---

[i]In the theory of probability and stochastic processes, the Chernoff bound provides exponentially decreasing bounds (asymptotic) on tail distributions of sums of independent random variables, compared to the first or second moment-based tail bounds such as Markov's inequality or Chebyshev inequality, which only yield power-law bounds on tail decay. Let $x_1, x_2,...$ x_n denote discrete and independent random variables with probability $p > 1/2$. Then, the probability of simultaneous occurrence of more than n/2 of the events x_k has an exact value P, where $P = \sum_{i=\lfloor n/2\rfloor}^{n}\binom{n}{i}p^i(1-p)^{n-i}$. The Chernoff bound provides a lower bound for P given by $P \geq 1 - \exp[-2n(p - 1/2)^2]$. The Chernoff bound applies to a class of random variables and provides an exponential fall-off of probability with distance from the mean. The critical condition that is needed for a Chernoff bound is that the random variable must be a sum of independent indicator random variables [39].

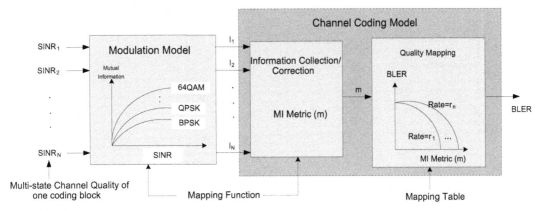

FIGURE 12-4

The procedure for an MI ESM link-to-system mapping method [8]

system-level simulation, denoted as $SINR_1$, $SINR_2$, $SINR_3$,… $SINR_N$, a mutual-information metric is calculated. Based on the computed MI metric, an equivalent SINR is obtained and used to lookup the BLER value [8].

12.5 IMT-ADVANCED TEST ENVIRONMENTS

For the evaluation of IMT-Advanced candidates, the ITU-R WP 5D defined several test environments where each test environment is characterized with certain user mobility, path loss and channel models, and system configuration parameters. The evaluation of candidate radio access technologies was performed in selected scenarios in the following test environments [6]:

- Indoor test environment models isolated cells at offices and/or in hotspots comprising stationary and pedestrian users. In this model, the emphasis is on very small cells, very high user throughputs, and high user density inside buildings.
- Urban micro-cellular test environment with higher user density was defined to model pedestrian and slow vehicular users. The micro-cellular test environment's focus is on small cells and high user densities which demonstrates typical traffic load in city centers and dense urban areas. The key characteristic of this test environment is high traffic loads, as well as outdoor and outdoor-to-indoor coverage. This scenario is interference-limited. A continuous cellular layout and the associated interference are assumed, and radio access points are located below rooftop level.
- Urban macro-cellular environment (base coverage urban) was defined to model coverage for pedestrian users up to fast-moving vehicular users. The base coverage urban test environment focuses on large cells and continuous coverage. The main characteristics of this test environment are continuous and ubiquitous coverage in urban areas. This scenario is interference-limited, using macro cells with radio access points above rooftop level. In the urban macro-cell test environment, mobile stations are located outdoor at street-level and fixed base station antennas are located clearly above surrounding building heights. Therefore, non-line-of-sight or obstructed sight can be considered in this scenario.

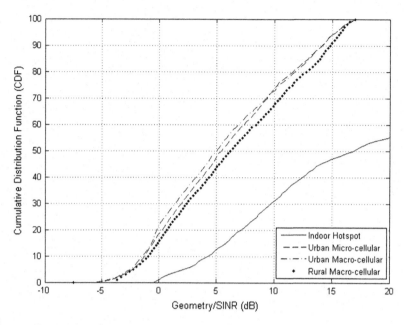

FIGURE 12-5

CDF of downlink SINR distribution under different test environments [82]

- High speed or rural test environment models high speed vehicules and trains. The high-speed test environment focuses on larger cells and continuous coverage. The key characteristics of this test environment are continuous wide area coverage supporting high speed vehicles. This scenario will therefore be noise-limited and/or interference-limited, and uses macro-sized cells.

An example downlink SINR distribution under the above test environments based on the IEEE 802.16m technology has been simulated and is shown in Figure 12-5.

12.6 NETWORK LAYOUT FOR SYSTEM-LEVEL SIMULATIONS

In rural or high-speed, macro- and micro-cellular environments, no specific topographical details are taken into consideration. Base stations are placed in regular grids conforming to a hexagonal layout. A basic hexagonal layout with three cells/sectors per site is shown in Figure 12-6 where antenna bore-sight, cell range, and inter-site distance are also illustrated. The system-level simulations use a wrap-around model with 19 sites each comprising 3 cells. The users are dropped uniformly over the entire coverage area (Monte Carlo method[ii]) in order to model the inter-cell interference. Depending on the

---

[ii]Monte Carlo methods are a class of computational algorithms that rely on repeated random sampling to compute their results. Monte Carlo methods are often used in simulating physical and mathematical systems. Because of their reliance on repeated computation of random or pseudo-random quantities, these methods are most suited to calculation by a computer and tend to be used when it is infeasible or impossible to find a closed-form result with a deterministic algorithm.

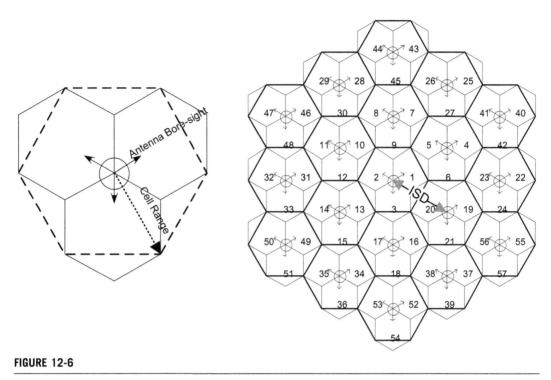

FIGURE 12-6

An illustration of a hexagonal cellular layout and basic antenna terms [6]

configuration being simulated and the required output, the effect of the surrounding outer-cells may be disregarded. In such cases, only 19 cells of the center cluster may be modeled. For the cases where modeling of the interfering outer-cells is necessary for accuracy of the results, the wrap-around structure with the 7-cluster network can be used. A cluster is defined as six displacements of the center hexagon. In the wrap-around inter-cell interference model, the network is extended to a cluster of networks consisting of seven replicas of the original hexagonal network, with the original hexagonal network in the middle while the other six copies are attached to it symmetrically on six sides. The number of mobile stations is predetermined for each sector/cell, where each MS location is randomly distributed with uniform distribution.

The serving cell of each MS is determined in two steps due to the wrap-around nature of the cell layout. The first step is to determine the 19 shortest-distanced cells for each MS from all 7 logical cell clusters, and the second step is to determine the serving cell/sector among the nearest 19 cells for each MS based on the strongest link according to path-loss and shadowing. To determine the shortest-distance cell for each MS, the distances between the MS and all logical cell clusters are calculated and the 19 cells with a shortest distance in all 7 cell clusters are selected. The serving cell for each MS provides the strongest link with a strongest received long-term signal power. It should be noted that the shadowing experienced on the link between the MS and cells located in different clusters is the same.

12.7 IMT-ADVANCED EVALUATION METHODOLOGY AND BASELINE CONFIGURATIONS

The IMT-Advanced candidate technologies were required to be evaluated under at least three test environments. The ITU-R WP 5D specified a common evaluation methodology, based on which the candidates were required to be configured and assessed against the IMT-Advanced minimum system requirements. Table 12-6 provides the common parameters for the evaluation of radio access technologies as specified by reference [6].

The BS antennas in each sector are assumed to have antenna patterns in azimuth and elevation directions that are given by:

$$A_e(\phi) = -\min\left[12\left(\frac{\phi - \phi_{tilt}}{\phi_{3dB}}\right)^2, A_m\right] \quad \text{Elevation Direction}$$

$$A(\theta) = -\min\left[12\left(\frac{\theta}{\theta_{3\,dB}}\right)^2, A_m\right] \quad \text{Azimuth Direction}$$

(12-14)

where $A(\theta)$ is the relative antenna gain (in dB) in the direction of $-180° \leq \theta \leq 180°$, θ_{3dB} is the 3 dB beamwidth ($\theta_{3dB} = 70°$), $A_m = 20$ dB is the maximum attenuation, $A_e(\phi)$ is the relative antenna gain (in dB) in the elevation direction $-90° \leq \phi \leq 90°$, ϕ_{3dB} is the elevation 3 dB beamwidth ($\phi_{3dB} = 15°$), and ϕ_{tilt} is the tilt angle. The value of the antenna tilt angle is assumed to be 0° for InH, 12° for UMi, 12° for UMa, and 6° for RMa test environments. The combined antenna pattern at angles off the cardinal axes is computed as $-\min[-A(\theta) + A_e(\phi)), A_m]$. The antenna bearing is defined as the angle between the main antenna lobe center and a line directed due east given in degrees. The bearing angle increases in a clockwise direction. The center directions of the main antenna lobe in each sector point to the corresponding side of the hexagon. For an indoor test environment, the BS is assumed to have an omni-directional antenna pattern. The mobile station is assumed to have omni-directional antennas.

12.8 LINK-LEVEL AND SYSTEM-LEVEL CHANNEL MODELS

The wireless communication channel is modeled to allow realistic modeling of the propagation conditions for radio transmissions in different environments [15–20,26–38]. The channel models were required to cover all test environments and scenarios related to the IMT-Advanced evaluations. The ITU-R WP 5D specified channel models for the evaluation of IMT-Advanced candidate technologies which consisted of primary and extension modules. The framework of the primary module is based on the WINNER II channel model,[iii] which applies the same approach as the 3GPP/3GPP2 Spatial Channel Model (SCM).[iv] The extension module extends the capabilities of the IMT-Advanced channel model to cover additional deployment scenarios beyond that of IMT-Advanced candidate technology evaluations, allowing the use of modified parameters to generate large-scale parameters. The ITU-R WP 5D-specified channel model is a geometric stochastic model. It does not explicitly specify the

---

[iii]Information on WINNER II channel models can be found at http://www.ist-winner.org/deliverables.html.
[iv]Information on 3GPP/3GPP2 Spatial Channel Model can be found in 3GPP TR 25.996 *Spatial Channel Model for Multiple Input Multiple Output* (MIMO), September 2003.

TABLE 12-6 Baseline Configuration Parameters for IMT-Advanced Evaluations [6]

Simulation Parameters	Indoor Hotspot (InH)	Urban Micro-cellular (UMi) Baseline Parameters	Urban Macro-cellular (UMa)	Rural Macro-cellular (RMa)
Base station Antenna Height (m)	6 (on the ceiling)	10 (below rooftop)	25 (above rooftop)	35 (above rooftop)
Number of BS Antennas (Receive/Transmit)	Up to 8/up to 8	Up to 8/up to 8	Up to 8/up to 8	Up to 8/up to 8
Base Station Transmit Power (dBm)	24 (40 MHz) 21 (20 MHz)	41 (10 MHz) 44 (20 MHz)	46 (10 MHz) 49 dBm (20 MHz)	46 (10 MHz) 49 (20 MHz)
Mobile Station Transmit Power (dBm)	21	24	24	24
Number of MS Antennas (Receive/Transmit)	Up to 2/up to 2	Up to 2/up to 2	Up to 2/up to 2	Up to 2/up to 2
Minimum Distance between MS and BS (m)	≥ 3	≥ 10	≥ 25	≥ 35
Parameters for Analytical Assessment of Peak Spectral Efficiency				
RF Carrier Frequency (GHz)	3.4	2.5	2	0.8
Outdoor to Indoor Penetration Loss	N/A	See Annex 1, Table A1-2 of [6]	N/A	N/A
Outdoor to In-Car Penetration Loss	N/A	N/A	9 dB (Lognormal, $\sigma = 5$ dB)	9 dB (Lognormal, $\sigma = 5$ dB)
Number of BS Antennas (Receive/Transmit)	Up to 4/up to 4	Up to 4/up to 4	Up to 4/up to 4	Up to 4/up to 4
Number of MS Antennas (Receive/Transmit)	Up to 4/up to 2	Up to 4/up to 2	Up to 4/up to 2	Up to 4/up to 2
Parameters for System-Level Simulations				
Network Layout	Indoor floor	Hexagonal grid	Hexagonal grid	Hexagonal grid
Inter-Site Distance (m)	60	200	500	1732
Channel Model	Indoor hotspot channel model	Urban micro-cellular channel model	Urban macro-cellular channel model	Rural macro-cellular channel model
User Distribution	Randomly and uniformly distributed over area	Randomly and uniformly distributed over area 50% pedestrian users and 50% users indoor	Randomly and uniformly distributed over area 100% vehicular users	Randomly and uniformly distributed over area 100% high speed vehicular users

(Continued)

TABLE 12-6 Baseline Configuration Parameters for IMT-Advanced Evaluations [6] *Continued*

Simulation Parameters	Indoor Hotspot (InH)	Urban Micro-cellular (UMi)	Urban Macro-cellular (UMa)	Rural Macro-cellular (RMa)
User Mobility	All mobile stations have fixed and identical speed with randomly and uniformly distributed direction	All mobile stations have fixed and identical speed with randomly and uniformly distributed direction	All mobile stations have fixed and identical speed with randomly and uniformly distributed direction	All mobile stations have fixed and identical speed with randomly and uniformly distributed direction
MS Speed (km/h)	3	3	30	120
Inter-Site Interference	Explicitly modeled	Explicitly modeled	Explicitly modeled	Explicitly modeled
BS Noise Figure (dB)	5	5	5	5
MS Noise Figure (dB)	7	7	7	7
BS Antenna Gain (Bore-Sight) (dBi)	0	17	17	17
MS Antenna Gain (dBi)	0	0	0	0
Thermal Noise Level (dBm/Hz)	−174	−174	−174	−174
Parameters for Assessment of Cell Spectral Efficiency and Cell Edge User Spectral Efficiency				
Traffic Model	Full buffer	Full buffer	Full buffer	Full buffer
System Bandwidth	2 × 20 MHz (FDD) 40 MHz (TDD)	2 × 10 MHz (FDD) 20 MHz (TDD)	2 × 10 MHz (FDD) 20 MHz (TDD)	2 × 10 MHz (FDD) 20 MHz (TDD)
Number of Users/Cell	10	10	10	10
Parameters for Evaluation of VoIP Capacity				
Traffic Model	VoIP	VoIP	VoIP	VoIP
System Bandwidth	2 × 5 MHz (FDD) 10 MHz (TDD)	2 × 5 MHz (FDD) 10 MHz (TDD)	2 × 5 MHz (FDD) 10 MHz (TDD)	2 × 5 MHz (FDD) 10 MHz (TDD)
Simulation Duration for a Single Drop (s)	20	20	20	20
Parameters for Link-Level Simulation (Mobility Requirement)				
Traffic Model	Full buffer	Full buffer	Full buffer	Full buffer
Channel Model	Indoor hotspot channel model	Urban micro-cellular channel model	Urban macro-cellular channel model	Rural macro-cellular channel model
System Bandwidth (MHz)	10	10	10	10
Number of Users/Cell	1	1	1	1

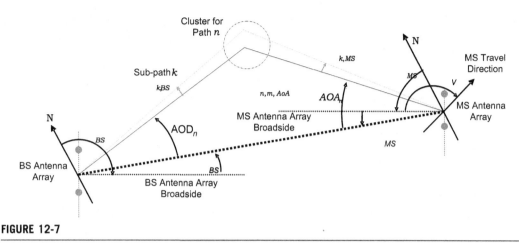

FIGURE 12-7

An illustration of angular parameters in the MIMO channels [6,8]

locations of the scatterers, rather the directions of the rays generated as a result of scattering objects. The geometric modeling of the radio channel enables separation of propagation parameters and antennas. The channel parameters for individual snapshots are determined stochastically based on statistical distributions extracted from channel measurements. Antenna geometries and radiation patterns can be defined properly by the model. As shown in Figure 12-7, channel realizations are generated through the application of the geometrical principles by summing the rays (plane waves) with specific small-scale parameters such as delay, power, Angle-of-Arrival (AoA), and Angle-of-Departure (AoD), and superposition of the results by taking into consideration the correlation between antenna elements and temporal fading with the geometry dependent Doppler spectrum. A number of rays constitute a cluster. A cluster and a propagation path diffused in space are equivalent in delay and angle domains. A generic MIMO channel model is applied to all test scenarios. The time-variant impulse response matrix of the MIMO channel can be written as (assuming N_t transmit antennas at the BS and N_r receive antennas at the MS) $\mathbf{H}(t; \tau) = \sum_{n=1}^{N} \mathbf{H}_n(t; \tau)$, where τ denotes the delay and $N = N_r \times N_t$ is the number of multi-paths. The channel impulse response is composed of the antenna array response matrices $\mathbf{F}_{tx}$ and $\mathbf{F}_{rx}$ for the transmitter and the receiver, respectively, as well as the dual-polarized propagation channel response matrix $\mathbf{h}_n$ for the nth cluster, hence:

$$\mathbf{H}_n(t; \tau) = \iint \mathbf{F}_{rx}(\phi) \; \mathbf{h}_n(t; \tau, \varphi, \phi) \, \mathbf{F}_{tx}^T(\varphi) \; d\varphi \, d\phi \tag{12-15}$$

The channel from transmit antenna element s to receive antenna element u for cluster n can be expressed as follows:

$$H_{u,s,n}(t; \tau) = \sum_{m=1}^{M} \begin{bmatrix} F_{rx,u,V}(\phi_{n,m}) \\ F_{rx,u,H}(\phi_{n,m}) \end{bmatrix}^T \begin{bmatrix} \alpha_{n,m,VV} & \alpha_{n,m,VH} \\ \alpha_{n,m,HV} & \alpha_{n,m,HH} \end{bmatrix} \begin{bmatrix} F_{tx,s,V}(\varphi_{n,m}) \\ F_{tx,s,H}(\varphi_{n,m}) \end{bmatrix} \times \exp(j2\pi\upsilon_{n,m}t)\delta(\tau - \tau_{n,m})$$
$$\times \exp(j2\pi\lambda_0^{-1}(\overline{\phi}_{n,m} \cdot \overline{r}_{rx,u})) \exp(j2\pi\lambda_0^{-1}(\overline{\varphi}_{n,m} \cdot \overline{r}_{tx,s}))$$

$$\tag{12-16}$$

where $F_{rx,u,V}$ and $F_{rx,u,H}$ denote the antenna element u field patterns for vertical and horizontal polarizations, respectively, $\alpha_{n,m,VV}$ and $\alpha_{n,m,VH}$ are the complex-valued gains of vertical-to-vertical and horizontal-to-vertical polarizations of ray n, m, respectively, λ_0 represents the wavelength of the RF carrier, $\overline{\phi}_{n.m}$ is the AoD unit vector, $\overline{\varphi}_{n.m}$ is the AoA unit vector, $\overline{r}_{tx,s}$ and $\overline{r}_{rx,u}$ are the location vectors of elements s and u, respectively, and $v_{n,m}$ denotes the Doppler frequency component of the ray n, m. [6]. If the radio channel is dynamically modeled, the above-mentioned small-scale parameters will be time-variant.

In Figure 12-7, the placement of the MS with respect to each BS is determined according to the cell layout. From this placement, the distance between the MS, the BS d, and the LoS directions with respect to the BS and MS, θ_{BS} and θ_{MS}, can be determined. Note that θ_{BS} and θ_{MS} are defined relative to the broadside directions. The MS antenna array orientations Ω_{MS} are independent and identically distributed, and are obtained from a uniform 0–360° distribution.

The generic MIMO channel model is a stochastic model with two levels of unpredictability; large-scale parameters such as shadow fading, delay, and angular spreads that are drawn randomly from tabulated distribution functions, and small-scale parameters such as delay, power, and direction of arrival and departure that are selected randomly according to tabulated distribution functions and large-scale parameters. An infinite number of different realizations of the channel model can be generated randomly by selecting different initial phases for the scattering objects.

The generic model is based on the drop concept. The system-level simulation is performed in a sequence of "drops," where a "drop" is defined as one simulation run over a certain time period when using the generic model. A drop (or snapshot of a channel segment) is a simulation instance where the random properties of the channel remain constant except for the fast-fading caused by the changing phases of the rays. The constant properties during a single drop are the power, delay, and direction of the rays. In a simulation, the number and the length of drops are selected based on the evaluation requirements and the deployment scenario. The generic model allows the simulation of several statistically-independent drops in order to obtain a statistical representation of system performance. The channel model parameters for evaluation of IMT-Advanced technologies in various test environments are shown in Table 12-7.

12.9 IEEE 802.16M LINK-LEVEL AND SYSTEM-LEVEL PERFORMANCE

In order to comply with the requirements of the ITU-R IMT-Advanced, the performance of the IEEE 802.16m standard was comprehensively evaluated based on the evaluation guidelines and methodology specified by the ITU-R Working Party 5D. The following section describes the simulation assumptions and configuration parameters that were used in the evaluation of the IEEE 802.16m technology using full-buffer data and VoIP traffic models. In addition to the common configuration parameters (technology agnostic) that were specified in Report ITU-R M.2135-1, technology-specific and test-environment-specific configuration parameters were assumed in the link-level and system-level simulations to maximize the system performance in the scenarios of interest; therefore, the configuration parameters may vary across different test environments. Note that the downlink to uplink ratio is 5:3, i.e., 5 downlink subframes and 3 uplink subframes, for the TDD mode for all full-buffer data simulations, and the cyclic prefix of 1/16 is used.

TABLE 12-7 Channel Model Parameters for IMT-Advanced Test Environments [6]

Channel Model Parameters		InH		UMi			UMa		RMa	
		Line-of-Sight	Non-Line-of-Sight	Line-of-Sight	Non-Line-of-Sight	Outdoor-to-Indoor	Line-of-Sight	Non-Line-of-Sight	Line-of-Sight	Non-Line-of-Sight
Delay Spread (DS) $\log_{10}$(s)	μ	−7.70	−7.41	−7.19	−6.89	−6.62	−7.03	−6.44	−7.49	−7.43
	σ	0.18	0.14	0.40	0.54	0.32	0.66	0.39	0.55	0.48
AoD Spread (ASD) $\log_{10}$ (degrees)	μ	1.60	1.62	1.20	1.41	1.25	1.15	1.41	0.90	0.95
	σ	0.18	0.25	0.43	0.17	0.42	0.28	0.28	0.38	0.45
AoA Spread (ASA) $\log_{10}$ (degrees)	μ	1.62	1.77	1.75	1.84	1.76	1.81	1.87	1.52	1.52
	σ	0.22	0.16	0.19	0.15	0.16	0.20	0.11	0.24	0.13
Shadow Fading (SF) (dB)	σ	3	4	3	4	7	4	6	4	8
K-Factor (K) (dB)	μ	7	N/A	9	N/A	N/A	9	N/A	7	N/A
	σ	4	N/A	5	N/A	N/A	3.5	N/A	4	N/A
Cross-Correlations	ASD vs. DS	0.6	0.4	0.5	0	0.4	0.4	0.4	0	−0.4
	ASA vs. DS	0.8	0	0.8	0.4	0.4	0.8	0.6	0	0
	ASA vs. SF	−0.5	−0.4	−0.4	−0.4	0	−0.5	0	0	0
	ASD vs. SF	−0.4	0	−0.5	0	0.2	−0.5	−0.6	0	0.6
	DS vs. SF	−0.8	−0.5	−0.4	−0.7	−0.5	−0.4	−0.4	−0.5	−0.5
	ASD vs. ASA	0.4	0	0.4	0	0	0	0.4	0	0
	ASD vs. K	0	N/A	−0.2	N/A	N/A	0	N/A	0	N/A
	ASA vs. K	0	N/A	−0.3	N/A	N/A	−0.2	N/A	0	N/A
	DS vs. K	−0.5	N/A	−0.7	N/A	N/A	−0.4	N/A	−0.5	N/A
	SF vs. K	0.5	N/A	0.5	N/A	N/A	0	N/A	0	N/A

(Continued)

TABLE 12-7 Channel Model Parameters for IMT-Advanced Test Environments [6] *Continued*

Channel Model Parameters		InH		UMi			UMa		RMa	
		Line-of-Sight	Non-Line-of-Sight	Line-of-Sight	Non-Line-of-Sight	Outdoor-to-Indoor	Line-of-Sight	Non-Line-of-Sight	Line-of-Sight	Non-Line-of-Sight
Delay Distribution		Exponential	Exponential	Exponential	Exponential	Exponential	Exponential	Exponential	Exponential	Exponential
AoD and AoA Distribution		Laplacian		Wrapped Gaussian			Wrapped Gaussian		Wrapped Gaussian	
Delay Scaling Parameter r_τ		3.6	3	3.2	3	2.2	2.5	2.3	3.8	1.7
Cross-Polarization Ratio (dB)	μ	11	10	9	8.0	9	8	7	12	7
Number of Clusters		15	19	12	19	12	12	20	11	10
Number of Rays per Cluster		20	20	20	20	20	20	20	20	20
Cluster ASD		5	5	3	10	5	5	2	2	2
Cluster ASA		8	11	17	22	8	11	15	3	3
Per Cluster Shadowing Std ζ (dB)		6	3	3	3	4	3	3	3	3
Correlation Distance (m)	DS	8	5	7	10	10	30	40	50	36
	ASD	7	3	8	10	11	18	50	25	30
	ASA	5	3	8	9	17	15	50	35	40
	SF	10	6	10	13	7	37	50	37	120
	K	4	N/A	15	N/A	N/A	12	N/A	40	N/A

In the downlink, and concerning the InH and UMi test cases, the sub-band CRU subchannelization scheme was utilized, whereas in the UMa and RMa scenarios, mini-band CRU subchannelization was utilized since the latter subchannelization scheme provides more frequency diversity gain compared to the former in higher mobility. Furthermore, in InH and UMi cases, the 6-bit transformed codebook-based MU-MIMO scheme with a 4×2 configuration was used, i.e., adaptive switching among rank-1, rank-2, rank-3, and rank-4 transmission, while in the UMa and RMa scenarios, MU-MIMO with long-term beamforming was utilized using a 4×2 antenna configuration, i.e., adaptive switching among rank-1, rank-2, rank-3, and rank-4 transmission. The MMSE receiver was employed in the MS for both channel estimation and data detection. The chase combining HARQ with maximum re-transmission delay of four frames and up to four re-transmissions was used in the simulations. The signaling error for the Assignment A-MAP and HARQ Feedback A-MAP were explicitly modeled in the simulations. The sounding estimation error has also been modeled for the midamble sequence. The dynamic overhead due to the Assignment A-MAP and HARQ Feedback A-MAP were modeled and the static overhead corresponding to non-user specific A-MAP, downlink synchronization channels, and superframe headers were taken into consideration in the simulations.

In the uplink, and with reference to the InH and UMi test cases, sub-band CRU subchannelization, as well as the 3-bit codebook-based MU-MIMO scheme with adaptive switching between single-user, and collaborative spatial multiplexing based on a 2×4 antenna configuration were used, whereas in UMa and RMa scenarios, mini-band CRU subchannelization in conjunction with MU-MIMO scheme with long-term beamforming and adaptive switching between single-user and collaborative spatial multiplexing with a 2×4 antenna configuration was utilized. The MMSE receiver was employed in the BS for both channel estimation and data detection. The chase combining HARQ scheme with a maximum re-transmission delay of four frames and up to four re-transmissions were assumed for the uplink simulations. Non-ideal channel estimation was assumed in the uplink and signaling errors were modeled for the primary fast-feedback channel (in UMa and RMa cases), the secondary fast-feedback channel (in InH and UMi scenarios), the HARQ feedback channel, and sounding estimation. Dynamic overhead due to the primary and secondary fast-feedback, as well as HARQ feedback channels, and fixed overhead corresponding to long-term covariance matrix, initial ranging, and bandwidth request channels were realistically modeled.

In the downlink, the Assignment A-MAP overhead dynamically varies as a function of the number of active users over the radio frame. The average DL/UL Assignment A-MAP overhead is taken into consideration in the calculation of the cell spectral and cell edge user spectral efficiencies. The overhead associated with the HARQ Feedback A-MAP is based on the number of required ACK/NACK derived from the uplink system-level simulations under each test environment. A fixed overhead of one OFDM symbol per frame is assumed for the primary and secondary preambles in both TDD and FDD duplex schemes. The fixed overhead due to the midamble is one OFDM symbol per frame in both TDD and FDD modes. There are 20 logical resource units over 5 OFDM symbols in every superframe that are used for the superframe headers. In the uplink, the overhead associated with the primary fast-feedback channel is calculated assuming a reporting period of 5 ms for the UMa and the RMa for all users/sectors. The CQI or PMI feedback is sent either through the secondary fast-feedback channel or MAC management messages. In both cases, the reporting period is 5 ms in both TDD and FDD modes. The overhead of the HARQ feedback channel was dynamically calculated based on the required ACK/NACK from the downlink system-level simulations for each test environment. Although the long-term covariance matrix is not fed back via dedicated uplink control

TABLE 12-8 Summary of Control Overhead for Various Test Environments [2]

Duplex Mode	Direction	InH	UMi	UMa	RMa
TDD	Downlink	9.19%	12.33%	11.17%	11.15%
	Uplink	7.85%	12.60%	9.23%	8.34%
FDD	Downlink	9.74%	16.28%	13.77%	13.63%
	Uplink	6.02%	10.58%	8.01%	6.51%

channel, rather through a regular MAC management message, its overhead is included in the calculation of the UL control overhead. A fixed overhead for the long-term covariance matrix feedback is taken into consideration assuming a reporting period of 20 ms for both the UMa and the RMa scenarios. Furthermore, a fixed overhead of one OFDM symbol per long TTI in both TDD and FDD modes is accounted for in the transmission of the sounding sequences in the uplink. The initial ranging channel is assumed to have occupied four LRUs per superframe in both TDD and FDD modes, while the overhead due to bandwidth request channel is four LRUs per superframe. Table 12-8 summarizes the downlink and uplink control overhead for various test environments.

12.9.1 Cell Spectral Efficiency and Cell Edge User Spectral Efficiency

Cell spectral efficiency SE_{cell} is defined as the aggregate throughput of all users, i.e., the number of correctly received bits delivered to the upper layers at the MAC SAP over a certain period of time, divided by the channel bandwidth divided by the number of cells. The channel bandwidth is defined as the effective bandwidth multiplied by the frequency reuse factor, where the effective bandwidth is the operating bandwidth that is appropriately scaled by the uplink/downlink ratio. The cell spectral efficiency is measured in bits/s/Hz/cell. If d_i denotes the number of correctly received bits by user i (in the downlink) or from user i (in the uplink) in a system comprising N active users and M cells, and if B denotes the channel bandwidth and T the time over which the data bits are received, the cell spectral efficiency can be expressed as follows:

$$SE_{cell} = \frac{\sum_{i=1}^{N} d_i}{TBM} \quad (12\text{-}17)$$

The (normalized) user throughput is defined as the average user throughput, i.e., the number of correctly received bits by users delivered to upper layers at the MAC SAP over a certain period of time, divided by the channel bandwidth and is measured in bits/s/Hz. The cell edge user spectral efficiency is defined as 5% point of the cumulative distribution function of the normalized user throughput. Let d_i, T_i, and B denote the number of correctly received bits by user i, the active session time for user i, and the channel bandwidth, respectively. The (normalized) user throughput of user i, is defined as:

$$\gamma_i = \frac{d_i}{T_i B} \quad (12\text{-}18)$$

The cell spectral efficiency and cell edge user spectral efficiency of IEEE 802.16m have been evaluated based on system-level simulations, and the results are shown in Table 12-9. It is shown that the IEEE 802.16m exceeds the IMT-Advanced requirements by a large margin.

TABLE 12-9 IEEE 802.16m Cell Spectral Efficiency and Cell Edge User Spectral Efficiency [2]

Requirements	Duplex Scheme	DL/UL	Test Environments			
			InH	UMi	UMa	RMa
Cell spectral efficiency (bits/s/Hz/cell) ITU-R requirement	TDD	DL	6.93 3.0	3.22 2.6	2.41 2.2	3.23 1.1
Cell edge user spectral efficiency (bits/s/Hz) ITU-R requirement			0.260 0.1	0.092 0.075	0.069 0.06	0.093 0.04
Cell spectral efficiency (bits/s/Hz/cell) ITU-R requirement	FDD		6.87 3.0	3.27 2.6	2.41 2.2	3.15 1.1
Cell edge user spectral efficiency (bits/s/Hz) ITU-R requirement			0.253 0.1	0.097 0.075	0.069 0.06	0.091 0.04
Cell spectral efficiency (bits/s/Hz/cell) ITU-R requirement	TDD	UL	5.99 2.25	2.58 1.8	2.57 1.4	2.66 0.7
Cell edge user spectral efficiency (bits/s/Hz) ITU-R requirement			0.426 0.07	0.111 0.05	0.109 0.03	0.119 0.015
Cell spectral efficiency (bits/s/Hz/cell) ITU-R requirement	FDD		6.23 2.25	2.72 1.8	2.69 1.4	2.77 0.7
Cell edge user spectral efficiency (bits/s/Hz) ITU-R requirement			0.444 0.07	0.119 0.05	0.114 0.03	0.124 0.015

12.9.2 VoIP Capacity

While most of the modern packet-switched air interface systems do not specify voice codecs, the calculation of VoIP capacity requires explicit assumptions about a particular voice codec and its parameters, such the frame rate, RTP payload size, DTX capability, etc. The VoIP capacity of the IEEE 802.16m was calculated assuming the use of the 3GPP AMR 12.2 kbps codec with a 50% speech activity factor such that the percentage of users in outage is less than 2%, where a user is defined to have experienced outage if less than 98% of the VoIP packets have been delivered successfully to the user within a one-way radio access delay bound of 50 ms. The assumptions and configuration parameters used in the system-level simulations for VoIP are as described in Table 12-4 and Table 12-5. Note that the downlink to uplink ratio is 4:4, i.e., four downlink subframes and four uplink subframes, for the TDD mode for all VoIP simulations. Furthermore, the cyclic prefix of 1/16 is used for all VoIP capacity calculations.

In the downlink, a combination of DRU and mini-band CRU subchannelization schemes is used in the simulations. Since the control channels are always allocated in the distributed resources, a fixed division is assumed in each subframe between the DRU and mini-band CRU resources. The SFBC mode with a 4×2 antenna configuration and non-adaptive precoding is used when DRU subchannelization is utilized, whereas with mini-band CRU, rank-1 transmission with wideband beamforming is employed. The MMSE receiver is modeled in the MS for channel estimation and data detection. The chase combining HARQ with a maximum re-transmission delay of four frames and up to four re-transmissions are used. A fixed overhead is assumed for non-user-specific and HARQ Feedback A-MAPs. The Assignment A-MAP overhead is explicitly modeled based on the scheduled allocations in each subframe in the downlink and uplink. It is assumed that two OFDM symbols per frame are consumed for the preambles and midamble, and the midamble estimation errors are taken into consideration in the results. The superframe headers consume 20 LRUs in the first subframe of every superframe and signaling errors are explicitly modeled in the VoIP simulations. Persistent scheduling for individual VoIP connections is used to reduce the control overhead. The MAC signaling and dynamic control overhead based on the Persistent Allocation A-MAP IE for initial transmissions is modeled; however, HARQ re-transmissions are not scheduled persistently. In general, dynamic overhead calculation model for the Assignment A-MAP is considered, whereas other downlink control channels are modeled with a fixed overhead.

In the uplink, DRU-based subchannelization was used to achieve frequency diversity. The SFBC mode with a 2×4 antenna configuration is further utilized. The MMSE receiver was simulated in the BS for both channel estimation and data detection. The chase combining HARQ with a re-transmission delay of four frames and up to four re-transmissions was used in the simulations. The primary fast-feedback channel is utilized for transmission of CQI feedback. Additionally, when wideband beam-forming is used, wideband PMI is also fed back using the primary fast-feedback channel. The control overhead due to the primary fast-feedback channel is assumed to be static according to the reporting period derived from the number of active users, voice activity factor, and the allocation interval for persistent scheduling. The overhead corresponding to the HARQ feedback channel assumes 30 ACK/NACK in each subframe derived from the downlink system-level VoIP simulation for each test environment. To accommodate sounding for 16 users per frame when the system is operating at the maximum VoIP capacity, a fixed overhead of two OFDM symbols per frame is assumed for the sounding channel. A fixed overhead of two LRUs per frame is assumed for the initial ranging and bandwidth request channels. Uplink channel signaling errors were explicitly modeled in the VoIP simulations for both FDD and TDD modes. Persistent scheduling for individual VoIP connections was modeled, and the MAC procedure and dynamic control overhead based on the Persistent Allocation A-MAP IE for initial transmissions was modeled, as well. The HARQ re-transmissions are not scheduled persistently. A fixed overhead model for the uplink control channels was used. The IEEE 802.16m VoIP capacity results in various test environments are given in Table 12-10. Since the downlink-to-uplink ratio is for the TDD and FDD modes, and from Table 12-8 the uplink control overhead is less than that of the downlink, the VoIP capacity in the uplink direction is higher than in the downlink direction; however, the system VoIP capacity is the least of the downlink and uplink capacities due to the symmetrical property of VoIP traffic.

12.9.3 Mobility

Following the methodology for evaluation of the mobility requirement [6] in the first step, the system-level simulation was performed for each test environment in the uplink to obtain the uplink SINR distribution. From the uplink SINR distribution, the 50th percentile value was obtained for each test environment. In the second step, link-level simulation was run to generate the spectral efficiency versus SINR curves for each test environment. The spectral efficiency values included the effect of the control channel overhead from the uplink system-level simulations for each test environment. The

TABLE 12-10 IEEE 802.16m VoIP Capacity in Various Test Scenarios [2]

Test Environment	Duplex Scheme	Downlink VoIP Capacity (Active Users/ MHz/Cell)	Uplink VoIP Capacity (Active Users/ MHz/Cell)	System Capacity (Active Users/ MHz/Cell)	ITU-R requirement (Active Users/ MHz/Cell)
InH	TDD	140	165	140	50
UMi		82	104	82	40
UMa		74	95	74	40
RMa		89	103	89	30
InH	FDD	139	166	139	50
UMi		77	102	77	40
UMa		72	95	72	40
RMa		90	101	90	30

mobility requirement is met, if the spectral efficiency values from the link-level curves, obtained for the median SINR value from the system-level simulations, are greater than the target values. These values were obtained assuming an antenna configuration of 4×2 in the downlink and of 2×4 in the uplink.

The statistical distribution of the uplink SINR for various test scenarios based on the IEEE 802.16m is simulated and is shown in Figure 12-8. The median uplink SINR is extracted from the distribution and is used in the calculation of the link-level spectral efficiencies for various test environments. The spectral efficiency for Line-of-Sight and Non-Line-of-Sight scenarios are calculated and provided in Table 12-11. It is shown that the IEEE 802.16m exceeds the ITU-R requirements in all test cases in both TDD and FDD duplex modes.

12.9.4 Peak Spectral Efficiency

The peak spectral efficiency is the highest theoretical data rate transmitted to a single mobile station normalized by bandwidth, assuming error-free transmission conditions, when all available radio resources for the corresponding link direction are fully utilized. The radio resources exclude those that are used for physical layer synchronization, reference signals or pilots, guard bands, and guard times. The peak spectral efficiency of the IEEE 802.16m was calculated based on the guidelines and definition provided in Report ITU-R M.2134 [5] and Report ITU-R M.2135-1 [6]. The FDD and TDD modes were considered separately, and channel bandwidth was assumed to be 20 MHz for a TDD and 2×20 MHz for an FDD duplex scheme with FFT size of 2048, and cyclic prefix length of 1/16 of the useful OFDM symbol length. In addition, four and two streams are considered in downlink and uplink, respectively. The physical layer overhead due to preambles, midamble, cyclic prefix, guard subcarriers, switching time, and idle time per radio frame were calculated and deducted from the total available time-frequency resources per frame. Since the peak spectral efficiency is calculated under error-free conditions, the coding rate is set to one and the maximum modulation scheme (64 QAM) is considered. It is further assumed that the downlink to uplink ratio is 1:1 in the TDD duplex scheme. Under the above assumptions, the peak spectral efficiency of the IEEE 802.16m is calculated as shown in Table 12-12. It is shown that the IEEE 802.16m exceeds ITU-R requirements.

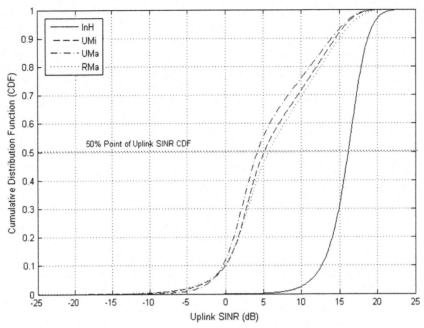

FIGURE 12-8

CDF of uplink SINR distribution for various test environments [2]

TABLE 12-11 Link-Level Results for Mobility [2]

Test Environment	Duplex Mode	Median SINR (dB)	LoS Spectral Efficiency (bits/s/Hz)	NLoS Spectral Efficiency (bits/s/Hz)	ITU-R Requirement (bits/s/Hz)
InH (10 km/h)	TDD	16.6	3.76	3.41	1.0
UMi (30 km/h)		5.0	1.81	1.50	0.75
UMa (120 km/h)		4.3	1.72	1.30	0.55
RMa (350 km/h)		5.6	1.70	1.23	0.25
InH (10 km/h)	FDD	16.6	3.86	3.56	1.0
UMi (30 km/h)		5.0	1.72	1.51	0.75
UMa (120 km/h)		4.3	1.63	1.34	0.55
RMa (350 km/h)		5.6	1.61	1.27	0.25

TABLE 12-12 Detail Calculation of the Peak Spectral Efficiency [2]

Parameter	TDD Mode		FDD Mode	
Bandwidth (MHz)	20		$2 \times 20 = 40$	
FFT Size	2048			
Cyclic Prefix Ratio	1/16			
Number of Spatial Streams	DL: $N_S = 4$ streams UL: $N_S = 2$ streams			
Subframe Types	6 Type-1 (6 Symbols) 2 Type-2 (7 Symbols)		5 Type-1 (6 Symbols) 3 Type-2 (7 Symbols)	
Number of Physical Resource Units	$N_{PRU} = 96$			
Number of Pilots per PRU	DL, 4 Streams, Subframe Type-1: 16 DL, 4 Streams, Subframe Type-2: 16 DL, 4 Streams, Subframe Type-3: 16 UL, 2 Streams, Subframe Type-1: 12 UL, 2 Streams, Subframe Type-2: 14			
Number of Data Sub-carriers per PRU	DL, 4 Streams, Type-1: $N_{data-T1} = 18 \times 6 - 16 = 92$ DL, 4 Streams, Type-2: $N_{data-T2} = 18 \times 7 - 16 = 110$ DL, 4 Streams, Type-3: $N_{data-T3} = 18 \times 5 - 16 = 74$ UL, 2 Streams, Type-1: $N_{data-T1} = 18 \times 6 - 12 = 96$ UL, 2 Streams, Type-2: $N_{data-T2} = 18 \times 7 - 14 = 112$			
Downlink Overhead Symbols	One OFDM Symbol for A-Preamble One OFDM Symbol for MIMO-Midamble			
DL:UL Ratio	1:1		N/A	
Number of Symbols per Subframe	DL: 5,7,5,6 UL: 6,6,6,7		DL: 5,7,6,6,7,6,5,7 UL: 6,7,6,6,7,6,6,7	
Number of Subframe Types	DL $N_{SF-T1} = 1$, Type-1 Subframes $N_{SF-T2} = 1$, Type-2 Subframes $N_{SF-T3} = 2$, Type-3 Subframes	UL $N_{SF-T1} = 3$, Type-1 Subframes $N_{SF-T2} = 1$, Type-2 Subframe	DL $N_{SF-T1} = 3$, Type-1 Subframes $N_{SF-T2} = 3$, Type-2 subframes $N_{SF-T3} = 2$, Type-3 Subframes	UL $N_{SF-T1} = 5$, Type-1 Subframes $N_{SF-T2} = 3$, Type-2 Subframes
Total Number of Data Sub-carriers per Frame	$N_{DATA} = N_{PRU} (N_{SF-T1}N_{Data-T1} + N_{SF-T2} N_{Data-T2} + N_{SF-T3}N_{Data-T3})$			
	DL: $N_{DATA} = 33600 = 96 \times (1 \times 92 + 1 \times 110 + 2 \times 74)$ UL: $N_{DATA} = 38400 = 96 \times (3 \times 96 + 1 \times 112)$		DL: $N_{DATA} = 72384 = 96 \times (3 \times 92 + 3 \times 110 + 2 \times 74)$ UL: $N_{DATA} = 78336 = 96 \times (5 \times 96 + 3 \times 112)$	
Coding Rate, FEC_{RATE}	1			
MCS 64QAM, N_b (bits/symbol)	6			
Frame Duration (ms)	5			
Peak Spectral Efficiency	$PEAK_{THR} = N_{DATA} N_S N_b FEC_{RATE}$			
Peak DL Throughput (Mbps)	161.28		347.44	
Peak UL Throughput (Mbps)	92.16		188.01	
DL Peak Spectral Efficiency (bits/s/Hz)	16.13		17.37	
UL Peak Spectral Efficiency (bits/s/Hz)	9.21		9.4	

12.9.5 User-Plane/Control-Plane Latency and Handover Interruption Time

The user-plane latency, alternatively known as transport delay, is defined as the one-way transit time between the availability of a packet at the IP layer at the source (user terminal/base station), and the availability of this packet at the IP layer at the destination (base station/user terminal). User-plane packet delay includes delay introduced by associated protocols and control signaling, assuming the user terminal is in the active state. The IMT-Advanced systems are required to have a user-plane latency of less than 10 ms in unloaded conditions (i.e., a single user with a single data stream) for small IP packets (e.g., 0 byte payload + IP header) in the downlink and uplink directions.

To calculate the user-plane and control-plane latencies, it is assumed that each 5 ms radio frame consists of eight subframes of 0.617 ms length, and the Transmission Time Interval (TTI) is equal to one subframe. The mobile station and base station processing times are assumed to take three subframes (3 × TTI). This analysis further assumes an intra-ASN handover mechanism where the serving and target base stations belong to the same ASN entity. Typical values of HARQ re-transmission probability or air-link error rate prior to HARQ processing range from 10% to 30%. The HARQ Round-Trip Time (RTT) or the time interval between two consecutive downlink or uplink data transmissions (or re-transmissions) is assumed to be one radio frame (8 × TTI) for the FDD mode. The HARQ RTT for the TDD mode, assuming a DL:UL ratio of 5:3 and one DL/UL switching point per radio frame, is 8 × TTI. Using the user-plane latency calculation model shown in Figure 12-9, the user-plane latency of the IEEE 802.16m has been analytically calculated, and the results are shown in Table 12-13.

Therefore, the IEEE 802.16m meets the ITU-R requirement for user-plane latency. The control-plane latency is typically measured as the transition time from idle state to active state. The IMT-Advanced systems are required to exhibit a transition time (excluding downlink paging delay and wireline network signaling delay) of less than 100 ms from an idle state to an active state in such a way that the user-plane can be established upon transition. Table 12-14 contains the assumptions and results for control-plane latency analysis for IEEE 802.16m.

As a result, the IEEE 802.16m meets the ITU-R requirement for control-plane latency.

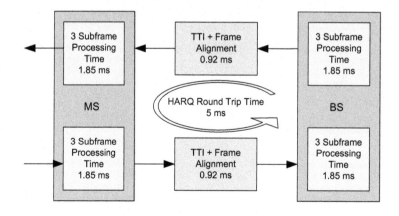

FIGURE 12-9

User-plane latency calculation model [2]

TABLE 12-13 User-Plane Latencies for 10% and 30% Probability of HARQ Re-transmissions [2]

Procedure	User-Plane Latency (10% HARQ Re-transmission Probability)	User-Plane Latency (30% HARQ Re-transmission Probability)
MS Wakeup Time	Implementation dependent	Implementation dependent
MS Processing Delay	$3 \times 0.617 = 1.85$ ms	$3 \times 0.617 = 1.85$ ms
Queuing/Frame Alignment	FDD: 0.31 TDD: 2.5 ms	FDD: 0.31 TDD: 2.5 ms
TTI For Uplink Data Packet (Piggy Back Scheduling Information)	0.617 ms	0.617 ms
HARQ Retransmission	0.1×5 ms	0.3×5 ms
BS Processing Delay	$3 \times 0.617 = 1.85$ ms	$3 \times 0.617 = 1.85$ ms
Total One-Way Access Delay	FDD: 5.13 ms TDD: 7.32 ms	FDD: 6.13 ms TDD: 8.32 ms

The handover interruption time is defined as the time duration during which a user terminal cannot exchange data packets with any base station. The handover interruption time includes the time required to execute any radio access network procedure, radio resource control signaling protocol, or other message exchanges between the user equipment and the radio access network. For the purpose of determining handover interruption time, the interactions with the core network (i.e., network entities beyond the radio access network) are assumed to occur in zero time. It is also assumed that all necessary attributes of the target channel, i.e., downlink synchronization is achieved and uplink access procedures are successfully completed, are known at initiation of the handover from the serving BS to the target BS.

As shown in Figure 12-10, the handover can be initiated by either the MS or the serving BS. The handover process is initiated when the MS issues a handover request to the serving BS, or when the serving BS issues a handover command to the MS. The handover request/command is issued following mobile-assisted handover procedures. The MS acquires the network topology through either serving BS broadcasts or unicast messages. The serving BS also provides the MS with the relevant trigger conditions to initiate or cancel neighbor-cell measurements, measurement reporting, and handover requests. Scanning intervals are provided for neighbor-cell measurements by the serving BS unilaterally or at the request of the MS. The handover request/command is an outcome of these measurements and reporting. The handover procedure is divided into four steps: handover initiation; handover preparation; handover execution; and handover cancellation procedure to allow the MS to cancel a handover procedure.

The following assumptions were made in the calculation of the handover interruption time: (1) the handover is of the intra-ASN type, i.e., the serving and target base stations belong to the same ASN; (2) this is a mobile-assisted handover; (3) the handover is a mobile-initiated type, which in terms of latency is the worst-case compared to BS-initiated type; (4) this is an optimized hard handover, which means that for intra-frequency, the MS is frame-synchronized with the serving BS and the target BS, and the MS context including security context is transferred to the target BS over the backhaul; and

TABLE 12-14 Control-Plane Latency for 10% and 30% Probability of HARQ Re-transmissions [2]

Procedure	Control-Plane Latency (10% HARQ Re-transmission Probability)	Control-Plane Latency (30% HARQ Re-transmission Probability)
MS Wakeup Time	Implementation dependent	Implementation dependent
DL scanning and synchronization + acquisition of the system configuration information for network re-entry Note: It can be assumed that MS might have updated system configuration prior to transition	40 ms Note that S-SFH SP1 that contains network re-entry information and is transmitted every 40 ms	40 ms Note that S-SFH SP1 that contains network re-entry information and is transmitted every 40 ms
Random Access Procedure (UL CDMA Code + BS Processing + DL CDMA_ALLOC_IE)	5 ms	5 ms
Initial Ranging (AAI_RNG-REQ + BS processing + AAI_RNG-RSP) + HARQ re-transmission of one message at 10% or 30%, only first-order estimation	6 ms Assuming the message will succeed in the first transmission with probability of 0.9 or in the second transmission with probability of 0.1	8 ms Assuming the message will succeed in the first transmission with probability of 0.7 or in the second transmission with probability of 0.3
Capability Negotiation (AAI_SBC-REQ + BS processing + AAI_SBC-RSP) + HARQ re-transmission	< 5 ms 1 × 5 ms for HARQ re-transmission)	< 5 ms 3 × 5 ms for HARQ re-transmission)
Authorization, Authentication, and Key Exchange (AAI_PKM-REQ + BS processing + AAI_PKM-RSP) + HARQ retransmission	< 5 ms 1 × 5 ms for HARQ re-transmission)	< 5 ms 3 × 5 ms for HARQ re-transmission)
BS Registration (AAI_REG-REQ + BS/ASN-GW processing + AAI_REG-RSP) + HARQ retransmission	< 5 ms 1 × 5 ms for HARQ re-transmission)	< 5 ms 3 × 5 ms for HARQ re-transmission)
RRC Connection Establishment (AAI_DSA-REQ + BS processing + AAI_DSA-RSP + AAI_DSA-ACK) + HARQ retransmission	< 5 ms1 × 5 ms for HARQ re-transmission)	< 5 ms3 × 5 ms for HARQ re-transmission)
C-Plane Connection Establishment Delay	< 31 ms	< 33 ms
IDLE_STATE → ACTIVE_ACTIVE Delay Note: It can be assumed that MS might have updated system configuration prior to transition	< 71 ms	< 73 ms

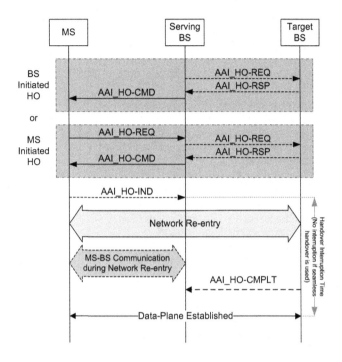

FIGURE 12-10

The IEEE 802.16m handover procedures [12]

(5) the seamless handover procedure is used [12]. All measurements are based on the synchronization channels. Table 12-15 summarizes the handover procedure delay budget.

Therefore, IEEE 802.16m meets the ITU-R requirement for handover interruption time.

12.9.6 Frequency Bands and Operation Bandwidth

The ITU-R requires IMT-Advanced systems to support a scalable bandwidth up to 100 MHz through aggregation of smaller frequency bands. The ITU-R further requires the IMT-Advanced systems to support the IMT bands that are designated by ITU-R for deployment of the 3rd and 4th generations of cellular systems. The IEEE 802.16m specifies multi-carrier techniques through which support of various contiguous or non-contiguous spectra are possible. The band classes specified by the WiMAX Forum allow deployment of the IEEE 802.16m in IMT bands in the form of TDD or FDD duplex schemes. Table 12-16 shows the frequency bands in which the IEEE 802.16m can be deployed.

12.9.7 IEEE 802.16m Link Budget

The IMT-Advanced submissions were required to provide a detailed link budget analysis under the ITU-R specified test environments. Report ITU-R M.2135-1 provided a methodology and common

TABLE 12-15 IEEE 802.16m Handover Interruption Time Analysis [2]

Step	Procedure	Estimated Latency (ms)
1	The MS initiates handover by sending an AAI_HO-REQ to the serving BS	20 to 35 ms
2	The serving-BS processes AAI_HO-REQ message and sends AAI_HO-REQ to one or more target base stations	5 ms
3	Target base stations reply to the serving BS with AAI_HO-RSP, which may include handover optimization related MAC update information	10 ms
4	The serving BS responds to MS with AAI_HO-CMD containing target BS list and the disconnect time	5 ms
5	MS acknowledges to the serving BS with AAI_HO-IND containing selected target BS and confirmation/rejection of the disconnect time (unsolicited uplink grant)	5 ms
6	At or after the Disconnect Time, the serving BS transfers the unacknowledged and the new data, if any, to the target BS to allow MS data continuity at target BS	0 to 10 ms (R8 interface latency)
7	MS switches to the target BS and acquires downlink broadcast overhead channels	5 ms
7.1	The MS waits for handover ranging opportunity to perform uplink synchronization with dedicated ranging code assigned by the target BS during handover preparation Note that initial ranging, uplink synchronization procedures are not counted into handover interruption time according to the definition	5 to 20 ms Note that 20 ms is the worst case when no dedicated ranging opportunity is provided for this handover. In most cases, target BS has prior knowledge of the MS capability and therefore target BS can prepare the ranging opportunity at the next frame, in that case the latency will be 5 ms
8	MS detects the downlink control channels for unsolicited uplink grant in order to send AAI_RNG-REQ message and data	10 ms
9	MS sends AAI_RNG-REQ to the target BS	5 ms
10	The target BS responds with AAI_RNG-RSP with the necessary information for the MS to perform uplink synchronization	10 ms
11	MS processes AAI_RNG-RSP	5 ms
12	If necessary, repeat steps 8 to 11 k times Note that the maximum value of k is calculated based on the number of times that steps 8 to 11 are repeated before expiration of a timer assigned by the serving BS	0 to $25k$ ms
13	The target BS and the MS continue data communication	0
Intra-FA handover Interruption Time (using seamless handover)		Sum of the processing times for Steps 6 and 7: 0 to 15 ms
Inter-FA handover Interruption Time (using seamless handover)		Sum of the processing times for Steps 6, 7, and 7.1: 5 to 35 ms

TABLE 12-16 Operation Bands of IEEE 802.16m [2]

Band Class	Uplink MS Transmission Frequency (MHz)	Downlink MS Receiving Frequency (MHz)	Duplex Mode
1	2300–2400	2300–2400	TDD
2	2305–2320, 2345–2360	2305–2320, 2345–2360	TDD
	2345–2360	2305–2320	FDD
3	2496–2690	2496–2690	TDD
	2496–2572	2614–2690	FDD
4	3300–3400	3300–3400	TDD
5L	3400–3600	3400–3600	TDD
	3400–3500	3500–3600	FDD
5H	3600–3800	3600–3800	TDD
6	1710–1770	2110–2170	FDD
	1920–1980	2110–2170	FDD
	1710–1755	2110–2155	FDD
	1710–1785	1805–1880	FDD
	1850–1910	1930–1990	FDD
	1710–1785, 1920–1980	1805–1880, 2110–2170	FDD
	1850–1910, 1710–1770	1930–1990, 2110–2170	FDD
7	698–862	698–862	TDD
	776–787	746–757	FDD
	788–793, 793–798	758–763, 763–768	FDD
	788–798	758–768	FDD
	698–862	698–862	TDD/FDD
	824–849	869–894	FDD
	880–915	925–960	FDD
	698–716, 776–793	728–746, 746–763	FDD
8	1785–1805, 1880–1920, 1910–1930, 2010–2025, 1900–1920	1785–1805, 1880–1920, 1910–1930, 2010–2025, 1900–1920	TDD
9	450–470	450–470	TDD
	450.0–457.5	462.5–470.0	FDD

parameters in order to calculate the downlink and uplink link budgets. In addition to the common parameters, the following assumptions were made in the calculations:

- There are four transmit antennas and four receive antennas in the BS, and there are two transmit antennas and two 2 receive antennas in the MS.
- The target packet error rate is 10% for initial transmission of data channels, and 1% for the control channels.
- The modulation and coding scheme for the DL Assignment A-MAP is QPSK 1/8.
- There are 6 bits transmitted over the uplink primary fast-feedback channel.

- The MIMO scheme for downlink data is rank-1 wideband beamforming, for downlink control is SFBC with non-adaptive precoder, for uplink data is rank-1 wideband beamforming, and for uplink control is single antenna transmission.
- The permutation (subchannelization) scheme used for DL/UL data is mini-band CRU and for DL control is DRU.
- There is 2 dB pilot-boosting over data tones in the downlink, and no pilot boosting in the uplink.
- No HARQ is assumed for the control channel.
- 0.5 dB HARQ combining gain for the data channel.

Shadowing fade margin is determined as a function of the cell edge coverage reliability and the standard deviation of the log-normal shadow-fading, including penetration loss. Cell edge coverage reliability is determined for the given area coverage reliability as a function of the shadow fading standard deviation and the path loss exponent obtained from the path loss model. The cell edge reliability can be determined using simulations or using traditional numerical methods. Cell area reliability is defined as the percentage of the cell area over which coverage can be guaranteed. It is obtained from the cell edge reliability, shadow fading standard deviation, and the path loss exponent. The latter two values are used to calculate a fade margin. Macro diversity gain may be considered explicitly in order to improve the system margin or implicitly by reducing the fade margin. The path loss models are summarized in Table A1-2 of Report ITU-R M.2135-1 [6]. The IEEE 802.16m link budget in the downlink and uplink under various deployment scenarios for the TDD mode is given in Table 12-17.

12.9.8 Additional Link-Level Simulation Results

The link-level performance of the IEEE 802.16m has been characterized under various conditions, including different mobility classes and beamforming schemes. The following assumptions have been made in the link-level simulations and the results are provided in this section. We assume the TDD mode of operation, transmission bandwidth of 10 MHz, FFT size of 1024, mini-band LRU, a modified ITU-PedB channel model at 3 km/h or a modified ITU-VehA channel model at 120 km/h, [8] a CTC channel coder with a code rate of $R = 0.30392$, QPSK modulation, uniform distribution of AoD, and $SIR = \infty$ dB. Furthermore, the packet error rate and spectral efficiency are simulated based on a different number of transmit and receive antennas in the downlink when antennas are either correlated, i.e., when inter-antenna distance $d = 0.5\ \lambda$, or when antennas are uncorrelated, i.e., when $d = 4\ \lambda$. Figures 12-11 through 12-14 show the PER and SE as a function of SNR for different antenna configurations and mobile speeds using adaptive beamforming based on a long-term covariance matrix. The channel state information in terms of a long-term covariance matrix is calculated and sent to the BS every four frames.

The effect of transmit diversity on the PER and SE in the low SNR region can be seen in the figures. Furthermore, it can be observed that the performance is better for correlated antennas compared with uncorrelated antennas, particularly in the low SNR region. The effect of low and high mobile speeds on the PER and SE have also been investigated and are illustrated in the figures. It is shown that at a given SNR, the increase in mobile speed would increase PER and decrease SE, as would theoretically be expected.

In order to investigate the effect of various feedback schemes on the link-level PER and SE, adaptive beamforming with wideband PMI was simulated and the results are shown in Figures 12-15 through 12-18.

TABLE 12-17 IEEE 802.16m Link Budget in Various Test Environments for the TDD Duplex Scheme [2]

Parameter	InH Downlink	InH Uplink	UMi Downlink	UMi Uplink	UMa Downlink	UMa Uplink	RMa Downlink	RMa Uplink
	System Configuration							
Carrier Frequency (GHz)	3.4	3.4	2.5	2.5	2	2	0.8	0.8
BS Antenna Heights (m)	6	6	10	10	25	25	35	35
MS Antenna Heights (m)	1.5	1.5	1.5	1.5	1.5	1.5	1.5	1.5
Cell Area Reliability	95%	95%	95%	95%	95%	95%	95%	95%
Transmission Bit Rate for Control Channel (kbps)	89.6	1.2	89.6	1.2	89.6	1.2	89.6	1.2
Transmission Bit Rate for Data Channel (Mbps)	20.23	0.98	7.53	0.21	7.53	0.21	7.53	0.21
Target Packet Error Rate for Control Channels	10^{-2}	10^{-2}	10^{-2}	10^{-2}	10^{-2}	10^{-2}	10^{-2}	10^{-2}
Target Packet Error Rate for Data Channels	10^{-1}	10^{-1}	10^{-1}	10^{-1}	10^{-1}	10^{-1}	10^{-1}	10^{-1}
Spectral Efficiency (bits/s/Hz) for Data	0.856	0.830	0.637	0.720	0.637	0.720	0.637	0.720
Pathloss Model	NLoS	NLoS	NLoS	NLoS	NLoS	NLoS	NLoS	NLoS
Mobile Speed (km/h)	3	3	3	3	30	30	120	120
Feeder Loss (dB)	2	2	2	2	2	2	2	2
Transmitter								
Number of Transmit Antennas	4	2	4	2	4	2	4	2
Maximum Transmit Power per Antenna (dBm)	18	18	38	21	43	21	43	21
Total Transmit Power (dBm)	24	21	44	24	49	24	49	24
Transmitter Antenna Gain (dBi)	0	0	17	0	17	0	17	0
Transmitter Array Gain (dB)	0	0	0	0	0	0	0	0
Control Channel Power Boosting Gain (dB)	0	0	0	0	0	0	0	0
Data Channel Power Loss Due to Pilot/Control Boosting (dB)	0.2734	0	0.2734	0	0.2734	0	0.2734	0

(Continued)

TABLE 12-17 IEEE 802.16m Link Budget in Various Test Environments for the TDD Duplex Scheme [2] *Continued*

Parameter	InH		UMi		UMa		RMa	
	Downlink	Uplink	Downlink	Uplink	Downlink	Uplink	Downlink	Uplink
System Configuration								
Cable, Connector, Combiner, Body Losses, etc. (dB)	3	1	3	1	3	1	3	1
Control Channel EIRP (dBm)	21	20	58	23	63	23	63	23
Data Channel EIRP (dBm)	20.73	20	57.73	23	62.73	23	62.73	23
Receiver								
Number of Receive Antennas	2	4	2	4	2	4	2	4
Receiver Antenna Gain (dBi)	0	0	0	17	0	17	0	17
Cable, Connector, Combiner, Body Losses, etc. (dB)	1	3	1	3	1	3	1	3
Receiver Noise Figure (dB)	7	5	7	5	7	5	7	5
Thermal Noise Density (dBm/Hz)	−174	−174	−174	−174	−174	−174	−174	−174
Receiver Interference Density (dBm/Hz)	−174	−174	−165	−166	−165	−166	−165	−166
Total Noise Plus Interference Density (dBm/Hz)	−166.21	−167.81	−162.88	−164.24	−162.88	−164.24	−162.88	−164.24
Occupied Channel Bandwidth (MHz)	37.81	3.15	18.90	0.79	18.90	0.79	18.90	0.79
Effective Noise Power (dBm)	−90.43	−102.82	−90.11	−105.27	−90.11	−105.27	−90.11	−105.27
Required SNR for the Control Channel (dB)	−0.56	−2.48	−1.57	−4.10	−1.95	−3.97	−1.19	−2.42
Required SNR for the Data Channel (dB)	1.41	−0.24	−1.05	−0.96	−0.21	−0.82	−0.70	0.37
Receiver Implementation Margin (dB)	2	2	2	2	2	2	2	2
HARQ Gain for Control Channel (dB)	0	0	0	0	0	0	0	0

HARQ Gain for Data Channel (dB)	0.5	0.5	0.5	0.5	0.5	0.5	0.5	0.5
Receiver Sensitivity for Control Channel (dBm)	−89.00	−103.30	−89.68	−107.37	−90.06	−107.24	−89.30	−105.69
Receiver Sensitivity for Data Channel (dBm)	−87.52	−101.57	−89.66	−104.73	−88.82	−104.59	−89.31	−103.40
Hardware Link Budget for Control Channel (dB)	110.00	123.30	147.68	147.37	153.06	147.24	152.30	145.69
Hardware Link Budget for Data Channel (dB)	108.25	121.57	147.39	144.73	151.55	144.59	152.04	143.40
Calculation of Available Path Loss								
Lognormal Shadow Fading Standard Deviation (dB)	4	4	4	4	7.8	7.8	9.4	9.4
Shadow Fading Margin (dB)	2.8	2.8	3.10	3.10	8.1	8.1	10.4	10.4
BS Selection/Macro-Diversity Gain (dB)	0	0	0	0	0	0	0	0
Penetration Loss (dB)	0	0	0	0	9	9	9	9
Other Gains (dB)	0	0	0	0	0	0	0	0
Available Path Loss for Control Channel (dB)	106.20	117.50	143.58	141.27	134.96	127.14	131.90	123.29
Available Path Loss for Data Channel (dB)	104.45	115.77	143.29	138.63	133.45	124.49	131.64	121.00
Range/Coverage Efficiency Calculation								
Maximum Range for Control Channel (m)	87.39	159.42	1030.29	891.56	891.93	562.90	2361.39	1412.39
Maximum Range for Data Channel (m)	79.65	145.37	1011.89	755.50	816.07	481.56	2323.96	1232.23
Coverage Area for Control Channel (km²/site)	0.02	0.08	3.33	2.50	2.50	0.99	17.52	6.27
Coverage Area for Data Channel (km²/site)	0.20	0.07	3.22	1.79	2.09	0.73	16.97	4.77

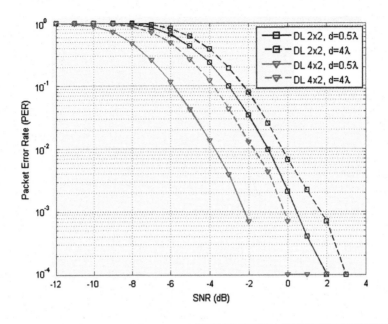

FIGURE 12-11

Link-level packet error rate versus SNR (modified ITU-PedB 3 km/h)

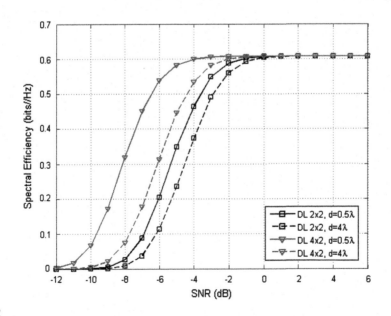

FIGURE 12-12

Link-level spectral efficiency versus SNR (modified ITU-PedB 3 km/h)

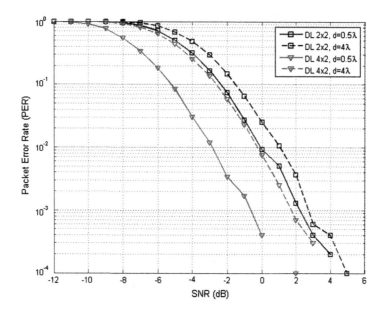

FIGURE 12-13

Link-level packet error rate versus SNR (modified ITU-VehA 120 km/h)

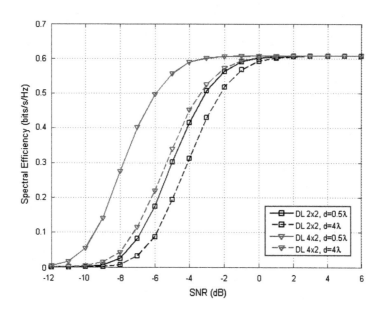

FIGURE 12-14

Link-level spectral efficiency versus SNR (modified ITU-VehA 120 km/h)

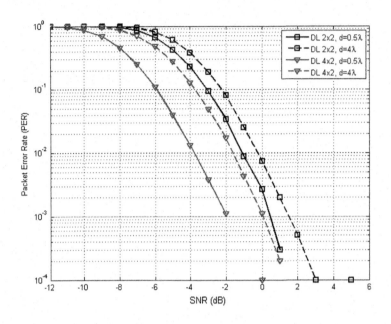

FIGURE 12-15

Link-level packet error rate versus SNR (modified ITU-PedB 3 km/h)

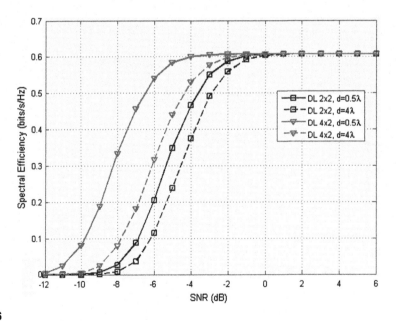

FIGURE 12-16

Link-level spectral efficiency versus SNR (modified ITU-PedB 3 km/h)

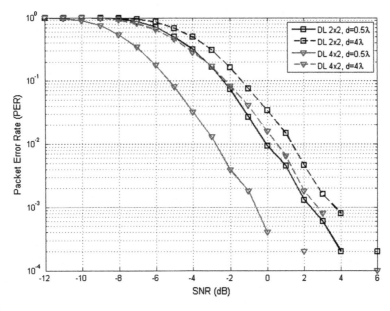

FIGURE 12-17

Link-level packet error rate versus SNR (modified ITU-VehA 120 km/h)

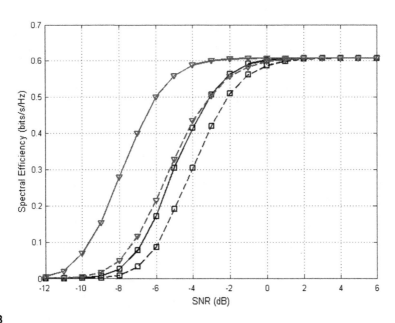

FIGURE 12-18

Link-level spectral efficiency versus SNR (modified ITU-VehA 120 km/h)

The results suggest that the performance with wideband PMI that is fed back to the BS every frame is slightly better than that of the long-term covariance matrix that is fed back every four frames.

12.10 3GPP LTE-ADVANCED LINK-LEVEL AND SYSTEM-LEVEL PERFORMANCE

The 3GPP LTE-Advanced is another candidate for IMT-Advanced whose performance was fully characterized based on the same evaluation methodology that was applied to the IEEE 802.16m. The following sections summarize the performance characterization of the 3GPP LTE-Advanced as part of the IMT-Advanced submission. Although the performance of the two candidate technologies, i.e., IEEE 802.16m and 3GPP LTE-Advanced, are not directly compared, it can be generally concluded that the two technologies perform similarly, and have the same level of functionality and performance in various test environments.

12.10.1 Cell Spectral Efficiency and Cell Edge Spectral Efficiency

Cell and cell edge spectral efficiencies were evaluated through extensive simulations conducted by a number of 3GPP member companies. The simulation results were reported based on specific 3GPP LTE-Advanced configurations, i.e., downlink and uplink MU-MIMO, downlink and uplink CoMP, and uplink SU-MIMO, for both FDD and TDD duplex schemes in various test scenarios. The performance differences among contributing sources can be explained by implementation-specific functionalities at the transmitter and the receiver that are not explicitly specified by the standard, such as receiver type, scheduling algorithms, etc. In the results reported for the downlink, the size of the control channel (L OFDM symbols) and the number of the MBSFN subframes are the factors that affect the overhead. Note that in the 3GPP submission, the control channel overhead is statically modeled, i.e., the downlink control channels occupy a fixed number of OFDM symbols irrespective of the number of users, although it is expected that the size of the PDCCH varies dynamically with the number of active users in the cell. In the MBSFN subframes, there are no common reference signals in data regions, which effectively reduce the overhead. Channel estimation and receiver types are the factors that affect the demodulation performance. The availability of CSI at the eNB is the assumption that impacts the transmit signal processing at the eNB for MU-MIMO and CoMP schemes. In the tables for the uplink, the PUCCH bandwidth is the factor that affects the overhead. Each value in Table 12-18 and Table 12-19 is obtained as an average of all the samples provided by different companies. The results suggest that the requirements are fulfilled with $L = 3$ corresponding to the largest overhead, except for the system bandwidth of 1.4 MHz where the requirements cannot be met. If the control overhead assumption is relaxed, that is the number of active users decreases, $L = 1$ and 2 can be considered and thereby the performance can be further improved. The tables show that the 3GPP LTE Rel-8 with SU-MIMO 4×2 can already fulfill the ITU-R requirements. The tables also show that further performance improvements can be achieved by using additional technical features.

In the following tables, various antenna configurations have been utilized. There are four or eight transmit antennas with the following configurations: (1) uncorrelated co-polarized, i.e., co-polarized antennas that are 4-wavelengths apart; (2) grouped co-polarized, i.e., two groups of co-polarized antennas where there is a 10-wavelength distance between the center of each group and a 0.5-

TABLE 12-18 3GPP LTE-Advanced Downlink Cell and Cell-edge Spectral Efficiencies [13]

Scheme and Antenna Configuration	Test Environment	Duplex Scheme	ITU-R Requirement Cell Spectral Efficiency/Cell-edge Spectral Efficiency	Cell Spectral Efficiency (bits/s/Hz/cell)			Cell-edge Spectral Efficiency (bits/s/Hz)		
				L=1	L=2	L=3	L=1	L=2	L=3
Rel-8 SU-MIMO 4 × 2 (A)	InH	FDD	3/0.1	4.8	4.5	4.1	0.23	0.21	0.19
MU-MIMO 4 × 2 (C)			3/0.1	6.6	6.1	5.5	0.26	0.24	0.22
Rel-8 SU-MIMO 4 × 2 (A)		TDD	3/0.1	4.7	4.4	4.1	0.22	0.20	0.19
MU-MIMO 4 × 2 (C)			3/0.1	6.7	6.1	5.6	0.24	0.22	0.20
MU-MIMO 4 × 2 (C)	UMi	FDD	2.6/0.075	3.5	3.2	2.9	0.10	0.096	0.087
MU-MIMO 4 × 2 (A)			2.6/0.075	3.4	3.1	2.8	0.12	0.11	0.099
CS/CB-CoMP 4 × 2 (C)			2.6/0.075	3.6	3.3	3.0	0.11	0.099	0.089
JP-CoMP 4 × 2 (C)			2.6/0.075	4.5	4.1	3.7	0.14	0.13	0.12
MU-MIMO 8 × 2 (C/E)			2.6/0.075	4.2	3.8	3.5	0.15	0.14	0.13
MU-MIMO 4 × 2 (C)		TDD	2.6/0.075	3.5	3.2	3.0	0.11	0.096	0.089
MU-MIMO 4 × 2 (A)			2.6/0.075	3.2	2.9	2.7	0.11	0.10	0.095
CS/CB-CoMP 4 × 2 (C)			2.6/0.075	3.6	3.3	3.1	0.10	0.092	0.086
JP-CoMP 4 × 2 (C)			2.6/0.075	4.6	4.2	3.9	0.10	0.092	0.085
MU-MIMO 8 × 2 (C/E)			2.6/0.075	4.2	3.9	3.6	0.12	0.11	0.099
MU-MIMO 4 × 2 (C)	UMa	FDD	2.2/0.06	2.8	2.6	2.4	0.079	0.073	0.066
CS/CB-CoMP 4 × 2 (C)			2.2/0.06	2.9	2.6	2.4	0.081	0.074	0.067
JP-CoMP 4 × 2 (A)			2.2/0.06	3.0	2.7	2.5	0.080	0.073	0.066
CS/CB-CoMP 8 × 2 (C)			2.2/0.06	3.8	3.5	3.2	0.10	0.093	0.084
MU-MIMO 4 × 2 (C)		TDD	2.2/0.06	2.9	2.6	2.4	0.079	0.071	0.067
CS/CB-CoMP 4 × 2 (C)			2.2/0.06	2.9	2.6	2.4	0.083	0.075	0.070
JP-CoMP 4 × 2 (C)			2.2/0.06	3.6	3.3	3.1	0.090	0.082	0.076
CS/CB-CoMP 8 × 2 (C/E)			2.2/0.06	3.7	3.3	3.1	0.10	0.093	0.087
Rel-8 SU-MIMO 4 × 2 (C)	RMa	FDD	1.1/0.04	2.3	2.1	1.9	0.081	0.076	0.069
Rel-8 SU-MIMO 4 × 2 (A)			1.1/0.04	2.1	2.0	1.8	0.067	0.063	0.057
MU-MIMO 4 × 2 (C)			1.1/0.04	3.9	3.5	3.2	0.11	0.099	0.090
MU-MIMO 8 × 2 (C)			1.1/0.04	4.1	3.7	3.4	0.13	0.12	0.11
Rel-8 SU-MIMO 4 × 2 (C)		TDD	1.1/0.04	2.0	1.9	1.8	0.072	0.067	0.063
Rel-8 SU-MIMO 4 × 2 (A)			1.1/0.04	1.9	1.7	1.6	0.057	0.053	0.049
MU-MIMO 4 × 2 (C)			1.1/0.04	3.5	3.2	3.0	0.098	0.089	0.083
MU-MIMO 8 × 2 (C/E)			1.1/0.04	4.0	3.6	3.4	0.12	0.11	0.10
Rel-8 Single-Layer BF 8 × 2 (E)				2.5	2.3	2.1	0.11	0.10	0.093

TABLE 12-19 3GPP LTE-Advanced Uplink Cell and Cell-edge Spectral Efficiencies [13]

Scheme and Antenna Configuration	Test Environment	Duplex Scheme	ITU-R Requirement Cell Spectral Efficiency/Cell-edge Spectral Efficiency	Cell Spectral Efficiency (bits/s/Hz/cell)	Cell-edge Spectral Efficiency (bits/s/Hz)
Rel-8 SIMO 1 × 4 (A)	InH	FDD	2.25/0.07	3.3	0.23
Rel-8 SIMO 1 × 4 (C)			2.25/0.07	3.3	0.24
Rel-8 MU-MIMO 1 × 4 (A)			2.25/0.07	5.8	0.42
SU-MIMO 2 × 4 (A)			2.25/0.07	4.3	0.25
Rel-8 SIMO 1 × 4 (A)		TDD	2.25/0.07	3.1	0.22
Rel-8 SIMO 1 × 4 (C)			2.25/0.07	3.1	0.23
Rel-8 MU-MIMO 1 × 4 (A)			2.25/0.07	5.5	0.39
SU-MIMO 2 × 4 (A)			2.25/0.07	3.9	0.25
Rel-8 SIMO 1 × 4 (C)	UMi	FDD	1.8/0.05	1.9	0.073
Rel-8 MU-MIMO 1 × 4 (A)			1.8/0.05	2.5	0.077
MU-MIMO 2 × 4 (A)			1.8/0.05	2.5	0.086
Rel-8 SIMO 1 × 4 (C)		TDD	1.8/0.05	1.9	0.070
Rel-8 MU-MIMO 1 × 4 (A)			1.8/0.05	2.3	0.071
MU-MIMO 2 × 4 (A)			1.8/0.05	2.8	0.068
MU-MIMO 1 × 8 (E)			1.8/0.05	3.0	0.079
Rel-8 SIMO 1 × 4(C)	UMa	FDD	1.4/0.03	1.5	0.062
CoMP 1 × 4 (A)			1.4/0.03	1.7	0.086
CoMP 2 × 4 (C)			1.4/0.03	2.1	0.099
Rel-8 SIMO 1 × 4 (C)		TDD	1.4/0.03	1.5	0.062
CoMP 1 × 4 (C)			1.4/0.03	1.9	0.090
CoMP 2 × 4 (C)			1.4/0.03	2.0	0.097
MU-MIMO 1 × 8 (E)			1.4/0.03	2.7	0.076
Rel-8 SIMO 1 × 4 (C)	RMa	FDD	0.7/0.015	1.8	0.082
Rel-8 MU-MIMO 1 × 4 (A)			0.7/0.015	2.2	0.097
CoMP 2 × 4 (A)			0.7/0.015	2.3	0.13
Rel-8 SIMO 1 × 4 (C)		TDD	0.7/0.015	1.8	0.080
Rel-8 MU-MIMO 1 × 4 (A)			0.7/0.015	2.1	0.093
CoMP 2 × 4 (A)			0.7/0.015	2.5	0.15
MU-MIMO 1 × 8 (E)			0.7/0.015	2.6	0.10

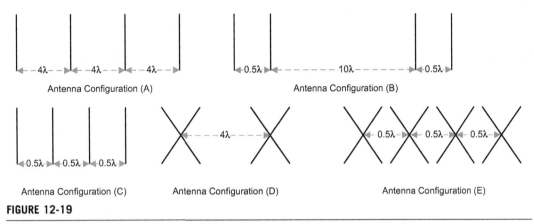

FIGURE 12-19

An illustration of antenna configurations that are used in 3GPP evaluations

wavelength separation between antennas within each group; (3) correlated and co-polarized, i.e., there is a 0.5-wavelength distance between antennas; (4) uncorrelated and cross-polarized, i.e., columns with ±45° linearly polarized antennas and separated by 4 wavelengths; and (5) correlated and cross-polarized, i.e., columns with ±45° linearly polarized antennas separated by 0.5 wavelengths. These configurations are illustrated in Figure 12-19.

The downlink and uplink cell and cell edge spectral efficiencies in various test environments for the TDD and FDD duplex schemes are shown in Tables 12-18 and 12-19, respectively. In these tables, the acronyms "JP-CoMP," "CS-CoMP," and "CB-CoMP" refer to joint processing, coordinated scheduling, and coordinated beamforming multi-point transmission schemes, respectively. The joint processing refers to a CoMP scheme where data is available at each of the geographically separated points and PDSCH transmission occurs from multiple points, whereas coordinated scheduling/beamforming is an alternative CoMP technique where data is only available at the serving cell and data transmission is from that point, but user scheduling/beamforming decisions are made by coordination between different cells.

12.10.2 VoIP Capacity

The number of active VoIP users was calculated through extensive system-level simulations that were conducted by a number of 3GPP member companies. Table 12-20 shows the VoIP capacity results in the indoor, micro-cellular, macro-cellular, and high speed test environments for the FDD and TDD duplex schemes. The results suggest that the 3GPP LTE Rel-8 can satisfy the ITU-R requirements in various deployment scenarios. The antenna configurations are illustrated in Figure 12-19.

12.10.3 Mobility

In order to evaluate the 3GPP LTE-Advanced candidate against the mobility requirement, the methodology that is set out in reference [6] is followed, where in the first step the system-level simulation is performed for each test environment in the uplink to obtain the uplink SINR distribution (see Figure 12-20). From the uplink SINR distribution the 50% point of the CDF is obtained for each test

TABLE 12-20 3GPP LTE VoIP Capacity [13]

Antenna Configuration	Duplex Scheme	Test Environment	ITU-R Requirement	VoIP Capacity (Active Users/MHz/ Cell)
Antenna configuration (A)	FDD	InH	50	140
		UMi	40	80
		UMa	40	68
		RMa	30	91
Antenna configuration (C)		InH	50	131
		UMi	40	75
		UMa	40	69
		RMa	30	94
Antenna configuration (A)	TDD	InH	50	137
		UMi	40	74
		UMa	40	65
		RMa	30	86
Antenna configuration (C)		InH	50	130
		UMi	40	74
		UMa	40	67
		RMa	30	92

environment. In the second step, link-level simulation is conducted to generate the spectral efficiency versus SINR curves for each test environment. The spectral efficiency values include the effect of the control channel overhead from the uplink system-level simulations for each test scenario. The mobility requirement is met, if the spectral efficiency values from the link-level curves, obtained for the median SINR value from the system-level simulations, are greater than the ITU-R targets. These values were obtained assuming different antenna configurations in the uplink, as shown in Table 12-21.

It can be concluded from Table 12-21 that any of the evaluated 3GPP LTE configurations can fulfill the ITU-R requirements concerning mobility for all test environments.

12.10.4 Peak Spectral Efficiency

The peak spectral efficiency of 3GPP LTE-Advanced is calculated assuming 20 MHz bandwidth, one OFDM symbol for downlink control signaling, cell-specific reference signals corresponding to one and four cell-specific antenna ports, UE-specific reference signals corresponding to 24 and zero resource elements per resource-block pair, and physical broadcast channel and synchronization sequences to occupy a total of 564 (Rel-10 MU-MIMO/CoMP with 6 MBSFN subframes per 10 ms) and 528 (Rel-8 downlink SU-MIMO) resource elements per radio frame for downlink eight- and four-layer spatial multiplexing, respectively. In addition, the UL/DL configuration 1 (2 DL subframes, 1 special-subframe, 2 uplink subframes) and special-subframe configuration 4 (12 DwPTS, 1 GP, 1 UpPTS, UpPTS is used for SRS transmission) are assumed for the TDD mode. The results shown in Table 12-22 suggest that the 3GPP LTE Rel-8 already fulfills the ITU-R requirements for downlink peak

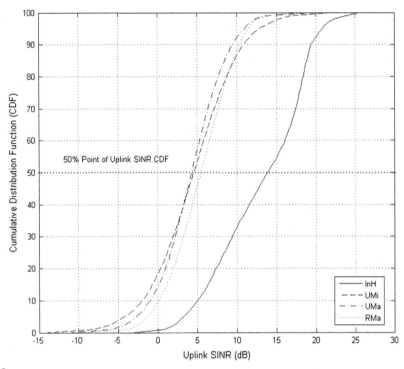

FIGURE 12-20

Statistical distribution of the uplink SINR in various test environments [42]

spectral efficiency. The table also shows the additional performance that can be achieved using advanced technical features, e.g., downlink eight-layer spatial multiplexing.

The uplink peak spectral efficiency is calculated assuming a 20 MHz bandwidth, PUCCH occupying two pairs of resource blocks per subframe, and PRACH consuming six resource block pairs per radio frame. The same UL/DL and special-subframe configurations are assumed for the downlink peak-spectral efficiency calculation. Table 12-22 shows that the extension of the 3GPP LTE Rel-8 with two-layer spatial multiplexing can fulfill the ITU-R requirement in the uplink.

12.10.5 User-Plane/Control-Plane Latency and Handover Interruption Time

The LTE user-plane one-way access latency for a scheduled UE consists of fixed node processing delays, which include radio frame alignment and 1 ms TTI duration. Using the latency calculation model shown in Figure 12-21, and assuming that the number of HARQ processes is eight for the FDD mode, the one-way latency is given as $T_{USER\text{-}PLANE} = 4 + 8p$ where p is the probability of HARQ re-transmissions or the error probability of the first HARQ re-transmission. While the minimum latency of $T_{USER\text{-}PLANE} = 4$ ms is achieved for $p = 0$, a more realistic value of $T_{USER\text{-}PLANE} = 4.8$ ms is obtained for $p = 0.1$.

TABLE 12-21 3GPP LTE Uplink Link-Level Simulation Results for the Mobility Requirements [13]

Path Loss Model (LoS/NLoS)	Duplex Scheme	Test Environment	ITU-RRequirements	Median SINR(dB)	Uplink Spectral Efficiency (bits/s/Hz)
Antenna Configuration 1 × 4 (NLoS)	FDD	InH	1.0	13.89	2.56
		UMi	0.75	4.54	1.21
		UMa	0.55	4.30	1.08
		RMa	0.25	5.42	1.22
Antenna Configuration 1 × 4 (LoS)		InH	1.0	13.89	3.15
		UMi	0.75	4.54	1.42
		UMa	0.55	4.30	1.36
		RMa	0.25	5.42	1.45
Antenna Configuration 1 × 4 (NLoS)	TDD	InH	1.0	13.89	2.63
		UMi	0.75	4.54	1.14
		UMa	0.55	4.30	0.95
		RMa	0.25	5.42	1.03
Antenna Configuration 1 × 4 (LoS)		InH	1.0	13.89	3.11
		UMi	0.75	4.54	1.48
		UMa	0.55	4.30	1.36
		RMa	0.25	5.42	1.38

TABLE 12-22 3GPP LTE-Advanced Peak Spectral Efficiency [3]

Scheme	Duplex Scheme	Direction	Spectral Efficiency (bits/s/Hz)
ITU-R Requirement	FDD	Downlink	15
Rel-8 4-Layer Spatial Multiplexing			16.3
8-Layer Spatial Multiplexing			30.6
ITU-R Requirement		Uplink	6.75
2-Layer Spatial Multiplexing			8.4
4-Layer Spatial Multiplexing			16.8
ITU-R Requirement	TDD	Downlink	15
Rel-8 4-Layer Spatial Multiplexing			16.0
8-Layer Spatial Multiplexing			30.0
ITU-R Requirement		Uplink	6.75
2-Layer Spatial Multiplexing			8.1
4-Layer Spatial Multiplexing			16.1

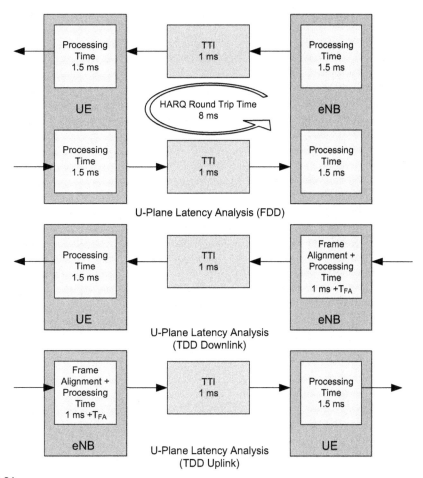

FIGURE 12-21

3GPP LTE user-plane latency calculation model [13]

The user-plane one-way latency for a scheduled UE in the TDD mode consists of fixed node processing delays, radio frame alignment, and TTI duration. The latency component can be seen in Figure 12-21 for the downlink and uplink. Using the latency model shown in Figure 12-21, the total one-way processing time is 2.5 ms, T_{FA} is the radio frame alignment which depends on various configurations of the TDD frame structure, and the TTI duration is 1 ms, hence, the user-plane latency of the TDD mode can be written as $T_{USER\text{-}PLANE} = 3.5 + T_{FA} + pT_{RTT}$ where T_{RTT} is the average HARQ round-trip time and p is the error probability of the first HARQ transmission.

Table 12-23 shows the user-plane latency component breakdown in downlink and uplink for different TDD UL/DL configurations when $p = 0.1$ is assumed. It is shown that in all cases, the 3GPP LTE can meet the ITU-R requirements for user-plane latency.

The above analysis further shows that the 5 ms user-plane latency requirement can be satisfied in the TDD mode in uplink and downlink using the UL/DL configuration 6 only when $p = 0$ is assumed.

TABLE 12-23 User-plane Latency Analysis with 10% HARQ Re-transmission Probability [13]

Step	Procedure	Direction	UL/DL Configuration						
			0	1	2	3	4	5	6
1	eNB Processing Delay (ms)	Downlink	1	1	1	1	1	1	1
2	Frame Alignment (ms)		1.7	1.1	0.7	1.1	0.8	0.6	1.4
3	TTI Duration (ms)		1	1	1	1	1	1	1
4	UE Processing Delay (ms)		1.5	1.5	1.5	1.5	1.5	1.5	1.5
5	HARQ Re-transmission (ms)		$0.1 \times$ 10	$0.1 \times$ 10.2	$0.1 \times$ 9.8	$0.1 \times$ 10.5	$0.1 \times$ 11.6	$0.1 \times$ 12.4	$0.1 \times$ 11.2
	Total One-Way Delay (ms)		6.2	5.62	5.18	5.65	5.46	5.34	6.02
1	UE Processing Delay (ms)	Uplink	1	1	1	1	1	1	1
2	Frame Alignment (ms)		1.1	1.7	2.5	3.3	4.1	5	1.4
3	TTI Duration (ms)		1	1	1	1	1	1	1
4	eNB Processing Delay (ms)		1.5	1.5	1.5	1.5	1.5	1.5	1.5
5	HARQ Re-transmission (ms)		$0.1 \times$ 11.6	$0.1 \times$ 10	$0.1 \times$ 10	$0.1 \times$ 10	$0.1 \times$ 10	$0.1 \times$ 10	$0.1 \times$ 11.5
	Total One-Way Delay (ms)		5.76	6.2	7	7.8	8.6	9.5	6.05

In order to calculate the control-plane latency, the model shown in Figure 12-22 is used. The transition from the RRC_IDLE to the RRC_CONNECTED state in the 3GPP LTE comprises several steps that are shown in procedural order in Table 12-24.

While the results of the analysis suggest that the 3GPP LTE Rel-8 satisfies the ITU-R requirements, using the improved features and protocols introduced in later releases would allow reduction of the control-plane latency to 50 ms. Note that since the NAS set-up process can be executed in parallel with the RRC set-up, it does not appear in the total latency calculation. The analysis in Table 12-24 shows that any of the 3GPP LTE configurations can fulfill the ITU-R requirements related to control-plane latency.

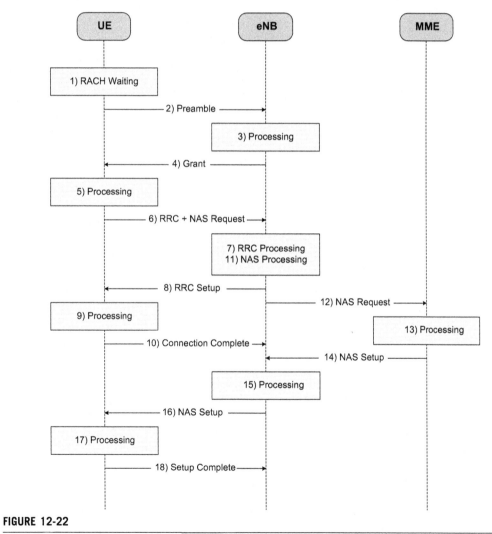

FIGURE 12-22

Breakdown of RRC_IDLE to RRC_CONNECTED state transition in 3GPP LTE [13]

TABLE 12-24 Break Down of C-Plane Latency Components [13]

Step	Procedure	Processing Time (ms)
1	Average delay due to RACH scheduling period (1ms RACH cycle)	0.5
2	RACH Preamble	1
3-4	Preamble detection and transmission of RA response (Time between the end RACH transmission and UE's reception of scheduling grant and timing adjustment)	3
5	UE Processing Delay (decoding of scheduling grant, timing alignment and C-RNTI assignment + L1 encoding of RRC Connection Request)	5
6	Transmission of RRC and NAS Request	1
7	Processing delay in eNB (L2 and RRC)	4
8	Transmission of RRC Connection Set-up and UL grant	1
9	Processing delay in the UE (L2 and RRC)	12
10	Transmission of RRC Connection Set-up complete	1
11	Processing delay in eNB (Uu→S1-C)	–
12	S1-C Transfer delay	–
13	MME Processing Delay (including UE context retrieval of 10ms)	–
14	S1-C Transfer delay	–
15	Processing delay in eNB (S1-C→Uu)	4
16	Transmission of RRC Security Mode Command and Connection Reconfiguration + TTI alignment	1.5
17	Processing delay in UE (L2 and RRC)	16
	C-Plane Delay	50

The calculation of the user-plane latency was carried out for synchronized (or pre-scheduled) UEs. In the case where the UE is not pre-scheduled or when the UE is required to synchronize in order to be allocated uplink resources, a reduced RACH scheduling period, a shorter PUCCH cycle, and reduced processing delays can be applied to satisfy the user-plane latency requirements.

The 3GPP LTE-Advanced general handover procedure is based on that of 3GPP LTE, which was described in Chapter 6, and is illustrated in Figure 12-23. The model shown in the figure has been used for calculation of the handover interruption time. As shown in Figure 12-14, once the handover command has been processed by the UE, the UE detaches from the source eNB and stops receiving data. This marks a point in time where the user-plane connectivity is interrupted. The UE then performs frequency synchronization with the target eNB, depending on whether the target cell is

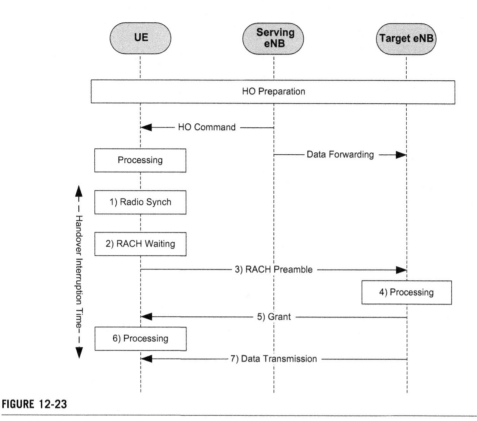

FIGURE 12-23

Model for calculation of handover interruption time [40]

operating on the same carrier frequency as the currently served frequency. Since the UE has already identified and measured the target cell, the corresponding delay will be negligible. The UE then performs downlink synchronization. Although baseband and RF timing alignments are part of the delay budget, since the UE has already acquired downlink synchronization with the target cell in conjunction with previous measurements and can relate the target cell downlink timing to the source cell downlink timing with a time offset, the corresponding delay is going to be less than 1 ms. The data forwarding between the two eNBs is initiated before the UE moves and establishes connection to the target cell, and because the backhaul is faster than the radio interface, forwarded data is already awaiting transmission in the target cell when the user-plane with the target cell is re-established. The latter delay component thereby does not affect the overall delay.

Based on the analysis shown in Table 12-25, the minimum handover interruption time is 10.5 ms for the FDD mode and 12.5 ms for the TDD mode. Note that this delay does not depend on the frequency of the target cell as long as the cell frequency has already been measured by the UE during the measurement gap.

As explained earlier, the minimum handover interruption time is 10.5 ms for FDD and 12.5 ms for TDD regardless of the frequency of the target cell and measurements. Note that in the FDD mode, the

TABLE 12-25 Breakdown of Handover Interruption Time [13]

Step	Procedure	FDD Mode Processing Time (ms)	TDD Mode Processing Time (ms)
1	Radio synchronization to the target cell	1	1
2	Average delay due to RACH scheduling period assuming 1 ms periodicity	0.5	2.5
3	RACH preamble	1	1
4-5	Preamble detection and transmission of Random Access (RA) response (time between the end RACH transmission and UE's reception of scheduling grant and timing adjustment)	5	5
6	Decoding of scheduling grant and timing alignment	2	2
7	Transmission of downlink data	1	1
	Handover Interruption Time	10.5	12.5

average delay due to RACH scheduling is 0.5 ms assuming 1 ms RACH periodicity, whereas in the TDD mode, the minimum delay is obtained with configuration 0 (i.e., six normal uplink subframes); however, when taking into account the average waiting time for a downlink subframe to receive the RA response and to initiate transmission of downlink data, the TDD configuration 1 (with random access preambles in special subframes) offers the shortest handover interruption time.

12.10.6 Estimation of the L1/L2 Overhead

The downlink L1/L2 overhead in the 3GPP LTE-Advanced includes different types of reference signals, i.e., Cell-Specific Reference Signals (CRS) or common pilots transmitted within each resource block, UE-specific Reference Signals (URS) or dedicated pilots, and the reference signals that are specifically utilized in the estimation of channel state information (CSI-RS). The overhead channels further include control signaling channels that are transmitted in the first L OFDM symbols ($L = 1, 2,$ or 3) of each subframe ($L = 4$ in the case of 1.4 MHz bandwidth) and the synchronization signals, as well as the physical broadcast control channel. The PDU headers corresponding to Layer 2 sub-layers (MAC/RLC/PDCP) are also included in the self-evaluation results. The overhead due to CRS and L1/L2 control signaling depends on the number of cell-specific antenna ports and the number of OFDM symbols used for the L1/L2 control signaling. The number of overhead resource elements per resource-block pair (168 resource elements in total) and the corresponding relative overhead are shown in Table 12-26 for non-MBSFN and MBSFN subframes. The overhead (due to L1/L2 control signaling and cell-specific reference signals) in Table 12-26 is shown in terms of the number of resource elements per resource-block pair in percentages.

In the case of MBSFN subframes, the overhead is reduced. All the subframes in a radio frame are assumed to be either regular or MBSFN type in overhead estimation. The overall overhead depends on the fraction of MBSFN subframes. In the self-evaluation results, a fraction of 0% and 60% has been assumed. The overhead due to UE-specific reference signals depends on the number of UE-specific

TABLE 12-26 3GPP LTE-Advanced L1/L2 Overhead [3]

Control-Region Size	Sub-frame Type	FDD Duplex Scheme			TDD Duplex Scheme		
		Number of Cell-Specific Antenna Ports			Number of Cell-Specific Antenna Ports		
		1	2	4	1	2	4
L = 1	Regular sub-frames	18 (10.7%)	24 (14.3%)	32 (19.0%)	18 (10.7%)	24 (14.3%)	32 (19.0%)
L = 2		30 (17.9%)	36 (21.4%)	40 (23.9%)	30 (17.9%)	36 (21.4%)	40 (23.9%)
L = 3		42 (25%)	48 (28.6%)	52 (31.0%)	42 (25%)	48 (28.6%)	52 (31.0%)
L = 1	MBSFN sub-frames	12 (7.1%)	12 (7.1%)	12 (7.1%)	12 (7.1%)	12 (7.1%)	12 (7.1%)
L = 2		24 (14.2%)	24 (14.2%)	24 (14.2%)	24 (14.2%)	24 (14.2%)	24 (14.2%)
L = 3		36 (21.4%)	36 (21.4%)	36 (21.4%)	36 (21.4%)	36 (21.4%)	36 (21.4%)

antenna-ports, and is 12 resource elements (7.1%) for one and two antenna ports, and 24 resource elements (14.3%) for three to eight antenna ports. Note that the overhead due to UE-specific reference signals is only present in resource blocks in which UE-specific reference signals are transmitted.

The relative overhead due to synchronization signals and physical broadcast channel depends on the operation bandwidth. A total of 528 resource elements per 10 ms radio frame are used for the synchronization signals and physical broadcast channel for four CRS ports, corresponding to approximately 0.6% and 0.3% overhead for 10 and 20 MHz operation bandwidth, respectively. The overhead due to PDU headers is proportional to data packet size and is approximately 2.7%, 0.51%, and 0.32% for physical-layer data rates of 1, 10, and 100 Mbits/s, respectively. This overhead is not included in the self-evaluation results. The relative overhead due to CSI-RS depends on the number of antennas and the periodicity. In a typical case, it is about 0.12% per antenna port (0.48% for four antenna ports and 0.96% for eight antenna ports). The relative overhead due to UE-specific RS is estimated at approximately 7% in the case of rank 1 and rank 2 transmission, and 14% for rank 3 to 8 transmission.

In the uplink, the L1/L2 overhead includes the Demodulation Reference Symbols (DM-RS) that are used in uplink channel estimation for coherent demodulation and that are transmitted once every 0.5 ms. It further includes the Sounding Reference Signal (SRS) used for uplink channel state estimation at the eNB, and the L1/L2 control signaling transmitted on configurable amounts of resource blocks, as well as L2 control overhead due to random access, uplink time-alignment control, power headroom reports, and buffer-status reports. The L2 overhead associated with the PDU headers (MAC/RLC/PDCP) is further taken into consideration in the overall overhead calculation. The amount of overhead due to DM-RS is approximately 14%, corresponding to one DFTS-OFDM symbol in each slot. The relative overhead is estimated independent of the rank of the transmission. The amount of SRS overhead depends on the sounding signal transmission interval and the bandwidth of the sounding signal. Using a 10 ms SRS transmission interval and full-band SRS, the relative overhead is approximately 0.7%. The amount of uplink resources reserved for random access depends on the PRACH configuration, e.g., a typical case with PRACH format 0 is six resource blocks per radio frame, resulting in a relative overhead of 0.6%, 1.2%, and 2.4% for a channel bandwidth of 20, 10, and 5 MHz respectively.

The relative overhead due to uplink timing-alignment control depends on the configuration and the number of active UEs within a cell. The absolute overhead is typically less than 32 bits/s per UE. The amount of overhead for buffer status reports depends on the configuration; assuming a continuous data and a reporting interval of 10 to 20 ms, the absolute overhead is 0.8–3.2 kbps. The amount of overhead due to PDU headers depends on the data packet size and is approximately 2.7%, 0.51%, and 0.32% for L1 data rates of 1, 10, and 100 Mbps, respectively. The above overhead calculations are based on normal CP length. In the case of MBSFN subframes in the TDD mode, all the subframes are assumed to be of the MBSFN type in the overhead estimation. The overall overhead depends on the fraction of MBSFN subframes. In self-evaluation results, a fraction of 0% and 33% has been assumed for a DL:UL ratio of 3:2 (i.e., UL/DL configuration 1).

In the TDD mode, the relative overhead due to synchronization signals and physical broadcast channels depends on the operation bandwidth and the DL/UL configuration. A total of 528 resource elements per 10 ms radio frame are used for the synchronization signals and the broadcast channel when using four CRS ports. For a DL:UL ratio of 3:2, the latter overhead translates into approximately 1.0% and 0.5%, for 10 and 20 MHz operation bandwidth, respectively. The amount of SRS overhead depends on the SRS transmission interval, the SRS bandwidth, and the usage of UpPTS in the special subframe. For a DL:UL ratio of 3:2 with four DL/UL switching points per 10 ms radio frame, the SRS transmission with the same interval as a TDD UL/DL transmission of 5 ms and full band SRS within UpPTS, the relative overhead is approximately 3.45%. The amount of uplink resources reserved for random access depends on the configuration of the random access channel. In a typical case with a DL:UL ratio of 3:2 and PRACH preamble format of 0, six resource blocks per radio frame are used for the physical random access channel, resulting in a relative overhead of 1.4%, 2.8%, 5.6% for a channel bandwidth of 20, 10, and 5 MHz, respectively.

12.10.7 Frequency Bands and Operation Bandwidth

The frequency bands in which the 3GPP LTE, 3GPP LTE-Advanced FDD, and TDD modes can be deployed are shown in Table 12-27. The operating bands of the 3GPP LTE-Advanced include E-UTRA operating bands, as well as possible new IMT bands to be identified by the ITU-R.

In both FDD and TDD modes, each component carrier supports a scalable bandwidth of 1.4, 3, 5, 10, 15, and 20 MHz. By aggregating multiple component carriers, wider transmission bandwidths up to 100 MHz are supported.

12.10.8 3GPP LTE-Advanced Link Budget

The 3GPP LTE-Advanced link budget under various deployment scenarios specified by the ITU-R was evaluated as part of the IMT-Advanced submission. Table 12-28 summarizes the link budget of the TDD mode. It is shown that the 3GPP LTE-Advanced control and data channel coverage meets the IMT-Advanced requirements.

12.10.9 Evaluation of 3GPP LTE-Advanced Against Release 10 Requirements

This section provides the self-evaluation results of the 3GPP LTE-Advanced against 3GPP Rel-10 requirements [40], where the target values for cell spectral efficiency and cell edge spectral efficiency

TABLE 12-27 Operating Bands of 3GPP LTE-Advanced [22,23]

Band Class	Uplink Operating Band eNB Receive/UE Transmit (MHz) F_{DL_low}–F_{DL_high}		Downlink Operating Band eNB Transmit /UE Receive (MHz) F_{DL_low}–F_{DL_high}		Duplex Mode
1	1920	– 1980	2110	– 2170	FDD
2	1850	– 1910	1930	– 1990	FDD
3	1710	– 1785	1805	– 1880	FDD
4	1710	– 1755	2110	– 2155	FDD
5	824	– 849	869	– 894	FDD
6	830	– 840	865	– 875	FDD
7	2500	– 2570	2620	– 2690	FDD
8	880	– 915	925	– 960	FDD
9	1749.9	– 1784.9	1844.9	– 1879.9	FDD
10	1710	– 1770	2110	– 2170	FDD
11	1427.9	– 1447.9	1475.9	– 1495.9	FDD
12	698	– 716	728	– 746	FDD
13	777	– 787	746	– 756	FDD
14	788	– 798	758	– 768	FDD
15	Reserved		Reserved		–
16	Reserved		Reserved		–
17	704	– 716	734	– 746	FDD
18	815	– 830	860	– 875	FDD
19	830	– 845	875	– 890	FDD
20	832	– 862	791	– 821	FDD
21	1447.9	– 1462.9	1495.9	– 1510.9	FDD
22	3410	– 3500	3510	– 3600	FDD
...					
33	1900	– 1920	1900	– 1920	TDD
34	2010	– 2025	2010	– 2025	TDD
35	1850	– 1910	1850	– 1910	TDD
36	1930	– 1990	1930	– 1990	TDD
37	1910	– 1930	1910	– 1930	TDD
38	2570	– 2620	2570	– 2620	TDD
39	1880	– 1920	1880	– 1920	TDD
40	2300	– 2400	2300	– 2400	TDD
41	3400	– 3600	3400	– 3600	TDD

TABLE 12-28 3GPP LTE-Advanced Link Budget in Various Deployment Scenarios for the TDD Mode [3]

Parameter	InH Downlink	InH Uplink	UMi Downlink	UMi Uplink	UMa Downlink	UMa Uplink	RMa Downlink	RMa Uplink
			System Configuration					
Carrier Frequency (GHz)	3.4	3.4	2.5	2.5	2.0	2.0	0.8	0.8
BS Antenna Heights (m)	6	6	10	10	25	25	35	35
MS Antenna Heights (m)	1.5	1.5	1.5	1.5	1.5	1.5	1.5	1.5
Cell Area Reliability for Control Channels	95%	95%	95%	95%	95%	95%	95%	95%
Cell Area Reliability for Control Channels	90%	90%	90%	90%	90%	90%	90%	90%
Transmission Bit Rate for Control Channel (kbps)	46.93	1.60	44.80	1.60	44.80	1.60	44.80	1.60
Transmission Bit Rate for Data Channel (Mbps)	4.65	0.37	2.25	0.074	2.25	0.074	2.25	0.074
Target Packet Error Rate for Control Channels	1%	1%	1%	1%	1%	1%	1%	1%
Target Packet Error Rate for Data Channels	10%	10%	10%	10%	10%	10%	10%	10%
Spectral Efficiency (bits/s/Hz) for Data	0.23	0.52	0.22	0.21	0.22	0.21	0.22	0.21
Pathloss Model	NLoS	NLoS	NLoS	NLoS	NLoS	NLoS	NLoS	NLoS
Mobile Speed (km/h)	3	3	3	3	30	30	120	120
Feeder Loss (dB)	2	2	2	2	2	2	2	2
Transmitter								
Number of Transmit Antennas	2	1	2	1	2	1	2	1
Maximum Transmit Power per Antenna (dBm)	21	21	41	24	46	24	46	24
Total Transmit Power (dBm)	24	21	44	24	49	24	49	24
Transmitter Antenna Gain (dBi)	0	0	17	0	17	0	17	0
Transmitter Array Gain (dB)	0	0	0	0	0	0	0	0
Control Channel Power Boosting Gain (dB)	0	0	0	0	0	0	0	0
Data Channel Power Loss Due to Pilot/Control Boosting (dB)	0	0	0	0	0	0	0	0
Cable, Connector, Combiner, Body Losses, etc. (dB)	3	1	3	1	3	1	3	1

Control Channel EIRP (dBm)	21	20	58	23	63	23	63	23
Data Channel EIRP (dBm)	21	20	58	23	63	23	63	23
Receiver								
Number of Receive Antennas	2	4	2	4	2	4	2	4
Receiver Antenna Gain (dBi)	0	0	0	17	0	17	0	17
Cable, Connector, Combiner, Body Losses, etc. (dB)	1	3	1	3	1	3	1	3
Receiver Noise Figure (dB)	7	5	7	5	7	5	7	5
Thermal Noise Density (dBm/Hz)	−174	−174	−174	−174	−174	−174	−174	−174
Receiver Interference Density for Control Channels (dBm/Hz)	−174	−174.9	−169.3	−161.7	−169.3	−161.7	−169.3	−161.7
Receiver Interference Density for Data Channels (dBm/Hz)	−174	−174.9	−169.3	−165.7	−169.3	−165.7	−169.3	−165.7
Total Noise Plus Interference Density for Control Channels (dBm/Hz)	−167	−168	−165	−161	−165	−161	−165	−161
Total Noise Plus Interference Density for Data Channels (dBm/Hz)	−167	−168	−165	−164	−165	−164	−165	−164
Occupied Channel Bandwidth for Control Channels (MHz)	36	0.18	18	0.18	18	0.18	18	0.18
Occupied Channel Bandwidth for Data Channels (MHz)	36	1.8	18	0.90	18	0.9	18	0.90
Effective Noise Power for Control Channels (dBm)	−91	−115	−92	−108	−92	−108	−92	−108
Effective Noise Power for Control Channels (dBm)	−91	−105	−92	−104	−92	−104	−92	−104
Required SNR for the Control Channel (dB)	−4.2	−10.6	−4.2	−10.5	−4.2	−10.1	−4.2	−9.9
Required SNR for the Data Channel (dB)	−1.6	−2.3	−1.7	−5.5	−1.7	−5.1	−1.7	−4.8
Receiver Implementation Margin (dB)	2	2	2	2	2	2	2	2
HARQ Gain for Control Channel (dB)	0	0	0	0	0	0	0	0
HARQ Gain for Data Channel (dB)	0.5	0.5	0.5	0.5	0.5	0.5	0.5	0.5
Receiver Sensitivity for Control Channel (dBm)	−94	−124	−95	−117	−95	−117	−95	−116

(Continued)

TABLE 12-28 3GPP LTE-Advanced Link Budget in Various Deployment Scenarios for the TDD Mode [3] *Continued*

Parameter	InH		UMi		UMa		RMa	
	Downlink	Uplink	Downlink	Uplink	Downlink	Uplink	Downlink	Uplink
System Configuration								
Receiver Sensitivity for Data Channel (dBm)	−91	−106	−93	−108	−93	−108	−93	−108
Hardware Link Budget for Control Channel (dB)	115	144	153	157	158	157	158	156
Hardware Link Budget for Data Channel (dB)	113	126	151	148	156	148	156	148
Calculation of Available Path Loss								
Lognormal Shadow Fading Standard Deviation (dB)	4	4	4	4	6	6	8	8
Shadow Fading Margin for Control Channels (dB)	2.8	2.8	3.1	3.1	5	5	5	5
Shadow Fading Margin for Data Channels (dB)	0.9	0.9	1.3	1.3	8.1	8.1	10.5	10.5
BS Selection/Macro-Diversity Gain (dB)	0	0	0	0	4.9	4.9	6.7	6.7
Penetration Loss (dB)	0	0	0	0	0	0	0	0
Other Gains (dB)	0	0	0	0	9	9	9	9
Available Path Loss for Control Channel (dB)	111	138	149	151	140	136	137	134
Available Path Loss for Data Channel (dB)	111	122	148	144	148	144	137	134
Range/Coverage Efficiency Calculation								
Maximum Range for Control Channel (m)	100.0	100.0	1405.2	1621.2	1175.6	978.0	3210.9	2634.3
Maximum Range for Data Channel (m)	100.0	100.0	1383.9	1062.7	1258.4	714.4	3565.5	1975.1
Coverage Area for Control Channel (km²/site)	0.031	0.031	6.20	8.26	4.34	3.01	32.4	21.8
Coverage Area for Data Channel (km²/site)	0.031	0.031	6.02	3.55	4.97	1.60	39.9	12.3

TABLE 12-29 Self-Evaluation of 3GPP LTE-Advanced against Release 10 Requirements [41]

Scheme and Antenna Configuration	Direction	Duplex Scheme	Release 10 Requirements(Cell Spectral Efficiency/Cell-edge Spectral Efficiency)	Cell Spectral Efficiency (bits/s/Hz/cell) L = 3	Cell-edge Spectral Efficiency(bits/s/Hz) L = 3
MU-MIMO 2 × 2 (C)	Downlink	FDD	2.4/0.07	2.69	0.090
JP-CoMP 2 × 2 (C)			2.4/0.07	2.70	0.104
MU-MIMO 4 × 2 (C)			2.6/0.09	3.43	0.118
CS/CB-CoMP 4 × 2 (C)			2.6/0.09	3.34	0.129
JP-CoMP 4 × 2 (C)			2.6/0.09	3.87	0.162
MU-MIMO 4 × 4 (C)			3.7/0.12	4.69	0.203
CS/CB-CoMP 4 × 4 (C)			3.7/0.12	4.66	0.205
JP-CoMP 4 × 4 (C)			3.7/0.12	5.19	0.269
MU-MIMO 2 × 2 (C)		TDD	2.4/0.07	2.88	0.113
JP-CoMP 2 × 2 (C)			2.4/0.07	3.15	0.130
MU-MIMO 4 × 2 (C)			2.6/0.09	3.76	0.151
JP-CoMP 4 × 2 (C)			2.6/0.09	4.64	0.199
MU-MIMO 4 × 4 (C)			3.7/0.12	4.97	0.209
CS/CB-CoMP 4 × 4 (C)			3.7/0.12	5.06	0.244
JP-CoMP 4 × 4 (C)			3.7/0.12	6.61	0.330
Rel-8 SIMO 1 × 2 (C)	Uplink	FDD	1.2/0.04	1.33	0.047
CoMP 1 × 2 (C)			1.2/0.04	1.40	0.051
SU-MIMO 2 × 4 (C)			2.0/0.07	2.27	0.091
Rel-8 SIMO 1 × 2 (C)		TDD	1.2/0.04	1.24	0.045
CoMP 1 × 2 (C)			1.2/0.04	1.51	0.051
SU-MIMO 2 × 4 (C)			2.0/0.07	2.15	0.090
MU-MIMO 2 × 4 (C)			2.0/0.07	2.59	0.079

are specified for the 3GPP case 1 channel model [41] with 2×2, 4×2, and 4×4 antenna configurations in the downlink, and 1×2 and 2×4 antenna configurations in the uplink. Table 12-29 shows the downlink/uplink cell and cell edge spectral efficiency results under the 3GPP case 1 configuration for FDD and TDD modes. The results suggest that, if 3GPP LTE Rel-8 is extended with MU-MIMO 4×2 in the downlink, the 3GPP Rel-10 targets can be fulfilled. The table also includes the advanced features by which the performance can be further improved. The uplink performance results indicate that the 3GPP LTE Rel-8 SIMO and 3GPP LTE-Advanced CoMP, SU-MIMO and MU-MIMO schemes can fulfill the 3GPP Rel-10 requirements. In the 3GPP case 1 scenario, the center frequency is 2.0 GHz, the transmission bandwidth is 2×10 MHz for the FDD mode and 20 MHz for the TDD mode, inter-site distance is 500 m, the path loss is 20 dB, and all users are assumed to be moving at 3 km/h.

References

[1] W.H. Tranter, K. Sam Shanmugan, T.S. Rappaport, K.L. Kosbar, Principles of Communication Systems Simulation with Wireless Applications, Prentice Hall, 2004.

[2] ITU-R Doc. IMT-ADV/4, Acknowledgement of Candidate Submission from IEEE under Step 3 of the IMT-Advanced Process (IEEE Technology), October 2009.

[3] ITU-R Doc. IMT-ADV/8, Acknowledgement of Candidate Submission from 3GPP Proponent (3GPP Organization Partners of ARIB, ATIS, CCSA, ETSI, TTA, and TTC) under Step 3 of the IMT-Advanced Process (3GPP Technology), October 2009.

[4] Report ITU-R M.2133, Requirements, Evaluation Criteria and Submission Templates for the Development of IMT-Advanced, November 2008.

[5] Report ITU-R M.2134, Requirements Related to Technical Performance for IMT-Advanced Radio Interface(s), November 2008.

[6] Report ITU-R M.2135-1, Guidelines for Evaluation of Radio Interface Technologies for IMT-Advanced, December 2009.

[7] L. Rade, Mathematics Handbook for Science and Engineering, Springer Berlin Heidelberg, (2010).

[8] IEEE 802.16m–08/004r5, IEEE 802.16m Evaluation Methodology Document (EMD), March 2009.

[9] 3GPP TS 26.101, AMR Speech Codec Frame Structure, June 2001.

[10] Ietf Rfc 3267, Real-Time Transport Protocol (RTP) Payload Format and File Storage Format for the Adaptive Multi-Rate (AMR) and Adaptive Multi-Rate Wideband (AMR-WB) Audio Codecs, J. Sjoberg et al, June 2002.

[11] S. Rodriguez-Herrera, D. McBeath, D. Pinckley andReed, "Link-To-System Mapping Techniques Using A Spatial Channel Model," IEEE 62nd Vehicular Technology Conference 2005, VTC-2005 Vol. 3, 2005.

[12] P.802.16m/D6, IEEE Standard for Local and Metropolitan Area Networks – Part 16: Air Interface for Broadband Wireless Access Systems, Advanced Air Interface, May 2010.

[13] 3GPP TR 36.912, Feasibility Study for Further Advancements for E-UTRA (LTE-Advanced), March 2010.

[14] Recommendation ITU-T G.729, Coding of Speech at 8 kbit/s Using Conjugate-Structure Algebraic-Code-Excited Linear Prediction (CS-ACELP), January 2007.

[15] V. Erceg, et al., "Channel models for fixed wireless applications,", IEEE 802.16.3c–01/29r4, (July 2001).

[16] Recommendation ITU-R M.1225, Guidelines for Evaluation of Radio Transmission Technologies for IMT-2000, (1997).

[17] 3GPP-3GPP2 Spatial Channel Ad-hoc Group, Spatial Channel Model Text Description, v7.0, August 2003.

[18] 3GPP TR 25.996, Spatial Channel Model for Multiple Input Multiple Output (MIMO), Simulations, June 2007.

[19] WINNER Project IST-4-027756, WINNER D1.1.2 v1.2, WINNER II Channel Models, September 2007 <https://www.ist-winner.org/>.

[20] WINNER Project IST-2003-507581, WINNER D5.4 v. 1.4, Final Report on Link Level and System Level Channel Models, November 2005 <https://www.ist-winner.org/>.

[21] WINNER Project, IST-2003–507581, WINNER D1.3 version 1.0, Final Usage Scenarios, June 2005.

[22] 3GPP TS 36.101, User Equipment Radio Transmission and Reception, March 2010.

[23] 3GPP TS 36.104, Base Station (BS) Radio Transmission and Reception, March 2010.

[24] 3GPP2 TSG-C30-20061204-062A, cdma2000 Evaluation Methodology (v6), December 2006.

[25] S. Tsai, A. Soong, "Effective-SNR Mapping for Modeling Frame Error Rates in Multiple-State Channels,", 3GPP2-C30-20030429-010, April 2003.

[26] Digital Mobile Radio Towards Future Generation Systems, COST Action 231 Final Report, EUR 18957, (1999).

[27] J.D. Parsons, "The Mobile Radio Propagation Channel,", John Wiley & Sons, 2000.

[28] Y. Oda, K. Tsunekawa, M. Hata, "Advanced LoS path-loss model in microcellular mobile communications," IEEE Transactions on Antennas and Propagation Vol. 51, May 2003.

[29] ETSI Technical Report 101 112 v3.2.0, "Universal Mobile Telecommunications System (UMTS); Selection Procedures for the Choice of Radio Transmission Technologies of the UMTS (UMTS 30.03 version 3.2.0)," April 1998.

[30] W.C. Jakes, Microwave Mobile Communications, Wiley, New York, 1974.

[31] M. Patzold, Mobile Fading Channels, John Wiley & Sons, 2002.

[32] 3GPP R1-061001, Ericsson, "LTE Channel Models and Link Simulations," March 2006.

[33] Source Code for a MATLAB/ANSI-C Implementation of the WINNER Phase I Channel Model, August 2006 <https://www.ist-winner.org/phase_model.html>.

[34] WINNER Project IST-WINNER II Deliverable D1.1.1 v1.0, "WINNER II Interim Channel Models," December 2006.

[35] M. Steinbauer, A.F. Molisch, E. Bonek, "The Double-Directional Radio Channel," IEEE Antennas and Propagation Magazine, August 2001.

[36] G.J. Foschini, M.J. Gans, "On Limits of Wireless Communications in a Fading Environment when Using Multiple Antennas," Wireless Personal Communications Vol. 6, February 1998.

[37] P. Almers, et al., "Survey of Channel and Radio Propagation Models for Wireless MIMO Systems," EURASIP Journal on Wireless Communication and Networking, January 2007.

[38] W. Dong, J. Zhang, X. Gao, P. Zhang, Y. Wu, "Cluster Identification and Properties of Outdoor Wideband MIMO Channel," IEEE 66th Vehicular Technology Conference, VTC-2007, September 2007.

[39] Athanasios Papoulis, Probability, Random Variables, and Stochastic Processes, fourth ed., McGraw Hill, 2002.

[40] 3GPP TR 36.913, Requirements for Further Advancements for Evolved Universal Terrestrial Radio Access (E-UTRA) (LTE-Advanced), March 2009.

[41] 3GPP TR 36.814, Further Advancements for E-UTRA Physical Layer Aspects, March 2010.

[42] 3GPP RP-090738, TR 36.912 Annex A3: Self-Evaluation Results, September 2009.

[43] S. Sesia, I. Toufik, M. Baker, LTE, the UMTS Long Term Evolution: From Theory to Practice, John Wiley & Sons, 2009.

Index

Printed in the United States
By Bookmasters